D1271095

Southern Forested Wetlands

Ecology and Management

edited by

Michael G. Messina
William H. Conner

Library of Congress Cataloging-in-Publication Data

Catalog record is available from the Libary of Congress

This book contains information obtained from authentic and highly regarded sources. Reprinted material is quoted with permission, and sources are indicated. A wide variety of references are listed. Reasonable efforts have been made to publish reliable data and information, but the author and the publisher cannot assume responsibility for the validity of all materials or for the consequences of their use.

Neither this book nor any part may be reproduced or transmitted in any form or by any means, electronic or mechanical, including photocopying, microfilming, and recording, or by any information storage or retrieval system, without prior permission in writing from the publisher.

All rights reserved. Authorization to photocopy items for internal or personal use, or the personal or internal use of specific clients, may be granted by CRC Press LLC, provided that $.50 per page photocopied is paid directly to Copyright Clearance Center, 27 Congress Street, Salem, MA 01970 USA. The fee code for users of the Transactional Reporting Service is ISBN 1-56670-228-3/97/$0.00+$.50. The fee is subject to change without notice. For organizations that have been granted a photocopy license by the CCC, a separate system of payment has been arranged.

The consent of CRC Press LLC does not extend to copying for general distribution, for promotion, for creating new works, or for resale. Specific permission must be obtained from CRC Press LLC for such copying.

Direct all inquiries to CRC Press LLC, 2000 Corporate Blvd., N.W., Boca Raton, Florida 33431.

Trademark Notice: Product or corporate names may be trademarks or registered trademarks, and are used only for identification and explanation, without intent to infringe.

© 1998 by CRC Press LLC.
Lewis Publishers is an imprint of CRC Press LLC

No claim to original U.S. Government works
International Standard Book Number 1-56670-228-3
Printed in the United States of America 2 3 4 5 6 7 8 9 0
Printed on acid-free paper

Editors

Michael G. Messina, Ph.D., is an Associate Professor in the Department of Forest Science, Texas A&M University, College Station, TX. Dr. Messina received his B.S. in forest science from The Pennsylvania State University and his Ph.D. in forestry from North Carolina State University. He worked for three years at the New Zealand Forest Service's Forest Research Institute in Rotorua on eucalypt silviculture and nutrition before accepting his current position.

His research involves a wide variety of topics ranging from the silviculture, nutrition, and ecophysiology of loblolly pine to the impacts of harvesting in bottomland hardwoods on vegetation recovery and diversity, and soil biology. Dr. Messina actively participates in the teaching and advising missions of his university. He is a member of numerous professional societies and has held leadership positions in several. He currently serves as an associate editor for the *Southern Journal of Applied Forestry*. He has authored or coauthored more than 30 technical papers in journals, proceedings, and other outlets.

William H. Conner, Ph. D., is an Associate Professor of Forestry in the Department of Forest Resources, Clemson University, and is stationed at the university's Baruch Forest Science Institute in Georgetown, SC. He received his B.S. in biology from Virginia Polytechnic Institute and State University and his M.S. in marine sciences and Ph.D. in forestry from Louisiana State University.

He team teaches a class on forested wetland ecology. He has taught a series of wetlands ecology classes at Nanjing Forestry Institute in China, is on the adjunct faculty of Louisiana State University and Auburn University, is an Associate Editor for *Wetlands*, and is a member of numerous professional societies.

Dr. Conner's research interests include the ecology and management of fresh- and saltwater forested wetlands, effects of natural and anthropogenic disturbance on forested wetland environments, use of wetlands for wastewater treatment, regeneration problems in forested wetlands, and estuarine/upland connections. He has authored or coauthored more than 80 technical papers in journals, proceedings, and other outlets.

Contributors

W. Michael Aust
College of Forestry
and Wildlife Resources
Virginia Polytechnic Institute
and State University
Blacksburg, VA

Stephen J. Brady
Natural Resources Inventory
and Analysis Institute
USDA Natural Resources
Conservation Service
Fort Collins, CO

Richard R. Braham
North Carolina State University
Deparment of Forestry
Raleigh, NC

Mark M. Brinson
Biology Department
East Carolina University
Greenville, NC

Marilyn A. Buford
USDA Forest Service
Forest Sciences Laboratory
Research Triangle Park, NC

James A. Burger
College of Forestry
and Wildlife Resources
Virginia Polytechnic Institute
and State University
Blacksburg, VA

William H. Conner
Belle W. Baruch Forest Science Institute
Clemson University
Georgetown, SC

Noel D. Cost
USDA Forest Service
Southern Research Station
Asheville, NC

Frederick W. Cubbage
North Carolina State University
Department of Forestry
Raleigh, NC

Katherine C. Ewel
USDA Forest Service
Pacific Southwest Research Station
Institute of Pacific Islands Forestry
Honolulu, HI

Deborah A. Gaddis
North Carolina State University
Department of Forestry
Raleigh, NC

Charles A. Gresham
Belle W. Baruch Forest Science Institute
Clemson University
Georgetown, SC

John M. Grymes III
Southern Regional Climate Center
Department of Geography
and Anthropology
Louisiana State University
Baton Rouge, LA

William. R. Harms
USDA Forest Service
Southern Research Station
Center for Forested Wetlands Research
Charleston, SC

John D. Hodges
Anderson–Tully
Memphis, TN

Donal D. Hook
Department of Forest Resources
Clemson University
Clemson, SC

Jan Jeffrey Hoover
U.S. Army Engineer Waterways
Experiment Station
Vicksburg, MS

Edwin J. Jones
North Carolina State University
Deparment of Forestry
Raleigh, NC

Robert C. Kellison
North Carolina State University
Deparment of Forestry
Raleigh, NC

K. Jack Killgore
U.S. Army Engineer Waterways
Experiment Station
Vicksburg, MS

Richard A. Lancia
Department of Forestry
North Carolina State University
Raleigh, NC

B. Graeme Lockaby
School of Forestry
Auburn University
Auburn, AL

Martha R. McKevlin
USDA Forest Service
Center for Forested Wetlands Research
Charleston, SC

Michael G. Messina
Texas A&M University
Department of Forest Science
College Station, TX

Kevin K. Moorhead
Environmental Studies
University of North Carolina
at Asheville
Asheville, NC

Robert A. Muller
Southern Regional Climate Center
Department of Geography
and Anthropology
Louisiana State University
Baton Rouge, LA

Richard D. Rheinhardt
Biology Department
East Carolina University
Greenville, NC

Irene M. Rossell
Environmental Studies
University of North Carolina
at Asheville
Asheville, NC

Amy A. Rozelle
South Carolina Forestry Commission
Sand Hills State Forest
Patrick, SC

Stephen Schoenholtz
Department of Forestry
College of Forest Resources
Mississippi State, MS

Rebecca R. Sharitz
Savannah River Ecology Laboratory
Aiken, SC

James P. Shepard
National Council of the Paper Industry
for Air and Stream Improvement, Inc.
Gainesville, FL

John A. Stanturf
USDA Forest Service
Center for Bottomland
Hardwoods Research
Stoneville, MS

Charles G. Storrs
National Wetlands Inventory
U.S. Fish and Wildlife Service
Atlanta, GA

Robert R. Twilley
Department of Biology
University of Southwestern Louisiana
Lafayette, LA

Mark R. Walbridge
Department of Biology
George Mason University
Fairfax, VA

T. Bently Wigley
National Council of the Paper Industry
for Air and Stream Improvement, Inc.
Department of Aquaculture
Fisheries & Wildlife
Clemson University
Clemson, SC

Thomas M. Williams
Belle W. Baruch Forest Science Institute
Clemson University
Georgetown, SC

Michael J. Young
North Carolina State University
Department of Forestry
Raleigh, NC

Preface

Forested wetlands have been an influential component of the economy and culture of the southern United States since the earliest settlements. These ecosystems have been used for transportation, food and fiber, flood control, wildlife habitat, recreation, and, all too often, as fertile sites for conversion to agriculture. The importance of wetlands to the southern economy and concern over their decreasing extent has lead to a surge of research activity as well as litigation and discord over competing wetland uses. Much of the research has been conducted by individuals or small research teams acting independently, with the nature of their work being largely dictated by geography and funding opportunities. Similarly, new information on managing forested wetlands is often developed in isolation of other activities occurring in the region. Although research findings, and sometimes gains in management technology, are usually disseminated in the traditional outlets of refereed journals and symposia proceedings, a need has developed for a concise compendium of information on the ecology and management of southern forested wetlands. This book has been written to that end.

The original idea for this book was developed by members of the Consortium for Research on Southern Forested Wetlands. The Consortium was formed in 1990 by an *ad hoc* group of scientists and research managers interested in the southwide coordination of research on the management and protection of forested wetlands. Consortium activities are organized within three areas: research coordination, science integration, and technology transfer. Membership in the Consortium includes individuals from forest products companies, state and federal agencies, universities, and environmental organizations. Consortium members organized this book, arranged for some of the authors, and selected the book's editors. A conference was held at Clemson University in March 1996, where authors presented their chapters in order to receive feedback from the audience while the book was being written.

The primary purpose of this book is to serve as a text for upper-level undergraduate and graduate forestry and ecology courses. The book also should serve as a first-source reference for those seeking information on particular wetland types or functions. The editors recognize that there are several volumes already published that deal with southern forested wetlands, but none that present both the ecological and management aspects of these important ecosystems. In other words, one of the major contributions of this text will be the presentation of information on current and past management practices, plus their potential impacts and appropriate mitigation.

Wetlands included in this book vary from mountain fens to mangroves. Types covered were selected based upon their prevalence and availability of information. In general, when choosing types to be included, the organizers of this book relied upon the Cowardin classification definition of wetlands; i.e., "Wetlands are lands transitional between terrestrial and aquatic systems where the water table is at or near the surface or the land is covered by shallow water. For purposes of this classification wetlands must have one or more of the following three attributes: (1) at least periodically, the land supports predominantly hydrophytes; (2) the substrate is predominantly undrained hydric soil; and (3) the substrate is nonsoil and is saturated with water or covered by shallow water at some time during the growing season of each year." The geographic extent of coverage by most authors ranges from West Virginia to Florida to Texas and inland north to Arkansas and Tennessee. The book is divided into three parts.

Part I represents an introduction to general southern forested wetlands issues with chapters on classification and inventory, functions and relations to societal values, and the policies, laws, and regulations concerning wetland management and protection.

Part II concerns general functions of southern forested wetlands and contains chapters on climate, hydrology, soils, biogeochemistry, plant adaptations, wildlife, and fish communities.

Part III is devoted to chapters on individual wetland types, including deepwater swamps, major alluvial floodplains, minor alluvial floodplains, pocosins and Carolina bays, mountain fens, pondcypress swamps, flatwoods wetlands, and mangroves. These chapters on wetland types are commonly organized to facilitate reading.

An **Appendix** of common and scientific names of plants and animals mentioned in the book is included to increase the book's usefulness as a reference source.

This book truly represents the collaboration of many scientists and managers working to conserve, study, or manage southern forested wetlands. We would like to thank those who devoted their energy to ensure a quality volume. Obviously, we include the authors who endured through the writing, responding to reviewers, chiding by the editors, presenting their chapters at the conference, checking scientific names, re-drawing figures and so forth.

Also, we relied upon many external reviewers who read and sometimes reread chapter drafts. The reviewers include Jim Allen, Sydney Bacchus, Craig Beals, Scott Bridgham, Marianne Burke, Norm Christensen, Malcolm Cleaveland, Neil Douglas, Malcolm Drew, Mitch Dubensky, Michael Duever, Kathy Ewel, Robert Fledderman, Tom Fox, Sue Grace, Jim Gregory, Paul Hamel, Ron Haynes, Jan Hoover, Wayne Hudnall, Wayne Inabinette, Bob Keeland, Sammy King, Kay Kirkman, Martin Lebo, Ariel Lugo, Bill McKee, Ken McLeod, Steve Meadows, Lee Norfleet, Wade Nutter, Tim Nuttle, Reza Pezeshki, John Stanturf, Robert Sutter, John Thorp, Ralph Tiner, John Toliver, Mark Walbridge, Charles Wax, Joe Yavitt, and Joy Young. We thank these people for volunteering their expertise to aid our goal of producing a comprehensive, factually correct textbook. The authors and reviewers of this text deserve any credit due; the editors accept any criticism for shortcomings.

Michael G. Messina
College Station, Texas

William H. Conner
Georgetown, South Carolina

Contents

PART I
INTRODUCTION

PART II
GENERAL FUNCTIONS OF SOUTHERN FORESTED WETLANDS

Part I

Introduction

1 Classification and Inventory

James P. Shepard, Stephen J. Brady, Noel D. Cost, and Charles G. Storrs

CONTENTS

1-56670-228-X/97/$0.00+$.50
© 1998 by CRC Press LLC

INTRODUCTION

Why classify and inventory forested wetlands? Classification is simply an agreement on naming conventions. It provides a common frame of reference for effective communication and organization of knowledge. Inventory follows closely behind classification. After naming forested wetlands and describing their distinctive characteristics, we then assess their frequency of occurrence, geographic distribution, and characteristics. Both classification and inventory are indispensable to understanding forested wetland ecology and devising intelligent management systems. Elucidation of forested wetland ecology proceeds in an inductive manner. Scientists gather ecological information from many individual wetland forests, then use classification to aid in finding generalities. Management is deductive. It begins with generalized knowledge and then prescribes techniques specific to individual types. There are substantial differences in ecological functions among forested wetlands. Classification formally acknowledges such differences and facilitates the use of management practices that are best-suited for specific wetland types. Inventory is important for both ecologists and managers. It is one way to establish priorities in the allocation of effort. One may favor research and management of commonly occurring taxa first since this results in the greatest reward. However, inventory also provides an assessment of scarcity and is thus a way of identifying rare or unique ecosystems.

The objectives of this chapter are to briefly discuss the historical development of forested wetland classification and inventory systems, to describe current classification systems, and to summarize information from recent inventories in the South (Alabama, Arkansas, Florida, Georgia, Kentucky, Louisiana, Mississippi, North Carolina, South Carolina, Tennessee, Texas, and Virginia). Some of the systems discussed are exclusively for wetlands, whereas others include wetlands only as one subset of forests ecosystems or land uses. These classification systems include a wide variety of attributes, from wetland area to timber volume, growth, and removals.

WETLAND DEFINITIONS

Before we can classify wetlands, we must define them. What is it that makes wetlands unique? Although there are many definitions of wetlands, the common element is water. Wetlands have unique hydrologic characteristics that usually result in soils and plant and animal communities that have special adaptations to live in or use the wetland environment. Wetlands are often transitional between aquatic and upland environments (Cowardin et al. 1979). As such, definitions based on hydrology are ultimately arbitrary. This is in contrast to classification of organisms, which has a firm phylogenetic base. For example, plant cells have cell walls and animal cells do not. Eukaryotic cells have nuclei whereas prokaryotes do not.

A thorough history of the development of wetland definitions was developed by the Committee on Wetland Characterization (National Research Council 1995). Two of the most widely used wetland definitions include the Clean Water Act Definition:

> Those areas that are inundated or saturated by surface or groundwater at a frequency and duration sufficient to support, and that under normal circumstances do support, a prevalence of vegetation typically adapted for life in saturated soil conditions. Wetlands generally include swamps, marshes, bogs, and similar areas (40 CFR 230.3).

and the Cowardin Classification Definition:

> Wetlands are lands transitional between terrestrial and aquatic systems where the water table is usually at or near the surface or the land is covered by shallow water. For purposes of this classification, wetlands must have one or more of the following three attributes: (1) at least periodically, the land supports predominantly hydrophytes; (2) the substrate is predominantly undrained hydric soil; and (3) the substrate is nonsoil and is saturated with water or covered by shallow water at some time during the growing season of each year (Cowardin et al. 1979).

EARLY CLASSIFICATION AND INVENTORY SYSTEMS

The first comprehensive wetland inventory was taken in 1906, when Congress directed the Department of Agriculture to gather information about the amount and location of wetlands that could be reclaimed for agriculture (Stegman 1976). This survey used a questionnaire sent to one or more persons in each county to supplement or verify existing data. It included only lands east of the 115th meridian (approximately the eastern boundary of the Pacific Time Zone) and excluded tidal wetlands. The classification was broken down into four categories related to crop production potential based on degree of flooding or ponding.

The second wetland inventory with a national scope was conducted in 1922 by the U.S. Department of Agriculture (USDA; Stegman 1976). It was based on data furnished by soil survey reports of the Bureau of Public Roads, topographic maps of the U.S. Geological Survey (USGS), various state reports, and the 1920 census of drainage. This survey reported acreage for tidal marsh, inland marsh, swamp and timbered overflow lands, and very deep peat. This was the most complete nationwide survey of wetlands ever conducted and, during the 1950s, was the basis for many estimates of reclaimable wetlands (Mader 1991).

The third extensive wetland inventory was conducted in 1954 by the U.S. Department of Interior Fish and Wildlife Service (USFWS) and marked a turning point in wetland inventory. Whereas previous inventories had assessed wetlands for the purpose of conversion to uplands, this inventory was conducted to facilitate wetland protection, primarily for waterfowl. The 1954 inventory developed maps that were kept in regional USFWS offices, but wetlands under 16 ha (40 acres) and wetlands outside of the major waterfowl flyways were excluded from the inventory. The classification system used for the inventory was the wetland classification of Martin et al. (1953) and is usually referred to as the Circular 39 system (Shaw and Fredine 1956). This classification included 20 wetland and deep water types in four major groupings: Inland Fresh, Inland Saline, Coastal Fresh, and Coastal Saline. Four of the 20 types included forested wetlands: Wooded Swamps, Shrub Swamps, Mangrove Swamps, and Bogs. Although the Circular 39 system was never officially adopted, it was used as a standard by other federal agencies.

Other assessments of southern forested wetlands have been compiled. Tansey and Cost (1990) used U.S. Forest Service (USFS) inventory data on dominant tree species, soil characteristics, and hydrology to estimate the extent of forested wetlands in Virginia, North Carolina, South Carolina, Georgia, and Florida. In their assessment, they displayed the geographic location of individual inventory plots that met various vegetation and hydrology criteria. McWilliams and Rosson (1990) used USFS inventory data to survey southern bottomland hardwood forests. Their assessment included all or part of Alabama, Arkansas, Louisiana, Mississippi, Oklahoma, Tennessee, and Texas. They divided bottomland hardwood forests into nine cover types and displayed their geographic distribution across the seven-state region. Cubbage and Flather (1993) compiled forest wetland and total wetland area information from the National Wetlands Inventory (NWI) and from the Soil Conservation Service (SCS) National Resource Inventory. They tabulated forested wetland area by cover type and sub-region. The NWI compiled information on wetland area in the 1980s for all wetlands in Alabama, Arkansas, Florida, Georgia, Kentucky, Louisiana, Mississippi, North Carolina, South Carolina, and Tennessee (Hefner et al. 1994). The assessment listed wetland area in all Cowardin et al. (1979) classes in the mid 1980s and changes since the 1970s. Keeland et al. (1995) assessed the status of southern forested wetlands and extent and causes of losses as part of a national assessment of ecosystems. For further information on the historical development of wetland classification systems, see Mader (1991) and Mitsch and Gosselink (1993).

CONVENTION ON WETLANDS OF INTERNATIONAL IMPORTANCE

There is no single, internationally accepted wetland classification system. The most prominent international treatment of wetlands is the Convention on Wetlands of International Importance, often referred to as the Ramsar Convention. This international treaty was signed in Ramsar, Iran, in 1971 (Navid 1988) and is dedicated to wetland conservation. Participating countries must designate at least one wetland for inclusion on the List of Wetlands of International Importance, consider responsibilities regarding migratory wildfowl, establish wetland nature reserves, and convene conferences on the conservation of wetlands and waterfowl. More than 60 nations participate in the Ramsar Convention. The United States ratified the treaty in 1986. Southern wetlands designated as Ramsar wetlands include the Great Dismal

Swamp in Virginia and North Carolina, Caddo Lake in Texas, Catahoula Lake in Louisiana, the Okefenokee Swamp in Georgia, and the Everglades National Park and Loxahatchie National Wildlife Refuge in Florida (Dugan 1993).

FEDERAL CLASSIFICATION AND INVENTORY SYSTEMS

NATIONAL WETLANDS INVENTORY

The NWI conducted by the USFWS is the fourth major inventory of the nation's wetlands. The NWI was undertaken to provide a more comprehensive wetland inventory than the 1954 Circular 39 inventory, which focused primarily on wetlands for waterfowl. A new system of classification, "Classification of Wetlands and Deepwater Habitats of the United States," was created for the NWI (Cowardin et al. 1979).

Legal Authority

Initially, justification for the NWI was based on broad interpretations of legislation. For example, the Fish and Wildlife Act of 1956 required the agency to "Conduct ... investigations, prepare and disseminate information" and report to the public, the President, and the Congress the "availability and abundance and the biological requirements of the fish and wildlife resources." As the project progressed, however, congressional legislation and funding mandated specific products. For example, the Clean Water Act of 1977 (33 U.S.C. 1288) required the Secretary of the Interior to "complete the National Wetlands Inventory of the United States ... and to provide information ... to states." The strongest directives came from the Emergency Wetland Resources Act of 1986 (P.L. 99-645): to complete NWI mapping of the 48 conterminous states by September 30, 1998.

Classification

Before beginning the NWI in the 1970s, the USFWS, along with other federal agencies, reviewed all existing inventories and classification systems. A new classification with a national scope was deemed necessary. The new classification system (Cowardin et al. 1979) was developed by a team of wetland ecologists with the assistance of local, state, and federal agencies, along with private groups and individuals. After extensive field testing and four major revisions, the classification was officially adopted by the USFWS in 1979 (Hefner et al. 1994). The Federal Geographic Data Committee has adopted the Cowardin classification as the federal standard for wetland data (Federal Geographic Data Committee 1994). Lists of wetland plants (Reed 1988) and hydric soils (SCS 1991) have been developed in support of the FWS's definition.

The Cowardin classification system is hierarchical and open-ended. It was developed to be applicable to remote sensing methods, as well as for use with ground-based information. The classification excludes uplands but classifies both terrestrial and aquatic habitats. At the most general level, wetlands and deepwater habitats are separated into five Systems: Marine, Estuarine, Riverine, Lacustrine, and Palustrine.

Each System groups wetlands and deepwater habitats according to hydrologic, geomorphic, chemical, and biological similarities. Forested wetlands are included in the Palustrine and Estuarine Systems.

At the next level of the hierarchy, Subsystems divide the Systems on the basis of water depth and other hydrologic characteristics. Next are Classes, followed by Subclasses. The 11 Classes are based on either vegetative life form or substrate and flooding regime. Classes describing vegetated wetlands include Aquatic Bed, Moss-Lichen Wetland, Emergent Wetland, Scrub–shrub Wetland, and Forested Wetland. Classes describing non-vegetated wetlands include Rock Bottom, Unconsolidated Bottom, Unconsolidated Shore, Rocky Shore, Streambed, and Reef. Subclasses provide additional life form detail (e.g., needle-leaved evergreen) or substrate information (e.g., sand). The Classes and Subclasses are easily recognized and can normally be identified using remote sensing techniques. The most detailed level of the classification is Dominance Type, named for the dominant plant species in vegetated wetlands or the predominant sedentary or sessile macroinvertebrate species in non-vegetated wetlands. At this level, the classification is open-ended, and Dominance Types can be identified and named as required. Dominance Types are not included on NWI maps, however.

As an example of the Cowardin classification, a partially drained slash pine plantation might be identified on an NWI map as PFO4Ad. This corresponds to:

SYSTEM: Palustrine (P)
SUBSYSTEM: None
CLASS: Forested (FO)
SUBCLASS: Needle-leaved Evergreen (4)
WATER REGIME: Temporarily flooded (A); partially drained (d)

The classification includes modifiers which describe hydrology, water chemistry, soil type, and the impact of beavers or man. Modifiers can be applied at the class, subclass, and dominance type levels. In the pine plantation example, the "temporarily flooded" water regime modifier (A) is added to indicate that the water is present for brief periods during the growing season, and the partially drained modifier (d) is added to show that the normal water level has been artificially lowered (but the stand remains a wetland).

The Cowardin classification system stresses the habitat function and life form as opposed to other factors such as hydrology or landform. An oxbow lake associated with a river floodplain, for example, is classified in the Lacustrine System because it is normally cut off from the main channel of the river. Trees are classified as shrubs when their height is less than 6 m (20 ft). Over time, a wetland forest regenerating following clearcut timber harvesting would change from the Emergent Subclass to the Scrub–shrub Subclass to the Forested Subclass as it progressed from seedling to maturity. For most forested wetlands in the South, there is no way to differentiate between geologic or landscape positions within a forest type using the Cowardin system. Thus an Atlantic white-cedar (*Chamaecyparis thyoides*) forest in the panhandle of Florida, a Carolina Bay in South Carolina, or a pocosin in North Carolina could all be classified PFO4B on NWI maps, corresponding to:

SYSTEM: Palustrine
SUBSYSTEM: None
CLASS: Forested
SUBCLASS: Needle-leaved evergreen
WATER REGIME: Saturated.

Mapping, Inventory, and Trend Analysis Strategies

The main goal of the NWI is to identify wetlands with a broad focus to include as many wetlands as possible without regard to wetland quality, value, or function. Cartographic delineation using remote sensing with limited field verification was deemed the best method to meet this ambitious goal, realizing that some types of wetlands are very difficult to recognize. The primary NWI product is map overlays registered to USGS 7.5′ quadrangle maps (1:24,000 scale) showing wetlands as polygons labeled according to the Cowardin classification system.

The NWI also conducts studies at 10-year intervals to estimate gains and losses of wetland area nationwide for use in wetlands policy. Similar to the wetland mapping program, these studies are conducted using remote sensing techniques on a permanent set of randomly selected plots located on USGS 7.5′ maps. Sets of aerial photography at 10-year intervals covering the plots are analyzed to detect changes in wetlands. The first study was completed in the early 1980s covering dates of photography for the 20-year interval from the 1950s to the 1970s (Frayer et al. 1983, Tiner 1984). The latest report covers the period from the 1970s to the 1980s (Dahl and Johnson 1991).

Wetland Mapping

Aerial photography was the first remote sensing platform used to test the NWI mapping process and remains the method of choice. Various scales and emulsions of aerial photography have been used. Color-infrared photography at a scale of 1:58,000 or 1:40,000 taken in late winter under normal water conditions is the best for mapping wetlands for most regions of the country. Most of the South has been mapped with 1:58,000 scale USGS National High Altitude Program (NHAP) color-infrared photography taken in the 1980s. Since the demise of the NHAP, 1:40,000 scale photos from the National Aerial Photography Program (NAPP) have been used. Portions of South Carolina, Virginia, and Louisiana have been and are presently being mapped using the 1990s NAPP photography. The change in scale from the smaller 1:58,000 to the larger 1:40,000 has significantly increased the cost of mapping but has improved map accuracy and detail.

The NWI periodically investigates the use of satellite imagery for monitoring wetland changes, updating NWI maps, and for producing maps in unmapped areas (Wilen and Pywell 1992). At present, satellite imagery is being used by some public agencies to update NWI maps; however, aerial photographs are the preferred tool for wetland mapping (Federal Geographic Data Committee 1992).

Skilled photointerpreters use stereoscopic interpretation and detailed guidance (mapping conventions) developed by the NWI (USFWS 1995) to identify and delineate wetlands on mylar overlays. Viewing the images in stereo provides a three-dimensional image that enables the interpreters to distinguish vegetation heights and to discern topographic relief that facilitates classification of wetland types and delineation of wetland boundaries. Colors, textures, tones, and relief visible on the photo, along with collateral data such as soil surveys and USGS quad maps, are used by photointerpreters to recognize characteristic wetland signatures. Field checks and quality control reviews are conducted at specific intervals throughout the interpretation process. Generally, the minimum mapping unit for wetlands in the South is 1 to 2 ha (2.5 to 5 acres) for polygons and around 10 m (33 ft) wide for linear features. Linear features less than about 0.5 km (0.3 mi) long are not mapped. However, smaller polygons less than 1 ha (2.5 acres) may occur occasionally on maps, but they are the exception and usually result from very distinct features on the aerial photographs.

The maps are prepared by transferring the line work on the photo overlay to an overlay of a 7.5′ USGS quad map using a Bausch & Lomb Zoom Transfer Scope. Recognizable features on the photo are aligned with corresponding features on the quad map, and the wetland boundaries and labels are inked on the map overlay. The NWI cartographic conventions (USFWS 1994a) include detailed guidance on the transfer process and numerous quality control steps to be carried out to check for proper alignment, drafting quality, legibility, and to make sure all wetland polygons were transferred from the photo overlay to the map overlay. The first iteration of each map is called a "draft map" which is distributed as an NWI product subject to review. Comments from map users are noted along with field review of the draft, and a final version of the map is produced.

The final manuscript map is converted to digital data following the digitizing conventions for the NWI (USFWS 1994b). Once digitized, maps may be digitally manipulated to produce an array of reports containing wetland acreage by System, Subsystem, Class, etc. For example, a study in South Carolina (Marshall 1993) focused on the Special Modifiers category to examine the percentage of wetlands that had been altered by man in the Edisto River Basin.

A noteworthy strength of the NWI is the accessibility of its products. Maps are routinely distributed to the U.S. Army Corps of Engineers, the U.S. Environmental Protection Agency, and the USDA Natural Resources Conservation Service (NRCS), as well as state agencies which have cooperated in their preparation. The FWS alone distributes about 150,000 copies of NWI maps annually throughout the 50 states. Maps are also available for purchase from USGS Earth Science Information Centers. The NWI has made available microfiche copies of NWI maps to the Library of Congress, as well as numerous regional and university libraries. For those NWI maps that have been digitized, direct transfer of the electronic files is available through the Internet. Digital maps are available for all of Florida and Virginia, most of North Carolina, and portions of coastal Louisiana, Georgia, and South Carolina. Since 1994, more than 150,000 maps have been transferred over the Internet.

Forested Wetland Areas in the South

Based on the most recent data from the NWI status and trends plots, Georgia had the greatest area of Palustrine Forested (PFO) wetlands among the southern states,

TABLE 1.1
Area (thousands of hectares) of National Wetlands Inventory forested wetlands in the 1970s and 1980s (Dahl 1996).

	Palustrine Forested		Palustrine Scrub–Shrub		Palustrine Forested plus Palustrine Scrub–Shrub	
State	mid 1980s	70s to 80s Change	mid 1980s	70s to 80s Change	mid 1980s	70s to 80s Change
Alabama	905.8	–39.4	76.2	10.5	982.0	–28.9
		(16.9)[1]	(24.8)	(18.9)	(63.7)	
Arkansas	1125.8	–85.1	77.4	–1.1	1203.2	–86.2
		(10.5)	(22.9)	(17.2)	0[2]	
Florida	2205.9	–74.5	470.5	1.7	2676.4	–72.8
		(11.8)	(19.7)	(11.8)	0[2]	
Georgia	2452.2	–184.0	267.7	61.9	2719.9	–122.0
		(11.2)	(19.5)	(11.5)	(29.4)	
Kentucky	110.9	–4.0	4.9	0.7	115.8	–3.2
		(58.0)	(42.1)	(65.2)	0[2]	
Louisiana	1973.3	–254.1	153.8	15.5	2127.1	–238.6
		(10.2)	(15.7)	(9.7)	(41.3)	
Mississippi	1487.3	–147.9	108.0	7.2	1595.3	–140.7
		(26.0)	(33.8)	(29.6)	0[2]	
North Carolina	1369.6	–399.0	522.5	–83.1	1892.1	–482.0
		(24.3)	(17.5)	(23.9)	(55.4)	
South Carolina	1456.4	–50.5	149.3	5.4	1605.7	–45.1
		(26.0)	(23.0)	(23.8)	0[2]	
Tennessee	205.2	–16.4	12.8	2.8	218.0	–13.6
		(29.4)	(49.0)	(36.0)	(92.2)	
Texas	972.5	–24.5	217.8	4.9	1190.3	–19.6
		(28.1)	(81.2)	(47.8)	0[2]	
Virginia	276.3	–7.9	45.4	2.8	321.7	–5.1
		(62.7)	(47.3)	(50.0)	0[2]	
Total	14,541.1	–1287.3	2106.2	29.4	16,647.5	–1257.9

[1] Numbers in parentheses are standard errors as a percent of the estimate, i.e., 100*((std. dev./n$^{1/2}$)/estimate).
[2] Standard error greater than estimate.

followed closely by Florida (Table 1.1). When all wetland types are considered, Florida had the greatest area in the 1980s, followed by Louisiana and then Georgia (Hefner et al. 1994). Kentucky, Tennessee, and Virginia were similar in that they had much less forested wetlands than the other southern states. While most forested wetlands fit the PFO Cowardin class, forests regenerating following clearcut harvesting, and young forests after restoration from non-forest land uses that are less than 6 m (20 ft) tall, are classified in the Palustrine Scrub–shrub (PSS) class by the Cowardin system. For this reason, Table 1.1 also lists PSS area. North Carolina had the greatest area of PSS, with Florida second (Table 1.1).

All southern states experienced loss of PFO wetlands during the decade ending in the mid 1980s. North Carolina experienced greater PFO loss rates than other states. In contrast, ten of the 12 states gained PSS area during the same period. It is likely that some declines in PFO area are caused by a change in classification between two Cowardin classes, rather than conversion of PFO wetlands to non-wetlands. Other taxonomic changes can also occur following disturbance, such as PFO to Palustrine Emergent.

The National Resources Inventory

The National Resources Inventory (NRI) is an inventory of natural resource conditions on non-federal lands in the United States. About 78% of the United States is in nonfederal ownership consisting of about 4% urban and 74% rural land. The NRI is conducted by the NRCS, formerly known as the SCS. The purpose of the NRI is to provide information that can be used for effectively formulating policy and developing natural resource conservation programs at both national and state levels. Congress has mandated that the NRI be conducted at 5-year intervals through the Rural Development Act of 1972, the Soil and Water Resources Conservation Act of 1977, and other supporting acts. The NRI is not exclusively a wetland inventory. It is a multiple resources inventory with information on land cover and use, soil erosion, prime farmland, wetlands, and other characteristics. Wetlands data were collected in 1982 and 1992 using an abbreviated form of the Cowardin et al. (1979) classification system.

The inventory is based upon a stratified clustered sample of 307,468 randomly located primary sample units (PSUs). The objective in determining sample sizes for the 1992 NRI was to have a coefficient of variation of less than 10% for any estimate of surface area with a particular resource condition on areas that constitute at least 10% of the surface area in the multi-county region defined by a Major Land Resource Area (MLRA) (SCS 1981, Nusser and Goebel, in press). MLRAs are physiographic regions defined by common soil associations, climatic factors, and major crops produced. There are about 176 MLRAs within the conterminous United States. PSUs are stratified geographically such that sampling intensity is greater where physiographic conditions were more heterogeneous. Typical PSUs are square parcels 65 ha (160 acres) in area, but range from 16 to 259 ha (40 to 640 acres). Within each PSU are one, two, or most often three random points where detailed resource data are collected. There are more than 800,000 such points in the inventory. Data are collected by making field visits and using aerial photographs, soil survey maps, NWI maps, and other ancillary materials. Wetland data from the NRI can be analyzed as simple totals of various categories or they can be combined into complex queries such as the extent of palustrine forested wetlands with a forest cover type of oak-gum-cypress.

Southern Wooded Wetlands Area from the 1992 National Resources Inventory

The 1992 NRI sampled 343,518 random points in the 12 southern states. Wooded wetlands are those in the Forested and Scrub–Shrub Cowardin Classes. The South

TABLE 1.2
Extent of palustrine wetlands (thousands of hectares) on non-Federal land[1] in 12 southern states in 1992 and changes since 1982 based on USDA National Resources Inventory data.

State	1992 Total Palustrine Wetlands (PW[2])	1992 PW Wetlands	PW Lost Since 1982	PW Gained Since 1982	PW Net Change 1982 to 1992
Alabama	1,494.7 ± 77.7[3]	1,235.5 ± 72.1	10.1 ± 5.1	1.1 ± 0.7	–8.9 ± 5.2
Arkansas	1,212.7 ± 89.5	1,003.5 ± 75.7	1.3 ± 0.7	0	–1.3 ± 0.7
Florida	3,162.8 ± 134.7	1,922.7 ± 99.2	44.2 ± 11.4	0	–44.2 ± 11.4
Georgia	2,614.2 ± 88.0	2,329.1 ± 85.1	31.9 ± 6.2	4	–31.0 ± 6.3
Kentucky	176.9 ± 21.9	79.5 ± 14.4	4	4	4
Louisiana	3,252.2 ± 87.8	2,129.2 ± 74.2	23.9 ± 5.3	2.8 ± 2.7	–21.1 ± 5.9
Mississippi	1,712.8 ± 74.7	1,370.3 ± 69.6	12.2 ± 4.9	24.5 ± 13.3	4
North Carolina	1,886.7 ± 79.1	1,812.6 ± 78.0	47.1 ± 11.1	9.1 ± 7.1	–38.0 ± 13.5
South Carolina	1,306.5 ± 59.2	1,238.2 ± 57.9	13.2 ± 3.9	0	–13.2 ± 3.9
Tennessee	270.4 ± 27.5	183.1 ± 22.4	2.3 ± 1.7	4	4
Texas	1,842.8 ± 95.6	957.2 ± 69.4	11.9 ± 4.4	4	–11.0 ± 4.6
Virginia	631.1 ± 39.7	484.7 ± 34.6	15.6 ± 4.0	4	–15.5 ± 4.0
Total	19,563.9 ± 273.8	14,745.6 ± 233.5	214.0 ± 20.6	40.7 ± 15.5	–173.3 ± 25.9

[1] Wetlands that went into or came out of federal ownership during the decade were not included in the lost, gained, or net change categories.

[2] Palustrine wooded wetlands = Palustrine Forested Wetlands plus Palustrine Scrub–Shrub Wetlands.

[3] 95% confidence interval.

[4] The confidence interval includes 0, therefore it cannot be concluded that there was a change in this category.

has approximately 20 million ha (49.4 million acres) of palustrine wooded wetlands in non-federal ownership (Table 1.2). The region has 48% of the total wetlands and about 81% of the estuarine wetlands on the non-federal lands in the conterminous United States. About 54% of the wooded palustrine wetlands in non-federal ownership in the conterminous United States occur in the 12-state region. Although not inventoried by the NRI, there are substantial federal holdings of wetlands in the region, for example, the Ramsar wetlands mentioned earlier.

More than 75% of the wetlands in the South are wooded. Wooded wetlands in North and South Carolina made up more than 90% of the palustrine wetlands in each of those states, whereas more than 80% of the palustrine wetlands in Georgia, Arkansas, Alabama, and Mississippi were wooded. Kentucky was the only state in the region where wooded wetlands made up less than half of the palustrine wetlands (45%).

In contrast, only about 3.3% of the estuarine wetlands in the region were wooded. More than 96% of the wooded estuarine wetlands in the region occurred in Florida (96,236 ± 26,548 ha [237,800 ± 65,600 acres]; 95% confidence interval), Louisiana

(13,072 ± 7,689 ha [32,300 ± 19,000 acres]), and Virginia (5,585 ± 3,642 ha [13,800 ± 9,000 acres]), while North Carolina, Alabama, Texas, South Carolina, and Mississippi combined had <4% of the region's estuarine wooded wetlands in non-federal ownership.

Oak-gum-cypress was the predominant forest cover type in palustrine wooded wetlands accounting for about 46% of the total area. Oak-pine was the next most abundant cover type representing 21% of the total, followed by loblolly-shortleaf pine (15%), oak-hickory (8.3%), longleaf-slash pine (7.1%), elm-ash-cottonwood (1.4%), and miscellaneous (1.4%). About 38% of the oak-gum-cypress forest type in the region occurs in Florida (1,258,519 ha [3,109,800 acres]) and Louisiana (1,207,892 ha [2,984,700 acres]). Forest cover types in estuarine wooded wetlands were 83% tropical hardwoods, followed by longleaf-slash pine (6.7%), oak-gum-cypress (6.0%), and other (4.1%).

Wooded Wetland Losses Based on the National Resources Inventory

About 55% of the total wetland loss in the nation from 1982 to 1992 occurred in these 12 southern states. There was a gross loss of 214,002 ha (528,800 acres) of wooded palustrine wetlands in this region during the decade from 1982 to 1992, or a loss of about 1.5%. This was the lowest recorded loss rate of wooded wetlands in recent decades. During this time there was also a gain of about 40,712 ha (100,600 acres) of wooded wetlands, making the net loss 173,290 ha (428,200 acres). Nearly all of the gain in wooded palustrine wetlands came from other palustrine system categories. Therefore, they do not represent new wetlands, but rather wetlands on which trees and shrubs have become the dominant vegetation during the last decade. About 64% of these new wooded palustrine wetlands were classified as being in pasture in 1982.

A closer look at the gross losses of palustrine wooded wetlands reveals that losses due to development accounted for 75% of the loss of wooded palustrine wetlands in these 12 southern states during the decade 1982 to 1992. About 77% of the losses due to development occurred in just five states: Florida (37,920 ± 10,441 ha [93,700 ± 25,800 acres]), Georgia (25,900 ± 5,625 [64,000 ± 13,900 acres]), North Carolina (24,848 ± 5,747 ha [61,400 ± 14,200 acres]), Louisiana (19,344 ± 5,059 ha [47,800 ± 12,501 acres]), and Virginia (15,378 ± 3,926 ha [38,000 ± 9,700]). While agriculture has historically been the principal reason for wetland loss in the United States, only about 15% of the 1982 to 1992 conversions of wooded wetlands were due to agriculture. Ninety percent of the wooded wetland to agriculture conversions occurred in just two states: North Carolina (21,570 ± 9,268 ha [53,000 ± 22,900 acres]) and Mississippi (6,880 ± 4,452 ha [17,000 ± 11,000 acres]). Certainly the interaction among agricultural economics, increased public awareness of wetland values, state and local regulations, and the "Swampbuster" provisions of the 1985 Food Security Act all contributed to the reduction in agriculture-related wetland conversions. The remaining losses were attributed to deepwater developments (8%) and to miscellaneous activities (2%).

Comparisons with other Wetland Estimates

Direct comparisons with other estimates of wetlands are difficult to make because of the dynamic nature of wetlands, differences in the time period in which estimates were made, differences in the geographic coverage of inventories, and differences in sampling and delineation protocols. The boundary between terrestrial and wetland habitats is often indistinct, especially in regions with low relief. In many cases wetlands represent the ecotone between terrestrial and aquatic habitats and typically exhibit great seasonal and annual variability in presence and depth of water. Inventories conducted during years with greater precipitation tend to report more wetlands than those conducted during years with less precipitation.

The 1992 NRI data compared very favorably to the national estimates made by Dahl and Johnson (1991). The estimates for palustrine wetlands, palustrine wooded, palustrine vegetated, and estuarine vegetated wetlands for the mid 1980s reported in Dahl and Johnson (1991) all fell within the 95% confidence interval for the 1982 estimates for those categories made by the 1992 NRI. Dahl and Johnson reported more estuarine wooded wetlands than did the NRI. Wooded estuarine wetlands represent a small portion of all estuarine wetlands, and it is probable that many are in federal ownership, and thus not inventoried by the NRI. It is important to recognize that both the NRI and the study by Dahl and Johnson were based on rigorous statistical designs. Implementation of these two inventories was slightly different, but both maintained high-quality inventory standards. The study by Dahl and Johnson utilized a stratified random sample of 3,629 sample units that were each 1,036 ha (2,560 acres) in area. Aerial photographs of the plots were sampled in the mid 1970s and again in the mid 1980s.

Forest Inventory and Analysis

The Forest Inventory and Analysis (FIA) is a USFS program which was initiated in 1928. The FIA is composed of five regional divisions, three being in the eastern United States and two in the western United States. These FIA regional divisions are concerned with resource issues, problems, and questions at the state and local level. At the national level, FIA is concerned with resource evaluations and assessments at the state, regional and national levels. Data and information needs are similar at all levels, so it is primarily a matter of aggregating data to the appropriate scale needed for a particular analysis.

The Southern Research Station's FIA unit is responsible for evaluating all renewable forest and rangeland resources in 13 southern states (includes Oklahoma in addition to the 12 states considered in this chapter). The evaluations require comprehensive state-by-state multi-purpose resource inventories of land use, timber, wildlife, range, recreation, water, and soils on a cycle of seven to ten years. Data presented in this chapter are taken from the latest state inventories: Alabama 1990, Arkansas 1988, Florida 1987, Georgia 1989, Kentucky 1988, Louisiana 1991, Mississippi 1994, North Carolina 1990, Oklahoma 1993, South Carolina 1993, Tennessee 1989, Texas 1992, and Virginia 1992.

FIA data do not include marshland, inland water, or agricultural and rangelands which may qualify as wetland. Only wet timberland area is included. Timberland, according to FIA definition, is land in tracts of at least 0.4 ha (1 acre) and sufficiently stocked by forest trees of any size (or formally having had such tree cover), not currently developed for non-forest uses, capable of producing a volume of industrial wood in excess of 1.4 m^3/ha/yr (20.0 ft^3/ac/yr), and not set aside from commercial timber production.

FIA has approximately 70,000 ground sample locations placed in both a systematic and systematic random fashion across all land uses in the South. Sample plots are proportionately distributed across all land uses to provide a measure of land use change. Approximately 44,000 plots are in timberland. At each sample location five to ten sub-plots are installed. Trees ≥ 12.7 cm (5 in) in diameter at breast height (DBH) and located within a radius determined using a basal-area factor of 8.6 m^2/ha (37.5 ft^2/acre) are selected for measurement. Trees less than 12.7 cm but ≥ 2.5 cm (1 in) DBH are tallied within a fixed radius of 2.1 m (6.9 ft) at each of the sub-plots.

The plot variables most frequently used to estimate wet timberland are forest cover type and physiographic class (Tansey and Cost 1990). Since consistent physiographic class classifications were unavailable across all 12 states, wet timberland was defined by forest cover type alone. At each sample location, a forest type classification is assigned based on the plurality of stocking of live dominant and codominant tree species. For purposes of this chapter, wet timberland is defined as the oak-gum-cypress, elm-ash-cottonwood, and pond pine forest types (Eyre 1980). From the inventories, plots dominated singly or in combination by tupelo (*Nyssa* spp.), sweetgum (*Liquidambar styraciflua*), wetland oaks (*Quercus* spp.), or cypress (*Taxodium distichum* and *Taxodium distichum* var. *nutans*) are classified as oak-gum-cypress. Plots dominated singly or in combination by elm (*Ulmus* spp.), ash (*Fraxinus* spp.), or eastern cottonwood (*Populus deltoides*) are classified in the elm-ash-cottonwood type. Any plots dominated singly or in combination by pond pine (*Pinus serotina*) are classified as pond pine forest types.

Extent and Ownership

Table 1.3 shows approximately 12.7 million ha (31.4 million acres) of wet timberland in the 12 southern states. This represents about 16% of the 78.5 million ha (194 million acres) of total timberland in the region. Oak-gum-cypress is the predominant wet timberland forest type in the region, accounting for 90% of the region's wet timberland area. Louisiana has the largest area of wet timberland, 1.9 million ha (4.7 million acres), with Florida second at 1.8 million ha (4.4 million acres). The wet timberland forests of the South are found largely on the Coastal Plain, but forested wetlands are also abundant in river terraces and stream margins in the Piedmont Plateau. Farmers and other private owners control about 8.6 million ha (21.2 million acres), or 68% of the wet timberland. Nearly 22% is owned or leased by forest industry. Public ownership is relatively small, only 10% of the wet timberland.

TABLE 1.3
Area of timberland (thousands of hectares) in oak-gum-cypress, elm-ash-cottonwood and pond pine forest types in the South, by state based on USFS forest inventory and analysis data.

State	All Classes	Oak-Gum Cypress	Elm-Ash-Cottonwood	Pond Pine
Alabama	918	912	7	0
Arkansas	1,107	1,043	64	0
Florida	1,826	1,729	34	64
Georgia	1,482	1,313	129	40
Kentucky	249	73	176	0
Louisiana	1,918	1,758	160	0
Mississippi	1,502	1,441	61	0
North Carolina	1,330	1,008	71	251
South Carolina	1,064	967	40	56
Tennessee	276	259	17	0
Texas	732	706	26	0
Virginia	258	159	98	1
Total	12,664	11,367	884	412

Volume of Growing Stock

The 12.7 million ha (31.4 million acres) of wet timberland contain 1.4 billion m^3 (49.4 billion ft^3) of growing stock (Table 1.4). This is 112 m^3/ha (1,600 ft^3/acre) averaged across the region. Of the 1.4 billion m^3, 54% is soft hardwood, 27% is hard hardwood, 13% is other softwood, and 6% is pine. The predominant species group, soft hardwood, is comprised primarily of blackgum (*Nyssa sylvatica* and *Nyssa sylvatica* var. *biflora*), water tupelo, sweetgum, and yellow-poplar (*Liriodendron tulipifera*). Louisiana and Florida have the largest total volume of growing stock, with 207 and 206 million m^3 (7,309 and 7,274 million ft^3), respectively. Georgia has the largest volume of soft hardwood growing stock, with 115 million m^3 (4,061 million ft^3). Florida accounted for 40% of other softwood growing stock volume, which is primarily cypress.

Net Growth and Removals

Net annual growth of growing stock is annual timber volume growth less the volume loss to mortality and cull volume. Table 1.5 shows that net annual growth of growing stock across the 12 southern states totaled about 39 million m^3 (1,377 million ft^3), which translates to an annual growth rate of 2.7%. On an area basis, net growth across all states averaged 3.0 m^3/ha/yr (42.9 ft^3/ac/yr). Eighty-four percent of all net annual growth was in hardwood growing stock. More than one-half of the total softwood growing stock growth occurred in North Carolina and Florida, where pond pine is most abundant. Alabama and North Carolina had the highest annual growing stock growth rates, at about 4.0 m^3/ha/yr (57.2 ft^3/ac/yr). South Carolina had the

TABLE 1.4
Volume of growing stock (million cubic meters) on oak-gum-cypress, elm-ash-cottonwood and pond pine forest types in the South, by state and species group based on USFS forest inventory and analysis data.

State	All Species	Pine	Other Softwood	Soft Hardwood	Hard Hardwood
Alabama	100.2	4.4	4.6	62.5	28.4
Arkansas	109.8	2.3	5.3	52.3	49.8
Florida	206.1	12.2	72.7	83.2	38.0
Georgia	185.4	12.4	20.8	114.5	37.8
Kentucky	23.6	0.0	0.3	18.8	4.4
Louisiana	207.1	4.4	41.2	101.9	59.6
Mississippi	147.2	6.3	6.0	77.3	57.6
North Carolina	176.7	25.5	13.6	106.5	31.0
South Carolina	142.3	8.7	12.3	86.4	34.9
Tennessee	31.0	0.0	2.3	20.1	8.6
Texas	57.5	3.5	3.0	24.1	26.9
Virginia	31.8	0.9	1.5	18.7	10.7
Total	1,418.8	80.6	183.8	766.5	387.7

lowest net growth rates; however, growth figures for South Carolina were based on 1992 and 1993 data and reflect damage from Hurricane Hugo in 1989.

Estimates of timber removals include timber cut during silvicultural operations and timber removed when clearing land during change to a non-forest land use. These estimates are derived from the remeasurement of permanent samples across the South and depict the situation for roughly the preceding decade.

Table 1.5 shows that annual timber removals of growing stock from wet timberland totaled 33 million m^3 (1,165 million ft^3). Sixty-five percent of all annual removals were hardwoods. For this latest round of inventories, net growth exceeded removals by 16%. In all states except South Carolina, hardwood growth exceeded hardwood removals by a fairly large margin. For softwoods, removals exceeded net growth by 5.25 million m^3 (185.4 million ft^3), which is 2% of the total softwood growing stock in wet timberland.

Jurisdictional Wetlands

For many landowners, there is another system for identifying wetlands that is far more important than the ones mentioned thus far. This is the federal Clean Water Act, which has regulated activities in wetlands since 1977 (Want 1996). Its classification system is quite simple and contains only two taxa: wetlands and non-wetlands. Wetlands identified for regulation by the Clean Water Act are often referred to as "jurisdictional wetlands." Delineation, the identification of the line separating jurisdictional wetlands from non-wetlands, has had a turbulent history. The Clean

TABLE 1.5
Net annual growth and removals (million cubic meters) of growing stock on oak-gum-cypress, elm-ash-cottonwood and pond pine forest types in the south, by state and species group based on USFS forest inventory and analysis data.

State	Growth			Removals		
	All Species	Softwood	Hardwood	All Species	Softwood	Hardwood
Alabama	3.64	0.28	3.36	2.02	0.35	1.67
Arkansas	3.25	0.19	3.06	1.73	0.23	1.49
Florida	4.82	2.09	2.73	3.74	2.63	1.11
Georgia	4.92	0.88	4.05	5.91	3.15	2.75
Kentucky	0.81	0.01	0.79	0.17	—	0.17
Louisiana	5.62	0.88	4.75	3.50	0.65	2.85
Mississippi	5.21	0.50	4.71	4.88	0.93	3.96
North Carolina	5.16	1.19	3.97	4.00	1.37	2.63
South Carolina	1.27	–0.17	1.44	4.30	1.69	2.61
Tennessee	0.92	0.03	0.89	0.33	0.00	0.32
Texas	2.08	0.32	1.76	1.28	0.45	0.83
Virginia	0.90	0.06	0.84	0.82	0.05	0.78
Total	38.59	6.25	32.34	32.66	11.50	21.16

Water Act definition of wetlands has not changed since written by Congress in 1977. However, guidance on how to distinguish between jurisdictional wetlands and non-wetlands has been revised several times by the agencies that administer the regulatory program, i.e., the Environmental Protection Agency and the Army Corps of Engineers. For a detailed history of the development of regulatory guidance for delineating jurisdictional wetlands, see National Research Council (1995).

Since 1991, the 1987 Corps of Engineers Wetlands Delineation Manual (USACE 1987) has been used for determination of jurisdictional wetlands. The 1987 manual requires that at least one positive wetland indicator be present for each of three criteria: hydric soils, hydrophytic vegetation, and wetland hydrology. NRCS, in conjunction with the National Technical Committee for Hydric Soils, is responsible for the definition and criteria of hydric soils and maintains a list of hydric soils of the United States (SCS 1991). Vegetation has been separated into five categories based on a species' estimated frequency of occurrence in wetlands (Reed 1988): Obligate Wetland (>99% probability), Facultative Wetland (67%-99%), Facultative (34%-66%), Facultative Upland (1%-33%), and Upland (<1%). To satisfy the vegetation criterion, the 1987 manual requires either that >50% of the dominant plant species are Obligate Wetland, Facultative Wetland, or Facultative; or that other indicators are present, such as visual observation of plant species growing in areas of prolonged inundation, technical literature, or morphological, physiological, or reproductive adaptations (National Research Council 1995). The 1987 manual's hydrology criterion requires inundation or saturation in the root zone (upper 30.5 cm [12 in]) long enough to create hydric soils and promote the growth and reproduction of hydrophytic vegetation. The manual includes guidance that areas inundated or saturated consecutively for less than 5% of the growing season are rarely wetlands, whereas areas inundated or saturated for more than 12.5% of the growing season are almost always wetlands. Areas that are inundated or saturated between 5 and 12.5% of the growing season are problematic and require greater reliance on soil and vegetative characteristics.

There is no inventory associated with the Clean Water Act jurisdictional wetlands. Wetlands are delineated on a site-specific basis, usually initiated by a landowner seeking a permit to conduct regulated activities in a wetland. Although the NWI is not designed to map jurisdictional wetlands, some comparisons have been made. Stolt and Baker (1995) compared NWI and jurisdictional wetlands in four 1,722 ha (4,255 acres) study areas in the Blue Ridge mountains of Virginia. They found that 91% of the wetlands mapped by NWI were deemed jurisdictional wetlands using the 1987 manual (USACE 1987). However, NWI mapped only a small percentage (0 to 16%) of the jurisdictional wetlands present. This discrepancy is likely due to wetlands in this area often being small or linear. Some types of forested wetlands are particularly difficult to identify from remotely sensed data because foliage obscures the ground (Tiner 1990, National Research Council 1995). This is especially true for wetlands in the drier end of the continuum.

The Clean Water Act requires a permit from the Corps of Engineers before activities involving a discharge of dredged or fill material can be conducted. Some activities, including agriculture, silviculture, and ranching are exempt from permit responsibilities. Permits are not required for normal, ongoing silvicultural operations such as disking, bedding, planting, minor drainage, and harvesting (33 CFR 323.4). Road building in wetlands as part of an ongoing silvicultural operation does not require a permit as long as 15 specific Best Management Practices (BMPs) are followed (USACE 1986). The Corps of Engineers and EPA have recently issued regulatory guidance clarifying the exemption when mechanical site preparation is used prior to planting pine (USACE 1996a). A permit is required for using mechanical site preparation to establish pine plantations in wetlands characterized by prolonged inundation such as: Cypress-Gum Swamps, Muck and Peat Swamps, and Cypress Strands and Domes. Wetland forests that are seasonally flooded, intermittently flooded, temporarily flooded (using Cowardin water regimes), and existing pine plantations do not require a permit for mechanical site preparation to establish pine.

Hydrogeomorphic Classification

A new federal wetland classification system is currently under development (Brinson 1993b, USACE 1996b). The "Hydrogeomorphic Classification for Wetlands" (HGM) is based on the premise that hydrology and geomorphology are the primary controls of wetland functions. HGM is not intended as a substitute for any existing classification systems, but rather as a means of organizing assessments of wetland functions in the Clean Water Act regulatory program. The system is hierarchical, with seven Classes at the highest level: Depressional, Slope, Mineral Soil Flats, Organic Soil Flats, Estuarine Fringe, Lacustrine Fringe, and Riverine. Lower-ordered taxa are being developed (Brinson and Rheinhardt 1996). No mapping or inventory systems are planned as part of the HGM. However, reference sites are required by the approach and are currently being developed. Reference sites are chosen to represent the range of wetland functions within a geographic region and are intended for use in comparison to sites being evaluated in the regulatory program.

STATE CLASSIFICATION AND INVENTORY SYSTEMS

Many states classify natural communities, including forested wetlands, in their natural heritage programs (Weakley et al. 1996). States often cooperate with the FWS to facilitate mapping for the NWI. Some states have digitized NWI maps for use in various state programs.

ALABAMA

A wetlands inventory is in progress in Alabama and is scheduled for completion in 1998 (Foster, pers. comm.).

ARKANSAS

Wetlands are included in a classification system of the natural vegetation of Arkansas. The system has approximately 215 community types, 75 of which are wetland (Foti et al. 1994). An inventory of the state's wetlands is also in progress and is scheduled to be complete in 2002 (Foti, pers. comm.).

FLORIDA

Florida recently published a state wetlands delineation manual used in regulatory programs by the state's five water management districts and by the Florida Department of Environmental Protection (Gilbert et al. 1996). The manual provides documentation for a system of 19 reference sites at 15 locations representing common Florida wetlands. Each reference site is located on public land and has the wetland/upland boundary marked. For each reference site, the manual includes information on the location, access, community characterization, and delineation specific to that wetland type, including descriptions and color photographs of the vegetation and soils.

Florida has also identified 26 ecological communities (Watts et al. 1993). Approximately half of these communities include forested wetlands; examples include Cypress Swamps, Swamp Hardwoods, and Wetland Hardwood Hammocks.

GEORGIA

Georgia included wetlands in its assessment of landcover using LANDSAT Thematic Mapper satellite imagery (Georgia Natural Heritage Program 1996). Landcover was separated into 15 classes, including Forested Wetlands (1,292,834 ha [3,193,300 acres]) and Scrub/Shrub Wetlands (156,938 ha [387,794 acres]). The landcover assessment is used for tax assessment purposes in conservation programs.

Kentucky

The state has cooperated with the NWI program. The NWI provided state-wide base maps, which the state Office of Information Services, Geographic Information digitized. From this information, the state estimates that there are 338,677 ha (836,871 acres) of wetlands with 32% forested (Balassa, pers. comm.). The state uses this information for water quality certification in Section 401 of the Clean Water Act and as a tool in preliminary stages of state wetland permitting.

Louisiana

NWI mapping is about 80% complete for Louisiana. Completion of the inventory is expected in 1999. Total state wetland area is estimated to be 4,486,847 ha (11,086,999 acres), with 62% forested (Hinds, pers. comm.).

Mississippi

Aside from the NWI, wetlands have been included in a state-wide landcover type survey based on 1:120,000 color infrared aerial photography. Forested wetland area in that survey was estimated as 240,000 ha (593,040 acres), 85% of all wetlands in the state. The state's landcover type inventory is currently being updated using LANDSAT Thematic Mapper satellite imagery (Belokon, pers. comm.).

North Carolina

The North Carolina Natural Heritage Program has a classification system for all natural communities in the state, including wetlands (Schafale and Weakley 1990). The lowest taxa in this classification system is the natural community. Communities are aggregated into 16 groups. Groups that include forested wetland communities in the Palustrine System are: River Floodplains, Nonalluvial Wetlands of the Mountains and Piedmont, Wet Nonalluvial Forests of the Coastal Plain, Pocosin and Peatland Communities of the Coastal Plain, Wet Savannahs of the Coastal Plain, Coastal Plain Depressions and Water Bodies, Nontidal Coastal Fringe Wetlands, and Freshwater Tidal Wetlands.

North Carolina is conducting an inventory of state wetlands that is scheduled for completion in late 1997. Wetland types that will be used in the inventory are: salt marsh/brackish marsh, freshwater marsh, ephemeral wetland, mountain bog, salt shrub wetland, pocosin, estuarine fringe forest, seep, wet flat, bog forest, headwater forest, bottomland hardwood forest, swamp forest, and transitional/disturbed site (Dorney, pers. comm.).

South Carolina

Other than the NWI, there is no state-wide wetland inventory. However, a survey of Carolina Bays based on black and white aerial photography from 29 coastal counties has been completed (Bennett and Nelson 1991). The survey found 2651 Carolina

Bays at least 0.8 ha (2.0 acres) in size. Wetlands and other cover types and land uses were addressed in detail in the Edisto River basin, an 808,000 ha (1,996,568 acres) area in southeastern South Carolina (Beasley et al. 1996). This assessment defined wetlands as the union of all wetlands mapped by the NWI and areas mapped as hydric soils in Natural Resource Conservation Service county soil surveys. Wetlands thus defined encompassed 277,000 ha (684,467 acres), about one-third of the Edisto River basin.

Tennessee

The Tennessee Wildlife Resources Agency is currently digitizing NWI maps (Eager, pers. comm.). A 1988 survey by the Tennessee Wetlands Program estimated that there were 258,671 ha (639,176 acres) of wetlands in Tennessee, forested and non-forested.

Texas

Texas is cooperating with the NWI, and the inventory of the state is complete. Total wetland area is estimated as 3,080,701 ha (7,612,412 ac; Dahl 1990) with 78% forested (Calnan, pers. comm.). The coastal area of Texas is being reinventoried to update the NWI inventory.

Virginia

There is no state-wide inventory of Virginia wetlands other than the NWI. The state cooperated with the NWI to complete wetlands mapping and a digital map database. NWI maps have been updated in several areas of the state as part of detailed wetland status and trends studies: the Chickahominy River watershed (Tiner and Foulis 1994a), the Norfolk/Hampton region (Tiner and Foulis 1994b), and selected areas of northern Virginia (Tiner and Foulis 1994c). Also, wetlands within watersheds draining into the Chesapeake Bay have received more attention in a multi-agency effort to restore and protect the bay (Tiner et al. 1994).

OTHER CLASSIFICATION SYSTEMS

Bottomland Hardwood Classification by Ecological Zones

An interdisciplinary workshop held at Lake Lanier, Georgia, in 1980 created a system for classifying bottomland hardwood wetlands into six zones (Clark and Benforado 1981). Zones were described with regard to hydrology, soil properties, and characteristic plant and animal communities. This system of classification has been used by some wetland scientists (Gosselink et al. 1990) as a useful way of organizing descriptions of wetland functions. It has little utility in the field because the six zones described are not recognizable on the ground, and patterns of zonation are more complex than the system depicts (Conner et al. 1990).

North Carolina State University Hardwood Research Cooperative Forest Site Definitions

The North Carolina State University Hardwood Research Cooperative created a simple classification of southern hardwood forests (Table 1.6) for use in forest management (Kellison et al. 1981). This system is commonly used by foresters, especially those employed by members of the Cooperative.

The Nature Conservancy

The Nature Conservancy (TNC) has an ecological classification system for the southeastern United States (Weakley et al. 1996). The system is a refinement of the previous southeastern classification system and has been made compatible with three other regional TNC systems. The new regional classification system is hierarchical based on vegetation structure at the upper levels and on floristics at lower levels. The system is complex, containing hundreds, perhaps thousands of taxa.

The Nature Conservancy classification system includes forested wetlands. For example, a southeastern cypress swamp could be classified as:

SYSTEM: Terrestrial
PHYSIOGNOMIC CLASS: Forest
PHYSIOGNOMIC SUBCLASS: Deciduous forest
FORMATION GROUP: Cold-deciduous forest
SUBGROUP: Natural/Semi-natural
FORMATION: Semipermanently flooded cold-deciduous forest
ALLIANCE: *Taxodium distichum* Semipermanently Flooded Forest Alliance
ASSOCIATION: *Taxodium distichum/Lemna minor* Forest

The classification system includes four freshwater hydrologic modifiers adapted from Cowardin et al. (1979): temporarily flooded, seasonally flooded, semipermanently flooded, and saturated.

SUMMARY

There is no consensus on one classification system for forested wetlands in the South. Classification systems are designed for specific purposes. A system that works well for one purpose may be ill-suited for another. However, this has led to a plethora of classification systems. Table 1.7 shows an approximate correspondence or "crosswalk" among several classification systems.

The Cowardin system is the most widely used classification system and has been mandated by Congress for mapping the wetlands of the entire country. Maps produced by the NWI are being used in a number of state programs in the region. Major drawbacks of the Cowardin system, however, are its technical taxonomic nomenclature and mapping codes. Few land managers or scientists routinely refer to a cypress dome as a PF06F. Wetland names such as cypress dome, pocosin, and alluvial floodplains are more commonly used.

TABLE 1.6
North Carolina State University Hardwood Research Cooperative forest site definitions (Kellison et al. 1981).

Muck Swamp	Very poorly drained area, usually with standing water. Broad expanses between tidewater and upstream runs and along blackwater rivers and branch bottom stands; also found in miniature in sloughs and old oxbows of redwater rivers and branch bottoms. Characterized by heavy accumulation of organic matter (amorphous, lacking structure). Soils range from silt loam through clay. Water tupelo (*Nyssa aquatica*) and baldcypress (*Taxodium distichum*) are common in deeply flooded areas and swamp blackgum (*Nyssa sylvatica* var. *biflora*) predominates toward the fringes.
Peat Swamp (Headwater Swamp)	Occupy broad interstream areas from which blackwater rivers and branch bottoms originate. These swamps are poorly drained, with heavy accumulations of raw organic matter. Soils resemble those of muck swamps but in general are heavier and of better site quality. Examples are: Dismal Swamp of Virginia-North Carolina; Green Swamp of North Carolina-South Carolina; Little Wambaw of South Carolina; and Okefenokee Swamp of Georgia-Florida. Swamp blackgum and red maple (*Acer rubrum*) predominate in mixture of many hardwood species; coniferous species found are loblolly (*Pinus taeda*) and pond (*Pinus serotina*) pines and Atlantic white-cedar (*Chamaecyparis thyoides*) to the north. This category includes the true pocosins of the Carolinas such as White Oak Pocosin in Hofmann Forest.
Wet Flat	Topographically similar to peat swamps and pocosins, they all lie in broad interstream areas where drainage systems are poorly developed. However, wet flats are better drained than their associates because of higher elevation. The nonalluvial soils may possess some accumulation of organic matter but fertility is superior to peat swamps and pocosins because of superior parent material. Abandoned rice fields of the southern lower Coastal Plain often fall in this category. Examples are: Jordan and J&W pocosins in North Carolina. Species generally encountered: (1) on wetter portion — sweetgum (*Liquidambar styraciflua*), red maple, water (*Quercus nigra*), swamp laurel (*Quercus laurifolia*) and willow (*Quercus phellos*) oaks, ashes (*Fraxinus* spp.), loblolly pine, elms (*Ulmus* spp.), and other species; (2) on "islands," with better drainage — cherrybark (*Quercus falcata* var. *pagodaefolia*), Shumard (*Quercus shumardii*) and swamp chestnut oaks (*Quercus michauxii*), yellow-poplar, hickories (*Carya* spp.), and occasionally American beech (*Fagus grandifolia*). (Locally, in the North Coastal Plain, wet flats are known as "pocosins," whereas in the South Coastal Plain they are called "hammocks.")
Red River Bottom	Floodplain of major drainage system originating in the Piedmont or mountains. Immediately adjacent to the drainage systems, sloughs and oxbows are commonly found; if of sufficient size, they are classified as muck swamps. Some organic matter may accumulate on the clay soils. Water tupelo predominates over cypress, red maple, swamp blackgum, swamp cottonwood (*Populus heterophylla*), swamp laurel oak, and others. Beyond the sloughs and oxbows are found first bottoms (low ridges) which flood periodically to considerable depths. However, drainage is fairly rapid because of high elevation. Soils are loam or silt loam. Species included sweetgum, green ash (*Fraxinus pennsylvanica*), water hickory (*Carya aquatica*), sycamore

TABLE 1.6 (continued)
North Carolina State University Hardwood Research Cooperative forest site definitions (Kellison et al. 1981).

Red River Bottom (continued)	(*Platanus occidentalis*), red maple, river birch (*Betula nigra*), elms, and willow, water, laurel, and overcup (*Quercus lyrata*) oaks. At still higher elevations, second bottoms and terraces (high ridges and fronts) are found. Here, flooding is infrequent or rare, and more mesophytic species of cherrybark, swamp chestnut, and white oaks, hickories, American beech, and occasionally yellow-poplar make their appearance. Examples of red river bottom are: Roanoke — Virginia, North Carolina; Santee — South Carolina; Oconee — Georgia; and Alabama River — Alabama.
Black River Bottom	Floodplain of major water system originating in Coastal Plain. Classification of minor site types and species similar to red river bottom, with exception of muck swamps being more prevalent and first and second bottoms and terraces being on a more modest scale. Examples of rivers where blackwater bottoms are found are: Blackwater — Virginia; Waccamaw — North Carolina, South Carolina; Black — South Carolina; and St. Marys — Georgia and Florida.
Branch Bottom	Relatively flat, alluvial land along minor drainage system which is subject to minor overflow. On wetter portions with heavier soils, the predominant species are willow, water and laurel oaks, swamp blackgum, sweetgum, red maple, and ash. The lighter soils of second bottoms and terraces support cherrybark, Shumard, swamp chestnut, and white (*Quercus alba*) oaks, sweetgum, hickory, yellow-poplar, and loblolly pine. Sloughs and oxbows of limited extent along the main channel support tupelo and swamp blackgum. Examples: Big Swamp — North Carolina.
Bottomland	In lower Piedmont, conditions identical to red river bottom are encountered. However, upstream, sloughs, oxbows, and first bottoms decrease in frequency and area until only well-drained bottomland (second bottom and terrace) is encountered. This type also includes the area adjacent to the tertiary streams of the red rivers. Species found here are sycamore, yellow-poplar, sweetgum, green ash, cottonwood (*Populus deltoides*), water and willow oak, loblolly pine, and others.
Coves, Gulfs and Lower Slopes Adjacent to Streams	This site type is characterized by uneroded, fertile mineral soils that are moist but well drained. It is common to upper Piedmont and mountain provinces. In the Cumberland Plateau range, valleys or hollows extending from the crest of the range to major drainage systems are known as gulfs. They are characterized by plateaus that are separated by steep terrain. Here the mixed mesophytic forest consisting of yellow-poplar, sweetgum, white ash (*Fraxinus americana*), northern red oak (*Quercus rubra*), black oak (*Quercus velutina*), black cherry (*Prunus serotina*) and black walnut (*Juglans nigra*) is at its best.
Upland Slopes and Ridges	Here the mineral soils are more shallow and less moist than the coves and lower slopes. White oak and hickory predominate, but yellow-poplar, black and scarlet oaks (*Quercus coccinea*), and blackgum are also common. Shortleaf (*Pinus echinata*) and Virginia (*Pinus virginiana*) pines are common coniferous associates.

TABLE 1.7
Approximate correspondence among wetland classification systems.

Book Chapter	National Wetlands Inventory and National Resources Inventory[1]	Forest Inventory and Analysis	Hydrogeomorphic Classification
Southern Deepwater Swamps	Palustrine Forested, Palustrine Scrub–shrub	Oak-Gum-Cypress	Riverine
Major Alluvial Floodplains	Palustrine Forested, Palustrine Scrub–shrub	Oak-Gum-Cypress Elm-Ash-Cottonwood	Riverine
Minor Alluvial Floodplains	Palustrine Forested, Palustrine Scrub–shrub	Oak-Gum-Cypress Elm-Ash-Cottonwood	Riverine
Pocosins and Carolina Bays	Palustrine Emergent, Palustrine Forested, Palustrine Scrub–shrub, Palustrine Open Water, Lacustrine Littoral Emergent, Lacustrine Littoral Open Water	Pond Pine	Depressional Organic Soil Flats
Mountain Bogs	Palustrine Emergent, Palustrine Forested, Palustrine Scrub–shrub	Not included	Depressional Organic Soil Flats
Cypress Domes	Palustrine Forested, Palustrine Scrub–shrub	Oak-Gum-Cypress	Depressional
Wet Flatwoods	Palustrine Forested, Palustrine Scrub–shrub	Oak-Gum-Cypress Pond Pine	Mineral Soil Flats Depressional
Mangroves	Estuarine Intertidal Forested, Estuarine Intertidal Scrub–shrub	Not included	Estuarine Fringe

[1] The National Resources Inventory uses an abbreviated form of the Cowardin classification used in the National Wetlands Inventory (i.e., without Cowardin subclasses and modifiers).

This chapter has briefly described the development of wetland classification and inventory systems from 1906 to the present. Although this history spans nearly a century, many improvements in wetland classification and inventory systems are still needed. Thus far, little attempt has been made to create correspondences between the taxa of different inventory or classification systems. For example, it would be useful to determine the Cowardin classification of all Forest Service FIA plots. In this way, the remotely-sensed Cowardin classification could be merged with individual tree data from ground-based measurements. Stand-level information such as tree species composition, standing volume, and growth and removals could be aggregated by Cowardin classes and landowner type. Such a merger of these two systems would take advantage of each system's strengths and yield a better understanding of the extent and geographic distribution of the different types of forest wetlands. Recent advances in Global Positioning Systems and Geographic Information Systems offer hope that such analyses can be accomplished in the near future.

Another need is a rigorous comparison between wetlands delineated for Clean Water Act regulatory purposes and wetlands mapped by the NWI. The NWI routinely conducts status and trends analyses for policy purposes, yet there have been few studies that quantitatively assess the NWI's suitability for this application. For example, a national analysis of wetland permit information revealed that 48% of the permits involved wetlands that were less than 0.4 ha (1.0 acres) in size. This is at or below NWI's minimum map unit size (Albrecht and Goode 1994), and thus the NWI status and trends analyses may not include these permitted wetland impacts.

Wetland classification systems are valuable for understanding the ecology of and the management of forested wetlands. Classification provides a way of organizing current knowledge of wetlands ecosystems and facilitates the acquisition of new knowledge to advance wetlands science, policy, and management.

2 Wetland Functions and Relations to Societal Values

Mark M. Brinson and Richard D. Rheinhardt

CONTENTS

INTRODUCTION

Wetlands perform functions that have environmental importance and social significance. Wetlands are often discussed in terms of their "functions and values," two terms that have been used interchangeably. In this chapter we will treat them as interrelated but separate for the purpose of more clearly distinguishing between ecological and socioeconomic issues. Functions of ecosystems, and hence wetlands, are the normal or characteristic activities that take place in wetland ecosystems or simply the things that wetlands do (Smith et al. 1995). Value, on the other hand, generally refers to the benefits derived from the goods and services that evolve from the functions that wetlands perform.

The depth of the social significance of wetland issues is reflected in the policy debates and regulations described in the next chapter. This high level of attention has inspired a number of approaches designed to measure the effects of altering wetlands. Most of these approaches are designed to estimate losses of functions when wetlands are changed or altered from their current status to some alternative use or condition. Examples include filling wetlands to create building sites, draining wetlands to grow annual crops, and flooding wetlands to create reservoirs for recreation and water supplies. Such alterations of wetlands are generally perceived as

1-56670-228-3/97/$0.00+$.50
© 1998 by CRC Press LLC

having negative effects on their functioning. Alterations that contribute to enhancement of wetlands functions, such as restoration of degraded wetlands, are generally perceived as positive.

Severe alterations reduce or totally eliminate the functioning of a wetland. While the effects of impacts may be viewed as negative by one sector of society, the same ones could be viewed as beneficial by another. Automobile drivers may benefit when a wetland is filled for a bridge crossing which reduces highway accidents and saves lives. The impact does not benefit the farmer or forester, however, who may have been displaced from land where he or she wanted to grow a field crop or tree plantation, or individuals living downstream who expect high standards for water quality. This means that alterations of wetlands can have both winners and losers in society. From an ecological perspective, however, alterations are negative in most cases because functions performed by wetlands are being reduced or impaired relative to their condition beforehand.

Besides losses of functions by alterations, it is useful to consider increases in functions that may be possible by restoring altered wetland ecosystems to a relatively unaltered condition. In order to determine the success of a restoration project, assessments can be made to determine whether functions at the restoration site increase over time. Since both increases and decreases in functions can be measured, it is better to refer to "changes in function" in wetlands as more neutral terminology than gains or losses.

We begin this chapter with a brief discussion of past approaches to assessing the functions and values of wetlands. We then discuss why it is necessary to classify wetlands before assessing their functions (classification is needed to determine whether differences in wetland characteristics between two sites are due simply to inherent differences among wetland types [i.e., one is a pocosin flat and the other a riverine cypress swamp] or to an alteration of one of the sites [i.e., the cypress swamp was drained]). We then examine how standards can be used to characterize a wetland class at the same time that the standards encompass the natural variation that occurs within the class. We close by discussing how functional assessment can be used in the decision-making process by relating assessments to both non-market and market-based economic assessments.

APPROACHES AND TECHNIQUES TO MEASURE FUNCTIONS AND VALUES

"Functions and values" is a phrase used repeatedly by those who become involved in scientific and policy issues that deal with wetlands. As pointed out in the next section, early efforts to evaluate functions and values did not distinguish between these two concepts. Much information has been written on the topic (Greeson et al. 1979, Lonard et al. 1981, Whigham and Brinson 1990, Richardson 1994, Smith et al. 1995), and techniques have been developed to estimate functions and values (Hollands and Magee 1986, Adamus et al. 1987, Maltby et al. 1994, Hruby et al. 1995), as well as the cumulative effects of impacts (Preston and Bedford 1988, Gosselink and Lee 1989, Leibowitz et al. 1992). Conceptual approaches fall into the following broad and sometimes overlapping categories that include but are not limited to (see Lugo and Brinson 1979):

(1) Qualitative statements of functions: A simple listing of functions for wetlands in general or for a specific group of wetlands that may include narrative that further describes or defines the functions.
(2) Value based on costs required to replace a service: One of the original papers on valuation calculated the wastewater treatment capacity of a tidal salt marsh (Gosselink et al. 1974). Other services may include the nursery function of producing finfish and shellfish and maintaining low levels of carbon dioxide in the atmosphere through peat accumulation.
(3) Value based on costs of replacing a wetland ecosystem: Costs of creation and restoration projects vary greatly (Kusler and Kentula 1990), but provide an alternative perspective on the value of self-sustaining ecosystems.
(4) Value based on energy flow: The proportion of total energy flowing through a wetland via primary production represents "natural" energy in contrast to fossil fuel energy to sustain society. When all energies are normalized nationwide to "gross domestic product" in dollars, a value per unit of energy can be calculated. The value can then be multiplied by the energy flow of the wetland in question to arrive at a monetary estimate of value.
(5) Value based on energy analysis adjusted for energy quality: Recognizing that different energies have different qualities, EMERGY analysis (not energy analysis; see Brown et al. 1995) has been developed to normalize data to solar equivalents of energy. This is perhaps the most holistic of approaches as it converts disparate types of energies to a common denominator based on embodied energy rather than dollars.
(6) Value based on combined ecological and economic approaches: This is a hybrid method known as "ecological economics" (Costanza et al. 1989). Such approaches struggle with the fact that many wetland functions occur outside of the marketplace where money is the standard currency for evaluation and exchange. Where market forces are lacking, the alternative is to employ non-market pricing techniques (Wolf 1979), often called shadow pricing or pseudo-market techniques. Such approaches standardize consumer demand with units of currency using "willingness-to-pay" and "willingness-to-accept" approaches (Costanza et al. 1989). The willingness approaches estimate the maximum amount an individual would pay for using a wetland or, conversely, the minimum amount an individual is willing to accept if a wetland were eliminated. Because many recognized goods and services of wetlands do not participate directly in the currency flow of economies, they are considered non-market factors or extramarket values.

A common property of most of these approaches is that they convert a wetland's attributes into the common monetary denominator, currency in dollars. This is an explicit use of a standard of price which is ultimately determined by a market system.

An assessment of ecological functions, in order to be consistent from one location and one time to another, however, must be separate from the consideration of the goods and services that society considers valuable. If there is no separation, an ecological assessment becomes an economic one, and is based largely on the opportunity that society will actually benefit from one or more of the functions. However, the fundamental properties of the ecosystem do not change just because society decides that a wetland is worth more or less money at a given point in time. The rate of photosynthesis, the quantity of nutrient cycling, and the capacity to support bird populations do not change when timber prices increase or decrease or when land speculators are able to market swampland to gullible investors. When timber is harvested or swampland filled, however, wetland functions are temporarily or permanently altered, and in some cases are completely eliminated.

When economics are not separated from ecological assessments, the results become highly contextual. This indicates that the assessment used is based on societal values rather than functions. Although the public interest review process (33 CRF 320.4) mandates that economic concerns be considered for activities that impact wetlands, the economic effect is just one of many issues that must be addressed. This "general balancing process" assesses cumulative effects including conservation, aesthetics, general environmental concerns, historic properties, fish and wildlife values, flood hazards, floodplain values, land use, navigation, shore erosion and accretion, and so forth. Consequently, ecological functions and economic values are simply two of an array of issues with which policy makers must deal in managing wetland resources. This book focuses on the ecology and management of southern forested wetlands in general. While some attention is correctly paid to a major objective of active forest management for fiber production, none of the chapters takes an explicitly economic perspective. For consistency, economics of forest practices will not be considered in any detail here, except as economics interfaces with ecological assessments.

DISTINCTION BETWEEN FUNCTIONS AND VALUES

One of the main difficulties encountered in using many functional assessment approaches available today is that they fail to separate functions from values. Without this separation, low assessment scores may be assigned to fully functional wetlands when assessed functions are naturally subdued due to inherent properties of wetland type and its landscape position, hydrology, or nutrient status. When values are the focus of an assessment (based on desired types or levels of functioning), and criteria are developed that allow this focus to be taken to extremes, favorable scores could be assigned to activities that cause severe alterations in functions or even destruction of a wetland itself. For example, the valued fertile soil of a wetland can be made more available for certain kinds of agriculture and silviculture by draining it. In maximizing the value of food and fiber production, several functions may be lost or severely reduced, such as nutrient removal and supporting a plant community of characteristic hydrophytes. The value for waterfowl hunting in a bottomland hardwood forest may be improved by increasing water depth by reducing drainage. By maximizing this use at the expense of forest structure, not only would the site be useless for another value, timber production, but the characteristic drying cycle important for seed germination and seedling establishment would be altered as well. The conversion of a floodplain to a reservoir to maximize the water storage function for flood control benefits can completely eliminate all other functions characteristic of floodplain forests.

Each of the above cases demonstrates how a particular wetland characteristic or function could be modified to maximize the outflow of goods and services that have value to society. However, if left unaltered, the wetland would still provide goods and services, but perhaps different ones. Moreover, even if society did not value wetlands at all, the functions of any given wetland would not change except in response to natural phenomena (e.g., hurricanes, fire, beaver activity, landslides, etc.). This is the fundamental reason for measuring impacts to wetlands using functional assessment rather than values assessment: functional assessment deals with properties of ecosystems and not vagaries of society's changing wants and needs. A wetland does not lose its functions by being a part of a wilderness area that has minimal contact with humans. In fact, most wetland scientists would argue that minimally impacted wetlands are those that function in a sustainable manner and at an appropriate level within a given set of landscape conditions.

The distinction between functions and values is further clarified in Table 2.1. This table provides several examples of wetland types that occur in the South and nearby regions, identifies a characteristic function for each class, describes the effects of the function, provides examples of goods and services associated with the function, and provides an indicator that the function exists. One example is a bottomland hardwood floodplain that stores water on its surface for short periods when the associated river overflows its banks. This water storage function reduces flood peaks downstream to levels below those that would occur if the floodplain did not store the water (e.g., if the bottomland were protected from flooding by levees). Society places value on this function because it provides the service or benefit of reducing downstream flood damages.

The separation of functions and values can be illustrated further by conceptualizing the relationship between a wetland and society (Figure 2.1). On the left-hand side of Figure 2.1, natural energies and materials (sun, water, and other imported materials such as nutrients) support processes such as photosynthesis, nutrient sorption, water supply, and so forth. In turn, these processes support functions which feed back to processes. An example would be the sediment retention function which provides soil fertility and a surface for the processes of plant establishment and growth. Plants provide feedback to the floodplain by stabilizing sediments and reducing water velocities for trapping more sediments. Natural stressors may exist, such as the inhibitory effects of saturated soils, that divert energy toward coping with stressors. For example, energy that otherwise might have been allocated to stem growth might instead be diverted to adventitious root growth to allow the plant to survive prolonged inundation (see Part II, Chapter 8).

On the right-hand side of the diagram, societal infrastructure encompasses goods and services derived from a wetland. Goods and services contribute to the life support and welfare of consumers in society. In addition, society depends on other ecosystems (agroecosystems and industrial forests) as well as fossil fuels to support its

TABLE 2.1
Functions for various classes of wetlands, related effects of the functions, corresponding goods and services of value to society, and relevant indicators of functions. The examples are based on several classes of forested wetlands common in the southern United States. Modified from National Research Council (1995).

Wetland Class (and example)	Function	Effects of Function	Goods and Services Valued by Society (Value)	Indicator of Function
Riverine (bottomland hardwood floodplain)	Stores surface water over the short-term	Reduced down-stream flood peaks	Reduced damage from floodwaters	Floodplain connected to stream channel
Slope (forested seep at foot-slope adjacent to floodplain)	Maintains a high water table	Maintenance of hydrophytic plant community	Maintenance of biodiversity through habitat heterogeneity	Presence of hydrophytes
Riverine (floodplain of low order stream)	Accumulates inorganic sediments	Retention of sediments and some nutrients	Maintenance of water quality	Increase in thickness of sediment
Peatland flat (sealevel controlled pocosin with deep peat)	Accumulates organic matter in soil	Maintains elevation relative to surrounding landscape	Maintains terrestrial condition in face of rising sea level	Presence of land in a coastal geomorphic context
Mineral soil flat (longleaf pine savanna)	Maintains characteristic plant community	Presence of longleaf pine with heart rot	Habitat for red-cockaded woodpecker and fire-dependent plant community	Presence of large trees with cavities
Depressional (Carolina bay wetland)	Supports characteristic vertebrate populations	Maintains black bear habitat	Production of individuals for hunting and viewing	Direct sign of bear
Estuarine fringe (red mangroves)	Maintains complex shallow water habitat	Supports nursery areas for finfish and shellfish	Recreational and commercial fisheries	Presence of viable mangrove swamps

own infrastructure. Values that consumers hold through culture, ethics, and religion (shown to the extreme right in Figure 2.1) can lead to actions that have either constructive or destructive effects on wetland ecosystems. If these feedbacks are destructive to wetlands, natural wetland functions will be diverted from their contribution to society. One example is building a levee which prevents the water storage function, but enhances the capacity for crop production on the site. Constructive feedbacks that support natural wetland functions act to reinforce natural processes that society enjoys. A common example is the restoration of altered wetlands.

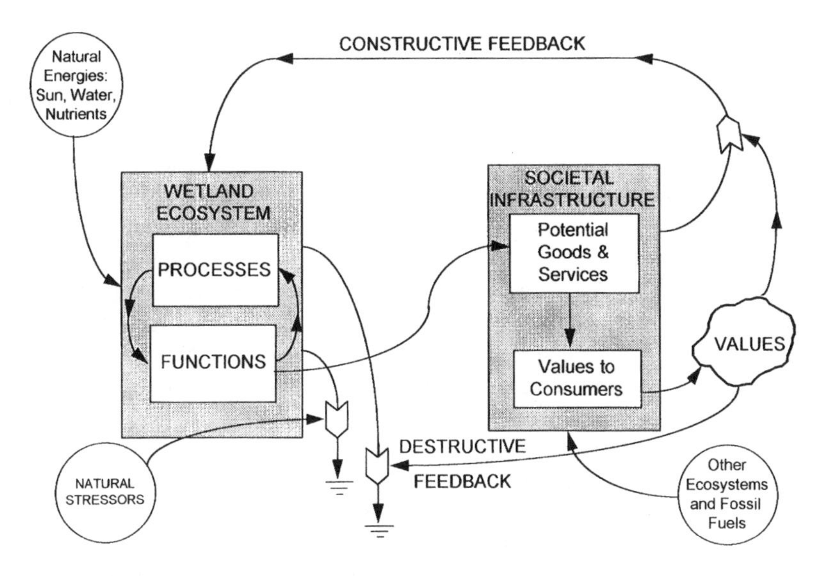

FIGURE 2.1 The relationship between functions and values of wetlands. Factors on the left-hand side of the figure represent wetland ecosystems with functions that may operate independently of society. Shown to the right is societal infrastructure which includes goods and services that may be extracted from a wetland by consumers. The values symbol is ultimately driven by consumers. Values can motivate actions that feed back either to destructively alter a wetland or constructively maintain its functions (adapted from Brinson 1993b).

APPROACHES TO FUNCTIONAL ASSESSMENT

The separation of functions from values in wetland functional assessment does not suggest that one should be chosen over the other; it merely allows the compartmentalization of assessments into ecological and economic phases. Consistent with other topics in this book, we focus on ecological issues rather than economic ones. However, we provide insight on how output from an ecological assessment can supply information that would be useful in an economic assessment phase.

There are several levels at which ecological assessments of impacts can be conducted. A common approach is to seek the best professional judgment of wetland scientists or ecologists to predict the impacts of a particular project on wetland functions. Such assessments are conducted routinely by personnel from the U.S. Army Corps of Engineers (Corps) and their counterparts in other agencies involved in Section 404 permitting. Assessment material submitted to the Corps is often prepared by consulting firms engaged by the party wishing to alter a wetland site. Information is commonly provided on wetland community type, hydrologic regime, and other pre-project characteristics. Post-project conditions are projected and probable changes in functions from the impacts are evaluated. Professional judgment of whether a project has significant impacts to wetland functioning may vary with the training of the regulator and his or her experience on other projects.

An alternative approach, or supplemental to the one just described, is to utilize a rapid method such as the Wetland Evaluation Technique (Adamus et al. 1987) or one of several other available methods (Larson and Mazzarese 1994). Most of these approaches fail to separate functions from values, however.

A more detailed level of functional assessment would be to conduct studies of the actual functions themselves and estimate how they would change after alteration. Hydrologic functions could be measured by employing water level recorders for several years in concert with topographic surveys to estimate hydroperiod. Biogeochemical functions may require the measurement of nutrient cycling and denitrification rates. Assessment of habitat and food web support functions could include the characterization of habitat structure and estimates of population densities over time. While such detailed studies may be justified for a project that is applying to alter large areas of wetlands, few studies of this magnitude are actually encountered in the public review process.

The approach that we describe here is ecosystem-based, and assesses functions irrespective of issues of values. We use the hydrogeomorphic (HGM) approach to functional assessment of wetlands (Smith et al. 1995) as an example of how alterations can be characterized in a semi-quantitative fashion, and the output can be placed into a decision-making framework for further assessment, including economic evaluations of value.

Classification: Grouping Wetlands by Functional Properties

Forested wetlands are highly variable. They include a spectrum ranging from massive stands of baldcypress (*Taxodium distichum*) in river swamps to shrub bogs that undergo cycles of wildfire. Due to natural differences in vegetation structure, hydrology, and landscape position, and degree of alteration, wetlands differ in their ability to perform functions. If the purpose of assessment is to detect changes or variations in functioning due to alterations, the natural variation among wetlands described earlier can interfere with the ability to detect change due to alteration. For example, if a shrub peat bog does not filter out sediments from runoff (because there is no upland as a source of sediments), should it be assessed by the same functional criteria as a streamside floodplain swamp that is a very effective filter? Of course not. The shrub bog is in the wrong landscape position to perform a sediment filtering function, so the wetland naturally lacks that function. Similarly, there are intact forested wetlands that do not support fish populations, shrub bogs that do not support migratory waterfowl, and peatlands that cycle nutrients at slow rates relative to other wetland types.

In order to avoid attributing functions to wetlands that may not naturally support them, it is appropriate to define them according to their innate functional characteristics and group them with other wetlands possessing the same characteristics (i.e., a functionally based classification). The purpose of classification is to enable one to determine changes in wetland functions appropriate to the wetland type being assessed. The classifications described in Chapter 1 are very useful for the purposes for which they were designed, but many are not functionally based and are of little value in categorizing wetlands prior to functional assessment. For example, the National Wetland Inventory classification (Cowardin et al. 1979) is based primarily on cover-type. Using vegetation cover alone to classify according to function can lead to problems. This is because similar cover-types may have very different geomorphic settings in the landscape and very different hydrologies. For example, forested wetland ecosystems are commonly characterized more by their plant communities than by their geomorphic or hydrologic conditions. By the same token, a variety of vegetation cover types may belong to the same hydrogeomorphic class. To illustrate, Table 2.2 shows commonly recognized vegetation communities listed by dominant species. Because species dominants can be found in two or more distinct geomorphic settings, vegetation alone is only marginally useful in determining functional classes or subclasses (subclasses can be thought of as regional hydrogeomorphic variations of the major classes of wetlands: riverine, depressional, organic soil flat, mineral soil flat, slope, lacustrine fringe, and estuarine fringe). Consequently, the hydrogeomorphic classification is a good starting point for categorizing wetlands prior to assessing functions. Extremely large wetlands or diverse wetland complexes, such as the Okefenokee Swamp or the Everglades, should obviously not be classified or assessed as just one wetland type. Large wetland areas likely contain several classes and thus require multiple assessments.

TABLE 2.2
Dominant tree species of common forested wetlands and examples of geomorphic classes where they occur.

Dominant Vegetation	Geomorphic Class
Atlantic white-cedar	1. Toe slope wetlands on coastal rivers
(*Chamaecyparis thyoides*)	2. Broad interstream organic flats
	3. Tidal swamps (New Jersey)
	4. Broad depression swamps (Lake Drummond, Va.)
	5. Streamheads in sandhill regions
Baldcypress	1. Bottomlands of low order streams
(*Taxodium distichum*)	2. Tidal swamps
Baldcypress/Water tupelo	1. Floodplains of high order streams
(*Taxodium distichum/Nyssa aquatica*)	2. Estuarine fringe swamps (tidal)
Ash/Swamp tupelo	1. Tidal swamps
(*Fraxinus* spp./*Nyssa sylvatica* var. *biflora*)	2. Second and third order floodplains of inner coastal plain
Mixed oak/swamp tupelo	1. Piedmont depressions
(*Quercus* spp./*Nyssa sylvatica* var. *biflora*)	2. Interstream wet flats protected from fire
	3. Low order stream floodplains
Longleaf pine	1. Broad interstream wet flats
(*Pinus palustris*)	2. Pleistocene sand ridges (non-wetland)

Using Reference Wetlands in the HGM Functional Assessment Approach

The purpose of assigning wetlands to their appropriate class or subclass before evaluating impacts is to prevent natural variation from confounding variation caused by alterations. Once a subclass is determined, the assessment of alterations in the HGM approach must be compared to some standard. How does one determine which wetlands in the class represent natural conditions against which impacts should be measured? A key word in this comparison is reference (Brinson and Rheinhardt 1996). The question can be posed as, "To what extent did a wetland lose or gain functions relative to some standard of comparison?" The standard used in the HGM approach is reference standard wetlands (i.e., those sites that sustain the overall highest level of functioning across the suite of functions characteristic of the class). These sites provide the information from which reference standards are developed. As a general rule, reference standard wetlands are mature, naturally occurring sites (forests, marshes, bogs, etc.) with little or no alteration of drainage patterns, species composition, and soil conditions.

The choice of reference standard wetlands is difficult because even the best classification will contain considerable natural variation. If we accept, according to the U.S. Fish and Wildlife Service (USFWS) definition, that wetlands are "transitional between terrestrial and aquatic environments ..." (Cowardin et al. 1979), and further, that wetland classes are transitional between one another, the importance of the initial grouping of wetlands into classes becomes central. How much do we lump, with the possibility of combining into one group the forests on major alluvial floodplains (Chapter 12) with the forests on minor alluvial floodplains (Chapter 13)? Alternatively, how much do we split the minor alluvial floodplains into separate classes as described in Chapter 13? There are no easy answers to these questions.

Besides the problem of determining or characterizing natural variation, it may be difficult to gain consensus on what constitutes natural variation for a specific wetland class. It is sometimes hard to find mature, minimally altered swamp forests in some areas even though they may have once been abundant. It is particularly difficult in some places to find enough of them to characterize natural variation. To some extent, the descriptive chapters in Part III of this book ("Ecology and Management of Specific Wetland Types") attempt to do this. Most of these chapters are oriented toward describing minimally altered conditions and dealing with the natural variation that occurs within each group. The chapters also attempt to separate natural variation from that due to active management practices that alter wetlands. The tacit assumption is that a relatively predictable wetland ecosystem and plant community assemblage can be regarded as a starting point for assessing functions, and that more passive forms of management will tend to perpetuate a wetland class within its range of natural variation. More active forms of management may alter both the structure and the functioning of an affected wetland. Seasonally flooded bottomland hardwood forests, Florida cypress domes, Atlantic white-cedar (*Chamaecyparis thyoides*) swamps, and red mangrove (*Rhizophora mangle*) swamps, even when completely unaltered, have successional stages or disturbance phases that can be compared as a stage or state of ecosystem development toward a single endpoint or a series of alternative endpoints. Reference or benchmark conditions are implicit for wetland categories described in each of the chapters in Part III.

What, then, are the conditions, variables, measurements, etc. that may serve as standards for comparing a wetland site before and after impact? Because the HGM approach is designed to be relatively rapid, criteria or standards consist of easily measurable attributes such as tree basal area, shrub and understory tree density, herbaceous cover, etc. If the assessment were done in more detail and time were not a constraint, more detailed measurements of function could be made. Measurements in Table 2.3 were determined from one or a small group of reference standard sites in each of two subclasses of wetlands — floodplains of low order streams in Virginia and a wet pine flat in North Carolina. Differences between the two hydrogeomorphic classes are readily apparent. Wet pine flats require periodic fire (frequency 1 to 5 yr) to maintain longleaf pine (*Pinus palustris*) and its characteristic savanna-like aspect (Boyer 1979, Ware et al. 1993). Fire leads to lower tree basal area, lower tree density, higher graminoid cover, and less coarse woody debris than would be found in hardwood-dominated floodplains of low order streams where fire is rare. This shows why classification is essential; it would be inappropriate to evaluate a stream floodplain wetland using reference standard conditions measured in unaltered pine flats.

TABLE 2.3
Reference standards for variables used in the "Maintains Characteristic Elemental Cycling" function for hardwood-dominated small floodplains of low order streams in Virginia and longleaf pine-dominated wet pine flats in North Carolina (unpublished data of authors).[a]

Names of Variables and Abbreviations	Units	Reference Standards for Floodplains of Low Order Streams (n = 7)	Reference Standards for Wet Pine Flats (n = 1)
Tree basal area, V_{BA}	m^2/ha	31.2	14.7
Shrub and understory tree density, V_{SUBC}	#/ha	4,474	12,550
Herbaceous cover, V_{HERB}	%	58	~100[b]
Forb cover, V_{FORB}	%	N/A	33.3
Graminoid cover, V_{GRAM}	%	N/A	66.8
Seedling cover, V_{SDLG}	%	36.0	60.3
Snag basal area, V_{SNAG}	m^2/ha	1.3	0.67
Fine woody debris cover, V_{FWD}	%	0.12	N/A
Leaf litter depth, V_{LITR}	cm	N/A	5.7
Coarse woody debris volume, V_{CWD}	m^3/ha	11.5	0.51
No. of decay classes of CWD, V_{DECY}	n	2–3	N/A

[a] All attributes were measured in the same manner. Natural variability is not characterized for wet pine flats in this table because only one site was measured. N/A = not measured.

[b] Sum of graminoid and forb cover.

Relationship Between Wetland Structure and Function

If alteration of wetlands can be determined by structural attributes alone, why deal with functions? Would knowledge of structural changes (i.e., changes in basal area, density, species composition, etc.) due to alteration be sufficient as a mode of assessment? The short answer is that a comparison of structural difference alone would not be functional assessment. It would lack interpretive power as to why an ecosystem might not meet reference standards. The more exact answer is that there is a close affinity between structure and function. By using structure as a method to depict function, the measurements can provide a higher level of insight and interpretation of how a given wetland-type works. Fields of academic study provide the basis for characterizing functions. They include biogeochemistry, plant physiology, plant ecology, hydrology, and population dynamics of vertebrates and invertebrates (see Chapters in Part II for details on these topics). We have long used these fields as perspectives of how ecosystems work. Since we obviously cannot conduct multiple research projects every time a wetland needs to be evaluated for impacts to functions, we can use reliable but rapidly measurable substitutes for estimating functions. In most cases, we must rely on ecosystem structure, which can be easily and rapidly measured in the field, and general ecological knowledge, which is based on a synthesis of many decades of research and publication. A connection between structure and function in assessment is similar to many decisions in resource management that must be made with "incomplete" information. This requires a "leap of faith" in which general ecological knowledge and common sense are applied to make management decisions.

There is an additional reason for using functions to depict how wetlands work. Functions are capable of being translated into goods and services that are valued by society. This may seem contrary to the point made previously on keeping functions and values separated. However, even if the analysis of functions of a project is done first without a societal context of values, context can be applied later at any time that economic or monetary analyses are initiated. Again, this allows the results of a functional analysis to be kept context free and separate from economic phases of decision making. Since the context of decisions on wetland impacts varies through time and among cultures, the functional assessment tool can remain stable over time if kept independent of economics. Of course, any method restricted to assessment of functions should be flexible enough to incorporate additional information about wetland functioning as it becomes available.

Here we provide an example of how the structure of ecosystems can be depicted or "modeled" to capture the essence of functions, and offer a semi-quantitative scale for the level of functioning. There are four steps to this process: identify functions, determine relevant variables, combine variables into equations, and scale the variables to field measurements. An example of a function is "Maintains Characteristic Elemental Cycling," defined as the uptake and return of elements (including nutrients) between vegetation and soil, on roughly an annual basis (other examples of functions are listed in Table 2.1). Research on elemental cycling in forested ecosystems generally regards the uptake of elements to be equivalent to annual accumulation in biomass as stem growth and litterfall as well as losses from plants by leaching (net throughfall and stemflow). Most commonly, only aboveground components are measured because of the difficulty of measuring root growth and other belowground processes. Return of elements is commonly measured as decomposition of litter and the woody components of the forest floor.

The second step determines which variables are relevant to the elemental cycling function. For example, a field measurement of biomass represents past nutrient uptake. Additional variables that can be measured rapidly in the field and are relevant to elemental cycling are given in Table 2.4.

TABLE 2.4
Variables used in Equation 2.1 for the function elemental cycling.

Variable	Definition and Measurement	Contribution to Process
Basal area (V_{BA})	Total cross-sectional stem area occupied by canopy trees >10 cm DBH.	Represents elements incorporated into woody biomass of canopy trees; elements may be immobilized for many decades.
Subcanopy vegetation (V_{SUBC})	Density of shrubs and understory tree species <10 cm DBH and >1 m tall.	Represents elements incorporated into subcanopy stratum; elements may be immobilized for several years or decades depending on mortality.
Herbaceous vegetation (V_{HERB})	Percent cover of annual, biennial, or perennial forbs and graminoids.	Represents elements incorporated into herbaceous biomass; elements are immobilized for normally less than 1 year.
Seedlings (V_{SDLG})	Percent cover of seedlings of canopy trees (not included as understory tree species).	Represents elements incorporated into seedling biomass; Immobilization period varies depending on mortality.
Snag (V_{SNAG})	Basal area of standing dead trees >10 cm DBH.	Represents elements available for mineralization over a period of decades.
Fine woody debris (V_{FWD})	Percent cover of small branches and twigs on forest floor <10 cm diameter.	Represents elements available for mineralization in less than a decade.
Leaf litter (V_{LITR})	Depth of leaf litter on forest floor.	Represents elements available for mineralization over several years.
Decay classes (V_{DECY})	Number of decay classes of coarse woody debris.	Indicates a mixture of decomposition stages of coarse woody debris.
Coarse woody debris (V_{CWD})	Volume of large woody tree trunks and branches on forest floor >10 cm diameter.	Represents elements available for mineralization over the long term.

The third step is to combine variables. The algorithm used for combining variables is to aggregate them in an equation in a format similar to that used by the Habitat Evaluation Procedure (USFWS 1981). Equation 2.1 shows how these biomass components are combined into logic equations to depict the Index of Function, where V_i is the index for a variable. For the elemental cycling function, the assumption is made that if biomass is present, then elements must have been taken up in the production of biomass. If dead biomass is present (as detritus), it is likely that the material is undergoing decomposition and thus releasing elements. Thus living biomass is a surrogate for elemental uptake and dead biomass is a surrogate for elemental release. Components of the equation utilize the variables listed in Table 2.4 along with definition, relationship to the function, and general methods for field measurements. These rapid measurements, then, supplant the need for actually measuring functioning. Instead, substitutes or indices of functioning are used as surrogates.

$$\text{Index of Function} = \left[\frac{\dfrac{\left(V_{BA} + V_{SUBC} + \left[\dfrac{V_{HERB} + V_{SDLG}}{2}\right]\right)}{3} + \dfrac{\left(V_{SNAG} + V_{FWD} + V_{LITR} + \left[\dfrac{V_{DECY} + V_{CWD}}{2}\right]\right)}{4}}{2} \right] \tag{2.1}$$

Variables in Equation 2.1 are weighted in very crude fashion according to perceived importance in contributing toward a particular function. For example, the indices for herbs (V_{HERB}) and seedlings (V_{SDLG}) are averaged before they are combined with the basal area and subcanopy variables. This simply reduces the weighting of the herb and seedling variables. Even so, the weighting is not in proportion to the biomass of trees in a mature stand. If data were available to indicate, for example, how important each of the nine variables in the equation is to the function, we could use coefficients to weight each variable. Without this information, best professional judgment must be invoked, with crude estimates ranging from most to least important. The default could be that each variable is equally important. In the equation presented above, groups of two to four variables are averaged to effectively reduce their importance relative to ones that are not averaged, and thus receive higher weighting. If data were available on the relative importance of each biomass category, then further refinements could be made.

The fourth and final part is to develop a scale for variables derived from field measurements. In each case the scale ranges from 0.0 to 1.0, with the highest score reflecting reference standard conditions. For example, basal area of canopy-sized trees in small stream bottoms average 31.2 m^2/ha (136 ft^2/ac; Table 2.3), so this basal area would receive an index of 1.0. If variables themselves, rather than indices, were plugged into the equations, we would inappropriately mix units, such as percent cover, with units proportional to biomass, such as basal area. This problem is partially avoided by first indexing or scaling all variables in Table 2.3 to an index of 1.0, defined by reference standard conditions. Thus, the Index of Function is 1.0 when all variables are assigned a 1.0 and the equation is solved. For example, a tree basal area of 15.6 m^2/ha (68 ft^2/acres) for a particular small stream bottom in Virginia would be one-half of the reference standard average of 31.2 m^2/ha (Table 2.3). The variable for this would be 0.5 on a scale of zero to 1.0 because it is one-half of the reference standard for floodplains of low order streams. Flexibility must be imbedded within the scoring so that variables reflect the natural variation found in reference standard sites. One approach (Brinson et al. 1995) uses discrete categories rather than continuous scoring. In this approach, reference standards for the stream bottom might range between 23.4 and 39.0 m^2/ha (101 and 170 ft^2/acres), or 75% to 125% of reference standards. Using a categorical scale, values within this range would be equivalent to an index of 1.0. An index of 0.5 might be 25% to 75% of reference standards, and so forth, as a way of capturing the variation that might exist in reference wetlands.

In the process of indexing all variables to reference standard conditions, we must account for situations wherein an index potentially exceeds 1.0 when variables are higher than reference standards, but clearly do not fall within the natural variation of the class. An example would be an immature stream bottom forest with a shrub and understory tree density of 8,000 stems/ha (3,238 stems/acres) in contrast to the reference standard that averages only 4,474 stems/ha (1,811 stems/acres). What should be done about this situation? In order to maintain stability in the Index of Function, variables may not exceed 1.0 in a reference-based assessment which assumes that the characteristic level of sustainable functioning occurs under least altered conditions. Thus, the 8,000/ha density would have to be assigned a score of either 1.0 or perhaps lower since the 8,000 level is clearly a departure from reference standards. In such cases, a rationale must be developed to logically deal with assigning indices to certain variables. In the example given, a density of 8,000 might be assigned a 0.5 since it exceeds reference standards by nearly twice. Other rules could be developed to satisfy assessment needs as long as assumptions and rationale are laid out in a logical and defensible manner. The point, however, is to deal with the fallacy that more is better. This fallacy is difficult to communicate for ecosystem functions, but one that must be revealed in an approach that establishes standards based on relatively unaltered, self-sustaining ecosystems. Some unfortunate consequences of "more is better" logic include excessive sedimentation rates (causing a wetland to become buried), excessive nutrient loading (causing changes in species composition to opportunistic weeds), high densities of stems (shrub wetlands score "better" than forested swamps), and extended hydroperiod (wetter is better).

Any ecological assessment must make allowances for natural disturbances. The assessment should separate natural cycles and disturbances from those induced by human activity. In fact, this separation is at the core of ecological assessment. While separation is not always straightforward, it usually is possible without being arbitrary. For example, Hurricane Hugo overturned and killed many trees in the Congaree Swamp National Monument in South Carolina, a relatively unaltered bottomland hardwood and swamp forest (Putz and Sharitz 1991). The consequence was greater amounts of coarse woody debris and greater cover of saplings and forbs in response to more light reaching the forest floor. Although infrequent, hurricanes are natural phenomena and the additional variation that they produce requires that reference standard conditions accommodate this situation. One alternative would be to broaden reference standard conditions to include hurricane-disturbed forests. Another would be to develop separate reference standards for time intervals following post-hurricane conditions. The need to encompass broad ranges of natural variability is particularly acute in wetlands dependent on fire cycles that result in stand replacement, such as pocosins. If a pocosin is assessed just one year following a surface fire, assessment measures of ecosystem functions should not become lower (i.e., the ecosystem is "being all it can be"). To do otherwise would fail to encompass natural variation that results from a type of disturbance (fire) to which the ecosystem not only is adapted, but is dependent upon. Thus, allowances could be made to include one or several ecosystem states, as appropriate, to represent reference standard conditions. Whether this is worthwhile is a practical matter to be handled on a case-by-case basis for each subclass.

USE OF ECOLOGICAL AND ECONOMIC ASSESSMENTS IN POLICY MAKING

Functional assessments of ecological conditions will not give answers on what policies and guidelines should be applied to the management of forested wetlands. Many other considerations must be taken into account, including whether the activity is regulated or exempt from regulation. With the public review process (33 CRF 320.4) as the basic guiding principle in deciding wetland regulation and policy, both ecological and economic assessments may be major components of any decision. Policy will determine whether an ecological assessment, an economic assessment, or some other factor (aesthetics, safety, etc.) dominates decision making, as well as how each may rank relative to other considerations. Decision-makers dealing with resource allocation in southern forested wetlands should be responsive to an array of matters regarding allocation of wetland resources. Functional assessment is just one tool to measure ecological conditions and change. In other words, issues of societal "value" are under the purview of the policy realm and cannot be claimed exclusively by either economic or ecological disciplines. Regardless, the ecological assessment component of an assessment should use the best information practically available so that policy-makers can make informed decisions.

If information from ecological assessments does not dictate policy choices, how might the information derived from them fit into the overall decision-making process? An example of how this might be approached is illustrated in Figure 2.2 which shows a forested wetland being assessed by three approaches: an ecological ledger and two economic ones, shadow pricing and market pricing. Before running specific alterations of wetlands through each of the ledgers, the logic of each needs to be explained.

The ecological ledger follows the approach outlined earlier in this chapter. A forested wetland is assessed for levels of functioning in its current condition (before impact or management), and after the alteration has occurred. The difference between the pre- and post-alteration assessments provides an estimate of losses or gains in functioning. That concludes the ecological ledger. If the alteration is temporary rather than permanent, another post-alteration assessment could be conducted in order to determine whether there were gains in functioning since the alteration took place (gains in functioning should also occur in restoration projects where degraded wetlands are used to achieve compensatory mitigation).

The economic ledger using shadow pricing may pick up from where ecological assessment ends, convert gained or lost functions into goods and services, and then estimate a dollar value for the changes (Figure 2.2). See also the relationship between

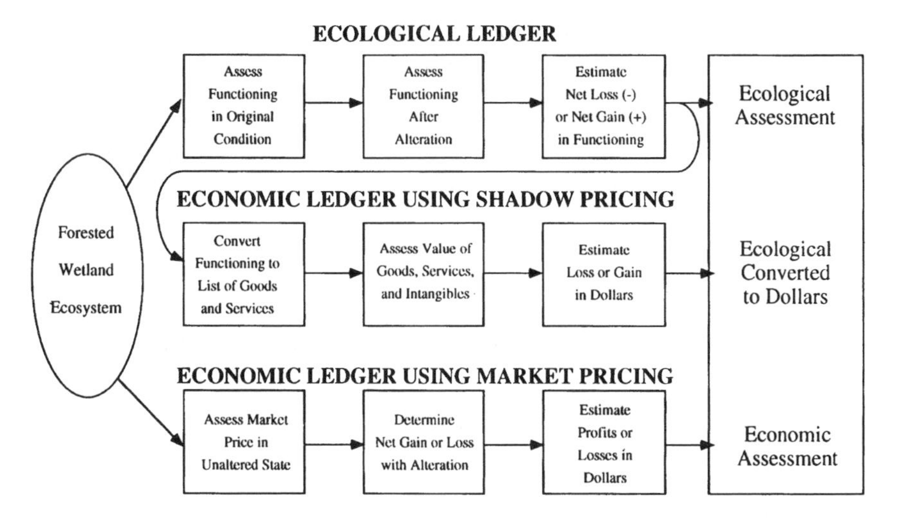

FIGURE 2.2 Three approaches to assessing wetland ecosystems. The ecological ledger estimates losses and gains in functioning using indices relative to reference standards derived from the least altered sites of a given wetland class. The economic ledger using shadow pricing converts functions to dollars based on the goods and services provided by functions in addition to other valued wetland attributes. The economic ledger using market pricing determines gains or losses of value that occur when the market responds to alterations of a wetland.

functions, goods and services, and values in Figure 2.1. This approach often focuses on goods and services related to non-market factors or the extramarket values mentioned earlier. When a wetland is altered, losses of goods and services are depicted as costs borne by sectors of society not involved in the benefits. Willingness-to-pay approaches can be used to broaden the scope of goods and services (shadow pricing is not typically applied to restoration efforts).

The economic ledger based on market pricing deals with the commodities and other features of wetlands that could result in profits or losses in real dollars (Figure 2.2). The procedure is to assess the monetary value of a wetland in its unaltered state, to determine the costs (marketing, overhead, production, etc.) for extracting resources and commodities during alteration, and finally to calculate the profits or losses due to the alteration. While this probably oversimplifies the economic ledger, the ecological and shadow pricing ledgers are equivalently simplified.

Two examples of wetland alterations are examined in a very qualitative manner to provide a comparison of ecological, non-market economic, and market-based economic approaches. A detailed analysis of each of these would be beyond the scope of this chapter, but they are offered to illustrate different perspectives on the same impact. The first example is an elevated bridge crossing (essentially no fill) over a stream bottom floodplain such as that described by the variables listed in Table 2.3. For the ecological ledger assessment, only two functions will be considered for simplicity, maintenance of characteristic surface water storage and maintenance of characteristic plant community. The surface water storage function would be virtually unaffected (assuming that the bridge design has not interfered with flows) while the plant community maintenance function would be reduced (trees permanently removed, shrubs temporarily removed, etc.). Thus there is minimal or no loss in surface water storage at the bridge site and a measurable reduction in the characteristic plant community where the bridge is built.

Upon running the economic ledger with shadow pricing, the result should be very similar to the ecological ledger. Various functions would be interpreted as having societal values (Table 2.1) and pricing could be applied to estimate economic losses of goods and services. The economic assessor could add intangibles such as aesthetics. Care should be taken so that the approach does not double count mutually exclusive goods and services. For example, bird hunters and bird watchers may have mutually exclusive benefits if they depend on the same species for their enjoyment.

Finally, the economic ledger estimated with market pricing assesses a wetland resource from society's perspective in terms of the value of the assessed land as a bridge site in comparison to an otherwise equivalent site not occupied by a bridge. According to standard acquisition procedures, the original landowner must be compensated at fair market value. If the wetland is worth $400/ha ($162/acre) on the real estate market when it was considered for a bridge crossing, and the bridge must occupy 2 ha (4.9 acres) of wetland area, the department of transportation would pay the landowner $800. Taxpayers would thus invest $800, plus transaction costs, to convert the site to their use as a bridge crossing. The cost to society of lost functions, such as flood water storage, plant community, and bird habitat, are not assessed because they are external to the market system.

The decision maker in this example has three sources of information available in addition to safety, public support, and other considerations. In each case, the information could be used to determine the costs of altering the wetland for the bridge, or could be used to compare the relative impacts for alternative locations for siting the bridge. The ecological assessment provides information on the magnitude of the impact, and what kind of compensatory mitigation would be appropriate to recover functions lost. While only two functions were discussed for simplicity, as many as 15 individual functions might be considered for riverine wetlands (Brinson et al. 1995). The shadow pricing approach would assign a dollar value to functions lost, which could be compared with the market pricing technique. Presently, there are no standardized guidelines for how to carry out assessments using shadow-pricing, but many examples have been published that could serve as models (Costanza et al. 1989). Extramarket pricing has been under-utilized in economic analyses of wetlands (Whitehead 1992–93). Finally, the market pricing approach requires no complex calculations, except for an estimate of transaction costs, because market signals are usually available. Because market pricing does not consider the "externalities" of lost functions due to the project, the costs are limited to transactions that occur within the confines of the project itself. With the above assessments and information at hand, the decision-maker can decide to stop the project, to continue the project without restoring environmental losses, or to continue the project and recover environmental losses through restoration.

The second example is that of converting a wet pine flat to a loblolly pine (*Pinus taeda*) plantation. The same functions will be treated (maintenance of characteristic surface water storage and maintenance of characteristic plant community) as were analyzed for the stream bottom floodplain. Surface water storage is altered temporarily, but not necessarily decreased, depending on the effectiveness of drainage ditches and the amount of soil compaction and organic horizon alteration that takes place during site preparation (Swindel et al. 1983). The characteristic plant community maintenance is highly altered, however, from the low basal area and high herbaceous cover typical of wet pine flats (Table 2.3) to a community with higher basal area and reduced ground cover. Species composition and structure are highly altered also, with closed canopy plantation loblolly suppressing shade intolerant grasses and shrubs that otherwise dominate ground cover in wetland pine flats. Thus, for the ecological ledger, there is alteration of surface water storage, but to a relatively small degree, and alteration in characteristic plant community to a relatively large degree.

As with the previous example, the economic ledger using shadow pricing would depend on output from the ecological ledger as its starting point. However, there may be overriding social issues that are not specifically addressed in the ecological assessment such as archeological sites and endangered species protection. Many mature pine flats are critical nesting habitat for the red-cockaded woodpecker (*Picoides borealis*), an endangered species. Moreover, mature wet pine flats are also relatively rare along the Atlantic and Gulf coasts. Therefore, shadow pricing should consider willingness-to-pay or other approaches in response to issues of endangered species and rareness of ecosystems.

The market pricing ledger would show the difference between costs of leaving the forest in an unaltered state (i.e., the opportunity cost of not extracting timber) and either profits or losses to be averaged over a silvicultural rotation. Land prices might also increase after improved drainage. Again, the decision maker is faced with information on whether to convert longleaf pine to plantation silviculture based on a variety of issues including costs due to loss in functions, costs to restore lost functions elsewhere, costs estimated by willingness to pay, and the difference between opportunity costs and potential value as plantation. Ecological assessments can provide information for the first two of these analyses.

Which of the three assessment approaches in Figure 2.2 is most useful? Each is useful to some extent because different perspectives are provided on altering a wetland. The ecological ledger is the most consistent with an ecological approach to forested wetlands emphasized in this book. The shadow pricing approach uses partial lists of functions and calculates what they might be worth through willingness to pay and other techniques. This approach is the most complex, because it depends upon either an ecological assessment to calculate differences in functioning before and after alteration or on perceived "willingness" estimates in the absence of such data. The market approach is the most straightforward. In the bridge crossing example, the market approach uses local price data to estimate the cost of making the wetland available for use. The cost of the bridge building and social benefits from traffic safety and convenience are outside of the scope of wetland resource economics. For the pine plantation example, market economics can be estimated based on projected market demand, rates of interest, production costs, extraction costs, and compensatory mitigation costs, if employed.

In the discipline of resource management, decision makers and policy makers can become better informed by evaluating results of the kinds of assessments described above. Based on broad principles of social welfare, they must decide whether a project, or policies regulating such activities, is worthwhile and should proceed. If all factors involved in the decision making could be truly converted to dollars, we would not need decision makers or policy. In such cases, sustainable social welfare (which presumably includes environmental living conditions for at least a tolerable quality of life) would be the item to optimize, and the translation of all factors to dollars would signal which alternative is highest, and thus the preferred alternative.

SUMMARY

The term "functions and values," when used to describe wetland ecosystems, has resulted in the mixing of ecological and economic standards of assessment without a distinction between the two. Functions should be used to refer only to ecosystem properties, and for forested wetlands these include such properties as surface water storage, plant community maintenance, nutrient cycling, and similar functions. Values, on the other hand, refer to the worth of a wetland to consumers and society. By recognizing the distinction between the two, ecologists can better assess changes in the functioning of forested wetlands due to management alterations. A functional assessment approach is briefly described that is based on recognizing characteristics of mature, unaltered wetlands, described in other chapters. Properties of these ecosystems can be used as standards for measuring changes in ecosystem functions. Simple algorithms using basic ecological principles and logic can be applied to make the transition between ecosystem structure, which can be quickly measured, and ecosystem function, which would require intensive, long-term research to measure. Outputs of an ecological assessment of hypothetical alterations to wetlands are compared with market and non-market approaches which may allow monetary estimates of a wetland's value. Decision makers and policy makers can use both ecological and economic assessments, in addition to other factors, in the allocation of natural resources. The assessment of functions can also contribute to determining whether restoration of wetland forests achieves desired goals.

ACKNOWLEDGMENTS

This manuscript benefited from the very helpful reviews of M. E. Lebo and Sammy King. David Knowles contributed numerous suggestions during manuscript development. The ideas and concepts expressed in this chapter are the result of workshops, discussions, training courses, and interactions with students. Principals in the development of the HGM functional assessment approach include L. C. Lee, D. Whigham, W. L. Nutter, G. Hollands, and R. D. Smith, in addition to the authors.

3 Wetlands Regulation: Development and Current Practices

Deborah A. Gaddis and Frederick W. Cubbage

CONTENTS

1-56670-228-3/97/$0.00+$.50
© 1998 by CRC Press LLC

WETLANDS REGULATION UNDER FEDERAL LAW

Wetlands regulation in the United States is a patchwork quilt of legislative, judicial, and institutional action pieced together over the last century. Starting with the Rivers and Harbors Act of 1899 and progressing through the various water pollution control acts of the 1970s and 1980s, the U.S. Army Corps of Engineers (Corps or USACE) gradually assumed regulatory control over wetlands in conjunction with the Environmental Protection Agency (EPA). Additionally, presidential directives, federal regulations, and judicial decisions directed the pattern of wetlands regulation. Knowledge of the incremental development of the regulatory process and the Corps' evolutionary role is necessary for the forest manager to understand the often inconsistent and ambiguous nature of wetlands regulation. This chapter will begin with a brief history of the regulatory process and conclude with information on the current situation in relation to forest management activities.

NAVIGATION TO WETLANDS PROTECTION

Wetlands regulation began as an offshoot of the Corps' mandate to protect navigation. In the 1800s Congress directed the Corps to build harbors and to maintain and improve the navigability of waters used for commerce. However, the Corps did not have specific regulatory authority to protect navigability as evidenced by the Supreme Court's decision in the *Willamette Iron Bridge Co. vs. Hatch* (125 U.S. 1, 1888). The court's decision gave the state of Oregon authority to control construction of dams, bridges, and other navigational obstructions in the absence of federal regulation of the area. As a result, Congress promptly passed the Rivers and Harbors Act (RHA) of 1890, assigning the Secretary of the Army regulatory responsibility for all construction activities, including the deposition of refuse into the navigable waters of the United States (Blumm and Zaleha 1989).

In 1899 Congress amended the RHA. Originally, Congress intended the amendments to provide funds for building, repairing, and preserving public works projects to maintain or improve navigation. On request from the Corps, Congress added ten sections to the bill to revise and clarify existing law concerning regulatory provisions (Adams 1993). For what was to become wetlands regulation, the relevant parts of the law were Sections 10 and 13. Section 10 gave the Secretary of the Army authority to regulate all dredging, filling, and construction in navigable waters. Section 13, informally known as the Refuse Act, prohibited the discharge or deposition of any refuse matter in navigable waters or their tributaries except street and sewer runoff (Stine 1983).

During the time between the 1899 RHA and the late 1950s, the Corps used Sections 10 and 13 of the RHA to maintain the navigability of streams and other waters in order to protect commerce as required under the U.S. Constitution. Routine activities regulated by the Corps were primarily construction activities or dredge and fill operations in navigable waters. The focus was on navigability — thus permit reviews did not consider environmental effects. Adverse effects on fish and wildlife from some of these operations influenced Congress to propose legislation to correct this problem (Stine 1983).

The Fish and Wildlife Coordination Act of 1958

The Federal Fish and Wildlife Coordination Act (FWCA) of 1934, as amended in 1958, required the Corps to consider how proposed water projects would affect wildlife and fish habitat and conservation. Under the law, any government or private agency proposing improvements to streams or other waters under federal authorization was required to consult with the U.S. Fish and Wildlife Service (USFWS) and the relevant state wildlife agency in order to conserve wildlife resources and to develop and improve habitat pursuant to such projects.

Public Interest Review

Despite the passage of the FWCA Amendments of 1958, the Corps continued to issue RHA permits based primarily on navigational concerns until July 1967. At that time the Secretaries of the Army and Interior signed a Memorandum of Understanding (MOU) whereby wildlife and fisheries' concerns would become a consideration for permitting activities. As a result of the MOU and the increasing public concern about environmental issues, the Corps revised the permit program in 1968 (Stine 1983). The Corps was required to consider the public interest during the permitting process, balancing resource utilization with resource protection. A total of seventeen factors affecting the general welfare of the public was weighted for each project. The factors were: conservation, economics, aesthetics, general environmental concerns, wetlands, historic properties, fish and wildlife values, flood hazards, floodplain values, land use, navigation, shore erosion, recreation, water supply and conservation, water quality, mineral needs, and property ownership concerns. The stated presumption of the regulation was that the permit would be granted unless contrary to public interest (Steinberg and Dowd 1988, Strand 1993).

The National Environmental Policy Act

In 1969 Congress passed the National Environmental Policy Act (NEPA) in response to public concerns over environmental degradation. With passage of the act, official government policy was "to create and maintain conditions under which man and nature can exist in productive harmony, and fulfill the social, economic, and other requirements of present and future generations of Americans." NEPA required analysis of impacts for major federal actions which significantly affected environmental quality. These actions included Corps projects and permitting functions. A preliminary environmental analysis (EA) was required to determine if an activity negatively affected the environment. A "finding of no significant impact" (FONSI) would allow the action to commence. If significant environmental impacts were likely, an environmental impact statement (EIS), including alternatives, would be required and alternatives considered.

The Corps Denies a Permit: *Zabel vs. Tabb 1970*

The scope of the revised permit program was not universally accepted by the Corps. Some Corps personnel felt that denying a Section 10 permit for environmental reasons was illegal. To test the new regulations, the Jacksonville, FL Corps office decided to deliberately deny a Section 10 permit for environmental reasons. The precedent-setting case of *Zabel vs. Tabb* (430 F. 2nd 199 [5th Cir. 1970] *cert. denied* and 401 [U.S. 910 1971]) was a major step forward in federal wetlands protection. Zabel and Russell, a development firm, wanted to dredge and fill 4.5 ha (11 acres) of tidelands under and next to Boca Ciega Bay in Jacksonville in order to create a trailer park. The resulting park would not affect navigation in any way, but would demonstrably harm the aquatic ecosystem. Col. Tabb, under the constraints of the Fish and Wildlife Coordination Act, denied the permit because the proposed dredge and fill needed to build the park would damage the fish and wildlife resources of Boca Ciega Bay (Stine 1983). The developers brought suit against the Corps, basing their case on the right of the Corps to reject a permit on grounds other than navigational constraint. The Court found for Zabel and Russell, ordering the Corps to grant the permit in their 1969 decision. The Corps appealed, and in 1970 the Fifth Circuit Court of Appeals reversed the decision. The Court found that Congress had the power to forbid such development because "the destruction of fish and wildlife in our estuarine waters does have a substantial, and in some cases, a devastating effect on interstate commerce" (1 Environmental Reporter 1451, 1452 as cited in Adams 1993).

The Court also cited NEPA in its support of the Corps action. Although NEPA had not been extant at the time of the permit denial, the Court stated that the decisions made must reflect the standards at the time of appeal. NEPA gave clear direction to federal agencies to consider ecological factors when an action was proposed under federal regulatory authority. Thus the findings obligated the Corps to consider environmental effects for Section 10 permits and respond accordingly (Adams 1993).

The Corps Assigned Pollution Control Responsibilities

The public continued to demand action to reduce and control pollution, but the only available avenue for pollution control was the Refuse Act which was created to maintain navigability. The Conservation and Natural Resources Subcommittee of the House Committee on Government Operations under the leadership of Wisconsin Representative Henry S. Ruess, recommended the expansion of the Refuse Act permit program in 1970 to cover pollution discharges. Partially as a result of Congressional and Justice Department interest in expanding the use of the Refuse Act, President Richard M. Nixon issued Executive Order 11574 requiring the Secretary of the Army to implement a Refuse Act permit program to control pollutant discharge into U.S. waters. The program required the Corps to consult with the Administrator of the EPA on water quality standards. Further consultation with the Secretaries of Interior and Commerce, along with state agency heads, was ordered to ensure consideration of wildlife and fisheries resources affected by such discharges. Compliance with NEPA, Federal Water Pollution Control Act of 1972 (FWPCA), and FWCA was required to reinforce a consideration of non-navigational concerns (Congressional Research Service 1973).

The Corps took quick action, issuing a notice of proposed rule making the next week. In April 1971 the final regulations outlined a permit program to regulate all discharges or deposits in navigable waters, tributaries, or wastewater treatment plants. Among the considerations required for permit issuance were impacts on water quality and fish and wildlife habitats. The EPA Administrator had final authority over water quality standards (Kalen 1993). With the proposed permit program, the Corps assumed a major role in the national efforts to stem pollution.

After the final regulations were established, permit applications flowed into Corps offices. Stine (1983) explained, "Nearly 20,000 permit applications were submitted to the Corps by December. Of the 11,000 processed and referred to EPA, only 21 were granted." Judicial action soon interrupted the program with a decision on *Kalur vs. Resor* (335 F. Supp. 1 [D.D.C. 1971]), which concerned the Grand River in Ohio. Entities wanting to use the river for disposal of material sued, alleging the Corps was not authorized under law to regulate disposal into non-navigable tributaries of navigable streams. The Court agreed with the plaintiffs, determining Section 13 specifically authorized permitting for discharges into navigable waters only. The regulation was held moot (Kalen 1993). Since the court decision left thousands of dischargers without any way to avoid violating the Refuse Act, the pressure increased on Congress to devise a solution.

FEDERAL WATER POLLUTION CONTROL ACT OF 1972

Congress responded to judicial action and public pressures to protect the nation's waters by passing the FWPCA in 1972. Its stated objective was "to restore and maintain the chemical, physical, and biological integrity of the Nation's waters" and included a national goal of elimination of pollutant discharges into navigable waters by 1985. One of the major innovations within the FWPCA was the creation of the National Pollution Discharge Elimination System (NPDES) which replaced the defunct Section 13 RHA regulation program. Each establishment desiring to discharge point source pollution had to apply for and receive a NPDES permit from the EPA (Greenfield 1985).

The FWPCA included a comprehensive water pollution control program. The aspects of water pollution addressed included sewage treatment, storm water discharges, industrial discharges and funding. Protection of wetlands was addressed only indirectly via nonpoint source (NPS) pollution control and permitting of dredge and fill activities.

NONPOINT SOURCE POLLUTION

There was no formal definition of NPS pollution in the FWPCA of 1972, but it may be understood to refer to an activity which generates pollution from a widespread land area, i.e., agricultural, developmental, or silvicultural practices. Congress focused on controlling nonpoint source pollution in Sections 201, 208, and 304. Section 201 provided grants for municipalities for construction of waste treatment works to control, as far as practicable, nonpoint pollution sources. Section 208 addressed both point and nonpoint source pollution with mandated area-wide waste treatment management plans. Under these guidelines, the governor of each state was to identify areas which had substantial water quality problems and designate their boundaries. Nonpoint source pollution, including that resulting from silvicultural activities, was to be controlled by the use of Best Management Practices (BMPs) rather than the more politically unpalatable land use planning or private land use regulations. BMPs were defined as:

> ...[T]hose methods, measures, or practices to prevent or reduce water pollution and include but are not limited to structural and nonstructural controls, and operation and maintenance procedures. BMPs can be applied before, during, and after pollution-producing activities to reduce or eliminate the introduction of pollutants into receiving waters. Economic, institutional, and technical factors shall be considered in developing BMPs. BMPs shall be developed in a continuing process of identifying control needs and evaluating and modifying the BMPs as necessary to achieve water quality goals ... To the extent practicable, BMPs should be set forth in a document which can be distributed widely in the planning area (12 CFR δ 35.1521-4(c) as cited in Greenfield 1985).

Section 304 addressed the need for agency guidance on NPS pollution activities. It directed the EPA Administrator to create guidelines for NPS identification, evaluation, and extent. In addition, guidelines were required for processes, procedures, and methods to control pollution from a variety of sources including agricultural and silvicultural activities (Greenfield 1985).

Section 404

The FWPCA switched the permitting system for discharges into navigable waters from the Refuse Act of 1899 to Section 404, thus eliminating the need for further action on President Nixon's permit program. In addition, the EPA Administrator was authorized to empower the states to assume responsibility for such permits if they were deemed capable of administrating a program (Congressional Research Service 1973).

Section 404 contained two basic regulatory elements — permit authority for dredge and fill operations and enforcement of the regulations. A permit was to be issued only if it "w[ould] not unreasonably degrade or endanger human health, welfare, or amenities, or the marine environment, ecological systems, or economic potentialities." Section 404 permit decisions were to be based on public interest, NEPA, and future guidelines to be created jointly with EPA. EPA was granted authority to disallow a Corps permit if the discharge would result in pollution of unique and valuable environmental resources.

IMPLEMENTATION OF THE FWPCA

Because the new law did not directly address wetlands, there was some disagreement over the range and scope of activities of both the Corps and EPA. The Corps had control over waters of the United States, but the extent of waters had not been clearly designated in law. The Corps' relationship with EPA was not clearly defined, and the EPA's attempts to fully respond to legislative mandates was impacted by this ambiguity.

The EPA Attempts To Regulate Forestry and Agriculture

Responding to the requirements of Section 208, the EPA attempted to regulate forestry and agricultural sources of pollution in December 1972. Agricultural and forestry sources were to be designated as point sources that would require a permit under the NPDES. As explained by Dana and Fairfax (1980), "This determination served mainly to illustrate EPA's lack of appreciation of the nature of land management and was a first indication to the forestry and range communities that their activities might be severely impacted by FWPCA implementation." Naturally, land managers objected and the EPA backtracked on their proposal. While agricultural and forestry sources were still considered by EPA as point sources, they were exempted from permitting requirements of Section 402. The Natural Resources Defense Council (NRDC) was not pleased with this exemption and sued EPA using the provision contained within the FWPCA for citizen suits (*NRDC vs. Train*. No. 1629-73 [D.D.C. 1975]).

EPA made an attempt in 1974 to control NPS pollution from forest management activities by drafting a model forest practices act that could be used by the states. The proposal contained requirements for regeneration, water quality protection, and amenity protection. The forestry community responded negatively, and the EPA abandoned its efforts for uniform state legislation (Cubbage and Harris 1988).

The Corps Begins Section 404 Implementation

In response to the newly passed legislation, the Corps proposed its initial 404 regulations in May 1973. The broad-reaching regulations were hailed by such environmental organizations as the Audubon Society. However, when the final regulations were issued in 1974, environmental groups were dismayed. The Corps had returned to its traditional definition of "navigable waters," limiting protection of wetlands in both coastal and freshwater sites. This decision was in spite of court cases such as *United States vs. Holland* (373 F. Supp. 665 [M.D. Fla. 1974]) which had extended federal jurisdiction to all waters that might be considered part of commerce — not just navigable waters . Environmental groups were outraged and began action to force the Corps to implement the law as written (Stine 1983, Blumm and Zaleha 1989).

The Environmental Defense Fund (EDF) began a program of challenging selected permit decisions through the judicial system hoping that a series of court decisions extending the range of permitting beyond traditionally navigable waters would influence the courts to order the Corps to revise the 404 permitting program. The NRDC took a different tack, filing suit directly against the Corps and Army officials in U.S. District Court, District of Columbia as *NRDC, Inc. vs. Callaway* (392 F. Supp. 685 [D.D.C. 1975]). The National Wildlife Federation (NWF) was joint plaintiff in the suit, which argued that regulations promulgated by the Corps did not meet the goals of the FWPCA . The case was decided in favor of the NRDC. Judge Aubrey E. Robinson, Jr. ordered the Corps to revise its regulations to extend the definition of navigable waters past mere water trafficking concerns, in accordance with the intent of Congress as spelled out in the FWPCA. The Corps responded promptly, putting forward four alternative regulations for public comment and review on May 6, 1975. Disappointed by the court's decision, the Corps Office of Chief Council prepared to appeal. However, the Justice Department declined to make the appeal, infuriating the Corps personnel involved with 404 revisions (Stine 1983).

Internal Resistance within the Corps

Gen. Kenneth E. McIntyre, deputy director of civil works, Col. Robert Hughes of the Public Affairs Office, and Jack Lankhorst of the Office of Chief Counsel plotted to bring the Section 404 issue to the attention of the public. Without formal approval of those higher in command, McIntyre and Hughes requested assistance from Locke Mouton, Deputy Chief of Public Affairs, in preparing an inflammatory press release to arouse public concern and action to restrict 404 jurisdiction. Mouton agreed to write the material but warned his peers of the consequences. He suggested that the release be cleared first with the Assistant Secretary of the Army for Civil Works. McIntyre refused to consult with the Assistant Secretary, and Mouton drafted a press release. After reading the draft, McIntyre and his staff decided it was not sufficiently inflammatory. Mouton revised his release by exaggerating the scope of the 404 program to incite maximum public outrage. One widely quoted section stated, "Federal permits may be required by the rancher who wants to enlarge his stock pond, or the farmer who wants to deepen an irrigation ditch or plow a field, or the mountaineer who wants to protect his land against stream erosion" (Corps of Engineers News Release 1975 as cited in Stine 1983).

McIntyre was successful and public opposition to the proposed expansion of 404 regulations exploded. Secretary of Agriculture Earl Butz added to the tumult with his own crusade against the program. Within the Army, the reaction was swift. Victor V. Veysey, Assistant Secretary of the Army for Civil Works demanded the firing of Mouton. Mouton was not fired, but Veysey made a public apology for the press release during the congressional hearings resulting from the excitement (Stine 1983).

The response from environmental groups was similarly condemnatory. The NRDC, the NWF, the Wilderness Society, the Audubon Society, and other groups publicly and privately protested the actions of the Corps. Mobilizing rapidly, environmental groups launched a counterattack with educational programs and outreach efforts to influence their group members and the general public to protest the evisceration of the Section 404 program (Stine 1983).

The public expressed intense feelings about the proposed new regulations. More than 4,500 written comments from politicians, federal and state agencies, agricultural and forestry groups, environmentalists, and other interested citizens were received and reviewed by the Washington Corps office. Together with the EPA, the Corps issued revisions on regulatory jurisdiction on July 25, 1975. Directives from top Corps officials made it clear that the Corps was to cooperate fully with the EPA and to comply without hesitation to the revised 404 regulations.

In order to regain the trust of the public, the Corps' Public Affairs Office launched a public relations initiative to improve relations with the EPA and improve the public conception of the Corps and its handling of the Section 404 program. As a result of the leadership from the top levels of the Corps, the regulations were implemented as written, and relationships between the Corps and environmental groups improved significantly (Stine 1983).

NPDES Permits Required for All Point Sources

While the Corps was in the midst of its internal dispute, *NRDC vs. Train* (1629-73 [D.D.C. 1975]) was decided. The case concerned the EPA's attempt to exempt small animal feedlot, agricultural, and silvicultural operations from the requirements of Section 402 for discharge permits. The court ruled EPA must issue regulations requiring NPDES permits for all point sources from concentrated animal feedlots, storm sewers, agriculture, and silviculture. In effect, this forced EPA to distinguish between nonpoint and point sources. Subsequent regulations categorized mills and woodyards as point sources (Cubbage and Siegel 1990). Rock crushing, gravel washing, log sorting, and storage facilities — all defined as silvicultural activities — were also classified as point sources (Dana and Fairfax 1980).

The Courts Address the Range of Section 208

Section 208 had required creation of area-wide waste treatment plans to control nonpoint source pollution. Originally, EPA regarded Section 208 planning as being limited to problem areas identified by the governor. The NRDC and the EDF sued over the limited interpretation. The courts decided in favor of the environmental groups, and EPA was required by the courts to extend Section 208 planning to cover all areas of a state in the decision of *NRDC vs. Train* (396 [F. Supp. 1386 1975] (Cubbage et al. 1993).

EPA Issues Guidelines on Section 404(B)

Under Section 404 (b), EPA was to promulgate guidelines for the Corps to use when making permit decisions. Due to the vast chasm that lay between the Corps' conception of the 404 program and that of the EPA, guidelines were not issued until September 5, 1975. Final action on these guidelines was prompted in part by the *Callaway* decision and the public outcry over the Corps' regulations. The EPA guidelines detailed concerns and goals for permit processing and criteria for evaluating proposed discharges. The guidelines also established a presumption against wetland filling unless the activity could not be performed on other sites, depended on certain methods of construction, or required proximity to water (Blumm and Zaleha 1989).

THE CLEAN WATER ACT

Congress amended the FWPCA as the Clean Water Act (CWA) of 1977. Although wetlands amendments were considered, none passed. Items of interest to wetland managers were the provisions for general permits and statutory exemptions. NPS programs were impacted by the passage of Section 303. Also included was a provision for citizen suits against violators of the CWA.

General Permits

The CWA authorized the Corps to continue their practice of issuing general permits covering states, regions, or the entire nation. Local general permits could be issued by district or division engineers for a five-year term, while nationwide general permits could be issued by the Corps headquarters. In essence, a general permit was a permit-by-rule as Section 404 would be satisfied by the landowner following the rules of the general permit system. In the subsequent 1982 Corps regulations, 26 nationwide permits were listed. Most of these were for isolated structures such as navigational aids or mooring buoys. Probably the most widely recognized was Nationwide 26, which authorized headwaters and isolated waters discharges (Strand 1993).

General Exemptions

Congress also exempted certain activities from regulation, giving as examples "Normal farming, silviculture, and ranching activities such as plowing, seeding, cultivating, minor drainage, harvesting…" Section 404(f)(1)(A), but other, similar activities are also exempt. Forest road construction is also exempted from the permit requirements under Section 404(f)(1)(E) as long as it is conducted in accordance with the 15 mandatory federal BMPs specified in Section 232.3(c)(6)(i-xv) of EPA's regulations. However, there are two important caveats in determining whether normal silvicultural activities are exempt. First, the EPA regulations stipulate that these activities must be part of an established (i.e., ongoing) operation. If they are not, a permit is required. Second, normal silvicultural activities must not result in a "new use," **and** cause a "reduction in reach/impairment of flow or circulation" of the waters. This two-part test is known as the Section 404(f)(2) recapture clause. The exemption for normal silvicultural activities was bound by the definitions of what constitutes normal silvicultural activities and the 404(f)(2) recapture clause. These constraints were imposed to prevent conversions of large acreages of wetlands into dry lands while allowing routine activities with only minor impacts to proceed (Strand 1993).

The Corps and EPA Work Out their Differences

The CWA failed to clarify whether the Corps or EPA had final jurisdictional authority to determine the extent of "navigable waters." The Secretary of the Army sought an opinion from Attorney Gen. Benjamin Civiletti to settle the matter. The Attorney General found that Congress had given EPA authority to make determinations of jurisdiction under the CWA including Section 404. The EPA also was deemed the ultimate authority for scope of Section 404(f) exemptions (Strand 1993).

After the Civiletti opinion was released, EPA issued guidelines as directed under Section 404(b) on December 24, 1980 (Kalen 1993). The new guidelines included the restriction for permitting activities that were not water dependent or could be constructed on an alternative site. They also expanded protection to areas important for fish and wildlife habitats or marine sanctuaries. There was a stipulation against discharging dredged or fill material unless it could be proven that the discharge would not create an "unacceptable adverse impact." The guidelines were declared by the EPA to be binding on the Corps and were not merely advisory (Blumm and Zaleha 1989).

Expansion of the Nationwide Permit Program

In July 1982 the Corps published an expanded nationwide permit program which immediately drew fire from environmental groups. The proposed regulations would have allowed general permits to be issued for isolated waters and waters above 'headwaters'. Previously, individual permits had been required in isolated waters greater than 4 ha (10 acres) in surface area. The USFWS estimated individual permit requirements would no longer be required for 283,400 to 364,400 ha (700,000 to 900,000 acres) of prairie potholes, up to two million lakes and wetlands in the upper Midwest, and 70% of Alaskan lakes. Sixteen environmental groups joined forces to file suit in *National Wildlife Federation vs. Marsh* (20262 [D.D.C. 1984]). The suit focused on Section 404(e), which allowed general permits to be issued for categories of activities when there were minimal cumulative impacts. There were no provisions within the law to issue general permits for classes of waters or wetlands (Blumm and Zaleha 1989).

The Corps settled the *National Wildlife Federation* suit in 1984 by agreeing to issue new regulations. These regulations required the Corps to:

1. Accept the 404(b) guidelines as mandatory.
2. Accept the presumption against wetland discharges.
3. Extend 404 coverage to agricultural clearing, draining, and channeling of wetlands across the entire nation.
4. Limit nationwide permits in isolated waters and headwaters to 4 ha (10 acres).
5. Require pre-discharge notification (PDN) on activities with significant impacts on 0.4 to 4 ha (1 to 10 acres).
6. Consult with EPA and FWS about proposed discharges on special aquatic sites of from 0.4 to 4 ha (1 to 10 acres) (Blumm and Zaleha 1989).

Adjacent Wetlands are Subject to Regulation

Judicial decisions continued to expand the reach of the Section 404 program. In 1985 the Supreme Court heard arguments on *United States vs. Riverside Bayview Homes, Inc.* (474 [U.S. 121 1985]). Riverside Bayview Homes was a development firm which owned approximately 32 ha (80 acres) near the Clinton River in Michigan. The normal drainage of the tract had been altered by a dike built by the Corps and Harrison Township, and construction on the property was impossible without some filling. Harrison Township issued Riverside a fill permit in 1976. When the developers did not begin fill operations, the Township declared the lands a public nuisance and threatened to fine the landowners. To avoid the fines, the developer began work. The Corps promptly issued a cease and desist order, alleging a violation of the FWPCA and the Section 404 permit program. The Supreme Court determined that the Riverside property was indeed wetland subject to regulation based on the vegetation present on the tract — vegetation which required saturated soil in order to flourish. The water that caused saturation of the soil was found to be ground water. In addition, the site was declared adjacent to navigable water. These three findings made it clear the site was a wetland. In their decision, the Court firmly stated that 'frequent flooding' of the area was not a prerequisite for wetlands delineation (Adams 1993).

The Corps Develops a Wetlands Delineation Manual

In January of 1987, the Wetlands Research Program of the U.S. Army Engineer Waterways Experiment Station at Vicksburg, Mississippi issued a manual for wetlands delineation replacing previous regional manuals. The manual contained technical guidelines and methods applicable for wetlands determination. A multi-parameter approach to determination was used with vegetation, soils, and hydrology as the parameters (USACE 1987).

The CWA is Amended

Congress amended the CWA in 1987 as the Water Quality Act. An important innovation in the legislation included Section 319 to control nonpoint source (NPS) pollution. This new language added implementation requirements to areawide waste treatment plans of Section 208 to control NPS pollution (Hohenstein 1987). A guide to the wetlands aspects of the CWA, as amended is found in Table 3.1.

Section 319 required the states to create detailed water quality management plans which identified water bodies not meeting water quality standards because of NPS pollution. Included within the plans were categories of NPS pollution and control mechanisms. These control mechanisms could be regulatory or voluntary, with regulatory mechanisms getting federal funding priority. Voluntary mechanisms were BMPs. The law mandated state water quality standards (WQS) as the criteria to measure success. The measurements had to respect EPA's antidegradation policy of maintaining at least the current water quality standards, even if they exceed state standards.

The General Accounting Office Studies 404 Regulation

The General Accounting Office (GAO) released "Wetlands: The Corps of Engineers' Administration of the Section 404 Program" in 1988. The House Committee on Public Works and Transportation had requested the GAO study coordination between federal agencies, Section 404 enforcement, sanctions on violators, and the general impact of the program on wetland resources (General Accounting Office 1988).

The GAO concluded the Corps did not have the authority to regulate most wetland losses which occur primarily as a result of normal farming and drainage activities. On individual project applications, the Corps considered resource agencies' comments but tended not to adopt the recommendations which would lead to permit denial or modifications. Appeal of Corps permit decisions by resource agencies was rare, largely due to the cumbersome appeals process. Monitoring wetland

TABLE 3.1
Quick guide to wetlands aspects of the Clean Water Act.

Section	Impact
Section 208	Area-wide waste treatment management. State or local government programs to identify agricultural and silvicultural related nonpoint sources of pollution and decide on feasible control procedures and methods.
Section 304	EPA Administrator to create guidelines for NPS identification, evaluation, and extent. Guidelines required for processes, procedures, and methods to control pollution from a variety of sources including agricultural and silvicultural activities.
Section 309	EPA may issue administrative compliance orders to stop illegal discharges and restore site. Authorization of civil judicial actions, criminal judicial actions, and penalties.
Section 319	NPS pollution control. State water quality management plans must include control of NPS pollution. BMPs are voluntary control methods.
Section 401	State water quality certification required for federal permitting.
Section 402	The National Pollution Discharge Elimination System (NPDES) to control point source pollution replaced Section 13 of RHA.
Section 404	Authority for dredge and fill permits given to the Corps. EPA given oversight authority over the permit granting process.
Section 404 (b)(1)	Spoil disposal sites for 404 permits must meet guidelines developed by the EPA and Corps.
Section 404 (c)	EPA given the power to prohibit discharges that could negatively affect municipal water supplies, marine or freshwater breeding areas, or wildlife habitat. Public notice and hearings were required along with consultation with the Corps.
Section 404 (e)	Authorization of general permits — nationwide, regional, or state.
Section 404 (f)(1)	Normal farming, silviculture, and ranching activities are exempted from regulation. The activity must be part of a continuing, ongoing operation. Conversion of wetlands for agricultural or forestry use is not exempt. Farm or stock ponds are exempt from regulation.
Section 404 (f)(2)	Recapture provision. An activity exempted in (f)(1) is subject to regulation if the purpose is to bring the area into a new use which would affect the flow or circulation of navigable water or reduce its reach.
Section 404(j) and Section 404(m)	Authority for DOI USFS to comment on permits.
Section 502(14)	Point source pollution defined as relating to pollution from a discrete source. Nonpoint sources were not defined but one legal definition was "disparate runoff caused primarily by rainfall around activities that employ or cause pollutants" (Whitman 1989).
Section 505	Authorization for citizen suits against violators, Corps, EPA, and governmental agencies.

activities was not a priority of either the Corps or EPA. Thus, the prevalence and extent of violations were unknown. When violations were observed, the Corps generally tried to obtain voluntary correction, seldom revoking permits or seeking the appropriate civil or criminal penalties (GAO 1988).

Other CWA Provisions Affecting Wetland Forest Management

The CWA granted standing for citizens to sue violators of the act including governmental agencies such as the Corps or EPA. The citizen had to notify the purported violator, the EPA Administrator, and relevant state officials at least 60 days prior to beginning the civil suit. Negligent violators of the CWA could be fined a criminal penalty of $2,500 to $25,000 or could receive a prison sentence of up to one year per day of violation under a first offense. Further violations increase the penalty to $50,000 and two years, respectively. Penalties were even more severe for deliberate violations or for knowingly endangering others. Civil penalties could be assessed when the offense was neither willful or negligent. In this case, the penalty could be as much as $25,000 for each day the CWA is violated (Adams 1993).

The CWA allowed states to establish their own water quality standards for point source pollution, but the EPA could impose its own standards if state standards were not as strict as federal standards. In Section 303, the act directed states to designate segments of streams or other watercourses which were "water quality limited." This term referred to areas within a watercourse which received so much point source pollution that discharge standards were not adequate to maintain water quality at their existing or designated beneficial uses (i.e., drinking, recreation, cold-water fishery, agriculture). States were to establish total maximum daily waste loads for such areas, including load allocations for NPS pollution and a wasteload allocation for point source discharges. Any NPS control program which was part of a statewide Section 303 program had to consider the impact of NPS pollution on wetlands. This was done with three requirements: (1) plans must be coordinated with state Section 404 programs; (2) discharge activities were performed according to BMPs; and (3) state groups with control of fish and wildlife resources had to be consulted (Greenfield 1985).

DEBATE OVER A MANUAL

In order to protect wetlands, there had to be a method of determining the extent and range of the regulated area. The Corps and EPA had jointly agreed upon a definition of wetlands which was used in the 1987 Manual:

> Those areas that are inundated or saturated by surface or ground water at a frequency and duration sufficient to support, and that under normal circumstances do support, a prevalence of vegetation typically adapted for life in saturated soil conditions. Wetlands generally include swamps, marshes, bogs, and similar areas (USACE 1987).

This definition and the delineation criteria from the 1987 Manual were not used by all federal agencies.

The 1989 Wetland Delineation Manual

Recognizing the need for a consistent method of delineating wetlands across federal agencies, the Corps, EPA, USFWS, and the Soil Conservation Service (SCS) jointly produced the Federal Manual for Identifying and Delineating Jurisdictional Wetlands (Federal Manual) in 1989. Tripartite criterion using hydrology, vegetation, and soils were used to determine if a site was a wetland under federal jurisdiction (Ogle 1993). The agencies were pleased with the new manual since delineation processes were simplified and the use of one manual across agencies increased the probability of coherence. However, the 1989 Federal Manual generated enormous public disapproval because landowners felt that the criteria vastly increased the amount of land subject to regulation under the CWA without opportunity for public review and comment prior to its use in the field. Agriculture, forestry, home developers, and other business interests, angered by a unilateral action they considered as an expansion of wetland area by the Administration, sought repeal of the Manual (Ogle 1993).

A New Manual is Proposed

A National Wetlands Policy Forum was held by the EPA in 1987 to make recommendations for national wetlands policy. The overriding principle of "no overall net loss of wetlands" was adopted as the centerpiece of the policy recommendations (The Conservation Foundation 1988). President George Bush adopted the "no net loss of wetlands" as national policy on August 9, 1991. As part of his wetlands policy initiative, on the advice of the Council on Competitiveness chaired by Vice President Dan Quayle, Bush also recommended the development of a new federal wetlands delineation manual. The resulting revised delineation manual was published on August 14, 1991. The new manual (Proposed Revisions) differed from the Federal Manual for all three factors required for a wetlands determination: hydrology, vegetation, and soils. The four federal agencies who developed the 1989 Manual claimed the net effect of the Proposed Revisions would have reduced the amount of land regulated under Section 404 (Ogle 1993). During the comment period on the Proposed Revisions, public interest was intense. more than 50,000 written comments were sent to EPA, and the EPA Wetlands Hotline received approximately 80,000 calls concerning the manual. No action was taken by EPA on the manual. Congress effectively nullified any attempt to implement the 1989 manual by the Energy and Water Development Appropriations Act of 1992, which specifically barred the Corps from using it or any other manual that did not follow the requirements of the Administrative Procedure Act. The Corps resumed use of the 1987 Manual, which continues to be the accepted manual as of 1997. A comparison of the three manuals is found in Table 3.2.

Recommendations of the National Research Council

As a result of the controversy over wetlands delineation, Congress sought further information on the science of wetlands characterization. Responding to Congressional concerns, EPA requested a study through the National Research Council (NRC). The NRC's Water Science and Technology Board created a committee to study wetlands delineation methods, consequences of the alternative methods, functions of wetlands, and regional variations. The resulting study was released by the NRC in the summer of 1995. The report contained a new reference definition of wetlands:

TABLE 3.2
Comparison of wetlands delineation manuals (modified from National Research Council 1995).

Working Title	1987 Manual	1989 Manual	1991 Proposed Revisions
Actual Title	Corps of Engineers Wetlands Delineation Manual	Federal Manual for Identifying and Delineating Jurisdictional Wetlands	Proposed Revisions to the 1989 Federal Manual for Identifying and Delineating Jurisdictional Wetlands
Sponsoring Agency	USACE Waterways Experiment Station, Vicksburg, MS	USACE, EPA, USFWS, SCS	Bush Administration & USACE, EPA
Hydrologic Criteria	Inundation or saturation in vegetation root zone (usually within 30 cm [1 ft] of surface) for more than 12.5% of the growing season; 5–12.5% with supporting evidence	Inundation or saturation of soil within 15–46 cm (6–18 in) of the surface for seven consecutive days	Soil inundated for at least 15 consecutive days or saturated to the surface 21 or more consecutive days
Vegetative Criteria	More than 50% of dominant plants obligate, facultative wetland, or facultative. Observed indicators of hydrophytic vegetation	More than 50% of dominant plants obligate, facultative wetland, or facultative. Frequency analysis of all species	Frequency analysis of all species; for non-vegetated tracts, criteria assumed to be met if soils and hydrology made growth of wetlands plants likely
Soils	National Technical Committee for Hydric Soils list	NTCHS criteria-same as hydrologic criteria	NTCHS criteria plus three other categories
Comments	Presence of fewer than three factors only for special cases; some conflicts with NRCS, USFWS, EPA manuals	Strong evidence of two factors can be used to infer presence of third.	Each factor considered separately; soil must have visible surface wetness; vegetative criteria more restrictive than 1987 manual
Present Status	Official Manual	Congress forbids use unless landowner consents	Not used

A wetland is an ecosystem that depends on constant or recurrent, shallow inundation or saturation at or near the surface of the substrate. The minimum essential characteristics of a wetland are recurrent, sustained inundation or saturation at or near the surface and the presence of physical, chemical, and biological features reflective of recurrent, sustained inundation or saturation. Common diagnostic features of wetlands are hydric soils and hydrophytic vegetation. These features will be present except where specific physiochemical, biotic, or anthropogenic factors have removed them or prevented their development.

The NRC explicitly recommended the creation of a new manual. Using existing manuals as a starting place, the study recommended that a draft be developed using the information gained by field use of previous manuals and the new knowledge gained by recent wetlands research. In addition, the NRC called for the use of regional supplements to "provide detailed criteria and indicators consistent with the federal manual" (National Research Council 1995). No action has been taken on these recommendations as of 1997.

RECENT CASE LAW AFFECTING FORESTED WETLANDS

While there have been many legal decisions which directed the development of wetlands regulations, three recent cases have particular importance to forest managers: *Avoyelles, Tulloch, and the Parker Tract.*

Avoyelles Sportsmen's League vs. Marsh

Avoyelles Sportsmen's League vs. Marsh (715 F.2nd 897, 910-13 [5th Cir. 1983]) had a definite impact on forest conversion activities. In the *Avoyelles* case, several Louisiana landowners with large acreages cleared their bottomland hardwood swamps of timber and prepared the ground for soybean production by shearing, piling, and burning. The Avoyelles Sportsmen's League filed suit as interested citizens against the Corps, EPA, and the landowners, claiming that the clearing constituted a violation of the Clean Water Act because the landowners did not have a 404 permit. The Court agreed, finding the clearing machinery caused point sources of pollution as fill material was being deposited via the leveling process. Not only was a permit required because of the filling involved, the clearing also qualified for regulation because it was intended to change the usage of the land from forestry to agriculture, triggering the recapture provision (Adams 1993).

The Tulloch Rule

Land clearing was also the focus of a more recent suit in North Carolina, *NCWF et al. vs. Tulloch* (C90 713 [E.D.N.C. 1992]). In the Tulloch case, the North Carolina Wildlife Federation and the NWF filed suit against the Corps over the development of an 729 ha (1800 acres) tract containing 283 ha (700 acres) determined by the Corps to be wetlands. The developer had originally applied for permitting, but withdrew the application after concerns were expressed over potential adverse effects. He then pursued the strategy of meeting with the Corps to determine ways to avoid the necessity of obtaining a Section 404 permit. His first step was to remove the vegetation without recontouring the land, which did not require a permit. Then the developer altered the drainage patterns by excavating new ponds and drainage ditches. The soil was removed with draglines or backhoes and all soil excavated was placed on uplands or in sealed containers. At the time, fill material was defined by the Corps as "any material used primarily for replacing an aquatic area with dry land or changing the bottom elevation of a body of water." Thus, the Wilmington District Corps determined that these activities did not deposit dredge material on the area.

Because of these actions, the water table dropped and the area was no longer considered wetland by the Corps. A golf course and residential development were built on the altered wetlands. The developer consulted with the Corps on a regular basis to make sure that each stage of the development could be implemented without need for a permit. The resulting successful lawsuit by the Wildlife Federation prompted the Corps to clarify the rules on discharges in order to prevent further evasions of this sort (Federal Register 1993).

In August 1993, the Corps released new regulations (commonly referred to as the Tulloch Rule) which explicitly stated that the term 'dredging for navigation' was not applicable to 'dredging activities in wetlands.' This definition made it clear that 'incidental movement' of dredged soil, excluded from Section 404 permitting, occurs only during normal dredging operations. Section 404 permits would be required for any activity such as mechanized land clearing that degrades or destroys wetlands. Any redeposition of dredged material is considered a discharge. Cutting of vegetation above the ground where roots are not disturbed is not considered a discharge, but pushing the roots up where soil is displaced would be. However, a permit may not be required if the landowner could prove in advance that proposed activities would not destroy or degrade waters of the United States, or would fall under the general exemptions (Federal Register 1993).

The Parker Tract

In 1991 the EDF, along with other environmental organizations, sued EPA, the Corps, and Weyerhaeuser Company alleging violations of the Clean Water Act in their management of land known as the Parker Tract (*EDF et al. vs. Hankinson,* 91467 [5th D. N.C. 1995, dismissed]). The Weyerhaeuser Company had purchased the Parker Tract in 1967 and 1969, which at the time was drained by major canals on three sides of the tract. Parallel drainage canals alongside the 39 km (24 mi) of forest roads bisected the tract. Acting under then current regulations, Weyerhaeuser performed some additional ditching on the tract and maintained the original ditches. Weyerhaeuser prepared and implemented written forest plans to convert the existing pine-mixed hardwood stands to managed pine plantations as part of their strategy of supplying wood to their nearby mill (USEPA 1994).

EDF claimed in its suit that converting a natural stand to a plantation was not part of a normal silvicultural operation and thus constituted a "use to which it was not previously subject" using the language of Section 404(f)(2). EPA asked the court to remand the case so EPA could make a formal determination of whether Section 404(f) would be applicable to this situation. After several years of debate, motions, and maneuvers, EPA determined the forestry activities conducted on the Parker tract were indeed part of Weyerhaeuser's normal silvicultural operation. In addition, they found that the plowing, bedding, cultivating, field ditching, and road construction occurring since 1978 was exempt under Section 404(f). Most importantly, EPA stated the conversion of the tract to a pine plantation did not represent a 'new use', thus no recapture had occurred (USEPA 1994).

Although this decision clearly supported Weyerhaeuser's managment activities, one question remained. EPA determined that the site preparation performed by

Weyerhaeuser was permissible under previous regulations, but questioned whether such future site preparation might be regulated as landclearing under the new Tulloch rule. EPA initiated a process to resolve this uncertainty. Participants included representatives from EPA and other federal agencies, state agencies, environmental groups, foresters, and other interested groups. The process concluded in 1995 with agency guidance that identified nine special wetland types in which mechnanical site preparation to establish pine plantations could not be initiated without a Section 404 permit: (1) permanently flooded, intermittently exposed, and semi-permanently flooded wetlands; (2) riverine bottomland hardwood wetlands; (3) white-cedar swamps; (4) Carolina bay wetlands; (5) non-riverine forest wetlands; (7) wet marl forests; (8) tidal freshwater marshes; and (9) maritime grasslands, shrub swamps, and swamp forests. Detailed descriptions of these wetlands are included in the guidance. For all other wetlands, no permit would be required as long as mechanical site preparation to establish pine adhered to the six BMPs listed in the guidance (Federal Register 1996a).

After the EPA guidance was issued on the Parker tract, the Court dismissed with prejudice the lawsuit against Weyerhaeuser, the Corps, and EPA. As part of this dismissal, a Conservation Partnership between the Weyerhaeuser Company and the environmental organizations involved in the lawsuit was established. Its goal "will be an interactive process to share scientific and economic information pertaining to Weyerhaeuser's management of water, wildlife, and plant resources on the properties included within the Conservation Partnership and to develop appropriate criteria for management of property to protect those values, consistent with Weyerhaeuser's forestry objectives. The Conservation Partnership will apply to Weyerhaeuser's 4,453 ha (11,000 acres) Parker Tract. In particular, the Conservation Partnership will focus on the following: natural heritage areas, water quality, unique wetland sites, and wildlife" (EDF 1995).

In January 1997, the District of Columbia federal court found that the Corps and EPA had unlawfully expanded their regulatory authority under the Tulloch Rule in *American Mining Congress vs. U.S. Army Corps of Engineers*, Civil No. 93-1754 (D.D.C.) Under the Parker Tract decision, the EPA had interpreted the general exemptions under Section 404(f)(1)(A) as being limited to those specifically listed (i.e., plowing, seeding, cultivation, minor drainage, [or] harvesting). Because of this interpretation, mechanized site preparation which had more than an inconsequential effect on the wetland would require a permit under the Tulloch rule. A suit was filed against this decision by the American Mining Congress, the National Association of Homebuilders, and the National Aggregate Association. The American Forest and Paper Association later joined the plaintiffs. Joining the government were the National Wildlife Federation, the North Carolina Wildlife Federation, the National Audubon Society, and the Sierra Club.

The government has requested that the court's judgment be limited to the plaintiffs in the case. The plaintiffs have objected, and no agreement has yet been reached on this limitation. It is expected that the government will appeal the court's finding of unlawful expansion of regulatory authority (AF&PA 1997).

Site Preparation Memorandum

Coinciding with the dismissal of the lawsuit by the environmental community, EPA and the Corps issued a clarifying Memorandum to the Field titled "Application of Best Management Practices to Mechanical Silvicultural Site Preparation Activities for the Establishment of Pine Plantations in the Southeast." This memorandum defines mechanical site preparation as including shearing, raking, ripping, chopping, windrowing, piling, and other similar physical methods used to cut, break apart, or move logging debris following harvest for the establishment of pine plantations. In essence, the Memorandum clarifies EPA's 1994 remand decision in the Parker Tract litigation by endorsing a BMPs approach to determine whether site preparation activities are within the scope of the silviculture exemption. The Memorandum has two components: (1) mechanical silvicultural site preparation activities conducted in mixed hardwood/pine sites and pine-dominated wetland types in accordance with BMPs that did not require a Section 404 permit; and (2) mechanical silvicultural site preparation activities that require a permit in certain of the wettest wetland types and special circumstance areas (BMP MOA 1995).

RELATED FEDERAL LEGISLATION

Although the Clean Water Act is the primary law affecting wetlands, other legislation exists which has implications for wetlands management. Coastal wetlands are affected by the Coastal Zone Management Act (CZMA) and the Coastal Wetlands Act. Agricultural wetlands are exempt from Section 404 regulation and must meet guidelines of the Food Security Act. Concerns for wetlands protection prompted passage of the Emergency Wetlands Resources Act. The North American Wetlands Conservation Act was designed to benefit migratory birds and other wildlife.

Coastal Zone Management Act

The CZMA was passed in 1972 in order to protect, preserve, develop, and restore or enhance the coastal lands of the United States. It called for the states to produce a voluntary coastal zone management program, meeting certain federal standards, to be approved by the federal authorities in the National Oceanic and Atmospheric Administration (NOAA). Financial assistance for the development of CZMA plans encouraged states to participate within the program. Another incentive was the requirement that federal activities within the affected areas must be consistent with the state CZMA plan. Despite the financial incentives, some states have chosen not to participate in the program.

CZMA includes federal wetlands permits under Section 404, which are subject to state approval within the realm of the coastal zone. Each state designates the boundaries of their coastal zones and determines which activities or resources are regulated within the coastal zone according to each state's plan. In some states, coastal wetlands are specifically regulated under the CZMA, but noncoastal wetlands are exempt from state regulation. Nonpoint source pollution control and wetlands protection are included in each state's program. The EPA has review authority along with NOAA over the NPS control plans. Forest nonpoint sources are to be controlled by the use of BMPs (Cubbage et al. 1993, Strand 1993, Want 1996).

Coastal Wetlands Act

The Coastal Wetlands Planning, Protection, and Restoration Act was passed in 1990. Although the act primarily focuses on Louisiana wetlands, there is a coastal wetlands restoration program which provides cost share assistance to coastal states. Seventy percent of the funds are designated for Louisiana, 15% for other coastal states, and 15% to the North American Waterfowl Management program for habitat protection (Strand 1993).

The Food Security Act

Agricultural wetlands are excluded from regulation under Section 404. The Food Security Act (FSA) of 1985 defined agricultural wetlands with three criteria, similar to the Corps' criteria: hydric soils, presence of surface or ground water that could support hydrophytic vegetation, and presence of such vegetation under normal conditions. The FSA required the USDA to develop criteria for hydric soils and hydrophytic vegetation and lists of soils and plants meeting the criteria. The Natural Resources Conservation Service (NRCS) used the National Technical Committee for Hydric Soils list to prepare soils maps that delineate wetlands. Vegetation lists were based on the national list of plants compiled by the USFWS (Strand 1993).

Under the FSA, the NRCS was charged with wetland identification and delineation on agricultural land, an analogous position to that of the Corps of Engineers on non-agricultural lands. The NRCS was to determine the status of lands in regard to USDA standards, i.e., degree of erodibility, presence of wetlands or converted wetlands, and existence and use of an approved conservation plan. Finally, they determined if crop production was allowed on converted wetlands based on the effect on wetland values (Adams 1993).

Unlike the CWA, the FSA listed quite a number of exceptions to wetland designation. Wetlands converted to cropland prior to 1985 were exempt. Also exempt were artificial lakes, ponds, wetlands used for agriculture, irrigated areas, wetlands that could be cropped with further alteration, and wetlands converted by acts of third parties unless there was a deliberate attempt to avoid regulation. In addition, where the NRCS determined that commodity production would only have a minimal impact, the wetland could be exempt. Finally, hardship exemptions could be granted by the Deputy Administrator, State and County Operations and Consolidated Farm Services Agency (formerly the ASCS), when debts were incurred prior to December 23, 1985 for the primary purpose of conversion (Strand 1993).

Swampbuster Provisions

The main provision of the Food Security Act affecting wetlands was the Swampbuster program, which linked the destruction of wetlands to overall farm management. Farmers were denied commodity program benefits such as crop insurance, disaster payments, and other federal benefits if they converted wetlands to agricultural production after December 23, 1985 (Strand 1993). Cubbage et al. (1993) reported the NRCS has applied Swampbuster provisions to the management of forested farm wetlands, advising landowners that federal incentives programs could not be used for mechanical site preparation where stumps would be removed and soil disturbed.

The Corps issued new regulations in 1993 which excluded prior converted (PC) agricultural wetlands from the definition of waters of the United States. Prior converted cropland has been modified by human activity so that it is no longer a wetland by hydrology or vegetation standards. This decision paralleled the National Food Security Act Manual. EPA also adopted the NRCS rules on abandonment of PC cropland, i.e., PC cropland had to be used for commodity production at least once every five years or be in a rotation with aquaculture, grasses, legumes, or pasture or it would be considered subject to regulation (Federal Register 1993). The 1996 Farm Bill made several changes in Swampbuster provisions. It expanded the areas used for mitigation and the options for mitigation. Under the bill, as long as PC wetlands are used for agriculture, the land will be designated as PC cropland. Farmers with Farmed Wetlands or Farmed Wetlands Pasture can allow an area to revert to wetlands and then convert it back to Farmed Wetlands without invoking the Swampbuster provision. Even if PC wetlands were abandoned, they would not be subject to regulation unless the wetlands hydrology became reestablished and all three wetlands criteria were met. The changes in Swampbuster also included giving the Secretary discretionary authority over penalties (USDA 1996).

The Emergency Wetlands Resources Act of 1986

The Land and Water Conservation Fund Act of 1965 created a fund to be used to purchase natural areas such as wetlands. This act was amended by the 1986 Emergency Wetlands Resources Act (EWRA). With this legislation, Congress specifically recognized the variety of benefits associated with wetlands rather than focusing on wildlife habitat or water quality issues. EWRA required states to consider wetlands within their overall outdoor recreation plans and continued the National Wetlands Inventory Project of the USFWS (Strand 1993). It also required the USFWS to develop a National Wetlands Priority Conservation Plan. The plan, which was approved on February 8, 1989, identified criteria for making public wetlands acquisition decisions. Such criteria included the presence of rare or declining wetland types, degree of public benefit, functions and values of wetlands, and pattern of wetlands losses (USFWS 1990). Under the 1996 Emergency Wetlands Resources Act, Regional Wetlands Concept Plans have been developed.

The North American Wetlands Conservation Act of 1989

Congress passed legislation in 1989 to protect and restore wetland ecosystems and other habitats for migratory birds and other wildlife. The North American Wetlands Conservation Act was designed to complement the North American Waterfowl Management Plan by encouraging partnerships between public agencies and other interests. Wetland projects with Canada and Mexico were to be funded through the act to obtain property rights to land or waters, restore habitats for migratory birds and other species, and to provide training and development of necessary infrastructure in Mexico to conserve wetlands (USFWS 1990).

FOREST MANAGEMENT IN WETLANDS

Landowners and foresters are often hesitant about making forest management decisions on wet areas. Although the specter of wetlands regulation is a daunting one, most forest management activities generally do not require permission from the Corps. The remainder of this chapter will cover what a landowner or forester must do to comply with federal regulations when making forest management decisions for wetlands. State regulations have not been covered and landowners are advised to consult with the state forestry agency to determine their responsibilities under state rules. This section will discuss wetlands delineation; subsequent sections will discuss the guidelines for forest management in wetlands.

PRELIMINARY WETLANDS JURISDICTIONAL DELINEATION

Regulation of land use under Section 404 depends on whether the area is considered as waters of the United States as determined by the legislative, judicial, and regulatory standards. Landscape/watershed position may be used to determine ecologically defined wetlands but jurisdictional wetlands definition depends on standards of law. Regulations issued in 1977 delineated four categories of jurisdictional waters:

Category 1. Coastal and inland waters, lakes, rivers, and navigable streams as well as adjacent wetlands.

Category 2. Tributaries of navigable waters, including adjacent wetlands. Nontidal drainage and irrigation ditches feeding into navigable waters were excluded.

Category 3. Interstate waters and their tributaries along with adjacent wetlands.

Category 4. Other United States waters where the destruction or degradation could affect interstate commerce. Examples included isolated lakes and wetlands, intermittent streams, and prairie potholes (Kalen 1993).

WETLAND CRITERIA

Wetlands within the four categories are delineated using the 1987 Corps Manual using the three-parameter criteria of soils, hydrology, and vegetation. Soils are considered to meet the criteria when there is either direct evidence that the soil is wet or it may be inferred that the soil developed under permanently wet or periodically saturated conditions. Some of the resources to aid in determining the presence of hydric soils include USDA Soil Surveys, National List of Hydric Soils from the NRCS, U.S. Geological Survey quadrangle maps, aerial photographs from the local Consolidated Farm Service Agency (CFSA), and the National Wetlands Inventory from the USFWS. Although these sources are helpful in preliminary delineation efforts, only evidence observed in the field can determine the presence of hydric soils. The soil should show evidence such as reduction and movement of iron or organic matter accumulation. Sulfudic odor and color can be used as indicators (Craft et al. 1995).

The vegetative criterion is met when the area is dominated by hydrophytes or plants which are adapted to survive saturated soil conditions during the growing season. Observations of adaptation to saturated conditions can indicate wetlands. The National Lists of Plant Species that Occur in Wetlands prepared by the USFWS categorizes plants by wetland indicator status such as obligate wetland plants, obligate upland plants, facultative wetland plants or facultative upland plants. Dominance can be measured by percent cover, stem density, frequency of occurrence, or basal area (Craft et al. 1995).

Hydrologic indicators such as visible evidence of soil saturation within 30 cm (12 in) of the surface, watermarks, sediment deposits, or evidence of drainage patterns can be used to determine a site's hydrology. Historical data, including aerial photos, groundwater well data, or stream gauge data can be used if available. Secondary data, such as oxidized root channels within 30 cm (12 in) of the soil surface, water marks, drift lines, or water stained leaves, can be used as well, if at least two secondary indicators are present (Craft et al. 1995).

Finally, the determination must be made in consideration of "normal circumstances." When human activities have altered the vegetation or drainage of an area, the criteria might not be met. In this case, the delineator needs to infer what conditions would be under normal circumstances. For agricultural land, crops are not considered normal circumstances (Strand 1993).

EXEMPTIONS AND EXCLUSIONS

Landowners and foresters who want to perform forest management activities in wetlands are sometimes daunted by the prospect of dealing with the Corps. If one depended on media reports, one might conclude that landowners are prohibited from managing their wetlands for any purpose. This is a misconception. All activities in wetlands are not regulated by the Corps. The following section will cover some of the exemptions and exclusions which are relevant to forest management of wetlands.

EXEMPTIONS

Section 404 (f)(1) contains statutory exemptions for certain activities in wetlands. Normal farming, forestry, and ranching activities are exempt from regulations when they are part of established, ongoing operations. Landowners are not required to contact the Corps or EPA prior to commencing activities which fall under these exemptions, although it would be wise to maintain records of such activities for proof of ongoing management. Exemptions apply only as long as the activities do not bring the area into a new use which might alter the natural flow of waters. Changing the current use of a wetland into farming, forestry, or ranching would require permitting because such a change would not be part of an established operation (Want 1996). One exception would be the conversion of PC wetlands into a new use, as these lands are exempted from regulation unless abandonment and wetland criteria are met.

The requirement that activities must be ongoing includes both a past and a future perspective. The land must have a history of forest management activities as well as a genuine intent for future timber management. Timber cannot be harvested under the forestry exemption in order to change to a new use without triggering a requirement for permitting.

Activities considered as normal forestry operations include harvesting timber, plowing, seedbed preparation, planting seedlings, cultivating, plowing fire lines, chemical applications, and even site preparation activities — following agency guidance (Federal Register 1996a). The primary purpose of any of these activities must be for forest management. Thus, fire lanes constructed for hunting access would require permitting while the same lanes created for timber protection under the general forestry exemption would not require a permit.

Farm and Forest Roads

Building and maintaining roads for farm and forest management are exempt from regulation, providing the BMPs and 15 baseline provisions listed under Section 404, Part 232.3 are followed. The number and size of roads must be kept to a minimum necessary to meet the specific purpose of the exempted activity. Water flows must not be restricted and fill must be stabilized to prevent discharges. Equipment travel within streamside management zones must be minimized. Migration and movement of aquatic life within waters must not be disrupted, and endangered species must be protected. Discharges should be avoided in migratory waterfowl areas and are prohibited in concentrated shellfish production areas. Temporary fills must be removed and the original contours restored (Strand 1993). The primary purpose of road construction must be for farming or forestry. Roads built primarily for hunting or recreation, even when used secondarily for forest management, are not exempt from regulation.

Minor Drainage and Farm Ponds

Irrigation and drainage ditches may be maintained under the 404(f) exemption for minor drainage. Levees, dikes, and ditches which were constructed prior to regulation may be maintained, but they may not be modified unless permitted. It is permissible to construct temporary sediment basins as well as perform minor drainage in order to continue a forestry or farming operation. Drainage activities which have significant impacts downstream or alter the flow of waters require a permit. The use of flashboard risers to control water levels may not require a permit if part of ongoing silvicultural activities.

Construction of farm or stock ponds which discharge dredge or fill materials into U.S. waters is allowed under Section 404(f)(1)(c). Ponds do not have to be part of an ongoing operation but they do have to be used for farming or ranching. They are exempt only to the size necessary for the farming or ranching operation. Ponds created for other uses such as recreation or aesthetics are not exempt from regulation. Dredging, rebuilding existing levees or dikes, and other maintenance activities are allowed to maintain existing ponds. Section 404 applies only if it brings a water into a new use and impairs the flow or circulation or reduces the reach of U.S. waters (Want 1996).

Nationwide Permits

The Corps has generated 39 nationwide permits (NWPS) which allow certain activities that pose minimal disruption to wetlands. These permits are subject to 15 general conditions on a variety of issues such as compliance with the Endangered Species Act (ESA) and the Wild and Scenic Rivers Act, erosion and siltation controls, and multiple use of the NWPS on a single project. Corps District Engineers have authority to modify or suspend nationwide permits where they deem it necessary to protect wetlands. For some of the permits, a pre-construction notification (PCN), formerly known as pre-discharge notification (PDN), is required along with a wetlands delineation. Some of the permits which may be of interest to forest managers include activities such as fish and wildlife harvesting or enhancement (NWP4), scientific measuring devices (NWP5), bank stabilization (NWP13), road crossings (NWP 14), minor discharges (NWP18), minor dredging (NWP19), headwaters and isolated wetlands discharges (NWP26), wetland and riparian restoration and creation activities (NWP 27), and moist soil management for wildlife (NWP 30). Specifics on these permits can be found in the Federal Register, the Corps offices, or on the Internet (Federal Register 1996b).

Nationwide 26 permits (NWP26), covering discharges into headwaters and isolated wetlands, has been the most controversial general permit because of its possible cumulative effects on wetlands destruction. This general permit was revised in December of 1996 and reissued for two years. The Corps plans to use the two year period to develop replacement permits for NWP26 which can be regionalized to reflect differences in wetlands functions and values due to geographic differences. NWP26 was the only general permit which based its application on wetlands position rather than on a specific activity and plans call for this, too, to be revised between 1997 and 1998. The Corps plans to replace NWP26 with several activity-specific permits which can be conditioned with regional requirements. Formerly, fills of up to 0.4 ha (1 acre) were allowed with no review, and up to 4 ha (10 acres) could be filled with minimal review. The interim permit changed the criteria to 0.13 ha (0.33 acre) and 1.2 ha (3 acres), respectively. The majority of projects over 0.13 ha will require a pre-construction notification (PCN). As before, District Engineers may require individual permits for specific cases and may exclude specific geographical areas from eligibility for NWP26 (Federal Register 1996b).

Regional General Permits

The Corps may issue general permits which apply only to specific states or regions. For example, some states have general permits for wildlife food plots which are not covered under the general forestry exemption. Foresters need to check with their local Corps office to see if local general permits may allow them to proceed with desired management activities without getting an individual permit.

AGRICULTURAL WETLANDS

Often agricultural landowners have forested acres included within their holdings and must consider their forest management activities within the context of their agricultural management. Landowners who wish to convert agricultural lands to forest lands need to consider regulation under the current Farm Bill as well as the Corps' Section 404 program. PC wetlands are exempt from regulation as long as they have not been abandoned according to current federal regulations and they have not recovered their wetlands characteristics. Regulation of agricultural wetlands under the 1996 Farm Bill is specified in 61 Federal Register, Vol. 61, No. 174, pp. 47019-47037. Because of the complexity of the rules, it would be wise for those dealing with agricultural wetlands to seek current information for specific projects.

STATE WATER QUALITY STANDARDS AND SILVICULTURAL BMPS

Activities requiring Section 404 permitting must also meet state WQS developed for Section 401 compliance. Landowners should contact the appropriate state agency for certification information. Under Section 319, each state must have a program in place to control NPS pollution. According to a report from the National Council of the Paper Industry for Air and Stream Improvement (NCASI), the twelve southern states have active NPS control programs which include the use of BMPs in forest management activities. Voluntary BMP programs are found in Alabama, Arkansas, Louisiana, Mississippi, Oklahoma, South Carolina, Tennessee, Texas, and Virginia. Florida and Georgia have programs with limited regulation, while North Carolina's program suggests use of BMPs to utilize the silvicultural exemption under the NC Sediment Pollution Control Act. Programs which monitor the compliance rate for silvicultural BMPs have been or are being developed in all southern states except for Mississippi (NCASI 1994).

In Arkansas, Louisiana, Mississippi, Oklahoma, and Texas, alleged infractions of WQS are regulated by the state water pollution management agency. However, these agencies have not invoked this authority for forest management activities. Alabama, Florida, Georgia, and South Carolina have water pollution management agencies which investigate and regulate water quality violations and assess fines if needed. In Virginia and North Carolina, the Department of Forestry regulates water quality violations from silvicultural activities (NCASI 1994). The Tennessee Division of Forestry investigates alleged silvicultural water quality infractions and the Tennessee Division of Water Pollution Control has regulatory authority when violators refuse to comply with BMPs (Applegate 1996).

THE SECTION 404 PERMITTING PROCESS

If landowners or forest managers determine that their proposed activity requires a permit, they must seek the individual permit before proceeding. The process for obtaining an individual Section 404 permit follows standard procedures found in the Corps regulations. Applicants are encouraged to meet with the Corps for a pre-application consultation and to determine what paperwork and field work are necessary.

Environmental Assessments or Environmental Impact Statements may be required to comply with NEPA. When the Corps determines the application is complete, a public notice is issued to let interested parties know of the proposed activity. All comments received must be considered by the district engineer and become part of the record kept on the application by the Corps. Hearings may be held if requested by the public and determined to be appropriate by the Corps.

Regulatory Guidance Letters

The evolutionary nature of wetlands regulation prompted the Corps to begin issuance of Regulatory Guidance Letters (RGLs). These documents provide field offices with specific guidance on a variety of topics. The letters expire five years from issuance. A complete listing of the current RGLs may be found in the Federal Register, Vol. 61, No. 118, pp. 30990–31002.

Commenting Authority

EPA may deny discharge permits under the authority of Section 404(c). Most permit denials by the EPA have been based on its interpretation of the Section 404(b)(1) guidelines. The USFWS was granted consultation rights over proposed permits by the FWCA and the ESA. The National Marine Fisheries Service may comment on permits proposed for areas within the coastal zone. The National Historic Preservation Act requires the Corps to consult with the Federal Advisory Council on Historic Preservation and state historic preservation officers to minimize harm to national historic properties and archaeological sites (Strand 1993). This statutory authority gives these agencies greater weight than other federal or state agencies. When these agencies appeal a permit decision, the Assistant Secretary of the Army must approve the elevation. Elevation of permit disputes is limited to situations with "substantial and unacceptable impacts to aquatic resources of national importance" (USACE 1992).

Corps Decision Criteria

Individual Section 404 permits may be granted by the Corps if certain conditions are met. The proposed activity must comply with the guidelines found in Section 404(b) which require denial of permits for discharges which have unacceptable adverse impact. The permit must meet the requirements of the Endangered Species Act, NEPA, and the National Historic Preservation Act. Public interest gives a presumption for granting permits to protect against erosion and against permits which would impede access to navigable waters. Permits cannot be granted where significant degradation to U.S. waters would occur. Table 3.3 gives criteria for Corps Section 404 permitting decisions (Want 1996).

Practicable Alternatives

The Corps may deny permits for discharges into wetlands or special aquatic sites where there is a practicable alternative. Non-water dependent projects are assumed to have practicable alternatives unless it is proven otherwise. Alternative sites which might be reasonably obtained by the permittee are considered practicable alternatives. Applicants can demonstrate the lack of alternatives by including an analysis of alternatives with the permit applications. Alternatives must be judged on the basis of broad public purpose rather than a narrow individual purpose (Want 1996).

Mitigation

Landowners may be required to perform mitigation in order to receive a Section 404 permit. EPA and the Corps have a Memorandum of Agreement (MOA) on Mitigation, issued February 7, 1990, which covers in detail the sequencing approach to mitigation to be used by the Corps in granting Section 404 permits, i.e., avoidance, minimization, and compensation (Mitigation MOA 1990). The Memorandum establishes a preference for on-site compensation over off-site compensation. In addition, functional values which are lost are generally to be replaced by in-kind compensation where possible. Mitigation banking and mitigation monitoring are covered in the MOA. Additional guidance on the use of wetland mitigation banks was provided by a later MOA between the Corps and EPA (Mitigation Banks MOA 1993).

After-the-Fact Permits

If a landowner performs an activity without obtaining a required permit, an after-the-fact permit may be required by the Corps. The application for an after-the fact permit follows the normal permitting process and may be denied. The District Engineer determines if legal action is appropriate and may seek penalties for the action. Generally, the Corps and EPA try to address violations by voluntary compliance or administrative enforcement. Although they have authority to seek criminal and civil penalties, the use of such authority has been limited to those cases involving flagrant and repeated violation of Section 404 (USEPA 1993b).

LANDOWNER OPTIONS FOR ASSISTANCE AND PROTECTION

There are several federal programs specifically created to assist private landowners with land management goals. Some of the programs apply only to wetlands, such as the Water Bank Program and the Wetlands Reserve Program (WRP). Other, more general programs which may be applied to wetlands include the Conservation Reserve Program (CRP), the Environmental Quality Incentives Program (EQIP), the Forestry Incentives Program (FIP), and the Stewardship Incentives Program (SIP). Landowners may be able to use conservation easements as a legal tool to give long-term protection to their wetlands and obtain some tax relief. The Forest Legacy Program was based on conservation easements between the government and landowners. The USFWS Partners for Wildlife Program offers technical and financial assistance for certain practices including restoring wetlands and other habitats. This program does not include assistance for mitigation requirements under any regulatory program.

TABLE 3.3
Corps criteria for issuing Section 404 permits.

Criteria	Effect	Source
Public Interest	Consideration of relevant factors including cumulative effects	33 CFR 320.4
Benefits of alteration outweigh the wetlands damage	Presumption against alteration of wetlands	33 CFR 320.4(b)(4)
Practicable alternative	If another activity/site can substitute without unreasonable costs and/or logistics, then permit will be denied	EPA 404(b)(1) guidelines found in 40 CFR 230
Water Dependent	Permittee must prove practicable alternatives do not exist to non water dependent activities	EPA 404(b)(1) guidelines found in 40 CFR 230
Significant degradation of waters	Permit can be denied even if there is no practicable alternative	EPA 404(b)(1) guidelines found in 40 CFR 230, RGL 87-2
Socioeconomic factors	Scope limited to socioeconomic effects from environmental alterations of the wetlands	RGL No. 88-11
Indirect effects	Gives lesser weight to effects which would result from a similar unpermitted project	RGL No. 88-11 RGL No. 88-13
Mitigation	Sequence: avoidance, minimize impacts, compensate	33 CFR 320.4 MOA on Mitigation between EPA and Corps, 2/7/1990
EPA, UAFWS, NMFS comments	Corps will consider agency comments but is responsible for final decisions	CWA, FWCA 33 CFR. 320.4 RGL 92-1
State Water Quality Standards (WQS)	States must certify permit compliance with WQS	CWA Sec. 401 RGL 92-4
Section 208 program requirements	BMPS must be used	EPA 404(b)(1) guidelines found in 40 CFR 230
Coastal Zone Management Act compliance	Permit applicant must certify compliance with state CZMA program. State can object	CZMA 33 CFR 325.2
Endangered Species Act compliance	DOI or DOC prepares jeopardy opinion if endangered species present. Corps makes final permit determination	ESA
National Environmental Policy Act compliance	Most projects require EA. Few projects require EIS. Mitigation can substitute for EIS	33 CFR 230, 53 Fed. Reg. 3127 (Feb. 3, 1988)
National Historic Preservation Act compliance	Corps must consider effect on historic properties on National Register; allow Advisory Council on Historic Preservation to comment	NHPA Sec. 106. 33 CFR 325, App. C

The Water Bank Program

Congress passed the Water Bank Act of 1970, which established the Water Bank Program to protect, restore, and improve private wetlands. By protecting wetlands, the program was expected to:

(1) conserve surface waters,
(2) preserve and improve habitat for migratory waterfowl and other wildlife resources,
(3) reduce runoff and soil and wind erosion,
(4) contribute to flood control,
(5) contribute to improved subsurface quality and reduce stream sedimentation,
(6) contribute to improved subsurface moisture,
(7) reduce acres of new land coming into production and to retire lands now in agricultural production,
(8) enhance the natural beauty of the landscape, and
(9) promote comprehensive and total water management planning (ASCS no date).

Ten-year contracts were established between cooperators and the Agricultural Stabilization and Conservation Service, now the Consolidated Farm Services Agency. Landowners received annual payments for the contracts as well as cost share payments for approved practices (Beck 1994). The first contracts were established in 1972, but enrollment has been very low. The Food, Agriculture, Conservation, and Trade Act of 1996 allows some Water Bank participants to enroll in the CRP during the final year of their Water Bank Program agreement.

Conservation Reserve Program

The CRP, administered by the CFSA, was designed to take erodible farmland out of production but has evolved over the years to become an environmental protection program. Its wetlands effects occurred primarily in the Midwest. Farmers agreed to put the land in some sort of permanent cover, usually pasture or trees, in exchange for annual rental payments for 10 to 15 years (Strand 1993). Land enrolled in the program must be subject to significant erosion and farmed as recently as three years before sign-up. Landowners receive cost-share assistance for approved conservation practices including tree planting. Rental payments are determined via a competitive bidding process. Preference is given to lands within Conservation Priority Areas, selected by state and federal agencies and state technical committees as being particularly environmentally sensitive. The federal farm agencies are to target these areas for enrollment in CRP in order to enhance water quality, encourage agricultural producers to comply with NPS requirements, and to improve wildlife habitat (Cubbage et al. 1993, USDA 1996).

A new CRP program under the 1996 Farm Bill is the Flood Risk Reduction program. For producers who crop land with high flood potential, the federal government will pay a one time, lump sum payment in return for the producer's exit from federal agricultural programs such a commodity loans, crop insurance, conservation programs, and disaster payments (USDA 1996).

Also under the 1996 bill is a Wildlife Habitat Incentives Program. Landowners can receive cost share assistance to develop habitat for wildlife. This program can be used to develop habitat for wetland species, fisheries, upland species, and endangered species and thus may be of interest to managers of private wetlands (USDA 1996).

Environmental Quality Incentives

The 1996 Farm Bill created a new program, EQIP, out of the Agricultural Conservation Program, Water Quality Incentives Program, Great Plains Conservation Program, and the Colorado River Basin Salinity Control Program. Fifty percent of the funding must be applied to livestock-related conservation practices. Conservation Priority Areas are to be established so that the EQIP program can be targeted toward areas where the most environmental benefit will be obtained. Land eligible for sign-up is agricultural land that affects soil, water, or related resources in a significant way. Technical assistance to landowners is provided with 5- to 10-year contracts, and financial assistance is available for up to 75% of the costs of certain conservation practices (USDA 1996).

Wetlands Reserve Program

The WRP was created in 1990. Lands which had been converted from wetlands and farmed wetlands were eligible if the area could be successfully re-created as a wetland. Landowners could sell a 30-year easement to the Department of Agriculture. In return, the landowner agreed to maintain or restore the wetland as directed by a Wetlands Reserve Plan of Operations (WRPO) prepared by the NRCS and approved by the USFWS. The landowner received payment for the easement as well as cost-share assistance for approved projects. Forest management, including harvesting, could be allowed if specifically stated in the contract. The purpose of the WRP was to restore wetlands functions and values to land altered for agriculture and contribute to the national goal of no net loss of wetlands (NCCES 1995). If the conservation plan was violated, repayment of the rentals could be required. This repayment did not, however, negate the easement. Wetlands Reserve Payments were not counted as 'benefits' subject to Swampbuster or Sodbuster repayment penalty provisions. Easement payments could continue even if the participant violated Swampbuster or Sodbuster. The 1996 Farm Bill requires that one-third of the area be in permanent easements, one-third in temporary easements, and the remainder in cost-share agreements for restoration only. Until at least 30,364 ha (75,000 acres) of temporary easements have been enrolled in the program, there will be a suspension of enrollment for permanent easements (Strand 1993, USDA 1996).

FIP, SIP, and State Forestry Incentives Programs

The FIP was created with the 1973 Agriculture and Consumer Protection Act and later expanded by the Cooperative Forestry Assistance Act of 1978. Cost-share payments were provided to landowners who chose to reforest their property under the guidance of federal farm agencies and state forestry agencies. The 1990 Farm Bill contained a new land management incentives program, SIP, which was designed to replace FIP with a more holistic management strategy. SIP provided cost share assistance for projects meeting a variety of objectives, from forest management to wildlife habitat improvement to recreation enhancement. Congress has continued to fund both FIP and SIP in the 1996 Farm Bill, and landowners may wish to apply for assistance under these programs when performing management activities on wetlands. State forestry agencies administer the program along with the CFSA (Cubbage et al. 1993).

Many states have their own incentives programs which can be used for forest management activities on wetlands. Landowners should contact their state forestry agency for information on program availability and criteria for assistance.

Technical Assistance

Landowners may receive technical advice on forested wetlands management from a variety of federal and state agencies. NRCS, CFSA, the Cooperative Extension Service, and state forestry agencies can provide information on wetlands management as well as information on assistance programs. State forestry agencies often have training programs on BMPs. The USFWS and state wildlife agencies provide technical assistance for wildlife habitat restoration and enhancement. Some environmental organizations, such as the EDF, can also provide information on wetlands protection and conservation easements.

Conservation Easements

Conservation easements can provide landowners with additional financial incentives to provide long-term wetlands protection. A conservation easement or restriction is a legal contract between a landowner and a government agency or conservation organization in which the landowner gives up some property rights, such as development or conversion rights, in order to receive certain benefits. The benefits received by the landowner may include lower property tax rates, reduced estate taxes, and income tax benefits. Contracts specify the rights, responsibilities, and duties required of both parties. Easements are usually permanent and transfer with ownership. Enforcement of the easement is the responsibility of the governmental agency, conservation organization, or land trust involved (Land Trust Alliance 1993).

Forest Legacy

Forest Legacy is a program geared to keeping important forest lands from being converted to nonforest uses. Landowners sign a perpetual conservation easement which restricts activities to those compatible with the goals for the particular tract, including timber harvesting and recreation. The objectives for maintaining the land as forest include protection of scenic values, cultural values, habitat for fish and wildlife, recreation, riparian areas, and other ecological values.

USFWS Partners for Wildlife Program

The USFWS Partners for Wildlife Program is a stewardship program focusing on assisting private landowners with habitat restoration and protection. Private landowners, including organizations, corporations, ranchers, and farmers, are eligible for cost sharing assistance of up to $10,000 per fiscal year for specific projects to restore habitat. The landowner may receive assistance of up to 100% of the cost of the project, though one of the program goals is to have 40% of total funding paid for by non-USFWS funds. Physical restoration activities must account for at least 70% of the federal cost share funds, with the remainder covering technical support, monitoring, and maintenance. Landowners must agree to voluntary contracts of at least 10 years. Usual terms of agreements include requirements for restoration with native plant species and communities, no agricultural use, and access for USFS personnel to verify compliance. Usually landowners must obtain needed permits and water rights. Even though the program has not been advertised widely, there is a nationwide waiting list of potential participants. From 1987 to 1995, the program restored more than 85,000 ha (310,000 acres) of wetlands, 373 km (600 mi) of riparian habitat, and 31 km (50 mi) of instream aquatic habitat. Information on the program may be obtained from the USFWS State Private Lands Coordinator for the landowner's state of residence (Mueller 1995, USFWS 1996).

SUMMARY

In the last two centuries, public wetlands policy has evolved from encouraging the draining of swamps to allow conversion to productive crop lands to protecting swamps, marshes, and other wetlands for a variety of environmental benefits. The nation's navigable waters, which were not defined to include wetlands, were first protected by legislation with the Rivers and Harbors Acts (RHAs) of 1890 and 1899. The RHAs controls on dredging and filling of rivers and disposal of refuse were the precursors of modern wetland law.

By the 1940s and 1950s, federal administrative and legal authorities began to extend to fish and wildlife protection and water quality. These modest initial efforts and increased public demands eventually led to passage of the Federal Water Pollution Control Act of 1972. As part of this comprehensive law, wetland protection was mandated under Section 404. This section required that individuals who perform dredge or fill activities in the nation's waters must receive a permit from the U.S. Department of the Army Corps of Engineers.

Section 404 has evolved over the last 22 years through a series of Congressional Amendments, court decisions, administrative regulations, agency memoranda of understanding, and individual implementation actions. Wetland protection under Section 404 jurisdiction has expanded to cover all the nation's waters and wetlands as defined by a complex federal wetlands delineation manual. "Normal" silvicultural activities remain exempt from permit requirements, but landowners must comply with the specific requirements enumerated in federal code to maintain exempt status. The exemption applies only to the extent that the "recapture provisions" of Section 404 do not apply.

Landowners have a variety of options to assist them in managing their forested wetlands. Federal incentives programs for forestry, stewardship, wildlife management, and wetlands protection may provide financial assistance to landowners. Conservation easements may be used to ensure a wetland will remain protected from development and destruction. Technical assistance may be obtained from a variety of federal and state agencies as well as from some environmental organizations.

Part II

General Functions of Southern Forested Wetlands

4 Regional Climates

Robert A. Muller and John M. Grymes III

CONTENTS

INTRODUCTION

Weather and climate are among the most significant environmental factors that impact and influence the viability of southern wetland forests. This chapter provides a brief overview of ways to describe and interpret climate and its potential interactions with vegetation. Special attention is directed toward synoptic weather types and water budget interpretations of climate, with some additional remarks about implications of extreme events, climatic variability and the potential of long-term climate change.

CLIMATIC CLASSIFICATION

Climates over the surface of the earth vary on geographical scales of continents, regions, landscapes and local areas, and even down to the internal microclimates of forest stands, agricultural fields, and city streets. Except at sharply defined boundaries normally related to the availability of water at the surface of the earth, this geographical variability is continuous and endless, and efforts to delineate and describe the regional climates of the earth often involve too much climatic detail.

Probably the most enduring regional climatic classification was developed initially in the late 19th century by Wladimer Köppen, a German climatologist and botanist. Köppen fit monthly average temperature and precipitation data to global regions of natural vegetation as perceived in the late 19th century, and he continued to revise the classification properties and geographical boundaries until his death in 1940. Even today, most introductory textbooks in climatology or physical geography use either the Köppen classification or some more recent modification for geographical

1-56670-228-3/97/$0.00+$.50
© 1998 by CRC Press LLC

differentiation of climates (Muller and Oberlander 1984, Strahler and Strahler 1992, McKnight 1996).

Within the Köppen climate system, properties of individual locations are described in terms of the annual and seasonal regimes of mean monthly temperatures and precipitation. Depending upon how subcategories are grouped, the Köppen global climatic classification is usually organized into 12 to 15 basic types. Almost all of the southern forested wetlands in the United States, except for southern Florida, are located within Köppen's Humid Subtropical type (Cfa), normally described as a humid "forest" climate with cool winters and warm to hot summers. In the United States, the humid subtropical climate region extends northward and westward from central Florida to New Jersey, southern Illinois, and eastern Texas. Locations of other humid subtropical climate regions on the other continents include much of southeastern China, most of the Japanese Islands, southern Brazil, Uruguay, and the Pampas of Argentina, small areas of South Africa, and southeastern Australia. Because of dynamic atmospheric controls on the delivery of climatic energy and precipitation, all are located in the subtropics in the southeastern sectors of the continents.

Wetland forests in southern Florida lie within the very northern margins of Köppen's Tropical Savanna (Aw) climate region, which is characterized by very mild to warm, dry winters, and rainy, hot summers. Hence, there is normally some degree of drought during the low-sun season in the tropical savanna regions, and the characteristic natural vegetation is grassland, with forest vegetation normally restricted to poorly-drained floodplains and swamps.

Figure 4.1 illustrates by means of climographs the mean monthly maximum and minimum temperatures and precipitation for the standard 30-year normals (1961 to 1990) of the National Weather Service at four locations that bracket the region of southern forested wetlands. Dallas (TX) and Raleigh (NC) are representative of the northern margins, and New Orleans (LA) and Miami (FL) the southern. Similarly, Dallas and New Orleans are to the west and Raleigh and Miami to the east. The scales are the same for geographical comparisons.

The climographs illustrate the increasing "continentality" of the annual temperature regimes toward the north and west, with much smaller differences in temperature between winter and summer at Miami as contrasted with Dallas. Temperatures at New Orleans are representative of forested wetlands in the vicinity of the Gulf of Mexico, but Raleigh, because of its more northern and inland location, is colder in winter than forested wetlands to the south across the Atlantic Coastal Plain. During summer average maximum temperatures everywhere are in the upper 20s to low 30s°C (upper 80s to low 90s°F), and minimum temperatures in the lower 20s°C (between 70 and 75°F). In contrast, there is a strong north-south thermal gradient in winter, with minimum temperatures at Miami very rarely as low as freezing, only occasionally below freezing at New Orleans, but near freezing on the average at Raleigh and Dallas.

The bar graphs of average monthly precipitation for Raleigh, New Orleans, and Dallas illustrate the remarkable regularity of "average" precipitation in the humid subtropical climate region, especially at Raleigh, where it averages 75 to 100 mm (3 to 4 in) per month year-round. Except for a drier season during fall at New Orleans, monthly precipitation there and along the central Gulf Coast averages 125

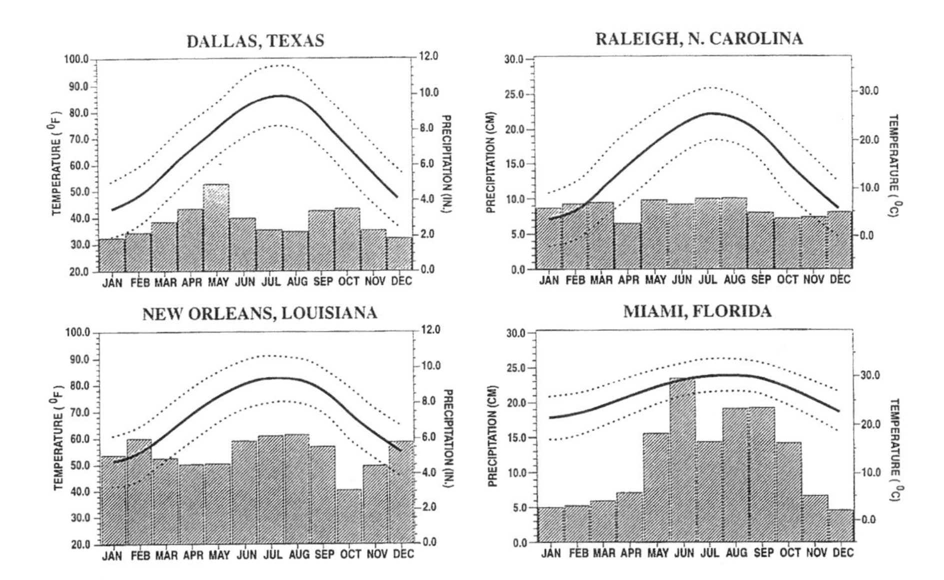

FIGURE 4.1 Annual regimes of mean monthly maximum and minimum temperature (line) and precipitation (bars) at Dallas, TX; Raleigh, NC; New Orleans, LA; and Miami , FL.

to 150 mm (5 to 6 in). The monthly averages at Dallas range between 50 to 100 mm (2 to 4 in), with the highest averages in spring and fall, and lower averages during late summer and winter. Miami, on the other hand, located at the northern margins of the tropical savanna climate, illustrates the great range between wet summers and dry winters typical of savanna climates.

ATMOSPHERIC CIRCULATION PATTERNS

The averages mask the variability and extremes that are characteristic of real-world climates. In the subtropical climate region of the United States, precipitation during late fall, winter, and spring is generated by midlatitude cyclones and their associated fronts that separate colder, drier polar or Arctic air to the north and west from warmer, moist tropical air to the east and south; the Gulf of Mexico and the western subtropical Atlantic are the moisture sources for precipitation over this region. Midlatitude cyclones move generally from west to east from 500 to 1,600 km (300 to 1,000 mi) per day, bringing warm and humid weather ahead of the storm systems, and colder and most often fair weather behind the systems. More widespread moderate precipitation is normally generated by the warm fronts to the east and northeast of midlatitude storm centers; thunderstorms, severe weather, and spotty very heavy rainfall are often associated with the trailing cold fronts "behind" the midlatitude storm centers. Much of the weather, therefore, consists of contrasting periods of fair and warmer weather, followed by stormy weather generated by the fronts, and then fair and much cooler or colder weather following the midlatitude cyclones.

The paths and frequencies of the midlatitude cyclones are controlled primarily by the patterns of airflow aloft (jet streams) more than 5,500 m (18,000 ft) above sea level. During fall, winter, and spring, this upper-air flow is a circumpolar vortex from west to east around the northern hemisphere extending equatorward from subpolar latitudes to the subtropics. This upper-level flow is rarely straight west-to-east parallel with latitude, but is most often in wave-like patterns, with large dips to the south (troughs) where cold Arctic and polar air is advected far to the south, and broad sweeps to the north (ridges) where warm, humid tropical air from the Gulf and adjacent Atlantic is advected over the region. The ridges tend to be associated with fair and mild weather, and the troughs with the development of midlatitude cyclones, stormy weather, and precipitation.

During summer, the circumpolar vortex weakens and retreats poleward, and not very many midlatitude cyclones reach the South. Instead, this humid subtropical climatic region is dominated by the effects of a huge subtropical high pressure system over the western Atlantic (the Bermuda High). The core of the subtropical high is associated with very warm, muggy weather but with a stable atmosphere aloft and reduced precipitation. Weak tropical disturbances, tropical storms, and great hurricanes are steered around the margins of the subtropical high, and they generate some of the summertime and early fall precipitation over Florida and other coastal areas adjacent to the Atlantic and Gulf; sea-breeze fronts also produce thundershowers and storms in these same coastal areas. Very heavy but localized summer convectional showers and thunderstorms are also common when the subtropical high weakens and there is less "vertical stability" in the atmosphere.

The annual regimes of observed daily temperatures and precipitation at the same four locations bracketing the southern forested wetlands are shown for calendar year 1995 in Figure 4.2. In contrast to the smooth averages of Figure 4.1, this figure illustrates the runs of warmer and colder weather and precipitation that are associated with the passage of midlatitude cyclones and associated fronts during most of fall, winter, and spring. Even at the southern location of Miami, the cooling effects of fronts are clearly visible in late December and again in early February, with milder episodes occurring as early as November and as late as April. During summer the limited number of weaker midlatitude cyclones and fronts rarely affect southern areas of the region, and temperatures there change very little from day to day. Unusual runs of extreme temperatures or precipitation occur when the upper-air ridges and troughs remain over the same region for extended periods of time, usually a week or more. These atmospheric circulation anomalies are responsible for most of the extreme climatic events (droughts, floods, freezes, and severe weather) that generate significant environmental and economic impacts (Lutgens and Tarbuck 1994).

SYNOPTIC CLIMATOLOGY

The analysis and interpretation of daily national weather map patterns of near-surface high and low pressure systems, fronts, and precipitation, are known as synoptic climatology. Climatologists have developed synoptic weather classification systems, within which daily weather patterns at a specific place or over a large region are organized into "types" based on circulation patterns (Yarnal 1993). Careful inspection of weather maps allows for manual classification of days into daily synoptic

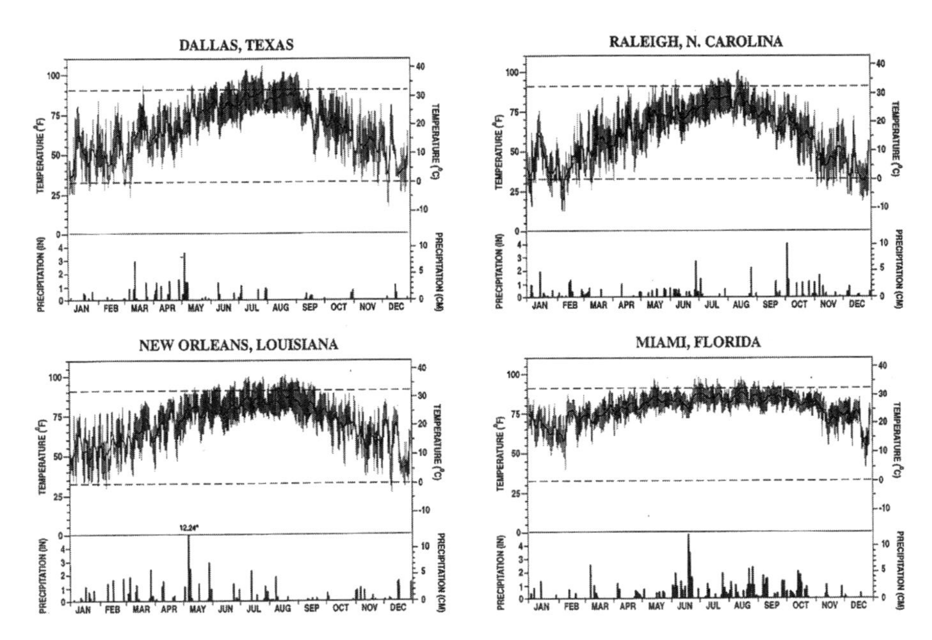

FIGURE 4.2 Annual regimes of daily maximum and minimum temperature and precipitation for 1995 at Dallas, TX; Raleigh, NC; New Orleans, LA; and Miami, FL.

weather-type calendars, which may be utilized to compare circulation frequencies through time, or for impact analyses. More recently the development of automated computer programs allows for the rapid creation of decades-long daily calendars that are especially useful for interpretations of atmospheric circulation changes through time.

Muller (1975) has developed a manual synoptic weather classification system consisting of eight all-inclusive types for Louisiana and adjacent states north of the Gulf of Mexico. Figure 4.3 shows typical examples of the eight weather types. For example, the Gulf Return (GR) type represents the sultry weather with southerly winds well ahead of an approaching cold front, and the Continental High (CH) type the fair and cool to cold weather behind a cold front. Pacific High (PH), Gulf High (GH), and Coastal Return (CR) weather types are additional fair weather types. There are three stormy weather types within which most of the precipitation occurs: Frontal Gulf Return (FGR), following GR weather and just ahead of approaching cold fronts; Frontal Overrunning (FOR), the cloudy and colder weather immediately following many cold fronts; and Gulf Tropical Disturbance (GTD) weather, including weak and poorly-defined tropical waves, tropical storms and hurricanes that occasionally plague the Atlantic and Gulf coastal regions from June through November.

For the New Orleans area, daily calendars for more than 35 years are available at the Louisiana Office of State Climatology at Louisiana State University, Baton Rouge, Louisiana. Average properties of the weather types for January at 0600 and 1500 hours, generally the coldest and warmest times of the day, are shown in Table 4.1. The data illustrate the contrasts of climatic properties among the types during the same month. The properties and calendars have proved to be very useful

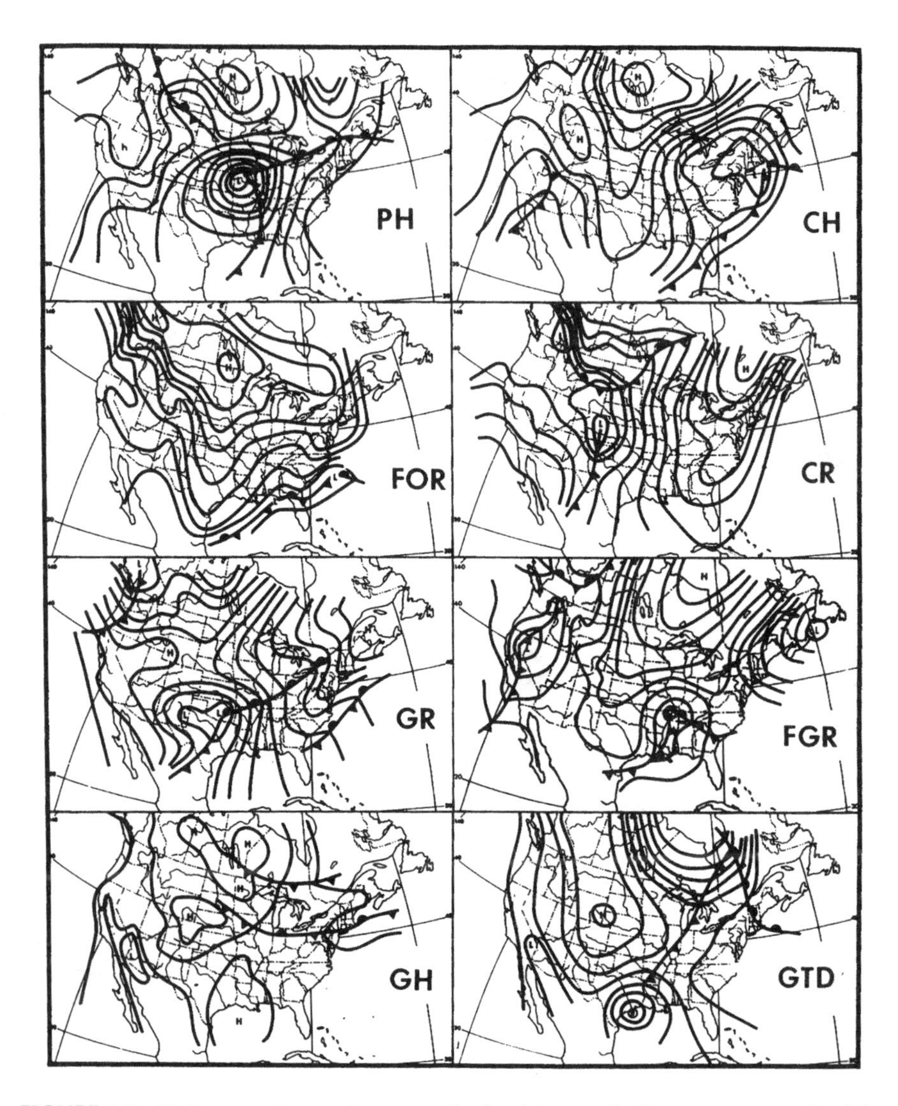

FIGURE 4.3 Eight synoptic weather types for Louisiana and adjacent states north of the Gulf of Mexico (Muller 1975). See text for definition of types.

for analyses of biological and ecological events related to weather and climate. For example, spring-season advances of corn earworm moths northeastward from Mexico across Texas and Louisiana have been clearly related to daily synoptic weather types (Muller and Tucker 1986). The weather-type names used by Muller are appropriate only for the northern Gulf Coastal region, but the synoptic weather patterns are similar all across the region of southern wetland forests. Therefore, the synoptic patterns can be identified as types 1, 2, and so on, and given local or regional names if necessary. The calendars and their average properties ought to be useful for study of weather-driven interactions among biological and environmental processes in forested wetlands.

TABLE 4.1
Average properties of synoptic weather types for New Orleans in January at 0600 and 1500 hours, 1961–80 (Muller and Willis 1983).

0600 CST	PH	CH	FOR	CR	GR	FGR	GH	GTD
Number of cases	22	143	246	40	67	75	27	0
Sky cover 0–10*	3	1	10	5	9	10	3	—
Visibility (km)	13	16	11	10	6	8	13	—
Air temperature (°C)	8	2	7	9	16	17	4	—
Dew point temperature (°C)	6	–2	4	8	15	16	2	—
Relative Humidity (%)	86	76	82	89	94	92	90	—
Wind direction†	32	02	02	09	15	18	23	—
Wind speed (knots)	4	7	8	5	6	7	3	—
1500 CST	**PH**	**CH**	**FOR**	**CR**	**GR**	**FGR**	**GH**	**GTD**
Number of cases	25	132	234	44	72	90	23	0
Sky cover 0–10*	2	2	10	6	7	9	3	—
Visibility (km)	16	18	11	13	13	11	16	—
Air temperature (°C)	18	11	11	17	22	21	17	—
Dew point temperature (°C)	7	–1	5	9	16	17	4	—
Relative Humidity (%)	51	46	73	61	70	77	44	—
Wind direction†	30	36	02	09	16	18	26	—
Wind speed (knots)	9	9	9	9	10	10	8	—

WATER-BUDGET ANALYSES

Forested wetlands, by definition, are areas distinguished by (1) the presence of water at or near the soil surface, (2) unique soil conditions, and (3) vegetation adopted to wet conditions (Mitsch and Gosselink 1993). This normal abundance of water results from several very different environmental conditions. One is location in a floodplain where river water is frequently available either by flooding or as near-surface groundwater. A second possibility is locally poor or impeded drainage so that rainwater remains stored at or very near the surface for extended periods instead of draining away. A third possibility is a very wet climate where precipitation regularly exceeds evapotranspiration rates.

Across the southern states from east Texas and Oklahoma to the Atlantic Coast, the climate is humid, and average annual precipitation far exceeds evapotranspiration. Climatic water-budget models evaluate the income, outgo, and storage of water at a site. Such budgets also exist for small and large drainage basins, for continents, ocean basins, and the entire earth, allowing hydrologic models to evaluate the movement and routing of water from place to place through drainage basins (Mather 1978, Muller and Thompson 1987).

The fundamental basis for the climatic water-budget model is potential evapotranspiration (PE), which was defined by Thornthwaite as the rates of evapotranspiration from a fully vegetated landscape with no shortages of water for the vegetation

(Thornthwaite and Mather 1955). Expressed in this form, PE is also an energy-budget term, and the rates of PE at a place are controlled by energy income, primarily solar radiation. At the time of the development of the models in the late 1940s, solar radiation data were essentially unavailable except at a small network of U.S. Weather Bureau stations and a few research sites. To estimate PE under these severe data constraints, Thornthwaite developed a series of equations using temperature, daylight hours, and a heat index (Thornthwaite and Mather 1957). Even at the present time, available "official" solar radiation data are very limited in the United States, and the Thornthwaite PE methodology is often used for geographical comparisons of PE and the water-budget components, as well as the basis for the series of Palmer drought indices used by numerous state and federal agencies for drought monitoring (Mather 1974). A number of studies have demonstrated that the monthly and annual estimates of PE by the Thornthwaite procedures are reasonably representative of potential water use by vegetation over most of the United States, but there are some seasonal lags. Daily estimates of PE, however, are not sufficiently sensitive to short-term meteorological variability to adequately reflect vegetative moisture demand on that time-scale (McCabe 1989).

Figure 4.4 shows the annual regime of average monthly PE and additional water-budget components at the same four places that bracket the region of southern forested wetlands in the United States. The graphs show that PE is low in winter and high in summer. At the two more northern places, Dallas and Raleigh, PE is less than 12 mm (0.5 in) per month in December and January, increasing southward to about 25 mm (1 in) at New Orleans and 50 mm (2 in) at Miami. At all four locations, PE peaks in July and August with nearly similar rates of 150 to 175 mm (6 to 7 in) per month.

Average monthly precipitation (P) is also shown in the average water-budget graphs in Figure 4.4. Precipitation is greatest at New Orleans and Miami, but seasonal patterns are very different. At Miami precipitation is very high from June through October, and low from November through March. At New Orleans, in contrast, monthly precipitation is much more evenly distributed through the year, but with a distinct autumn minimum, especially during October. To the northeast at Raleigh, average monthly precipitation is much less, but remarkably evenly distributed throughout the year. Precipitation is lowest at Dallas, with strong bimodal maxima in spring and fall.

In the climatic water-budget model, P is treated as "income" and budgeted against PE, the climatic demand for water in the landscape. When P is greater than PE, precipitation is first allocated to meet the PE demand, and actual evapotranspiration (AE) is estimated to be the same as the PE; it is important to note that AE in the climatic water budget is equivalent to evapotranspiration (ET), the term normally used by hydrologists, agricultural scientists, foresters, and ecologists. The second priority for precipitation in the model is to recharge soil moisture storage if soil moisture is below capacity. Finally, any additional excess precipitation goes into surplus (S), representing water that becomes available for groundwater recharge and runoff into streams and rivers.

When P is less than the PE, the model assumes that all of the precipitation will be used for evapotranspiration. Soil moisture will also be utilized to meet the PE

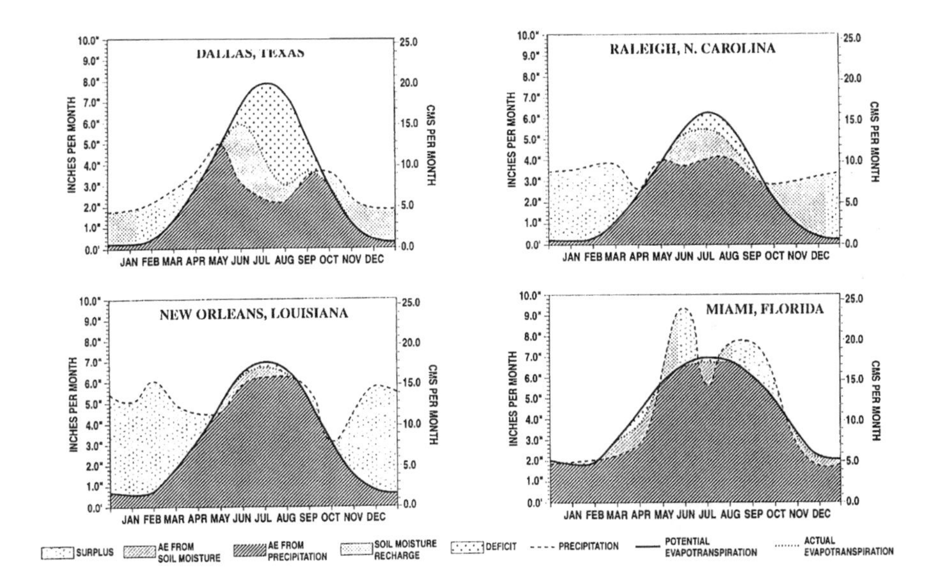

FIGURE 4.4 Mean monthly PE and water-budget components at Dallas, Raleigh, New Orleans, and Miami for 1961 to 1990.

demand. The soil moisture storage capacity can be set to represent soil texture and typical rooting depths of the vegetation, and the available soil moisture for ET is the difference between field capacity and the wilting point. A general worldwide average for available soil moisture storage capacity is often taken as 150 mm (6 in), but capacities can range from only 25 mm (1 in) for very shallow soils above bedrock to 300 mm (12 in) or more in deep soils with extended root systems. In many wetland situations, it should be recognized that water is usually available at or very near the surface, and that ET will essentially be equal to PE.

At more well-drained sites, the model assumes decreasing availability of soil moisture, and rates of withdrawal will be proportional to the degree to which the soil storage capacity is full. For example, when soil water is at or near capacity, soil moisture withdrawal rates will allow vegetation to utilize all that is needed. When soil moisture has been depleted to 25% of capacity, vegetation will be able to withdraw only about 25% of the demand. In the climatic water-budget model, ET will be less than the PE, and the site or region experiences a deficit (D), which is an index of water shortage to the vegetation. Therefore, the average climatic water-budget graphs show the monthly regimes of AE (ET) directly from precipitation and from soil moisture reserves, soil moisture recharge, deficits, and surplus, as well as PE, P, and AE.

The four graphs in Figure 4.4 illustrate regional patterns of average water-budget components across the region. At Raleigh, despite the evenly-distributed average monthly precipitation, some deficits occur during summer and early fall, largely because of the strong PE demand at these times. Soil moisture recharge occurs later in the fall, and generation of surplus water and runoff is usually restricted to winter and spring. At Dallas, because of the lower precipitation, summer-fall deficits are

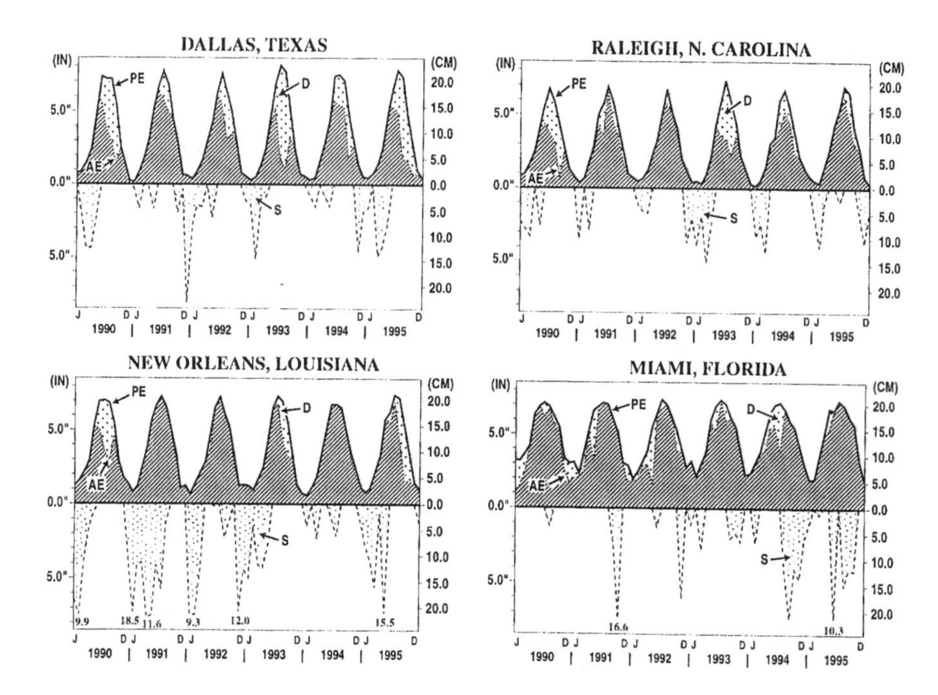

FIGURE 4.5 Continuous monthly water-budget components at Dallas, Raleigh, New Orleans, and Miami, for 1990 to 1995.

much greater and soil-moisture recharge persists into midwinter, with smaller surpluses restricted to late winter and early spring. At New Orleans, the patterns are similar, but because of much greater rainfalls, summer deficits are very small, with large surpluses from October into April. The patterns at Miami and across southern Florida are different, however. Because of low rainfall in winter and spring, no surplus water is generated on average, and although summer PE is high, heavy rainfalls keep deficits to a minimum.

While the water-budgets of Figure 4.4 show long-term climatic averages at these locations, they are not very representative of "real-world" climate because they exclude the natural variability of atmospheric circulation patterns. Some of the effects of climatic variability can be seen in Figure 4.5 by means of continuous monthly water budgets for the six most recent years at the same four locations. The continuous monthly water budget model is based on monthly temperatures and precipitation through time. For simplicity, only the four curves of PE, AE, D, and S on monthly time scales are shown. Similar continuous water budgets can be calculated for weekly or even daily data.

The graphs indicate that PE does not vary greatly at each place annually, suggesting that energy income, especially solar radiation, does not vary to the same degree as precipitation. Actual evapotranspiration is generally the same as PE during winter, but usually less than PE during summers and autumns.

Despite the overall classification of the region as having a humid climate, summer-autumn deficits occur frequently at all four locations. At Raleigh and Dallas, there were deficits every summer-autumn. Deficits were very large at Raleigh for the drought conditions during 1990 and 1993, but 1992 experienced only a very small seasonal deficit. At Dallas the deficits were much larger. New Orleans is much wetter with only insignificant deficits during most years, but with a large deficit in 1990. In the tropical savanna climate at Miami, relatively large deficits can occur any time of the year, but not consistently from one year to another.

Figure 4.5 also shows that generation of surplus water tends to be a late fall-winter-spring phenomenon across the region, with the exception of southern Florida (Miami). A small summer-season surplus was generated only once (1995) during the six years at Raleigh, and never at Dallas. Even at New Orleans, it was very uncommon for surplus water to be generated during summer. At Miami, rainfall can be so great during some summers that surpluses are generated; 1995 is a good example.

An additional point is that the surplus is the "residual" in the climatic water-budget model in terms of allocations of monthly precipitation. Hence, the variability of surplus water and runoff is much greater than the variability of precipitation, a matter of great significance for monitoring and potential management of southern wetland forests. At Raleigh, the contrasts between the winter-springs of 1991–92 and 1992–93 should be obvious. The Miami data illustrate the truly "feast or famine" generation of freshwater across southern Florida, especially the Everglades, with very minimal surpluses during 1990 through 1992, but very large surpluses in 1994 and 1995.

Comparisons of the temporal sequences of surplus at the four locations suggests that atmospheric circulation patterns generating precipitation produce runs of wetter and drier seasons that occur on spatial scales much smaller than the region encompassing the southern wetland forests. Stated another way, a wetter or drier summer or winter in Raleigh is not necessarily the same in Dallas, New Orleans, or Miami.

Figures 4.6, 4.7, and 4.8 summarize this water-budget section by showing the mean annual precipitation, deficits, and surpluses, respectively, across the region of southern forested wetlands for the recent 30-year normal period (1961–1990) of the National Weather Service. The data are organized by the climatic divisions of the National Climatic Data Center and represent the arithmetic averages of monthly temperature and precipitation data from roughly five to ten stations within each division. The annual averages represent the means of continuous monthly water-budget analyses, so that much of the effects of monthly and seasonal climatic variability are taken into account for the estimation of annual deficits and surpluses.

Figure 4.6 shows that mean annual divisional precipitation ranges from a high of 1,830 mm (72 in) over extreme northwestern South Carolina to a low of 330 mm (13 in) in western Texas. This single small division in northwestern South Carolina consists mostly of mountain and upland climate stations where orographic precipitation is very high. Other climatic divisions in the Appalachian Mountains from northcentral Alabama northeastward into Georgia, Tennessee, South Carolina, and North Carolina include localized areas with precipitation as high or even higher than for the small division in South Carolina, but these larger climatic divisions also include sites located in "rainshadow" valleys where precipitation is significantly lower, resulting in lower divisional averages. Mean annual precipitation is more than

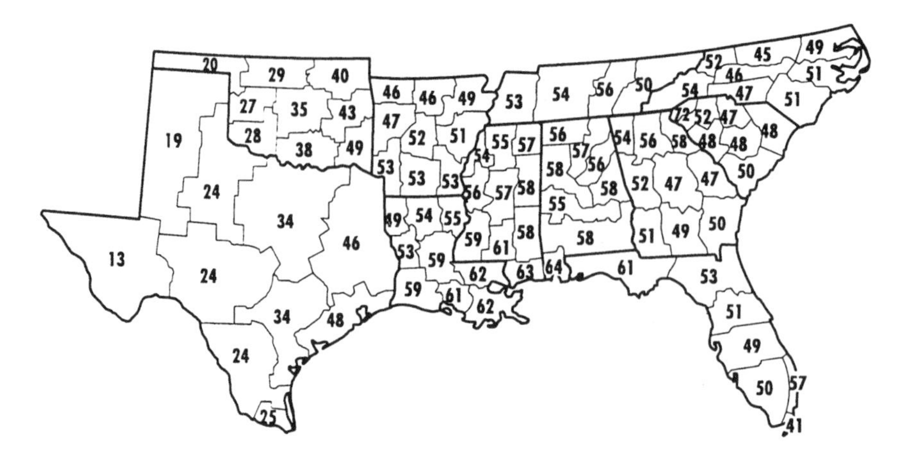

FIGURE 4.6 Mean annual precipitation (inches) by climatic divisions, 1961 to 1990.

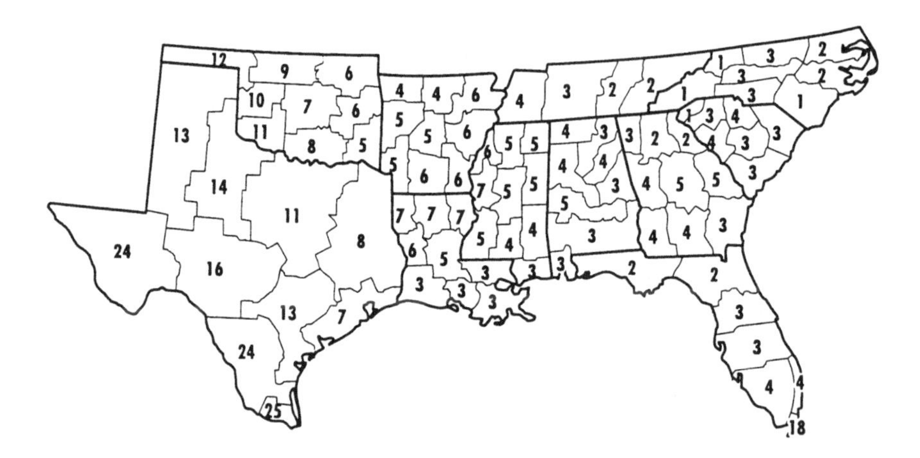

FIGURE 4.7 Mean annual deficits (inches) by climatic divisions, 1961 to 1990.

1,140 mm (45 in) everywhere east of eastern Texas and Oklahoma, except in extreme southern Florida where rainfall over the Florida Keys is much lower than over the mainland. Precipitation tends to be highest, on the order of 1270 to 1524 mm (50 to 60 in), in a broad swath from southern Louisiana, Mississippi, and Alabama, northeastward into Tennessee and western North Carolina.

The geographical pattern of mean annual divisional deficits is shown in Figure 4.7. Mean annual deficits are smallest in the east, ranging from as little as 25 mm (1 in) in sections of North Carolina to 75 mm (3 in) across much of Tennessee, South Carolina, Georgia, Florida, eastern Alabama, and westward across southern Mississippi and Louisiana. Deficits increase gradually westward to 125 mm (5 in) over the northern half of Mississippi and Arkansas, and even southeastern Oklahoma, and 125 to 250 mm (5 to 10 in) across western Mississippi, southern Arkansas and

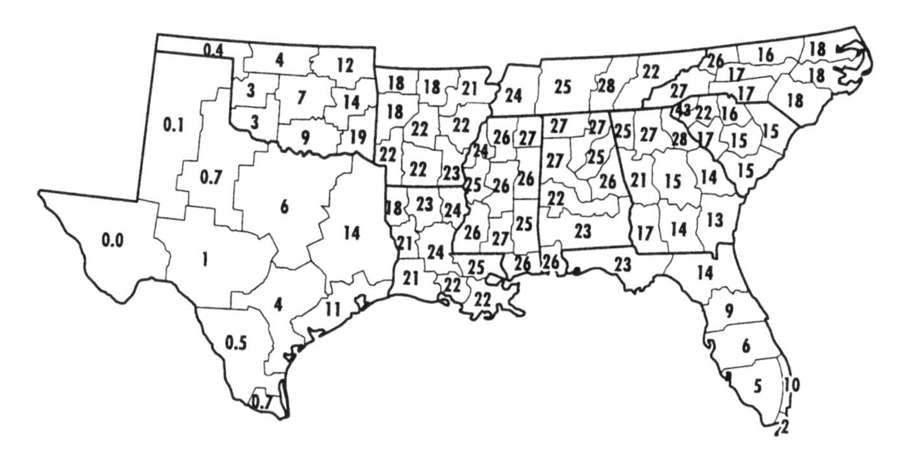

FIGURE 4.8 Mean annual surpluses (inches) by climatic divisions, 1961 to 1990.

northern Louisiana, and then westward across Oklahoma and eastern Texas. Further west, deficits increase rapidly to nearly 630 mm (25 in) over western and southern Texas, and to nearly 500 mm (20 in) over extreme southern Florida in a division that is representative only of the much drier weather over the Florida Keys, where frontal and convectional rainfalls are much less than over the Florida mainland.

Mean annual divisional surpluses are shown in Figure 4.8. This figure highlights the steep gradients of surpluses, representing water available for runoff, across the region. There is again a broad swath with a southwest-northeast axis of 500 mm (20 in) or more per year on average from Louisiana, Mississippi, Alabama, and the Florida Panhandle to the north and northeast across the eastern half of Arkansas, Tennessee, the northern half of Georgia and the mountainous western sections of the two Carolinas. Surpluses averaging 750 to 1000 mm (30 to 40 in) and more occur over the mountainous terrain of the southern Appalachians, as indexed again by the upland division positioned in northwestern South Carolina. Surpluses decrease towards the south regularly across southern Georgia and Florida, down to only 50 mm (2 in) per year for the southern Florida division representing conditions mostly over the Florida Keys. Surpluses also decrease rapidly and regularly to the west and south over Oklahoma and Texas so that surpluses across the western sections of Texas and Oklahoma and the southern sections of Texas average less than 25 mm (1 in). Hence, there are only limited riparian opportunities for forested wetland environments in these areas.

MICRO-CLIMATOLOGY

In addition to geographic patterns of regional climates outlined above, it has been recognized that climates can vary considerably over very small distances. For example, minimum overnight temperatures with clear and calm conditions often vary greatly across hilly terrain, with coldest temperatures in frost and freeze pockets in the lowest positions in the landscapes. Under these conditions, temperatures within the first 1 to 2 m (3.3 to 6.6 ft) above the surface will vary greatly from temperatures at the standard climatological level of 1.5 m (4.9 ft); near surface temperatures will be much warmer above warm water in a swamp, cooler over soil, and much colder over snowcover.

Geiger has been recognized at the "Father of Microclimatology." Microclimatology is the small-scale climate perspective of environmental responses to regional climates. Geiger wrote the most comprehensive and respected international summary of classical microclimatic research up to midcentury, and has been updated by contemporary climatologists (Geiger et al. 1995). The very empirical measurements of climatic elements in microclimatic settings have been partially replaced with a research approach known as boundary-layer climatology. Oke (1987) and Rosenberg et al. (1983) have published popular overview and summary texts of microclimatological methodologies. In brief, the focus of this approach is the measurement and subsequent modeling of the exchanges of energy and moisture and related airflow patterns, between the underlying surfaces and the lower atmosphere below about 3,048 m (10,000 ft), attempting to explain the how and why of spatial and temporal microclimatic differences.

EXTREME WEATHER EVENTS

Extreme events can be especially destructive from the perspective of the natural succession of southern wetland forests. Extremely low temperatures and extended periods of very low temperatures, known as "Arctic Outbreaks," can be associated with die-back and losses of temperature-sensitive species. During these episodes, most attention focuses on commercial crops, such as the citrus industry in Florida. Despite warnings about the probable onset of global warming, "Arctic Outbreaks" have continued to plague Florida's citrus industry during recent decades, driving the heartland of the industry farther southward from north to central Florida (Rogers and Rohli 1991).

Ice storms or glaze are occasionally very destructive to forest vegetation across the southern region. Ice storms occur because of very special atmospheric and surface temperature conditions during the passage of midlatitude cyclones, with trees losing much of their crowns during extreme events. Much of the midsouth is vulnerable to ice storms, and the most recent extensive ice storm in February, 1994, swept across the region from eastern Texas and Oklahoma to the Carolinas, causing estimated economic losses of up to $3 billion (National Climatic Data Center 1996).

Severe weather, including thunderstorms, lightning, damaging winds, tornadoes, and flooding rains, also impact southern wetland forests in destructive ways. Lightning often kills major limbs or entire trees, sometimes making them vulnerable to infestations of destructive insects. Very strong "straight-line" winds and tornadoes, associated with severe thunderstorms, can result in swaths of forest blow-downs. Tornadoes are not uncommon across this region, with the greatest average frequencies of about nine tornadoes per year per 26,000 km^2 (10,000 mi^2) in the "tornado belt" from central Texas to Kansas. Frequencies average about five per year inland from the Gulf Coast from Mississippi to Florida, and even higher in central Florida (Lutgins and Tarbuck 1994).

Just about every year, during the season between June and November, tropical storms and hurricanes cut swaths of destruction somewhere along the coastline between Virginia and Texas. When the coastline is divided into 80 km (50 mile) segments, historical records dating back more than 100 years suggest that average hurricane recurrence intervals range between 6 and 85 years (Simpson and Riehl 1981). The shortest recurrence intervals are along the southern half of the east coast of Florida, and along the Gulf Coast from Pensacola west to Brownsville. The longest intervals are from Savannah south to Daytona Beach.

For southern wetland forests not tolerant of saline conditions, storm surges, when ocean water levels typically rise at least 1.8 to 2.7 m (6 to 9 ft), and as much as 6.4 m (21 ft) during very extreme events such as Hurricane Camille, can be especially destructive on flood plains immediately inland from coastal estuaries. Hurricane and gale-force winds and small tornadoes can also produce "blow-downs" on floodplains upstream from the coasts. Flooding generated by heavy storm rains is not normally a significant hazard.

GLOBAL CLIMATE CHANGE

The potential impact of global climate change on southern forested wetlands is an issue which must be considered. Most of the leaders of the global climate change community continue to predict warming in association with increasing atmospheric concentrations of carbon dioxide and other trace gases. The estimated rates of warming, however, tend to be less than those of about a decade ago (Henderson-Sellers 1994). In very recent years, it has become recognized that some of the projected warming due to carbon dioxide has been offset by the cooling effects of sulfur compounds in the lower atmosphere because of increasing industrialization, and in the upper atmosphere by irregular loadings from volcanic eruptions. The empirical analyses of global and Northern Hemisphere temperatures over the last 150 years suggest some overall warming, in the Northern Hemisphere especially from about 1920 to about 1950, and more rapidly again since about 1980 (Hulme 1994). For the Northern Hemisphere, average temperatures now are about 0.5°C (1°F) higher than those recorded about 150 years ago (Henderson-Sellers 1994).

There is also some evidence that global precipitation has increased during this same time period (Balling 1992), and analyses of more than 100 years of climate data for Louisiana suggest an increase of more than 10 percent. At this time, however, the spatial resolutions of most global climate models are too coarse to yield reliable estimates of regional patterns of temperature and precipitation changes with a significant degree of accuracy. Slightly warmer temperatures are likely, but precipitation may or may not continue to increase. Hence, slightly higher rates of PE demand might be offset by increasing precipitation. If precipitation were to remain static, decrease slightly, or become increasingly variable, resultant regional climate could become less favorable for some forested wetlands. A final point is that the coastal and estuarine forested wetlands would also be further impacted by projections of global sea level rise, but the current estimates of rates are much lower than the nearly catastrophic rates projected initially by some researchers 10 to 20 years ago (Tooley 1994).

5 Hydrology

Thomas M. Williams

CONTENTS

INTRODUCTION

The chief aim of this chapter is to explain how hydrologic processes produce various southern forested wetlands. In Chapter 2, wetland functions, and how they related to societal value, were defined. Hydrologic processes (why wetlands form) determines wetland functions (the things wetland do). Hydrogeomorphic (HGM) functional analysis classifies wetlands based on geomorphic setting, water source, and dominant flow direction (Brinson 1995). A major strength of this classification is the relationship of classes with hydrologic processes that produce wetlands. This chapter will define three types of hydrology, based on the water balance, that can create wetlands.

The water balance equation is the basis of hydrology and is usually written as:

$$P = E + T + R_s + R_g \pm \Delta S \tag{5.1}$$

where: P = precipitation
E = evaporation
T = transpiration
R_s = surface runoff
R_g = subsurface runoff
ΔS = change in moisture storage

1-56670-228-3/97/$0.00+$.50
© 1998 by CRC Press LLC

General forest hydrology texts (Hewlett 1982, Brooks 1991) provide complete discussions of the water balance equation for well-drained forest land. In those texts, the water balance equation is generally considered over an entire watershed for a relatively long time period, often over one annual cycle. Wetlands are areas that are flooded or saturated long enough in the growing season to support plants adapted to anaerobic conditions. Precise definitions of water table depths to meet the hydrology criteria for jurisdictional wetlands are contained in wetland delineation manuals (USACE 1987, Federal Interagency Committee for Delineation 1989). To be useful to an explanation of wetland formation, the order of terms can be changed in the equation, a more limited geographic extent can be used, and a shorter time period can be considered.

The geographic extent considered in this chapter will be analogous to the pedon in soil taxonomy (Soil Survey Staff 1975), i.e., an area large enough to contain a normal range of variation associated with a process, yet small enough that the process does not differ across the area. For example, a pond could be a single unit from deepest to shallowest part. However, a river floodplain would contain several units such as terraces, channels, and oxbow lakes. A single process causes the pond to fill, while several processes are active on the floodplain.

The time frame considered will also be determined by the way we define a wetland. In the 1987 delineation manual (USACE 1987), wetland hydrology requires soil saturation for at least five percent of the growing season, or approximately 14 days in the Southeast. An appropriate time scale for the water balance equation may be a week in this case.

The water balance equation can be rewritten in the form:

$$\Delta S = P - E - T \pm R_s \pm R_g \qquad (5.2)$$

In this form, the equation provides a mathematical prediction of wetland hydrology. Soil saturation depends on the sign and magnitude of the storage factor in the water balance equation. If ΔS is positive and larger than the volume of unsaturated soil porosity, the soil will become saturated. Since this equation can be used on areas much smaller than an entire watershed and for time periods much shorter than a year, the two runoff terms can have either sign.

On most sites, the four right-most terms are all negative and are generally referred to as loss terms; i.e., water is removed by evaporation, transpiration, and surface and subsurface runoff. Evaporation and transpiration are always water losses from a site and are therefore negative. Across the landscape, there are positions where surface water or groundwater accumulate (primarily where the land surface has a concave upward profile), and these terms can be of either sign. Wetland site hydrology can be classified into groups based on these four terms of the water balance equation. Where all four terms are negative, wetlands occur where precipitation is greater than the sum of all four loss terms for a considerable portion of the year. The term "rainwater hydrology" will be used for sites where all four terms are negative, since rain is the direct cause of site wetness. Sites where the R_g term is positive will have groundwater hydrology. In these areas, subsurface flows supplement rainfall to result in wetland hydrology. Finally, a site can be flooded by surface water (R_s is positive).

COMPONENTS OF THE WATER BALANCE EQUATION

PRECIPITATION

Precipitation is the basic water input to the hydrologic system. For southern forested wetlands, rainfall is the dominant form of precipitation, occurring primarily as winter frontal storms and summer convective thunderstorms with orographic effects limited to the higher portions of the Appalachians. In addition, tropical cyclones often contribute large amounts of rainfall to coastal areas. Rainfall is measured by standardized gauges, and daily data are published for the region in the NOAA publication "Climatological Data." Compilations of these publications are also available in CD-ROM format for all stations in multi-state regions (National Climatic Data Center 1994).

Although standardized measurement devices are available, estimation of precipitation input to forested areas is still difficult due to spatial and temporal variability. General estimates of expected extreme events are available as atlases for the United States. Rain gauge networks can be designed with spatial statistical techniques to obtain optimal unbiased estimates of point rainfall and areal averages (O'Connel 1982). In addition to gauge networks, radar imaging can be used to improve estimation of spatial variability in rainfall (Fan et al. 1996).

EVAPORATION AND TRANSPIRATION

Evaporation and transpiration are second only to precipitation in importance to the water balance equation and together may utilize 80 to 90% of available solar energy (Priestley and Taylor 1972). Evaporation and transpiration are the mechanisms by which water is returned to the atmosphere as vapor. These processes require three preconditions: available liquid water, available energy, and an atmosphere not saturated with water vapor. Chapter 4 presents a detailed discussion of the major climatic factors driving evaporation and transpiration in southern forested wetlands. Hewlett (1982) presented a thorough discussion of these factors in relation to forested environments. In well-drained temperate forests, all rainfall that reaches the soil infiltrates (Hewlett and Hibbert 1967, Bonnel 1993). In this case, only rainfall intercepted on vegetation is evaporated, and most water is returned to the atmosphere through transpiration. Since it is difficult to measure evaporation of intercepted water independently of transpiration, the two processes are usually combined as evapotranspiration (ET). ET in well-drained temperate forests is proportional to the amount of forest cover. A 10% change in evergreen forest cover results in about a 4 cm (1.6 in) change in ET per year, while the same change in deciduous forest cover changes ET by 2.5 cm (1 in) (Bosch and Hewlett 1982). The difference is due to smaller interception during the dormant season in deciduous forests.

Most forest hydrology research on which the above reviews were based was conducted on entire watersheds with moderate topography. On most of those watersheds, forested wetlands were only a minor portion of the watershed. On areas of reduced relief, the position of the water table is a reflection of the change in storage. Decreased ET following logging resulted in elevated water tables in wet mineral soils (Trousdell and Hoover 1955). Langdon and Trousdell (1978) followed the same stand during regrowth and found water table level to decrease (corrected for differences in rainfall and potential evaporation) in proportion to stand density. However, on hydric soils, with water table close to the surface under full canopies, the water table rise was much smaller following cutting (Williams and Lipscomb 1981). Ewel and Smith (1992) found no change in water levels following cutting of cypress domes in Florida. On very wet sites, Lockaby et al. (1994) found a drop in the water table following logging.

In well-drained forests, evaporation and transpiration decrease with forest cutting due to reduced interception and transpiration. Most upland sites experience periods of limited water availability (Chapter 4). Deep-rooted trees continue to transpire using water from the soil during these periods. After the forest is cut, water deep in the soil is no longer available for evaporation or transpiration. On flooded or very wet sites, however, evaporation is not limited by availability of water. In this case, evaporation or transpiration are limited by energy and atmospheric moisture. Transpiration requires energy to overcome resistance to water movement (primarily, soil-to-root and stomatal resistance). Therefore, cutting a very wet site will result in a shift from transpiration to evaporation. Since evaporation uses energy more efficiently, the total loss of water to the atmosphere can increase.

Surface Runoff

Flow of larger rivers is published by the U.S. Geologic Survey (USGS 1994), while flow in streams draining smaller watersheds can be measured by devices described by Bos (1978). Surface water runoff from forested areas has been a subject of considerable research during the last three decades. Hewlett and Hibbert's (1967) forceful arguments and subsequent data (Hewlett et al. 1977) dispelled the idea that forest soils acted as described by Horton (1933). Horton's theory that surface flow originates when rainfall exceeds infiltration capacity is valid only for highly disturbed forest soil, like logging decks and skid trails. In temperate regions, rainfall rates do not exceed the infiltration capacity of undisturbed forested soils (Bonnel 1993).

There are a number of mechanisms by which forested watersheds produce streamflow quickly (called stormflow) after rain begins (Dunne 1983). The variable source area concept (Hewlett and Hibbert 1967, Hibbert and Troendle 1988) proposes that stormflow comes from two components, subsurface flow from saturated areas near the stream bank and expansion of the stream channel into concave landforms. Between storms, water moves downslope in the unsaturated soil profile, creating near saturated conditions in concave landscape positions. With rain, the upper ends of perennial streams expand into these concave areas as the soil saturates. With increasing rain, the streams continue to expand both laterally and longitudinally. In this theory, stormflow comes entirely from rain on this expanding source area, and upslope areas do not contribute directly to stormflow.

Others have found clear evidence of water flowing in large soil pores and bypassing the unsaturated profile (Bouma and Decker 1978, Leany et al. 1993). These workers have argued that stormflow results from rain entering large soil pores and running downhill into the stream channel. Column studies have shown that water flows through macropores, even in unsaturated soils (Radulovich et al. 1992, Bootlink et al. 1993), and the soil matrix has limited influence on concentrations of materials in this water (Wilson et al. 1991). However, workers have also found that chemical composition and isotope ratios show stormflow to contain a much higher percentage of pre-event water (water that has interacted with the soil matrix) than column macropore studies would suggest (Sklash and Farvolden 1979, Neal et al. 1992). McDonnell (1990) could account for all stormflow with subsurface macropore flow. He also found that water in macropores quickly acquired chemical characteristics of pre-event water and could account for the pre-event character of stormflow. Mulholland (1993) found geochemical evidence of three separate flow paths involving both macropores and groundwater in stormflow from Walker Branch in Tennessee.

On watersheds with less relief, as those found on the lower Coastal Plain, there are no steep slopes to cause water to travel to streams in the unsaturated profile. Likewise, macropore flow is downward to the water table. On low-gradient watersheds, water tables near streams are quite close to the soil surface, with a capillary fringe above the water table. Rain added to such a water table will result in rapid and large increases in water table height (Abdul and Gillham 1984, Gillham 1984) which will cause a rapid increase in groundwater gradient to the stream (Abdul and Gillham 1989) and delivery of groundwater to stormflow. Waddington et al. (1993) found that water flowing from saturated areas accounted for stormflow from low-gradient watersheds. They also found that rain quickly mixed with pre-event water in saturated areas so that stormflow from saturated surfaces contained a large percentage of pre-event water.

Subsurface Runoff

Subsurface runoff of water is of primary importance to forest hydrology. All rain that is not intercepted by vegetation or litter enters and flows within the soil. Water moves through the soil in both unsaturated and saturated conditions. The rate of subsurface flow depends on the hydraulic gradient and hydraulic conductivity of the substrate, most often described by the Darcy equation:

$$V = k\ dh/dx \tag{5.3}$$

where: V = apparent velocity of flow
k = saturated hydraulic conductivity
dh/dx = change of hydraulic head per unit distance along the flow path

The Darcy equation has been the only function used to predict the rate of water movement in saturated soil for well over 100 years. When combined with continuity (short-term water balance) and expressed in all three dimensions, it can be written in the differential form:

$$d(K_x dh/dx)/dx + d(K_y dh/dy)/dy + d(K_z dh/dz)/dz = S\ dh/dt - R \quad (5.4)$$

where: $K_{x,y,z}$ = hydraulic conductivity in the x, y, z directions
S = the storage coefficient (amount of water released from storage per unit change in head
R = water added or withdrawn (depending on sign).

Equation 5.4 also may be modified by allowing K to vary with soil moisture for flow in unsaturated profiles, usually known as the Richard's equation.

In general, Equation 5.4 cannot be solved by analytic techniques without making assumptions of homogeneity of K and S and allowing flow in only one or two dimensions. The earliest solutions were developed to estimate flow to wells (Theim 1906, Theis 1935). Later work included methods to estimate hydraulic conductivity and storage coefficients from test pumping of a well (Chow 1952, Hantush 1966, Kruseman and De Ridder 1979). Pumping test analytic techniques are available for a wide variety of conditions represented by different simplifying assumptions.

Equation 5.4 can be approximated using one of several numerical methods, the most common being finite difference and finite element techniques (Anderson and Woessner 1992). Computer programs have been written to calculate these numeric techniques. One of the more commonly used programs, MODFLOW (McDonald and Harbaugh 1988), uses a finite difference technique to represent the equation as a series of difference equations for a three-dimensional mesh of nodes representing the aquifer. The computer program then solves these equations with matrix algebra, yielding a solution of head distributions and estimates of flow across the mesh for an increment of time.

Computerized representations of groundwater flow suffer from mathematical and physical limitations. For example, MODFLOW uses an iterative procedure to solve the matrix algebra. Since the solution of the matrix equations is not known, the iteration cannot proceed to some predetermined error but only to some predetermined amount of change between iterations which converge on the solution. If the difference equations were solved exactly, convergence on the exact solution of the differential equation occurs only with infinitely small mesh interval and time step. Finally, the differential equation is simply a representation of the actual flow field and is only as accurate as the estimates of hydraulic conductivity and storage coefficient at each grid point.

A large variety of prediction models is available (Van Genutchen and Alves 1982, Wang and Anderson 1982, Bear and Verruijt 1987) with alternative mathematical approximations. However, field heterogeneity in the hydraulic conductivity and storage coefficient most seriously limits our ability to quantitatively estimate groundwater flow. Estimation of these parameters can be done by laboratory analysis of intact core material. Cores do not reproduce larger scale structure. Slug tests (Hvorslev 1951, Cooper et al. 1967, Bouwer and Rice 1976) produce *in situ* estimates of these values but must be repeated to estimate large-scale structure of the aquifer. Bore hole meters (Molz et al. 1989) can produce large-sale, *in situ* estimates of the vertical distribution of average horizontal conductivities. Although measurement of spatial heterogeneity is still not completely possible, several stochastic techniques have been developed to estimate heterogeneity of conductivity and storage (Dagan 1987, Gelhar 1993) and tested with tracer experiments (Garabedian et al. 1991, Rehfield et al. 1992).

Change in Moisture Storage

Water can be stored as surface ponds or within the soil matrix, often reflected by a change in the water table on wetland sites. Water within the soil may or may not exist in hydrostatic equilibrium. Hydrostatic equilibrium means that the water pressure in soil pores is the same as in a free-standing column of water at the same elevation. The water table is the point at which water pressure in soil pores equals atmospheric pressure and air is driven out of the soil below that level. In small pores, water is held at pressure less than atmospheric by adhesion and cohesion. These forces are the same as those in capillary tubes, leading to the term capillary fringe for water in this zone. This water is also in hydrostatic equilibrium with a suction equal to the weight of a water column extending down to the water table. As in capillary tubes, the force of gravity overcomes cohesion in large soil pores allowing air to enter, and water is then held in layers around soil particles. This water is no longer in hydrostatic equilibrium, the normal condition for water in most upland soils. Factors important in water storage in unsaturated soil are well known and can be found in basic soils texts (Hillel 1971, Brady 1974).

Soil water in the capillary fringe is very important to understanding storage in wetland soils. The storage coefficient (S in Equation 5.3) determines the rate a water table will fall per unit discharged as drainage or the rate it will rise per unit of rain. It is determined by the unsaturated pore space in the capillary fringe and is sometimes called drainable porosity. It can be estimated from pumping tests (Kruseman and De Ridder 1979), slug tests (Cooper et al. 1967), measures on drainage ditches (Skaggs 1976), or measures of water table rise after rain (Williams 1978). On coastal sediments it can vary from 5% to 13.5%, which translates to a water table rise of from 7.5 to 20 units for each unit of rain (Williams 1978, Skaggs et al. 1991).

WATER BALANCE OF FORESTED WETLANDS

As shown in Equation 5.2, the components of the water balance equation can be combined to predict hydrology of a wetland. In this way, hydrologic processes cause a wetland to exist in particular areas. This is the most basic evaluation of hydrologic function. It is a building block for functions that can be used to set values. For example, a forested wetland can be caused by inflow of groundwater. Such wetlands have been found to remove nitrate from contaminated groundwater (Peterjohn and Correl 1984). For water quality improvement (value) (Odum 1989) to occur, both groundwater flow into the wetland and contamination of groundwater must occur.

The first step to defining the ecological value of a forested wetland will be to define the hydrologic processes in that particular wetland.

The HGM classification system is now a standard method for functional analysis of wetlands (Smith et al. 1995). In the HGM classification there are seven hydrogeomorphic classes: Riverine, depressional, lacustrine fringe, tidal fringe, slope, wet mineral flats, and wet organic flats. Each of these categories has a unique set of functions related to landscape position and hydrologic process. In this section the three basic hydrologic processes will be illustrated though a series of examples.

RAINWATER HYDROLOGY

Rainwater creates wetlands by an interaction of climate and geomorphology. Rainwater hydrology is defined by a negative sign for all the loss terms in the water balance equation (5.2). By this process, wetlands occur where rainfall is greater than evaporation and transpiration, and storage accumulates because both surface and subsurface flows are relatively ineffective in removal of excess rainfall. Rainwater hydrology dominates where excess rainfall is high, and landform reduces the rate of surface and subsurface drainage.

The following example calculation can be used to elucidate the factors important in determining the presence of rainwater wetlands. For this example, we will assume a simplified section of the Coastal Plain in northeastern South Carolina. This simplified section is a symmetrical forested interfluve between two rivers that are 1 km (0.62 mi) apart. We will assume a thick impermeable, horizontal layer below the entire section, the section is homogeneous fine sand, the saturated thickness is 10 m (33 ft), and both rivers completely penetrate the saturated thickness but do not penetrate the underlying impermeable horizon. We will use the 30-year average surplus in this region (38 cm [15 in]) from Figure 4.8. In Equation 5.2, (P-E-T) equals 38 cm (15 in), DS = 0 when averaged more than 30 years, $R_s = 0$ for forested land without channels, and, therefore, R_g is –38 cm (–15 in). If we assume water flows horizontally in our section, and we are interested in an equilibrium water table, then we can reduce Equation 5.4 to the simple linear equation:

$$-R = Kx\ dh/dx \qquad (5.5)$$

Figure 5.1a shows the simplified section from one river to the watershed divide. By considering a 1-m (3.3-ft) wide section, we can convert the rate equation (5.5) into quantities:

$$Q = 365(W*D*Kx\ dh/dt) \qquad (5.6)$$

where: Q is quantity of water (m^3/year)
W = width in meters
D = depth of saturated profile in meters
Kx = hydraulic conductivity in m/day.

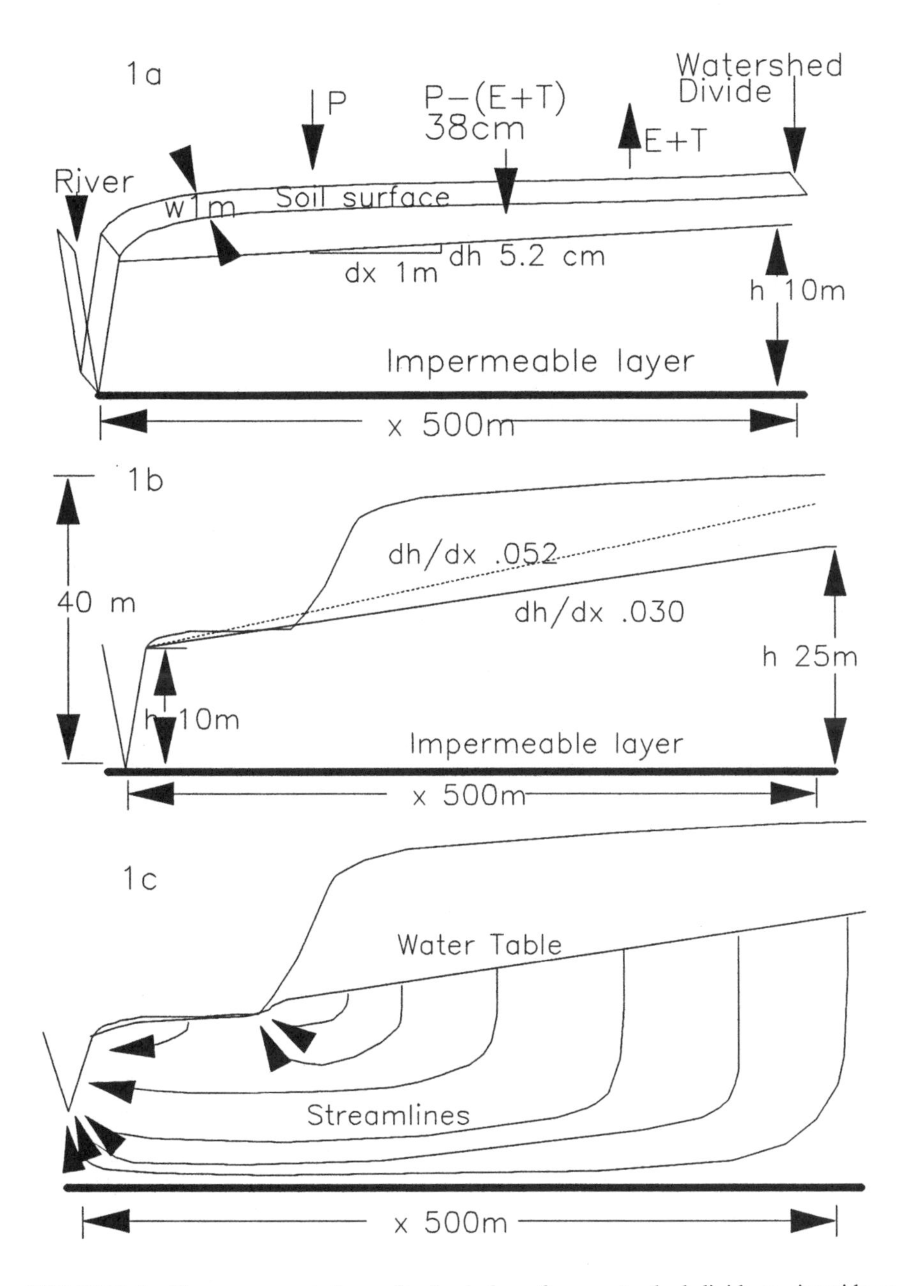

FIGURE 5.1 Three representations of a land slope from watershed divide to riverside as examples of groundwater movement in forested watersheds. (a) Representation of most simplistic flow model used with Equation 5.5 to estimate rainwater wetlands. (b) A more realistic representation of a coastal watershed. In this case, the actual slope of the equilibrium in (a) is modified to include influence of increasing saturated thickness with higher water table position. (c) More realistic depiction of (b) with streamlines.

For this example, we will use a value of 1 m/day (3.3 ft/d) for Kx (Kruseman and De Ridder 1979). Therefore,

$$Q = -(-0.38 \text{ m/yr})* 500 \text{ m} *1 \text{ m} = 190 \text{ m}^3/\text{year}$$

and

$$Q = 365 \text{ days/year}* 1 \text{ m}* 10 \text{ m} * 1 \text{ m/day} *dh/dx$$
$$190 \text{ m}^3/\text{year} = 3{,}650 \text{ m}^3/\text{year} * dh/dx$$
$$dh/dx = 0.052$$

For these conditions, an equilibrium water table will slope at 5.2% from the river. Consequently, there will be soil saturation on a portion of the interfluve if the average slope from river level to watershed divide is less than 5.2%.

Climate is one of the chief variables in development of rainwater hydrology. Figure 4.8 presents mean annual surplus for climate regions in the South. This figure can be viewed as the climatic potential for rainwater hydrology. If we maintain the hydraulic characteristics of the example constant, wetlands would form due to rainwater hydrology at slopes below 6% in eastern North Carolina, 4.5% in Coastal Georgia, 2% near Orlando, Florida, and 9% near Mobile, Alabama. Of course, these numbers represent a simplified example and should be treated only as a relative comparison of the climatic differences across the South.

Quantity of subsurface flow is the product of the rainfall excess and the area of input determined by the distance between streams. In the above example, streams were 1,000 m (3,280 ft) apart resulting in 190 m^3/y/m length (2,040 ft^3/y/ft length). Had streams been 100 m (328 ft) apart the subsurface flow would have been 19 m^3/y/m length (204 ft^3/y/ft length) with an equilibrium water table slope of 0.52%. Drainage system development is complex, but drainage density, or length of stream per unit area (e.g., m/ha, km/km^2, ft/ac, mi/mi^2), is best related to elevation, age, erodibility, and excess rainfall (Leopold et al. 1964). High elevation provides the energy for rapid erosion and headward migration of streams into interstream depressions. Since erosion is a slow process, only sites that are geologically old or are extremely erodible have highly developed stream systems. The southern Appalachian mountains are a prime example of stream development. Relatively high mountains have been exposed to a warm, damp climate for several hundred million years. Although lower in elevation, the Piedmont has had a similar history of prolonged weathering and erosion. In both provinces the density of streams is quite high, such that for a given rainfall excess, the quantity of subsurface flow (Q) need not be very large. Also, erosion in these provinces has resulted in landslopes often greater than 10%. With both higher drainage density and steeper slopes, these provinces are unlikely to have wetlands with rainwater hydrology.

On the Coastal Plain, elevations are generally less than 100 meters (328 ft) and the land surface has been exposed to erosion less than 50 million years. The Coastal Plain is composed of a series of marine terraces that represent ancient sea levels (Cooke 1936). Each terrace contains landforms similar to the present beach: offshore sands, beaches, and salt marsh plains (Colquhoun 1974). As one approaches the coast, these marine terraces become progressively younger, and drainage density decreases.

Equation 5.5 also indicates that equilibrium water table varies inversely with both hydraulic conductivity and aquifer depth (the term transmissivity is used for the product of these two). Each terrace contains sediment similar to the present coast. On former beaches, sediments are often relatively thick sands and have relatively high transmissivity. In contrast, former salt marshes have clays in the soil profile or subsoil. These soils have lower hydraulic conductivity (Natural Resources Conservation Service 1994) and, therefore, will have steeper equilibrium water tables. In addition, salt marshes tend to have very little relief, and land slopes tend to be very small on old salt marsh plains.

Rainwater hydrology is determined by climate, drainage density, soil permeability, and land slope. In the HGM classification, rainwater hydrology is dominant in wet flats with mineral and organic soils. Wet flats in the Southeast are most likely to occur in the Atlantic Coastal Plain in eastern North Carolina and in the Gulf Coastal Plain from Tallahassee, Florida to the Texas border. Within the Coastal Plain the younger terraces near the coast are most likely to have wet flats. Within any terrace, rainwater hydrology is most likely to form wetlands on former salt marsh deposits and least likely on former beach deposits. In all cases, wetlands are most likely along interstream divides in positions most distant from streams. Since organic soils form with prolonged soil saturation, this same list would explain the likelihood of organic soil wet flats.

Rainwater hydrology is determined by large-scale phenomena. Rainfall excess, land slope, and drainage density interact in a relatively simple way to determine where wet flats are likely to be found. A reasonably useful example could be developed by reducing Equation 5.3 to a simple linear equation using simplified geology and a long-term climatic average. However, as stated earlier, determination of jurisdictional wetlands requires a specific length of flooding during the growing season. More realistic evaluation of wet flats has been done using the model DRAINMOD (Skaggs 1978) which has been modified to model drained pine plantations (McCarthy and Skaggs 1992, Richardson and McCarthy 1994, Amatya et al. 1995).

Groundwater Hydrology

Wetlands where the groundwater runoff term in Equation 5.2 is positive have groundwater hydrology. In discussions of rainwater hydrology, large-scale topography was most important. In consideration of groundwater hydrology, topography close to the point is usually the most important. If we modify the rainwater example slightly, as shown in Figure 5.1b, the overall transmissivity and land slope will move the rainfall excess to the river, and no wetland with rainwater hydrology would exist. In the example, we ignored the fact that the saturated thickness increases with higher water tables farther from the river. If we modify the calculation to determine dh/dx assuming h increases, an iterative solution can be found with a mean thickness of 17.5 m (57.4 ft) and an equilibrium slope of 3% (the solid line in Figure 5.1b).

It is obvious that saturated thickness is a very important parameter in the calculation of groundwater flow. Figure 5.1b presents a common landscape with a break in slope and a floodplain along the stream. This example represents one of the most common instances of groundwater hydrology. If the aquifer sediment

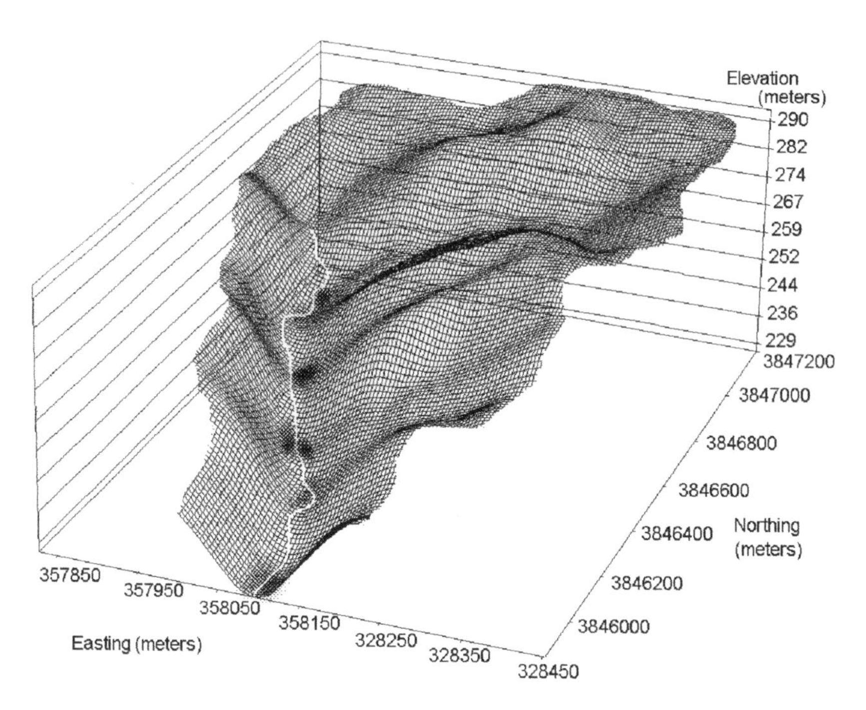

FIGURE 5.2 Wire mesh representation of watershed in South Carolina Piedmont. Groundwater wetlands (dark gray areas) are primarily along eastern side of stream (white line) where small valleys intersect main valley. Note size of area (about 72 ha [178 acres]) and total relief (65 m [213 ft]).

thickness decreases, the transmissivity of the aquifer declines, and groundwater flow from a site is less than groundwater flow into the site. The actual groundwater flow pattern resembles streamlines in Figure 5.1c. The HGM classification calls these sites slope wetlands. Parsons et al. (1991) have developed two-dimensional models that predict water tables in such a two-dimensional flow field.

Real landscapes are not limited to two-dimensional flow fields and simple mathematical representations (see Marsily [1986] for quantitative discussions of groundwater flow in real systems). However, qualitative understanding of groundwater wetlands can often be determined from the surface topography and understanding of underlying geology. The following examples show how slope wetlands might be identified in more complex landscapes.

The mountains and Piedmont are characterized by soils developed by weathering of underlying bedrock. In these provinces, the water table aquifer is usually in the saprolite over the unaltered bedrock. Topography is well dissected with many small streams and ephemeral streams in valleys created by weathering of less-resistant rock. Figure 5.2 is the surface representation of a small watershed in the northwestern South Carolina Piedmont demonstrating distribution of slope wetlands in these provinces.

Northwestern South Carolina has large rainfall excesses (Figure 4.8). However, weathering has produced a relatively steep and dissected landscape. In this example, slopes are only about 100 m (328 ft) long so the quantity of groundwater flow to any point along the stream or ephemeral is relatively small. On the western side of the watershed, the stream channel is incised into the toe of the slope and groundwater flows directly into the channel. All the valley bottoms have accumulated colluvial material, and the main valley has some alluvial fill. In the preceding two-dimensional analysis, we considered only the flow of groundwater from hillslope into the valley fill, however, water also flows in the colluvial or alluvial fill. In this watershed, small wetlands exist where the eastern ephemeral channels join the main channel. The distribution along the main channel also demonstrates the concept of thinning of aquifer depth very well. Topography of this watershed is dominated by east-west bands of resistant rock that form the small ridges. At each point where the channel meets these bands of resistant rock, there is little alluvial fill and transmissivity in the downstream direction is minimal. Small slope wetland areas are just upstream of each band of resistant rock.

The Piedmont and mountains have steep terrain and high drainage density. Wetlands occur where groundwater flows accumulate or resistant rocks result in thinning of the alluvial fill. In most of the Coastal Plain, aquifers are bounded by marine clays that are nearly horizontal. On the lower Coastal Plain, drainage systems are poorly developed and wet flats are widespread. In the middle Coastal Plain, drainage is more completely developed, often in relict marine topography.

A section of Dillon County, SC demonstrates the development of wetlands by both rainwater and groundwater hydrology in this province (Figure 5.3). There are three types of wetlands depicted in this 3 × 4 km (2 × 2.5 mi) section of topography. The basic landform is a portion of old beach that now forms the northern bank of the Little Pee Dee river floodplain. Small streams have formed in the lower landforms between relict dune lines. Although the entire area is formed of beach sand, the topography creates a complex three-dimensional flow field (Figure 5.4). The Little Pee Dee, about 10 m (33 ft) below the dune lines, creates a regional flow to the south, thus producing a large slope wetland along the edge of the floodplain (A1). The two smaller streams between the dune lines also create local regions of aquifer thinning and slope wetlands (A2 & A3). The distinct dune lines are not present in the western portion of the section, and a wet flat (C) occurs in this area roughly 3 km (2 mi) from the Little Pee Dee. Although the major regional slope (0.003) is to the south, there is also a minor slope to the east (0.001) which is intercepted by the small streams (A2 & A3), producing local groundwater discharge in this direction also. In fact, the gradient from C to A2 (0.004) is greater than toward the Little Pee Dee, resulting in more flow to the east than to the south. In this direction the aquifer thins, producing three small slope wetlands (B1, B2, and B3).

Although groundwater hydrology can often be predicted by topography, aquifer materials are seldom homogeneous and boundaries are seldom either parallel to the surface topography or completely horizontal as assumed in the previous examples. In general, groundwater hydrology produces wetlands where transmissivity decreases. Transmissivity will decrease where the surface declines, where a less permeable layer comes closer to the surface, or where the aquifer material becomes less permeable.

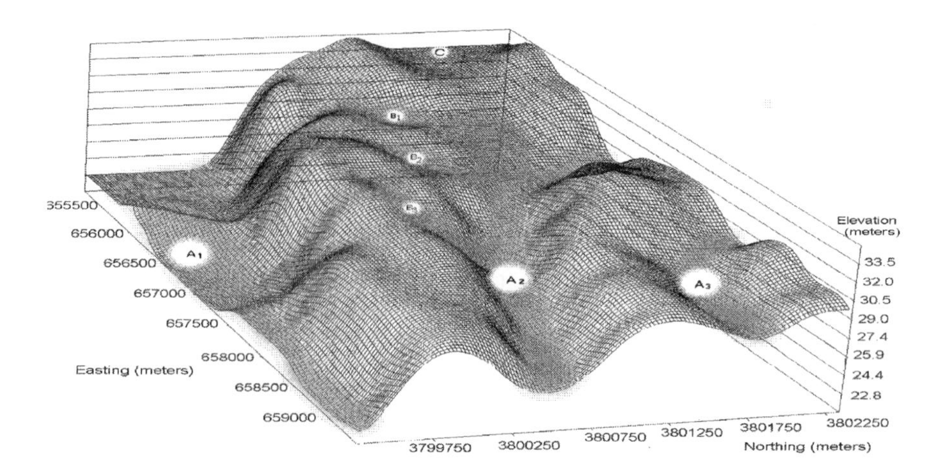

FIGURE 5.3 Wire mesh representation of middle Coastal Plain. (Viewed from above southeast corner). Wetlands are represented in gray and are of three types: (A) slope wetlands at slope toes beside small streams developed between ancient sand dunes and along valley of Little Pee Dee River; (B) slope wetlands where aquifer thins due to lower elevations; and (C) a wet flat about 2 km (1.2 mi) distant from river valley where topography does not include ancient sand dunes.

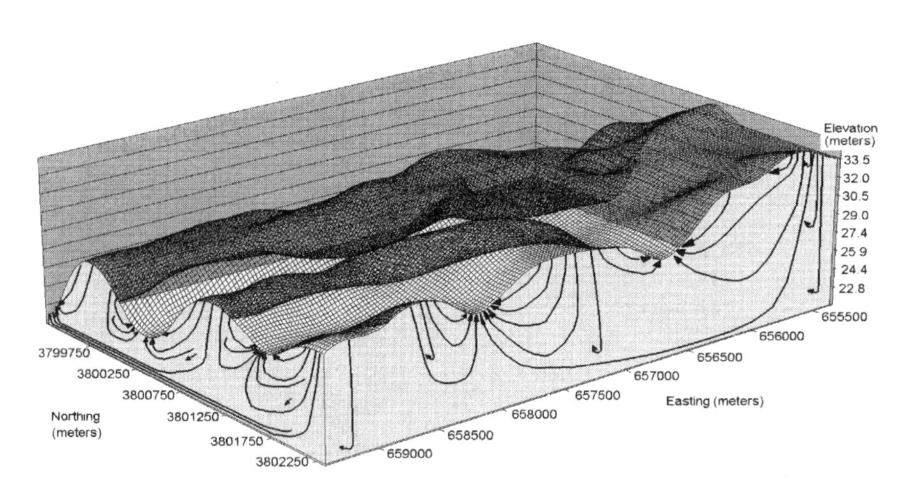

FIGURE 5.4 Block diagram of Figure 5.3 (viewed from above northeast corner) with a representation of the complex three dimensional flow developed by surface topography. Arrows with points into block on north side represent southward flow. Arrows originating within the block on the east side represent eastward flow.

In much of coastal Georgia and Florida, the Floridian Aquifer is artesian with pressure often above the local water table. This aquifer is separated from the surface by the relatively impermeable Hawthorn Formation in most places. However, where the Floridian Aquifer is artesian and the Hawthorn Formation is thin or slightly permeable, groundwater hydrology may form wetlands. Location of such wetlands is difficult to predict. Also, water levels in such wetlands are determined by the artesian level in the Floridian Aquifer and permeability of the Hawthorn Formation. Wetland boundaries will change with changes in Floridian Aquifer artesian level.

SURFACE WATER HYDROLOGY

Given the sizes of rainfall excess and deficits in the Southeast (Figures 4.7 and 4.8) any closed depression without groundwater or surface water outlet will fill with water to form lakes or ponds. However, forested floodplains are the most important forested wetlands with surface water hydrology. These wetlands receive water by overbank flooding from adjacent rivers. Forested floodplains have long been recognized as mosaics of various species (Van Doren 1955), with species occupying sites based on length of inundation (Dollar et al. 1992). These wetlands have been stratified into several levels of flood depth and duration (Mitsch and Gosselink 1993). However, studies of flooding tolerance of tree species have been based on empirical measures of inundation (McKnight et al. 1981, Hook 1984b) rather than predicted flood elevations based on flow data.

Determining the wetland characteristics of any particular floodplain forest requires knowledge of the river flooding regime. The flows of all the major river systems in the United States are measured and published by the U.S. Geologic Survey. Average daily flows are published for each state and are also available on CD ROM media for computerized access (USGS 1994). From these data, a flow-duration curve can be determined for the river in question (Lindsey et al. 1958). A flow duration curve presents the percentage of time a specific flow could be expected to be exceeded. A flow duration curve can also be produced for a portion of the year to relate flows to a particular physiological activity.

In an effort to predict regeneration in the Congaree Swamp National Monument (CSNM), Rikard (1988) constructed a flow duration curve to coincide with germination and early growth (March to June) using flow records of the Congaree River at Columbia, SC, from 1940 to 1985 (Figure 5.5). This curve indicates that there is a 95% chance that flow will exceed 100 m^3/sec (3,500 ft^3/sec) and a 95% chance flow will be below 850 m^3/sec (30,000 ft^3/sec). A more intuitive view of these data is as a histogram of occurrences scaled to the 122 days of the germination period (Figure 5.6). The most likely flow is between 147 m^3/sec (5,190 ft^3/sec) and 173 m^3/sec (6,100 ft^3/sec) which is likely to occur for 15 days in the period. Rikard (1988) then chose four ranges of discharge based on the shape of Figure 5.6: 147 to 283 m^3/sec (5,190 to 9,990 ft^3/sec), 283 to 538 m^3/sec (9,990 to 18,900 ft^3/sec), 538 to 1,161 m^3/sec (18,900 to 41,000 ft^3/sec), and >1,161m^3/sec (41,000 ft^3/sec). These ranges represent flooding for 82.1, 28.7,12.8, and 1.4 days of the growing season, respectively (flows below 147 m^3/sec [5,190 ft^3/sec] were within the main channel and do not represent flooding).

Since the Congaree River has no tributaries between Columbia and the CSNM, these flow rates are equally valid there. However, flow rates must be translated into water elevations to determine which land elevations will be flooded. Water flows in

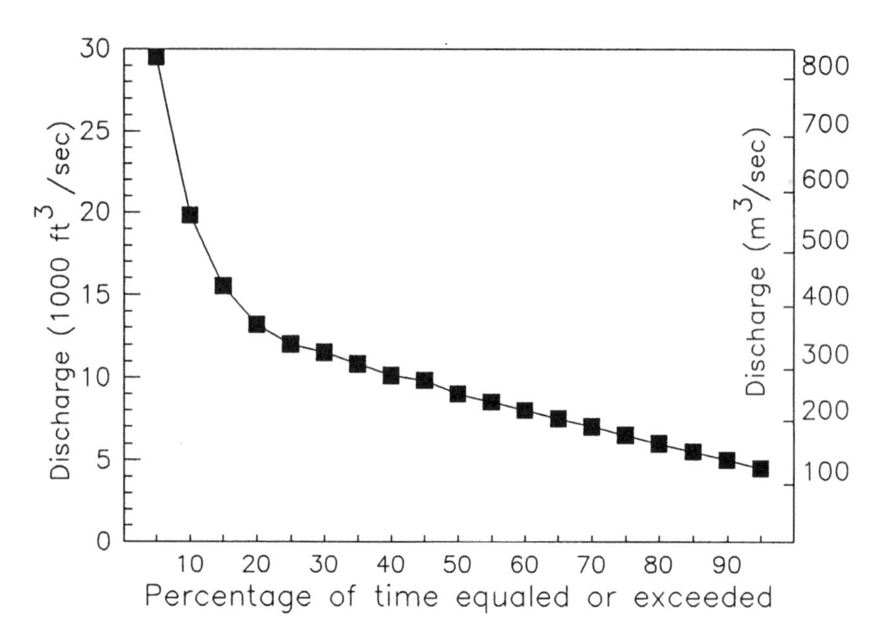

FIGURE 5.5 Flow duration curve for the Congaree River at Columbia, SC, for March through June, based on records from 1940–1985. (adapted from Rikard 1988).

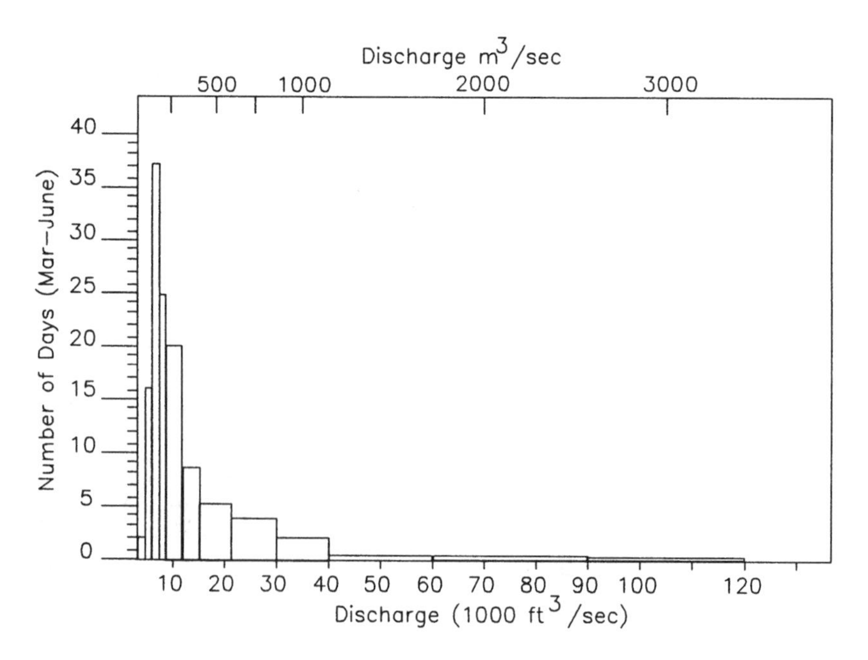

FIGURE 5.6 Histogram of data used in Figure 5.5 scaled as days of a possible 122 for the period March through June.

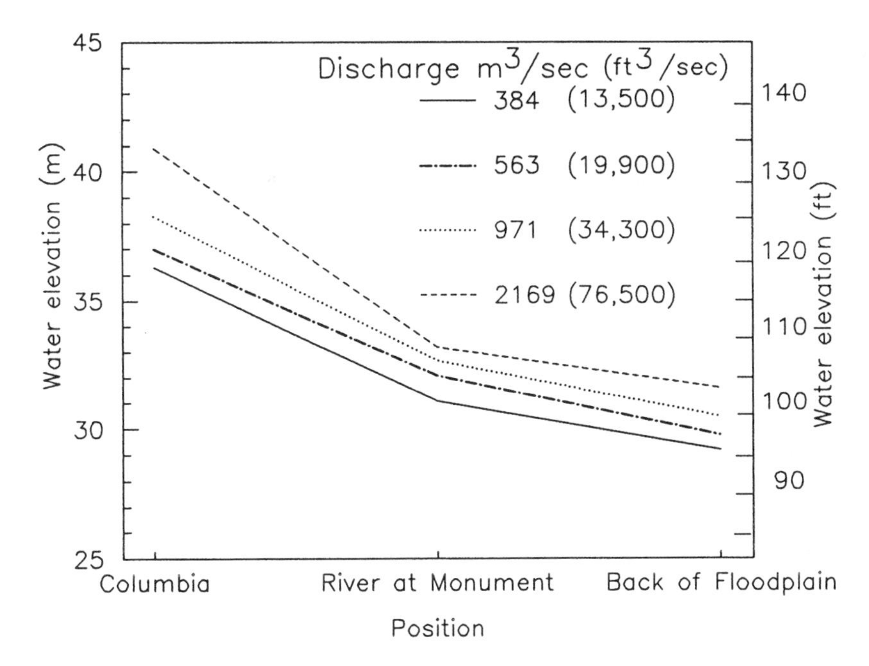

FIGURE 5.7 Changes in river elevation along the Congaree River from Columbia to and into the Congaree Swamp National Monument (based on data presented in Patterson et al. 1985).

open channels by converting potential energy (height above a datum) into kinetic energy as kinetic energy is dissipated by turbulence. The slope of the water surface is proportional to the rate of energy dissipation (Henderson 1966). Patterson et al. (1985) determined an empirical relationship from Columbia to the CSNM and across the floodplain to Cedar Creek (Figure 5.7). At lower flow, energy loss from Columbia to the Monument (32 km [20 mi]) is represented by a drop in the surface of 5.2 m (17 ft) and flow across the floodplain (4 km [2.5 mi]) by 1.9 m (6.2 ft). The energy loss across the swamp increases to 2.3 m (7.5 ft) when the distributary channels are flowing full, but declines to 1.6 m (5.2 ft) when the entire swamp is flooded. At the river edge, Rikard's four flooding ranges correspond to elevations of <30.9 m (101.3 ft), 30.9 m to 32.1 m (101.3 to 105.3 ft), 32.1 m to 32.7 m (105.3 to 107.2 ft), and >32.7 m (107.2 ft) . However, near the bluff the same flow ranges corresponded to elevations of <29.0 m (95 ft), 29.0 to 29.8 m (95.1 to 97.7 ft), 29.8 to 30.6 m (97.7 to 100.3 ft) and >30.6 m (100.3 ft).

When Rikard (1988) analyzed tree species found on the floodplain in these elevation ranges, the lowest range corresponded very closely to the species cited as most tolerant of flooding and the highest zone had species listed as only slightly tolerant (McKnight et al. 1981).

These same techniques were used to examine the Ogeechee river floodplain near where Meyer and co-workers studied river ecology (Findley et al. 1986, Meyer and Edwards 1990, Pulliam 1993). However, on many of the sites (especially near the

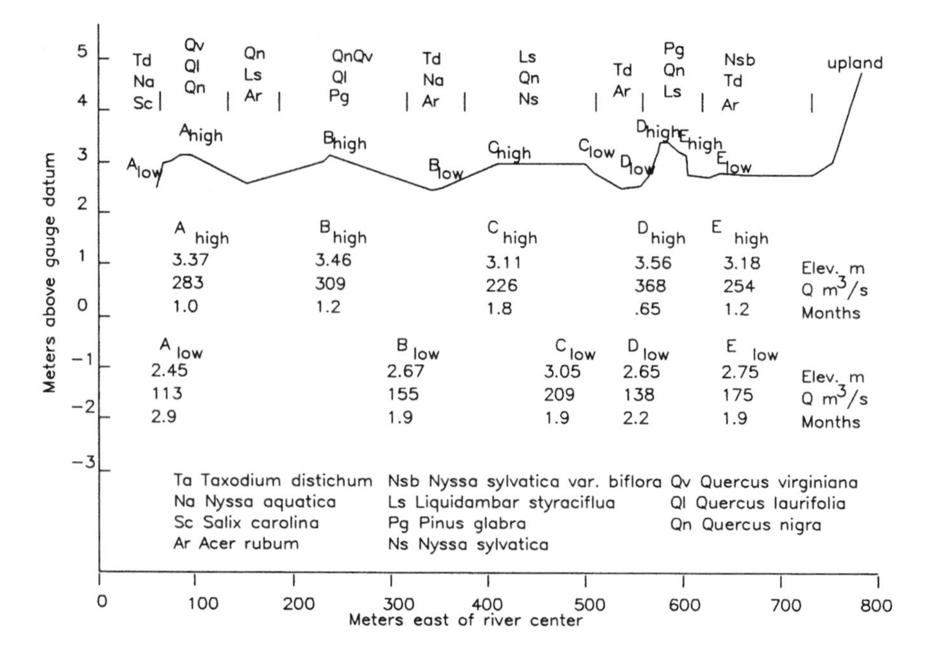

FIGURE 5.8 Cross section of Ogeechee River near Eden GA. Land elevation (line), sampling points (A-E), distribution of dominant overstory tree species (vertical lines and abbreviations), elevations of sampling points, discharge which would flood that point, and number of months that discharge would be expected to occur (calculated as in the Congaree example).

bluff), species were more flood tolerant than would be expected from the flow records (Figure 5.8). This study also included continuous water table recorders on five sites. During periods without rain, transpiration lowered the water table across the floodplain, as could be clearly seen by water table drops during the day. On the two sites closest to the bluff, a groundwater subsidy was clearly evident by a rise in the water table each night (Figure 5.9). On this cross section a groundwater hydrology created wetlands in about 250 m (820 ft) of an 800 m (2600 ft) wide floodplain.

On a moderately large, generally free-flowing river (only 1 large dam in the watershed), flow records could be used to determine forest species composition. However, a given flow rate does not flood the entire floodplain to the same elevation. Hydraulics of the floodplain must also be considered. At present hydraulic models capable of estimating flooding across a forested floodplain are limited by the inability to measure or model the spatial heterogeneity of both surface elevations and flow resistance across real floodplains. Only empirical methods (Patterson et al. 1985) are available, and these are site specific.

DETERMINATION OF WETLAND HYDROLOGY AT A SPECIFIC SITE

Forested wetland hydrology can be classified based on the water balance equation. On convex upward landscape positions, all the loss terms in the equation (5.2) will

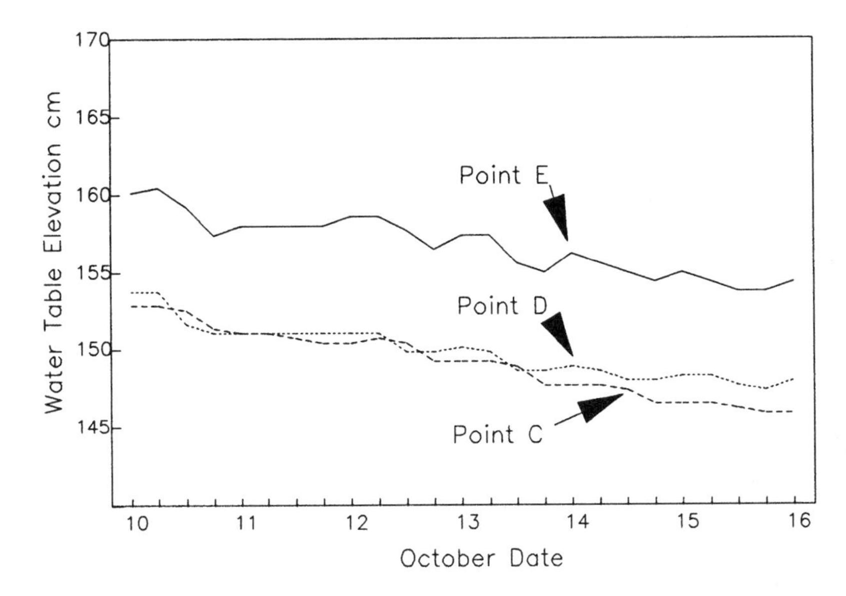

FIGURE 5.9 Water table elevations of wells at sample points C, D, and E in Figure **5.8.**

be negative, and wetlands will be formed by rainwater hydrology. These wetlands will most likely be pocosins or wet pine flats. In the same general regions (lower Coastal Plain of NC, AL, LA, and MS) any concave landscape position will also receive a groundwater subsidy . Where rainfall excess is smaller (coastal SC, GA, and FL), wetlands are more likely to occur in concave landscape positions. Hardwood flats, Carolina bays, cypress domes, and small blackwater swamps are all likely to have groundwater hydrology. Hydrology of cypress domes (Ewel 1976) and Carolina bays (Schalles 1989) shows interaction of both rain and groundwater.

Finally, forested wetlands on floodplains can be influenced by all three. On a small piedmont watershed, Hook et al. (1994) reported no relation between mature vegetation and flooding. They found hydric soils and soil anaerobisis were related only to landscape position (concave toe of steeper slope). On the Ogeechee River, mature vegetation was not well related to flooding, especially near the bluff where groundwater input maintained a wetland independent of flooding. However, on the Congaree River, flood durations determined from flow records corresponded well with empirically derived flood tolerance.

The stream Hook et al. (1994) studied has a 100 m (328 ft) wide floodplain and maximum flood under 20 m^3/sec (700 ft^3/sec), the Ogeechee River has a floodplain about 800 m (2600 ft) wide and a maximum flood about 1,000 m^3/sec (35,000 ft^3/sec), whereas the Congaree floodplain is roughly 4 km (13,000 ft) wide and a maximum flood of 11,400 m^3/sec (400,000 ft^3/sec) (before the construction of the Lake Murray dam [Whetstone 1982]). One might speculate that surface water hydrology is more likely to determine wetland characteristics where river flows are large in relation to floodplain width.

The southeastern United States has a humid climate which results in rainfall that exceeds evaporation and transpiration on an annual basis. Wetlands will occur where the excess is not removed by surface or subsurface runoff. In the Piedmont and mountains, slopes are steep and subsurface flow moves water into numerous streams. In these regions, wetlands with groundwater hydrology predominate. The HGM slope and riverine classes are most prevalent. In the upper Coastal Plain, older landscapes have a number of small streams in both erosional and relict marine depressions. Wetlands of the slope, depressional, and riverine classes will occur. Groundwater hydrology is dominant in several wetlands in this province, but floodplains of large rivers will also have surface water hydrology. On the lower Coastal Plain, elevation and erosion are least, and rainwater hydrology will create wetlands on interstream divides. Large wet flats, with both mineral and organic soils, can be found here. Conditions are most favorable on the lowest marine terraces in eastern North Carolina and from Florida to East Texas along the Gulf Coast. Depressional and riverine wetlands are also common. Groundwater hydrology is important to both classes, including wide floodplains of smaller blackwater rivers. Along the immediate coast surface hydrology dominates tidal fringe wetlands.

6 Soils and Landforms

John A. Stanturf and Stephen H. Schoenholtz

CONTENTS

1-56670-228-3/97/$0.00+$.50
© 1998 by CRC Press LLC

INTRODUCTION

Wetland soils have varying biological, physical, and chemical characteristics, although all are saturated with water and depleted of oxygen for portions of the growing season. Water is the dominant cause of, or influence on, soil transformations, additions, removals, and transfers in wetland soils. These changes include immobilization of soluble soil components, decomposition and accumulation of organic matter, mineral dissolution and synthesis, fluctuations in redox potential and pH, and translocation of solutes, sediments, colloids, and products of physical and chemical weathering (Mausbach and Richardson 1994).

In this chapter we concentrate on the distribution and characteristics of forested wetland soils of the Atlantic and Gulf Coastal Plains and Mississippi River Valley regions. Wet soils dominate these areas, and forested wetlands are widespread. Although forested wetlands do occur within upland areas of the mountains and plateaus, they are less extensive there. Many riparian areas in uplands are similar to minor stream floodplains in the Coastal Plain, at least in terms of soil characteristics. Soils of relatively rare wetlands such as mountain bogs have been studied little, and what is known about them will be presented in other chapters in this book.

SOIL TERMINOLOGY

Soils are dynamic natural bodies consisting of horizons (layers) of predominantly organic or mineral material of variable thickness that are generally parallel to the land surface. Horizons differ from their parent material (the geologic material in which soils develop) in their morphological, physical, chemical, mineralogical, and biological characteristics (Birkeland 1984). A vertical arrangement of horizons is called a soil profile and can be described by standardized methods (Soil Survey Staff 1993). Properties of horizons change laterally, as a result of differences in environmental conditions. Such spatial variation is generally gradual, but abrupt differences in parent material do sometimes occur (e.g., on floodplains where depositional environments have varied over time). Especially in the Coastal Plain, the origin of sediments determines mineralogy and texture (Daniels and Gamble 1978).

Similar profiles make up a named soil unit called a series. Series have been placed within a hierarchical classification system called Soil Taxonomy (Soil Survey Staff 1975). Wet soils are not recognized at the highest level of Soil Taxonomy (the Order level) but are separated initially at the next level, the Suborder level (Table 6.1). The important distinction between soils that receive their wetness from regional ground water and those that receive it from perched water is not explicit in Soil Taxonomy or in series either (Dudal 1992), but changes in both classification and series definition have been proposed that will better define wet conditions in soils (Vepraskas 1995).

Moisture regime within a soil profile controls the chemical and physical characteristics of the profile. Anything that affects retention or movement of water on the regional or local landscape will affect moisture regime within a soil profile (Daniels and Gamble 1978). Two patterns of wetness can occur in soils: continuous saturation of the profile from lower horizons to the surface horizons, caused by

TABLE 6.1
Taxonomic classification of forested wetland soils in the southern United States (after Broadfoot 1964, 1976, Daniels et al. 1984, Brown et al. 1990, Shepard 1993, Schoeneberger 1996).

Order and Suborder	Description
Entisols	Soils with little profile development
Aquents	*Wet Entisols*
Fluvents	*Flood-prone Entisols*
Inceptisols	Soils with some profile development
Aquepts	*Wet Inceptisols*
Ochrepts	*Drier than Aquepts, occur on floodplains and terraces that flood periodically.*
Mollisols	Soils with moderate organic matter accumulation giving rise to a thick, dark surface horizon; base saturation 50% or more
Aquolls	*Wet Mollisols*
Udolls	*Moist but freely drained Mollisols*
Alfisols	Soils with an argillic horizon (denoting clay movement and redeposition) and moderate base saturation (35% or more)
Aqualfs	*Wet Alfisols, saturated for an extended period annually*
Udalfs	*Drier than Aqualfs but still moist*
Ultisols	Soils with an argillic horizon but more soil development than Alfisols, hence base saturation is less than 35%
Aquults	*Wet Ultisols, wet usually as a consequence of high groundwater*
Spodosols	Soils with a spodic subsurface horizon where organic material in combination with aluminum or iron (or both) has accumulated after leaching from surface horizons
Aquods	*Wet Spodosols*
Humods	*Drier than Aquods, usually because depth to a water table is greater*
Vertisols	Soils with shrink and swell characteristics resulting from moderate to high content of expanding clays, with moderate to high base status
Uderts	*Vertisols with cracks at the surface for part of the year*
Histosols	Organic soils, peat or muck deposits over mineral soil or bedrock
Hemists	*Wet Histosols in which the organic material has mostly but not entirely decomposed beyond recognition*
Saprists	*Wet Histosols consisting of almost completely disintegrated plant material*

groundwater or water of the capillary fringe; or wetness of horizons at or near the surface caused by a slowly permeable horizon at depth that "perches" water (Dudal 1992). Both patterns can exist within a profile, e.g., in heavy clay soils where water perches at the surface following heavy rain and is separated from a true water table at 3 m (10 ft) or more by soil that is without free water. Sources of wetness include groundwater, seepage water, perched water, and floodwater.

The movement of water into and through soil is controlled by soil properties at the surface and within the soil volume by the amount and supply of water and by other environmental factors. Relevant soil properties include surface roughness, water repellency, slope, cracks, size and continuity of pores, total pore space, and the antecedent moisture content of the soil.

TABLE 6.2
Drainage class definitions for soils in forested wetlands in the Coastal Plain of the southern United States (modified from Collins 1982).

Well drained — Sloping to rolling topography, with little or no run-off received from adjacent areas. Surface water may perch for short periods. Bright colors, and mottling, if present, occurs as few, faint, or distinct mottles at depth. Depth to permanent water table is greater than 2 m (6 ft). Soil has adequate moisture for plant growth, and wetness during the growing season does not limit plant growth.

Moderately well drained — Gently sloping to rolling topography but with low gradient and generally lower elevation; some seepage from adjacent areas. Soil absorbs much of the water it receives. Groundwater, if present, seldom rises above 50 cm (20 in) below the soil surface. May have subsoil horizon with slow permeability, and this horizon may impede internal drainage. Perched water table sometimes occurs. Soil properties such that moisture retention is generally adequate for plant growth. Some evidence of moisture accumulation or water table fluctuation within 75 cm (30 in) of the surface. Mottling occurs at depths greater than 50 cm (20 in), and gleying, if present, is 1 m (3 ft) or deeper. Subsoil has some gray mottles in pale-red and yellowish-grey subsoil matrix.

Somewhat poorly drained — Relatively flat land. Receives and absorbs both surface run-off and seepage from adjacent land or has a seasonally high ground-water table, or both. Sometimes has water standing on the surface. Surface soil usually has moderate to rapid permeability but is often underlain by layer with slow permeability; perched water table common. Mottling and gleying present in the surface 25 to 75 cm (10 to 30 in). Subsoil matrix yellowish-gray with pale red and strong brown mottles.

Poorly drained — Generally flat land or land concave in cross section. Ground water table, if present, generally high throughout the season; land often has water standing on the surface. Receives surface run-off and seepage from adjacent land. May be subject to flooding. Many poorly drained soils have fine-textured horizons with slow permeability within 50 cm (20 in) of the surface that seriously restrict internal drainage and may cause ponding or perched water tables. Mottling and gleying are quite pronounced even in the surface 50 cm (20 in) of the profile.

Very poorly drained — Ponded areas and depressions, or low areas subject to frequent flooding. Surface soil colors are black to dark-gray and gleying occurs within 25 cm (10 in) of the surface.

Runoff is the movement of precipitation water across the soil surface. In floodplain soils, a related term is sheetflow for the movement of water across the surface. Infiltration is the entry of water into the soil at its surface. Hydraulic conductivity is the rate of movement of water within a soil and varies depending on soil moisture status (saturated or unsaturated conditions). Hydraulic conductivity can refer either to horizontal or to vertical movement. Soil drainage class is an important if somewhat vague concept; it refers to the removal of water from the soil, which affects the degree and duration of wetness. While seven drainage classes are recognized, only five are relevant to discussion of forested wetlands. Drainage classes for Coastal Plain forested wetland soils are defined in Table 6.2.

The wetland classification created by Cowardin et al. (1979) introduced the term hydric soil. An interagency National Technical Committee for Hydric Soils determines the criteria for classifying soils as hydric, publishes a list of hydric soils and their taxonomic classification (SCS 1991), and provides descriptions of field indicators (NRCS 1995a). The presence of hydric soils is one of three indicators used to define jurisdictional wetlands which are subject to federal and state regulation (see Chapter 3).

TABLE 6.3
Criteria for hydric soils
(National Technical Committee for Hydric Soils, memo dated 9/27/90).

1. All Histosols except Folists, or
2. Soils in Aquic suborder, Aquic subgroups, Albolls suborder, Salorthids great group, Pell great groups of Vertisols, Pachic subgroups, or Cumulic subgroups that are:
   a. Somewhat poorly drained and have a frequently occurring[a] water table at less than 15 cm (6 in) from the surface for a significant period (usually more than 2 weeks)[b] during the growing season, or
   b. poorly drained or very poorly drained and have either:
      (1) a frequently occurring water table at less than 15 cm (6 in) from the surface for a significant period (usually more than 2 weeks) during the growing season if textures are coarse sand, sand, or fine sand in all layers within 50 cm (20 in), or for other soils.
      (2) a frequently occurring water table at less than 30 cm (12 in) from the surface for a significant period (usually more than 2 weeks) during the growing season if permeability is equal to or greater than 15 cm/hr (6 in/hr) in all layers within 50 cm (20 in), or
      (3) a frequently occurring water table at less than 45 cm (18 in) from the surface for a significant period (usually more than 2 weeks) during the growing season if permeability is less than 15 cm/hr (6 in/hr) in any layer within 50 cm (20 in), or
3. Soils that are frequently ponded for long duration or very long duration during the growing season, or
4. Soils that are frequently flooded for long duration or very long duration during the growing season.

[a] "Frequently occurring" means the condition occurs more than 50 years in 100 years.
[b] The period of soil saturation is 14 consecutive days or more during the growing season.

A hydric soil forms where persistent high water table, flooding, or ponding during the growing season causes anaerobic conditions to develop in the upper part of the soil. Anaerobic conditions occur in soils when molecular oxygen is virtually absent. The most recent technical criteria for hydric soils are listed in Table 6.3. It is important to remember that not all hydric soils are in jurisdictional wetlands, and that not all soils within wetland areas meet the criteria for hydric soils.

Hydric soils are widespread in the South, which has more hydric soils and wetlands than any other region of the United States except Alaska (Shepard 1993). The distribution of hydric soils among soil orders is shown in Figure 6.1. Hydric soils in the South are most commonly Entisols; the most common drainage classes are poorly or very poorly drained.

Organic Soils

Organic soils, or Histosols, develop under conditions of almost continuous saturation where drainage is poor, anaerobic conditions prevail, and litter production exceeds decomposition (Clymo 1984). If soils become deficient in oxygen because they are waterlogged, oxidation of organic matter is limited and rate of decomposition slows because anaerobic decomposition processes are much less efficient than aerobic ones (Tate 1980).

Organic soils are composed primarily of plant remains in various stages of decomposition. Botanical origin and degree of decomposition are two characteristics

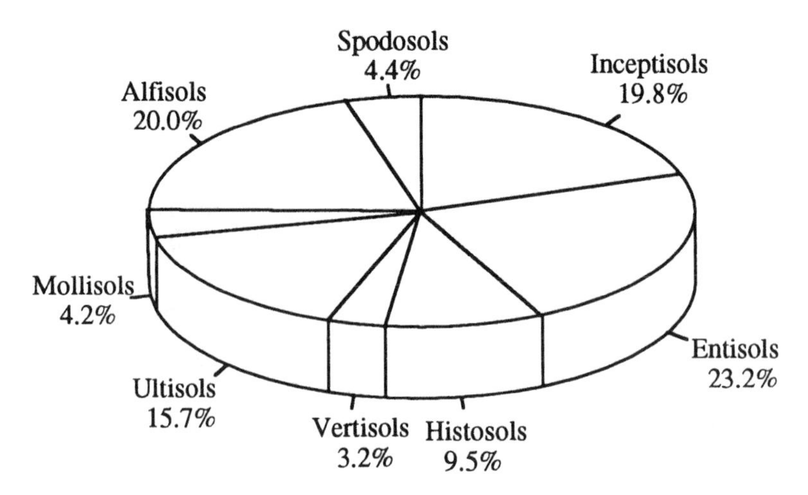

FIGURE 6.1 Percentages of hydric soils in the southern United States in each soil order (data from Shepard 1993).

TABLE 6.4
Criteria for organic soil (after Soil Survey Staff 1975).

Saturated	Clay Content (%)	Organic Carbon Content (%)
Yes	>60	>18
Yes	0	>12
Yes	0–60	12–18
No	— [1]	>20

[1] No clay content criterion.

in determining organic soil classification. These factors also influence physical and chemical soil properties including bulk density, hydraulic conductivity, porosity, and cation exchange capacity (Boelter and Veery 1977, Gafni and Brooks 1990). In Histosols, 40 cm (16 in) or more of the upper 80 cm (32 in) is organic soil material. Organic soils have an organic carbon (C) content of at least 12% by weight, depending on soil clay content (Table 6.4). Soils with higher clay content require higher levels of organic C to be considered organic. All organic soils other than Folists are hydric. Folists are freely drained soils with excess litter over shallow bedrock.

Organic hydric soils are called Fibrists or peats if >2/3 of the plant fibers they contain are identifiable and <1/3 decomposed, Saprists or mucks if <1/3 of the plant fibers they contain are >2/3 decomposed, and Hemists or mucky peats if they have decomposition intermediate between that of peats and that of mucks (Soil Survey Staff 1975). Fibers are fragments of plant tissue, excluding live roots, that are retained on a 100-mesh sieve and have recognizable plant cell structure (Soil Survey Staff 1975). In the southern United States, most Histosols are saprists or hemists because of high temperature and rapid decomposition (Leighty and Buol 1983).

Histosols often contain two zones. The aerobic upper zone near the soil surface is called the acrotelm and is characterized by relatively high biological activity and rate of decay, moderately decomposed plant material, and a fluctuating water table. The lower zone, or catotelm, is permanently saturated, usually anaerobic, highly decomposed, and has little biological activity and low decay rates (Clymo 1984, Moore and Bellamy 1974). The boundary between these two layers is approximately at the mean depth of the minimum water table level (Clymo 1984). Physical and chemical properties of each layer are determined largely by the degree of decomposition of organic matter.

Examples of organic soils found in forested wetlands of the southeastern United States include organic mucks and peats associated with pocosins, Carolina bays, and mountain bogs and fens. These wetlands receive water inputs primarily during periods when precipitation exceeds evapotranspiration. They are usually flooded or saturated during the cool seasons and dry during summer months. Overland flow and, in some cases, groundwater seepage can also contribute to their hydrologic budgets. Because of their positions in the landscape, these forested wetlands are characterized by restricted water movement, limited export of organic materials, and accumulation of organic matter (Brown 1990). The predominant water flux in these systems tends to be vertical rather than lateral (Brown 1990). Soil cation exchange sites are frequently dominated by H^+ ions causing low pH and low phosphorus (P) availability (Richardson and Gibbons 1993). Pocosins and Carolina bays overlie mineral sands and clays, but plant roots are isolated from these substrates by peat deposits. Organic soils are also scattered within other types of southern forested wetlands, including cypress domes and deepwater swamps, where pockets or bands of organic matter accumulation occur in sufficient concentrations and depths to qualify as mucks (Maki 1974).

Mineral Soils

Any soil that does not have an adequate level of organic C to qualify as an organic soil (Table 6.4) is classified as a mineral soil. Mineral hydric soils are sufficiently saturated to have chemical and physical properties indicative of an anaerobic, reducing environment. They often have aquic (seasonally saturated) or peraquic (continuously saturated) moisture regimes and exhibit gray or mottled horizons near the soil surface. Most hydric soils develop distinctive morphologies that reflect oxygen depletion resulting from microbiological decomposition during periods of saturation. Biogeochemical processes resulting from this anaerobiosis cause changes in organic matter accumulation and decomposition and alterations in the oxidation state and combined form of iron (Fe), manganese (Mn), sulfur (S), nitrogen (N), P, and other elements.

Surface horizons of many hydric mineral soils have relatively high organic matter content because of inhibitory effects of anaerobic conditions on decomposition. Mineral soils with histic epipedons (surface layers) are characterized by surface

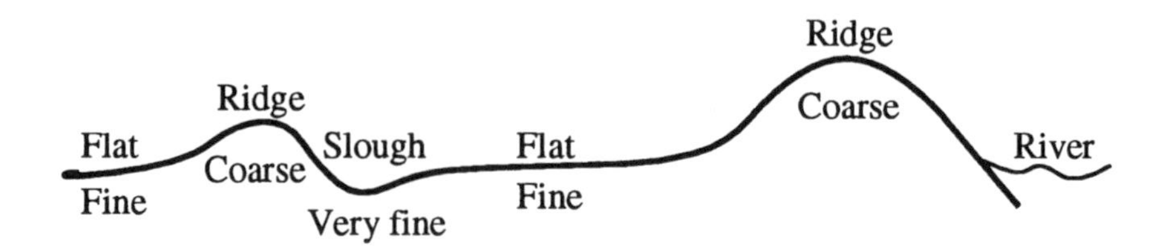

FIGURE 6.2 Floodplain soil texture variation.

horizons with >20 cm (8 in) thickness of organic soil material (see Table 6.4). Organic pans or spodic horizons can also form in sandy soils when organic matter, Fe, and aluminum (Al) move downward through surface horizons and accumulate in the zone of a fluctuating water table. Spodic horizons are most frequently observed 30 to 75 cm (12 to 30 in) below the soil surface and are relatively impervious to water. Wet spodic soils are characterized by dark surface horizons with high organic matter content and leached, dull gray horizons between the surface horizons and the spodic horizon (Brasfield et al. 1983). Spodosols are common features of pine flatwoods which have developed on thick sand beds deposited by coastal waters along former shorelines (Stout and Marion 1993). Seasonally saturated soils and seasonal flooding are characteristic of pine flatwood forests.

Mineral soils also predominate in riverine forests, but levels of organic matter are usually higher in these soils than in upland forests because productivity is high and because decomposition is inhibited periodically when riverine soils are flooded and anaerobic (Wharton et al. 1982). However, in contrast to water movement in wetlands that are isolated from lateral water flow, floodplain water movement is dominated by strong lateral fluxes which provide import and export of organic matter and maintain aerated soil conditions that promote decomposition at sufficient levels to minimize accumulation of organic matter.

Riverine forest soils are often weakly developed and classified as Entisols or Inceptisols because they are relatively recent in origin and are exposed to fluvial erosion and deposition. These processes strongly influence soil textural patterns, particularly in large floodplains which have heterogeneous topography (Figure 6.2). Localized ridges that form along current or former river banks in large floodplains have coarse-textured soils that contain large proportions of sand and silt because these areas have or formerly had rapidly moving floodwaters which deposit heavier sediments. Relatively low-lying backwater flats and former river channels or sloughs retain floodwater longer and are characterized by deposition of fine-textured soils that contain much clay. Resultant gradients in soil texture within floodplain forests influence soil aeration, porosity, permeability, structure, organic matter content, nutrient content, and ultimately vegetative composition.

Physical and Chemical Properties of Wetland Soils

Organic and mineral wetland soils can be distinguished by numerous properties (Table 6.5). Organic soils develop from organic sediments that originate within the ecosystem through accretion of organic matter, whereas mineral wetland soils are

TABLE 6.5
Comparison of mineral and organic wetland soils (After Mitsch and Gosselink 1993).

Property	Organic soils	Mineral soils
Sediment source	Autogenic	Allogenic
Organic matter (%)	>20–35	<20–30
Organic C (%)	>12–20	<12–20
pH	Acid	Approximately neutral
Nutrient availability	Often low	Generally high
Cation exchange capacity	High, dominated by H^+	Low, dominated by base cations (Ca^{++}, Mg^{++}, K^+, Na^+)
Bulk density (Mg/m^3)	Low, 0.2–0.3	High, 1.0–1.8
Porosity (%)	High, 80	Low, 45–55
Water holding capacity	High	Low
Forested wetland type	Pocosin	Floodplain forest

dominated by sediments from flowing waters originating outside the ecosystem (Kangas 1990). The higher organic matter content of organic soils results in high cation exchange capacity (CEC). However, CEC of organic soils is dominated by H^+ ions, and mineral nutrients are predominantly in organic forms, so nutrient availability is generally lower in organic than in mineral soils (Mitsch and Gosselink 1993). Finally, organic soils have higher porosity than mineral soils and, therefore, have lower bulk densities and higher rates of hydraulic conductivity. As organic matter decomposition increases with increasing soil depth in organic soils, macroporosity decreases, bulk density increases, and hydraulic conductivity decreases.

Physical Properties

Origin, amount, and degree of decomposition of organic matter are the dominant influences on physical properties of organic soils. The physical structure of organic matter affects soil porosity which in turn determines bulk density, hydraulic conductivity, infiltration capacity, and water holding capacity (Boelter and Veery 1977, Gafni and Brooks 1990). Although organic soils have a high total porosity, the nature of the porosity varies with decomposition status of the soil. Fibric soil materials are the least decomposed and have high total porosity, low bulk density, and high water holding capacity (Boelter 1974). As plant materials decay in the aerobic surfaces of organic soils, mass is lost and structural elements of the plant materials lose their integrity and collapse (Clymo 1984). Highly decomposed sapric soil materials have less total porosity, higher bulk density, and relatively low hydraulic conductivity.

Pore size distribution significantly affects water movement and retention in soils (Figure 6.3). Undecomposed organic soils and mineral soils dominated by sand-size particles contain many large pores that allow rapid water movement and are easily drained at low suctions. In contrast, well-decomposed organic soils and clay-dominated

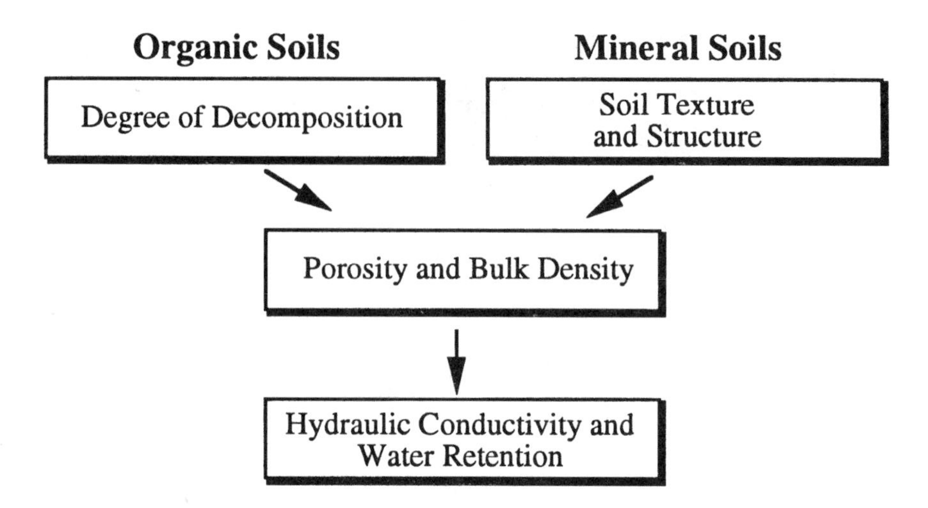

FIGURE 6.3 Organic and mineral soil properties controlling water movement and retention.

mineral soils have finer pores and relatively slow water movement, and they do not drain at low suction (Boelter 1974, Brady 1990). As suction increases, water content of soil materials with fine pores is higher than that of soil materials with coarse pores. However, when soils are drained, those with large pores actually retain less water than do soils with small pores.

Chemical Properties

Oxygen (O_2)and carbon dioxide (CO_2) in the gas phase are transported through soil by diffusion. When soil pores are filled with water, O_2 diffusion is very slow (about 1/10,000th the rate of aerated pores), and anaerobic conditions can result within hours of saturation as dissolved O_2 in waterlogged soils is consumed by microbial activity. The rate of O_2 depletion depends on ambient temperature, availability of organic substrates, and chemical O_2 demand for reductants. Soil aeration is also influenced by soil texture, degree of decomposition, permeability, and degree of compaction. Relatively undecomposed organic soils and sandy mineral soils have more rapid water movement and drainage and, therefore, have better aeration than more decomposed organic soils and clay mineral soils.

Lack of O_2 in flooded or saturated wetland soils greatly influences soil chemistry. In the absence of O_2, the electrochemical status of soils can change rapidly. Most soil reactions of N, Fe, Mn, S, and C involve electron transfer from one ion or molecule to another (Bohn et al. 1985). Oxidizing compounds accept electrons from other substances, whereas reducing compounds donate electrons to other substances. Oxygen is the strongest electron acceptor in soils and yields the most energy to microorganisms from oxidation. The major electron donor in soil is organic matter. Once O_2 in soil is depleted, other electron acceptors including N, Fe, Mn, S, and C are utilized but yield less energy and may yield compounds toxic to plants. Redox potential is a measurement of the tendency of soil solution to oxidize or reduce substances; it is a useful measure of the degree of reduction once O_2 is depleted. Aerated soils have redox potentials in the range of +400 mV to +700 mV, whereas waterlogged soils have redox potentials as low as –250 mV to –350 mV (Gambrell and Patrick 1978). Redox potential decreases as a sequence of reductions takes place and often decreases with increasing depth in wetland soils (Kangas 1990). However, transport of O_2 from plant roots to surrounding soil can create oxidized microsites in an otherwise reducing environment, and anaerobic microsites may develop in wetland soils that otherwise give readings of relatively high redox potential. Redox conditions are more homogeneous in saturated or flooded soils than in drier ones. As soils dry, large pores become aerobic while soil aggregate interiors may remain anaerobic (Bohn et al. 1985).

Depending on the duration of anaerobic conditions, compositional forms and oxidation states of N, Fe, Mn, S, and C will change. Once O_2 is depleted, the next strongest electron acceptor in soils is NO_3, followed by Fe and Mn hydroxyoxides, SO_4, and finally by organic matter itself during fermentation (Ponnamperuma 1972, Gambrell and Patrick 1978, Bohn et al. 1985, Mitsch and Gosselink 1993).

Changes in oxidation state cause corresponding alterations in physical and chemical properties of wetland soil elements. Some ions have oxidation states that make them volatile (e.g., CO_2, CH_4, N_2O, N_2, NH_3, SO_2, H_2S), others are easily soluble (e.g., NO_3^-, Fe_2^+, Mn_2^+, SO_4^-), and others are insoluble (e.g., Fe_3^+, Mn_4^+).

Nitrogen is frequently a limiting nutrient in forested wetland ecosystems, where it is not abundant in forms available to plants. Nitrogen availability is a function of inputs from floodwater, precipitation, and microbial N fixation, microbial transformations from organic forms to plant-available inorganic forms, and losses via denitrification, leaching, volatilization, and in flow-through wetlands, surface water export. Nitrogen is characterized by relatively high chemical versatility (Bowden 1987). It is one of the few nutrient elements that exists in both soluble and gaseous phases. Most of the N in wetland soils is in organic forms, with much lower levels occurring as inorganic N . Low levels of inorganic N are the result of a close balance between plant uptake and the mineralization of organic N to inorganic N. Mineralization of organic N produces ammonium (NH_4^+), which is the dominant form of inorganic N because of the prevalence of anaerobic conditions. Ammonium can be immobilized on soil particles by cation exchange or can be taken up by microorganisms and plants and immobilized into organic compounds. In oxidized zones, nitrification (the aerobic microbial conversion of ammonium to nitrate [NO_3^-]) will occur. Nitrate can be assimilated by plants and microorganism or lost via leaching since it is a soluble, mobile anion. However, shortly after saturation and the subsequent onset of anaerobic conditions, NO_3^- is reduced to NH_4^+. Denitrification also takes place under anaerobic conditions and results in the microbial conversion of nitrate to gaseous nitrous oxide (N_2O) and molecular nitrogen (N_2).

The form of Fe, Mn, and S in forested wetland soils is controlled directly by redox conditions. Under aerobic conditions, insoluble Fe_3^+ and Mn_4^+ occur. Manganese and Fe_3^+ are reduced to more soluble forms (Mn_2^+ and Fe_2^+) under anaerobic conditions and thereby become more available for plant uptake. Reduced Fe and Mn can reach toxic concentrations in wetland soils (Mitsch and Gosselink 1993).

Reduction of Fe and Mn leads to development of soil color that is often indicative of hydric soil status. Many mineral soils develop gray-green or gray-blue coloration (gleying) as a result of the chemical reduction of Fe. If oxidized zones occur within a gleyed soil matrix, oxidized Fe, oxidized Mn, or both will cause red-brown or black mottles. Gleying and mottle formation are controlled by redox potential, availability of organic substrates for microbial activity, and sufficient soil temperature to promote microbial activity. Sulfur transformations are important because hydrogen sulfide (H_2S), which can form under reduced conditions, can be toxic to plants and microbes. However, sulfides often combine with Fe_2^+ or Mn_2^+ to form nontoxic insoluble ferrous or manganous sulfides.

Phosphorus, too, is important in the chemistry of forested wetlands. Unlike N, it does not occur in a gaseous phase and is not altered directly by changes in redox potential. Phosphorus occurs in both organic and inorganic forms and is either soluble or insoluble depending on its associated compounds and the redox conditions of the wetland soil. Mineral wetland soils are characterized by a predominance of inorganic P attached to mineral soil sediments, whereas organic wetland soils have a predominance of organic P immobilized in organic matter. Phosphorus is indirectly influenced by changes in redox potential, becoming more soluble under reduced conditions through the hydrolysis and reduction of ferrous, aluminum, or calcium phosphates to more soluble compounds. The resultant soluble inorganic P occurs as orthophosphate (PO_4^{-3}, HPO_4^{-2}, or $H_2PO_4^-$). Under aerobic conditions, P precipitates as insoluble Fe, Al, or Ca phosphates. Phosphorus additions to riparian and floodplain forests often occur through deposition of clay particles in floodwater sediment. These adsorb the negatively charged phosphates.

The redox potential at which soil components become electron acceptors is affected by pH. This relationship can be demonstrated by pH-redox stability diagrams (Bohn et al. 1985). Basic soils require lower redox potentials for reduction of soil components (Bohn et al. 1985). Generally, pH tends towards neutrality in reduced soils. Thus, pH will increase in an acid soil undergoing reduction but decrease in an alkaline soil. These changes are controlled by soil Fe and organic matter content. In acid soil, reduction of Fe consumes H^+ ions and increases pH. In alkaline soils, production of CO_2 is greater than reduction of Fe and pH decreases (Bohn et al. 1985). Wetland soils have a wide range of pH; organic soils are often acidic, and mineral soils are often closer to neutral or alkaline (Mitsch and Gosselink 1993).

Hydrologic Classification of Wetlands

All wetlands can be classified as riverine, basin, or fringe wetlands, based upon their hydrology (Kangas 1990, Lugo 1990b). Some common terms for forested wetlands in the South are given in Table 6.6 and grouped by these classes. Many of these terms refer to wetlands defined by their distinctive vegetation and landscape position. Some of these forested wetland communities are described more fully in other chapters.

TABLE 6.6
Wetlands by hydrologic type.

Riverine	Basin	Fringe
Major bottom	Pocosins	Mangrove
Minor bottom	Flatwoods	Tidal swamps
Branch bottom	Wet mineral flat	Fringe hardwoods
Upland drain	Savannas	
Alluvial river bottom	Pitcher plant flat	
Red river bottom	Wet depression	
Black river bottom	Carolina bay	
Loess bluff bottom	Cypress dome	
Piedmont bottomland	Head, bay or cypress	
Blackland bottom	Pond, cypress or gum	
Cypress strand or stringer	Slough	
Hardwood run	Oxbows	
Terrace	Swale	
	Mountain bog	
	White-cedar swamp	
	Maritime swamp	
	Peat swamp	
	Headwater swamp	
	Muck swamp	
	Basin mangroves	

Riverine Wetlands

In the South, riverine wetlands occupy a larger area than do wetlands of any other kind. These forests are dominated by river hydroperiod, sedimentation, and nutrient import and export (Brinson 1990). Terminology for riverine wetlands in the South generally recognizes differences in floodplain size — the distinctions between major bottom, minor bottom, and upland drain. The source of alluvial deposits provides another distinction, that between alluvial or red river bottoms where deposits are derived from rocks of the mountains or Piedmont, and blackwater bottoms with deposits derived from the sedimentary rocks of the Coastal Plain. Many features within floodplains (Figure 6.2) vary in frequency and extent depending on the size of the floodplain (Chapters 12 and 13).

Basin Wetlands

Basin wetlands are found in depressions where water accumulates (Brown 1990). Inflows may come from the surrounding landscape as surface runoff or interflow, from precipitation, or from groundwater. In the South there are many basin wetlands with a variety of vegetation communities. Wet mineral flats, savannas, pitcher plant flats, and prairies which occur along the Atlantic and Gulf coasts are wetlands that have fine-textured, poorly and very poorly drained soils (Ultisols and Inceptisols)

(Pritchett and Fisher 1987). The flatwoods of southern Georgia and northern Florida resemble basin wetlands, but they developed on coarse-textured soils characterized by subsoil accumulations of organic material and minerals rich in Fe and Al (Spodosols). Pocosins, found in the Carolinas and Virginia, occur on domed peats (Histosols and Inceptisols) on broad interstream divides and other areas where runoff is blocked by recent dunes or old beach ridges (Daniels et al. 1984). Carolina bays (Chapter 14) are oval depressions with poorly drained Ultisols and Histosols in the center of the bay and a rim of well-drained sandy or loamy soil (Daniels et al. 1984). The common hydrological characteristic of all these features is stagnant or very slow sheetflow. The major flux is vertical and is often restricted to precipitation and evapotranspiration. Basins are rarely flushed completely, and organic matter such as peats or mucks accumulates.

Fringe Wetlands

Fringe wetlands occur along the edges of oceans, estuaries, and lakes (Lugo 1990a). Water flux is bidirectional and perpendicular to the wetland such as from river flow and tidal action in an estuarine fringe tidal swamp. That is, water flows out of the swamp into the river at low tide, and into the swamp from the river at high. In fringe mangrove communities, at least periodic overflow of brackish water is required to favor mangroves over potentially competing vegetation that is not adapted to saline soils (Odum and McIvor 1990). Fringe wetlands usually have simple vegetative structure and may support only a single tree species as in some mangrove forests, or they may exhibit a zoned vegetation pattern (Chapter 18). There is usually a clear zonation of species or productivity along moisture, aeration, or salinity gradients, or all three.

Physiographic Setting

Forested wetlands and related riparian zones occur in all regions of the southern United States but are most extensive in the Coastal Plain and Mississippi River Valley (Martin and Boyce 1993, Schoeneberger 1996). The Coastal Plain is a broad lowland stretching from Virginia to Texas. It is underlain by alluvial and marine sediments of mostly Cretaceous (oldest), Tertiary, and Quaternary (youngest) age. Within the Coastal Plain, the age of surficial deposits (soil parent material) decreases from the Piedmont boundary seaward. The sediments that form the Coastal Plain were laid down in various onshore, near shore, and offshore environments since the Cretaceous. Deposition began about 130 million years ago and continues to the present. Some areas have been influenced by aeolian (wind-blown) materials.

Atlantic Coastal Plain

The Atlantic Coastal Plain along the coast from Virginia to northern Florida is mostly level land with extensive areas of poor drainage and striking offshore features such as barrier islands and bars. The Atlantic Coastal Plain can be divided into the Upper (Inner, Hilly), Middle, and Lower Coastal Plain.

The Lower Coastal Plain is a level lowland of elevation less than 30 m (98 ft) above mean sea level. Because the Lower Coastal Plain has minimal relief and a poorly integrated drainage network, its soils are predominantly somewhat poorly to very poorly drained except along drainage ways. Lower Coastal Plain soils include Spodosols in sandy Quaternary or older low-lying deposits along the coast, Ultisols and Inceptisols with kaolinite mineralogy inland, formed in older Cretaceous and Tertiary sediments, and in some sections, extensive deposits of Histosols. In most of the Lower Coastal Plain, a slightly dissected belt of younger terraces with extensive swampy and marshy areas occurs near the coast, and broad flats (mineral flats and flatwoods) are found inland. Maritime plant communities on the offshore barrier islands include some small forested wetlands (Bellis and Keogh 1995).

The Middle Coastal Plain is more dissected and has more relief than the Lower Coastal Plain. Elevation ranges from 30 to 90 m (98 to 295 ft). Very poorly and poorly drained soils are less common, although Fluvaquents and Fluvaquepts are found in valleys along streams. A striking feature of the Middle Coastal Plain is the common occurrence of basin wetlands called Carolina bays.

The Upper Coastal Plain is a transition to the Piedmont uplands. It is older than either the Middle Coastal Plain or the Lower Coastal Plain. The Upper Coastal Plain consists chiefly of broad uplands or low-lying plateaus that gently slope seaward. Relief is greater, up to 200 m (656 ft). The dominant soils are the older Paleudults and Paleaquults. Riverine forested wetlands occur in valleys. These are typically minor bottoms, but large swamps, such as the Congaree Swamp in South Carolina, have formed where major rivers leave the Piedmont and enter the Upper Coastal Plain (Wharton et al. 1982).

Peninsular Florida

Peninsular Florida is the most recently emerged land in the Coastal Plain. The Peninsular Arch or Central Florida Ridge is surrounded on three sides by marine terraces. Carbonate rock underlies most of the ridge, and sinkholes are common and widespread. In the south of peninsular Florida, large basins contain lakes, marshes, and some forested wetlands such as the Everglades, Lake Okeechobee, and Big Cypress Swamp. Coastal areas of extreme south Florida support fringe and basin mangrove wetlands (Chapter 18; Odum and McIvor 1990).

Gulf Coastal Plain

The Gulf Coastal Plain extends from Georgia and the panhandle of north Florida on the east to Texas on the west. It is split into eastern and western portions by the Mississippi River Valley. Gulf Coastal Plain sediments are thicker, and upland soils there are more influenced by aeolian additions than those of the Atlantic Coastal Plain. Extensive loess deposits blanket Coastal Plain sediments to the east and west of the Mississippi Embayment.

The East Gulf Coastal Plain Section in its greater part is a belted coastal plain, with lowlands on weaker rock members and bounded seaward by cuesta scarps and dip slopes on more resistant rock, typically sandstone. Adjacent to the coast are recent marine terraces. Inland, distinctive Vertisols with shrink and swell properties are developed on a band of outcropping Selma Chalk. This band is known locally as the Black Belt or Black Prairie and extends across central Alabama and northeastern Mississippi. Other blackland belts occur in Texas, Louisiana, and Arkansas.

The West Gulf Coastal Plain Section is similar to the East Gulf Coastal Plain Section, although it is considerably wider and more faulted. The section's major rivers have very extensive drainage basins and have formed large deltaic deposits. In the West Gulf, soils are more neutral to alkaline than in the East Gulf and have mixed and smectite mineralogies.

SOILS OF RIVERINE WETLANDS

River systems supporting riverine wetlands abound in the South. Floodplain widths vary from several km to narrow strips of vegetation (Sharitz and Mitsch 1993). Alluvial (redwater) rivers originate in the mountains and the Piedmont. These rivers have many tributaries and relatively large watersheds, and so have sustained high flows in the winter and spring. Blackwater rivers and their tributaries originate entirely in the Coastal Plain. They receive most of their water from local precipitation, and baseflow (groundwater seepage) is a large component of low flows. Blackwater rivers are often tributaries of alluvial rivers. High water levels in alluvial rivers can impede discharge from, and cause flooding of, blackwater rivers. Blackwater rivers get their name from their dark color, a legacy of humic (organic) substances flushed from swamps. Because of their low gradient, blackwater systems tend to have less sediment transport and deposition than alluvial rivers, and hence have less well-developed floodplain features.

We now recognize that the present southern floodplains owe at least some of their features to Pleistocene glaciation and periglacial climate (Wharton et al. 1982). Solifluction, the slow downslope movement of saturated soil masses, during the Pleistocene caused Piedmont valley filling and subsequent incision of Piedmont rivers. The rising and lowering of sea level during interglacial and glacial periods caused changes in gradients of rivers originating in the mountains and Piedmont. One consequence is that most rivers presently occupy floodplains too large to have been formed by their present discharge volume and meander dimensions (Wharton et al. 1982). Another consequence is the occurrence of relict floodplain features. There are two or more relict terraces along many alluvial rivers and relict braided channels in the Mississippi River system. Post-glacial sea-level rise has drowned the mouths of Atlantic Coastal Plain rivers such as the Roanoke and Chowan and caused the formation of Holocene meander belts in the Mississippi system (Saucier 1994).

Although major rivers were covered by the sea in their lower reaches as a result of sea-level rise, their tributaries often have extensive floodplains. Examples are the James, Chowan, and Roanoke Rivers in Virginia and northern North Carolina and the Escambia and Choctawhatchee Rivers in Florida (Wharton et al. 1982). Rivers with narrow floodplains include the Pamlico and Neuse Rivers in North Carolina.

Floodplain Features

Landforms typical of floodplains (Figure 6.2) develop as a result of the interplay between hydroperiod and energy relationships in riverine systems. While a thorough explanation of the origin of floodplain features is beyond the scope of this chapter, brief descriptions of typical landforms will help to organize the presentation of soils information. Although floodplains may appear to be flat and featureless, elevation differences of just a few centimeters can create quite distinct soil and vegetation zones.

Natural levees form adjacent to the river channel. They are often high points on the floodplain, and all are well-drained due to their coarser textured soils. Old levees that indicate the previous location of a river channel may be far from the present channel, in a broad floodplain. Oxbows are formed when a river in flood seeks a shorter path and cuts off a meander. Deepwater alluvial swamps often form in oxbow lakes. Meander scrolls form as depressions and rises as the river migrates laterally across a floodplain. Sloughs form in meander scrolls when water ponds above deposited clay. If they are large enough and permanently flooded, sloughs support deepwater swamps. Backwater deposits of fine sediments occur between natural levees and higher ground.

Deposition processes in riverine systems give rise to distinctive landforms and soil patterns. Rivers in flood generally drop the heaviest particles first, forming the fronts on the river side of the levee. Slightly lighter particles drop next, on the levee, while finer particles drop on the inland side of the levee. Thus many natural levees have steep slopes on their river side. In still or slackwater areas behind the levee, the finest particles drop.

The terms first bottom, second bottom, and terrace are sometimes confused, presumably because it is difficult to name floodplain features that differ regionally. As Mader (1991) points out, most foresters and ecologists working in the Mississippi and Gulf Coastal Plain systems have adopted the usage introduced by Putnam (Putnam and Bull 1932, Putnam 1951, Putnam et al. 1960), who used first bottom to denote the sites between the river channel and its natural levee. In the Mississippi and other large systems, this extensive area is also called batture, front land, or river front. The first bottoms may also extend beyond the levee if land beyond the levee is flooded frequently. In Putnam's terminology, the area beyond the active floodplain is the first terrace. The first terrace is frontland of a former river front (Mader 1991). Second terraces are even older floodplain features. Saucier (1994) recognized varying numbers of terraces in the different constituent river basins of the Mississippi Valley (four in the lowlands west of Crowley's Ridge and two in the St. Francis basin).

In the Atlantic Coastal Plain, terraces have two meanings: marine terraces, representing old (pre-Holocene) shorelines, and floodplain terraces. Thornbury (1965) provided descriptions of the marine terraces, as well as a correlation among local names in several states based on characteristic altitude. Daniels et al. (1984) described these marine terraces in the Coastal Plain of North Carolina. They called these features plains, listing three in the Middle Coastal Plain: Brandywine, Coharie, and Sunderland. They listed three slightly eroded plains in the Lower Coastal Plain:

Wicomico, Talbot, and Pamlico. These plains are separated by scarps that mark former high sea levels. Proceeding inland from the coast, these scarps are named the Suffolk, Walterboro, Surry, Kenly, Wilson Mills, and Coats.

Wharton et al. (1982) described three floodplain terraces in Atlantic Coastal Plain floodplains. The most recent is the Holocene terrace, often called the first bottom (Braun 1972, Mader 1991). The next terrace is called variously Terrace I, Deweyville, second bottom, or backswamp (Wharton et al. 1982, Mader 1991). Daniels et al. (1984) recognized two or three terraces above the modern floodplain and distinguished between Coastal Plain and Piedmont river terrace soils. Coastal Plain river terrace soils formed in sediments derived from local sources only, by blackwater rivers. Piedmont river terrace soils occur in the Coastal Plain but are developed in sediments from both the Piedmont and Coastal Plain. These sediments tend to have more variable mineralogy and more than 10% weatherable minerals in the subsurface B horizon. Soils developed in Coastal Plain terrace sediments have less than 10% weatherable minerals and usually siliceous mineralogy (Daniels et al. 1984).

Plant growth depends on four major soil factors (Baker and Broadfoot 1979): soil physical properties, moisture availability during the growing season, nutrient availability, and aeration. Profile data for riverine wetland soils are limited, but four properties that are sometimes reported are percentage organic matter, pH, percentage clay (or texture determined in the field), and less frequently, extractable P. At the landscape level, many soil physical properties are related to texture, and nutrition may be inferred from organic matter, pH, and P levels. Phosphorus in particular appears to be a good indicator of nutrition, as many soils developed in material that was already weathered. Extreme P deficiency has been demonstrated in some soils (e.g., Pritchett and Comerford 1982). Brown et al. (1990) suggested that species composition of many wet acid soils may be influenced dramatically by levels of available phosphorus. While many riverine wetlands appear to benefit from nutrient additions from sedimentation (Mitsch and Gosselink 1993), minor bottoms in particular may show extreme deficiencies (e.g., Lockaby et al. 1997b).

Alluvial Rivers — Atlantic Coastal Plain Soils

Floodplains of major alluvial rivers with drainages originating in the Piedmont or the mountains are commonly called red river, redwater, or alluvial river bottoms. They may exhibit the full range of floodplain features including levees or low ridges, flats, and meander scrolls, and these features are typically broad. These rivers often have muck swamps associated with, or adjacent to, the active floodplain, especially in their Lower Coastal Plain reach. Smaller sloughs and oxbows are considered as part of the bottomland. Beyond these usually saturated soils are the first bottoms, which periodically flood to considerable depths. Relict floodplain features such as terraces may be associated with these floodplains, and in some locations these are called second bottoms. Soils texture ranges from sandy loam to clay loam. Aquults, Fluvaquents, Udifluvents, and Humaquepts predominate. Mineralogies are mostly kaolinitic in the Atlantic Coastal Plain, with mixed and smectite mineralogies more common to the west. Examples of red river bottoms are the bottoms of the Roanoke

River in Virginia and North Carolina, the Santee River in South Carolina, the Oconee and Flint Rivers in Georgia, and the Alabama River in Alabama.

Soils of alluvial rivers in the Atlantic Coastal Plain are generally acidic (Table 6.7). Organic matter accumulates in depressions, especially in muck swamps that frequently occur along the river channels. Clay percentages are moderate, as sources of alluvium are the clay-rich soils of the mountains and reworking of Coastal Plain deposits (Phillips 1992). Phosphorus values in soils of the surrounding wet mineral flats and flatwoods are lower than those in alluvial river soils. Some common series include Chewacla, Wehadkee, Portsmouth, Roanoke, and Nimmo.

Blackwater Rivers — Atlantic Coastal Plain Soils

Floodplains of major river systems originating in the Coastal Plain are termed blackwater bottoms. Features of first bottoms and terraces in blackwater bottoms are less developed than in alluvial river bottoms. Muck swamps are more prevalent because relief and energy are low. Soils tend to have higher organic matter accumulations than in alluvial river bottoms, generally more than 5% (Wharton et al. 1982). Blackwater bottoms receive most of their sediment from local sources and receive proportionally more groundwater than do the main alluvial river bottoms. For these reasons, blackwater bottom soils are often closer in their nutrient status to their associated uplands than to the soils of the main alluvial river bottoms. Examples of river systems with blackwater bottoms are the Blackwater River in Virginia, the Waccamaw in North Carolina and South Carolina, the Black River in South Carolina, and the St. Mary's in Georgia and Florida.

Typical soils in blackwater bottoms are acidic except where calcareous sediments outcrop or are near the surface (Table 6.7). Clay content may be quite low if source materials are predominantly beach sands. Phosphorus tends to be lower than in alluvial rivers and to be quite deficient. Some common soil series include Bibb, Chastain, Chowan, Masontown, and Muckalee. Terrace soils in blackwater bottomlands may be of the Lumbee and Paxville series.

Piedmont Bottomland

Extensive erosion in the Piedmont since European settlement has caused infilling of smaller Piedmont streams with subsequent downcutting (Markewich et al. 1990, Phillips 1992). The active floodplains of these streams are narrow, and features such as sloughs and oxbows are fewer and smaller than in the Coastal Plain. In the lower Piedmont, these bottoms resemble redriver bottoms. Upstream, only well-drained soils on terraces are found. Some examples of Piedmont bottomland can be found in the Meherrin River in Virginia, Neuse River in North Carolina, Saluda River in South Carolina, Oconee River in Georgia, and Sipsey River in Alabama (Kellison et al. 1988).

Red And Arkansas Rivers — Gulf Coastal Plain

The distinction between sediment sources inside and outside the Coastal Plain is generally less meaningful in the Gulf Coastal Plain than in the Atlantic Coastal Plain. In the Red River and Arkansas River bottoms however, reddish-brown soils

TABLE 6.7
Generalized soil properties of some major southern forested wetland types.

Location/Landform	pH	Organic Matter (%)	Clay (%)	Phosphorus (mg/kg)
Alluvial Rivers in the Atlantic Coastal Plain[1]				
Swamp	5.3	35.0	45	10
Low flat	5.4	4.4	44	10
High flat	6.2	2.5	35	6
Ridge	6.2	3.0	18	4
Blackwater Rivers in the Atlantic Coastal Plain[1]				
Swamp	5.5	17.1	19	11
Flat	5.1	7.9	12	10
Red and Arkansas Rivers in the Gulf Coastal Plain[2]				
Mixed bottom	4.6	2.4	50	16
Red river bottom	5.8	3.7	59	19
Terrace	4.6	2.0	18	18
Minor Bottoms in the Gulf Coastal Plain[3]				
Coastal plain alluvium	5.0	1.0	12	6
Blackland alluvium	4.8	1.9	32	12
Blackland alluvium, acidic	4.0	2.1	35	4
Blackland alluvium, calcareous	6.0	2.5	40	20
Mississippi River Alluvium[4]				
Recent levee	6.8	2.5	38	20
Old levee	5.1	2.2	40	22
Slackwater	5.6	2.4	60	18
Loess Bluff Bottoms in the Gulf Coastal Plain[1]				
Acid	5.2	1.8	22	14
Neutral	6.0	1.2	11	11

[1] Kellison et al. 1982, Wharton et al. 1982, Daniels et al. 1984
[2] Broadfoot 1964, 1976
[3] Broadfoot 1964, 1976, Southern Forest Soils Council 1990, Alabama Forestry Planning Committee 1993
[4] Broadfoot 1964, 1976, Brown et al. no date, Patrick et al. 1981

are typical, and these are formed in alluvium mainly derived from the Western Plains. More alkaline soils occur in the floodplains of the Red River, and more acid soils in the Arkansas River bottoms (Table 6.7). Distinct older terraces are found in these systems, and soils of these terraces have coarser textures than those of the active floodplain. Alfisols, Mollisols, and Vertisols are common. Common soils in these systems include Portland, Perry, Buxin, and Roebuck in the bottom and Asa, Morse, and Gore on terraces.

Minor Bottoms

Minor bottoms have floodplains from less than 1 km (0.6 mi) to several km in width, and their streams flood frequently but for short duration (Hodges, pers. comm.).

Sediments in minor bottoms reflect local sources. Many soils of minor bottoms in the Gulf Coastal Plain contain aeolian sediments (loess). Distinctive outcroppings occur in bands in the East Gulf Coastal Plain, and this is reflected in terminology such as Blackland bottoms and loess bottoms (Broadfoot 1964). Most bottomland soils outside the loess influence are sandy, acid, lacking in natural fertility, and may be severely P deficient (Table 6.7). Typical coarser-textured soils in these bottoms are Iuka, Menatchie, Johnston, Kinston, and Chennaby series. Terrace soils include the Izagora, Stough, Myatt, and Leaf series. Where alluvium is derived from local outcroppings of marls and soft limestone such as the Selma Chalk, however, soils are more fertile and finer in texture. These Blackland or prairie soils are found extensively in Mississippi and Alabama and occur to a lesser extent in Louisiana and Arkansas. Blackland bottom soils may have weathered to an acid surface or retain the calcareous nature of their parent material (Table 6.7). Typical series with acidic surfaces include Houlka and Una; if soils are calcareous, typical series include Catalpa, Leeper, and Tuscumbia.

In the Atlantic Coastal Plain, minor bottoms may be called branch bottoms (Kellison et al. 1982). These areas along minor drainages are flat and are subject to some overflow. Examples include Big Swamp in North Carolina and the Wambaw in South Carolina. Soils vary from almost organic in sloughs and oxbows to those typical of other alluvial or blackwater systems. Soils in these systems include Bibb, Kinston, and Johnston.

Mississippi River Alluvium

The Mississippi Alluvial Plain Region bisects the Gulf Coastal Plain section. It is divided into five relatively flat river basins (Saucier 1994) and the basin of the Mississippi River itself. The five additional basins are the Atchafalaya and Boeuf-Tensas Basins in Louisiana, to the west of the Mississippi River; the Yazoo Basin in Mississippi, east of the river; and in Arkansas, the Black and St. Francis basins. Intervalley ridges such as Crowley's Ridge and Macon Ridge parallel the river and loessal bluff highlands border the valley. The present delta of the Mississippi River in Louisiana is characterized by extensive marshes and coastal prairies (Lytle and Sturgis 1962) in the Deltaic Plain east of the Atchafalaya Basin, sometimes grading into deepwater swamps.

All of the river basins that make up the plain are in places lower than the Mississippi River due to depositional features in the plain and faulting of underlying rocks (Hunt 1974). The depositional history of the plain was determined by two interacting forces, the tremendous sediment loads carried by meltwater from the retreating Laurentide ice sheets in the northern United States and Canada and concomitant fluctuations in sea level during glacial and interglacial periods. Valley filling occurred when sea level was high during interglacials, and valley erosion during glacial periods (Thornbury 1965). In the late Wisconsin chronology suggested by Saucier (1994), outwash deposition ended 10,000 years before present (BP), and the river then changed from a braided to a meandering pattern.

Soils formed on young natural levees tend to have alkaline or neutral pH. Soils on older levees were leached, hence they tend to be acidic. Slackwater or backswamp deposits of fine textured materials on almost level or gently sloping areas occur away from the current or former channels of the Mississippi River. Soils formed in slackwater deposits typically have very high clay content, some of the highest in the United States (Pettry, pers. comm.), and poor internal drainage. The lowest-lying areas are in depressions such as swales and old, partly filled river channels. Soils here developed in alluvium washed from adjacent areas (Broadfoot 1964). Mississippi River alluvium originated in the northern Appalachians, upper Midwest, or the Rocky Mountains. Soils developed in this alluvium tend to be very fertile, but differences in depositional environments and drainage produce great variety (Table 6.7). Soils formed on recent natural levees show only the beginnings of profile development and tend to be neutral or alkaline, and low in organic matter. These soils are typically Fluvaquents. Typical series on recent levees are Crevasse, Robinsonville, Commerce, and Mhoon. Older natural levee soils are acid and have greater profile development, including the formation of an argillic horizon. These soils may be Alfisols such as Ochraqualfs. Typical series include Bosket, Dubbs, Dundee, and Forestdale.

Slackwater deposits are from almost level to gently sloping, with high clay content (sometimes greater than 90%). These soils have different thicknesses of fine-textured sediments overlying the sandier sediments at depth. The Sharkey series, the most extensive hydric soil in the United States, is Vertic Haplaquepts developed in slackwater deposits of Mississippi River alluvium. Other series found in this position are Bowdre, Tunica, Alligator, and Amagon.

Depressions occur in old, partly filled river channels throughout the floodplain. Soils in these depressions formed in alluvium washed from adjacent soils. All have weak profile development and are mottled or gleyed. Often organic matter accumulation is high. Several older series found in depressions, such as Dowling, are now regarded as depressional phases of the Sharkey series.

Loess Bluff Bottoms

Associated with the Mississippi River floodplain are extensive areas of loess, wind-blown deposits predominantly of silt-sized particles. While these occur on both the eastern and western boundaries of the floodplain, they reach their greatest expression on the east. A number of river floodplains in the loess area border the Mississippi River floodplain on the east, and on the west near Crowley's Ridge and Macon Ridge. Soils in loess bluff bottoms may be acid or neutral and are usually Aquic or Typic Udifluvents (Table 6.7). Soils developed in acid bottoms include the Collins, Falaya, Arkabutla, and Waverly series. Typical series in nonacidic bottoms include Morganfield, Adler, and Wakeland.

SOILS OF BASIN WETLANDS

In the Atlantic Coastal Plain from Chesapeake Bay in Virginia south to the Neuse River in North Carolina, estuarine embayments divide the Coastal Plain into broad peninsular tracts. Depressions on the surface are similar to Carolina bays. Extensive areas of organic soils are present in this section, especially in North Carolina (Daniels et al. 1984, Campbell and Hughes 1991). These domed peats occur on interstream divides and broad flats, obscuring the original surface. Organic soils are generally less than 2.5 m (8 ft) thick, although some in old stream channels or Carolina bays may be as much as 4.5 m (15 ft) thick (Daniels et al. 1984). Major basin wetlands occur close to the coast, in association with blackwater rivers and their estuaries. Pocosins in Virginia and North Carolina are examples of major basin wetlands. The sources of many blackwater rivers are basin wetlands called peat swamps or headwater swamps. The wetter portions of the flatwoods and savannas can be basin wetlands. Other basin wetlands of note include the karstic features such as cypress ponds and domes in Florida and the large basins of South Florida such as the Everglades, Lake Okeechobee, and Big Cypress Swamp which support some forested wetlands.

Muck Swamps

Muck swamps have already been mentioned in conjunction with alluvial river systems. In some places, such as at the contact between the Piedmont and the Upper Coastal Plain, muck swamps are large enough to be named separately, as is the Congaree Swamp in South Carolina. These are very poorly drained areas, usually perennially saturated or flooded. They typically have deep accumulations of organic matter over silt loam to clay soils. Muck swamps are found along blackwater rivers, branch bottom streams, and in old oxbows of redwater rivers (Kellison et al. 1988). Typical soils include Chastain, Wehadkee, and Dorovan.

Peat Swamps

Also called headwater swamps, these basins occur on broad interstream divides and are source areas for blackwater rivers and branch bottoms. Peat swamps resemble muck swamps in that they are poorly drained, with deep accumulations of organic matter. Peat swamps, however, tend to have more mineral soils of greater fertility. Some examples include the Great Dismal Swamp on the Virginia-North Carolina border, the Green Swamp of North and South Carolina, Little Wambaw in South Carolina, and the Okefenokee Swamp on the Georgia-Florida border. According to Kellison et al. (1988), the true pocosins of the Carolinas are peat swamps.

Pocosins

Pocosins are most extensive in the Embayed Section (Virginia and North Carolina) of the Lower Coastal Plain. They occur on broad interstream divides where runoff is restricted and organic material has accumulated to depths of as much as 2.5 m (8 ft). Their domed peat soils are classified as Histosols, primarily saprists (Daniels et al. 1984). Histosols have low hydraulic conductivity and are generally low in available plant nutrients (Richardson and Gibbons 1993). In many pocosins, precipitation may be the only source of plant nutrients. Where the organic matter is shallower, plant roots may contact mineral soil. Wet mineral flats are often associated with the Histosols of pocosins. A cross section of a typical pocosin would show the Portsmouth series (mineral soil) grading into the Ponzer and Dare series (organic soils) as the thickness of the organic layer increased. Other Histosols in the peat or muck areas of North Carolina include Belhaven, Scuppernong, Mattamuskeet, Croatan, Pungo, and Pamlico (Daniels et al. 1984).

Flatwoods and Savannas

True flatwoods in south Georgia and north Florida occur on Spodosols (predominantly Haplaquods), poorly to somewhat poorly drained soils developed in coarse-textured sediments (Abrahamson and Hartnett 1990, Morris and Campbell 1991). More widespread are the savannas, which are also known as wet mineral flats or pitcher plant (*Sarracenia* spp.) flats. These areas are found in both the Atlantic and Gulf Coastal Plains. Savannas are found on finer-textured soils, usually Ultisols (Paleaquults and Haplaquults) and Inceptisols (Humaquepts). They developed in slack-water deposits and are poorly to very poorly drained. Many of these areas are now in pine plantations. Typical soils include Portsmouth, Bladen, Rutlege, Plummer, Mascotte, and Sapelo series.

Cypress Domes

Cypress domes or ponds are isolated depressions filled with acid, peaty, mineral soils. In Florida, they often occur within pine flatwoods (Ewel 1990b). A related feature, the cypress strand, has some flow and an outlet, and thus is a riverine wetland, although the difference between domes and strands is problematic. Cypress domes are moist or inundated for long periods but dry out at the surface, especially in droughty summers. They may have clay pans or lenses beneath their sandy surface soils or occur in depressions in marl and limestone bedrock (Coultas and Deuver 1984, Brown et al. 1990, Ewel 1990b).

Carolina Bays

Carolina bays are curious, ovate depressions that occur in the Coastal Plain (Richardson and Gibbons 1993), most abundantly in the Middle Coastal Plain. They are seldom found in areas of clay soils or where silt content is high in North Carolina (Daniels et al. 1984), but many South Carolina bays have clayey soils of the Coxville, Rembert, or McColl series (McKee, pers. comm.). The longer axes of these features trend from northwest to southeast. On the northwest, the transition to the surrounding upland is gradual, but around the remainder of the oval, a distinct rim is common. The soil of the rim is sandy and may rise 1 to 3 m (3 to 10 ft) above the bay floor. The bay floor may have organic or wet mineral soils (Daniels et al. 1984).

Soils of Carolina bays are organic (Croatan, Mattamuskeet, and Pamlico) or mineral in the interior (Byars, Pantego, Torhunta, Coxville, McColl, and Rains). Soils of the sandy rim are well to moderately well drained series such as Norfolk, Bonneau, Wakulla, Lakeland, or Chipley.

SOILS OF FRINGE WETLANDS

Coastal areas also support fringe and basin mangroves in Florida and extensive marshes and deepwater swamps along the Gulf Coast. Mangroves in south Florida are best developed in subtropical conditions, along low-energy coastlines (Gilmore and Snedaker 1993). Fringing mangroves occur mostly in estuarine environments (bays, lagoons, coves, tidal rivers, and tidal creeks), where they are limited to areas that are periodically inundated with brackish water that excludes competitors (Odum and McIvor 1990). Mangroves are known for their ability to trap inorganic sediments and may modify soils by peat formation (Odum and McIvor 1990). Pore water in mangrove soils is almost always anaerobic. Upon drainage, some mangrove soils produce acidic drainage as a result of microbial oxidation of sulphidic materials.

7 Biogeochemistry

B. Graeme Lockaby and Mark R. Walbridge

CONTENTS

INTRODUCTION

Biogeochemical cycles of wetland forests are among the most complex and difficult to study of those associated with any forest ecosystem type. Numerous biogeochemical interactions (as well as high spatial and temporal variability associated with each) occur at the interface of terrestrial and aquatic systems. In addition, because of the landscape position of floodplain forests, some of the biogeochemical transformations which may occur there are critical to the maintenance of an acceptable quality of life in our society. Consequently, scientific interest in wetland biogeochemistry remains very high.

Much of the complexity inherent to biogeochemical processes in forested wetlands stems from hydrologic linkages with associated streams and uplands. The influence of these linkages on nutrient cycling cannot be overemphasized and dominates at the geochemical, biogeochemical, and biochemical levels. It is the influence of flooding and belowground saturation which chiefly distinguishes the processes controlling the biogeochemistry of wetland forests from those of uplands and justifies separate consideration of wetland processes. In many cases, hydrologic influences create environments which require the development of unique research methodologies

1-56670-228-3/97/$0.00+$.50
© 1998 by CRC Press LLC

as well. As a result of their role as the primary determinants of ecological function, the mechanisms through which hydroperiod defines wetland forests in a functional sense receives and deserves much research attention.

Hydroperiod reflects a composite of geomorphic, climatic, and anthropogenic influences which affect the nature and magnitude of geochemical inputs of elements moving to riparian forests in flood- and ground-water. As noted by Brinson (1993a), the consideration of hydroperiod solely as an index of wetness severely restricts its utility as an explanatory tool. Consequently, the initial biogeochemical status (i.e., the degree to which a system is naturally oligotrophic or eutrophic) of a floodplain forest is determined by its landscape position and the geomorphology and mineralogy of the upstream portion of its watershed.

The specific nature of the integration between hydrology and biogeochemical status may vary greatly among floodplain forests and through time and, as a result, continues to elude a thorough understanding on the part of wetland biogeochemists. However, it is this integration which results in the "biogeochemical behavior" which is characteristic of particular riparian forests. Other processes which are intimately linked with biogeochemistry, such as those controlling primary productivity, are also driven by this integration.

Even at a very general scale, several categories of input-output balance behavior may be distinguishable (Brinson 1993a), and many subtle variations on each can be expected. These variations include temporal influences as in the case of geochemical budgets in undisturbed systems shifting in terms of source-sink activity between flooding seasons (Lockaby et al. 1997c). Input-output budgets also reflect the cumulative effect of spatial variation in nutrient processing among floodplain communities with different hydroperiods. Consequently, it is useful to view floodplain forests biogeochemically as heterogeneous assemblages of nutrient processing units.

This chapter will explore the direct and indirect influences of hydrology on nutrient processing within floodplain systems. Direct influences include the nature and magnitude of alluvial inputs, definition of the microenvironment in which decomposition occurs, and changes in soil chemistry. An example of an indirect influence of hydrology is determination of vegetation composition/productivity and associated changes in nutrient circulation. Also, variation in these processes among floodplain forests resulting from differences in landscape position, geomorphology, or biogeochemical status will be examined.

IMPORT–EXPORT DYNAMICS

Effects of Landscape Position on Biogeochemical Function

Because of their unique landscape position as ecotones between terrestrial and aquatic ecosystems, floodplain forests are biogeochemically linked to neighboring uplands and streams. Floodplain wetlands receive chemical inputs from surrounding uplands via surface, subsurface, and groundwater flow (termed riparian transport), and from adjacent streams via floodwaters, termed overbank transport (Brinson 1993a). Overbank transport also provides a mechanism for floodplain wetlands to receive chemical inputs from uplands outside of their immediate subwatershed.

Riparian transport is a unidirectional linkage — materials flow from uplands to wetlands, but not vice versa. These material and chemical inputs are then stored, transformed, and/or exported to either the atmosphere or to neighboring aquatic ecosystems. Overbank transport can be a bi-directional pathway of material exchange between floodplain and stream ecosystems: floodwaters bring chemical inputs from streams to the floodplain surface during flooding, and also transport materials from the floodplain surface to adjacent streams. Brinson (1993a) hypothesized that the ratio of riparian to overbank transport would decrease with increasing stream order.

The import-export dynamics of southern forested wetlands have frequently been examined from the perspective of elucidating their nutrient retention and transformation functions (cf., Walbridge and Lockaby 1994). Southern forested wetlands can remove and/or retain significant amounts of nitrogen (N) and phosphorus (P) from floodwaters. Kuenzler (1989) reviewed removal efficiencies of forested riparian systems of the southeastern Coastal Plain receiving agricultural and mixed inputs, and found efficiencies ranging from 22 to 89% for total N, and 20 to 80% for total P. Removal efficiencies of forested wetlands receiving wastewater inputs were frequently higher, often approaching 100% (Kuenzler 1989).

Walbridge and Lockaby (1994) reviewed the mechanisms associated with N removal in southern forested wetlands, and found rates for individual processes ranging from 0.5 to 350 kg/ha/yr (0.45 to 312 lbs/ac/yr; Table 7.1). Denitrification, a transformation function (*sensu* Richardson 1989) in which nitrate is converted to gaseous N (N_2O and N_2) and exported to the atmosphere, was the dominant mechanism associated with N removal. When N inputs occur in forms other than nitrate, the internal processes that lead to nitrate production and denitrification may be complex (Brinson et al. 1984). The potential for nitrification to be inhibited under acid conditions (cf., Hemond 1983) has led to some speculation concerning the importance of denitrification in floodplain forests that occur on acid soils (Walbridge and Lockaby 1994). Few direct measurements of denitrification exist for forested floodplains in the southern United States (cf., Bowden 1987, Walbridge and Lockaby 1994).

Sediments often provide significant deposition of P in forested floodplains, while adsorption by soil minerals (aluminum [Al] and iron [Fe] in acid soils, calcium [Ca] in alkaline soils) appears to be the dominant mechanism for removal of dissolved phosphate (Table 7.2). Studies that have examined input/output budgets of southern forested wetlands during flooding events have usually found either a net retention of P by the wetland ecosystem (cf., Kuenzler 1989, Walbridge and Lockaby 1994), or a transformation of inorganic P to organic form, with little net storage in some cases (e.g., Elder 1985). The details of this transformation have not been explained. For example, it is not known whether inorganic P is converted to organic form immediately upon entering the floodplain and then exported, or alternatively, is sequestered in the floodplain ecosystem, with floodwater inputs balanced by the export of similar amounts of organic P produced by the floodplain's biogeochemical system at some time prior to the flooding event. Similarly, the forms of organic P in export waters have not been identified or characterized — are they living organisms (i.e., microbial biomass P), simple components of living organisms (e.g., RNA-P, lipid-P), more complex polymers formed during the decomposition process, or P adsorbed to organic matter/metal complexes?

TABLE 7.1
Mechanisms associated with nitrogen removal in floodplain and riparian forests in the southern United States (with permission, from Walbridge and Lockaby 1994).

Mechanism	Study	Location	N Removal Rate (kg/ha/yr)
Sediment/Particulate Deposition	Peterjohn and Correll 1984	Maryland	11.0
Denitrification (potential estimates)	Engler and Patrick 1974	Bayou Sorrel, LA	350.1[1]
	Lowrance et al. 1984	Little River, GA	31.5
	Ambus and Lowrance 1991	Coastal Plain, GA	2–224
Denitrification (mass balance estimates)	Brinson et al. 1984	Tar River, NC	130.0[2]
	Peterjohn and Correll 1984	Maryland	47.7[3]
	Jacobs and Gilliam 1985	Coastal Plain, GA	0.5–2.5
NH_4^+ adsorption	Brinson et al. 1984	Tar River, NC	64.2[2]
	Peterjohn and Correll 1984	Maryland	0.8[4]
Microbial Immobilization[5]	Brinson et al. 1984	Tar River, NC	16.2[2]
	Qualls 1984	Coastal Plain, NC	87.0
Plant Uptake	Brinson et al. 1984	Tar River, NC	15.5[2]
	Lowrance et al. 1984	Little River, GA	51.8
	Peterjohn and Correll 1984	Maryland	15.0

[1] Potential estimate under conditions favoring denitrification (cf. Bowden 1987).
[2] Estimated over a 10-month period under an addition rate of 10 kg/ha/mo.
[3] Nitrate removal from both surface (2.7) and subsurface (45.0) flow; at least two thirds attributed to denitrification.
[4] Estimate based on NH_4^+ removal.
[5] Associated with leaf litter.

Nutrient retention and transformation are two important biogeochemical functions that floodplain forests perform, providing value to society by improving water quality (Walbridge 1993). Transformations that convert inorganic nutrients (N and P) to organic form may also be important in regulating the productivity of downstream aquatic ecosystems. High concentrations of inorganic N and P can trigger aquatic eutrophication. Organic forms of these nutrients operate more like 'slow release' fertilizers, with inorganic nutrient release (mineralization) controlled by enzyme activity (cf., Wetzel 1991).

INTRASYSTEM CYCLING

Vegetation Processes

Vegetation processes (uptake, storage, reabsorption, and above- and belowground litterfall) are important linkages in the floodplain forest's biogeochemical system. Plants utilize primarily inorganic forms of N and P, converting them to organic form

TABLE 7.2
Mechanisms associated with phosphorus removal in floodplain and riparian forests in the southern United States (with permission, from Walbridge and Lockaby 1994).

Mechanism	Study	Location	P Removal Rate (kg/ha/yr)
Sediment/Particulate Deposition	Brown 1978	Prairie Creek, FL	32.5
	Mitsch et al. 1979a	Cache River, IL	36.0
	Mitsch et al. 1979b	Kankakee River, IL	13.6
	Yarbro 1983	Creeping Swamp, NC	1.6–1.7
	Peterjohn and Correll 1984	Maryland	3.0
Adsorption	Brinson et al. 1984	Tar River, NC	199.0[1]
	Richardson et al. 1988	Little River, GA	130.0[2]
Microbial	Brinson et al. 1984	Tar River, NC	6.6[1]
Immobilization	Richardson et al. 1988	Clarkton, NC	35.0–40.0[2,3]
Woody Plants	Schlesinger 1978	Okefenokee, GA	0.2
	Mitsch et al. 1979a	Cache River, IL	1.0
	Yarbro 1983	Creeping Swamp, NC	0.6–1.2
	Brinson et al. 1984	Tar River, NC	1.3[1]
	Lowrance et al. 1984	Little River, GA	3.8
	Peterjohn and Correll 1984	Maryland	1.6
Non-woody Plants	Mitsch et al. 1979a	Cache River, IL	3.3
	Yarbro 1983	Creeping Swamp, NC	0.4

[1] Estimated over a 10-month period under an addition rate of 10 kg/ha/mo.
[2] Sewage-enriched system.
[3] Based on increase in $CHCl_3$–labile (microbial) P.

following uptake. Reported annual rates of plant uptake and storage in southern forested wetlands range from 15.0 to 51.8 kg/ha (13.4 to 46.2 lbs/acre) for N, and from 0.4 to 3.8 kg/ha (0.4 to 3.4 lbs/acre) for P (Tables 7.1 and 7.2). Plant uptake can be significant with respect to overall rates of N and P retention in southern forested wetlands, but will vary as a function of stand age. Rates may be high in young aggrading stands, but will decline as forests mature. At steady state, although large amounts of N and P may circulate annually within the forest ecosystem, retention of external nutrient inputs will be minimal (cf., Vitousek and Reiners 1975). Successful utilization of plant uptake as a mechanism of nutrient removal and/or retention in southern forested wetlands would require periodic harvests to sustain high uptake rates (Walbridge and Lockaby 1994).

Reabsorption of nutrients from senescing foliage (retranslocation) and nutrient return in litterfall are also important biogeochemical processes that are intimately linked in floodplain forests. Numerous studies have examined plant retranslocation efficiencies in a variety of systems worldwide (see Chapin and Kedrowski 1983 and Walbridge 1991 for partial listings), but few studies exist for forested wetlands in the southern United States. In general, retranslocation efficiency (the percentage of foliar N or P withdrawn prior to leaf abscission) appears to vary as a function of habitat N and P availability; nutrient deficient habitats are more likely to be dominated by plants with comparatively high retranslocation efficiencies. However, Schlesinger (1978) found that an Okefenokee cypress swamp appeared to conserve nutrients by minimizing foliar mass, retranslocating only potassium (K) in significant amounts. Whether individual plant species are capable of altering retranslocation efficiencies in response to changes in nutrient availability is more questionable. Walbridge (1991) did not find significant differences in P retranslocation efficiencies of native plants along a natural P availability gradient of pocosin and bay forest wetlands in North Carolina, although the retranslocation efficiencies of some species did decline significantly in response to P fertilization in pine plantations established on sites previously dominated by forested wetlands. Under equilibrium conditions in relatively undisturbed stands, most species probably function at or near their maximum retranslocation efficiency, although disturbances that lead to dramatic increases in nutrient availability could cause retranslocation efficiencies to decline.

Detrital Processing

The decomposition process in both terrestrial and aquatic systems is controlled by the integration of three sets of factors: (1) the "quality" of the litter substrate, (2) the nature of the microenvironment in which decomposition occurs, and (3) microbial populations inherent to (2) above (Tenney and Waksman 1929, Swift et al. 1979). These may shift order in terms of relative importance as rate controlling factors during early decomposition which is dominated by losses of soluble constituents vs. later stages where recalcitrant substances dominate (Berg 1986, Melillo et al. 1989). When attempting to identify to what extent decomposition in a floodplain is unique and/or differs from that described generally for terrestrial systems, a useful approach is to ask if factors (1) to (3) operate differently in the former. Due to a lack of information regarding belowground decomposition in riverine forests, the discussion of factors (1) to (3) will focus only on decay of aboveground detritus.

In the temperate zone, decomposition rates of foliar litter in riverine forests are more rapid than those of deciduous forests in general. Decomposition rates are often described by a constant percentage weight loss per unit time using:

$$\% \text{ original remaining} = e^{-kt}$$

where: t = time

k = litter-specific decomposition rate change.

The average k from temperate riverine forests is 1.01 (calulated from Brinson 1990) while the mean k for temperate deciduous forests is 0.77 (Swift et al. 1979). However, it is important to realize that rates measured in floodplains may include a larger "export" component and, thus, be inflated relative to those of non-flooded systems. Thus, "decomposition" as measured there includes a large component of mechanical disintegration as well as the metabolism of organic to inorganic constituents (Boulton and Boon 1991).

Soils of floodplain forests are sometimes thought to contain greater concentrations of organic matter than do uplands (Sharitz and Mitsch 1993). However, this observation seems to be accurate only for poorly drained floodplain soils. A comparison of organic matter percentages in surface mineral soils of southeastern floodplains indicates levels of 5, 3.4, 2.8, and 3.0 (calculated from Wharton et al. 1982) for Clark and Benforado's (1981) zones II-V, respectively, values which fall within a general range for forest soils. Organic matter levels may be much higher (i.e., 10 to 50%; Wharton et al. 1982) in those floodplain soils which are grading toward Histosols due to poor drainage.

Decomposition of coarse woody debris (CWD) in forested floodplains has received far less attention than foliage (Chueng and Brown 1995). Some authors have suggested that CWD degradation rates are inversely related to the degree of wetness along a terrestrial — aquatic continuum (Harmon et al. 1986). However, there are indications that rates in floodplains may exceed those in non-flooded systems by a wider margin than is the case for foliar litter. A range of k values between 0.089 and 0.182 has been reported in forested floodplains (Chamie and Richardson 1978, Chueng and Brown 1995) while those associated with log fragmentation in temperate forests generally fall between 0 and 0.06 (Harmon et al. 1986).

As has been emphasized repeatedly, one of the primary influences within floodplains which does not occur in other terrestrial systems is that of hydroperiod. All aspects of decomposition from the timing, magnitude, and quality of above- and belowground detrital inputs as well as definition of the microenvironment in which decomposition takes place are subject to the influence of hydrology. Flooding is probably the largest single influence on the spatial distribution of the forest floor and ultimately is the main determinant of detrital import — export relationships in floodplain systems. Thus, examination of the many interactions between hydroperiod and the three factors which drive decomposition is warranted.

Influence of Litter Quality

Gradations in vegetation species composition and microsite characteristics due to spatial variability in hydroperiod and/or soil saturation have often been described in connection with floodplain forests (Wharton 1978). At a range of microsite scales, hydroperiod will vary within localized areas and may give rise to distinct changes in species composition. The result is often a mosaic of communities within floodplain ecosystems. Since vegetation composition may change markedly among these communities, it is reasonable to assume that litter quality will change measurably as well.

In general, long-term decomposition rates (i.e., those encompassing stages where recalcitrant materials dominate) on individual sites are controlled primarily by the quality of the substrate (Johansson et al. 1995) since the microenvironment is homogeneous spatially. This truism may be applied to particular communities within a floodplain where hydroperiod (and, consequently, the overall microenvironment) is similar and intrasite variation in litter quality is minimal.

Generally, as spatial scales expand, both microenvironment and litter quality increase in variation with microenvironment increasing in importance relative to and apart from its influence on quality (Meentemeyer 1978, Meentemeyer and Berg 1986, Johansson et al. 1995). This is particularly true of rates during early decomposition

(Johansson et al. 1995). However, late stages are still heavily influenced by quality. As an example, across a 2,000 km (1,243 mi) transect with average temperature and precipitation ranges of 8.5°C (47.3° F) and 645 mm (25 mm) respectively, Johansson et al. (1995) found that quality (i.e., lignin:N ratios, which ranged between 23 and 114), was the dominant influence on rates during later stages in Scotch pine (*Pinus sylvestris*) litter. However, a major difference between upland and riverine systems may be associated with the much smaller spatial scales over which these effects are manifested in the latter.

Thus, a key question is the nature of the relationship between litter quality and hydroperiod. One of the few investigations in forested wetlands which evaluated the roles of litter quality vs. microenvironment on decomposition is that of Day (1982). Day compared rates among four communities within a depressional swamp in Virginia which differed substantially in term of species and flooding regime. Although decay was more rapid in extensively flooded communities where tupelo (*Nyssa* spp.) species dominated, litter originating there was also less resistant to degradation. Lignin:N and C:N ratios from the wetter communities averaged 12 and 28, respectively, whereas those from less frequently flooded sites averaged 22 and 43. The differential in quality influenced loss rates after two years to a greater degree than did the direct effect of flooding regime.

However, increasing "wetness" does not always reflect higher litter quality within particular sites. In the ACE (Ashepoo, Combahee, and Edisto Rivers) Basin of South Carolina, litter quality improved as extent of flooding decreased. Four communities there exhibited decreasing lignin:N (from 86 to 49) and lignin:P ratios (from 176 to 112) in deciduous species as flooding tendency lessened. The inverse relationship between wetness and quality in the ACE is even clearer when C:N ratios of a single species, sweetgum (*Liquidambar styraciflua*), are compared across the three communities in which it occurred (i.e., 89 to 60 as flooding tendency decreased; Lockaby and Burke unpublished data). In the ACE gradient, tupelo species exhibited the highest lignin:N ratio (86) while in Day's study, the genus exhibited the narrowest ratio (10) relative to others present.

There are suggestions from data collected in some basin wetlands that an interaction between water deficits and low nutrient availability may be a factor contributing to reduced quality of foliage (Lugo et al. 1990a). Some insight regarding this interaction may be gained by contrasting quality indices between two blackwater riverine systems in the southeastern United States which have been well-characterized in terms of both hydroperiod and nutrient availability. Vegetation productivity on the Ogeechee River floodplain in southeastern Georgia is probably limited by N availability primarily and, to a marginal degree, by that of P. The floodplain is moderately drained for the most part and foliar litter there exhibits N:P, lignin:N, and lignin:P ratios of 8, 20, and 154, respectively (Lockaby et al. 1996). In contrast, the Little Escambia floodplain in southwest Alabama is saturated to the soil surface most of the year and is deficient in N and particularly in P (Lockaby et al. 1994). In the latter system, the respective indices are 15, 18, and 240, which may indicate lower quality there driven heavily by P deficiency and, to an unknown degree, by excessive wetness.

Litter quality, as measured by N:P ratios, is thought to provide an indication of how these elements may behave as decomposition progresses. Vogt et al. (1986) have suggested that N:P ratios wider than 15 imply P limitation to the decomposition process. Consequently, there exists a stronger tendency for P immobilization in those situations. However, relations in soils may not be the same as in freshly abcised litter. As N:P ratios in organic wetland soils decreased from 57 to 18, Walbridge (1991) observed three-fold increases in microbial P.

In the Great Dismal Swamp, Day (1982) reported that both N and P were conserved throughout the year in litter which exhibited N:P ratios of 15 to 17. In a North Carolina riverine swamp, foliar litter with a very wide ratio displayed a tendency to immobilize both N and P initially followed by mineralization after 16 and 38 weeks, respectively (Brinson 1977). During the 60-week investigation, N behavior tended to reflect mineralization while that of P was dominated by immobilization.

Comparisons of N and P dynamics in litter from the Ogeechee River floodplain (N:P = 8) with those from the Little Escambia (N:P = 15) provide a useful contrast between theoretically P non-limited vs. limited scenarios. During decomposition on unharvested plots of the Little Escambia, P content increased during a 68-week period while that of N changed little (Figure 7.1; Lockaby et al. 1996). In contrast, Ogeechee litter P showed an early immobilization phase followed by mineralization thereafter (Figure 7.2; Lockaby et al. 1996) as would be suggested by Vogt et al. (1986). In addition, N in the latter system exhibited a strong mineralization tendency throughout a 106-week period. Thus, narrow ratios may promote release of both elements.

Changes in N and P content in CWD during decomposition may behave similarly. In two instances where logs exhibited narrow N:P ratios of 7 and 2, respectively, a strong tendency toward P mineralization occurred (Chueng and Brown 1995, Rice et al. in press). However, N was generally immobilized in logs with a C:N ratio of approximately 300 (Chueng and Brown 1995) and exhibited slight mineralization in an instance with a C:N ratio of 204 (Rice et al. in press). However, discussions of N and P behavior during decomposition in different systems are made more complex by variation among systems in hydroperiod and/or nutrient inflow. As an example, it is possible that prolonged saturation of the Little Escambia foliar litter (rather than a wide N:P ratio) may have reduced N mineralization.

Influence of Microenvironment

Decomposition rates and/or masses of steady state forest floors differ among microsites within forested floodplains (Bell and Sipp 1975, Brinson 1977, Day 1982, Peterson and Rolfe 1982, Briggs and Maher 1983, Shure et al. 1986, Cuffney 1988). As previously noted, the observed variation is probably attributable to differences in litter quality if microsites are sufficiently large to entail changes in vegetation species composition.

However, the role of microenvironment variation may be pronounced as well and has probably attracted more research attention in floodplains than has litter quality. Prior to consideration of the role of microenvironment, the concept must be

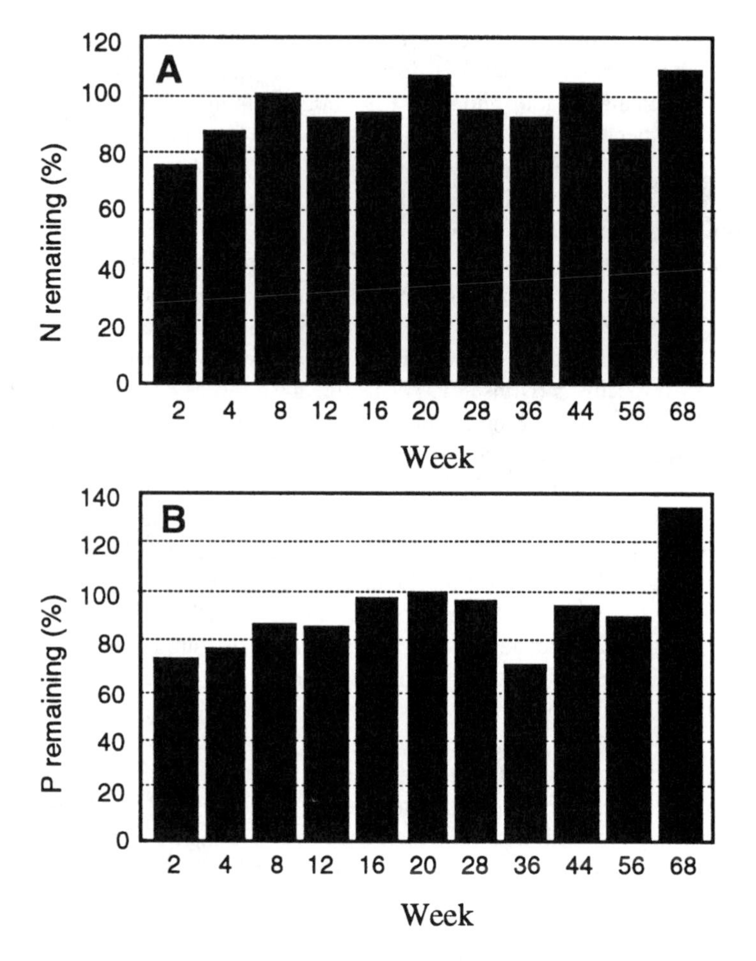

FIGURE 7.1 Effect of wide N:P ratio on immobilization/mineralization patterns for N (a) and P (b) in litter of the Little Escambia floodplain, AL.

expanded through consideration of the timing of detrital inputs and the duration of exposure to a given microenvironment.

In order to facilitate logistical and statistical considerations associated with litterbag studies, comparisons of decomposition among communities often are initiated simultaneously. Although the reasons for simultaneous initiation are important, this probably represents a moderate departure from natural processes in many cases. This is illustrated by comparing the phenology of litterfall inputs between two adjacent communities (differing in terms of hydroperiod and species) on the Flint River floodplain in central Georgia (Figure 7.3). The community which is inundated longer each year is composed primarily of tupelo species and completes litterfall by late autumn. In contrast, litterfall in the more mesic community (composed of oaks [*Quercus* spp.] and sweetgum) extends well into late winter.

The significance of these two patterns relative to decomposition is that the two litters are exposed to microenvironments which are not only spatially different but

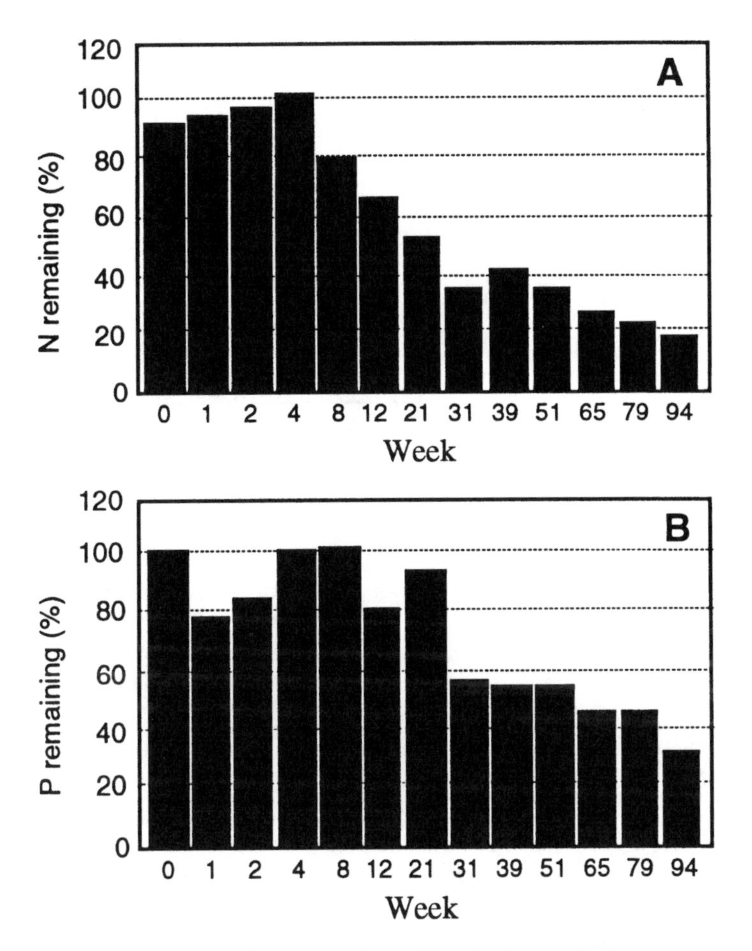

Figure 7. 2 Effect of narrow N:P ratio on immobilization/mineralization patterns for N (a) and P(b) in litter of the Ogeechee floodplain, GA.

are offset temporally as well. This results in greater variation between the two in terms of abiotic factors such as temperature and may contribute measurably to temporal offsets in immobilization/mineralization patterns as well (Lockaby et al. 1996). These offsets may be significant if they reflect differential synchroneities between nutrient availability, vegetative uptake, and/or geochemical leakage within different communities.

In addition, annual amounts of aboveground litter are likely to differ markedly among microenvironments on a floodplain. An extensive data set compiled by Megonigal et al. (1997) for forested wetlands in the southeast United States indicates that substantial differences in litterfall quantities exist between wet and drier floodplain sites.

Confusion exists concerning the nature of the direct influence of hydroperiod on decomposition (Brown 1990, Mitsch and Gosselink 1993). Much of the uncertainty stems from the use of relative terms such as "increased wetness" or "flooded

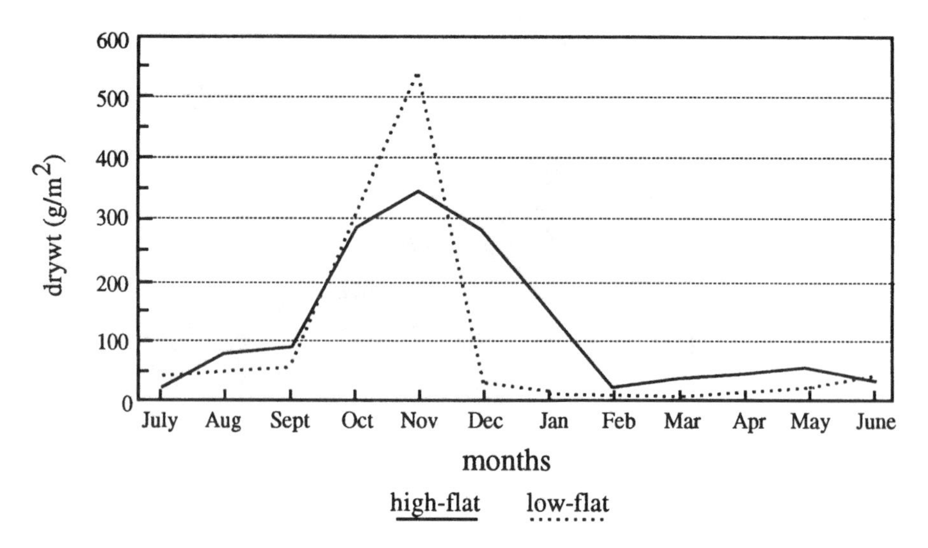

FIGURE 7.3 Temporal patterns of litterfall within communities on a well-drained (solid line) vs. a poorly-drained site (dashed line) of the Flint River floodplain, GA.

to a greater extent," which have little utility apart from intrasystem comparisons. However, detailed information on duration, velocity, aeration, depth, and chemistry is often lacking with regard to hydroperiod and, consequently, the use of relative terms is difficult to avoid.

In general, flooded conditions are thought to promote decomposition, although exceptions have been reported (Irmler and Furch 1980, Duever et al. 1984a). Mitsch and Gosselink (1993) suggest that aerobic/moist conditions promote decomposition to the greatest extent followed by dry and anaerobic conditions in that order. The stimulatory influence of wetting-drying cycles on decomposition has been emphasized in non-flooded systems (Swift et al. 1979), and Brinson et al. (1980) have suggested that these cycles are probably equally important in wetlands as well.

The effect of flow rate on decomposition has been emphasized by Brinson (1990) and Conner and Day (1991). These authors indicate that rates of mass loss and N and P immobilization may be suppressed and stimulated respectively in situations with minimal flow. In contrast, they suggest that rapid mass loss and mineralization are more likely to be characteristic of wetlands with higher flows.

An attempt to examine flooding influences in a controlled environment on the Ogeechee floodplain was made by Lockaby et al. (1996). Hydroperiod variations included variable durations with velocity and aeration being maintained at similar levels. Single episodes of flooding stimulated mass loss to the greatest extent. It was thought that the failure of the alternating flooded-non-flooded treatment to promote the most rapid mass loss was perhaps attributable to the differences between a wet-dry cycle in an upland vs. the latter treatment in the floodplain. In the latter, moisture may be less limiting during the "dry" portion of the cycle than in a system which never floods.

TABLE 7.3
Physical and chemical characteristics of forested wetland soils in the southern United States.

Site/Location	pH (flooded)	pH (non-flooded)	Sand	Silt	Clay	OM	Reference
Big Thicket, TX	5.3	4.5	22	31	47	2.8	Walbridge et al. 1992
Ogeechee, GA	5.6	5.0	48	28	24	7.7	Walbridge unpublished
Brown Marsh, NC	n.d.	4.1	n.d.	n.d.	n.d.	17.1	Richardson et al. 1988
Cashie, NC	n.d.	4.8	n.d.	n.d.	n.d.	17.6	ibid
Wahtom, NC	n.d.	4.2	n.d.	n.d.	n.d.	3.2	ibid
Short Pocosin, NC	n.d.	3.9	n.d.	n.d.	n.d.	95.2	Walbridge 1991
Tall Pocosin, NC	n.d.	3.7	n.d.	n.d.	n.d.	94.4	ibid
Bay Forest, NC	n.d.	3.7	n.d.	n.d.	n.d.	76.2	ibid
Bottomland Hardwoods (VA Coastal Plain)	n.d.	4.7	53	28	18	5.1	Fuad 1995
Bottomland Hardwoods (VA Coastal Plain)	n.d.	4.4	52	23	25	10.6	Axt and Walbridge 1997
Bottomland Hardwoods (VA Piedmont)	n.d.	4.9	43	26	31	7.5	Axt and Walbridge 1997

The influences of variation in N and P inflow on decomposition were also investigated in the same study on the Ogeechee. Elevated N slowed decomposition by reducing lignin degradation (Lockaby et al. in press) as has been previously reported in non-flooded systems (Fog 1988). However, Dierberg and Ewel (1984) observed a stimulation of decomposition as a result of influx of N-enriched sewage into a cypress (*Taxodium* spp.) dome. It is possible that this case reflects the effects of improved litter quality since plant production was stimulated as well. In contrast, P elevation generally has little effect on decomposition rates (Howarth and Fisher 1976, Lockaby et al. 1996), although some exceptions have occurred in which a stimulatory effect was recorded (Elwood et al. 1981).

SOIL CHEMISTRY

General Physical and Chemical Characteristics

Soils of southern forested wetlands are generally neutral to acidic, with reported pH's ranging from 3.7 to 3.9 in pocosins and bay forests on highly organic soils, to 5.6 in the Ogeechee River floodplain under flooded conditions (Table 7.3). Flooding may cause a temporary pH increase of 0.5 to 1.0 unit or more. Textural composition can vary from >90% organic matter in Histosols characteristic of pocosin wetlands in coastal North Carolina, to nearly 50% clay in the Neches River floodplain (The Big Thicket), Texas, to >50% sand in bottomland hardwood forests in the Virginia Coastal Plain. The organic matter content of surface horizons of southern forested wetlands on mineral soils can range from 2.8 to 17.6%.

Phosphorus Retention

Sediment deposition and adsorption by soil minerals appear to be the two primary mechanisms associated with P retention in southern forested wetlands (Table 7.2). Adsorption by soil minerals will be particularly important in wetlands receiving significant amounts of dissolved phosphate in input waters. Sources of dissolved phosphate would include commercial fertilizers, wastewater and/or septic systems, and phosphate detergents, and could come from both point and nonpoint sources. Since all of these dissolved phosphate sources are linked to human activities, such activities are likely to increase dissolved phosphate loadings to wetlands downslope/downstream.

Richardson (1985) used an isotherm approach to compare wetland soil P sorption capacities and found significant variation as a function of wetland type. Similar variations are observed when soil P sorption capacities of southern forested wetlands are compared (Figure 7.4). Both Fuad (1995) and Axt and Walbridge (1997) observed that on average, the P sorption capacities of wetland soils in Caroline (n = 8) and Spotsylvania (n = 6) Counties, Virginia, were higher than those of adjacent uplands (Figure 7.4). However, individual site comparisons of wetlands and adjacent uplands did not show consistently higher P sorption capacities in wetland soils, due to variations in upland soil clay content (Fuad 1995, Axt and Walbridge 1997). When upland soils are sandy and associated wetland soils are significantly higher in clay content, as can frequently occur in the southern Coastal Plain (e.g., The Big Thicket, Texas; Walbridge et al. 1992), wetland soils do exhibit higher P sorption capacities than surrounding uplands. The P sorption capacities of upland soils bordering the Ogeechee River floodplain, Georgia, are also higher that those of adjacent wetlands. Darke (1997) hypothesized that because large amounts of P had already been adsorbed by upland soils, wetlands in this situation represented a second line of defense against the eutrophication of downstream aquatic ecosystems.

Soil clay and organic matter content, and concentrations of Al and Fe oxides and hydroxides, may all control P sorption capacity in the predominantly acid soils of southern forested wetlands. Oxalate-extractable (amorphous) Al exhibits a strong correlative relationship with P sorption capacity in a wide range of both wetland and upland soils (Richardson 1985, Richardson et al. 1988, Walbridge et al. 1991, Walbridge and Struthers 1993, but see Fuad 1995), and appears to be a good predictor of soil P sorption capacity in southern forested wetlands (Figure 7.5). The stronger correlations of oxalate-extractable Al vs. Fe with P sorption capacity frequently observed in wetland soils could be due to the susceptibility of Fe to reduction and dissolution under flooded conditions (cf., Walbridge and Struthers 1993), which can release associated phosphate. Although all forms of Al and Fe oxides and hydroxides sorb P by the same mechanism, amorphous forms tend to dominate soil P sorption reactions when present in significant amounts because of their greater adsorptive surface area per unit mass/volume than more highly ordered crystalline forms (Parfitt and Smart 1978, Hsu 1989, Schwertmann and Taylor 1989). Some studies have suggested that flooding can favor the formation and/or persistence of amorphous Fe and Al hydroxides, thereby increasing soil P sorption capacity (Willett et al. 1978, Kuo and Mikkelsen 1979, Kuo and Baker 1982, Sah and Mikkelsen 1986).

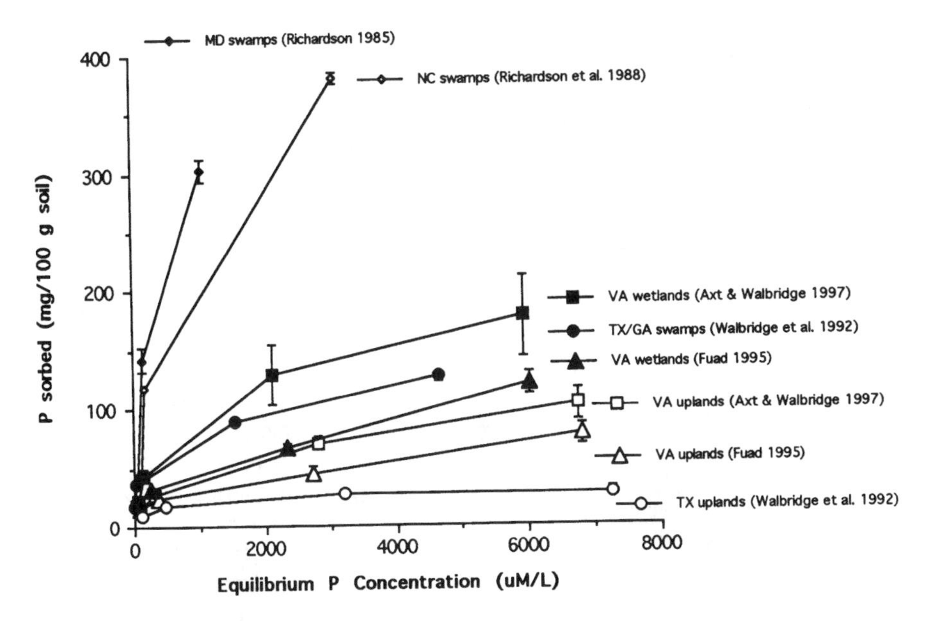

FIGURE 7.4 Phosphorus adsorption isotherms for a range of forested wetlands in the southern United States. Values represent means of P sorbed during 24-hour laboratory equilibrations from initial solution concentrations ranging from 0.31 to 260 mg PO_4–P/L, plotted as a function of the equilibrium concentration of PO_4–P remaining in solution, for individual sites: n = 10 for Maryland (MD) swamps (Richardson 1985); n = 3 for North Carolina (NC) swamps (Richardson et al. 1988); n = 6 for Virginia (VA) wetlands (bottomland hardwoods), Spotsylvania Co. — 3 each in Piedmont and Coastal Plain physiographic regions (Axt and Walbridge 1997); n = 2 for Texas (TX) (The Big Thicket)/Georgia (GA) (Ogeechee) swamps (Walbridge et al. 1992); n = 8 for Virginia (VA) wetlands (Coastal Plain bottomland hardwoods), Caroline County (Fuad 1995). Upland values are means for sites adjacent to wetland sites by study, with the exception of Texas uplands, which represent a mean of two upland sites adjacent to the Texas wetland site. All studies used equivalent methods of isotherm determination.

More recently, strong correlative relationships have been observed between soil organic matter content and P sorption capacity in some southern forested wetlands, and between oxalate-extractable Al and percent organic matter (Axt and Walbridge 1997, Darke 1997). The correlative relationship observed (r = 0.854) when oxalate-extractable Al and percent organic matter were compared across a range of southern forested wetland sites (Figure 7.6) was significantly improved (r = 0.964) by deleting three outlier sites, suggesting that the relationship may be somewhat site specific. While it has been generally assumed that the Al and Fe extracted by acid ammonium oxalate in wetland soils represents primarily amorphous inorganic Al and Fe oxides and hydroxides, these new studies suggest that in some systems, oxalate-extractable Al may represent Al complexed by organic ligands (oxalate also extracts organically-bound Al and Fe) (cf., Walbridge et al. 1991). It is interesting to speculate on the mechanism(s) that might cause soluble Al to be available to bind with organic matter. One possible source would be the solubilization of Al from low base saturation upland soils subjected to high salt concentrations, such as occurs during fertilization

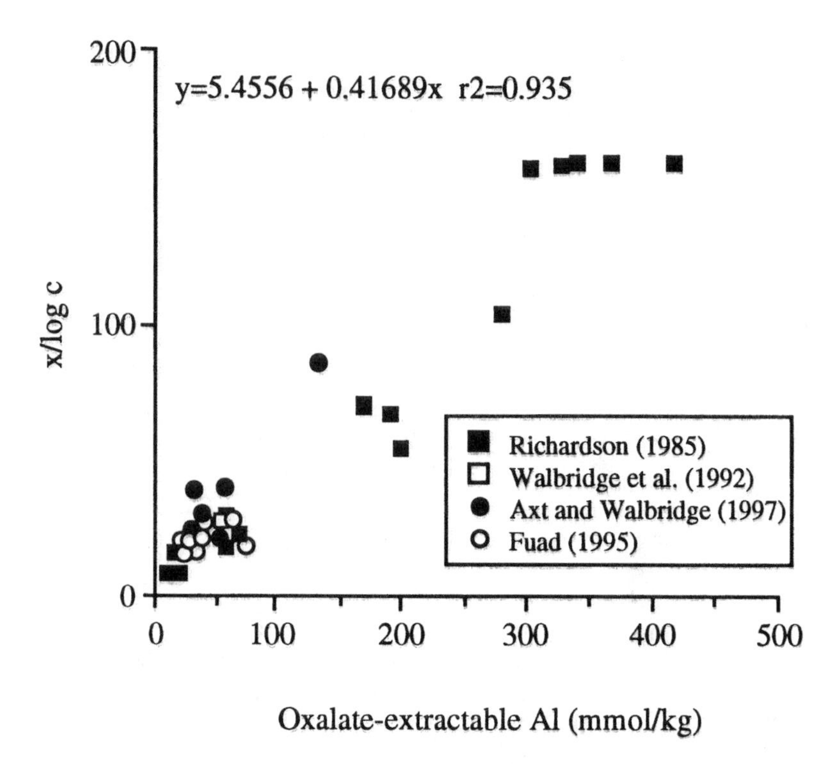

FIGURE 7.5 P sorption capacity, indexed by x/log c, where x = P sorbed (mg P/100 g soil) from a solution containing 130 mg PO_4–P/L during a 24 h equilibration at a soil:solution ratio of 1:12.5, and c = the corresponding equilibrium P concentration (μM/L), as predicted by oxalate-extractable (noncrystalline + organically-bound) Al in a range of southern forested wetland soils. Sites are identical to those of Figure 7.4, with the addition of six North Carolina (NC) pocosin sites from Richardson (1985).

(Egli and Fitze 1995). Such a process of Al solubilization in upland soils, and subsequent complexation by organic matter in wetlands downslope, would have important implications for biogeochemical cycling in southern forested wetlands.

One such implication relates to the potential importance of upland groundwater discharge in floodplains along higher order streams — a revision of Brinson's (1993a) hypothesis. While the volume of floodwaters and mass of materials deposited probably do, in general, follow Brinson's hypothesized trend (i.e., the significance of overbank transport increases with increasing stream order), elemental dynamics will follow this rule only to the extent that input concentrations are roughly equivalent. For example, if soluble Al concentrations were extremely low in floodwaters, and much higher in groundwaters, then Al inputs might be dominated by riparian transport in floodplains along higher order streams despite the fact the overbank transport was the dominant vector of hydrologic and material inputs. This type of pattern is a distinct possibility in the Ogeechee River floodplain where 1) low base saturation upland soils are fertilized periodically to enhance the productivity of pasture species, 2) groundwater discharge has been found to represent a significant

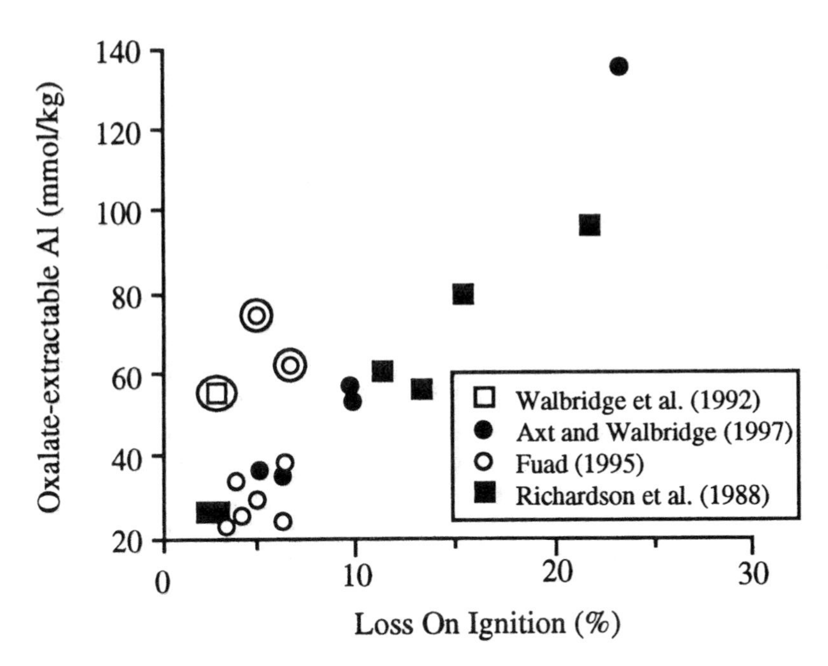

FIGURE 7.6 Oxalate-extractable (noncrystalline + organically-bound) Al as predicted by loss on ignition for a range of southern forested wetland soils. Sites are identical to those of Figure 7.4, with the exception of Richardson et al. (1988), where points represent averages for separate locations (n = 2 per site) at each of three different sites (Brown Marsh, Cashie, and Wahtom swamps, North Carolina). The overall relationship ($y = 16.156 + 4.15261x$; $r^2 = 0.729$) was significantly improved ($y = 6.4258 + 4.7346x$; $r^2 = 0.930$) by the deletion of four outliers, circled on the figure.

component of the Ogeechee floodplain's hydrologic budget (T. M. Williams, pers. comm.), and 3) preliminary data suggest that Al deposition in floodwater sediments is low (Darke 1997).

Spatial Variability

Spatial variability may be extremely important in controlling soil chemistry and biogeochemical cycling in floodplain forests. At the coarsest level of resolution, the composition of flood-deposited materials will vary systematically along a gradient perpendicular to the direction of streamflow as distance from the river margin increases, because larger particles (e.g., sands) settle out rapidly as flood waters enter the floodplain (i.e., levee formation), while smaller particles (e.g., clays) settle out more slowly, with higher deposition rates farther away from the river margin (in backswamps or sloughs) (Leopold et al. 1964).

At a slightly finer scale, the floodplain surface itself is characterized by a series of convex (ridge) and concave (swale) surface features, which vary in soil composition and hydroperiod. Ridges are frequently characterized by sandy surface horizons, while swales have higher clay and organic matter contents and longer hydroperiods; both have potential effects on biogeochemical cycling and nutrient retention. Added to these potential variations are variations in the quality and quantity of litterfall (discussed above) and the tendency for litter to accumulate in depressions (swales).

When upland surface runoff and groundwater contributions are included, one can envision a gradient through the floodplain from river margin to upland, perpendicular to the direction of streamflow, along which potential contributions from overbank transport decline, while potential contributions from riparian transport increase with increasing distance from the stream channel.

RELATIONSHIPS BETWEEN VEGETATION PRODUCTIVITY AND BIOGEOCHEMISTRY

Some authors have suggested that "high" levels of available nutrients in soils of floodplains combined with the opportunities for geochemical inputs of both N and P in general (i.e., in contrast to basin wetlands) imply that deficiencies of those elements are not a critical limitation for vegetation in riverine forests (Lugo et al. 1990a, Sharitz and Mitsch 1993). While the contention that vegetation productivity is not severely limited by lack of nutrients in many riverine forests is probably correct, there are relatively few natural forests worldwide where nutrient deficiencies might be termed severe (Pritchett and Fisher 1987). However, if we are to improve our understanding of processes controlling productivity in these systems, some attention should be devoted to the occurrence of more subtle levels of nutrient limitation.

The nutrient cycling efficiency indices suggested by Lugo et al. (1990a) are highly useful as a means to compare systems and to develop hypotheses regarding interactions of hydrology, nutrient cycling, and productivity. Unfortunately, in most studies, data required for calculation of biogeochemical indices and productivity have not been jointly estimated and, consequently, scrutiny of comprehensive relationships for riverine forests are not possible. However, some speculation regarding the degree to which growth may be restricted due to nutrient deficiency can be derived from reports of biological (as opposed to economically viable) growth responses following additions of some nutrients.

Stimulation of productivity in some riverine forests due to influx of sewage effluent has been observed (Brinson 1990). In addition, during the 1960s and 70s, an active bottomland hardwood fertilization program existed in the southeastern United States. Researchers involved with that program noted that measurable growth responses following application of N, P, and/or K could be observed with a variety of floodplain hardwood species, particularly in floodplains that had been formerly cultivated (Francis 1985). Cultivation during the 19th and/or early 20th centuries was very common in many floodplain forests of the southeastern United States (Walker 1991).

Suggestions that many riverine forests functioned as nutrient sinks (Kitchens et al. 1975, Nelson et al. 1987) gave rise to the notion that floodplain forest soils were generally fertile. However, as is the case with most generalities in nature, there are many exceptions. Levels of available P reported from numerous floodplain forest soils of the southeastern United States range between 2 to 18 mg/kg (3.2×10^{-5} to 2.9×10^{-4} oz/lb) with a few higher exceptions (Wharton et al. 1982). A similar range of available P was reported among floodplain soils in southwest Alabama by White and Carter (1970) and in the midsouth by Broadfoot (1976). These ranges may be compared to a crude deficiency level for available soil P of approximately 7 mg/kg (Kennedy, pers. comm.) for hardwoods of southern floodplains in general. While this deficiency level and reported ranges of extractable P should not be used in an absolute sense (due to minimal fertilizer response data and variations in extraction methods), it is apparent that considerable variation exists among floodplain soils in terms of relative P availability.

Most (i.e., 90%) of the floodplains surveyed by Wharton et al. (1982) exhibited organic matter percentages above 2.0, which is the value below which Baker and Broadfoot (1979) have suggested nutritional limitations occur for a variety of floodplain deciduous species. However, all sites surveyed by Wharton et al. (1982) had not been recently disturbed and probably would reflect lower levels of organic matter (and correspondingly greater tendencies to be deficient) if recently cultivated (Francis 1985). In the midsouth, 44% of the floodplain forest soil series described by Broadfoot (1976) exhibited surface soil organic matter percentages less than 2.0.

Thus, there is evidence in some riverine forests of the southeastern United States, such as those associated with low P availability and/or those which have been recently cultivated, that macronutrient availability may represent a constraint on vegetation productivity. These nutrient limitations may be the result of historical human activities and/or may reflect the natural status of a site. This suggests that net primary production in such systems may, consequently, be responsive to shifts in the rates and/or magnitude of nutrient circulation mechanisms. However, the nature of the relationship between nutrient circulation and productivity, as well as how the strength of this relationship varies among floodplains, remains unclear at present due to insufficient data. The general topic of biogeochemical controls on vegetation productivity in floodplain forests remains one of the least understood facets of the ecology of those systems.

Circulation of N and P: Riverine vs. Depressional Systems

Some insights into biogeochemical relationships to vegetation productivity may be gained by examining patterns of N and P restitution among forested wetland communities. Theoretically, we should expect clarity of these patterns to increase as N and/or P deficiency increases. As an example, we should see more definitive patterns between P circulation and net primary productivity (NPP) in depressional than in riverine wetlands due to the lack of geochemical input mechanisms for P in the former (Brinson 1993a). However, this is not the case.

Relations between litterfall (used as an index of aboveground NPP) and N or P content of litterfall within southeastern forested wetland communities indicate that patterns are expressed more strongly for N than P in both riverine and depressional system types (Figures 7.7a-d). The relationship for N within depressional systems is particularly strong (Figure 7.7c). The strength of N relationships is not surprising since this element is widely documented to represent the primary limiting element to forest productivity on a global scale (Kimmins 1987).

Given the limited potential for P geochemical input mechanisms within depressional wetlands, it is somewhat unexpected to see a stronger relationship between litterfall and P circulation in riverine systems (Figure 7.7b vs. 7.7d). However, this may be due to the higher productivity (as indicated by higher litterfall) in floodplain forests which possibly might induce greater annual requirements of P therein. In addition, in terms of position on Figure 7.7b, a separation of oligotrophic blackwater (i.e., communities 10, 5, 12) vs. euthrophic redwater (i.e., 9, 8, 7) floodplain systems is apparent.

The efficiency of N and P circulation via litterfall can be examined using ratios suggested by Vitousek (1982) and Lugo et al. (1990a). Clear patterns emerge for circulation of both elements with a wider range of efficiencies being apparent for riverine forest communities (Figures 7.8 and 7.9). As expected (Waring and Schlesinger 1985), those communities with highest productivities and larger amounts of N or P in circulation are less efficient. As in the case of Figure 7.7b, this generality is particularly apparent within the riverine category when comparing graph positions for blackwater (i.e., highly efficient — low productivity) vs. redwater systems (i.e., low efficiency — high productivity).

The degree to which the balance between N and P within a forested wetland relates to productivity is explored in Figure 7.10. A pattern (i.e., a normal distribution, $p > 0.05$) emerges which is broadly suggestive of a bell curve. The apex of litterfall is associated with a N/P ratio mean of approximately 12, which may imply that those systems with ratios below this (i.e., redwater, mean = 7.8) are primarily N deficient while those above (i.e., blackwater, mean = 13) are mainly lacking in P. As an example, system 6a (Flint River floodplain) exhibits a very narrow N/P ratio relative to its associated litterfall. This community is apparently very inefficient in cycling P (Figure 7.9) and has large amounts of P in circulation (Figure 7.7b). These data suggest that this particular Flint community may exhibit luxury consumption of P due to P enrichment stemming from large geochemical inputs from unknown sources.

Effects of Forest Management on Biogeochemical Functions

The current state of knowledge regarding the influence of forest management on wetland forest biogeochemistry in the southern United States has been discussed in detail in Walbridge and Lockaby (1994) and Lockaby et al. (1997a). Most research describes influences resulting from clearcut — natural regeneration combinations, although some assessments of logging road impacts exist as well (Rummer et al. 1997). As pointed out in Lockaby et al. (1997c), the nature, magnitude, and longevity of shifts in some biogeochemical functions will be strongly affected by the species composition and productivity of vegetation communities which occur following silvicultural activities. However, at present, no long-term data (i.e., of sufficient age to compare directly with pre-harvest conditions) are available and, consequently, considerable uncertainty exists regarding potential influences of harvesting on nutrient circulation within wetland forests.

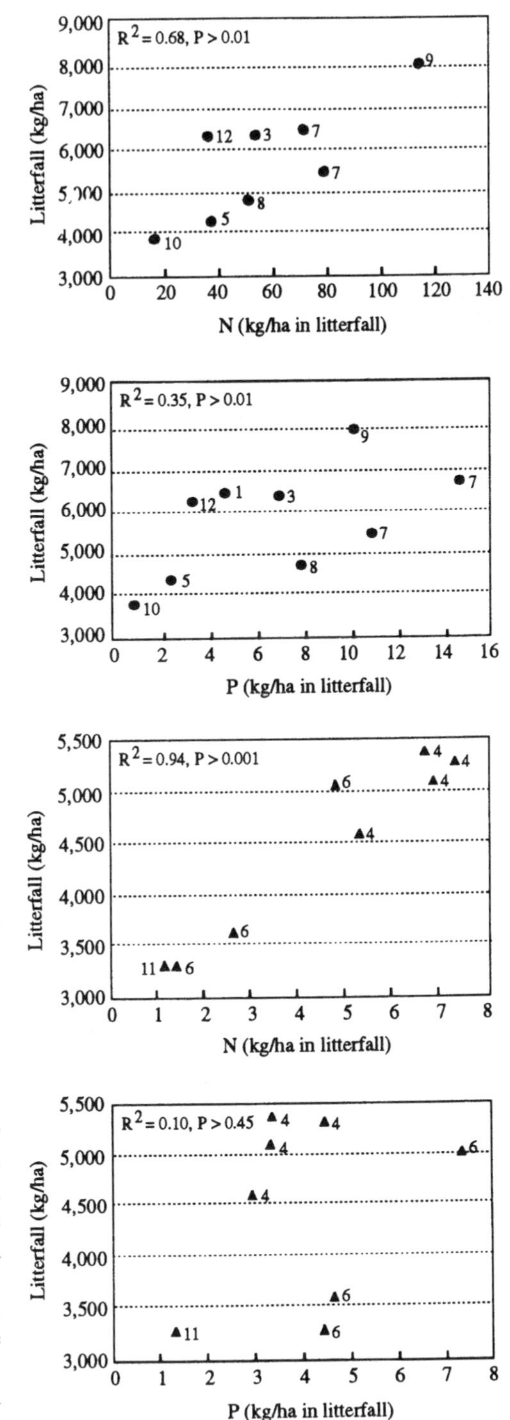

FIGURE 7.7 Relationships between litterfall and N or P content of litterfall for riverine (•) and depressional (>) wetland forest communities in the southeastern United States. Source: 1 = Brinson et al. (1980); 2 = Conner and Day (1992b), N/P ratio from same authors; 3 = Cuffney (1988), N and P data from Lockaby (unpublished data); 4 = Gomez and Day (1982); 5 = Jones et al. (1996), N and P data from Lockaby (unpublished data); 6 = Lockaby and Burke (unpublished data from ACE Basin, Charleston, SC); 7 = Lockaby and Clawson (unpublished data from Flint River, GA); 8 = Nelson et al. (1987), 9 = Peterson and Rolfe (1982); 10 = Post and de la Cruz (1977); 11 = Schlesinger (1978); 12 = Shure et al. (1986), P data from Jones (unpublished data).

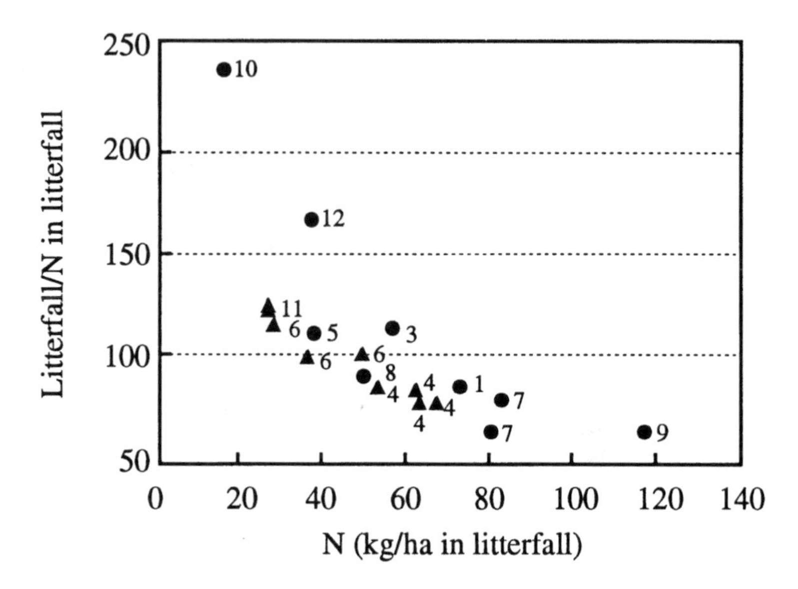

FIGURE 7.8 Nitrogen circulation efficiency in litterfall for forested wetland communities (> = depressional, • = riverine) in the southeastern United States. See Figure 7.7 for explanation of location numbers.

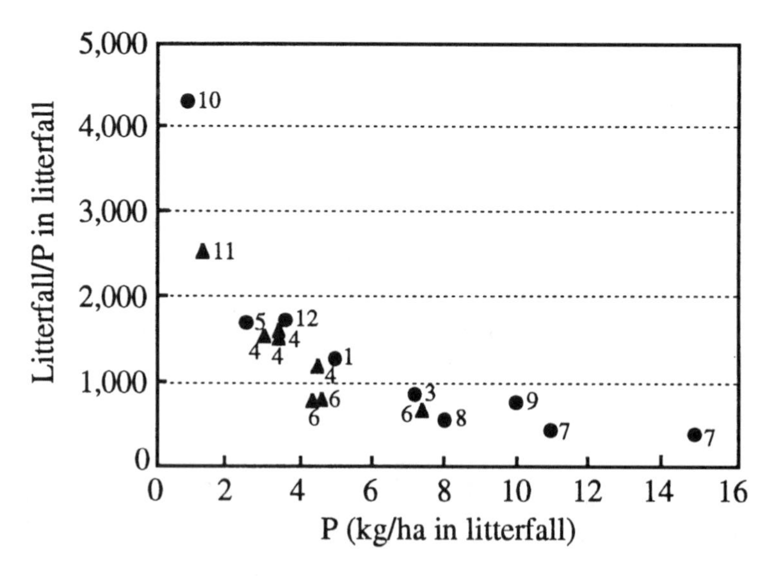

FIGURE 7.9 Phosphorus circulation efficiency in litterfall for forested wetland communities (> = depressional, • = riverine) in the southeastern United States. See Figure 7.7 for explanation of location numbers.

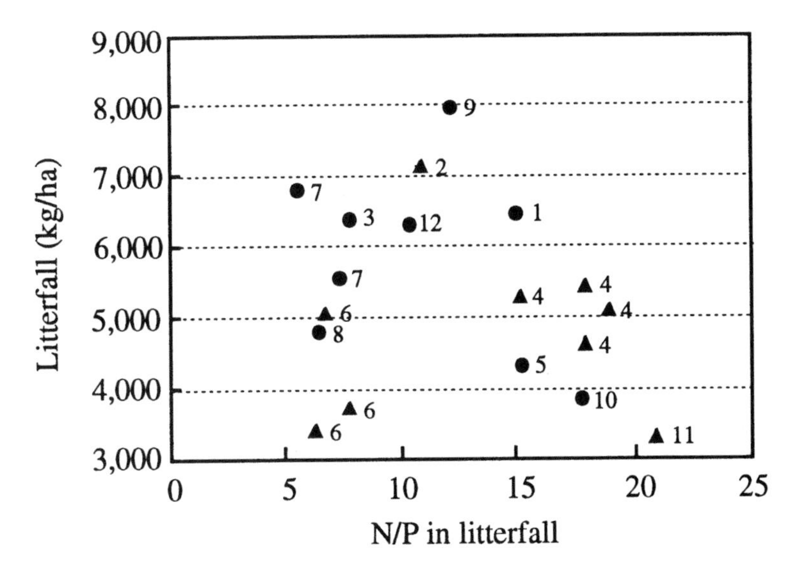

FIGURE 7.10 Relationship between litterfall and N/P ratio in litter for forested wetland communities (> = depressional, • = riverine) in the southeastern United States. See Figure 7.7 for explanation of location numbers.

The short-term influence of harvesting on decomposition rates may be categorized as either (1) no effect or (2) an acceleration (Lockaby et al. 1997a). The degree to which (1) or (2) occurs is a function of the depth of the soil water table. Where water tables are at or near the soil surface following harvest (Lockaby et al. 1994), little change in rates may occur. However, where moist but non-inundated conditions prevail, an acceleration may be noted (Mader et al. 1989). There apparently is less potential for rate reduction to occur as is sometimes the case in upland systems and this is likely a function of lower potential for moisture to limit decomposition (even with increased temperatures) in wetland forests.

The long-term influence of harvesting on detrital processing is a function of the potential for changes to occur in the quantity and quality of detrital inputs as new stands regenerate. Consequently, our ability to anticipate the nature of such changes is severely limited by a lack of data regarding rotation-length vegetation dynamics.

Some documentation exists regarding impacts of natural regeneration methods on biogeochemical source-sink relationships in floodplain forests (Lockaby et al. 1997c). During the first year following harvest, sink behavior was accentuated for K and P, while inorganic N increased in floodwaters. Also, during the initial post-harvest period, suspended solids were generated from harvested sites. However, there are indications from other studies that sedimentation rates increase in harvested areas once complete re-occupancy of growing space is achieved by vegetation (Aust et al. 1997).

CONCLUSIONS

While a considerable amount of information has been compiled regarding the biogeochemistry of forested wetlands in the southeastern United States, a strong need exists for identification of "unifying" concepts which can help predict and explain the variations that are observed in biogeochemical functions among wetland forests. Also, there has been insufficient effort toward integrating biogeochemistry with other processes in wetlands such as hydrology and/or vegetation productivity. While linkages between aboveground vegetation and biogeochemical processes remain insufficiently understood, the greatest information "void" in this regard is certainly belowground. In this vein, one wishes for data sets comparable to the excellent above- and belowground work of Frank Day and his colleagues in the Great Dismal Swamp to be available for many wetland systems.

Similarly, there remain gaps in our knowledge of human influences on forested wetlands. While some work has evaluated forest and agricultural management effects on forested wetland functions, little has been done to examine effects of other human activities such as highway construction which have greater potential to alter hydrology and, consequently, might cause significant modifications in biogeochemical functions.

8 Adaptations of Plants to Flooding and Soil Waterlogging

Martha R. McKevlin, Donal D. Hook, and Amy A. Rozelle

CONTENTS

1-56670-228-3/97/$0.00+$.50
© 1998 by CRC Press LLC

INTRODUCTION

Over the last 30 years, numerous reviews, book chapters, and entire books have been written on the effects of flooding on plant growth, both in terms of those species tolerant and those intolerant of flooding (Gill 1970, Hook et al. 1972, Coutts and Armstrong 1976, Hook and Crawford 1978, Yoshida and Tadano 1978, Drew and Lynch 1980, Crawford 1982, Kozlowski 1982, 1984, Morisset et al. 1982, Ernst 1990, Jackson 1990, Jackson et al. 1991, Rozema and Verkleij 1991, Schaffer et al. 1993). This chapter summarizes information on the ecology and management of plant species common to southern forested wetlands. However, much of the research done in this field has involved commercially important agronomic crops and horticultural species, as well as timber species, and those plant species native to the tropics. When appropriate, information obtained from species other than those native to forested wetlands of the Southeast will be presented to provide examples of common adaptations.

ADAPTATION VS. ACCLIMATION

The terms "adaptation" and "acclimation" are often used interchangeably. The subtle differences in their meanings are important when considering plants in relation to a changing environment. An adaptation is normally a species characteristic resulting from natural selection during evolution. Such a characteristic imparts an assumed increased level of fitness permitting survival, growth, and reproduction, hence allowing the transfer of the characteristic to the next generation. Therefore, adaptations are often viewed on an evolutionary scale. Immediate responses of individual organisms to some recent change in conditions over short time scales are considered acclimations. However, the ability to acclimate may be an adaptation arrived at over many generations. The subject of this chapter deals with plant characteristics affording survival under flooded or hydric conditions. Some of those characteristics may be present continually and could be considered true adaptations. However, many characteristics and responses of flooded plants are induced and represent the ability of an individual or species to acclimate to new conditions. Because it is genetically conferred, the ability to acclimate can also be considered an adaptation.

Tolerant vs. Intolerant

The capacity to survive under reduced or hypoxic (less than optimum oxygen) soil conditions varies greatly, even among broad groups of species considered to be flood-tolerant. In general, broad-leafed trees are more tolerant than conifers (Kozlowski 1986); some of the most tolerant southern forest species include bald-cypress (*Taxodium distichum*), water tupelo (*Nyssa aquatica*), green ash (*Fraxinus pennsylvanica*), black willow (*Salix nigra*), and black-mangrove (*Avicennia germinans*) (Kozlowski 1982, Hook 1984b). Different degrees of tolerance can be arbitrarily divided into four classes: very tolerant, tolerant, moderately tolerant, and intolerant plant species, based upon the ability to survive, grow, and reproduce under flooded conditions. However, tolerance varies among families, individuals, and clones within a species depending upon individual genotype (Hook et al. 1987, Hallgren 1989). Keeley (1979) found very different responses to flooding in black-gum (*Nyssa sylvatica*) seedlings among swamp, floodplain, and upland populations over a one-year period. Thus, placing plants into specific categories such as hydrophyte or "facultative wet" becomes difficult when accounting for within-species variation (Tiner 1991).

Plant responses to flooding do not always provide a benefit to the individual. Some responses are simply the result of hypoxic stress and may be a precursor to impending death. Hence, the responses of intolerant plants will also be discussed in this chapter, especially when the **lack** of some response is associated with flooding intolerance.

Why Must Plants Adapt?

Plant adaptations to flooding or soil waterlogging are found when soil conditions differ from those normally encountered in mesic environments. It is to this below-optimum root environment that wetland plants have adapted in order to survive, grow, and reproduce. Flooding or soil saturation to the surface virtually eliminates gas-filled pores, and gas exchange becomes limited to molecular diffusion of gases in soil water. The supply of oxygen (O_2) from the atmosphere to the soil matrix, therefore, is diminished greatly, and soil fauna and plant roots utilize the remaining O_2 in respiration. Anaerobes replace aerobes, and the soil becomes reduced as alternate electron acceptors (e.g., nitrate [NO_3^-], manganese [Mn^{+4}], and iron [Fe^{+3}]) are utilized in respiration (Ponnamperuma 1984a). These activities result in changes in the bioavailability of soil chemical constituents (Ponnamperuma 1984a) and the accumulation of soil gases (Arshad and Frankenberger 1990). Inherent variability among soils in soil chemical and organic constituents leads to differing degrees of soil reduction; therefore, the soil series is an important factor in determining the nature of the root environment under flooded conditions.

In order to survive, plant roots must not only be able to respire in an O_2–deficient environment, they must also be able to maintain selective nutrient acquisition and tolerate or avoid potentially toxic products of soil reduction and anaerobic respiration (Drew and Lynch 1980). Such responses result from changes both in plant function and structure. However, every adaptive response requires additional energy input, hence trade-offs between survival and production must be considered (Wiedenroth 1993).

FUNCTION

METABOLIC ADAPTATIONS

Two contrasting schools of thought on metabolic adaptations of plants to flooding currently exist in the literature (Keeley 1979, McKee et al. 1989). One theory proposes that ethanol (ETOH) accumulation is greater in the roots of intolerant species whereas ETOH accumulation in tolerant species is avoided by limiting alcoholic fermentation and activating alternate pathways that generate non-toxic organic acids such as malate (Crawford 1967, Crawford and McManmon 1968). The second theory argues that alcoholic fermentation is stimulated in tolerant species upon flooding (Hook et al. 1971, John and Greenway 1976, Keeley 1979, Good and Patrick 1987, Van Toai et al. 1987). Keeley (1979) argues that the acceleration of alcoholic fermentation is imperative in preventing a cessation of glucose catabolism by inhibiting accumulation of pyruvate and the reduced form of nicotinamide adenine dinucleotide (NADH). Support for the second theory is further evidenced by greatly enhanced alcohol dehydrogenase (ADH) activity in flood-tolerant maize seed (Van Toai et al. 1987).

Anaerobic Respiration

Plant roots typically respond to anaerobiosis by increasing the rate of glycolysis and fermentation (Figure 8.1). As O_2 supply to root tissue declines, the electron transport chain ceases to function due to the lack of available O_2 as the terminal electron acceptor. As a direct result, NADH accumulates momentarily. A negative feedback to the Tricarboxylic Acid (TCA) Cycle ensues, suppressing the further metabolism of pyruvate. As pyruvate accumulates along with NADH, glycolysis is impaired, and there is no net production of adenosine triphosphate (ATP), one of the major energy carriers involved in cell physiology. The NADH produced in glycolysis during the oxidation of 3-phosphoglyceraldehyde must be re-oxidized to NAD^+ in order for glycolysis to proceed. This reoxidation occurs by the transfer of electrons and protons to pyruvate. Several biochemical pathways capable of facilitating this reaction exist. The function and importance of each varies by species (Joly 1994). Pyruvate produced by glycolysis is either converted to acetaldehyde and ETOH utilizing pyruvic acid decarboxylase (PDC) and ADH, or to lactate via lactate dehydrogenase (LDH).

Ethanolic fermentation is the most commonly encountered form of fermentation; however, the production of lactic acid and malic acid have been shown to occur in some species under some circumstances (Crawford 1967, Hook and Brown 1967, Hook et al. 1971, Keeley 1979, Smith and ap Rees 1979, Joly and Crawford 1982, DeBell et al. 1984, Roberts et al. 1984, Hook and Denslow 1987, Van Toai et al. 1987, McKee et al. 1989, Muench et al. 1993, Joly 1994). Ethanolic fermentation

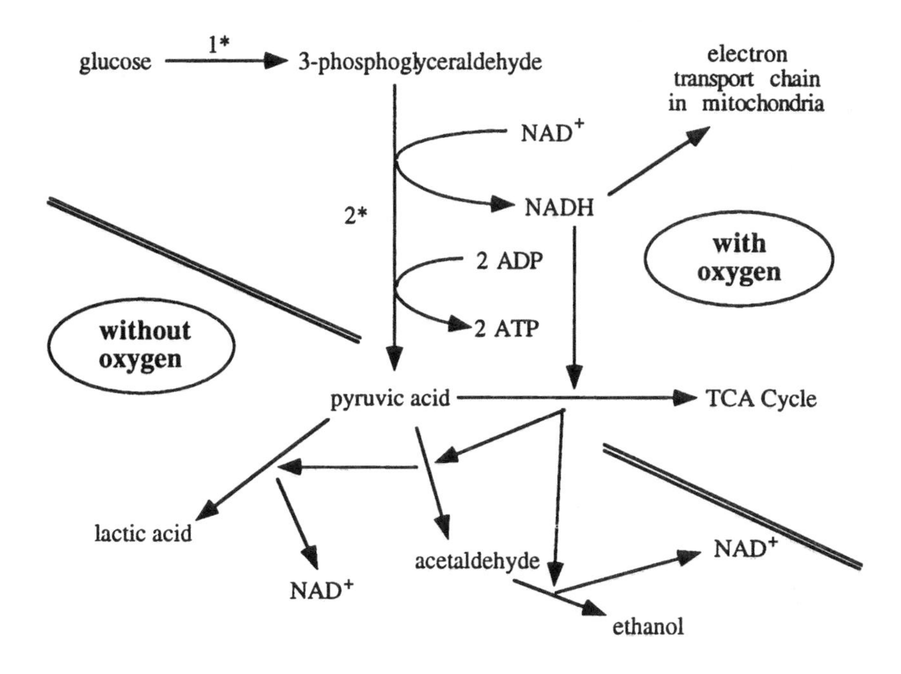

FIGURE 8.1 Simplified diagram of anaerobic respiration pathways including glycolysis and fermentation. In the presence of oxygen, pyruvic acid from glycolysis would move through the Tricarboxylic Acid (TCA) Cycle (also known as Kreb's Cycle), and NADH would be oxidized to NAD^+ via the electron transport chain where oxygen is the terminal electron acceptor. In the absence of oxygen, the electron transport chain and TCA cycle cease to function. NADH is oxidized via fermentation.

is characterized by the decarboxylation of pyruvate to carbon dioxide (CO_2) and acetaldehyde which is further reduced to ETOH, resulting in the oxidation of NADH to NAD^+, which is a necessary cofactor in glycolysis. The potential for ethanolic fermentation is commonly demonstrated in the laboratory by measuring the activity of the enzyme, ADH, which is responsible for converting acetaldehyde to ETOH. In the presence of NAD^+, the glycolytic pathway will proceed and 2 ATP per molecule of glucose (net) are produced. Although this quantity is small in comparison to that produced in the electron transport chain, the reaction is quite rapid, and over time, large quantities can be produced provided that sufficient carbohydrates are available.

Ethanolic fermentation is ubiquitous in plant tissues (Harry and Kimmerer 1991). Alcohol dehydrogenase activity has been observed in the stem cambium, and acetaldehyde and ETOH have been detected in the cambium and transpiration stream of several tree species (Kimmerer and Stringer 1988, Crawford and Finegan 1989, MacDonald and Kimmerer 1991). Alcohol dehydrogenase activity has also been observed in the leaves of eastern cottonwood (*Populus deltoides*) (Kimmerer 1987). Roots generally exhibit some degree of ethanolic fermentation regardless of root environment

(Hook et al. 1971, Wiedenroth 1993), and root tips grown in normoxic conditions may show higher rates of ADH activity than those grown in hypoxic conditions (McKevlin and Hook 1996).

The importance of elevated levels of fermentation to waterlogging tolerance appears to depend on the plant species, and the amount of increase in ADH activity depends on species as well as the length and seasonality of flooding (Rozelle 1997). Most of the recent literature involving ADH proposes that an increase in ADH activity upon flooding is a short-term adaptation (Chow 1984, Van Toai et al. 1987, Kimmerer and Stringer 1988). Also, several isozymes of ADH have been characterized, and subtle differences in their induction mechanisms could partially explain the contradictory information regarding ADH activity and waterlogging tolerance (Harry and Kimmerer 1991, Newman and Van Toai 1991, Muench et al. 1993, Johnson et al. 1994). Expression of certain ADH genes changes during the development of some species. Several genes including ADH show increasing levels of transcripts (mRNA) during anoxia. This is evidenced by studies showing increases in particular mRNAs following anaerobiosis (Harry et al. 1988, Newman and Van Toai 1991, Peschke and Sachs 1994). Current research is attempting to identify DNA sequences important in regulating ADH expression. Alcohol dehydrogenase activity is also sensitive to zinc (Zn) concentration so that Zn deficiency could result in low activity levels as reported under flooded conditions in some studies (Moore and Patrick 1988). This is not surprising since Zn is a cofactor of ADH.

Many studies show increases in ADH activity for both tolerant and intolerant species immediately following inundation, with some decrease after a few months (Keeley 1979, Pedrazzini and McKee 1984). Chow (1984), Van Toai et al. (1987), and numerous others claim that high ADH content and rapid increases of ADH activity following flooding were characteristic patterns of flood-tolerant species. Pezeshki et al. (1993) showed increased levels of ADH activity in hypoxic roots of marshhay cordgrass (*Spartina patens*). Tripepi and Mitchell (1984) showed that both flood-tolerant river birch (*Betula nigra*) and intolerant European white birch (*Betula pendula*) had higher levels of ADH activity after one day of hypoxia. After six days, ADH activity levels in the intolerant birch returned to control levels whereas activity in the tolerant birch remained elevated. Adenylate energy charge and O_2 consumption capacity decreased for both species over the course of the 18-day experiment.

Rice (*Oryza sativa*), considered to be very tolerant of flooding, exhibited increased ADH activity in roots after 21 days under reduced conditions in the laboratory and after four weeks in the field; however, values declined after 12 weeks in the field, suggesting a return to aerobic respiration (Tolley et al. 1986). This suggests that internal transport of oxygen to the roots replaced anaerobic conditions that stimulated ADH activity. Pedrazzini and McKee (1984) showed similar results with rice. Rice roots growing in a flooded environment experienced initial increases in ADH activity, but activity declined to control levels after 47 days. The authors attributed the decline to the development of aerenchyma tissue (see section on Structure). Fermentation of glucose in anaerobic rice roots is primarily by the ethanolic pathway (Mayne and Kende 1986).

Chow (1984) experimented with several cultivars of maize (*Zea mays*) and concluded that rapid increases in ADH activity and high concentrations of ADH in primary root tissue were related to improved waterlogging tolerance. Pearson and Havill (1988) noted that although both tolerant and intolerant herbaceous species showed increased rates of ADH activity with hypoxia, the increase in the intolerant species was much higher than that in the tolerant species. They also suggested that tolerance was related more to the maintenance of high levels of cytochrome oxidase activity than ADH activity. McKee et al. (1989) also observed differential increases in ADH activity with five freshwater marsh species upon flooding, with the least tolerant species showing the highest activity.

Working with flood-tolerant and intolerant cultivars of sweet potato (*Ipomoea batatas*), Chang et al. (1983) observed higher ETOH accumulation in the intolerant cultivar. Both cultivars exhibited an initial period of electrolyte leakage from tissue after onset of anaerobic conditions; however, leakage declined for the tolerant cultivar after three days. Electrolyte leakage was not a direct result of ETOH accumulation.

Crawford (1976) noted that excised roots from several tree species intolerant of flooding showed an increase in the rate of anaerobic CO_2 evolution when flooded and had higher rates of CO_2 evolution when compared to roots from flood-tolerant tree species when flooded. Working with pea (*Pisum sativum*), Barclay and Crawford (1981) showed increased ETOH production with increased length of anaerobic conditions up to four days. High concentrations of ETOH were associated with mortality. Laboratory experiments using several species suggest that high accumulations of ETOH may have detrimental effects on plant tissue, either directly (Crawford and Zochowski 1984, Perata et al. 1986) or indirectly during post-anoxic recovery (Monk et al. 1987). However, ETOH may be leaked from root tissue or transported to the shoot via the transpiration stream (Fulton and Erickson 1964, Bolton and Erickson 1970) and released via lenticels and stomata or metabolized in the shoot (Crawford and Finegan 1989, MacDonald and Kimmerer 1991).

Cytoplasmic Acidosis

Plants lacking flood tolerance may rely on lactic acid production during short periods of hypoxia; however, during extended periods, cytoplasmic acidosis and eventual death may result. Roberts et al. (1984) found that root tips with less than 1% of normal ADH activity were unable to regulate cytoplasmic pH, resulting in cytoplasmic acidosis and rapid cell death. The experimenters hypothesized that low ADH levels retarded ETOH production and that the lactic acid pathway was induced. Johnson et al. (1994) also observed cytoplasmic acidosis and temporary increases in lactic acid production during the first 20 minutes of anoxia in two mutants of maize. ADH_1 and ADH_2 are isozymes of ADH. ADH_1 and ADH_2 are two genes expressed in the roots of maize that encode ADH enzymes. ADH_1 is the primary enzyme involved in ethanolic fermentation in wild-type maize roots and has 10 to 20 times greater specific activity than ADH_2. ADH_{1-} mutants of maize lack active ADH_1 protein and are very sensitive to anoxia but still have ADH_2 activity, whereas ADH_{1+} wild type contain the ADH_1 protein.

In both of the above studies, the ADH_{1-} seedlings failed to stabilize the pH of the cytoplasm, whereas the ADH_{1+} seedlings recovered. The ADH_{1-} seedlings could survive anoxia for a greater time when hypoxically pretreated by gassing with 4% O_2 for 18 hours before sparging with O_2-free N_2 gas. This pretreatment enhanced expression of the ADH_2 gene in the ADH_{1-} mutants. Even though the ADH_2 gene typically makes less of a contribution than the ADH_1 gene, fermentation is enhanced by induction of ADH_2 in the absence of ADH_1 (Johnson et al. 1994). Flood tolerance therefore may be related to the expression of additional ADH isozymes (Kennedy et al. 1987, Harry et al. 1988, Johnson et al. 1994).

Other metabolic pathways aside from fermentation may be involved as a means of transferring the O_2 debt from the roots to the shoot. The use of nitrate as an electron acceptor in plants grown under hypoxic conditions has been documented by several workers (Garcia-Novo and Crawford 1973, Bertani et al. 1982). Production of glycerol, shikimate, the amino acids alanine and glutamate, and other unidentified acidic metabolites has been documented in terrestrial plants (Crawford 1976, DeBell et al. 1984, Jackson and Drew 1984, Jenkin and ap Rees 1986), and the production of formate has been observed in unicellular green algae (Kreuzberg 1984). These products, as well as malate, may also be transported to the shoot via the transpiration stream and metabolized in the above-ground components of the plant.

Post-anoxic injury, which may occur in the tissues of intolerant species when aerobic conditions return, results from the generation of toxic free oxygen radicals (Crawford 1993). An increase in superoxide dismutase (SOD) activity, which protects plants from post-anoxic injury, has been observed in root tissues of some flood-tolerant species (Grosse et al. 1993).

Adenylate Energy Charge

Changes in the biochemical pathways of a plant cell due to hypoxia can be expected to result in a cascade of effects on other physiological processes aside from respiration per se. One of the most important is the reduction in cellular levels of ATP, the ratio of ATP to adenosine diphosphate (ADP), and adenylate energy charge ratio ([ATP] + 0.5 [ADP]/[ATP] + [ADP] + [AMP {adenosine monophosphate}]) of root tissue under hypoxic conditions (Saglio et al. 1980, Mocquot et al. 1981, Tripepi and Mitchell 1984, Drew et al. 1985). Mendelssohn and McKee (1981), working with two *Spartina* species, showed that the adenylate energy charge ratio could be used as an index of anoxia stress. Dependence on anaerobic respiration often leads to a decline in the energy status of the cell resulting from the small quantity of ATP produced per unit of glucose metabolized (Pradet and Raymond 1983). The ensuing energy deficit impacts virtually all energy-requiring processes within the cell.

Nutrient Acquisition

Nutrient acquisition by roots is an active process requiring the input of energy. However, several other factors related to soil conditions are also involved. Inherent fertility of the soil and soil acidity are critical aspects of the overall nutrient supplying ability of a site. Reduced soil conditions, which develop with flooding, alter nutrient availability depending on the mineralization and electrochemistry of the specific nutrient involved (Ponnamperuma 1972, 1984a, Ernst 1990, McKee and McKevlin 1993). Forested wetlands in the Southeast can vary from nutrient-rich alluvial river swamps to nutrient-poor, highly acidic scrub–shrub bays. These conditions will have a strong influence on nutrient acquisition regardless of the energy status of root tissue cells.

The transport of most nutrients across cell membranes from the soil solution of the rhizosphere and apoplastic space into the symplasm requires energy because the movement is usually against a concentration or an electrochemical gradient. Although a complete discussion of nutrient uptake is beyond the scope of this chapter, the energy contained in the phosphate bonds of ATP is used to drive hydrogen (H^+) ion pumps and to activate ion channels which are involved in nutrient transport across membranes (Marschner 1995). Root tissue that experiences a reduction in cellular energy charge due to anaerobiosis may lose the ability to absorb nutrients selectively because the energy required to absorb nutrients and retain them within the symplasm against a concentration gradient is insufficient. Aside from low adenylate energy charge, lipid composition of cellular membranes may be altered under hypoxic conditions (see section on Membrane Biosynthesis). Lipid composition is strongly associated with ion selectivity.

In the waterlogging tolerant species baldcypress and water tupelo, Dickson et al. (1972) noted that concentrations of phosphorus (P), potassium (K), calcium (Ca), and magnesium (Mg) were generally highest in both species when grown on a saturated-aerated soil treatment and lowest when grown on an unsaturated soil. Nitrogen (N) levels were low, and the authors attributed this to soil denitrification. Nutrient uptake by baldcypress grown in saturated-unaerated soil was less restricted than that of tupelo, suggesting a greater ability to maintain adequate nutrient uptake. McKevlin et al. (1995), also working with water tupelo, showed that waterlogging resulted in decreased concentrations of N in all components and K in the foliage and increased concentrations of P in the roots and stem and K in the roots. Water tupelo transports O_2 from the atmosphere to roots and rhizosphere via internal aeration, consequently it can take up nutrients selectively in waterlogged soils.

Morris and Dacey (1984) showed that uptake rate of ammonium (NH_4^+) by smooth cordgrass (*Spartina alterniflora*) was less from an anoxic root environment than from an oxidized root environment. Working with Scotch pine (*Pinus sylvestris*), Heiskanen (1995) noted that under waterlogged conditions, foliar concentration of N was reduced whereas concentrations of P, Ca, sulfur (S), Mn, and sodium (Na) were increased compared to those from seedlings grown in well-drained medium. Smith et al. (1989) found decreased N and P in the shoot and increased N and P in the root of pecan (*Carya illinoensis*), a bottomland species, when grown under reduced soil aeration conditions. Working with two-year-old loblolly pine (*Pinus taeda*), a species common on southeast wet flats, McKee et al. (1984) also found that the N concentration in the foliage was less with flooding, whereas the concentrations of N and P in the roots were higher with flooding. Topa and McLeod (1986b), working with several pine species common to the Southeast, also observed increased P concentrations in the roots and decreased P concentrations in the shoots of seedlings after 8 weeks in anaerobic solution culture. However, in a separate study using the same species, Topa and McLeod (1986b, 1986d) observed decreases in the concentrations of N, P, K, Fe, and Mn in both the roots and the shoots when grown under anaerobic solution culture conditions for 15 and 30 days. Nutrient acquisition response is apparently very sensitive to the conditions under which the test material is grown (i.e., flooded soil vs. anaerobic nutrient solution culture) and to the duration of the experiment. With short-term anoxia, the internal aeration system may not develop sufficiently to influence uptake.

Potentially toxic nutrients which become more available under reduced soil conditions (e.g., Fe, Mn, and S) can accumulate in plant tissue due to passive uptake and transport to the shoot (Jones and Etherington 1970, Jones 1971a, 1971b, 1972, Hook et al. 1983). Many studies have shown the accumulation of both Fe^{+3} and Mn^{+4} in and on roots of flooded plants (Armstrong 1967, Green and Etherington 1977, Good and Patrick 1987, McKevlin et al. 1987a, Gries et al. 1990, Van Diggelen 1991). Concentrations of Fe in and on the roots of water tupelo and swamp tupelo (*Nyssa sylvatica* var. *biflora*) were an order of magnitude greater with flooding; however, levels of Fe were not increased in the foliage (Hook et al. 1983, McKevlin et al. 1995). The ability to immobilize Fe within the root system, preventing its transport to the shoot via the transpiration stream, is associated with internal ventilation of the root system and rhizosphere oxidation (Green and Etherington 1977). However, depending on the location of aerenchyma, antagonistic reactions may occur such as the co-precipitation of other nutrients (e.g., P with Fe in the form of ferric phosphate) (Armstrong and Boatman 1967, Van Diggelen 1991). This scenario is often associated with P deficiency in the shoot (McKevlin et al. 1987b). Accumulation of Mn in and on roots has also been documented with moderately tolerant individuals (Keeley 1979). On the other hand, Fe and Mn does not always accumulate in the leaf tissue of intolerant species even when soil redox potentials and chemical concentrations are consistent with the potential for increased uptake (Etherington 1984).

The addition of nutrients via fertilization can often compensate for the effects of flooding stress (Hook et al. 1983, DeBell et al. 1984). For example, McKee and Wilhite (1986) showed that the application of P on loblolly pine growing on poorly drained soils improved survival and growth after 10 years to a level greater than that achieved by bedding alone. Day (1987a) also showed that applications of N and P compensated for flood stress in red maple (*Acer rubrum*) seedlings in a greenhouse experiment.

Membrane Biosynthesis

The biosynthesis of fatty acids used in the production of cellular membranes is sensitive to O_2 deficiency, and many researchers report effects of hypoxia on membrane properties. Membrane fluidity and permeability can be altered. Aside from impacting the cell plasmalemma, mitochondrial membranes and the membranes of other organelles are also altered by hypoxia, often resulting in changes in ultrastructure and function (Vartapetian et al. 1978, Gindin et al. 1989).

Another process involving altered membrane properties is the transport and unloading of sucrose from phloem. Drew and Lynch (1980) report that phloem transport from shoot to roots is inhibited when the roots are subjected to hypoxic conditions. Schumacher and Smucker (1985) showed that anoxia of the root system of bean (*Phaseolus vulgaris*) resulted in reduced transport of carbon (C) into the roots. Giaquinta et al. (1983) showed that the pathway of sucrose unloading from the phloem in maize roots is via the symplasm, although an apoplastic pathway may exist for some species. Saglio (1985) also showed that phloem unloading in maize root tips was inhibited by root anoxia and suggested that the inhibition was associated with modifications of the endoplasmic reticulum involved in plasmodesmata structure as opposed to effects of anoxia on respiration and cellular energy status. Regardless of the mechanism, a reduction in the transport of C from the shoot to the roots could lead to a decline in the availability of substrate for ethanolic fermentation and a reduction in anaerobic respiration (Saglio and Pradet 1980). However, Atwell and Greenway (1987), working with rice seedlings, observed that endogenous sugar levels were adequate for coleoptile growth in anaerobic solution culture.

Stomatal Response, Gas Exchange, and Moisture Stress

Hypoxic conditions in the root environment are responsible for other modifications in both the C and water budgets of the plant. Reduced soil conditions associated with flooding have been shown to decrease stomatal conductance, transpiration, and net photosynthesis in a variety of both flood-tolerant and intolerant species (Blake and Reid 1981, Pezeshki and Chambers 1985a, 1985b, 1986, Pezeshki et al. 1986, 1993, Smith and Ager 1988, Van der Moezel et al. 1989, Smith et al. 1990, Kludze et al. 1994), and in some cases, xylem pressure potential is also reduced (Coutts 1981, Syvertsen et al. 1983, Zaerr 1983, Grossenickle 1987, Harrington 1987, Osonubi and Osundina 1989, Lewty 1990).

Davison and Tay (1985), working with eucalyptus (*Eucalyptus marginata*) seedlings, observed an increase in the resistance to water movement through the seedlings due to the blockage of xylem vessels by tyloses. The transpiration rate of the seedlings was negatively correlated with the proportion of occluded vessels. Declines in the xylem pressure potential in sunflower (*Helianthus annuus*) were also believed to be related to occlusion of the xylem and subsequent reductions in hydraulic conductivity (Everard and Drew 1989).

Stomatal closure in relation to anoxia has been observed in at least 58 species (Sojka 1992). In many cases, flood-tolerant species return to near pre-flood levels of stomatal conductance and photosynthesis (Kozlowski 1982, Tang and Kozlowski 1984a, Pezeshki et al. 1986, 1996, Sojka 1992, Pezeshki 1993). These positive responses may result from the development of morphological adaptations such as adventitious roots (Tang and Kozlowski 1984a, Van der Moezel et al. 1989, Topa and Cheeseman 1992a). It appears that in species that produce adventitious roots, stomatal closure and severe shoot desiccation are prevented (Sena Gomes and Kozlowski 1980a, 1980b, 1980c).

On the other hand, highly flood-tolerant species such as mangroves (*Avicennia germinans, Laguncularia racemosa,* and *Rhizophora mangle*) exhibited no reductions in leaf conductance or net carbon assimilation per unit of leaf area when subjected to low soil redox potentials (Pezeshki et al. 1990). A reduction in total leaf area per plant was observed, however, which could result in an overall reduction in net C gain per plant. Harrington (1987) also noted the absence of stomatal closure during the flooding of red alder (*Alnus rubra*) and black cottonwood (*Populus trichocarpa*).

Marshhay cordgrass exhibited reduced stomatal conductance and net C assimilation in response to hypoxia although chlorophyll content was not affected (Pezeshki et al. 1993). On the other hand, chlorophyll content has been shown to decrease with flooding stress for some species (e.g., rubber tree [*Hevea brasiliensis*]) (Sena Gomes and Kozlowski 1988), loblolly pine (McKevlin 1984), and water tupelo (McKevlin et al. 1995).

Factors affecting stomatal closure include sensitivity to CO_2 (Bradford 1983a), available N (Liu and Dickman 1992), salinity (Pezeshki et al. 1986), soil redox potential (Pezeshki et al. 1993, Kludze et al. 1994), species differences (Pezeshki and Chambers 1986, Pezeshki et al. 1993), age of plants (Sojka 1992, Kludze et al. 1994), leaf diffusional resistance (Sena Gomes and Kozlowski 1980a, 1980b, 1980c), water temperature (McLeod et al. 1986, Foster 1992), light (Foster 1992), and amount and duration of flooding (Liu and Dickman 1993, Sojka 1992).

Abscisic acid (ABA) produced in the roots is the primary stimulator of stomatal closure under O_2 stress (Drew and Lynch 1980, Sojka 1992, Van Toai 1993), while cytokinins prevent stomatal closure (Bradford 1983b, Pezeshki et al. 1993). The interaction between the two hormones regulates stomatal behavior. Lower leaf water potentials appear to increase ABA movement to the shoot, and N deficiency can stimulate the ABA response (Bradford 1983a). Species response to supplemental N differs greatly (Liu and Dickman 1993, Sojka 1992).

Stomatal closure after inundation impacts photosynthesis by limiting gas exchange (Pezeshki et al. 1993, Sojka 1992); however, in some species, the effects on photosynthesis were separate from those on stomatal aperture, suggesting non-stomatal control of photosynthetic rate (Bradford 1983a, Pezeshki and Chambers 1985a, 1985b, Foster 1992). Flooding-induced changes in levels of photosynthetic enzymes such as rubisco could be involved (Pezeshki et al. 1993).

Other effects on the above-ground components of flooded plants include reductions in leaf expansion. Flooding has been shown to reduce leaf size in cottonwood (Liu and Dickman 1993) and in oaks (Angelov et al. 1996). However, flooded swamp tupelo had normal leaf size (Anglelov et al. 1996). Smit et al. (1989), in an earlier study with hybrid cottonwood, found reduced leaf expansion and showed that the reduction was due to decreases in cell wall extensibility as opposed to decreases in bulk leaf water potential.

Plant Growth Regulators

Processes involved in protein synthesis, the transcription of DNA, and translation of RNA rely heavily on membrane structures and also require an input of energy. The possible impacts of flooding on cellular function become enormous when viewed in this light. The production of plant growth regulators such as ethylene and auxins constitutes several of those processes known to be altered by soil flooding.

The role of ethylene as a phytohormone is well documented. Its effects include promotion and inhibition of root extension, leaf chlorosis, leaf epinasty, leaf senescence and abscission, tissue hypertrophy, formation of aerenchyma tissue, and production of adventitious roots (Blake and Reid 1981, Tang and Kozlowski 1984b, Yamamoto and Kozlowski 1987a, 1987c, 1987d, Sena Gomes and Kozlowski 1988, Topa and McLeod 1988, Gindin et al. 1989, Jackson 1990, Voesenek et al. 1993). Ethylene seems to be one of the most important hormones involved in acclimation to flooding (Blom et al. 1994). In many species, both leaves and roots show greater ethylene accumulation or production (or both) under anoxic conditions than aerated conditions.

Ethylene synthesis increases upon inundation as a direct result of low O_2 concentration (Tang and Kozlowski 1984b). The precursor to ethylene is amino cyclopropane-1-carboxylic acid (ACC). Its production is stimulated under anaerobic conditions, and it is easily transported in xylem sap (Yamamoto and Kozlowski 1987a). The conversion of ACC to ethylene requires O_2. When the O_2 concentration within the tissue is sufficiently high, ACC is converted to ethylene (Yang and Hoffman 1984). Flooding has been shown to stimulate ACC production in the roots and ethylene production in the stem of woody seedlings (Yamamoto and Kozlowski 1987a, 1987b). Ethylene may also become entrapped within the tissues of flooded plants (Pezeshki et al. 1993, Voesenek et al. 1993). Water surrounding the plant inhibits the diffusion of ethylene out of the tissue. Entrapment of ethylene is influenced by temperature, root circumference, and whether the surrounding water is stagnant or moving (Jackson 1990).

Ethylene concentration is very important in its effect upon plants. A high concentration of ethylene can produce an opposite effect than that of a lower concentration. For example, high concentrations in a flood-intolerant species of dock (*Rumex acetosa*) caused growth reduction in petioles (Voesenek et al. 1993, Blom et al. 1994); growth reduction was prevented when ethylene (5 nL/ml) was applied in combination with subambient O_2 partial pressures (3 kPa) and higher CO_2 partial pressures (5 kPa). Thus, the response of petioles of *R. acetosa* to ethylene depends on the specific balance of the gaseous products of respiration and photosynthesis under flooded conditions. Also in this study, a flood-tolerant species of dock (*R. palustris)* showed increased leaf elongation. This and other studies show differences in species responsiveness to ethylene (Pezeshki et al. 1993). Plant species that are more flood tolerant may entrap less ethylene and, therefore, accumulate the optimum concentration for increased growth, while some intolerant plants may entrap larger amounts of ethylene in roots, resulting in an inhibitory effect (Jackson 1985). For example, in rice, ethylene accelerates root extension (Drew and Lynch 1980, Jackson 1985, Voesenek et al. 1993).

The involvement of ethylene in the formation of aerenchyma tissue has been well established (Drew et al. 1981, Kawase 1981a, Jackson 1985, 1990, Kozlowski 1986) and contradicts an older hypothesis suggesting that the breakdown and lysis of cells in aerenchyma development was caused strictly by anoxia. In order for ethylene to promote breakdown of cortical cells resulting in development of air space, small internal O_2 partial pressures are necessary (Jackson 1985, Topa and McLeod 1988). Aerenchyma development can also be stimulated by applying ethylene exogenously. One ppm ethylene caused aerenchyma development in the stem cortex of sunflower, tomato (*Lycopersicon esculentum)*, and bean when applied to part of the stem and root system (Kawase 1981a). Abscisic acid, a known inhibitor of ethylene, has been shown to reduce gas space formation in maize (Jackson 1985).

Ethylene's effects on aerenchyma development may be through increases in cellulase activity which can also be observed soon after flooding (Kawase 1979, 1981a, 1981b). Cellulase is an enzyme that hydrolyzes cellulose, which results in the degrading of cell walls and eventual dissolution of parenchyma tissue (Salisbury and Ross 1978). The formation of adventitious roots and hypertrophy of stem and lenticels has also been attributed to an increase in cellulase activity (Kawase 1981b, Jackson 1985, Kozlowski 1986, Topa and McLeod 1988).

Auxins may also be involved in morphological responses to flooding (Wample and Reid 1979). It has been suggested that ethylene production causes an accumulation of indole-acetic acid (IAA) (Jackson 1985). On the other hand, flooding may interrupt the transport of auxin down the shoot causing accumulations at the shoot base. Auxin accumulations in the shoot may stimulate ethylene biosynthesis and eventual adventitious root formation (Haissig 1974, Jackson 1990). Regardless of the mechanism, these two hormones appear to work together in producing stem and lenticel hypertrophy and other morphological responses. Pezeshki et al. (1993) showed that the formation of large intercellular spaces within stem tissue and lacunae within roots was the result of the combined actions of ethylene, auxin, and cellulase.

STRUCTURE

The initial stimulation of ADH activity may provide a temporary means to support essential metabolic functions. The anaerobic pathway is much less efficient than the aerobic pathway in terms of energy production (39 moles ATP/mole hexose vs. 3 moles ATP/mole hexose) but may provide an energy source while anatomical changes (i.e., aerenchyma development) are occurring, which will allow the eventual return to near normal aerobic activities. Thus, long-term flood tolerance may be related to the plant's ability to quickly make these anatomical changes enabling aerobic metabolism.

Plants may acclimate or adapt to wetland conditions on several temporal scales. Changes in physiological processes can be immediate upon the onset of an O_2 deficiency to the roots. Changes in root physiology can result in changes in morphology and anatomy over time. Individuals and species which undergo these changes tend to be selected for on frequently flooded sites because of the benefits associated with those changes, specifically, the development of a continuous internal ventilation system that allows for the transport of gases from the shoot to the roots and vice versa. Plants that have evolved to live in wetland conditions possess characteristics that allow them to tolerate the initial problems associated with O_2 deficiency and avoid hypoxic conditions by changes in their anatomy and morphology. They also possess characteristics that aid in reproduction under periodically flooded conditions.

Plants that do not tolerate flooded conditions but acclimate to short-term anoxia may also exhibit a specific set of characteristics including reduced shoot and leaf extension rate, yellowing of foliage, wilting, leaf epinasty, leaf abscission, and cytoplasmic acidosis (Drew and Lynch 1980, Etherington 1984, Roberts et al. 1984,

Terazawa and Kikuzawa 1994). Prolonged waterlogging or flooding usually results in death in species that do not possess adaptations (Terazawa and Kikuzawa 1994).

Morphological Adaptations

Biomass allocation patterns are often indicative of species' tolerance to waterlogging or lack thereof. Tolerant species or individuals tend to allocate less biomass to roots under waterlogged conditions when compared to other components (Keeley 1979). The duration and periodicity of the flooding regime can also result in different biomass allocation patterns. Working with red maple, Day (1987a) observed that continuously flooded treatments resulted in the greatest decline in root-to-shoot ratios compared to periodically flooded and drained treatments. Megonigal and Day (1992) also showed that periodically flooded baldcypress allocated more C to roots than did continuously flooded trees, but total productivity did not differ between the two treatments after three years.

Rapid Early Shoot Growth

Rapid stem elongation is an important attribute of flood-tolerant species, allowing growth of the shoot above the water surface. Several herbaceous and woody species are known to exhibit this characteristic. Raskin and Kende (1984) and Stunzi and Kende (1989) observed rapid internodal elongation in deep-water rice upon submergence. The wetland herbaceous vine, climbing hempweed (*Mikania scandens*), also exhibits rapid stem elongation under flooded conditions compared to drained conditions (Moon et al. 1993). McKevlin et al. (1995) noted similar responses with the swamp species, water tupelo. Blom et al. (1990) and Laan and Blom (1990), working with several species of *Rumex*, suggested that the ability to elongate petioles and stems was an important aspect of flooding tolerance in this genus.

Stem and Lenticel Hypertrophy

Hypertrophy is the enlargement of an organ without an increase in the number of constituent cells. Stem hypertrophy is more commonly known as butt swell (or buttressing) and is characterized by an increase in the diameter of the basal portion of the stem (Figure 8.2). Stem hypertrophy has been observed as a response to flooding in many herbaceous and woody species including tomato, sunflower, corn, Mediterranean pine (*Pinus halepensis*), bur oak (*Quercus macrocarpa*), baldcypress, green ash, swamp tupelo, water tupelo, and loblolly pine (Hook 1984a, Tang and Kozlowski 1984b, Yamamoto and Kozlowski 1987c, McKevlin et al. 1987a).

The apparent role of hypertrophy is to increase air space which allows increased movement of gases within the plant. Increased air space has been correlated with low bulk density. In seedlings, stem hypertrophy is characterized by an increase in the intercellular space of the cortex and phloem and also in the phelloderm after the initiation of secondary growth (McKevlin et al. 1987a). Yamamoto and Kozlowski (1987c) report that stems of Mediterranean pine seedlings had more parenchymatous tissue consisting of more xylem rays, larger ray cells, many epithelial cells, and more phloem and parenchyma cells resulting in a lower bulk density than nonflooded

FIGURE 8.2 Swamp tupelo tree with stem hypertrophy (butt swell), adventitious water roots, and hummock around the base of the tree.

seedlings. However, production of false growth rings has been reported in water tupelo by Hook et al. (1973) and in loblolly pine by McKevlin et al. (1987a) as a contributor to stem hypertrophy.

Several studies confirm the regulatory role of ethylene in altering growth and stem anatomy of woody plants (Wample and Reid 1979, Jackson 1985, Yamamoto and Kozlowski 1987a, 1987b, 1987c, 1987d). Tang and Kozlowski (1984a) found

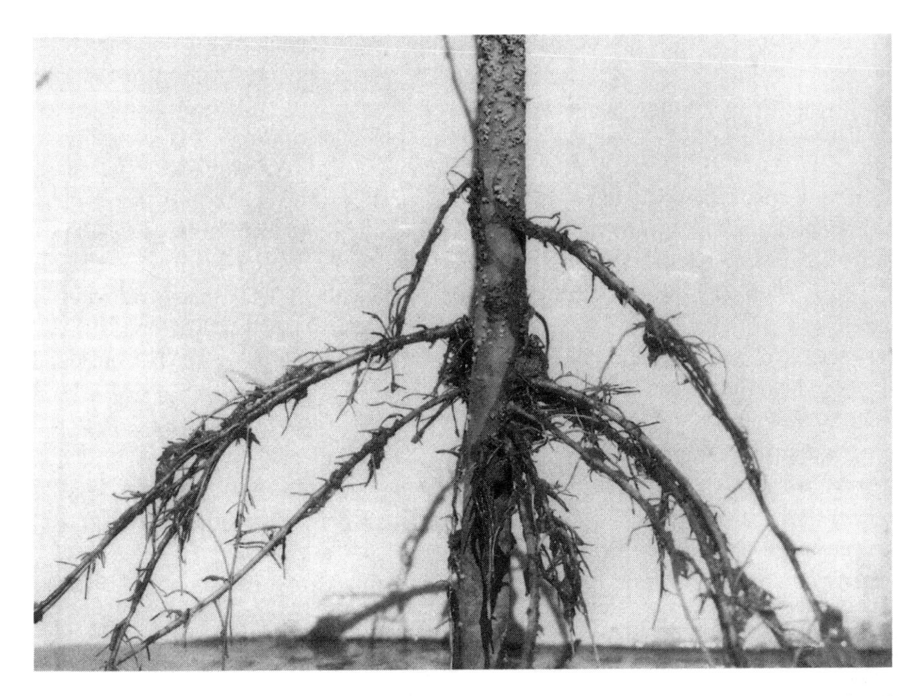

FIGURE 8.3 Swamp tupelo seedling with prolific lenticel and adventitious water root development on flooded section of stem.

that ethylene production was higher in stems with well-developed hypertrophy than those without stem hypertrophy in flooded green ash seedlings. Stem hypertrophy has been induced by exogenous ethylene application (Blake and Reid 1981).

Lenticels are small pores located on the stems of woody plant species and are the major points of atmospheric O_2 entry. Lenticels aid in ventilation by providing continuity of intercellular space in the lenticel tissues with that in the cortex and phloem (Hook and Scholtens 1978, Kozlowski 1982, Kozlowski 1986, Topa and McLeod 1986c, Hook and McKevlin 1988). Lenticel hypertrophy (Figure 8.3) has long been associated with flooding as a response which acts to increase internal gas transport from the stem to the roots (Hook et al. 1970a). Flood tolerance in many species is correlated with the plant's ability to either form lenticels or produce hypertrophied lenticels on submerged portions of the stem. Lenticel hypertrophy is often associated with stem hypertrophy.

Lenticels also serve as excretory sites for toxic substances such as ETOH and acetaldehyde (Kozlowski 1986, Larson et al. 1991). Small quantities of ETOH have been observed emanating from lenticels of flooded lodgepole pine (*Pinus contorta*) roots (Crawford and Finegan 1989). The gaseous release of ETOH was suggested by the authors as a sensitive indicator of root hypoxia but limited as a means of ETOH removal from the plant.

Production of lenticels is a result of increased phellogen activity, cell enlargement, and cell elongation (Kozlowski 1986, Larson et al. 1991). Duration of flooding apparently does not affect the number of lenticels formed but does affect the size.

Enlarged lenticels allow greater influx of O_2 into the stem and release of toxins from the stem because of the continuity between large, intercellular spaces among the complementary cells and breaks in the closing layers of the lenticel structure (Hook and Brown 1970, Philipson and Coutts 1980, McKevlin et al. 1987a). According to Larson et al. (1991), nonhypertrophied lenticels become hypertrophied when periclinal divisions take place to produce radially arranged rows of cells in the phellem and phelloderm. These formative processes require species-specific soil temperature and moisture conditions (Kozlowski 1986).

In some species, the formation of hypertrophied lenticels under anoxic conditions appears to be induced by ethylene (Kozlowski 1982, Tang and Kozlowski 1982, Tang and Kozlowski 1984a, 1984b). Floodwater O_2 concentrations have been found to stimulate ethylene production in shoot tissues above the submerged root (Jackson et al. 1978, Larson et al. 1993). Ethylene production is also temperature dependent, with the most favorable temperature at about 30°C (86°F) (Yang 1980). For mango (*Mangifera indica*) and green ash, increased phellogen activity was observed at temperatures around the 30°C (86°F) range, further suggesting a relationship between phellogen activity and ethylene (Borger and Kozlowski 1972).

In a comparison of waterlogging tolerant slash pine (*Pinus elliottii*) and intolerant Honduras Caribbean pine (*Pinus caribaea* var. *hondurensis*), Lewty (1990) noted that slash pine exhibited stem and lenticel hypertrophy in response to soil waterlogging whereas the Caribbean pine did not. Topa and McLeod (1986a), working with three additional species of *Pinus*, noted lenticel hypertrophy on stems of pond pine (*P. serotina*) and wet- and dry-site loblolly pine, but none on stems of sand pine (*P. clausa*). However, lenticels are often produced on the roots of tree species that are considered to be intolerant of flooding (Aronen and Haggman 1994).

Other means of gas exchange both with the surrounding atmosphere and within the plant have also been reported. Herbaceous species often exhibit an increase in the number of stomata on the stem surface as opposed to lenticels as a means of gas exchange (Moon et al. 1993). The wetland tree species water tupelo and green ash have a highly permeable cambium allowing the movement of gases from the phloem to the xylem and vice versa (Hook and Brown 1972).

Buttress Roots, Cypress Knees, and Pneumatophores

Other commonly observed morphological features of woody species adapted to life in a frequently flooded environment include buttress roots, cypress knees, and pneumatophores. Buttress roots appear as fluted projections at the base of mature trunks extending several feet from the trunk outward and down into the soil. They have the general appearance of the gothic architectural structures known as "flying buttresses" common on many European cathedrals. Often, root systems of many species common to wetlands are shallow because of the high water tables and frequent flooding. It is believed that buttress roots provide additional support (Sutton and Tinus 1983). The development of buttress roots is probably related to differential growth activity of the stem cambium of the lower trunk and upper woody roots in that it resembles a form of hypertrophy. Ethylene and auxin are probably involved, but there is little evidence other than anecdotal to support this speculation.

Cypress knees are knobby, vertical projections on the upper surfaces of roots protruding above the soil and flood water surface and also result from localized cambial activity. Current thinking suggests that the production of cypress knees has no specific adaptive significance and that knees develop as a consequence of fluctuating water tables and the subsequent effects on ACC production and conversion to ethylene in conjunction with auxin in a manner similar to stem hypertrophy (Yamamoto 1992). Hook (1984a) suggests that knee roots probably are beneficial if they occur but are not necessary for survival.

Mangroves are one of the best examples of a tree species that can tolerate permanently flooded soils. All species of mangroves have a highly specialized root system designed for support made up of laterally spreading cable roots and vertically descending anchor roots. Species differ somewhat as to the exact nature of the system used to ventilate the below-ground portion of the root system, but most rely on some form of above-ground root such as pneumatophores or stilt roots (Stewart and Popp 1987). Pneumatophores are negatively geotropic roots covered in lenticels with well-developed aerenchyma tissue. Stilt roots are aerial roots also rich in lenticels (Sutton and Tinus 1983).

Other Structures

In some wetland species, it is the ability of the rhizome and stolons to tolerate hypoxic conditions that imparts waterlogging tolerance to the species (Braendle and Crawford 1987). In bulrush (*Scripus lacustris*) and cattail (*Typha* spp.), a true tolerance to anoxia exists, as opposed to an avoidance of anoxia by the development of an internal ventilation system such as with common rush (*Juncus* spp.) (Crawford 1993). Rhizomes of these truly tolerant species can over-winter in a dormant state and produce new shoot growth in an entirely O_2-deficient environment, although these eventually extend above the water or mud surface into an aerobic environment, so that O_2 may be transported to the rhizome to support aerobic metabolism.

Root System

Low redox potentials associated with waterlogged soil conditions have been shown to inhibit root elongation, and the degree of inhibition differs according to the apparent tolerance of the species to waterlogging (Coutts and Philipson 1978a, Pezeshki 1991). Most plants that have developed under drained conditions experience some degree of root dieback upon flooding, regardless of their ability to tolerate flooded conditions (Hook and Brown 1973, Sena Gomes and Kozlowski 1980c, Syvertsen et al. 1983, Topa and McLeod 1986a). Lateral roots may either die back to the tap root or to the primary lateral from which they originated. Thick, succulent, relatively unbranched roots may be initiated from the point of dieback in root systems undergoing acclimation to flooding. These roots are often referred to as soil water roots (Hook et al. 1970b). Hook et al. (1971) noted that in swamp tupelo, these roots had less suberization in the epidermis and that casparian strips were less evident.

Adventitious roots are often produced in response to flooding after initial root dieback. Adventitious roots are formed on the stem above the level of the soil, usually at the surface of the flood water (Hook 1984a). Adventitious roots are not necessary for survival but significantly increase the capacity to tolerate flooded conditions (Kozlowski 1986). A correlation exists between adventitious root formation and stomatal reopening (Sena Gomes and Kozlowski 1980a, 1980b, Kozlowski 1982). This adaptation is more prevalent in flood-tolerant species such as water tupelo, baldcypress, green ash, and pond pine (Hook 1984a). Also, herbaceous species such as tomato, maize, and sunflower produce adventitious roots under flooded conditions (Hook 1984a).

Adventitious roots have a greater capacity for oxidizing the rhizosphere (Blom et al. 1994, Hook 1984a). Reoxidation of soil around these roots stimulates bacterial nitrification which generates available NO_3^- for the plant and may result in conversion of divalent metals like Mn to a less soluble, higher valency state (Osundina and Osonubi 1987). This conversion may allow greater selectivity of nutrient uptake and prevent toxic accumulation of metals in various plant parts. Osundina and Osonubi (1987) suggested that plants which did not produce or had limited production of adventitious roots had high concentrations of Mn in leaves causing increased leaf abscission.

Newly formed roots are often more succulent and contain more aerenchymous tissue than roots of the original root system (Hook et al. 1971, Keeley 1979, Rogers and West 1993, Blom et al. 1994). This change in root structure may be short-term. Roots that appear succulent after flooding in short-term experiments often appear much like those of drained plants in long-term studies (Keeley 1979).

Ethylene has also been shown to induce formation of adventitious roots upon flooding (Tang and Kozlowski 1982, 1984b, Blom et al. 1994). Increases in ethylene concentration may be due to increased ethylene production upon inundation and/or reduction of the diffusion of ethylene out of tissue because of entrapment by water (Pezeshki et al. 1993). Ethylene may not be the only substance involved. It has been suggested that production of adventitious roots involves a balance of substances including carbohydrates, nitrogenous compounds, auxins, and cofactors (Tang and Kozlowski 1984b, Kozlowski 1986). It is possible that failure of some species to form adventitious roots may result from a deficiency of certain enzyme activators (Kozlowski 1986).

Anatomical Adaptations

The formation of large cavities or lacunae has been observed in the roots of many herbaceous and woody species when grown under flooded conditions (Figure 8.4) (Chirkova and Soldatenkov 1965, Green and Etherington 1977, Topa and McLeod 1986c, 1988, Justin and Armstrong 1987, McKevlin et al. 1987a, Pezeshki et al. 1993, Blom et al. 1994, Kludze et al. 1994). These lacunae are in most cases longitudinally continuous with the tap root and stem intercellular spaces and have been given the general name of aerenchyma tissue. This adaptation is characteristic of the more flood-tolerant species, with the time frame for development of aerenchyma varying among species (Hook and Brown 1973, Pezeshki et al. 1993, Rogers and West 1993). In baldcypress, aerenchyma formation occurs in both flooded conditions and to a lesser extent in aerated conditions, suggesting that this is a genetic trait of this species (i.e., an adaptation) (Kludze et al. 1994). Aerenchyma tissue has been shown to form in either the root cortex or within the stele. Many herbaceous species (rice, sedges [*Carex* spp.], rushes, dock, and cordgrass to name a few) (Justin and Armstrong 1987), and at least one woody species (water tupelo) (McKevlin and Hook, unpublished), develop aerenchyma within the cortex. However, most conifers develop aerenchyma within the stele (Coutts and Philipson 1978b, Philipson and Coutts 1978, Topa and McLeod 1986c, McKevlin et al. 1987a).

The development of aerenchyma tissue is induced by flooding or reduced soil conditions. As with many of the other morphological changes, increases in ethylene concentration and cellulase activity are believed to be responsible for aerenchyma formation (Kawase 1981a 1981b, Jackson 1985, Topa and McLeod 1988, Blom et al. 1994). Aerenchyma formation that is a result of cell lysis is termed lysigenous, and that which results from the separation of cells is termed schizogenous.

Collapse of stems and roots with extensive aerenchyma tissue is prevented by the production of a cylinder of stone cells and sclerenchyma fibers which provide support; these structures have been observed in palm (*Astrocaryum jauari*) (Schluter et al. 1993), swamp tupelo (Hook et al. 1970b), and water tupelo (McKevlin and Hook, unpublished).

Age of the plant at the onset of flooding is important to the plant's ability to carry out developmental changes (i.e., aerenchyma formation) which will increase O_2 transport to the roots, thus increasing long-term tolerance (Keeley 1979, Tang and Kozlowski 1983, Melick 1990, Kludze et al. 1994). This is consistent with the observation that mature, flood-tolerant trees suddenly exposed to flooded conditions often have a lower survival rate than the seedlings (Keeley 1979).

Aside from visual observations of aerenchyma tissue, measuring root porosity is another means to estimate a root's ability to facilitate the transport of gases. Armstrong (1972), in a series of laboratory experiments, showed that root porosity was strongly associated with radial O_2 loss. Roots of waterlogging tolerant species grown under hypoxic conditions have increased porosity, which results from cubic and radial arrangement of cells as opposed to hexagonal arrangement in combination with aerenchyma production (Justin and Armstrong 1987). Increased root porosity along with aerenchyma production has also been associated with ethylene accumulation (Pezeshki et al. 1993). Baldcypress, a well-known inhabitant of southern swamps, exhibits an increase in root porosity and radial O_2 loss when grown under reducing conditions (Kludze et al. 1994). Fisher and Stone (1990) noted that green wood from sinker and tap roots of slash pine had high air contents and was capable of supporting mass air flow through the woody root tissue. Oxygen transport in woody roots of two conifers was also reported by Philipson and Coutts (1980).

Benefits of Morphological and Anatomical Responses

Internal Ventilation

In conjunction with the change in root system morphology and anatomy over the long term is an increase in internal O_2 transport to roots and a decrease in alcoholic fermentation (Keeley 1979, Drew and Lynch 1980). This transformation suggests that anatomical changes such as aerenchyma formation, adventitious rooting, and hypertrophied lenticel formation eventually result in greater internal ventilation

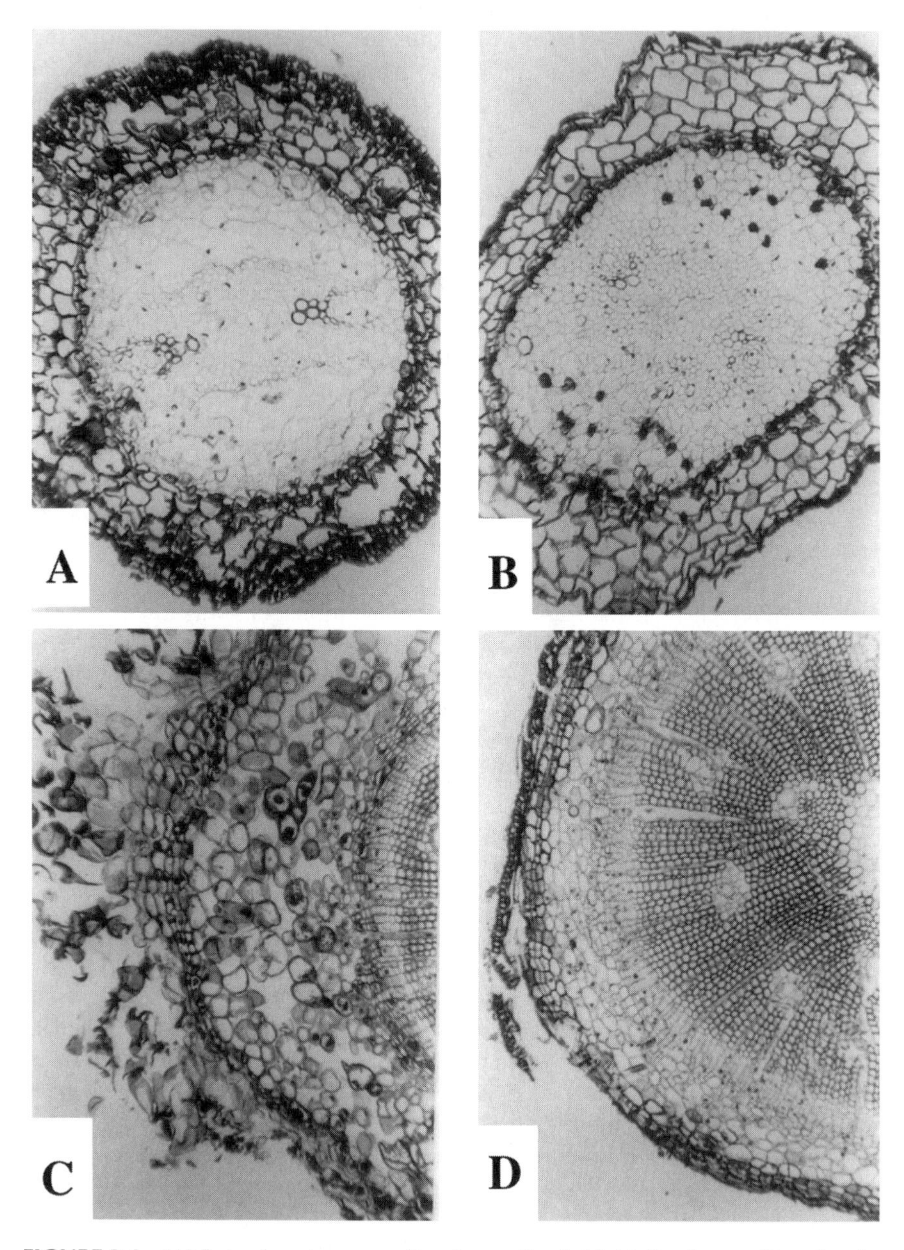

FIGURE 8.4 (A) Lateral root cross section from a flooded loblolly pine seedling showing large lacunae within the stele on either side of the xylem poles; (B) Lateral root cross section from a non-flooded loblolly pine seedling showing lack of aerenchyma tissue within the stele between the xylem poles; (C) Cross section of stem from a flooded loblolly pine seedling showing hypertrophied lenticel and large areas of intercellular space within the phloem; (D) Cross section of stem from a non-flooded loblolly pine seedling showing minimal intercellular space within the phloem.

leading to increased capacity for aerobic respiration. In an early study, Barber et al. (1962), using radio-labeled O_2, showed gaseous diffusion from the shoot to the root in rice and barley. It is generally accepted that the development of aerenchyma in roots, increased intercellular space in stem sections, and hypertrophied lenticels act in concert to provide a pathway for gaseous flux between the aerated shoot and the hypoxic root system. Moon et al. (1993) noted that climbing hempweed produced extensive zones of aerenchyma in the cortex of the lower stem and main root upon flooding and that concentrations of ETOH and malate were undetectable. The authors suggested that aerenchyma tissue allowed sufficient gas exchange to support aerobic respiration. In marshhay cordgrass, flooding resulted in decreased root specific gravity (i.e., increased porosity) associated with a decline in ADH activity after an initial increase in ADH activity immediately upon flooding (Burdick and Mendelssohn 1990). The authors attributed the decline in ADH activity to the increased root aeration provided by aerenchyma. Hook and McKevlin (1988) were able to detect changes in the partial pressure of O_2 within aerenchymatous roots of loblolly pine depending on the atmosphere surrounding the shoot.

Rhizosphere Oxidation

The increase in O_2 transport to the roots results in an increase in radial O_2 loss from roots into the surrounding rhizosphere (Figure 8.5) (Armstrong 1968, Philipson and Coutts 1978, Good and Patrick 1987, Kludze et al. 1994). The soil water roots of the flood-tolerant swamp tupelo have been shown to oxidize their rhizosphere, whereas roots from non-flooded seedlings were not capable of doing so (Hook et al. 1971). Armstrong and Read (1972) observed O_2 flux from the roots of conifer seedlings (especially pine) while maintained in an anaerobic solution. The flux rate varied depending on the atmosphere surrounding the shoot and the rate of root respiration as influenced by temperature of the solution culture. Topa and McLeod (1986c) showed rhizosphere oxidation in pine but only under conditions of either low temperature or high O_2 partial pressure, neither of which would be considered as ambient growing season conditions. It has been hypothesized that O_2 diffusion from roots into a reducing environment could potentially remove toxic substances from the rhizosphere by oxidation and subsequent precipitation. Precipitation of iron plaques on the roots of common reed (*Phragmites* spp.) was taken by Gries et al. (1990) as an indicator of rhizosphere oxidation. Bedford et al. (1991), working with five wetland species including rice and cattail, failed to detect oxygenation of nutrient solution cultures in which the plants were growing. Oxygen was transported from the shoots to root tissue but was consumed by respiration within the plant or by microbes supported by C exudates from the roots. However, Sorrell and Armstrong (1994) criticized Bedford's et al. (1991) techniques and raised doubt that their conclusions were entirely valid. Bedford and Bouldin (1994) responded stating that they were valid for the soil mass but not for individual rhizospheres. They reiterated that assessing plant effects on soil biogeochemistry in real systems is difficult but important. Both Sorrell and Armstrong (1994) and Bedford and Bouldin (1994) agreed that much more work is needed in this area.

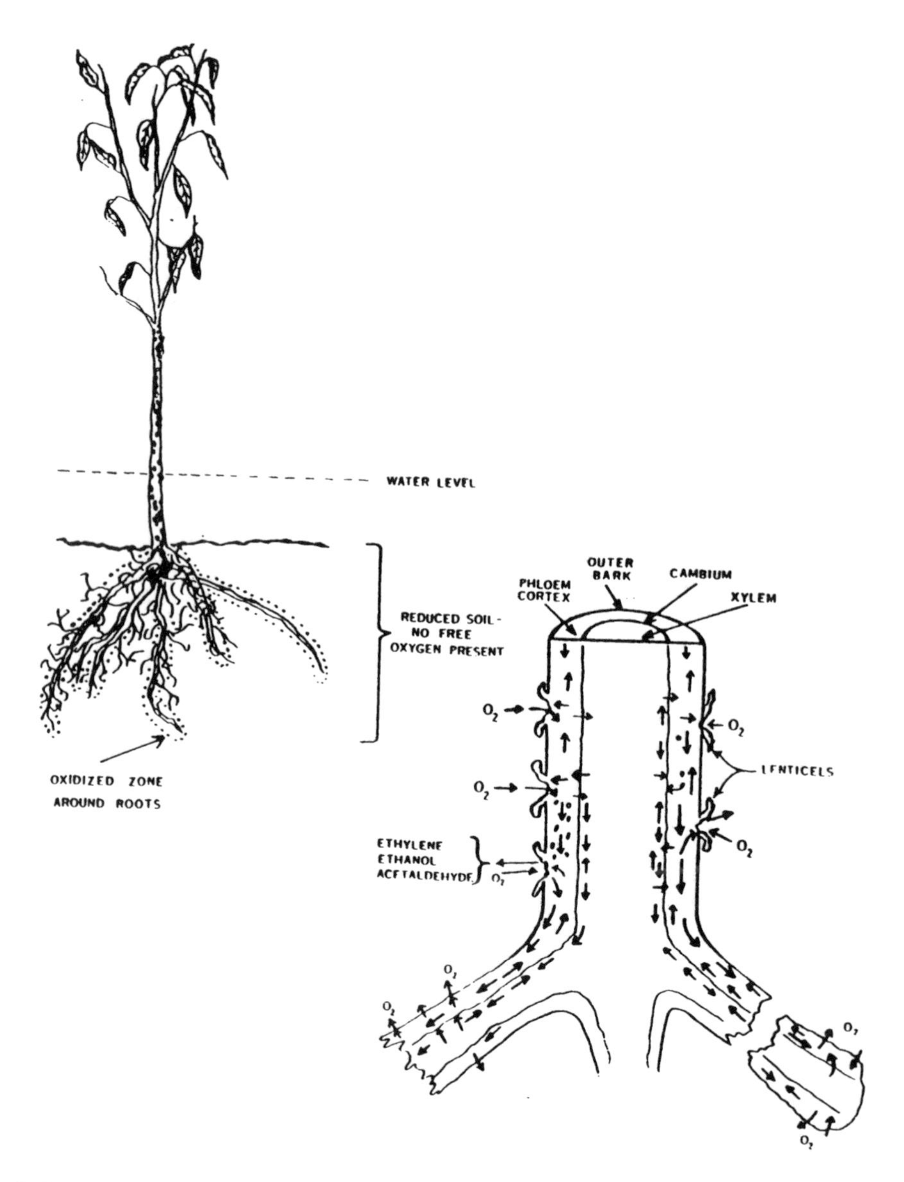

FIGURE 8.5 Schematic diagram of the pathway for internal ventilation and rhizosphere oxidation. Gases diffuse into the stem via lenticels and move within the intercellular space in the phloem to the tap root and into the aerenchyma tissue of the laterals. Products from anaerobic respiration may also be removed from the plant via the ventilation system (with permission, Hook and Scholtens 1978).

One immediate and often overlooked benefit of the development of aerenchyma tissue is the reduction in the volume of respiring tissue in the root. The formation of aerenchyma has been described as an energy efficient adaptation in that resistance to O_2 movement is lowered and amount of respiring tissue decreases without loss of structural support (Pezeshki 1991). Theoretically, O_2 demand along the root path of diffusion is lowered, thus increasing the amount of O_2 available to the meristematic tissue and actively absorbing regions of the root distal to the apex.

Return to Aerobic Respiration and Improved Energy Status

Aerenchymatous roots have been shown to have higher values for ATP content, adenylate energy charge, and ATP/ADP ratios than roots not forming aerenchyma tissue (Drew et al. 1985). The higher energy status was attributed to the internal transport of O_2 via the aerenchyma tissue. In some cases, ADH activities return to pre-flood levels after the development of aerenchyma tissue (Burdick and Mendelssohn 1990). Once aerobic respiration has resumed and O_2 partial pressures are sufficient to support other O_2 requiring processes, many structures and their functions return to near pre-flood conditions. For example, root elongation growth in wheat growing in an anoxic medium was related to O_2 transport from the shoot, and rate of elongation was proportional to O_2 concentration (Thompson et al. 1990).

Variations in Ventilation

Passive Diffusion

Passive diffusion has long been accepted as the primary means for gaseous movement within plant tissues. Many early studies concluded that O_2 diffused from the leaves to the roots and that the kinetics of movement were consistent with diffusion (Barber et al. 1962). However, diffusion of O_2 as a means of transport is effective only over short distances; passive diffusion of O_2 through root aerenchyma tissue has been estimated to be effective for a distance of approximately only 6 cm (2.4 in) (Armstrong 1979, Drew et al. 1985). However, Philipson and Coutts (1978) measured O_2 "diffusion" over distances of more than 30 cm (12 in). The effectiveness of passive diffusion as a means for transporting O_2 depends on the length of the diffusion path, radial O_2 loss, and O_2 consumption rate (Rogers and West 1993, Blom et al. 1994).

The concept that simple diffusion is the primary means of gas movement for many hydrophytic species has recently been challenged (Crawford 1993), and mass flow and pressurized transport mechanisms have been suggested. Stunzi and Kende (1989) determined that mass flow was responsible for the transport of air from the atmosphere to the root system of deep water rice. Mass flow and pressurized transport mechanisms may explain some of the discrepancies in measured distances of O_2 diffusion.

Mass Flow and Pressurized Transport

A gas mixture can be driven by gradients, such as a gradient in total pressure. Dacey (1981), working with water lilies, found that methane (CH_4) produced by anaerobic bacteria was a major constituent in lacunar gas within the plant. His earlier work showed that pressure could form in the root from CH_4 that had formed in sediments. After observing that CH_4 occurred in the petioles of older leaves during daylight but was not present in the petioles of younger, floating leaves, Dacey decided to investigate mass flow. He discovered that the pressurized gases in young leaves cause a mass flow to the rhizome through the petiole. Total pressure is lowered in the lacunae of the younger leaves when this mass flow occurs causing partial pressure of all gases to drop. As a result, a gradient is set up causing gases (specifically O_2 and N_2) to diffuse into the leaf. Diffusion of a gas through a porous membrane is facilitated by a gradient in partial pressure of the gas across the membrane. Because the partial pressure of O_2 and N_2 is greater on the outside of the leaf while the partial pressure of CH_4 is greater inside the leaf, CH_4 will diffuse out in an attempt to reach equilibrium. At the same time, partial pressure of water vapor is kept near saturation due to continual evaporation of water into the lacunae. Thus a constant mass flow is present because of continuous leaf pressurization. That is, the molecules of O_2 and N_2 leaving the lacunae by mass flow are replaced by molecules of water vapor produced by evaporation, hence mass flow is possible without loss of total pressure in the lacunae. There must be molecules left behind to prevent lower pressure at the top of the plant so that the mass flow continues towards the rhizome.

Various factors affect pressurized ventilation, such as transpirational cooling, light, humidity, and internal resistance (Grosse et al. 1991). Grosse and Schroeder (1984, 1985) found that in European alder (*Alnus glutinosa*), a temperature gradient created by the absorption of solar radiation by the dark-colored bark of the stem established a pressure gradient between air spaces within the stem tissues and the surrounding atmosphere. According to the principles of thermo-osmosis, a porous partition between inner tissues of the stem and the atmosphere must be present. The temperature gradient causes a one-way diffusion of gas molecules across the partition into the warmer tissues resulting in a pressure-induced mass flow of gas within the plant. Grosse and Schroeder (1985) went on to suggest that the phellogen is the porous membrane that is responsible for the pressure differential resulting in gas flow to the roots. This is evidenced by the eight-fold increase in O_2 supply to the roots when hypertrophied lenticel formation is present.

Another study on European alder examined the mechanism involved in thermo-osmotic pressurization of the trunk (Buchel and Grosse 1990). A porous partition between the two compartments is required to support the development of a pressure gradient. To produce a thermo-osmotic flux, the pores of the partition must be smaller than the mean free path length of the gas molecules on either side of the partition (Atkins 1987 as cited by Buchel and Grosse 1990). Results show that the pores of the phellogen layer under the lenticels in the alder species observed had diameters that were less than the mean free path length of O_2, allowing the development of a temperature-driven pressure gradient.

Mass and pressurized flow of gases have been described as an adaptation to anoxia which aids in ventilation of submerged organs and in removal of toxic substances produced by anaerobiosis (Buchel and Grosse 1990). The mass flow of gas to the roots as a result of pressurized transport greatly increases the rate of O_2 loss into the rhizosphere (Grosse 1989, Schroder 1989). Mass flow of gases has been observed in many other species including yellow floating-heart water lily (*Nymphoides peltata*) (Grosse and Mevi-Schutz 1987), South American elodea (*Egeria densa*) (Sorrell 1991), white water lily (*Nymphaea alba*) (Grosse et al. 1991), and marsh reed (Armstrong and Armstrong 1988). It has also been shown to improve the internal ventilation of the wetland tree species baldcypress, white birch (*Betula pubescens*), and quaking aspen (*Populus tremuloides*) (Grosse et al. 1992). Pressurized gas transport was also observed in the wetland species Japanese alder (*Alnus japonica*) but not in the more upland species *Alnus hirsuta* (Grosse et al. 1993).

External Transport

Deep-water rice is able to trap air between the hydrophobic, corrugated surface of the leaf blades (Raskin and Kende 1983). Air movement down the shoot is via diffusion and mass flow created by changes in CO_2 concentrations during dark respiration and O_2 concentrations during photosynthesis. Plants without air layers because of contact with surfactants exhibited no growth and experienced deterioration of submerged tissue.

Oxygen from Photosynthesis

Oxygen produced during photosynthesis may also be important in whole plant O_2 consumption and flux. Gries et al. (1990), working with marsh reed, observed increased transport of O_2 to the roots during periods of illumination compared to periods of darkness. They attributed this increase to a steeper O_2 gradient between shoots and roots when photosynthesis was occurring. Underwater photosynthesis was shown to aid in survival of submerged dock (*Rumex maritimus*) by contributing more than 50 percent to root O_2 consumption when levels of dissolved CO_2 present in the water were adequate to support photosynthesis (Laan and Blom 1990).

Reproductive Adaptations

Seed Germination and Dormancy

The ability of seed to germinate while inundated has both advantages and disadvantages depending on the conditions of the inundation. Those species whose seeds are adapted to germination and growth under flooded conditions experience less competition and have an early advantage in capturing the site provided that the flood water is sufficiently shallow to allow for emergence of the shoot (Hook 1984a). Rice is one such species capable of germinating and initiating coleoptile growth under conditions of anoxia, although root growth is inhibited (Metraux and Kende 1983, Rumpho et al. 1984). This ability enables rice to outcompete species that are incapable of growth under anoxic conditions. Trees and shrubs common to southeastern forested wetlands whose seeds have been shown to germinate under water include buttonbush (*Cephalanthus occidentalis*), American elm (*Ulmus americana*), black willow, and cottonwood (Hook 1984a).

On the other hand, the ability of individual seeds to survive long periods of flooded conditions while buried in sediments in a dormant state could also be advantageous. Infrequent droughts would expose soil surfaces, and those surviving seeds would have the opportunity to germinate. Dominant tree species common to deep swamps, such as baldcypress, pond cypress (*Taxodium distichum* var. *nutans*), water tupelo, and swamp tupelo, do not germinate under water (Hook 1984a). Therefore, seed longevity is important in natural regeneration of these species.

Hydrochory

Frequent flooding can result in the dispersal of seeds by water, especially those seeds with morphological features allowing them to float, such as the samaras of red maple and elm (*Ulmus* spp.). Seed flotation devices prevent seeds from becoming waterlogged and sinking into the hypoxic sediments (Schneider and Sharitz 1988). Seed dispersal in association with litter displacement is an important determinant of community composition (Sharitz and Lee 1985a, Schneider and Sharitz 1986, Nilsson and Grelsson 1990, Skoglund 1990).

Reproductive Phase Strategies

Another mechanism that enables species to survive flooding is to change their timing of reproduction. Some species delay their flowering and seed production during periods of flooding and survive as vegetative plants, whereas others accelerate flowering during short dry periods, utilizing the opportunity to reproduce between successive floods (Blom et al. 1990).

MANAGEMENT IMPLICATIONS

As the twenty-first century approaches, management of southern forested wetlands will become a necessity to ensure sustainable productivity of our natural resources. Knowledge of the ecophysiology of those plants, both herbaceous and woody, native to southern forested wetlands is a valuable tool in planning and designing management scenarios.

Genetic Material

Knowledge of relative species flood tolerance is important in choosing appropriate species to include in restoration and regeneration projects. Equally important is information regarding the genetic variation within a species as it influences overall waterlogging tolerance. Knowledge of genotypes containing more adaptive characteristics will be most valuable for planting and reseeding sites likely to be flooded. Using families selected for their ability to tolerate wet sites will minimize the need for minor drainage and bedding. Restrictions in the use of these silvicultural tools may become a reality in the future, so steps taken now to reduce reliance on these and other site preparation methods could ensure the means for future management of wet sites.

Water Management

Water management techniques may be required to return disturbed wetlands to conditions similar to those in existence prior to disturbance. In some situations, the water table will rise after removal of site vegetation and will remain high until the restoration of the evapotranspiration gradient. In other situations, the natural hydrology of a site has been disrupted by man, resulting in flooding during the growing season as opposed to the early spring, which has resulted in a decline in seedling survival (Sharitz et al. 1990). A more natural hydroperiod must be temporarily established to permit restoration or regeneration of a manipulated site (Sharitz and Lee 1985a, Schneider and Sharitz 1988). For example, Conner and Muller (1986) showed that in a Louisiana swamp, the years of cypress regeneration corresponded with the years of low water conditions.

Greentree Reservoirs

Proper management of greentree reservoirs requires balancing the hydrologic requirements of the plant species present along with those of the wildlife inhabiting the reservoir (Fredrickson 1991). This could mean drawing down the reservoir earlier every four to five years to allow for growth and recovery of the tree species present (Wigley and Filer 1989, Moorhead et al. 1991). Imposing a "drought" on a regular cycle would also be beneficial for woody species reproduction (Moorhead et al. 1991). Changes in the hydroperiod may not be optimum for wood duck (*Aix sponsa*) reproduction but are necessary to maintain the health of the trees inhabiting the reservoir and a desirable community structure (Karr et al. 1990), which over the long term, benefits the waterfowl population (Allen 1980). Modifying draw-down dates and reflood depths can also be used to manage recruitment of herbaceous species and to eliminate undesirable species (Merendino and Smith 1991).

Fertilization

Other techniques available to minimize the harshness of periodically flooded sites include fertilization using N, P, and lime (Langdon and McKee 1981, Hook et al. 1983, McKee et al. 1984, DeBell et al. 1984, McKee and Wilhite 1986). Fertilization with the appropriate nutrients at the time of planting could decrease mortality and increase early growth, allowing seedlings to become established.

Planting Stock

Planting stock is also critical to successful management of forested wetlands. Using containerized seedlings may prove to be more successful than bare-root seedlings on wet sites (Yeiser and Paschke 1987). With the development in techniques for mass production of clonal material such as rooted cuttings, managers need to be aware that cuttings may not be as tolerant of flooded conditions as seedlings of the same parent stock due to the morphology of the shoot-root interface (Osundina and Osonubi 1987).

Physiology Research Needs

Vascular plants in nature tend to arrange their density and composition along moisture gradients from dry to wet based on their ability to adapt to soil hypoxia or anoxia (Hook and Scholtens 1978, Tiner 1991). As a result of this general relationship, at least in the United States, many thousands of plant species have been classified according to their relative flood tolerance (Reed 1988) for the purpose of aiding in the delineation of wetlands (i.e., defining the physical boundary). Since plants tend to segregate along moisture gradients in a general relationship to flood tolerance, physiological and biochemical research has been aimed at developing a unified theory of flood tolerance (McManmon and Crawford 1971, Roberts et al. 1984). Although natural or wild plants tend to conform roughly to the moisture gradient pattern, domesticated crop and horticultural plants, with a few exceptions (rice and cranberries to name two), tend to be very intolerant (sensitive) to even hours or days of root hypoxia or anoxia. However, maize has been used for about two decades or more as an experimental plant in this area. It was selected for two very good reasons. First, strains of the species show varying degrees of root anoxia tolerance up to about four days, and second, genetic strains or mutants of the species (with known metabolic traits) can be developed quickly. Thus, Roberts et al. (1984) have used the highly effective technique of nuclear magnetic resonance (NMR) for testing wild and mutant strains of maize subjected to short-term anoxia (6 to 96 hrs) for assessing cytoplasmic acidosis. Some very informative and exciting findings have resulted from this research. However, we feel that too much of the effort in this area is being focused on maize and that the results from maize may be applicable only to those species that are very sensitive to root anoxia (Menegus et al. 1971, Kennedy et al. 1992, Vartapetian and Jackson 1997).

There is a real need to explore the range of metabolic acclimations and adaptations of tree and other wetland species to root and shoot anoxia. Do flood intolerant tree species succumb because of cytoplasmic acidosis or other physiological or morphological limitations in their ability to colonize wet sites? Increased understanding in this area could aid in wetland delineation and functional assessment and in designing crop plants to fit various moisture regimes as has been done with rice. For instance, population expansions and accompanying demands drive the development of wetlands (Hook 1993). Consequently, these forces tend to reduce wetland areas at rapid rates and alter the natural hydrodynamics of the remaining wetlands. More sophisticated information on the physiology and genetics of tree species is needed than currently exists in order to match species, families, clones, or genetically engineered varieties to sites with altered hydrodynamics to: a) maintain or increase forest productivity, b) provide diverse faunal and floral habitat, and c) enhance wetland mitigation efforts.

SUMMARY

Plants exhibit a range of responses when experiencing saturated and flooded soil conditions. Immediate but often short-term responses include changes in cellular respiration resulting from a lack of O_2. In turn, other processes, both cellular and tissue related, are altered which may result in changes in morphology and anatomy over an extended time-frame. When such responses to flooding impart a degree of fitness to the individual or species exhibiting them, we term the responses adaptations. Not all responses result in beneficial changes in structure and function. Many plants exhibit altered processes and yet cannot survive even short periods of flooding.

Most adaptations involve developing the means to avoid hypoxic conditions and providing an initial energy level sufficient to meet that goal. However, other variations do exist within the vast collection of plant species that can survive the harsh conditions of life in a reduced soil environment. Highly tolerant species usually possess a suite of responses, and it is this combination of multiple adaptations that makes them successful. There can be much variation both among species and within a single species.

Genotypic variation provides a major tool for the forest manager. The use of highly adapted species and clonal material within a species can make management a reality in forested wetlands with a minimum of water control structures and site preparation. However, timber is only one of many resources managed in southern forested wetlands, and knowledge of how different plant species respond to hydrologic conditions, both natural and man-made, can provide the means to develop management strategies for a healthy, sustainable, forested wetland ecosystem for the future of the South.

9 Wildlife Communities

T. Bently Wigley and Richard A. Lancia

CONTENTS

INTRODUCTION

Wildlife species and communities of southern forested wetlands have many values. Historically, native Americans and early European settlers made extensive use of wetland-associated animals (Hudson 1976). White-tailed deer (*Odocoileus virginianus*), wild turkeys (*Meleagris gallopavo*), rabbits (*Sylvilagus* spp.), squirrels (*Sciurus* spp.), and waterfowl were highly favored as foods. Bones and hides of these and other species were used for clothing, shelter, and implements. Wildlife played a role in the social and spiritual lives of some native Americans; animal parts such as feathers of bald eagles (*Haliaeetus leucocephalus*) and the bills and crests of ivory-billed woodpeckers (*Campephilus principalis*) were highly prized. Pursuit of wetland-associated furbearers such as American beavers (*Castor canadensis*), mink (*Mustela vison*), bobcats (*Lynx rufus*), and northern river otters (*Lutra canadensis*) was a major impetus for early European settlers moving westward across North America. The beauty and recreational values of wildlife in wetlands continue to add

1-56670-228-3/97/$0.00+$.50
© 1998 by CRC Press LLC

to human enjoyment of the South. Large numbers of Southerners enjoy hunting, feeding, photographing, or simply observing wildlife in these areas.

Ecologically, forested wetlands are transitional habitats between uplands and standing and flowing bodies of water. They are characterized by detritus-based food webs that are enriched with nutrient and energy inputs from uplands. These inputs are largely a function of hydroperiod (the timing, frequency, and duration of flooding). Periods of flooding followed by drying necessitate transitions between aquatic and terrestrial-based food webs. Some wildlife species associated with southern forested wetlands reflect the transitional nature of wetlands in their life histories. Many amphibians, such as the gopher frog (*Rana capito*), breed in wetlands and live the remainder of their adult lives in uplands. Other species do just the opposite. Aquatic turtles, such as snapping turtles (*Chelydra serpentina*), live their adult lives predominately in water and lay eggs in sandy, upland habitats. Other transitions are evident when wildlife move between uplands and lowlands while foraging, migrating, or dispersing.

In this chapter, we discuss ecological functions and relationships of wildlife, physical and biological factors affecting their presence and abundance, wildlife that occur in southern forested wetlands, and considerations related to accommodating wildlife communities in forested wetlands that are managed. Although the term "wildlife" includes a wide variety of taxa, we limit most of our discussion to terrestrial vertebrates.

ECOLOGICAL FUNCTIONS AND RELATIONSHIPS OF WILDLIFE

Wildlife has many ecological roles (Table 9.1) and help maintain important processes within forested wetlands. One important function of wildlife is facilitating the dispersal of plants and other organisms (van der Pijl 1982). Through their caching behavior, squirrels, some small mammals, and blue jays (*Cyanocitta cristata*) aid in dispersing hard mast such as acorns, beech nuts, and hickory nuts. White-tailed deer, black bears (*Ursus americanus*), common raccoons (*Procyon lotor*), and other wildlife scarify the seeds of soft mast (e.g., berries) during digestion and disperse them through defecation. Cedar waxwings (*Bombycilla cedrorum*) and American robins (*Turdus migratorius*) are major dispersers of soft mast among ecosystems. Some birds may also disperse eggs of amphibians and fish. Wetland mammals help some herbaceous plants disperse when seeds adhere to or become entangled in their fur. Southern flying squirrels (*Glaucomys volans*) and other small mammals may play a significant ecological role by dispersing mycorrhizae-forming fungi.

Wildlife communities alter the physical characteristics of forests and,subsequently, ecological processes. By creating impoundments and felling trees, beavers create snags and change the age, structure, and composition of forest vegetation. Herbivores such as white-tailed deer, rabbits, and small mammals have a similar influence when they consume hard mast and seedlings. Such changes affect succession, the susceptibility of wetland forests to disturbance (e.g., ice, wind, insects), and the abiotic component of ecosystems. Hydroperiod may be changed when beaver

TABLE 9.1
Examples of ecological functions provided by wildlife in forested wetlands, behaviors or activities that provide the functions, and examples of species or species groups that exhibit those behaviors/activities.

Ecological function of wildlife	Behavior or activity	Example of species
Dispersal of plants	Caching hard mast	Small mammals, squirrels, blue jays
	Scarification and transportation of seeds	Cedar waxwing, American robin
Alteration of forest structure and composition	Herbivory	Swamp rabbit, white-tailed deer
	Creating impoundments	Beaver
Alteration of soil characteristics and forest productivity	Predation of soil macro-invertebrates	Amphibians
	Burrowing	Crayfish, moles, fossorial rodents
Support of food webs	Predation of detritivores	Amphibians
	Primary consumption	Eastern woodrat
Transport of energy to uplands and other ecosystems	Migration	Waterfowl, Neotropical migratory songbirds
	Reproduction	Flatwoods salamander, gopher frog, snapping turtle
Recolonization of adjacent habitats	Dispersal	Cavity-nesting birds, White-tailed deer

impoundments moderate stormflow and locally elevate the water table. Biogeochemical cycles and soil characteristics are affected when wildlife (e.g., amphibians and some small mammals) burrow and feed on soil macroinvertebrates.

When wildlife alter their environment, they sometimes enhance or diminish habitat for other wildlife species. Beaver burrows, lodges, and dams provide cover or resting sites for other wildlife (e.g., snakes, mink, waterfowl) and enhance local diversity. Browsing by white-tailed deer can reduce shrub cover, thereby reducing habitat for shrub-nesting birds (DeCalesta 1994). Crayfish (e.g., *Cambarus* spp.) tunnels that extend downward to the water table sometimes are used by amphibians and reptiles to escape droughts (e.g., water snakes [*Nerodia* spp.], flatwoods salamanders [*Ambystoma cingulatum*], lesser sirens [*Siren intermedia*], crawfish frogs [*Rana areolata*], gopher frogs, mud salamanders [*Pseudotriton montanus floridanus*], and queen snakes [*Regina septemvittata*]). Burrows and wallows created by vertebrates such as gopher tortoises (*Gopherus polyphemus*) and alligators (*Alligator mississippiensis*) are important refugia for other species.

Wildlife species play an important role in food webs. The wildlife community of forested wetlands includes primary consumers that engage in all forms of herbivory including granivory, frugivory, nectarivory, and foliovory. Detrital food webs, characterized by a large community of invertebrates that shred detritus, add stability to forested wetland ecosystems by using energy seasonally fixed through primary production. Amphibians, which often comprise a substantial portion of the standing animal biomass in wetland communities, are predators of invertebrate detritivores and form an important link in the food chain upon which higher-level consumers depend (deMaynadier and Hunter 1995).

Predators and vertebrate decomposers (i.e., scavengers) are involved in both detritus- and primary production-based food webs in forested wetlands. Some predators, such as bobcats, barred owls (*Strix varia*), red-shouldered hawks (*Buteo lineatus*), and timber rattlesnakes (*Crotalus horridus*) feed mostly on terrestrial vertebrates such as rabbits, birds, and small mammals. Others feed mostly on aquatic organisms. The glossy crayfish snake (*Regina rigida*) eats crayfish, salamanders, frogs, fish, and insects (Behler and King 1979). Many predators such as raccoons and mink feed on both terrestrial and aquatic organisms. Vertebrate decomposers in forested wetlands include bald eagles, black vultures (*Coragyps atratus*), turkey vultures (*Cathartes aura*), and Virginia opossums (*Didelphis virginiana*).

Invertebrates are a major food item for many predators and are available in both detritus- and production-based webs. Oligochaetes, molluscs, isopods, amphipods, and crustaceans are found throughout many forested wetland systems and serve as prey to many predators. Many birds, reptiles, salamanders, and some small mammals (e.g., Soricidae) consume large volumes of insects. Insects such as caterpillars are especially important to forest birds during periods of egg-laying and brood-rearing.

Wildlife provide an ecological link between forested wetlands and other ecosystems and among ecological zones within forested wetlands. Some amphibians and other species such as beavers, river otters, water snakes, and turtles transfer energy between forested wetlands and aquatic ecosystems. Others transfer energy between forested wetlands and uplands, or between drier and wetter hydrologic zones. During winter, waterfowl commonly divide their time between forested wetlands, open water (e.g., lakes, rivers), and agricultural lands. Migratory birds provide an energetic link between southern forested wetlands and ecosystems on other continents and in other portions of North America.

FACTORS AFFECTING WILDLIFE PRESENCE AND ABUNDANCE

A combination of many biotic and abiotic factors determines the inherent productivity of a forested wetland site and its capacity to support a community of wildlife species (Table 9.2). For many years, biologists have recognized that inherent site fertility is reflected throughout the entire food web (e.g., Allen 1954), and wetland forests tend to be highly productive because of deep soils and abundant water. Although they are discussed separately, abiotic and biotic factors are interrelated.

The abundance of wildlife species varies with the temporal context of factors affecting populations. In many cases, controlling or limiting factors will have different short- (months, seasons) and long-term (years, decades) effects. Clearly, historical events continue to influence present-day wildlife populations and the range of practical management options. Settling the South in the past 200 years caused great reduction in the extent of wetland (and upland) forests and increasingly fragmented the remnants into smaller, more isolated tracts. As a result, several species including the ivory-billed woodpecker and Bachman's warbler (*Vermivora bachmanii*) were extirpated or severely reduced in abundance and distribution. Equally clear, present-day management decisions and land-use practices will determine the future potential for fauna of forested wetlands.

TABLE 9.2
Principal ecological factors affecting wildlife presence and abundance in southern forested wetlands.

Factor	Component
Abiotic Factors	
Soil characteristics	
Topography	
Hydrological regime	
Disturbance regime	
Climate	
Biotic Factors	
Stand-level vegetation	
	Presence of micro-habitat features, e.g., snags, down wood, cane patches
	Structure and diversity of vegetation in vertical plane
	Structure and diversity of vegetation in horizontal plane
	Plant species composition and distribution
	Detritus composition and distribution
Landscape-level vegetation	
	Composition of habitats
	Arrangement and juxtaposition of habitats
	Shape and area of habitats
	Extent and character of edges between habitats
Interactions within wildlife community	
	Inter- and intra-specific competition
	Predation
	Dispersal
	Migration
Human-related factors	
	Hunting and trapping
	Forest conversion
	Alteration of forest structure and composition

Abiotic Factors

Principal abiotic factors in forested wetlands are soil, water, weather, topography, and disturbance. Of these, soils have a major influence on the inherent fertility of a site and reflect other considerations such as the predominant historical role of climate and hydrology. Abiotic factors are an important determinant of a site's ability to function as wildlife habitat. Soil type and hydrology can influence the extent of burrowing by crayfish, herpetofauna, and mammals. The water capacity of a soil can affect whether vernal pools will be present for amphibian breeding, just as it reflects the historical patterns of deposition and erosion.

Abiotic factors have both direct and indirect effects on wildlife populations. Direct effects include mortality caused by natural events such as fires, storms, drought, unusual temperatures, and flooding. Indirect effects include adverse impacts on reproduction and survival. For example, Coulter and Bryan (1995) found that predation on wood storks (*Mycteria americana*) by raccoons increased in years with low rainfall. The dry ground allowed raccoons to climb nest trees and prey on nestlings. They also found fledgling success was positively related to prey density for wood storks, which in turn is proximately related to water levels in wetlands and ultimately to rainfall and runoff.

Flooding is a major abiotic factor in many forested wetlands. Over the long-term, each species of flora and fauna has adapted to a particular water regime. Just as the reproductive cycles and growth of obligate wetland trees (e.g., baldcypress [*Taxodium distichum*], water tupelo [*Nyssa aquatica*]) and moist-soil, herbaceous vegetation are geared to patterns of flooding, so too are the life histories of many reptiles and amphibians. Thus, maintaining or restoring a natural water regime is an important management consideration if forested wetlands are to support a full complement of wetland forest and wildlife species (Drayton and Hook 1989).

The wood stork is an excellent example of a species greatly affected by water regime. As described earlier, high water during nesting season helps reduce egg and nest predation by raccoons. However, because wood storks are "touch" feeders, they have much greater hunting success when water levels drop and concentrate their prey. When prey are concentrated, hunting is easier and brood survival greater. Thus, wood storks are most benefited by high water levels early in the reproductive season and lower water levels later, a natural pattern for many wetlands of the South.

Although flooding can provide windfalls of abundant food, it also can temporarily disrupt breeding cycles, displace animals from their habitat, and alter the composition and structure of understory and overstory vegetation. Many species of reptiles, amphibians, and small mammals seek shelter from floods by climbing above the high water, whereas larger species may temporarily move to drier sites. In large alluvial systems, however, escape may not be possible. If flooding occurs during the breeding season, reproductive output of ground-nesting birds, especially Neotropical migrants that may not attempt to renest, may be severely reduced. Other species, such as wild turkeys, will readily renest if the disturbance occurs early during egg laying or incubation. Some, including the yellow-crowned night-heron (*Nyctanassa violacea*), take advantage of periodic flooding by nesting over temporarily flooded forest.

Although not readily apparent, fire is a powerful abiotic factor that ecologically shapes wetlands in several ways (Cerulean and Engstrom 1993). During droughts, fire can set back plant succession in ordinarily wet forests and begin the cycle of renewal and regrowth as much as a 100-year flood could. Wet forests provide refugia for wildlife when fire consumes drier, upland forests. Species of plants and animals harbored in wetlands become a source to recolonize uplands. The exclusion of fire from uplands can allow lowland species to invade uplands. Loblolly-bay (*Gordonia lasianthus*) normally is restricted to wet sites, but will invade more xeric pine sites in the absence of fire. Historically, American beech (*Fagus grandifolia*), a thin-barked tree and important producer of hard mast, also probably was confined to lowlands. However, in the absence of fire, it has colonized many upland sites enhancing habitat for wildlife species that consume hard mast.

Changes in topography associated with wetlands are critical abiotic features. High-order river bottoms have a complicated pattern of berms and levees with intervening sloughs, flooded depressions, and wet benches. These features reflect patterns in elevation, soils, and the water table, and create vegetative diversity which leads to diversity in animals (Sallabanks et al. in preparation, Wakeley and Roberts 1996). Mitchell and Lancia (1990) found that species such as white-eyed vireos (*Vireo griseus*), northern cardinals (*Cardinalis cardinalis*), and American redstarts (*Setophaga ruticilla*) reached peak abundance on berms adjacent to the Mobile/Tensaw River. Berm habitats can be particularly diverse because they form a "double edge," with water on one side and lower elevation wetland forests on the other (Howard and Allen 1989). Steep banks, another topographic feature, provide nesting substrate for bank nesting species like belted kingfishers (*Ceryle alcyon*) and bank swallows (*Riparia riparia*).

Biotic Factors

Stand- and Micro-habitat Scales

An important stand-scale biotic feature affecting wildlife abundance and diversity in forested wetlands is the structural diversity of vegetation in vertical and horizontal dimensions. As a general rule, the taller and more stratified the vegetative profile from the forest floor to the top of the canopy, the greater the diversity of forest-dwelling birds (MacArthur and MacArthur 1961, Dickson et al. 1995). More vertical structural diversity means more opportunities to forage, nest, and escape from predators.

Vertical structure is important to other arboreal vertebrates. Some snakes, e.g., rat snake (*Elaphe obsoleta*), and lizards may be found throughout the forest profile. The broadhead skink (*Eumeces laticeps*) and green anole (*Anolis carolinensis*) climb readily, hunting and taking refuge throughout the forest canopy. Interestingly, some species such as the five-lined skink (*Eumeces fasciatus*) are terrestrial in some parts of their range and largely arboreal in others (Conant and Collins 1991).

Plant succession is a major ecological process that shapes wildlife habitat. As plant succession proceeds, forest canopy trees increase in height, diameter, and volume, but decrease in density. Biomass accumulates asymptotically. In older stands, less energy is channeled into net growth, and more energy goes to maintenance of existing biomass. Older deciduous trees, a feature common in later successional stages, serve as sources of food and den or nest sites and are a key habitat component.

Although vertical diversity increases during succession, this pattern may be altered by hydroperiod such that little or no midstory or understory develops. Thus, wildlife species associated with the understory and midstory layers, e.g., the wood thrush (*Hylocichla mustelina*), hooded warbler (*Wilsonia citrina*), and Swainson's warbler (*Limnothlypis swainsonii*), can be absent in wetlands subjected to prolonged flooding (Howard and Allen 1989). In contrast, prolonged flooding causes tree mortality which creates snag habitat for other species such as the prothonotary warbler (*Protonotaria citrea*). Midstory and overstory vegetation also increase or decrease as a direct function of sunlight availability through the overstory canopy.

During succession, the animal community inhabiting the forest also changes. Bird diversity in many ecosystems tends to increase as succession progresses

(Johnson and Odum 1956), but in some forested wetlands, bird abundance is high in both very young and very old stands (Mitchell et al. 1991). Young forests usually support large numbers of species that prefer edge and shrub habitat. Older forests often support large numbers of primary and secondary cavity-nesting species (Mitchell et al. 1991), which contribute to high species diversity. These species will also use younger forests if residual trees and snags are present.

Horizontal habitat diversity at the stand scale is an important habitat attribute. Patches of residual vegetation or the presence of other structural features (e.g., snags, a dense midstory) create horizontal diversity in young successional stages and provide habitat for species sometimes considered associates of older forest. Acadian flycatchers (*Empidonax virescens*) in coastal South Carolina frequently breed in pine plantations (on wet flats) as young as 15–20 years of age where a hardwood midstory is present. Conversely, gaps in the forest canopy create pockets of early successional habitat in an otherwise mature stand. The size of the gap in relation to the height of the forest largely determines the amount of sunlight that penetrates to the forest floor, which in turn stimulates growth of advance regeneration and understory species. Edge-associated wildlife, such as indigo buntings (*Passerina cyanea*), can sometimes be found singing in the smallest, most isolated gaps. Gaps in old stands usually are created by tree falls from the canopy, or they can be created through management.

Distribution and composition of overstory, midstory, and understory tree species are important horizontal habitat features. Because mast production varies considerably among years and by species, a diversity of mast-producing species is important. If production is poor for one species, it tends to be complemented by good production in another species or group. Therefore, the presence of a variety of hard and soft mast producers can ensure available nuts, fruits, and berries throughout large portions of the year (Harris and Vickers 1984). How resources such as patches of soft mast are distributed within habitat can heavily influence the distribution and activities of animals that depend upon them (e.g., black bears).

The presence of micro-habitat features within stands also can increase horizontal diversity. Snags, particularly those of large diameter, provide cover and opportunities for nesting, foraging, perching, hunting, displaying, and other activities. Snags are used by a large number of primary and secondary cavity-nesting species, such as pileated (*Dryocopus pileatus*) and red-headed woodpeckers (*Melanerpes erythrocephalus*) in the former group and prothonotary warblers and wood ducks (*Aix sponsa*) in the latter. Large, hollow snags provide roost sites for chimney swifts (*Chaetura pelagica*) and den sites for black bears (Weaver and Pelton 1994). Bats such as Rafinesque's big-eared bat (*Plecotus rafinesquii*), currently rare throughout its range in the southeastern Coastal Plain, use old, hollow baldcypress trees in swamps for roosting and maternal colony sites. Even snakes such as the scarlet kingsnakes (*Lampropeltis triangulum elapsoides*) sometimes appear above ground in snags with the bark still intact.

Other micro-habitat features in forested wetlands are critical to certain species. Swainson's warbler frequents patches of cane (*Arundinaria gigantea*) while the northern parula's (*Parula americana*) abundance in old stands may reflect its preference for Spanish moss (*Tillandsia usneoides*), an epiphyte used for nesting material. Down woody debris is important for many amphibians, reptiles, small mammals, and snails, and is used for foraging, travel, feeding, hunting, display, nesting, and other purposes. Small mammals often travel along logs while predators such as timber rattlesnakes may focus their hunting activities near down wood. Down wood is especially important as cover. Many species such as ground skinks (*Scincella lateralis*) use rotting logs and stumps as shelter and as cover to escape predators (Conant and Collins 1991).

Inter-stand and Landscape-level Habitat

Edges are created by differences in habitats between forest communities or between forests and other land uses. The degree of contrast between abutting habitats determines the nature of the edge. Edges have various consequences. They often increase species diversity, which potentially can increase predation (Paton 1994), increase brood parasitism by brown-headed cowbirds (*Molothrus ater*) (Brittingham and Temple 1983), and add exotic species. Usually, the wildlife community in edges is composed of species found in both abutting habitats, plus edge specialists. Along clearcut-forest edges of baldcypress-water tupelo, Mitchell and Lancia (1990) found that 11 bird species attained peak abundance, including some shrub species (e.g., white-eyed vireo, northern cardinal), some open woods species (e.g., summer tanager [*Piranga rubra*], great crested flycatcher [*Myiarchus crinitus*]), and some permanent resident, generalist species (e.g., Carolina chickadee [*Parus carolinensis*], tufted titmouse [*Parus bicolor*]).

Forested wetlands often are an integral component in landscape diversity. In agricultural or semi-urban landscapes, wetland forests may be the largest contiguous blocks of forest land. Wet forests are more difficult to access and convert to other uses than are upland sites, so they often are the last remaining forested habitats in a landscape. Because the process of fragmenting forests into smaller and more isolated stands diminishes habitat for many species (Burdick et al. 1989, Kress et al. 1996), remaining wetland forests often provide refugia for area-sensitive species such as Kentucky warblers (*Oporornis formosus*), cerulean warblers (*Dendroica cerulea*), and scarlet tanagers (*Piranga olivacea*).

Because of the dendritic pattern of riparian habitat across the southern landscape, some wetland forests potentially function as corridors. Large alluvial floodplains serve as regional and continental migration corridors for waterfowl, raptors, and songbirds. At a smaller scale, corridors of wetland forest might facilitate dispersal and other movements among otherwise isolated forest stands in agricultural-dominated landscapes or similar stand types in forested landscapes. Like edges, corridors can have a variety of other consequences (Hess 1994) such as increasing edge for predation or brood parasitism (some predators travel along edges), spreading disease, predators, and parasites, reducing inbreeding effects and reducing population isolation. However, few of these effects have been documented experimentally (Lindenmayer 1994).

A mosaic of habitat patches within a landscape is important to species that require several habitats to meet their life requisites. This is particularly obvious for large birds that nest in forested wetlands and forage in open wetlands. Wood storks nest in trees and forage in shallow water where prey is concentrated. These foraging areas change with water levels over the four months required to fledge young. Fleming et al. (1994) showed that spatial extent and heterogeneity of wetland habitat are landscape properties that significantly impact nesting success of wood storks. Large decreases in the extent of short-hydroperiod wetlands delayed nest initiation so that young could not be fledged before the start of the rainy season. Because prey no longer was available in the rainy season, nestlings starved and reproductive success plummeted.

Relationships within Wildlife Communities

In addition to habitat and abiotic considerations, interactions within the wildlife community can affect the relative abundance of species. Birds may not successfully reproduce each year due to brood parasitism, predation, competition for preferred nest sites or brood-rearing habitat, or low abundance of prey items such as caterpillars. Some interactions such as predation can alter the age distribution and sex ratio, as well as density, of a population.

Sometimes, habitat attributes are closely tied to relationships within and among species. Habitat area and "edge effects" are thought to be important influences on the dynamics of many wildlife populations. Theoretically, populations in habitats with high edge-to-interior ratios will be susceptible to high levels of predation or forced to compete with invading species. Alluvial floodplain forests often are long and narrow, or adjacent to urban or agricultural habitats that support predators. Isolated wetlands such as baldcypress or water tupelo ponds often are small with large amounts of edge per unit area. Therefore, forested wetlands may be especially susceptible to edge effects.

Forested wetland habitats may serve as either sources or sinks (Pulliam 1988). Sink habitats are those where natality (birth rate) is insufficient to balance local mortality (death rate). In source habitats, natality exceeds mortality. Populations in sink habitats persist through immigration from nearby sources.

The source/sink paradigm may describe wildlife populations in southern forested wetlands that have strong zonation. Sometimes, however, source habitat for one species is sink habitat for others. For example, wetter zones of alluvial floodplains may be source habitat for green frogs (*Rana clamitans*) and northern cricket frogs (*Acris crepitans*), but sink habitat for timber rattlesnakes. The source/sink paradigm also may describe the relationship between populations associated with both forested wetland and adjacent upland habitats. Because alluvial floodplains often have abundant hardwoods which readily form cavities (Melchiors and Cicero 1987), floodplains may be source habitats for some cavity-nesting species that are found in both uplands and wetlands. Because of such considerations, wildlife communities in forested wetlands likely are a mixture of populations, some of which are self-maintaining and some of which are not (Pulliam 1988).

The ability of a species to disperse also affects its abundance in forested wetlands. Although species differ in this ability, dispersal also is affected by the arrangement of habitats in the landscape. Isolation of similar habitats can make dispersal and genetic interchange more difficult (Templeton et al. 1990). If this occurs, the population may be susceptible to inbreeding, restricted gene flow, genetic drift, and even local extinction (Templeton et al. 1990). Local extinction may ultimately be caused by some stochastic environmental event such as habitat alteration, predation, competition, or natural disturbance (Soule 1983). Or, it can be caused by stochastic fluctuations in population size due to variation in demographic factors such as sex ratio or age structure.

Because of isolation, some populations in forested wetlands may exist as metapopulations. A metapopulation has been defined as a group of populations which interact through dispersal (Hanski 1991, Hanski and Gilpin 1991). Supposedly, movements of dispersing individuals would be infrequent and across habitat types not suitable for feeding and breeding activities, and dispersers would face substantial risk of failing to locate another suitable habitat in which to settle (Hanski and Gilpin 1991). Colonizing organisms would come from the set of local populations making up the metapopulation. Because local populations and the metapopulation itself may become extinct, successful dispersal would be critical to metapopulation maintenance. Wildlife populations may exist as metapopulations where forested wetlands are separated by urban or agricultural land. For example, many amphibians are highly adapted to the moist conditions of forested wetland habitat and face substantial risks such as dessication or predation when dispersing across agricultural land, urban areas, or even upland forest.

Competition and predation are other processes within wildlife communities that affect the abundance of species. Competition may be defined as the active demand by two or more organisms for a common resource that is actually or potentially limiting (Miller 1967, Wilson 1975, Aarssen 1984). Exploitative competition occurs when an individual uses a resource thus reducing its availability to other individuals (Schoener 1983). Examples might include aerial-feeding birds such as common nighthawks (*Chordeiles minor*) competing with bats for nocturnal insect prey, or martins and swallows (Hirundinidae) foraging on the same mosquito (Culicidae) prey base as dragonflies (Odonata). Foliage-gleaning birds and insectivorous lizards such as the white-eyed vireo and green anole, respectively, also compete for insects.

Another form of competition, interference competition, occurs when an individual is denied access to a resource by the actions of another individual. Interference (and exploitative) competition can occur at several levels including the individual and population levels. Male black bears sometimes exclude female bears from high quality food sources, e.g., berry patches. Interference competition also occurs among birds, mammals, reptiles, and amphibians for micro-habitat features such as cavities or down woody debris. Southern flying squirrels, for example, are particularly aggressive competitors for cavities, and can exclude other species locally.

Predation influences the abundance of species by directly reducing prey populations and can limit the distribution of some species. Bullfrogs (*Rana catesbeiana*) prey upon other frogs such as southern leopard frogs (*Rana utricularia*) and limit their distribution. In addition to competing with aquatic salamanders for food, fish prey upon them and can locally eliminate them. Predation effects are difficult to predict in wildlife populations and can vary with habitat characteristics, climate, elevation, availability of buffer prey, and predator density. Although there have been few studies in forested wetlands, recent research from other ecosystems (J. P. Hoover et al. 1995) suggests that predation can severely reduce reproductive success of some birds. Nest predation appears to be highly correlated with forest and core area and may be worst where forests are fragmented by agriculture or suburban development (J. P. Hoover et al. 1995). Predation also can affect other population interactions such as competition by reducing densities of competing species below levels at which resources are limiting.

Human-Related Factors

Humans have influenced wildlife communities in forested wetlands in a variety of ways. They directly harvest wildlife, convert forested wetlands to other uses, and alter forest structure and composition through silvicultural practices, water-level regulation, and firewood harvesting. Humans will continue to depend upon and alter forested wetland ecosystems. Therefore, it is important to understand human-wildlife interactions in forested wetlands and develop strategies for integrating wildlife considerations with human objectives.

Hunting and trapping are popular in the South, where there were about 5.2 million hunters (about 8% of the population >16 years old) during 1991 (USFWS 1992). Hunting and trapping alter the density and demographics (e.g., sex ratio, age structure) of wildlife populations. Bucks-only hunting has skewed the sex ratio in many deer populations heavily towards females, reduced the average age of males, increased population densities, and perhaps altered some aspects of reproduction (Miller et al. 1995). Variation in deer numbers (and other browsers) affects the understory of wetland forests and other wildlife species (e.g., shrub-nesting birds) that are associated with that habitat feature.

Extensive areas of southern forested wetlands have been converted to other uses, particularly agricultural and urban uses. For many species, forest conversion renders once suitable habitat unsuitable. Some birds, such as the yellow-throated vireo (*Vireo flavifrons*) and Swainson's warbler, are associated with extensive tracts of forest or forest-interior conditions (Hamel 1992), and are unable to exploit agricultural or urbanized habitats. Many migratory birds and other wildlife species such as the Louisiana black bear (*Ursus americanus luteolus*) also have been affected by reductions in area of alluvial floodplain forests (Smith et al. 1993, Burdick et al. 1989). For such species, lands converted to agricultural or urban uses may become barriers to dispersal, and once continuous populations may become metapopulations. A once continuous population of Louisiana black bears in the Tensas River Basin is now divided into two populations separated by agricultural lands and a four-lane interstate highway. During four years of recent radio-telemetry studies, very limited dispersal has been observed between the two populations (Pelton, pers. comm.).

For some species, however, agricultural lands can be an important habitat component when food crops are grown. Wintering waterfowl in the Mississippi Alluvial Valley make extensive use of food crops that have replaced forested wetlands (Delnicki and Reinecke 1986). Black bears also feed heavily on corn, winter wheat, and sugarcane in the Tensas and Atchafalaya River basins of Louisiana and in coastal North Carolina (Pelton, pers. comm.).

Wildlife are affected when humans alter the structure and composition of forests. For thousands of years, native Americans greatly influenced forest structure and composition through their activities (Day 1953, Martin 1973, Hudson 1976). They foraged long distances from their villages for plants and animals that were used for food, fiber, and medicine, and collected wood and bark for utensils, weapons, canoes, houses, and fuelwood. Native Americans used stone axes and fire to fell living trees for fuelwood, to create cropland, and for other purposes. When resources became limiting, villages were moved to new locations. Fire was used by native Americans for many other purposes, such as hunting and war. Although the full extent of its use in forested wetlands is unclear, fires started by Native Americans undoubtedly burned wetlands during dry periods.

European settlers continued and expanded this tradition of using forests. Many eastern forests, including forested wetlands, have a history of prior conversion to agriculture by European settlers. Even forests that have never been converted to other uses usually have been altered through historical timber harvesting. Birch (1996) reports that even today, about 70% of privately owned forests in the South are controlled by landowners that have harvested their forests in the past. In earlier times, timber harvesting often was conducted without regard for residual stand characteristics or regeneration. Most (87%) private, nonindustrial forest landowners today do not have formal plans describing their future management activities (Birch 1996).

As a result of human activities, forested wetlands are different in composition and structure from those that existed on this continent before settlement by Europeans (Palik and Pregitzer 1992). Present-day wetland forest landscapes can have less forest area (Burdick et al. 1989, Kress et al. 1996), greater diversity, and more forest patches that are smaller and more simply shaped than pre-settlement landscapes (Mladenoff et al. 1993). In addition, there may be fewer large patches of forest, fewer patches of old forest, and altered ecosystem juxtapositions (Mladenoff et al. 1993).

Because of historical alterations of forests and changes in land use, some present-day management options for forested wetlands have been largely foreclosed. The option of maintaining pre-settlement conditions across large areas is not feasible across most of the East. Perhaps the only certainty regarding southern forested wetlands is that they will continue to change and be heavily influenced by humans.

WILDLIFE COMMUNITIES AND HABITAT ATTRIBUTES OF WETLAND TYPES

Wildlife communities of forested wetlands are as diverse as the wetland habitats they inhabit and often are difficult to characterize. Species such as the eastern cottonmouth (*Agkistrodon piscivorus piscivorus*), bullfrog, belted kingfisher, prothonotary warbler, water shrew (*Sorex palustris*), and mink are closely linked to wetlands. Of vertebrate taxa, however, amphibians may be most closely associated with forested wetlands. Forested wetlands often are excellent sources of moist microhabitats (e.g., pools, down woody debris) important to terrestrial amphibians. When conditions are moist enough, wetland forests even provide temporary habitat for some aquatic amphibians such as the two-toed amphiuma (*Amphiuma means*).

Many wildlife species that occur in wetland forests are not found there exclusively. However, they may fulfill part of their life requisites in forested wetlands or reach peak densities there (Schitoskey and Linder 1978). The eastern bluebird (*Sialia sialis*) often is described as an associate of upland pine habitats. During winter, however, they will forage on berries and insects in bottomland hardwoods and beaver ponds (Allen and Sweeney 1990). Leonard (1994) described only five of 69 birds in his study in Florida as restricted to forested wetlands. The eastern box turtle (*Terrapene carolina carolina*) feeds in forested wetlands on slugs, earthworms, berries, and mushrooms, especially during dry periods. Raccoons use a variety of habitats, but their densities may be 13 to 23 times higher in wet forests than in upland sites (Leberg and Kennedy 1988).

Some species primarily associated with aquatic ecosystems also use forested wetlands. Birds such as brown pelicans (*Pelecanus occidentalis*), anhingas (*Anhinga anhinga*), and green herons (*Butorides virescens*) use wetland forests to rest, roost, nest, and feed. Aquatic turtles such as common musk turtles (*Sternotherus odoratus*) commonly bury their eggs in drier sites within forested wetlands.

In the following sections we discuss wildlife communities and habitat attributes of major forested wetland types. Some characteristic birds, mammals, and herpetofauna documented in several forested wetland types are described in Tables 9.3, 9.4, and 9.5, respectively.

Alluvial Floodplain Forests

Alluvial floodplains often have many important habitat features such as abundant detritus, hard mast, soft mast, ground-level vegetation, aquatic and terrestrial invertebrates, snags, arboreal cavities, and down woody debris. In many floodplain forests, the vertical profile is multi-layered and dominated by deciduous trees, some of which reach extremely large diameters. This variety usually supports rich and diverse wildlife communities. Larger alluvial floodplains have an easily distinguished zonation of habitats where hydrological regimes range from sloughs to the transition communities between floodplain and upland habitats (Wharton et al. 1982). In small floodplains, however, habitat zonation is less distinct.

In alluvial floodplains, many birds are found across the hydrologic gradient and show no particular affinity for cover types or hydrological regimes (Mitchell and Lancia 1990, Mitchell et al. 1991, Wakeley and Roberts 1996). Some species, however, are most numerous in specific zones. In Arkansas, Wakeley and Roberts (1996) found the chimney swift, prothonotary warbler, and great crested flycatcher to be most dense in wetter zones. Water birds such as wood ducks and hooded mergansers (*Lophodytes cucullatus*) also are found primarily within wetter zones of alluvial forests (Fredrickson 1979).

TABLE 9.3
Selected birds associated with southern forested wetland habitats.

Taxa[2]	Common name	Habitat Type[1]				
		AF	PCB	MW	FWM	MAN
Pelecaniformes: Totipalmate Swimmers						
Pelecanidae						
Pelecanus occidentalis	Brown pelican					•
Anhingidae						
Anhinga anhinga	Anhinga	•				
Fregatidae						
Fregata magnificens	Magnificent frigatebird					•
Ciconiiformes: Herons, Ibises, Storks and Allies						
Ardeidae						
Ardea herodias	Great blue heron	•		•		•
Casmerodius albus	Great egret				•	•
Egretta thula	Snowy egret					•
Egretta caerulea	Little blue heron				•	•
Bubulcus ibis	Cattle egret	•				•
Butorides virescens	Green heron	•	•	•	•	•
Nyctanassa violacea	Yellow-crowned night heron	•			•	•
Threskiornithidae						
Eudocimus albus	White ibis	•			•	•
Ajaia ajaja	Roseate spoonbill					•
Ciconiidae						
Mycteria americana	Wood stork					•
Anseriformes: Swans, Geese, and Ducks						
Anatidae						
Aix sponsa	Wood duck	•	•		•	
Anas platyrhynchos	Mallard	•				
Lophodytes cucullatus	Hooded merganser	•				
Falconiformes: Diurnal Birds of Prey						
Cathartidae						
Coragyps atratus	Black vulture	•	•			•
Cathartes aura	Turkey vulture	•	•	•	•	•
Accipitridae						
Pandion haliaetus	Osprey					•
Elanoides forficatus	American swallow-tailed kite	•			•	•
Ictinia mississippiensis	Mississippi kite	•				
Buteo lineatus	Red-shouldered hawk	•	•	•	•	•
Buteo brachyurus	Short-tailed hawk					•
Galliformes: Gallinaceous Birds						
Phasianidae						
Bonasa umbellus	Ruffed grouse			•		
Meleagris gallopavo	Wild turkey	•			•	•
Colinus virginianus	Northern bobwhite	•	•		•	•

TABLE 9.3 (continued)
Selected birds associated with southern forested wetland habitats.

Taxa[2]	Common name	Habitat Type[1]				
		AF	PCB	MW	FWM	MAN
Charadriiformes: Shorebirds, Gulls, Auks and Allies						
Scolopacidae						
Scolopax minor	American woodcock	•	•			
Columbiformes: Sandgrouse, Pigeons and Doves						
Columbidae						
Zenaida macroura	Mourning dove	•	•		•	•
Cuculiformes: Cuckoos and Allies						
Cuculidae						
Coccyzus americanus	Yellow-billed cuckoo		•		•	•
Coccyzus minor	Mangrove cuckoo					•
Strigiformes: Owls						
Strigidae						
Strix varia	Barred owl	•	•		•	•
Caprimulgiformes: Goatsuckers, Oilbirds, and Allies						
Caprimulgidae						
Caprimulgus carolinensis	Chuck-will's-widow		•			•
Apodiformes: Swifts and Hummingbirds						
Apodidae						
Chaetura pelagica	Chimney swift	•	•			
Trochilidae						
Archilochus colubris	Ruby-throated hummingbird	•	•		•	•
Piciformes: Puffbirds, Toucans, Woodpeckers and Allies						
Picidae						
Melanerpes erythrocephalus	Red-headed woodpecker		•		•	•
Melanerpes carolinus	Red-bellied woodpecker	•	•		•	•
Picoides pubescens	Downy woodpecker	•	•		•	
Picoides villosus	Hairy woodpecker	•	•		•	•
Colaptes auratus	Northern flicker		•		•	•
Dryocopus pileatus	Pileated woodpecker	•	•		•	•
Passeriformes: Passerine Birds						
Tyrannidae						
Contopus borealis	Olive-sided flycatcher			•		
Contopus virens	Eastern wood-pewee	•	•		•	•
Empidonax virescens	Acadian flycatcher	•	•	•		•
Empidonax traillii	Willow flycatcher			•		
Myiarchus crinitus	Great crested flycatcher	•	•		•	•
Tyrannus tyrannus	Eastern kingbird	•	•			•

TABLE 9.3 (continued)
Selected birds associated with southern forested wetland habitats.

Taxa[2]	Common name	Habitat Type[1]				
		AF	PCB	MW	FWM	MAN
Corvidae						
Cyanocitta cristata	Blue jay	•	•	•	•	•
Corvus brachyrhynchos	Common crow	•	•	•	•	
Corvus ossifragus	Fish crow	•	•			•
Paridae						
Parus carolinensis	Carolina chickadee	•	•		•	
Parus bicolor	Tufted titmouse	•	•		•	•
Sittidae						
Sitta carolinensis	White-breasted nuthatch	•				
Sitta pusilla	Brown-headed nuthatch	•	•		•	
Troglodytidae						
Thryothorus ludovicianus	Carolina wren	•	•		•	•
Muscicapidae, Subfamily Sylviinae						
Polioptila caerulea	Blue-gray gnatcatcher	•	•	•	•	•
Muscicapidae, Subfamily Turdinae						
Sialia sialis	Eastern bluebird		•		•	
Catharus fuscescens	Veery			•		
Catharus guttatus	Hermit thrush			•		
Hylocichla mustelina	Wood thrush	•	•			
Mimidae						
Dumetella carolinensis	Gray catbird	•	•			•
Mimus polyglottos	Northern mockingbird				•	
Toxostoma rufum	Brown thrasher		•		•	•
Vireonidae						
Vireo griseus	White-eyed vireo	•	•		•	•
Vireo flavifrons	Yellow-throated vireo	•	•		•	•
Vireo olivaceus	Red-eyed vireo	•	•	•	•	•
Emberizidae, Subfamily Parulinae						
Parula americana	Northern parula	•	•	•	•	•
Dendroica virens	Black-throated green warbler		•	•		•
Dendroica dominica	Yellow-throated warbler	•	•	•	•	•
Dendroica pinus	Pine warbler	•	•		•	
Dendroica discolor	Prairie warbler		•		•	•
Dendroica cerulea	Cerulean warbler			•		
Mniotilta varia	Black-and-white warbler	•	•	•		•
Setophaga ruticilla	American redstart	•	•			•
Protonotaria citrea	Prothonotary warbler	•	•		•	•
Helmitheros vermivorus	Worm-eating warbler		•		•	•
Limnothlypis swainsonii	Swainson's warbler	•	•	•	•	
Seiurus aurocapillus	Ovenbird		•			•
Seiurus motacilla	Louisiana waterthrush	•	•	•		
Oporornis formosus	Kentucky warbler	•	•	•	•	•
Geothlypis trichas	Common yellowthroat	•	•	•	•	•

TABLE 9.3 (continued)
Selected birds associated with southern forested wetland habitats.

Taxa[2]	Common name	Habitat Type[1]				
		AF	PCB	MW	FWM	MAN
Wilsonia citrina	Hooded warbler	•	•	•	•	
Wilsonia canadensis	Canada warbler			•		
Icteria virens	Yellow-breasted chat	•	•	•	•	•
Emberizidae, Subfamily Thraupinae						
Piranga rubra	Summer tanager	•	•		•	•
Emberizidae, Subfamily Cardinalinae						
Cardinalis cardinalis	Northern cardinal	•	•	•	•	•
Pheucticus ludovicianus	Rose-breasted grosbeak			•		
Guiraca caerulea	Blue grosbeak	•			•	
Passerina cyanea	Indigo bunting	•	•		•	•
Emberizidae, Subfamily Emberizinae						
Pipilo erythrophthalmus	Rufous-sided towhee	•	•		•	•
Aimophila aestivalus	Bachman's sparrow				•	
Spizella passerina	Chipping sparrow		•			
Spizella pusilla	Field sparrow		•			
Emberizidae, Subfamily Icterinae						
Agelaius phoeniceus	Red-winged blackbird	•				•
Sturnella magna	Eastern meadowlark		•		•	
Quiscalus quiscula	Common grackle	•	•		•	•
Molothrus ater	Brown-headed cowbird	•	•	•		
Icterus galbula	Northern oriole	•				
Fringillidae						
Carpodacus purpureus	Purple finch			•		
Carduelis tristis	American goldfinch	•	•			•

[1] AF = alluvial floodplain forests, FWM = flatwoods/wetlands mosaic, MW = mountain wetlands, MAN = mangroves, PCB = pocosins/Carolina bays.

[2] Partial species lists taken from: AF = Sharitz and Mitsch (1993), Mitchell et al. (1991), Mitchell and Lancia (1990), Hamel (1989), Dickson (1978); FWM = O'Meara (1984), Johnson and Landers (1982), Harris and McElveen (1981); MW = Boynton (1994); PCB = Karriker (1993), Lee (1986, 1987); MAN = Hamel (1992), Odum et al. (1982).

Some birds are closely associated with specific micro-habitat features most common in certain zones. The Acadian flycatcher is most common near streams and rivers where the overstory and midstory are well developed and the ground layer is relatively open (Murray 1992, Murray and Stauffer 1995). Prothonotary warblers are common near standing water with scattered snags and stumps (Hamel 1992). Wakeley and Roberts (1996) speculated that chimney swifts were favored by the open canopy, sparse midstory, and large, hollow snags characteristic of wetter sites on their study area.

During winter, alluvial floodplain forests support a large number of migratory bird species. Over-wintering waterfowl are locally common to abundant where hard

TABLE 9.4
Selected mammal species associated with southern forested wetland habitats.

Taxa[2]	Common name	Habitat Type[1]				
		AF	PCB	MW	FWM	MAN
Marsupalia: Pouched Mammals						
Didelphidae						
Didelphis virginianus	Virginia opossum	•	•	•	•	•
Insectivora: Shrews and Moles						
Soricidae						
Sorex cinereus	Masked shrew			•		
Sorex longirostris	Southeastern shrew	•	•	•		
Sorex palustris	Water shrew			•		
Sorex fumeus	Smoky shrew					
Blarina brevicauda	Northern short-tailed shrew	•	•	•	•	•
Cryptotis parva	Least shrew		•		•	
Talpidae						
Scalopus aquaticus	Eastern mole		•			
Condylura cristata	Star-nosed mole		•	•		
Chiroptera: Bats						
Vespertilionidae						
Pipistrellus subflavus	Eastern pipistrelle		•	•		
Eptesicus fuscus	Big brown bat		•			
Lasiurus borealis	Eastern red bat		•	•		
Nycticeius humeralis	Evening bat		•			
Plecotus rafinesquii	Rafinesque's big-eared bat		•			
Lagomorpha: Pikas, Rabbits and Hares						
Leporidae						
Sylvilagus palustris	Marsh rabbit	•	•		•	•
Sylvilagus floridanus	Eastern cottontail rabbit	•	•	•	•	
Sylvilagus aquaticus	Swamp rabbit	•	•			
Rodentia: Rodents						
Sciuridae						
Sciurus carolinensis	Eastern gray squirrel	•	•			•
Sciurus niger	Eastern fox squirrel	•	•		•	•
Glaucomys volans	Southern flying squirrel	•				
Castoridae						
Castor canadensis	American beaver	•	•			
Cricetidae						
Oryzomys palustris	Marsh rice rat	•	•		•	•
Reithrodontomys humulus	Eastern harvest mouse		•			
Reithrodontomys fulvescens	Fulvous harvest mouse	•				
Peromyscus polionotus	Oldfield mouse		•			
Peromyscus leucopus	White-footed mouse		•			
Peromyscus gossypinus	Cotton mouse	•	•		•	

TABLE 9.4 (continued)
Selected mammal species associated with southern forested wetland habitats.

Taxa[2]	Common name	Habitat Type[1]				
		AF	PCB	MW	FWM	MAN
Ochrotomys nuttalli	Golden mouse	•	•	•	•	
Sigmodon hispidus	Hispid cotton rat		•		•	•
Neotoma floridana	Florida woodrat	•	•		•	
Clethrionomys gapperi	Southern red-backed vole			•		
Microtus chrotorrhinus	Rock vole			•		
Microtus pinetorum	Pine vole		•			
Ondatra zibethicus	Common muskrat	•	•			
Synaptomys cooperi	Southern bog lemming		•	•		
Muridae						
Mus musculus	House mouse	•	•		•	
Zapodidae						
Napaeozapus insignis	Woodland jumping mouse			•		
Carnivora: Carnivores						
Canidae						
Urocyon cinereoargenteus	Eastern gray fox		•		•	•
Ursidae						
Ursus americanus	Black bear	•	•		•	•
Procyonidae						
Procyon lotor	Raccoon	•	•		•	•
Mustelidae						
Mustela frenata	Longtail weasel		•		•	
Mustela vison	Mink	•	•			•
Lutra canadensis	River otter	•	•		•	•
Felidae						
Felis concolor coryi	Florida panther				•	•
Lynx rufus	Bobcat		•		•	•
Sirenia: Manatees and Sea Cows						
Trichechidae						
Trichechus manatus	Manatee					•
Artiodactyla: Even-toed Hoofed Mammals						
Suidae						
Sus scrofa	Wild pig			•	•	
Cervidae						
Odocoileus virginianus	White-tailed deer	•	•		•	•

[1] AF = alluvial floodplain forests, MW = mountain wetland, FWM = flatwoods/wetlands mosaic, MAN = mangroves, PCB = pocosins/Carolina bays.

[2] Partial species lists taken from: AF = Dickson and Warren (1994), Sharitz and Mitsch (1993); FWM = Stout and Marion (1993), Abrahamson and Hartnett (1990), Labisky and Hovis (1987), Harris and Vickers (1984); MAN = Gilmore and Snedaker (1993), Odum et al. (1982); MW = Boynton (1994); PCB = Mitchell (1994), Clark et al. (1985), Gibbons and Semlitsch (1981).

TABLE 9.5
Selected amphibians and reptiles associated with southern forested wetlands habitats.

Taxa[2]	Common name	Habitat type[1]				
		AF	PCB	MW	FWM	MAN
Caudata: Salamanders						
Sirenidae						
Pseudobranchus striatus	Dwarf siren				•	
Siren intermedia	Lesser siren	•			•	
Siren lacertina	Greater siren				•	
Salamandridae						
Notophthalmus viridescens	Red-spotted newt		•		•	
Amphiumidae						
Amphiuma means	Two-toed amphiuma				•	
Ambystomidae						
Ambystoma cingulatum	Flatwoods salamander				•	
Ambystoma maculatum	Spotted salamander			•		
Ambystoma opacum	Marbled salamander	•	•	•		
Ambystoma talpoideum	Mole salamander		•	•		
Ambystoma texanum	Smallmouth salamander	•				
Ambystoma tigrinum	Eastern tiger salamander		•			
Plethodontidae						
Desmognathus auriculatis	Southern dusky salamander	•			•	
Desmognathus fuscus	Northern dusky salamander	•		•		
Desmognathus ochrophaeus	Mountain dusky salamander			•		
Eurycea bislineata	Two-lined salamander	•	•	•		
Eurycea longicauda	Longtail salamander	•		•		
Eurycea quadridigitata	Dwarf salamander	•	•		•	
Gyrinophilus porphyriticus	Spring salamander			•		
Plethodon glutinosus	Slimy salamander	•	•			
Pseudotriton montanus	Rusty mud salamander		•		•	
Pseudotriton ruber	Southern red salamander		•			
Salientia: Frogs and Toads						
Pelobatidae						
Scaphiopus holbrooki	Eastern spadefoot		•		•	
Ranidae						
Rana areolata	Crawfish frog		•		•	
Rana catesbeiana	Bullfrog	•	•			
Rana clamitans	Bronze frog	•	•			
Rana grylio	Pig frog				•	
Rana heckscheri	River frog				•	
Rana utricularia	Southern leopard frog	•	•		•	
Microhylidae						
Gastrophryne carolinensis	Eastern narrowmouth frog	•	•		•	
Bufonidae						
Bufo marinus	Giant toad					•
Bufo quercicus	Oak toad	•	•		•	

TABLE 9.5 (continued)
Selected amphibians and reptiles associated with southern forested wetlands habitats.

Taxa[2]	Common name	Habitat type[1]				
		AF	PCB	MW	FWM	MAN
Bufo terrestris	Southern toad	•	•		•	
Bufo valliceps	Gulf Coast toad	•				
Bufo woodhousii	Woodhouse's toad	•				
Hylidae						
Acris gryllus	Southern cricket frog	•	•		•	
Hyla andersoni	Pine barrens treefrog		•			
Hyla chysoscelis	Gray treefrog	•	•			
Hyla cinerea	Green treefrog	•	•			
Hyla femoralis	Pine woods treefrog		•		•	
Hyla gratiosa	Barking treefrog		•		•	
Hyla squirella	Squirrel treefrog	•			•	•
Hyla versicolor	Gray treefrog	•				
Limnaoedus ocularis	Little grass frog				•	
Osteopilus septentrionalis	Cuban treefrog					•
Pseudacris crucifer	Spring peeper	•	•			
Pseudacris nigrita	Southern chorus frog		•		•	
Pseudacris ornata	Ornate chorus frog		•		•	
Pseudacris triseriata	Boreal chorus frog	•				
Crocodylia: Crocodilians						
Alligator mississippiensis	American alligator		•			•
Crocodylus acutus	American crocodile					•
Testudines: Turtles						
Chelydridae						
Chelydra serpentina	Snapping turtle	•	•		•	
Kinosternidae						
Kinosternon baurii	Striped mud turtle				•	•
Kinosternon subrubrum	Eastern mud turtle	•	•		•	•
Sternotherus minor	Loggerhead musk turtle			•		
Sternotherus odoratus	Common musk turtle	•	•		•	
Emydidae						
Clemmys guttata	Spotted turtle	•				
Deirochelys reticularia	Chicken turtle		•		•	•
Malaclemys terrapin	Northern diamondback terrapin					•
Pseudemys nelsoni	Florida redbelly turtle					•
Terrapene carolina	Eastern box turtle	•	•		•	
Trachemys scripta	Red-eared slider	•	•			
Chelonidae						
Chelonia mydas	Green turtle					•
Trionychidae						
Apalone ferox	Florida softshell					•

TABLE 9.5 (continued)
Selected amphibians and reptiles associated with southern forested wetlands habitats.

Taxa[2]	Common name	Habitat type[1]				
		AF	PCB	MW	FWM	MAN
Squamata, suborder Lacertilia: Lizards						
Iguanidae						
Anolis carolinensis	Green anole	•	•		•	•
Anolis distichus	Bark anole					•
Anolis sagrei	Brown anole					•
Sceloporus undulatus	Fence lizard		•			
Anguidae						
Ophisaurus attenuatus	Slender glass lizard		•			
Ophisaurus ventralis	Eastern glass lizard		•		•	
Teiidae						
Cnemidophorus sexlineatus	Six-lined racerunner		•		•	
Scincidae						
Eumeces anthracinus	Coal skink			•		
Eumeces fasciatus	Five-lined skink	•	•		•	
Eumeces inexpectatus	Southeastern five-lined skink		•	•	•	•
Eumeces laticeps	Broadhead skink	•	•	•	•	
Scincella lateralis	Ground skink	•	•		•	
Squamata, suborder Serpentes: Snakes						
Colubridae						
Carphophis amoenus	Eastern worm snake	•				
Cemophora coccinea	Scarlet snake		•		•	
Coluber constrictor	Southern black racer	•	•		•	
Diadophis punctatus	Southern ringneck snake	•	•		•	
Elaphe guttata	Corn snake				•	•
Elaphe obsoleta	Yellow rat snake	•	•	•		•
Farancia abacura	Mud snake	•	•		•	
Heterodon platirhinos	Eastern hognose snake	•	•		•	
Heterodon simus	Southern hognose snake		•			
Lampropeltis getula	Eastern kingsnake	•	•	•		
Lampropeltis triangulum	Eastern milk snake	•			•	
Masticophis flagellum	Eastern coachwhip	•				
Nerodia cyclopion	Mississippi green water snake					•
Nerodia erythrogaster	Yellowbelly watersnake	•	•			
Nerodia fasciata	Southern water snake	•	•		•	•
Nerodia rhombifer	Diamondback water snake	•				
Nerodia taxispilota	Brown water snake	•				
Opheodrys aestivus	Rough green snake	•			•	
Pituophis melanoleucus	Northern pine snake		•			
Regina rigida	Glossy crayfish snake	•			•	
Rhadinaea flavilata	Pine woods snake				•	

TABLE 9.5 (continued)
Selected amphibians and reptiles associated with southern forested wetlands habitats.

Taxa[2]	Common name	Habitat type[1]				
		AF	PCB	MW	FWM	MAN
Seminatrix pygaea	Swamp snake				•	
Storeria dekayi	Northern brown snake	•	•			
Storeria occipitomaculata	Redbelly snake		•	•		
Tantilla coronata	Southeastern crowned snake		•			
Thamnophis proximus	Western ribbon snake	•				
Thamnophis sauritus	Eastern ribbon snake		•		•	
Thamnophis sirtalis	Chicago garter snake		•		•	
Virginia striatula	Rough earth snake		•		•	
Virginia valeriae	Smooth earth snake		•			
Agkistrodon contortix	Southern copperhead	•				
Viperidae						
Agkistrodon piscivorus	Eastern cottonmouth	•	•		•	•
Crotalus adamanteus	Eastern diamondback rattlesnake				•	
Crotalus horridus	Timber rattlesnake	•	•	•	•	

[1] AF = alluvial floodplain forests, FWM = flatwoods/wetlands mosaic, PCB = pocosins/Carolina bays, MW = mountain wetlands, MAN = mangroves.
[2] Partial species lists taken from: AF = Phelps and Lancia (1995), Foley (1994), Clawson et al. (1997); FWM = Abrahamson and Hartnett (1990), Enge and Marion (1986), Vickers et al. (1985); MAN = Gilmore and Snedaker (1993), Odum et al. (1982); MW = Boynton (1994); PCB = Richardson and Gibbons (1993), Gibbons and Semlitsch (1981), Wilbur (1981).

mast and cover are available. Ground- and bark-foraging birds are particularly common. Winter bird communities may not differ greatly among hydrological zones (Wakeley and Roberts 1996).

Small mammal communities in alluvial floodplains often are dominated by a few species which sometimes vary among hydrologic zones. Some small mammals are semi-arboreal and escape seasonal flooding by climbing. Although many medium- and large-sized mammals are found throughout alluvial floodplains, some such as marsh rabbits (*Sylvilagus palustris*), nutria (*Myocastor coypus*), and mink are closely associated with wetter zones. Most larger mammals associated with wetter sites are capable of climbing or swimming. When flooding occurs, species less proficient at swimming or climbing will move to drier sites, seek refuge in brush tops or on other emergent structures, or perish.

Herpetofauna in alluvial floodplains have been described by Wharton et al. (1982), Foley (1994), Clawson et al. (1997), Phelps and Lancia (1995), and others. Standing water following floods and heavy rains is important for reproduction of many amphibians in alluvial floodplains. Amphibians in floodplains often are most abundant in moist conditions provided by a closed canopy and abundant leaf litter (Rudolph and Dickson 1990). Reptiles usually are most abundant where understory vegetation and associated prey are dense.

POCOSINS

Pocosins often are categorized as "short" or "tall." Structurally, short pocosins are characterized by scattered pond pine (*Pinus serotina*) with a short and very dense shrub layer. Tall pocosins typically have a nearly closed canopy of pond pine and hardwoods such as loblolly-bay, redbay (*Persia borbonia*), and sweetbay (*Magnolia virginiana*). The shrub layer is taller but less dense than in short pocosins.

Wildlife communities of pocosins are composed of few endemic species (Sharitz and Gibbons 1982, Richardson and Gibbons 1993). Breeding and winter bird communities of short pocosins are relatively low in diversity and richness, and species often occur at low densities (Lee 1986, Karriker 1993). Scattered clumps of trees in short pocosins appear to be an important habitat attribute for some birds such as eastern wood-peewees (*Contopus virens*) and brown-headed nuthatches (*Sitta pusilla*). Karriker (1993) speculated that a lack of thermal cover, overstory structure, and seeds from autumn-blooming herbs limit habitat quality of short pocosins during winter. Breeding birds associated with mature forests are more common in tall pocosins due to abundant broadleaf midstory and overstory trees, substantial overstory canopy closure, and snags (Lynch 1982, Lee 1986, Karriker 1993). These habitat attributes also appear to provide winter birds with foraging sites, fruits, and thermal cover.

Currently, concern exists over possible declines in some early-successional birds found in short pocosins. Sauer and Droege (1992) estimated that between 1966 and 1988, eight early-successional species including indigo buntings, painted buntings (*Passerina ciris*), prairie warblers (*Dendroica discolor*), and yellow-breasted chats (*Icteria virens*) had experienced significant declines in abundance. It is unclear whether these declines are due to alteration of breeding habitat, changes in winter habitat, or some other factor.

Small mammal communities in pocosins are greatly affected by understory composition and structure. High levels of herbaceous cover, low overstory density, and minimal litter depth appear to enhance small mammal richness and density (Mitchell et al. 1995). Because of greater structural diversity, semi-arboreal species (e.g., cotton mice [*Peromyscus gossypinus*] and golden mice [*Ochrotomys nuttalli*]) sometimes are more abundant in tall pocosins than in short. Limited scent station data suggest that richness and abundance of medium-sized mammals also may be greater in tall pocosin than in short (Mitchell 1994).

Little is known about herpetofaunal communities of pocosins. Permanently flooded ponds in small pocosins are important habitat for herpetofauna such as spotted turtles (*Clemmys guttata*). One amphibian possibly endemic to pocosins is the pine barrens treefrog (*Hyla andersonii*) (Wilbur 1981).

CAROLINA BAYS

Plant communities within Carolina bays often occur in concentric zones and usually offer habitat attributes that are much less common in the surrounding landscape. Many species primarily associated with uplands use bays for some life requisite (e.g., food, breeding sites). Bays are especially important to amphibians for reproductive purposes (Gibbons and Semlitsch 1981). Thus, even though few wildlife species are endemic (Sharitz and Gibbons 1982, Richardson and Gibbons 1993), wildlife diversity associated with bays can be quite high (Clark et al. 1985).

Many mammals using Carolina bays are opportunistic associates of early-successional habitat (Clark et al. 1985). Some, such as cotton mice and eastern harvest mice (*Reithrodonomys humulus*), are found across hydrologic zones. Others, such as eastern moles (*Scalopus aquaticus*), are restricted to outer portions of bays where soil moisture is lower and more conducive to burrowing. Larger, more mobile mammals such as raccoons, black bears, and white-tailed deer often include bays as part of their home ranges, but also use upland habitats.

Carolina bays support over 100 species of breeding birds (Lee 1987). However, most are found in other habitats, and no single group of birds can be considered characteristic of bays (Lee 1987). Several bay-associated birds such as the worm-eating warbler (*Helmitheros vermivorus*), Louisiana waterthrush (*Seiurus motacilla*), and chipping sparrow (*Spizella passerina*) are on the periphery of their distribution where Carolina bays occur. Aquatic birds such as American coots (*Fulica americana*) and great blue herons (*Ardea herodias*) often use bays with permanent water for feeding and breeding. The sand rim vegetation surrounding many Carolina bays also supports a variety of forest- and shrub-associated birds.

Mountain Wetlands

Mountain wetlands are relatively isolated, rare, small, and scattered unevenly across the landscape. Beyond Boynton's (1994) list of species potentially using these habitats, wildlife communities of mountain wetlands are poorly understood. Mountain wetlands support an unusually large proportion of rare species such as the star-nosed mole (*Condylura cristata*), bog turtle (*Clemmys muhlenbergii*), mole salamander (*Ambystoma talpoideum*), four-toed salamander (*Hemidactylium scutatum*), and southern bog lemming (*Synaptomys cooperi*) (Murdock 1994). Boynton (1994) lists Cooper's hawk (*Accipiter cooperii*), mountain chorus frog (*Pseudacris brachyphona*), olive-sided flycatcher (*Contopus borealis*), and pygmy shrew (*Microsorex hoyi*) as species of special concern.

Several wildlife species found in mountain wetlands are at the southern periphery of their range. For example, the rose-breasted grosbeak (*Pheucticus ludovicianus*), woodland jumping mouse (*Napaeozapus insignis*), and red squirrel (*Tamiasciurus hudsonicus*) are most common in the northern United States or Canada (Bull and Farrand 1994, Whitaker 1995). However, they also are found in the southern Appalachians at high elevations.

Flatwoods/Wetlands Mosaic

The pine flatwoods landform often has depressional wetlands dominated by baldcypress, pondcypress (*Taxodium distichum* var. *nutans*) (collectively referred to hereafter as cypress), or other species (e.g., water tupelo) distributed within a matrix of pine flatwoods. Depressional wetlands and flatwoods usually differ in hydrology.

Pine flatwoods sometimes have poorly drained soils and may experience shallow water tables or shallow inundation for brief periods, especially early in the growing season or after large storms. However, these sites are commonly quite dry for much of the year with water tables 1 m (3 ft) or more below the surface. In contrast, depressional wetlands are commonly inundated for long periods and, even when surface water is absent, the water table is close to the surface.

Many wildlife species commonly use both dry and wet habitats within the flatwoods landform (e.g., Harris and McElveen 1981, O'Meara 1984). However, cypress ponds provide habitat attributes unique within the landscape and support some species not commonly found in drier sites. Open water and abundant snags in the center of ponds provide habitats for aquatic and cavity-nesting birds and mammals. Around the periphery of ponds, understory vegetation and the evergreen midstory provide abundant fruit and a substrate for arthropod activity. Thus, there are many foraging opportunities for small mammals and birds (Harris and Vickers 1984). Because a variety of species use cypress ponds, species richness usually is high and actually can be greatest in small ponds with large amounts of edge per unit area (O'Meara 1984).

Bird and mammal communities in drier flatwoods sites dominated by pines often are described as depauperate. The hispid cotton rat (*Sigmodon hispidus*), cotton mouse, least shrew (*Cryptotis parva*), northern short-tailed shrew (*Blarina brevicauda*), and house mouse (*Mus musculus*) were the only small mammals captured by Labisky and Hovis (1987). This low number of species may reflect the relatively simple structure of fire-influenced habitat or biogeographic considerations.

Because of the interspersion of wet and dry sites, amphibian and reptile communities within the flatwoods landform generally are rich in species. Depressional wetlands are especially important to amphibians for reproductive purposes and as refugia during dry periods, while drier sites are used for other activities. In cypress ponds, amphibians can outnumber reptiles by as much as five to one (Vickers et al. 1985), with ranids being especially common (Harris and Vickers 1984). Amphibians usually increase in abundance in ponds as water depth increases. In drier sites, burrows and decaying stumps and logs are important habitat features for herpetofauna, especially reptiles. Burrows of the gopher tortoise also provide habitat for green anoles, squirrel treefrogs (*Hyla squirella*), greenhouse frogs (*Eleutherodactylus planirostris*), ringneck snakes (*Diadophis punctatus*), indigo snakes (*Drymarchon corais*), eastern diamondback rattlesnakes (*Crotalus adamanteus*), and eastern box turtles (Brandt et al. 1993).

Mangrove Forests

Wildlife communities in mangrove (*Rhizophora* spp.) forests vary greatly among hydrologic zones (Gilmore and Snedaker 1993). Faunal communities associated with the aquatic prop root zone of mangroves usually are dominated by fish and invertebrates such as sessile or mobile tunicates, crustaceans, and molluscs. Invertebrates such as molluscs, nematodes, amphipods, shrimp, and crabs characterize benthic communities associated with mangrove fringe habitats and ponds in basin forests of mangrove when they are flooded.

Birds are the most obvious vertebrate taxa associated with the arboreal canopy of mangrove forests. Some such as the mangrove cuckoo (*Coccyzus minor*), brown pelican, and roseate spoonbill (*Ajaia ajaja*) usually are not found in other southern forest types, although many others are. A variety of wading birds and birds of prey (e.g., roseate spoonbills, bald eagles) nest, roost, or perch in mangrove forests but feed in nearby sublittoral/littoral areas, tidal creeks, ditches, and basins. Mangrove forests also are important stopover habitat for neotropical migratory birds during spring and fall migrations.

Few small mammals have been documented in mangrove forests, probably because of a lack of intensive surveys. Odum et al. (1982) list only five species: the marsh rice rat (*Orzyomys palustris*), key rice rat (*Oryzomys argentatus*), hispid cotton rat, black rat (*Rattus rattus*), and short-tailed shrew. With the exception of the insectivorous short-tailed shrew, these species are omnivores that feed on vegetation, fungi, and aquatic invertebrates such as crayfish and crabs. Medium- and large-sized mammals such as the river otter, raccoon, white-tailed deer, marsh rabbit, and bobcat are associated with tidal creeks, ditches, and basins where they feed on a variety of molluscs, crustaceans, and vertebrates, or on vegetation.

Few amphibians are associated with mangroves, presumably because they are unable to osmoregulate in salt water, and few comprehensive surveys have been conducted. Odum et al. (1982) list only three amphibians in mangroves: the giant toad (*Bufo marinus*), squirrel treefrog, and Cuban treefrog (*Osteopilus septentrionalis*). The Cuban treefrog, an exotic species introduced by humans, is North America's largest treefrog. The giant toad, which can reach 24 cm (9.4 in) in length, also is an exotic. Odum et al. (1982) suggest that several other species of toads and frogs, including the eastern narrow-mouthed toad (*Gastrophryne carolinensis*), eastern spadefoot (*Scaphiopus holbrooki holbrooki*), southern cricket frog (*Acris gryllus gryllus*), green treefrog (*Hyla cinerea*), and southern leopard frog, may use mangrove forests.

Aquatic ecosystems associated with mangrove forests support several reptiles not commonly found in other forested wetland systems. These include the American crocodile (*Crocodylus acutus*) and several species of marine turtles. However, reptiles more common to other forested wetland ecosystems (e.g., cottonmouths, eastern mud turtles [*Kinosternon subrubrum subrubrum*]) also occur in mangrove forests. Only three species of lizards, all arboreal *Anolis*, have been documented in mangroves (Odum et al. 1982).

ACCOMMODATING WILDLIFE IN MANAGED WETLAND FORESTS

It is not unusual for forest landowners to have wildlife community objectives such as a sustained annual harvest of a game species or providing habitat for a threatened, endangered, or rare species. Thus, many forested wetlands in the South will continue to be managed for both economic and ecological objectives. Because of this consideration, landowners and managers should understand how wildlife respond to stand- and landscape-level management of forested wetlands. Because most existing wetland forests have been altered at least once during their history, it should be possible to maintain current wildlife communities in managed forests under most conditions.

Stand-level Management

Individual wildlife species typically respond differently to the immediate consequences of management. The response will vary according to the type, extent, and intensity of management within the forest community and to the sensitivity of the wildlife species to specific habitat alterations. The landscape context of the management event likely will affect wildlife response at the stand level. Other factors such as aspect, topographic position, flooding regime, site quality, structure and dynamics of a species' population, and wildlife community interactions (e.g., competition) also can affect wildlife response to forest management. For any stand-level management practice, populations of some species may remain relatively unaffected, some may decrease in abundance, and some may increase. Thus, responses of wildlife to forest management in wetlands are not simple and cannot be characterized as simply "good" or "bad."

At the local scale, harvesting usually immediately reduces canopy cover, simplifies stand structure, reduces the availability of some micro-habitat features (e.g., cavities, hard mast), and enhances the growth of understory vegetation. The availability of some habitat features such as foraging substrates (e.g., multiple foliage layers, boles, snags, limbs), dense "interior" conditions (if the stand is large enough), hard mast, and arboreal cavities are directly related to the amount and character of residual overstory canopy. Thus, even-aged silvicultural systems, especially clearcutting, typically have a greater immediate, stand-level impact on habitat attributes than do uneven-aged systems. Because of these considerations, habitat suitability for wildlife species associated with these overstory features usually will decline on freshly harvested sites (McComb and Noble 1980, Mitchell 1989, Mitchell and Lancia 1990, Thompson and Fritzell 1990, deMaynadier and Hunter 1995). Enge and Marion (1986) found that clearcutting in pine flatwoods habitats immediately reduced stand-level populations of arboreal lizards and amphibians, species groups favored by closed canopy conditions.

Conversely, some habitat features increase in abundance in newly harvested stands. Understory vegetation generally increases in diversity and abundance. Stumps and logging slash enhance cover and harbor beetles, grubs, and other invertebrates that are protein-rich food sources for many species of wildlife (Weaver et al. 1990). Ruts made by skidders and other logging equipment hold water and serve as breeding locations for amphibians (Clawson et al. 1997). Thus, species associated with ground-level habitat often increase in abundance because of enhanced food and cover resources (USFS 1980), including some small mammals (McComb and Noble 1980, Wesley et al. 1981), amphibians (Enge and Marion 1986, Clawson et al. 1997), and large vertebrates such as black bear and white-tailed deer (Weaver et al. 1990).

Young, regenerating stands support a number of birds that prefer dense thickets of saplings, shrubs, or vines. Songbirds such as yellow-breasted chats, prairie warblers, indigo buntings, white-eyed vireos, hooded warblers, and golden-winged warblers (*Vermivora chrysoptera*) usually benefit from activities that open the canopy. A large percentage of these songbirds are Neotropical migrants that may be declining (Peterjohn and Sauer 1993). Game birds such as wild turkeys often nest in or near openings (Speake et al. 1975, Wesley et al. 1981) and feed on insects associated with herbaceous vegetation found there (Hurst and Stringer 1975, Healy and Nenno 1983). Wintering waterfowl feed on native moist-soil plants that grow in forest openings (Heitmeyer 1985).

It is important to remember that many habitat alterations due to forest management activities are relatively temporary and must be considered at several temporal and spatial scales. Short- and long-term responses usually differ, as do responses at various spatial scales. Many habitat features that are immediately reduced in abundance by harvesting usually become more abundant as the stand matures. Availability of hard mast and arboreal cavities usually increases as a harvested stand matures. Depending upon structure and composition of the regenerated stand, some habitat features may eventually be more abundant than prior to harvesting.

In many ecosystems, wildlife communities regain approximate pre-harvest structure and composition as harvested stands mature. Pough et al. (1987) and Petranka et al. (1993) found that amphibian communities approximate pre-harvest structure and composition within 50–70 years in upland forests. Amphibian communities in bottomland hardwood ecosystems may approximate their pre-harvest structure and composition much sooner following clearcutting (Foley 1994, Clawson et al. 1997). Similarly, many birds normally associated with older forests can reach high densities in stands as soon as 25–30 years following harvest (Mitchell 1989, Mitchell and Lancia 1990). Other late-successional species may require much longer to reach peak density.

Landscape-level Management

Although there is much interest in accommodating wildlife at the landscape level, there are few data related to responses of wildlife species and communities to specific landscape configurations. At present, a major concern among resource managers is the possibility that forest management results in "fragmentation" which leads to dysfunctional wildlife communities. Harris and Silva-Lopez (1992) referred to a fragmented forest as "a landscape that formerly was forested but now consists of forested tracts that are segregated and sometimes isolated in a matrix of nonforested habitat." The concept of forest fragmentation is directly drawn from theories regarding island biogeography. However, the equilibrium theory of island biogeography remains largely untested in managed forest landscapes, is not universally accepted, and may be overemphasized as a conservation tool (Reed 1983, Woolhouse 1983).

Fragmentation often is viewed as having a number of consequences such as reduction in the total area of forest, the creation of patches or islands of remnant forest, and alteration of microclimate conditions within patches which may affect the biological community in many ways (Harris and O'Meara 1989, Wilcove 1989). Non-forest habitat surrounding forest patches might present barriers to successful dispersal if the habitats are unsuitable to a species due to factors such as high predation rates, a lack of food items, or if other species normally serving as cues are absent. Fragmentation also might reduce the amount of suitable micro-habitats in a landscape, increase amounts of edge, and exacerbate problems sometimes suggested to afflict small populations, e.g., demographic fluctuations, genetic deterioration, and susceptibility to environmental perturbations.

One of the most popular proposals among conservationists is to use corridors as mitigation for fragmentation in managed landscapes. In wetland forest landscapes, this would mean leaving unharvested or lightly harvested strips, such as streamside management zones, to connect older forests. This notion is a natural and logical consequence of the equilibrium theory (Wilson and Willis 1975). If a habitat is isolated from other such habitats, it is intuitive that connecting these patches might mitigate for some adverse effects of fragmentation.

Corridors would be most effective if they actually were used for immigration and if those immigrations served to mitigate local extinctions which were occurring at a predictable rate (Simberloff and Cox 1987). Apparently, there are reasonable doubts about whether either of these conditions are met (Gilbert 1980, Williamson 1989, Simberloff et al. 1992). Additionally, there are few data substantiating that immigrating animals actually contribute much to the fitness in small populations (Noss 1987), and small, isolated populations do not always go extinct (Simberloff et al. 1992). Corridors also can have negative effects such as the introduction of disease or exotic organisms (Hess 1994). For these reasons, the application of corridors within managed forest landscapes probably should be determined on a site-specific basis, considering the ecological factors pertinent to the animals of concern (Simberloff et al. 1992).

Because of these uncertainties, it is important to realize that our understanding of how forest management activities affect biotic communities at the landscape scale is incomplete. Most landscape-scale studies to-date have been conducted in areas permanently fragmented by agriculture or urbanization and with large amounts of high-contrast edges. Conversely, managed forested wetland landscapes usually are a mosaic of heterogeneous forest types, differing in age, species composition, structure, area, and shape. Many stands in managed forest landscapes will have some characteristics of preferred habitats and may be suitable for some uses such as dispersal. Probably because of these differences, recent studies in forest-dominated landscapes (Thompson et al. 1992, Hamel et al. 1993, Welsh and Healy 1993, McGarigal and McComb 1994, Hagan et al. 1995) indicate that these landscapes function differently from landscapes fragmented by agriculture or urbanization. More such research is needed in southern forested wetland landscapes.

Despite the lack of information regarding responses of wildlife to specific landscape configurations, managers should consider landscape characteristics when planning management activities. In many cases, this means that managers will only be able to consider the landscape context of stand-level management activities. For example, harvesting the only mature stand in a landscape dominated by young successional stages could increase stand-level diversity but could reduce diversity at the landscape scale. In most cases, a landscape composed of a variety of forest types and successional stages is most diverse and supports the largest complement of species (Hunter 1990). Enge and Marion (1986) suggested that herpetofaunal diversity would be greatest in a landscape that is a mosaic of small clearcuts, cypress domes, and different-aged pine stands.

Often wildlife-related objectives are best met by planning at the landscape scale. Interestingly, planning at the landscape scale may actually increase management options at the stand scale. For example, habitat requirements of black bears cannot be met fully within an agricultural field, but an agricultural field can contribute to the support of black bears at the landscape scale. Similarly, arboreal cavities and hard mast-producing trees usually are not abundant in young forest stands. Thus, it is difficult to feature wildlife species associated with cavities or hard mast in newly regenerated stands. However, streamside management zones and other buffer-type stands often support tree species that readily form cavities (Melchiors and Cicero 1987) and sometimes are managed on longer rotations and with different silvicultural systems than other stands. By planning the maintenance of such important habitat components within the landscape, forest management activities can contribute to maintaining diversity at larger scales.

10 Fish Communities

Jan Jeffrey Hoover and K. Jack Killgore

CONTENTS

1-56670-228-3/97/$0.00+$.50
© 1998 by CRC Press LLC

INTRODUCTION

Fish communities of southern forested wetlands were first characterized by William Bartram in 1791 (Van Doren 1955). He described wetland geo-morphometry, plant communities, hydrology, and fish. Bartram suggested that wetlands provide fish with high-water avenues for migration, low-water refugia, and areas of intensive fish production and predation. Ecology of fish in southern forested wetlands, however, is not extensively documented (e.g., Ross 1987). This is partly attributable to difficulty in sampling structurally complex aquatic habitats, but is more likely due to emphasis on stream habitats by ichthyologists and to pervasive, long-term reductions in wetlands (Junk et al. 1989). Insufficient study also reflects public attitude toward wetland fish such as pirate perch (*Aphredoderus sayanus*), gar (Lepisosteidae), and bowfin (*Amia calva*). Forested wetlands are viewed as "unfamiliar and primitive" because fish inhabiting them are "secretive and primitive" (Clark 1978). Described herein are fish of southern forested wetlands, their natural history, and the environmental factors that influence populations.

ICHTHYOFAUNA

ZOOGEOGRAPHY

The Mississippi Basin, the Appalachians, and near-shore waters of the continental shelf are important conduits of fish dispersal and adaptive radiation, so the fish fauna of the Southeast is extremely rich. Ichthyofauna consist of more than 450 species (Lee et al. 1980). There are 36–55 species/1° quadrat in this region, compared with <35 species/1° quadrat for the rest of North America (Hocutt and Wiley 1986). Most species, including wetland fish, have extensive geographic distributions defined by broad physiographic boundaries, such as the lower Mississippi Basin, Atlantic and Gulf Coastal Plains, or some combination thereof (Lee et al. 1980). Florida, particularly the Apalachicola Basin, is a notable "contact zone," in which ranges of closely-related species and subspecies overlap, and intergrades (hybrids) occur. Because of relative isolation of some wetland systems, geographic clines and races exist for several broadly distributed wetland fish; e.g., chubsuckers (*Erimyzon* spp.), sailfin mollies (*Poecilia latipinna*), mud sunfish (*Acantharcus pomotis*), and swamp darters (*Etheostoma* [*Hololepis*] spp.) (Hubbs 1941, Frey 1951, Collette 1962, Cashner et al. 1989).

Two centers of endemism exist (Hocutt and Wiley 1986). Lake Waccamaw, a Carolina bay, is inhabited by the endemic Waccamaw killifish (*Fundulus waccamensis*), Waccamaw silverside (*Menidia extensa*), and Waccamaw darter (*Etheostoma perlongum*), and by a distinctive form of coastal shiner (*Notropis petersoni*) (Hubbs and Raney 1946). Freshwater wetlands in peninsular Florida are inhabited by an endemic gar (*Lepisosteus platyrhincus*) and several killifish (*Lucania goodei, Jordanella floridae, Fundulus seminolis, F. confluentus*) (Hunt 1953, Barnett 1972, Lee et al. 1980); mangrove forests in southernmost Florida are inhabited by a mosquitofish (*Gambusia rhizophorae*), silverside (*Menidia conchorum*), and rivulus (*Rivulus marmoratus*) restricted to that area and parts of the Caribbean (Gilbert

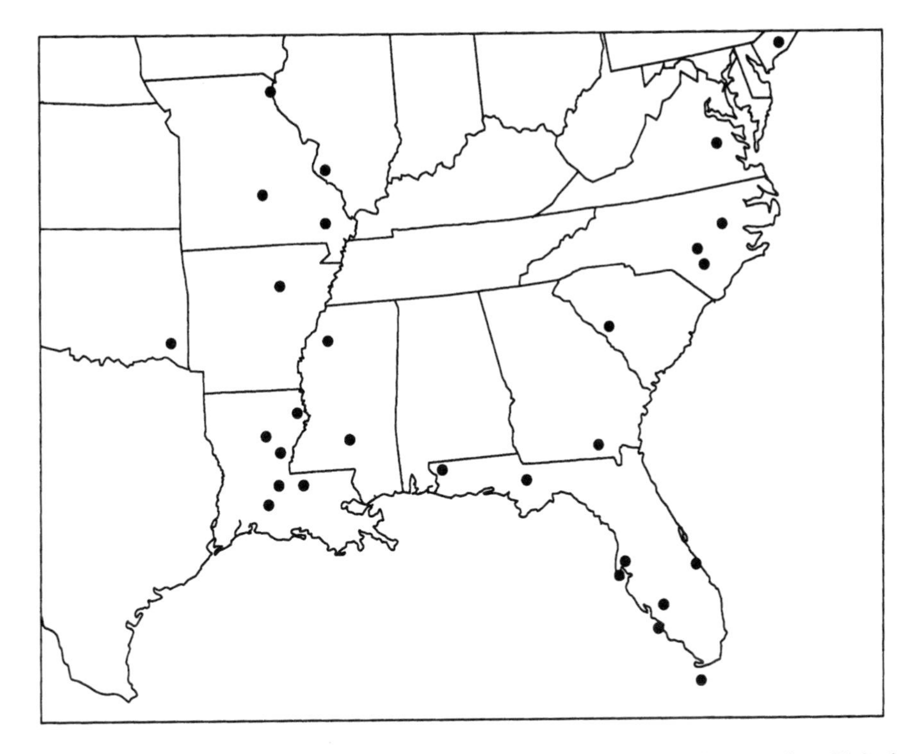

FIGURE 10.1 Locations of field studies of fish in forested wetlands of the southern United States.

1992). Evolution of these unusual fauna is attributed to a stable, benign physical environment in Lake Waccamaw (Hubbs and Raney 1946, Frey 1948) and, in Florida, to insular isolation of fish populations during Pliocene sea level changes (Neill 1957).

Field studies of fish in southern forested wetlands encompass ranges of the broadly-distributed and endemic forms (Figure 10.1). They are conducted throughout the Gulf and Atlantic Coastal Plains and in the lower Mississippi Basin. More than 80 species of wetland fish occur (Table 10.1). These include transient, riverine species that seasonally exploit and occasionally reside in wetlands, such as buffalo (*Ictiobus* spp.) and freshwater drum (*Aplodinotus grunniens*), and resident, slack water species that permanently inhabit wetlands. This latter group is dominated taxonomically by killifish (*Fundulus* spp.), sunfish (*Lepomis* spp.), oligotypic taxa such as gar, pickerel (*Esox* spp.), and pygmy sunfish (*Elassoma* spp.), and monotypic taxa such as bowfin (*Amia calva*) and pirate perch. Darters (Percidae) and minnows (Cyprinidae), speciose (>100 spp) in southern streams, do not dominate taxonomically in wetlands.

Fish Diversity and Wetland Habitats

Fish species richness is high in bottomland hardwood systems and some mangrove forests (Table 10.2). In very large drainages dominated by bottomland hardwoods,

TABLE 10.1
Native fish characteristic of southern forested wetlands: bottomland hardwoods (BLH), bays and pocosins (POC), mangroves (MAN), and pinewoods (PIN). Characteristic species are not ubiquitous and are more abundant in wetland habitats than in adjacent water bodies.

	BLH	POC	MAN	PIN
Paddlefish, Polyodontidae				
Paddlefish, *Polyodon spathula*	•			
Gars, Lepisosteidae				
Spotted gar, *Lepisosteus oculatus*	•			
Longnose gar, *L. osseus*	•			
Shortnose gar, *L. platostomus*	•			
Florida gar, *L. platyrhincus*	•			
Bowfin, Amiidae				
Bowfin, *Amia calva*	•			
Tarpons, Elopidae				
Tarpon, *Megalops atlanticus*			•	
Eels, Anguillidae				
American eel, *Anguilla rostrata*	•	•		
Minnows, Cyprinidae				
Blacktail shiner, *Cyprinella venusta*	•			
Bluehead shiner, *Notropis hubbsi*	•			
Taillight shiner, *N. maculatus*	•			
Blackmouth shiner, *N. melanostomus*	•			
Weed shiner, *N. texanus*	•			•
Pugnose minnow, *Opsopoeodus emiliae*	•			
Fathead minnow, *Pimephales promelas*	•			
Bluenose shiner, *Pteronotropis welaka*				•
Suckers, Catostomidae				
Smallmouth buffalo, *Ictiobus bubalus*	•			
Black buffalo, *I. niger*	•			
Creek chubsucker, *Erimyzon oblongus*	•			•
Lake chubsucker, *E. succetta*	•	•		
Spotted sucker, *Minytrema melanops*	•			
Bullhead catfish, Ictaluridae				
Black bullhead, *Ameiurus melas*	•			
Yellow bullhead, *A. natalis*	•	•		
Brown bullhead, *A. nebulosus*	•			
Tadpole madtom, *Noturus gyrinus*	•	•		
Pikes, Esocidae				
Redfin pickerel, *Esox americanus americanus*	•	•		
Grass pickerel, *E. a. vermiculatus*	•	•		
Chain pickerel, *E. niger*	•	•		
Mudminnows, Umbridae				
Central mudminnow, *Umbra limi*				•
Eastern mudminnow, *U. pygmaea*		•		•

TABLE 10.1 (continued)
Native fish characteristic of southern forested wetlands: bottomland hardwoods (BLH), bays and pocosins (POC), mangroves (MAN), and pinewoods (PIN). Characteristic species are not ubiquitous and are more abundant in wetland habitats than in adjacent water bodies.

	BLH	POC	MAN	PIN
Pirate perches, Aphredoderidae				
Pirate perch, *Aphredoderus sayanus*	•			•
Cavefish, Amblyopsidae				
Swampfish, *Chologaster cornuta*				•
Needlefish, Belonidae				
Atlantic needlefish, *Strongylura marina*			•	
Redfin needlefish, *S. notata*			•	
Rivulins, Aplocheilidae				
Mangrove rivulus, *Rivulus marmoratus*			•	
Killifish, Cyprinodontidae				
Sheepshead Minnow, *Cyprinodon variegatus*			•	
Goldspotted killifish, *Floridichthys carpio*			•	
Golden topminnow, *Fundulus chrysotus*	•			
Marsh killifish, *F. confluentus*	•		•	
Starhead topminnow, *F. dispar*	•			•
Gulf killifish, *F. grandis*			•	
Mummichog, *F. heteroclitus*	•			
Striped killifish, *F. majalis*			•	
Blackspotted topminnow, *F. olivaceus*	•			•
Blackstripe topminnow, *F. notatus*	•			
Bayou topminnow, *F. notti*	•			•
Longnose killifish, *F. similis*			•	
Waccamaw killifish, *F. waccamensis*		•		
Flagfish, *Jordanella floridae*	•			
Pygmy killfish, *Leptolucania ommata*	•			
Bluefin killifish, *Lucania goodei*	•			
Rainwater killifish, *L. parva*			•	
Livebearers, Poeciliidae				
Western mosquitofish, *Gambusia affinis*	•			•
Eastern mosquitofish, *G. holbrooki*	•	•	•	
Mangrove gambusia, *G. rhizophorae*			•	
Least Killifish, *Heterandria formosa*	•			
Sailfin molly, *Poecilia latipinna*	•		•	
Silversides, Atherinidae				
Hardhead silverside, *Atherinomorus stipes*			•	
Brook silverside, *Labidesthes sicculus*	•			
Inland silverside, *Menidia beryllina*	•		•	
Waccamaw silverside, *Menidia extensa*		•		
Tidewater silverside, *M. peninsulae*			•	

TABLE 10.1 (continued)
Native fish characteristic of southern forested wetlands: bottomland hardwoods (BLH), bays and pocosins (POC), mangroves (MAN), and pinewoods (PIN). Characteristic species are not ubiquitous and are more abundant in wetland habitats than in adjacent water bodies.

	BLH	POC	MAN	PIN
Snooks, Centropomidae				
Common snook, *Centropomus undecimalis*			•	
Pygmy sunfish, Elassomatidae				
Carolina pygmy sunfish, *Elassoma boehlkei*	•			
Everglades pygmy sunfish, *E. evergladei*	•			
Bluebarred pygmy sunfish, *E. okatie*	•			
Okefenokee pygmy sunfish, *E. okefenokee*	•			
Banded pygmy sunfish, *E. zonatum*	•			•
Sunfish, Centrarchidae				
Mud sunfish, *Acantharchus pomotis*	•	•		
Flier, *Centrarchus macropterus*	•	•		•
Blackbanded sunfish, *Enneacanthus chaetodon*	•			
Bluespotted sunfish, *E. gloriosus*	•	•		
Banded sunfish, *E. obesus*	•	•		
Warmouth, *Lepomis gulosus*	•	•		
Dollar sunfish, *L. marginatus*	•	•		•
Redspotted sunfish, *L. miniatus*	•			
Blackspotted sunfish, *L. punctatus*	•			
Bantam sunfish, *L. symmetricus*	•			•
Black crappie, *Pomoxis nigromaculatus*	•			
Largemouth bass, *Micropterus salmoides*	•			
Perches, Percidae				
Bluntnose darter, *Etheostoma chlorosomum*	•			
Swamp darter, *E. fusiforme*	•	•		
Waccamaw darter, *E. perlongum*		•		
Cypress darter, *E. proeliare*	•			
Gulf darter, *E. swaini*				•
Yellow perch, *Perca flavescens*		•		
Mojarras, Gerreidae				
Spotfin mojarra, *Eucinostomus argenteus*			•	
Silver jenny, *E. gula*			•	
Drums, Sciaenidae				
Freshwater drum, *Aplodinotus grunniens*	•			
Spot, *Leiostomus xanthurus*		•		
Mullets, Mugilidae				
Striped mullet, *Mugil cephalus*	•		•	
Total number of characteristic species	45	20	22	16

TABLE 10.2
Species richness of southern forested wetland fish communities: number of species recorded for wetlands/number of species known for locality (percentage of local fauna known to occur in wetlands).

Habitat, Location	Wetland/Total	Reference
Bottomland Hardwoods		
Mississippi River, LA	62/91 (68%)	Guillory 1979
Atchafalya River, LA	53/95 (56%)	Lambou 1990
Ochlockonee River, LA	37/48 (77%)	Leitman et al. 1991
Appalachicola River, FL	51/91 (56%)	Light et al. 1995
Hillsborough River, FL	31/42 (74%)	Barnett 1972
Escambia River, FL	29/58 (50%)	Beecher et al. 1977
Creeping Swamp, NC	23/23 (100%)	Walker 1985
Mangrove Forests		
Tampa Bay, FL	19/31 (62%)	Mullin 1995
Indian River, FL	88/397 (22%)	Gilmore 1995
Rookery Bay, FL	13/35 (37%)	Sheridan 1992
Florida Bay, FL	64/87 (74%)	Thayer et al. 1987
Pine Forests		
Pine Hills, IL	23/82 (28%)	Gunning and Lewis 1955
Black Creek, MS	26/42 (62%)	Ross and Baker 1983
Bays and pocosins		
Lake Waccamaw, NC	25/51 (49%)	Frey 1951
Cape Fear Bays, NC	18/71 (25%)	Frey 1951
Savannah Bays, SC	12/88 (14%)	Snodgrass et al. 1996

most fish present will occur in wetlands, although some only rarely (Baker et al. 1991). Field surveys of 16 different wetland systems range from 12 to 88 species and represent 14 to 100% of all fish species known from those localities. Bottomland hardwood communities consist of 23 to 62 species, 50 to 100% of local fish fauna. Mangrove forest fish communities consist of 19 to 88 species, 22 to 74% of local fish fauna. Fish communities in pine forests, bays and pocosins consist of 12 to 26 species, 14 to 62% of local fish fauna. Isolated bays are particularly depauperate, individual water bodies typically harboring only 1 to 5 species (Snodgrass et al. 1996).

Fish communities in wetlands exhibit log-normal patterns of species abundance. Two or three species are numerically dominant (cumulatively >70% of individuals), several species are extremely rare (each $\leqslant$1% individuals), and the majority of species are intermediate in abundance. Diversity of different communities, however, may be compared by quantifying number of species expected from random subsamples of the fish community or "rarefaction" (Ludwig and Reynolds 1988). Higher numbers of species expected from subsamples of equal size indicate higher species diversity or community complexity.

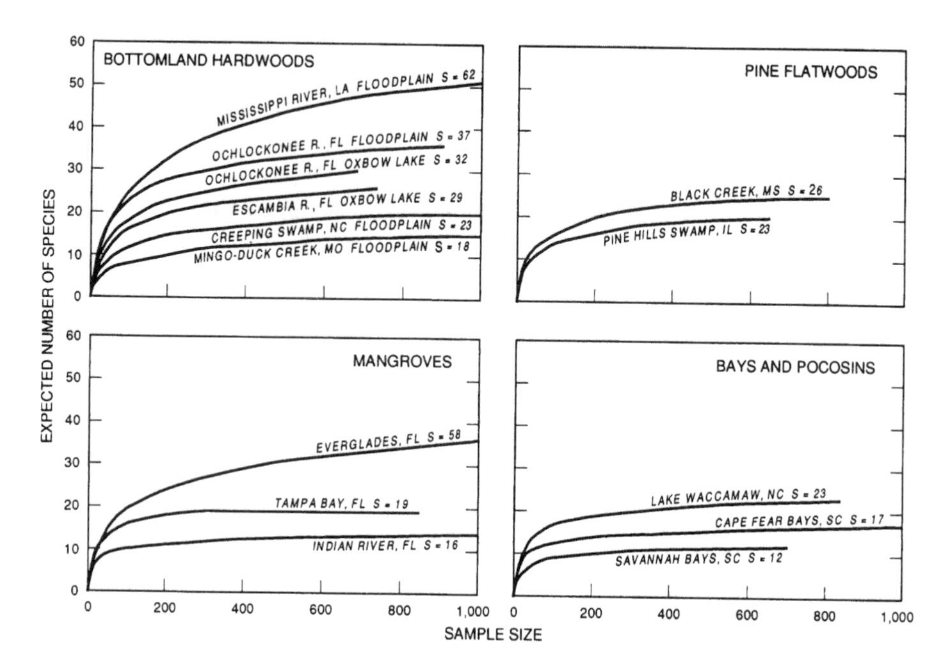

FIGURE 10.2 Rarefaction curves for 14 published studies of forested wetland fish communities (references listed in text). S = number of species collected during survey (total number of species known may be higher).

Rarefaction curves illustrate fish diversity for 14 wetland communities (Figure 10.2). Diversity is highest for bottomland hardwood floodplains of the Mississippi and Ochlockonee Rivers: 32 and 41 spp/600 fish (Guillory 1979, Leitman et al. 1991). Moderate diversity exists in Florida oxbow lakes, Lake Waccamaw, pine woods of Black Creek, Mississippi, and Everglades mangrove forests: 20 to 29 spp/600 fish (Frey 1951, Ross and Baker 1983, Thayer et al. 1987, Leitman et al. 1991). Comparatively low diversity exists in small South Carolina bays and in mangrove forests and bottomland hardwoods of higher latitudes: 11 to 19 spp/600 fish (Harrington and Harrington 1961, Walker 1985, Finger and Stewart 1987, Mullin 1995, Snodgrass et al. 1996).

Wetland fish may be considered as a guild, or complex of guilds, since resident species share several morphological, physiological, and reproductive traits. Most species are small, elongate, surface-dwelling forms, with auxiliary methods of respiration, and high reproductive investment (Figure 10.3).

MORPHOLOGICAL ADAPTATIONS OF WETLAND FISH

Gross Morphology

Two body forms predominate, each associated with distinctive behavior (Hubbs 1941, Webb 1988, 1994). "Accelerators" are elongate, terete, streamlined, with

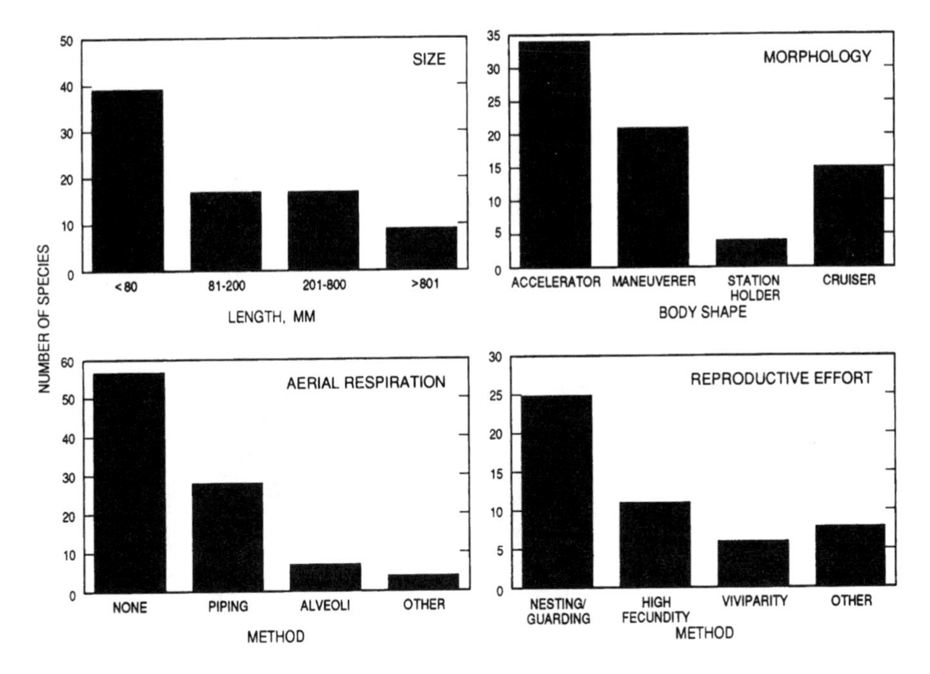

FIGURE 10.3 Wetland adaptations in 85 fish (species listed in Table 10.1). Data compiled from Lee et al. (1980), Breder and Rosen (1966), Pflieger (1975), Robison and Buchanan (1988), and Rohde et al. (1994).

flexible bodies and large tails (deep peduncle and low, expansive caudal fin); dorsal and anal fins are typically small, situated posteriorly (Figure 10.4). Propulsion, by body and caudal fin, maximizes thrust for fast starts, especially through open water. Accelerators include small, pelagic surface-dwellers such as livebearers (Poeciliidae), killifish (Cyprinodontidae), and silversides (Atherinidae). Accelerators also include large, littoral fish that inhabit submersed cover, such as gars (Lepisosteidae) and pickerels (Esocidae). "Maneuverers" are deep-bodied, compressed, gibbose, with symmetrical dorsal-ventral contours; large paired fins are situated anteriorly (Figure 10.4). Propulsion, by median and paired fins, allows precise, low-speed swimming (including hovering) in structurally complex habitats. Most maneuverers are found near submersed wood and vegetation at middepth. The group includes fliers (*Centrarchus macropterus*), dwarf sunfish (*Enneacanthus* spp.), and bantam sunfish (*Lepomis symmetricus*).

Two other forms are less common in wetlands (Figures 10.3 and 10.4). "Stationholders" have low, elongate bodies adapted for small, low velocity microhabitats; they have large pectoral fins and lack swim bladders creating negative lift. Stationholders also have rough body surfaces and large pelvic fins used to grasp substrate. Adaptations allow fish to maintain benthic position in flowing water. Primary mode of locomotion is by darting — short bursts of swimming close to the bottom, then resting on sediments or submersed cover. This group includes the swamp (*Etheostoma fusiforme*), cypress (*E. proeliare*), bluntnose (*E. chlorosomum*) and slough (*E. gracile*) darters. "Cruisers" have various body shapes but peduncles are narrow

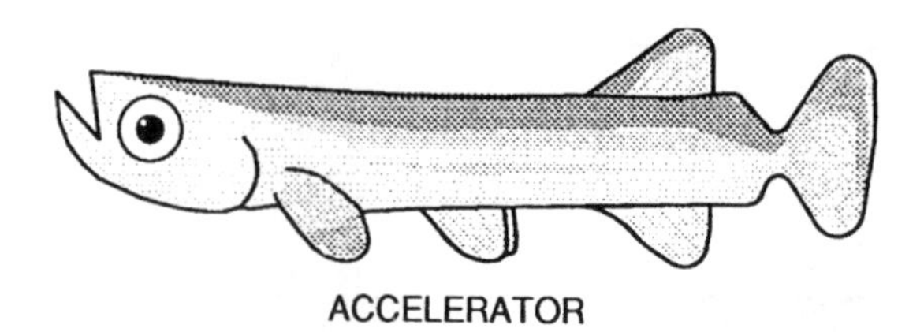

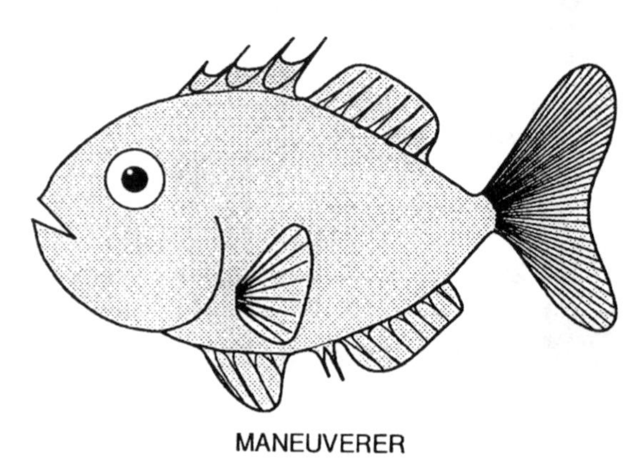

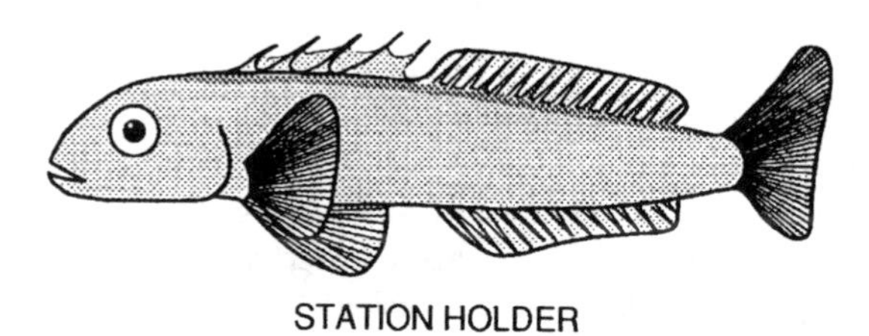

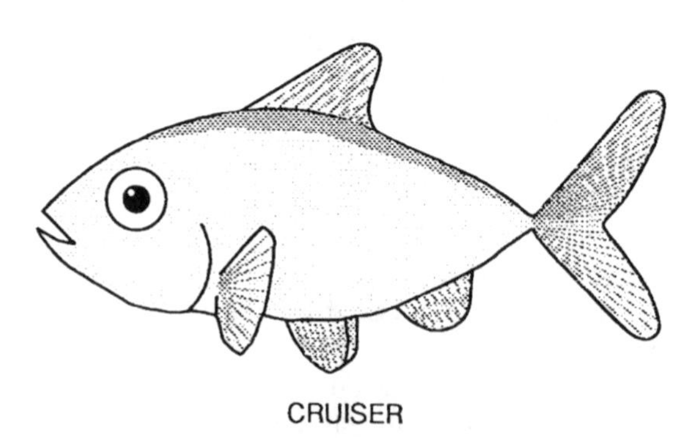

FIGURE 10.4 Morphological forms of wetland fish.

and caudal fins high, deeply forked with relatively small area. Their morphology enhances sustained swimming. This group includes minnows (Cyprinidae) and suckers (Catostomidae).

Two species have distinctive morphologies. Paddlefish (*Polyodon spathula*) form is intermediate between that of an accelerator and a cruiser, allowing this species to inhabit the surface of deep, flowing water. Bowfin are unique among freshwater fish, having a long, undulatory dorsal fin that provides locomotion (Webb 1994). Both species, together with gar, have primitive heterocercal tails (caudal fins with a muscular extension of the body in the upper lobe). Such tails generate lift, and may facilitate surfacing.

Pigmentation

Hues of wetland fish are subdued, typically drab blends of white, brown, green, and black. Bright colors, usually associated with reproduction, are conspicuous in only a few species (e.g., taillight shiner [*Notropis maculatus*], golden topminnow [*Fundulus chrysotus*], and dollar sunfish [*Lepomis marginatus*]). Conservative coloration of most fish provides crypsis in tannin-stained waters with fine sediments and leaf litter.

Markings such as lateral bands in weed shiners (*Notropis texanus*) and chubsuckers, and posterior spots in blacktail shiner (*Cyprinella venusta*), taillight shiner and bantam sunfish (*Lepomis symmetricus*), may deflect attacks by predators. The prominent "eyespot" on the tail of the adult male bowfin, however, provides sexual dimorphism and allows paternal recognition by juveniles (Breder and Rosen 1966). Function of a similar marking in mangrove rivulus is unknown, although that species occurs in habitats free of predatory fish and is a self-fertilizing hermaphrodite unknown to provide parental care (Gilbert 1992).

Melanism is commonly reported for five wetland species: spotted gar, sailfin molly (*Poecilia latipinna*), eastern mosquitofish (*Gambusia holbrooki*), western mosquitofish (*Gambusia affinis*), and golden topminnow (Inness 1948, Suttkus 1963, Barnett 1972). Numbers of melanistic individuals are low (<<5% population), however, and records are more likely in colored waters of cypress and flatwood swamps than from turbid systems. Functional significance of melanism is unknown except for the eastern mosquitofish (Martin 1977). In that species, melanistic males exhibit greater aggression towards normally pigmented males, greater sexual activity, and greater avoidance of open water.

PHYSIOLOGICAL ADAPTATIONS OF WETLAND FISH

Respiration

In flowing streams, dissolved oxygen (DO) concentrations are high and fish absorb sufficient oxygen from water through their gills. Southern forested wetlands, however, frequently experience hypoxia (DO <5.0 mg/l) and anoxia (DO <3.0 mg/l), especially in systems that are heavily shaded and wind-sheltered, and in which high levels of turbidity, nutrients, and decaying material occur. In those habitats, dissolved oxygen is invoked as a limiting factor for fish (Baker et al. 1991), but suffocation of resident species is rare. Coping mechanisms of fish include: 1) shifts in microhabitat; 2) tolerance of anaerobic metabolism; 3) changes in activity; 4) extraction of low concentrations of oxygen; 5) increased breathing at the water's surface; and 6) increased breathing of atmospheric air (Bevelander 1934, Kramer 1987). Such mechanisms not only reduce (or prevent) suffocation, but also increase energy available for locomotion, growth, and reproduction.

Fish experiencing respiratory distress or detecting low concentrations of DO move laterally or higher into the water column. This has been demonstrated experimentally for spotted gar (*Lepisosteus oculatus*), which preferentially inhabit DO concentrations of 6 to 10 mg/l (Hill et al. 1973). Anaerobic metabolism, suggested by prolonged tolerance of low oxygen, is documented for bowfin, grass pickerel (*Esox americanus vermiculatus*), chubsuckers, several minnows, bullheads (*Ameiurus* spp.), and sunfish (Moore 1942, Odum and Caldwell 1955, Lewis 1970, Smale and Rabeni 1995). Some of these species reduce movements (and oxygen demand), some exhibit special anatomical adaptations that permit extraction of oxygen under very low concentrations (e.g., sieve-shaped gills of the bowfin) (Bevelander 1934).

Killifish and live-bearers are small (<80 mm [3.5 in] SL) and dorsally-flattened, with small, oblique, superior mouths. This allows exploitation of the oxygen-rich surface film of water that exists even during extreme anoxia (Lewis 1970). These fish break the surface film and draw water across their gills, a behavior called "piping." In mangrove forests, larger species, such as striped mullet (*Mugil cephalus*) and common snook (*Centropomus unidecimalis*), also employ aquatic surface respiration (Hoese 1985, Peterson et al. 1991).

Modified gas bladders that allow fish to breathe air occur in three taxa: gars, bowfin, and mudminnows (*Umbra* spp.) (Renfro and Hill 1970, Horn and Riggs 1973, Gee 1980). These species are physostomous; the gas bladder connects to the roof of the foregut via the pneumatic duct. Fish can survive complete anoxia, by surfacing and swallowing air, which passes into the gas bladder, and oxygen can be absorbed through "alveoli." Air breathing is performed in conjunction with branchial respiration but increases at higher temperatures and lower DO concentrations. Synchronous aerial respiration by mudminnows and young gar is presumed to reduce predation by birds (Hill 1972, Gee 1980). Surface-skimming fish, such as livebearers and mullet, also gulp air, absorbing oxygen through highly vascularized structures in the pharynx called pseudobranchs (Odum and Caldwell 1955, Hoese 1985).

Temperature Preference/Tolerance

Wetland habitats are typically warmer than river channels (Baker et al. 1991), and wetland fish apparently prefer and tolerate high water temperatures. Final preferenda (a genetically fixed measure of temperature preference) is 28.4 to 34.0°C (83.1 to 93.2°F) for yellow bullhead (*Ameiurus natalis*) (Reynolds and Casterlin 1978), bluespotted sunfish (*Enneacanthus gloriosus*) (Casterlin and Reynolds 1979), bowfin (Reynolds et al. 1978), and lake chubsucker (*Erimyzon succetta*) (Negus et al. 1987). Values for these fish are higher than final preferenda indicated for riverine and lacustrine fish (Coutant 1977). Wetland fish also exhibit higher lethal temperatures than riverine species (Brett 1956). Lethal temperature for most southern stream fish is approximately 35°C (95°F), but for yellow and black bullheads (*Ameiurus melas*), blackstripe (*Fundulus notatus*), and blackspotted topminnows (*F. olivaceus*), it is 38 to 39°C (100 to 102°F) (Smale and Rabeni 1995).

Tolerance of Extreme Water Quality

Most wetland fish are moderately tolerant of degraded water quality (Jester et al. 1992). Acidic water (pH <5.0) impairs osmoregulation, respiration, and reproduction, and is associated with low species richness of the Okefenokee Swamp and some Carolina bays (Frey 1951, Laerm et al. 1980). Sensitive species include several minnows; tolerant species include pickerels, mudminnows, dwarf sunfish, and swamp darters (Robison and Buchanan 1988, Rohde et al. 1994). Turbid water (>35 NTU's) impairs fish movement, vision, feeding, and survival of eggs and larvae (Wilber 1983). Sensitive species include taillight shiner and swamp darters; tolerant species include buffalo, several sunfish, and blacktail shiner (Ewing 1991, Hoover et al. 1995).

REPRODUCTION

Spawning

Egg placement in shallow water, nest construction and guarding, "fanning" of eggs by parents, and oral manipulation of eggs are observed among disparate groups such as bowfin, bullheads, and sunfish and are believed to be adaptive responses to hypoxia and siltation (Breder 1936, Hubbs 1941, Breder and Rosen 1966). Viviparity (live-bearing), also believed to be hypoxia-adaptive, permits variation in embryo size allowing rapid response to fluctuating temperature, food availability, and predation (Hubbs 1941, Meffe 1990).

Ovaries are well-developed in wetland species. Gonadal weights of 6 to 12% of body weight are typical for minnows and silversides, as are multiple clutches (Cowell and Barnett 1974, Hubbs 1982, Heins and Dorsett 1986, Heins and Rabito 1988, Taylor and Norris 1992). Fecundities of 5,000 to 40,000 eggs occur in gars and bowfin (Carlander 1969). Also, gar eggs are believed to be toxic to potential predators such as birds and mammals (Netsch and Witt 1961). Eggs of bowfin are protected in a large (45 to 75 cm [18 to 30 in]), sheltered nest, guarded by the highly aggressive male; after hatching, young are "herded" in compact, ball-shaped schools by circular movements of the male, who continues to protect them until they are at least 100 mm (4 in) total length (Breder and Rosen 1966).

Spawning Chronology

Collectively, wetland fish spawn throughout most of the year, principally during spring and summer. Spawning among co-occurring species is staggered so that reproduction by different species is temporally segregated. Early spawners include buffalo, darters, and black basses (*Micropterus* spp.); late spawners include sunfish and minnows.

Temperature is strongly associated with spawning (Hontela and Stacy 1990). Widespread species are presumed to spawn over similar ranges of temperatures in different parts of their range. A few species (e.g., mud sunfish, flier, bowfin, and pirate perch) spawn at temperatures <15°C (59°F) (Shepherd and Huish 1978, Murdy and Wortham 1980, Wharton et al. 1981, Pardue 1993). Most wetland fish spawn at 15 to 25°C (59 to 77°F). Taxa spawning at warmer temperatures include taillight and blacktail shiners, and sunfish (Cowell and Barnett 1974, Heins and Dorsett 1986, Turner et al. 1994). These data are consistent with spawning chronologies documented from individual forested wetlands in the lower Mississippi Basin (Turner et al. 1994, Hoover et al. 1995, Killgore and Baker 1996). Other environmental factors influencing reproduction include photoperiod, lunar cycles, nutrient availability, and river stage.

Forested Wetlands as Nurseries

For larval fish, wetlands provide abundant food that promotes rapid growth and structurally complex habitats used as refugia from predators (Paller 1987, Lambou 1990). Larval fish abundance in the Big Sunflower River Basin of Mississippi demonstrates spawning and rearing value of forested wetlands, especially for perciform fish (Figure 10.5). Freshwater drum spawn in open water of streams, but maximum numbers of larvae are observed in inundated forests fringing the river. Shad and common carp, invasive and ubiquitous, readily exploit non-forested land. Buffalo also spawn on flooded agricultural fields, but maximum abundance of larvae is in forested oxbow lakes. Densities of sunfish and crappie are low in non-forested habitats, high in forested habitats. Black bass and darters occur in measurable numbers only in backwaters of a large national forest.

HYDRODYNAMICS

Hydraulics

Most forested wetlands are lentic. However, hydraulics can affect habitat use and dispersal of fish in side channels and oxbows connected to rivers and in fringing floodplain wetlands. Velocity influences abundances of larval fish within different habitats (Turner et al. 1994) and contributes to their downstream movement (Harvey 1987, Copp and Cellot 1988). Hydraulic mixing of larval fish on floodplains, however, may obscure habitat-use patterns (Killgore and Baker 1996). Velocity is associated with increased species richness of invertebrates and fish, possibly due to increased levels of dissolved oxygen and allochthonous nutrient sources (Ellis et al. 1979, Thorpe et al. 1985).

Hydrology

Most wetland fish are littoral spawners, and lateral movement of the "aquatic-terrestrial transition zone" during flooding provides expanded habitat for reproduction (Junk et al. 1989). Timing of floods, however, is critical, as rising waters must coincide with rising temperatures and lengthening daylight to be of maximum benefit to fish, as

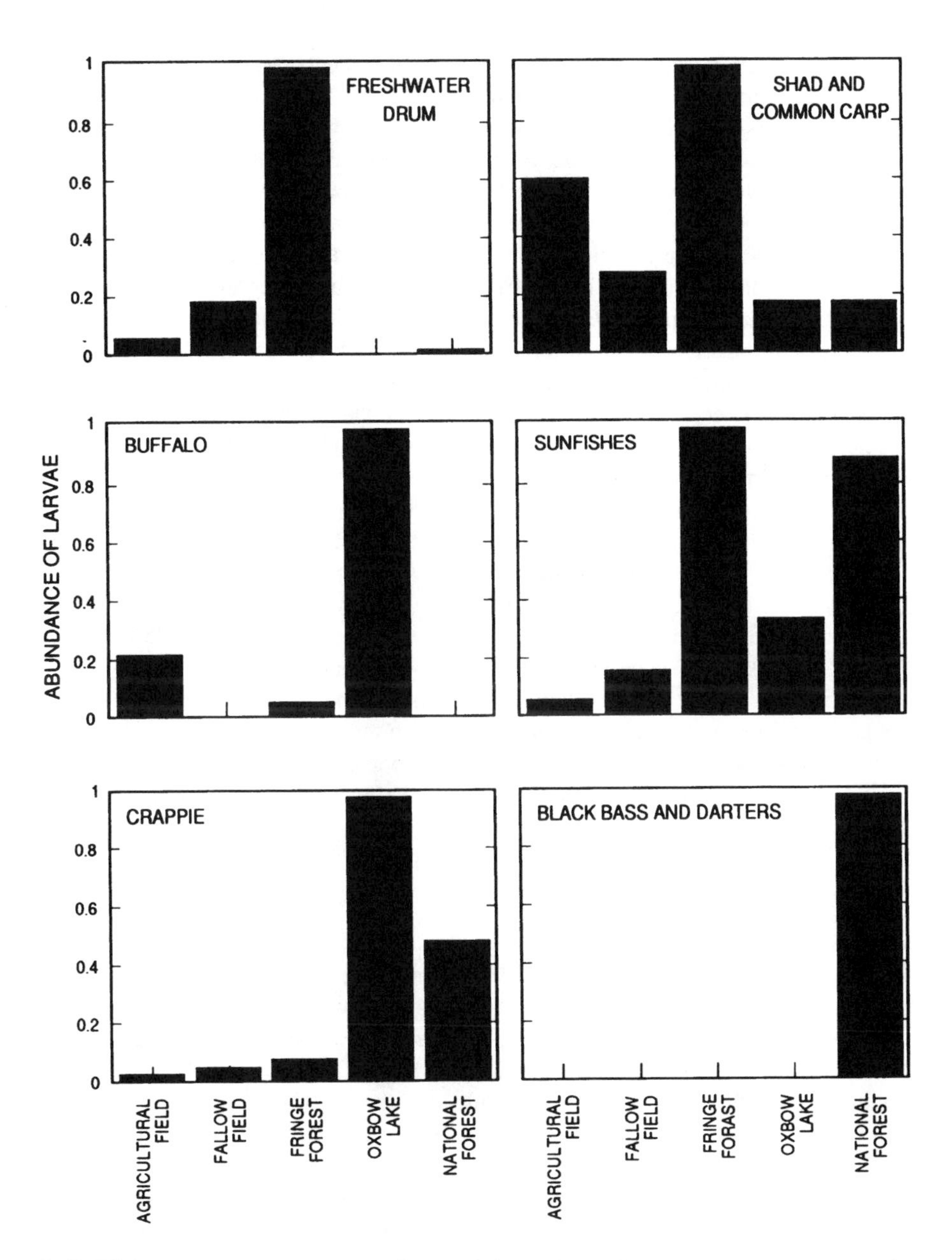

FIGURE 10.5 Relative abundance of larval fish in Big Sunflower River Basin, MS; x-axis represents a gradient of increasingly forested habitats; y-axis is standardized fish abundance (spring-summer density for that habitat divided by maximum spring-summer density for any habitat).

this is the period of gonadal maturation. If flooding takes place too early or too late in the season, benefits to fish are diminished.

Temporal changes in fish abundance coinciding with hydrologic cycles are documented. Pirate perch are twice as abundant during a year of high discharge in early spring as they are when discharge is low and peaks in late winter (Killgore and Baker 1996). Weed shiners, a flood exploitative species, are significantly more abundant during high water years (Ross and Baker 1983). Abundances of western mosquitofish, banded pygmy sunfish (*Elassoma zonatum*), eastern mosquitofish, flag fish (*Jordanella floridae*), and Everglades pygmy sunfish (*Elassoma evergladei*) are also associated with hydrology (Barney and Anson 1920, 1921, Streever and Crisman 1993).

Flood-associated increase in fish numbers is attributable to migrations of riverine fish onto the floodplain (Guillory 1979). It is also attributed to enhanced food resources that occur via nutrient releases, increased photosynthesis, and increased availability of detritus, terrestrial insects, and aquatic invertebrates, including crayfish (Lantz 1970a, Sklar 1985, Gladden and Smock 1990, Leitman et al. 1991).

When water recedes, food again becomes abundant for predatory fish as prey species are concentrated (Kushlan 1974, Carlson and Duever 1977, Kushlan and Kushlan 1980). Densities of smaller fish subsequently decline (Lambou 1959). Without predators, numbers of small fish remain high, food and space become limiting, and competition is likely. Intraspecific competition may be intense, resulting in lower growth rates, establishment of territories, and increased combat (Rubenstein 1981a, 1981b, 1981c). Interspecific competition, however, is reduced by habitat and food partitioning among sympatric, closely-related species (Holloway 1954, Bohlke 1956, Barnett 1972, Graham and Hastings 1984, Graham 1989).

Occurrence of fish in individual wetlands is determined by distribution of wetlands and by the extent and frequency with which they dry (Snodgrass et al. 1996). Community structure results from sequential colonization (high water) and extirpations (low water). A drying wetland, however, is not necessarily a death sentence for the fish living there.

Dessication

Tropical fish are well-known to survive dessication, but many sub-tropical and temperate wetland fish also survive drying. Mangrove rivulus can survive 66 days out of water (Taylor 1990). Bowfin are found in muddy soil and reportedly aestivate in spheres of dried mud (Bevelander 1934, Neill 1950). Marsh killifish (*Fundulus confluentus*) survive on damp mud for more than 24 hours (Kushlan 1973), and eggs remain viable after more than three months of exposure (Harrington 1959). Aerial incubation has also been confirmed for mummichog (*Fundulus heteroclitus*) and is likely for other residents of mangrove forests: gulf (*F. grandis*), longnose (*F. similis*), and striped killifish (*F. majalis*) (Taylor 1990). It has also been suggested for flagfish, juveniles of which appear in large numbers and of uniform size in previously dry, isolated ponds; also, flagfish exhibit relatively high rates of hatching failure and ranges of hatching time when cultured under completely aquatic conditions (Nilsson 1973). Dried eggs of shortnose gar (*Lepisosteus platostomus*) can be hatched in the laboratory, and dessication of their floodplain spawning habitats is believed frequent, with rains reconstituting eggs (Breder and Rosen 1966).

fish that cannot resist direct dessication may still survive drying wetlands by escaping to small, aquatic refugia. They may survive in puddles (depth <6 cm [2.4 in]), as do some invertebrates (White 1985). Subterranean streams may be utilized by fish, although apparently not by marine-derived species such as sailfin molly (Van Doren 1955, Hellier 1966). Crayfish burrows are abundant in bottomland hardwoods (Lambou 1990), providing refugia for many small fish. Bluefin killifish (*Lucania goodei*), pirate perch, banded pygmy sunfish, tadpole madtoms (*Noturus gyrinus*), and, possibly, mosquitofish and grass pickerel occupy burrows when floodplain ponds dry for 3 to 4 months (Creaser 1931, Neill 1951).

PREDATION

Short, efficient food chains and high densities of predators are observed in bottomland hardwoods (Hunt 1953, Hansen et al. 1971, Wharton et al. 1981, Lambou 1990), bays (Frey 1951, Schalles and Shure 1989), and mangrove forests (Harrington and Harrington 1961, Thayer et al. 1987). Top-level carnivores, such as gar, typically feed on herbivores such as flagfish, crayfish, and shrimp, or on planktivores such as silversides (Hunt 1953, Echelle and Riggs 1972). Omnivory, however, is not uncommon, especially among small species, but even these fish exhibit selective feeding in specific microhabitats or on individual taxa (Beach 1974, Cowell and Barnett 1974, Lorman and Magnuson 1978, Graham 1989). Productivity of these small fish is high because of their rapid life cycles and ability to respond to algae and zooplankton pulses (Freeman and Freeman 1985).

Most wetland predators are solitary (or occur in very small groups) and sedentary. Examples include gar, bowfin, mud sunfish, pirate perch, and pygmy sunfish (Holloway 1954, Barnett 1972, Parker and Simco 1975, Rubenstein 1981b, Pardue 1993). Limited movement, especially among elongate fish, may confer substantial energetic benefits (Hunt 1960, Netsch and Witt 1961). Active predators eat large quantities, digest food quickly, and grow slowly, but sedentary predators, like gar, eat small quantities, digest food slowly, and grow rapidly. Longnose gar (*Lepisosteus osseus*) attain lengths >50 cm (20 in) in their first year (Netsch and Witt 1961). Spotted gar and bowfin attain lengths of 22 (9 in) and >40 cm (15.7 in), respectively, during their first four months (Carlander 1969).

Predatory fish affect appearance and demography of wetland fish and impact invertebrate communities. Attenuated bodies, smaller scales, and reduced coloration and lateral lines of the Lake Waccamaw endemic fish probably evolved as responses to intense predation by white catfish (*Ictalurus catus*), white bass (*Morone americana*), bowfin, and yellow perch (*Perca flavescens*) (Hubbs and Raney 1946). Predation by mosquitofish on least killifish (*Heterandria formosa*) results in smaller populations, with fewer and larger adults and fewer males (Belk and Lydeard 1994). The fathead minnow reduces invertebrate abundance, biomass, and richness in prairie marshes (Hanson and Riggs 1995); it and other invertivorous species probably have similar effects in forested wetlands.

ANTHROPOGENIC DISTURBANCE

Deforestation

Most wetland fish spawn in flooded forests (Breder and Rosen 1966). Elimination of woody debris and erosion of once stable substrates result in direct loss of habitat for spawning and larval fish. In the Atchafalaya River basin, deforestation is associated with reduced biomass of some sport fish, but increased biomass of common carp (*Cyprinus carpio*) and shad (*Dorosoma* spp.) (Wharton et al. 1981). The 40% reduction in forest around Lake Pontchartrain, Louisiana, is coincident with profound (52 to 84%) changes in species composition of lower trophic levels and reduced species richness of benthic invertebrates and fish (Stone et al. 1982). In contrast, although nearly 80% of the Cache River, Arkansas, basin has been deforested (Graves et al. 1995), diversity and abundance of larval fish are high (Killgore and Baker 1996), presumably because of its connection to the speciose White River and movement of fish between the two water bodies.

Floodplain Reduction

Long-term reductions in floodplain are believed to shorten food chains and eliminate higher trophic levels (Power et al. 1995). Floodplain habitats in the lower Mississippi Basin, particularly ponds and seasonally inundated floodplain, are reduced 80% from historic areas (Baker et al. 1991), and presumably, so are populations of fish abundant in them: taillight shiner, pirate perch, flier, warmouth, pygmy sunfish. Declines in sport fisheries and commercial fishing are reported for the Atchafalaya River basin following levee construction (Lambou 1963). Levees in southeastern Missouri are also associated with reductions in fish diversity and abundance of characteristic species like starhead topminnow (*Fundulus dispar*), banded pygmy sunfish, and bantam sunfish (Finger and Stewart 1987). Certain recreationally important species, however, such as largemouth bass, sunfish, and bullheads spawn successfully in leveed waters (Finger and Stewart 1987, Hoover et al. 1995).

MANAGEMENT OF WETLAND FISH COMMUNITIES

Wetland Fisheries

Fisheries are the exploitable components of fish and crustacean communities. In southern forested wetlands, sport fisheries exist for sunfish, largemouth bass, and crappie (*Pomoxis* spp.); commercial fisheries exist for crayfish (*Procambarus* spp.), buffalo, freshwater drum, and gar (Lambou 1963). Commercial fisheries for paddlefish formerly existed in oxbow lakes and backwaters of the Mississippi River (Stockard 1907), but habitat destruction, decreasing paddlefish abundance, and government restrictions on harvest now confine these fisheries mostly to riverine environments. Most sport and commercial species, however, are broadly distributed, relatively abundant, and possibly underexploited (Lantz 1970b, Huner 1978, Lambou 1990).

Wetland fisheries are "managed" like other fisheries: by limiting catch, by imposing size limits, by stocking, and by delineating a legal "fishing season" (Holloway 1954, Lambou 1963, Huner 1978). Another technique, fish removal, is also practiced. Theoretically, the reduction of predators (e.g., gars) or competitors (e.g., buffalo) enhances populations of remaining species, but empirical demonstrations are few. Data from case studies are equivocal (Holloway 1954, Lantz 1974), and the wisdom of predator eradication programs is being re-evaluated. Gar and bowfin, for example, may provide substantial benefits to sport fisheries by selectively feeding on small, diseased, parasitized fish (Herting and Witt 1967, Scarnecchia 1992).

Other Fish Resources

Non-fishery components of the fish community also merit consideration for management. Numerous wetland species are valuable as study organisms, biocontrol agents, and as indicators of water pollution (Clark 1978). Grass shrimp (*Palaemonetes* spp.), crayfish, bowfin, and gar are used as specimens in laboratory courses. Mosquitofish and killifish consume mosquitos and midges providing biological control of nuisance emergences of aquatic insects. Numerous wetland species, with specific habitat tolerances and preferences, are readily maintained and propagated in the laboratory and could be used as bioassay organisms: sailfin molly, least killifish, flagfish, taillight shiner, bluehead shiner (*Notropis hubbsi*), pygmy sunfish, and bantam sunfish (Innes 1948, Cowell and Barnett 1974, Quinn 1990, Jester et al. 1992).

Persistent interest and trade in ornamental wetland fish has existed since the 1860's (Klee 1968). Mosquitofish, sailfin molly, least killifish, golden topminnow, bluefin killifish, flagfish, pygmy sunfish, and dwarf sunfish were popular among early aquarists, frequently depicted in aquarium literature, and are still kept in home aquaria (Innes 1948, Quinn 1990). Demand among aquarists is currently satisfied by commercial European breeders, sporadic availability from domestic fish farmers, and exchanges among collector/hobbyists, but interest in native wetland fish is growing.

Few of the above taxa are regularly "harvested" so that traditional fishery management techniques are of limited value. Threatened fish require habitat preservation, population monitoring, and field surveys for additional populations. Other groups may be effectively managed by enhancing wetland habitats.

Imperiled Fish

None of the fish characteristic of southern forested wetlands are federally listed as "endangered," "threatened," or "of special concern," but 42 species are listed by one or more states (Schmidt 1996). Of eighteen states, two list no wetland species, ten list one to five species, and six list six to 12 species (Table 10.3). Number of fish species listed may be politically influenced but also reflects zoogeography. States listing larger numbers of fish are those with: (1) diverse wetland ichthyofauna, such as Arkansas and Missouri (Pflieger 1975, Robison and Buchanan 1988); (2) multiple

TABLE 10.3
Imperiled fish of southern forested wetlands: extinct/extirpated (X), endangered (E), threatened (T), special concern (S) according to surveys of state resource agencies (Schmidt 1996).

	AL	AR	DE	FL	GA	IL	LA	MD	MO	MS	NC	NJ	OK	SC	TN	TX	VA	WV
Paddlefish	T	S							S		E		S			E	T	S
Longnose gar			S															
Shortnose gar	S																	
Blacktail shiner						S												
Bluehead shiner		S				E	T						S			T		
Taillight shiner		S				E			E				S					
Blackmouth shiner				E														
Weed shiner						E												
Bluenose shiner				S			S											
Pugnose minnow									S									S
Black buffalo													S					S
Creek chubsucker																T		
Lake chubsucker		S							T									
Yellow bullhead			S															
Brown bullhead									T				X					
Grass pickerel																		S
Chain pickerel													S					
Central mudminnow		X							E									
Mangrove rivulus				S														
Golden topminnow									X						S			
Marsh killifish	S																	
Starhead topminnow						S			S									

Gulf killifish	S																	
Waccamaw killifish											S							
Bluefin killifish	E				S						S			S				
Least Killifish	E										S							
Brook silverside																	S	
Waccamaw silverside											T							
Common snook				S														
Carolina pygmy sunfish											T			T				
Bluebarred pygmy sunfish														S				
Mud sunfish			T															
Flier									S									
Blackbanded sunfish					S			S									E	
Banded sunfish	S		T															
Blackspotted sunfish						T												
Bantam sunfish						T			T									
Swamp darters		S							E									
Waccamaw darter											T							
Cypress darter						S												
Freshwater drum											T							
Striped mullet		S							T									
Extirpated	**0**	**1**	**0**	**0**	**0**	**0**	**0**	**0**	**1**	**0**	**0**	**0**	**1**	**0**	**0**	**0**	**0**	**0**
Endangered	**2**	**0**	**1**	**1**	**0**	**3**	**0**	**0**	**3**	**0**	**1**	**0**	**0**	**0**	**0**	**1**	**1**	**0**
Threatened	**1**	**0**	**2**	**0**	**0**	**2**	**1**	**0**	**4**	**0**	**4**	**0**	**0**	**1**	**0**	**2**	**1**	**0**
Special Concern	**4**	**6**	**2**	**3**	**2**	**3**	**1**	**1**	**4**	**0**	**3**	**0**	**5**	**2**	**1**	**0**	**1**	**4**
Total	**7**	**7**	**5**	**4**	**2**	**8**	**2**	**1**	**12**	**0**	**8**	**0**	**6**	**3**	**1**	**3**	**3**	**4**

endemic species, such as North Carolina (Rohde et al. 1994); and (3) multiple species occurring at edges of their range, such as Illinois (Lee et al. 1980).

Several fish with broad geographic distributions are imperiled, either throughout their range or at specific locations. Paddlefish, bluehead shiner, and taillight shiner are listed in four or more states (Table 10.3); all species are considered "intolerant" of degraded habitats (Jester et al. 1992). Central mudminnow, golden topminnow, and brown bullhead are apparently extirpated in Arkansas, Missouri, and Oklahoma, respectively (Table 10.3); in each case, however, these populations occurred at periphery of geographic ranges (Lee et al. 1980). State records for the central mudminnow and golden topminnow were based on one or two collections made more than 50 years ago, and brown bullhead is naturally sporadic or rare throughout much of its range (Pflieger 1975, Robison and Buchanan 1988).

Apparent rarity of some species may reflect lack of scientific collecting or insufficient habitat data rather than fragmented or declining populations. Taillight shiner, for example, only recently discovered in Kentucky, is common there in oxbow lakes with cypress swamp margins (Burr and Page 1975); the same species, once considered, "rare and vulnerable" in Arkansas, is now known from numerous sites (Robison 1978). Blackmouth shiner (*Notropis melanostomus*), known from a single location in Florida, went undetected during 35 years of intensive collecting, but was later found to be abundant in specific backwater habitats inhabited by brook silverside (*Labidesthes sicculus*) (Gilbert 1992). Some species not listed by resource agencies may also be in jeopardy. Mud sunfish are broadly distributed, but occurrence, population size, and survivorship are typically low and movement to other areas limited (Tarplee 1979, Cashner et al. 1989, Pardue 1993). Conservation of these and other wetland fish necessitates habitat preservation or habitat enhancement.

Habitat Enhancement

Invertebrate and fish production are maximized by floods of long duration (White 1985, Brown and Coon 1994, Killgore and Baker 1996) followed by periods of low water in which increased temperature and zooplankton production facilitate growth and feeding (Lantz 1974, Finger and Stewart 1987, Rutherford et al. 1995). Operating water control structures (weirs, flap-gates) so that water rises slowly in late winter-early spring, remains elevated with minor stage fluctuations (to reduce hypoxia) throughout spring, and then is slowly decreased in summer, will provide substantial benefits. Likewise, reopening side channels or backwaters blocked by sedimentation or levees will provide expanded spawning and rearing grounds for many species of migratory riverine fish (Ellis et al. 1979, Hoover et al. 1995). Ideally, managed hydrologic regimes should mimic natural cycles of flooding. Unregulated mangrove forests, for example, exhibit higher fish densities and species richness than those "managed" for mosquito control by impounding water during the summer (Lin and Beal 1995).

Floodplain reforestation provides habitat benefits for fish that spawn on seasonally inundated floodplains. Typically, mixed assemblages of hardwoods are planted as saplings on agricultural land. Maximum habitat benefits to forest-spawning fish are assumed to exist after 25 to 50 years.

Altering wetland geomorphometry, especially those water bodies not typically considered "wetlands," can also provide measurable benefits. Many forested backwaters of large rivers have been lost, but mouths of tributaries apparently provide comparable habitats (Brown and Coon 1994). Protecting fringe forests in these areas and maintaining or controlling flow through these mouths will benefit different species dependent on hydrologic connection to structurally complex slack water (Rey et al. 1990). Losses of seasonally inundated floodplain are high, but borrow pits harbor a similar fauna (Baker et al. 1991), and fish standing crops are high, suggesting substantial contribution to riverine communities when pits are inundated during floods (Sabo et al. 1991, Sabo and Kelso 1991). Borrow pits are man-made water bodies and thus can be excavated so that morphometry is optimal for fishery benefits (i.e., long, sinuous shorelines, high water volume, and variable depth).

Analytical Techniques and Models

Predator/prey ratios and growth/mortality models of wetland fish are difficult to apply due to extensive feeding by piscivores on decapods, seasonal cycles of hyperdispersion and overproduction during floods, and catastrophic mortality during drought (Penn 1950, Neill 1951, Holloway 1954, Lambou 1961, Kushlan 1974, Lambou 1990).

Habitat-based approaches use correlations between physical and biological data to predict responses of fish to wetland perturbations. Associations between multivariate ordinations of physical habitat and the occurrence (or abundance) of fish are used to objectively describe and quantify habitats (Meffe and Sheldon 1988, Pyron and Taylor 1993, Snodgrass et al. 1996). Path analysis provides working models describing causal, directional effects of physical variables on fish communities (Sheldon and Meffe 1995). "Anoxic factor," an index relating area of water bodies to dissolved oxygen, is significantly correlated with fish species richness (Nurnberg 1995). "Drawdown ratio," an index relating maximum water area during flooding to minimum water area during dry season, is correlated with fish biomass and rate of exploitation, presumably because intensity of flooding dictates abundance and individual weight of fish, and because water level during drought determines between-year survivorship (Welcomme and Hagborg 1977). A model based on landscape-level variables, trophic relationships among fish, and hydraulic geometry of streams is being used to predict effects of water regulation and floodplain reduction on floodplain food webs and fish communities (Power et al. 1995). Some of the above techniques have yet to be applied to southern forested wetlands, but all have been used to describe ecology of wetland species.

CONCLUSIONS

Fish communities of southern forested wetlands are comprised of taxonomically disparate species derived from riverine fauna. Many of these fish share morphological, physiological and reproductive characteristics. Most are small, elongate surface-dwellers or deep-bodied, vegetation dwellers. They are tolerant of hypoxia and resist dessication. Modes of reproduction vary, but parental investment is high (e.g., mul-

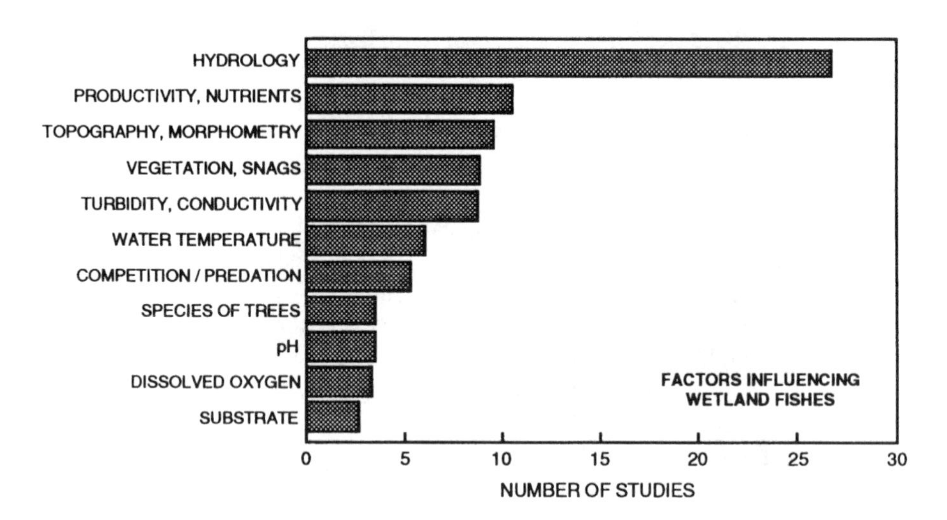

FIGURE 10.6 Factors important to fish (or fish food organisms) in 52 published population- and community-level studies (field studies listed in References). Cumulative frequency is >52 because multiple factors were invoked in several papers.

tiple clutches, high fecundity, nesting, guarding, or viviparity). These adaptations are comparable to those of marine decapods that have invaded freshwater (i.e., small size, habitat specialization, high tolerance of physico-chemical stress and large, brooded offspring) (Anger 1995). Such adaptations suggest environments that are food-limited and in which intense predation dictates export of offspring. In wetland fish communities, particularly speciose systems proximate to large rivers (bottomland hardwoods) or the open ocean (mangrove forests), this export strategy is coupled with hydrologic cycles of flooding. The importance of hydrology in structuring fish communities is indicated by the frequency with which it is discussed in field and laboratory studies of wetland fish (Figure 10.6). More than half of the wetland studies cited in this chapter invoke hydrology as a principal factor influencing fish occurrence, abundance, and demography.

ACKNOWLEDGMENTS

Assistance with this manuscript was provided by C. Dickerson, Jr., S. G. George, C. Murphy, A. Murphy, M. Vitaya-Odum, and T. Robinson. S. R. Adams, D. D. Dickerson and N. H. Douglas reviewed earlier drafts of this chapter. Funding was provided by the United States Army Corps of Engineers Wetlands Research Program. Permission to publish this document was granted by the Chief of Engineers.

Part III

Ecology and Management of Specific Wetland Types

11 Southern Deepwater Swamps

William H. Conner and Marilyn A. Buford

CONTENTS

1-56670-228-3/97/$0.00+$.50
© 1998 by CRC Press LLC

INTRODUCTION

Geographical Extent

Deepwater swamps, primarily baldcypress-water tupelo (*Taxodium distichum-Nyssa aquatica*), pondcypress-swamp tupelo (*Taxodium distichum* var. *nutans-Nyssa sylvatica* var. *biflora*), or Atlantic white-cedar (*Chamaecyparis thyoides*), are freshwater systems with standing water for most or all of the year (Penfound 1952). Cypress-tupelo swamps are generally found along rivers and streams of the Atlantic Coastal Plain from Delaware to Florida, along the Gulf Coastal Plain to southeastern Texas, and up the Mississippi River to southern Illinois (Johnson 1990, Wilhite and Toliver 1990). Baldcypress was a very dominant tree in the coastal plain of the southern United States when settlers first arrived (Mattoon 1915), and this resource seemed inexhaustible to these early settlers, with more than 35 million m^3 (15 billion board feet) of timber estimated in the Louisiana delta swamps alone (Kerr 1981). Other deepwater swamp types include cypress domes (see Chapter 16) and depressional swamps like the Okefenokee and Dismal Swamps. Atlantic white-cedar swamps occur in a narrow band (80 to 210 km [50 to 130 mi] wide) from southern Maine to northern Florida and west to southern Mississippi (Little and Garrett 1990). For further information on deepwater swamps, the reader is referred to Lugo et al. (1990b), Ewel and Odum (1989), Kirk (1979), Cohen et al. (1984), Dennis (1988), Thomas (1976), and Laderman (1989).

Classification

Probably the first classification of deepwater swamp forests was Shaler (1890), who used both physical and vegetational characteristics to identify two types of forested wetlands: freshwater swamps and estuarine swamps. Freshwater swamps were located in lowlands where drainage was hindered by barriers made by the rivers, were composed of freshwater plants, and contained alluvium supplied by freshwater streams. Estuarine swamps occurred where water levels fluctuated from tidal action and were composed of grasses and river alluvium.

In the 1950s, the U.S. Fish and Wildlife Service (USFWS) recognized 20 types of wetlands in relation to wildlife habitat value (Shaw and Fredine 1956). Wooded swamps were characterized by waterlogged soil and frequent inundation with 30 cm (12 in) of water, and they were found along sluggish streams, flat uplands, and shallow lake basins. This was the most widely used classification system in the United States until the National Wetlands Inventory classification system (Cowardin et al. 1979) was adopted.

Other classification systems developed in the 1950s were based on vegetation, habitat, and the quantity, depth, and duration of water as diagnostic criteria (Putnam 1951, Penfound 1952). Penfound (1952) classified swamps of the Atlantic and Gulf Coastal Plains and the Mississippi River Alluvial Valley on the basis of water depth and duration of flooding. Deep swamps were identified as freshwater, woody areas with surface water throughout all or most of the growing season. Shallow swamps and peaty swamps were inundated for only part of the growing season. The terminology of Putnam (1951) has probably been the most widely accepted by the forestry profession. He recognized first bottoms as areas developed from recent deposits as a result of frequent flooding by the present drainage system. Secondary classifications within this site type included ridges, flats, sloughs, and swamps. These site types generally apply to the floodplain of all major streams (see Chapters 12 and 13).

Küchler (1964) compiled a map of the United States with 116 different plant community types. His system was based primarily on vegetation and unique in that he described the potential vegetation at one point in community succession if human influence were removed. Of the 10 types he identified as inland wetlands, only one pertains to deepwater swamps: southern floodplain forest. Unfortunately, this classification system is difficult to apply on the ground (Bailey et al. 1978), and the units are too large to be of practical value other than for broad land-use and resource planning (Mader 1991).

Wharton et al. (1976) classified Florida freshwater swamps into four types based on hydrologic inputs. Stillwater cypress domes are poorly drained to permanently flooded depressions dominated by pondcypress. Lake edge swamps are found on the margins of many lakes and isolated sloughs of the southeastern United States. Both of these swamp systems experience little water inflow except through precipitation and, in some cases, groundwater. Slow-flowing cypress strands are shallow forested depressions on a gently sloping plain, with seasonal wet and dry cycles, although deep peat deposits retain moisture even in extremely dry periods. Alluvial river swamps and floodplains are confined to permanently flooded depressions on floodplains such as abandoned river channels or elongated swamps paralleling the river.

The USFWS National Wetland Inventory includes deepwater swamps within the class of Palustrine Forested Wetlands (Cowardin et al. 1979), which is characterized by woody vegetation at least 6 m (20 ft) tall. This class is further subdivided into five subclasses, of which four are representative of southern deepwater swamps: (1) broad-leaved deciduous where water tupelo, swamp tupelo, and ash (*Fraxinus* spp.) are represented; (2) needle-leaved deciduous where baldcypress dominates; (3) broad-leaved evergreen where bays are prevalent; and (4) needle-leaved evergreen where Atlantic white-cedar is dominant.

A national system of forest cover types was developed by the Society of American Foresters (Eyre 1980) based solely on species dominance, with the name of each forest type usually limited to one or two species. Deepwater swamp types under this classification include baldcypress (Type 101), baldcypress-tupelo (Type 102), water tupelo-swamp tupelo (Type 103), Atlantic white-cedar (Type 97), and pondcypress (Type 100). This classification system is not intended for intensive land management.

The U.S. Forest Service (USFS) conducts periodic inventories of the nation's forests and uses a system of 20 forest type groups based on overstory species composition (USFS 1967). These type groups are further broken down into local groups but are seldom used when reporting the data. Wetland hardwoods are equivalent to the combined oak-gum-cypress and elm-ash-cottonwood type groups in the southern United States (Boyce and Cost 1974).

Economic Importance

Both the Spanish in Florida and French in Louisiana found Indians using cypress, which the Seminoles called "hatch-in-e-haw," meaning everlasting (Neubrech 1939). The Europeans quickly recognized that cypress (refers to both baldcypress and pondcypress when used like this) wood was very rot resistant, strong, and easily worked, and efforts to establish a timber trade with Louisiana began around 1700 (Mancil 1980). Harvesting in these wet swamps was seasonal in nature until the invention of the pullboat in 1889. Pullboats and the expansion of the railroad system (Sternitzke 1972), combined with a massive national campaign by cypress dealers (Burns 1980), resulted in a logging boom during the period 1890 to 1925. Production of cypress lumber increased from 1.17 million m^3 (495 million board feet) in 1899 to more than 2.36 million m^3 (1 billion board feet) in 1913 (Mattoon 1915, Betts 1938), with the majority of the early commercial trade being baldcypress (Brown 1934). By 1925, nearly all of the virgin timber had been cut and most of the mills closed. In 1933, only about 10% of the original standing stock of cypress remained (Brandt and Ewel 1989), but some cypress harvesting continued throughout the southern United States on a smaller scale.

Atlantic white-cedar logging began as early as 1700 in North and South Carolina (Frost 1987) and 1749 in New Jersey (Little 1950). Frost's (1987) analysis of Ashe's (1894) figures suggest that up to 50% of the Atlantic white-cedar area in North Carolina was cut between 1870 and 1890. As in other parts of the southern United States, the rate at which Atlantic white-cedar swamps were logged greatly increased following the introduction of railroads, steam logging technology, portable sawmills, and dredging technology (Earley 1987, Frost 1987). Another period of logging Atlantic white-cedar occurred during 1970–1980 (Baines 1990).

Although the majority of cypress/tupelo swamps were cut over during the late 1800s and early 1900s and there has been a general decline in land area of this forest type (Dahl et al. 1991), there are currently between 1.2 and 2 million ha (2.9 and 4.9 million acres) of second-growth timber (Williston et al. 1980, Kennedy 1982). Standing stock volumes continue to increase (Brandt and Ewel 1989, Conner and Toliver 1990), with the greatest concentration of baldcypress in Louisiana, pondcypress in Florida, and Atlantic white-cedar in Florida and North Carolina (Table 11.1). Tupelo growing stock is more widespread among the states.

Second-growth cypress is much less decay resistant than old-growth trees (Campbell and Clark 1960, Choong et al. 1986). Harvesting conditions, wetlands legislation, and confusion over the durability of cypress have resulted in an erratic market (Marsinko et al. 1991). Use of cypress wood remains popular, however. Baldcypress has many uses including fencing, boat planking, river pilings, furniture, interior trim, cabinetry, siding, flooring, and shingles (Brown and Montz 1986). Pondcypress is primarily used for fenceposts, mulch, and pulp (Terwilliger and Ewel 1986), although fence posts last less than five years (Applequist 1957). Second-growth Atlantic white-cedar has the same properties and durability of old-growth Atlantic white-cedar (Baines 1990, Earley 1987). There continues to be a high demand for all dimensions of Atlantic white-cedar for boat building, shingles, pilings, posts, furniture, and industrial millwork (Schroeder and Taras 1985, Little and

TABLE 11.1
Growing stock volume (million m³) of major deepwater swamp species in the southern United States based on most recent U.S. Forest Service surveys.

State	Year of Survey	Baldcypress	Pondcypress	Tupelo	Atlantic white-cedar
Alabama	1990	4.53	14.40	0.30	
Arkansas	1988	5.43	2.58		
Florida	1987	16.81	58.50	42.74	2.21
Georgia	1989	6.53	17.82	63.50	
Louisiana	1991	42.46	22.23		
Mississippi	1994	5.74	2.72	0.04	
North Carolina	1990	9.20	3.14	54.83	2.08
South Carolina	1993	9.34	3.41	40.64	0.04
Tennessee	1989	2.30	0.35		
Texas[a]	1992	2.60	0.99		
Virginia	1992	1.13	0.29	12.99	2.55[b]

[a] East Texas only

[b] Value calculated from information provided by D. Brownlie, U.S. Department of Interior, Fish and Wildlife Service, Great Dismal Swamp National Wildlife Refuge, Suffolk, VA.

Garret 1990). With reasonable management, these forests may once again become significant sources of wood products (Sternitzke 1972, Williston et al. 1980).

PHYSICAL ENVIRONMENT

CLIMATE

Although cypress and tupelo species are found across a wide climatic range, they grow best in the southern United States. Average annual precipitation ranges from 1120 to 1630 mm (44 to 64 in) across this range, although southeast Texas receives only 760 mm/year (30 in). The growing season ranges from 190 days in southern Illinois to 365 days in southern Florida. Temperatures average 27°C (81°F) in summer and 7°C (45°F) in winter across the range of these species. Baldcypress has been planted in the northern U.S. and southern Canada, where it survives minimum winter temperatures of –29 to –34°C (-20 to –29°F) (Harlow and Harrar 1979), but few seeds mature in these extreme conditions (Fowells 1965).

Atlantic white-cedar also grows in a humid climate with average annual precipitation ranging from 1020 to 1630 mm (40 to 64 in). The frost-free season for this species ranges from 140 days at its northern limit to 305 days at its southern-most location. Temperatures can range from extremes of –38°C (-36°F) in the winter to 38°C (100°F) in the summer (Fowells 1965).

GEOLOGY, GEOMORPHOLOGY, AND SOILS

Deepwater swamps occur in a wide variety of geomorphic situations ranging from broad, flat floodplains to isolated basins. Major features of these floodplain systems include meandering river channels, natural levees adjacent to the rivers, meander scrolls created as the rivers change course (ridge and swale topography), oxbow lakes created as meanders become separated from the main channel, and sloughs which represent areas of ponded water in meander scrolls and backwater swamps (Leopold et al. 1964, Bedinger 1981, Brinson et al. 1981b, Mitsch and Gosselink 1986). In the Mississippi River drainage, large loads of meltwater and soil material were carried by the river during glacial intrusions and recessions. Soil from the western plains, the midwest, and the Allegheny and Appalachian mountains was worked and reworked in the Mississippi floodplain, creating isolated backwater swamps and bayous and building ridges and natural levees (McKnight et al. 1981). On the Atlantic Coastal Plain, development occurred through cycles of continental submergence and emergence. Soils are generally of Appalachian and coastal origin, and rivers have modified the landscape to a lesser degree than in the Mississippi drainage (Sharitz and Mitsch 1993). Backwater swamps are less extensive and the bottoms are less dissected on the Atlantic Coastal Plain (Braun 1950).

Even though these swamps are areas of very low topographic relief, slight changes in elevation (a few cm) produce quite different hydrologic conditions, soils, and plant communities (Brown 1972). Peat deposition is characteristic of these systems because of slow decomposition rates, and the thickness of the peat decreases towards the shallow edges of the swamps (Dennison and Berry 1993). Baldcypress and water tupelo commonly grow on soils ranging from mucks and clays to silts and sands (Alfisols, Entisols, Histosols, and Inceptisols). These soils are moderately to strongly acidic with a subsoil that is rather pervious (Johnson 1990, Wilhite and Toliver 1990). Although Atlantic white-cedar sometimes grows on sandy soils, it mainly grows on muck (peat) soils of the orders Spodosols and Histosols. The muck ranges in depth from a few centimeters to 12 m (39 ft) and is generally acid with a pH between 3.5 to 5.5 (Little and Garrett 1990). Atlantic white-cedar is not normally found in areas where the muck is underlain by clay or contains appreciable amounts of silt or clay (Fowells 1965). It also occurs sporadically along blackwater streams from the Sandhills region in North and South Carolina to blackwater streams in Florida and Alabama. Small stands occur along these streams where organic matter overlays or is layered with clay or stream bed alluvium (Moore and Carter 1987, Laderman 1989).

Soil oxygen content is one of the most important characteristics in these flooded soils. Anaerobic conditions are created rapidly upon flooding and can persist for long periods in deepwater swamps, especially during low flow or stagnant conditions. Soils high in clay content (small pore size) hinder drainage more than do sandy or loamy soils, and are thus more likely to be poorly aerated. High organic matter content can both increase and deplete soil oxygen. Organic matter can improve soil structure in clayey soils, increasing soil aeration, but decomposing organic matter creates an oxygen demand (Sharitz and Mitsch 1993). While the organic matter content of upland soils is low (0.4–1.5%), organic levels in deepwater swamps can reach 36% (Wharton et al. 1982).

Hydrology

Hydrologic inflows are dominated by runoff from surrounding uplands and by overflow from flooding rivers (Mitsch and Gosselink 1993). Topographic features may impound water and cause flooding from rainfall rather than from stream overflow. Examples where this type of flooding is common are oxbows, backswamp depressions, and swales between relict levees where drainage patterns are poorly developed (Brinson 1990). Even though deepwater swamps are usually flooded, water levels vary seasonally and annually. High water levels coincide with winter-spring rains and melting snow runoff. Low levels occur in the summer from high evapotranspiration and low rainfall (Wharton and Brinson 1979). During extreme droughts, like those of 1924 and 1960, even deepwater forests may lack surface water for extended periods (Mancil 1969).

Biogeochemistry

Soils of deepwater swamps generally have ample nutrients. High clay content results in higher concentrations of phosphorus, and the relatively high organic matter content results in higher concentrations of nitrogen (Sharitz and Mitsch 1993). Soils tend to be highly reduced, thereby increasing mobilization of minerals such as phosphorus (P), nitrogen (N), magnesium (Mg), sulfur (S), iron (Fe), manganese (Mn), boron (B), copper (Cu), and zinc (Zn) (Mitsch and Gosselink 1986). In addition, low oxygen may foster accumulation of potentially toxic soil compounds (Sharitz and Mitsch 1993). Low oxygen also causes a shift in the redox state of several nutrients to more reduced states, making them unavailable to plants (Wharton et al. 1982).

Biogeochemical processes are strongly linked to hydrologic characteristics. The depth and duration of flooding, as well as whether floodwaters are flowing or stagnant, affect whether these wetlands serve as sources, sinks, or transformers of nutrients. Flooding in these swamps causes a lateral transport of elements. Dissolved and particulate forms are carried into the wetland from upstream or are transported downstream (Brinson 1990). Phosphorus inputs, based on sedimentation rates, range from 0.17 g $P/m^2/yr$ (6×10^{-4} $oz/ft^2/yr$) in North Carolina swamps (Yarbro 1979) to 3.6 g $P/m^2/yr$ (1×10^{-2} $oz/ft^2/yr$) in southern Illinois swamps (Mitsch et al. 1979a). Deepwater swamps can also serve as sinks for nutrients. Kitchens et al. (1975) reported a 50% reduction in P as overflow waters passed through a South Carolina swamp. Day et al. (1977) found a 48% reduction in N and a 45% reduction in P as water passed through a swamp/lake complex in Louisiana.

Some wetlands serve as sinks for nutrients for a number of years. A Florida cypress strand was found to be an effective sink even after 50 years of enrichment with partially treated wastewater (Nessel 1978a, b, Nessel and Bayley 1984, DeBusk and Reddy 1987). Floodplain forests along the Apalachicola River in Florida are nutrient transformers rather than sinks (Elder and Mattraw 1982, Elder 1985). While inputs and outputs of total N and P are similar, there are net increases in particulate organic N, dissolved organic N, particulate P, and dissolved P along with decreases in dissolved inorganic P and soluble reactive P. These transformations of inorganic forms to organic forms may be important for secondary productivity in downstream ecosystems (Sharitz and Mitsch 1993).

Major flows of nutrients most frequently measured in these swamps include decomposition, wood accumulation, sedimentation, and return of nutrients from the forest canopy as litterfall (Brinson 1990). Nutrient return from the canopy to the forest floor is high in riverine forests compared to upland forests, suggesting that fluvial processes are important in maintaining the relative high fertility of these systems (Brinson et al. 1980). When floodplain soils are nutrient poor, resorption from leaves prior to abscission may be important in conserving nutrients (Adis et al. 1979). Decomposition rates of leaf litter vary greatly (Duever et al. 1975, Brinson 1977, Burns 1978, Nessel 1978b, Brinson et al. 1981a, Kemp et al. 1985, Conner and Day 1991), with peat accumulation occurring in areas with long hydroperiods (Brinson 1990). Flowing water alters litter through transport, concentration, sorting, physical destruction, siltation, and increased moisture regime (Bell and Sipp 1975). Debris piles accumulate on the upstream side of trees (Hardin and Wistendahl 1983). Decomposing litter in slowly flowing situations, however, may immobilize N and P, providing a mechanism for nutrient conservation during the dormant season (Brinson 1977). Annual P accumulations in the wood of trees is quite low when compared with recycling in litterfall. Uptake of P by stem wood does respond to supply, however. The rate of P accumulation in the stem wood increased three-fold when nutrient rich sewage effluent was released into a cypress strand in Florida (Nessel 1978a). Sedimentation rates vary greatly in deepwater swamps, and only a few studies have reported P deposition rates (Mitsch et al. 1979a, Yarbro 1983). The proportion of the sediment P that is available for plant uptake has not been determined (Brinson 1990).

In Atlantic white-cedar stands, acid conditions generally occur with soil and water pH ranging from 2.5 to 6.7 (Day 1984, Golet and Lowry 1987, Schneider and Ehrenfeld 1987, Whigham and Richardson 1988, Laderman 1989). Available information indicates that Histosols under Atlantic white-cedar stands are generally high in organic matter content (20 to 30%) and cation exchange capacity. They have relatively high levels of Ca, Mg, and Al and relatively low levels of P (Bandle and Day 1985, Whigham and Richardson 1988, Laderman 1989). Day (1984, 1987b) determined that litter accumulation rates in Atlantic white-cedar stands exceeded those of associated cypress and mixed hardwood communities, and that decomposition rates were generally lower in cedar-dominated communities than in associated cypress or mixed hardwood-dominated communities in the Great Dismal Swamp.

Fire

Fire is generally infrequent in natural deepwater forests of the southern United States because of the continuously moist conditions. During droughts, or after drainage, however, fire can have a significant effect on these forests. Mature cypress trees are seldom killed by fire, although tree vigor may be reduced (Duever et al. 1986). Broad-leaved species, on the other hand, are killed, thereby helping maintain cypress dominance (Ewel and Mitsch 1978, Schlesinger 1978). Even-aged cypress stands developed after fire in Georgia's Okefenokee Swamp (Duever and Riopelle 1984). Although cypress returned to most sites in the Florida Everglades after a series of fires in 1937, there was little regeneration after a 1962 fire (Craighead 1971); this was probably due to the lowering of the water table in the area by drainage (Brandt and Ewel 1989). Atlantic white-cedar has thin bark and flammable foliage, making it extremely susceptible to injury and death by fire. Living trees with fire scars are rare, as most of the impacted trees are killed (Akerman 1923, Korstian and Brush 1931, Little 1950).

WIND

In deepwater swamp forests, hurricanes are capable of defoliating, topping, and overturning trees, but it is usually the defective and hollow trees that break, and windthrow of these wetland species is generally rare (Craighead and Gilbert 1962, Duever et al. 1984a, Hook et al. 1991a). Windthrow of bottomland species is more common and may be related to shallow rooting in moist, soft soil (Hedlund 1969, Gunter and Eleuteris 1973). In south Florida, new leaf growth was unusually rapid for several species of trees, and many species flowered a second time immediately following Hurricane Donna in 1960 (Vogel 1980). The major short-term effect to bottomland and swamp species during Hurricane Hugo was the loss of foliage and small branches (Gresham et al. 1991, Putz and Sharitz 1991). Putz and Sharitz (1991) also reported that trees that had previously suffered wind damage were more susceptible to new damage.

VEGETATIONAL COMMUNITIES

TYPES

Based on field observations and study in four southern states, Wharton et al. (1982) described 30 dominance types in deepwater swamps. These dominance types can be broken down into four broad categories: cypress/tupelo gum forests, bay swamps and shrub bogs, tidal forests, and Atlantic white-cedar forests. In the cypress/tupelo gum forests, subtle differences determine the relative dominance of baldcypress, water tupelo, swamp tupelo, and Ogeechee tupelo (*Nyssa ogeche*). Water tupelo occurs primarily in alluvial floodplains of the coastal plain and tolerates deeper and longer flooding than swamp tupelo. Swamp tupelo is also common in coastal plain floodplains as well as in upland swamps and ponds and in brackish water fringing coastal estuaries (Penfound 1952). Ogeechee tupelo occurs only in Florida and Georgia but is found in both alluvial and blackwater systems. Baldcypress is found throughout the southern United States, but is often replaced by water tupelo because of erratic reproduction, slower growth rates, and poor stump and root sprouting (Wharton et al. 1982). Frequent disturbance, such as logging, also favors water tupelo dominance (Penfound 1952, Putnam et al. 1960, Eyre 1980). Pondcypress is codominant with water tupelo and swamp tupelo on some Florida blackwater floodplains and depressional wetlands (Wharton et al. 1982).

Bay swamps and shrub bogs are comparatively rare in the floodplain environment (Wharton et al. 1982). For more information on these upland counterparts, see Chapter 14.

Tidal forest types can be found within the zone of tidal influence of all of our coastal floodplains and may extend a considerable distance inland. Soils are peaty and tightly bound by interwoven root mats. The water table is continuously high and the forest floor covered up to twice a day as a result of tidal action (Wharton et al. 1982, Rheinhardt and Herschner 1992).

Atlantic white-cedar seems to be a disturbance-adapted successional species and is found in bog stream swamps on peat overlying sandy soils or in acid backswamps (Wharton et al. 1982). Fire is a common precursor to Atlantic white-cedar forests developing, although logging, flooding, or windthrow can yield similar results (Korstian and Brush 1931, Little 1950, Frost 1987).

Dominant Species within Types

Southern deepwater swamps have unique plant communities that either depend on or adapt to the almost continuously wet conditions. The dominant canopy species found in these swamps include baldcypress, water tupelo, swamp tupelo, and Atlantic white-cedar. These species grow together or in pure stands. Other species include red maple (*Acer rubrum*), black willow (*Salix nigra*), swamp cottonwood (*Populus heterophylla*), green and pumpkin ash (*Fraxinus pennsylvanica* and *F. profunda*), pondcypress, Atlantic white-cedar, pond pine (*Pinus serotina*), and loblolly pine (*Pinus taeda*) (Barry 1980, Eyre 1980, Wharton et al. 1982, Sharitz and Mitsch 1993). Along the shallow margins of deepwater swamps, species diversity is greater and may include overcup oak (*Quercus lyrata*), water hickory (*Carya aquatica*), waterlocust (*Gleditsia aquatica*), American elm (*Ulmus americana*), persimmon (*Diospyros virginiana*), sweetbay (*Magnolia virginiana*), and redbay (*Persea borbonia*). Ogeechee tupelo occurs in southwest Georgia and northern Florida. The understory in deepwater swamps is generally sparse because of low light conditions and long periods of flooding. Small tree and shrub associates include buttonbush (*Cephalanthus occidentalis*), redbay, swamp-privet (*Forestiera acuminata*), water-elm (*Planera aquatica*), sweetbay, swamp dogwood (*Cornus* spp.), poison sumac (*Rhus vernix*), Virginia willow (*Itea virginica*), swamp cyrilla (*Cyrilla racemiflora*), fetterbush (*Lyonia* spp.), swamp leucothoe (*Leucothoe racemosa*), hollies (*Ilex* spp.), swamp rose (*Rosa palustris*), Carolina ash (*Fraxinus caroliniana*), southern bayberry (*Myrica cerifera*), and viburnums (*Viburnum* spp.) (Eyre 1980, Johnson 1990, Wilhite and Toliver 1990).

Because Atlantic white-cedar grows over such a broad latitudinal range, a variety of species are associated with it. In the southern United States, these include red maple, swamp tupelo, baldcypress, loblolly bay (*Gordonia lasianthus*), redbay, sweetbay, pond pine, and slash pine (*Pinus elliottii*). Common fetterbush (*Lyonia lucida*), greenbriar (*Smilax* spp.), and southern bayberry are the most common associated shrubs (Korstian and Brush 1931, Buell and Cain 1943, Laderman 1989).

Adaptations

In order to survive standing water for most of the year, plants have developed many morphological and physiological adaptations. For a detailed description of these features, see Chapter 8 or Kozlowski (1984). Common features in deepwater swamps are knees and buttressed tree trunks. Knees are produced by baldcypress, pondcypress, water tupelo, and swamp tupelo, and represent extensions of the root systems to well above the average water level. On baldcypress, the knees are conical in shape and typically less than a meter in height, although some knees reach 3 to 4 m (10 to 13 ft) (Hook and Scholtens 1978, Brown and Montz 1986). Water tupelo produce fewer knees than baldcypress (Hall and Penfound 1939a), and the knees of swamp tupelo are actually arching roots that approximate the appearance of cypress knees (Mitsch and Gosselink 1993). Atlantic white-cedar possesses neither knees nor arching roots (Kearney 1901).

The exact function of these knees has not been resolved. It is commonly believed that they function in gas exchange for the root systems, but Kraemer et al. (1952) concluded that they do not provide aeration for the rest of the tree. Even though some CO_2 evolves from knees (Cowles 1975, Brown 1981), this does not prove that oxygen transport is occurring through the knees (Mitsch and Gosselink 1993). Another possible function is an adaptation for anchoring trees in unstable soils. Under each knee, there is a secondary root system similar to and smaller than the main root system (Mattoon 1915, Brown 1984).

Another adaptation in flooded swamps is buttressing of the lower part of the tree trunk. The height of the buttress varies depending on the water depth. Swelling generally occurs along the part of the tree where there is a frequent wetting and soaking of the tree trunk but where the trunk is also above the normal water level (Kurz and Demaree 1934). The value of this swelling to survivability is also unknown (Mitsch and Gosselink 1993) but has been proposed as an aid in support.

Atlantic white-cedar has developed reproductive strategies that are likely an adaptation to the hydrologic and fire disturbance regimes in cedar habitats. Cones and viable seeds are produced at a very early age (Little 1950, C. G. Williams, pers. comm.). Seeds are generally wind or water disseminated. Poor seed years are rare, and seeds can germinate at rates exceeding 2.5 million seedlings/ha (1 million/acre) (Korstian and Brush 1931). Seedlings and saplings can produce shoots from lateral branches or dormant buds when injured (Little 1950).

Successional Patterns

On permanently flooded sites with little sediment deposition, succession tends to be stalled, and changes in composition may not occur for hundreds of years without disturbance. Cypress-tupelo forests may be 200–300 years old before canopy trees begin to die (Hodges 1994a). In some poorly drained areas, deposition of fine-textured material on the floodplain eventually creates better drained conditions. The pioneer tree species on such sites is generally black willow, a short-lived species. Black willow stands start to deteriorate as early as age 30 and few survive to age

60 (Johnson and Shropshire 1983). Further compositional changes depend on the degree of sedimentation. Low sedimentation rates usually result in an association of swamp privet, water-elm, and buttonbush which may eventually be replaced by baldcypress. High rates of sedimentation (or drainage) may foster a replacement of the cypress-tupelo forest with species like water hickory-overcup oak association (Hodges 1994a).

According to Korstian and Brush (1931), the best conditions for establishing Atlantic white-cedar stands are open, warm conditions following recent burns in wet swamps, recently cut-over lands, and clearings. The species aggressively regenerates from seed under favorable conditions, often developing in dense, pure stands. In the South, it appears to be naturally limited to those sites where return intervals for catastrophic fires vary from 25 to 250 years and that occupy the relatively narrow moisture gradient bounded by high and stable water tables of deep swamps and the seasonally variable water table of pocosins. Alterations in the fire/hydrology regime shift regenerating stand composition toward a cedar-hardwood mixture and ultimately to a mixed hardwood stand (Frost 1987).

Productivity

Aboveground primary productivity values for cypress/tupelo forests are among the highest reported for forest ecosystems, due largely to fluctuating water levels and nutrient inflows (Brinson et al. 1981a, Brown 1981, Conner and Day 1982, Brinson 1990, Lugo et al. 1990c, Conner 1994). Aboveground biomass production of forests with unaltered seasonal water flow frequently exceeds 10 t/ha/yr (8,900 lbs/acre) in these forests, with a maximum of nearly 20 t/ha/yr (17,800 lbs/acre) being reported for an undisturbed cypress/tupelo forest in South Carolina (Table 11.2). Litterfall accounts for an average of 39% of the aboveground primary production in wetland forests. Very little is known about belowground processes, although there is evidence that roots contribute as much or more to the detrital pool than does litterfall (Symbula and Day 1988). Powell and Day (1991) reported that belowground productivity in a frequently flooded Atlantic white-cedar swamp (3.66 t/ha/yr or 3,266 lbs/acre) and a cypress swamp (3.08 t/ha/yr or 2,748 lbs/acre) was much lower than in a mixed hardwood swamp (9.89 t/ha/yr or 8,906 lbs/acre), suggesting that the allocation of carbon to the root system decreases with increased flooding.

Brinson et al. (1981a) suggest that the amount and frequency of water passing into and through a wetland are the most important determinants of potential primary productivity. Periodic inundation subsidizes the forested wetland with nutrients and sediments that stimulate plant production (Gosselink et al. 1981). Forested wetlands with stagnant or sluggish waters are usually less productive, but not always (Brown and Peterson 1983). Communities with permanently impounded conditions or on sites with poor drainage leading to continuously high water tables and the accumulation of acidic peat soils have lower productivity, primarily because of low nutrient turnover under anoxic conditions, N limitations, and low pH (Brown et al. 1979). This change in productivity with respect to flooding has been discussed by several authors (e.g., Conner and Day 1976, 1982, Odum 1978) and is illustrated in Figure 11.1.

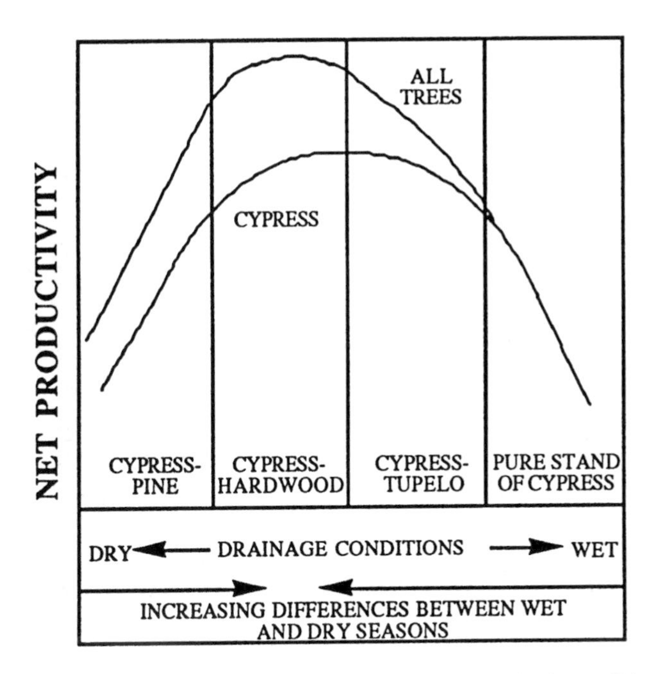

FIGURE 11.1 The relationship between productivity and hydrologic conditions for deepwater swamps (from Mitsch and Ewel 1979, copyright © by American Midland Naturalist, reprinted with permission).

Mature cypress and tupelo prosper under flooded conditions (Kennedy 1970, Dickson et al. 1972), but when changes in the natural regime occur, tree growth can be affected. Cypress has tolerated flood depths of 3 m (10 ft) or more (Wilhite and Toliver 1990). In Florida, Harms et al. (1980) found that 0 to 16% of the cypress trees died within seven years in water from 20 to 100 cm (8 to 39 in) deep. In water more than 120 cm (47 in) deep, 50% of the cypress died after four years. A long-term study of cypress survival was conducted near Lake Chicot, Louisiana (Penfound 1949, Eggler and Moore 1961). After four years of flooding with water 60 to 300 cm (24 to 118 in) deep, 97% of the cypress survived. Eighteen years after flooding, 50% of the cypress were still alive. However, most of the living trees in the deep water had dead tops (Eggler and Moore 1961). Conner and Day (1992a) found that growth of both baldcypress and water tupelo was greater in a permanently flooded swamp than in a natural cypress-tupelo forest. Keeland and Sharitz (1995) found that water tupelo and swamp tupelo achieved best growth under deep periodic flooding while maximum growth of baldcypress occurred with shallow permanent flooding. Stahle et al. (1992) and Young et al. (1995b) reported that increased flooding resulted in a short-term increase in baldcypress growth rate followed by a long-term decline in both South Carolina and Tennessee. Although Atlantic white-cedar is found in a variety of hydrologic regimes from nontidal coastal wetlands to permanently flooded areas, it generally grows on slightly elevated hummocks surrounded by water up to 1 m (3.3 ft) deep. When the boles are under water, cedars are stressed, and while surviving, they do not grow well (Laderman 1989).

TABLE 11.2
Aboveground biomass production of deepwater swamps in the southern United States.

Location/type	Flood periodicity	Water	Leaf litterfall (t/ha/yr)	Stem growth (t/ha/yr)	Reference
Florida					
baldcypress-pop ash	seasonal[a]	flowing	4.76	6.10	Brown 1981
tupelo[b]-cypress[c]-ash	semi-permanent	poor drainage	4.81	—	Elder & Cairns 1982
water tupelo-baldcypress	semi-permanent	poor drainage	4.76	—	ibid
baldcypress	seasonal	undrained	3.45	7.72	Burns 1978
cypress	permanent	stagnant	—	1.54	Mitsch and Ewel 1979
Georgia					
pondcypress	permanent	stagnant	3.28	3.53	Schlesinger 1978
baldcypress-sweetgum-oak-blackgum-tupelo	seasonal, 3–6 mo.	flowing	6.50–8.50	—	Cuffney 1988
Illinois					
baldcypress-water tupelo	permanent	stagnant	2.35	—	Middleton 1994
Kentucky					
green ash-baldcypress	semi-permanent	slowly flowing	1.36	4.98	Mitsch et al. 1991
baldcypress	permanent	slowly flowing	2.53	2.71	ibid
baldcypress	semi-permanent	stagnant	0.63	1.42	ibid
Louisiana					
baldcypress-water tupelo	semi-permanent	slowly flowing	6.20	5.00	Conner and Day 1976
baldcypress-water tupelo	semi-permanent	slowly flowing	4.17	7.49	Conner et al. 1981
baldcypress-water tupelo	permanent	stagnant	3.30	5.60	ibid
baldcypress-water tupelo	semi-permanent	slowly flowing	4.88	3.38	Megonigal et al. 1997
baldcypress-water tupelo	seasonal	slowly flowing	7.25	4.30	ibid
baldcypress	permanent	slowly flowing	3.33	3.30	ibid

North Carolina					
water tupelo	seasonal	flowing	5.52–6.77	—	Brinson 1977 Brinson et al. 1980
South Carolina					
baldcypress-water tupelo	frequent flooding	flowing	4.66	2.93	Muzika et al. 1987
baldcypress-red maple	frequent flooding	flowing	5.44	13.43	ibid
black willow-red maple-baldcypress (recovering)	seasonal	flowing	4.35	9.09	Bates 1989
water tupelo-baldcypress	permanent	flowing	4.38	2.16	Megonigal et al. 1997
Virginia					
Atlantic white-cedar	seasonal, 4 mo.	stagnant	5.69	—	Gomez and Day 1982
Atlantic white-cedar	seasonal		5.06	1.68[d]	Day 1984
baldcypress-red maple-blackgum	seasonal, 6 mo.	stagnant	5.68	—	ibid
green ash-blackgum-bluebeech-red maple	daily tidal	flowing	2.52	4.92	Fowler & Hershner 1989

[a] Prior to dam construction, flooding was year-round

[b] Includes water tupelo, swamp tupelo, and Ogeechee tupelo

[c] No distinction made as to whether baldcypress or pondcypress

[d] Average value over 57 years

Drainage of swamp forests can also affect primary productivity rates. Drainage of a cypress swamp in Florida led to a thinning of the overstory canopy and a reduction in biomass production of the trees, litterfall, and herbaceous plants (Carter et al. 1973). Productivity of a drained cypress stand in Florida was 3.87 t/ha/yr (3,453 lbs/acre) compared to 8.58 t/ha/yr (7,656 lbs/acre) for an undrained stand.

ANIMAL COMMUNITIES

MAMMALS

The most common mammals found in permanently-flooded swamps include American beaver (*Castor canadensis*) and northern river otter (*Lutra canadensis*). Beavers often change local hydrologic conditions through their dam building activities and cause tree mortality through removal as well as flooding. Baldcypress distribution along small drainages in Missouri appears to be related to beaver activity. In Louisiana swamps, nutria (*Myocastor coypus*) imported into the state during the 1930s have severely limited cypress regeneration attempts by feeding on planted seedlings (Blair and Langlinais 1960, Conner and Toliver 1990). Mink (*Mustela vison*) concentrate their activities adjacent to permanent water while raccoons (*Procyon lotor*) utilize both wetland and upland areas. Occasionally, white-tailed deer (*Odocoileus virginianus*), bobcats (*Lynx rufus*), and swamp rabbits (*Sylvilagus aquaticus*) can be found in these swampy areas. Eastern gray squirrels (*Sciurus carolinensis*) and southern flying squirrels (*Glaucomys volans*) are common throughout the floodplain forests (Sharitz and Mitsch 1993).

Few studies have quantified mammal use of Atlantic white-cedar swamps. White-tailed deer can be a significant damaging agent in regenerating stands (Little 1950, Laderman 1989, Little and Garret 1990). Some mammals known to use Atlantic white-cedar on the Atlantic Coastal Plain are gray squirrel, eastern red bat (*Lasiurus borealis*), common muskrat (*Ondatra zibethicus*), red fox (*Vulpes vulpes*), black bear (*Ursus americanus*), and bobcat (Laderman 1989).

REPTILES AND AMPHIBIANS

Wharton et al. (1982) report that only a few reptiles and amphibians are locally abundant in deepwater swamps. Major species include mud turtles (*Kinosternon subrubrum subrubrum* and *K. baurii*), glossy crayfish snakes (*Regina rigida*), mud snakes (*Farancia abacura*), plainbelly water snakes (*Nerodia erythrogaster*), and eastern cottonmouth snakes (*Agkistrodon piscivorus*). Water snakes are more common in swamps, in both numbers and biomass, and are commonly misidentified as cottonmouths. Another reptile of deepwater swamps is the American alligator (*Alligator mississippiensis*), found from North Carolina to Louisiana and making a tremendous comeback after being hunted almost to extinction.

Dominant amphibia of deepwater swamps include the lesser siren (*Siren intermedia*) and two-towed amphiuma (*Amphiuma means*). Amphibious salamanders include the dusky (*Desmognathus* spp.), the many-lined (*Stereochilus marginatus*), and the dwarf (*Eurycea quadridigitata*). The rusty mud salamander (*Pseudotriton montanus floridanus*) and the northern two-lined salamander (*E. bislineata*) occur on the edge of the permanently flooded zone (Wharton et al. 1982). Frogs are less specific in this forest type, but include the green frog (*Rana clamitans melanota*), southern leopard frog (*Rana utricularia*), southern cricket frog (*Acris gryllus gryllus*), and the bird-voiced treefrog (*Hyla avivoca*) (Wharton et al. 1982).

Detailed information on reptiles and amphibians associated with Atlantic white-cedar swamps in the South is limited. Reports from the Great Dismal Swamp and from Dare County, NC, include the five-lined skink (*Eumeces inexpectatus*), redback salamander (*Plethodon cinereus*), carpenter frog (*Rana virgatipes*), southern copperhead snake (*Agkistrodon contortrix contortrix*), and timber rattlesnake (*Crotalus horridus*) (Laderman 1989).

Birds

Deepwater swamps are used by birds for nesting and summer and winter foraging (Fredrickson 1979). In open areas within the forest, shorebirds are attracted to areas that are muddy or of shallow depth. Dabbling ducks are attracted best when water depths are 30 cm (12 in) or less (Taylor 1977), while wading birds and other deepwater foragers exploit deeper waters (Fredrickson 1979). Characteristic passerine birds include the prothonotary warbler (*Protonotaria citrea*), tufted titmouse (*Parus bicolor*), northern parula warbler (*Parula americana*), and common grackle (*Quiscalus quiscula*) (Wharton et al. 1982). Prothonotary warblers nest in cavities within the swamp and forage in the vicinity of their nests. Parula warblers nest in Spanish moss (Bent 1953). Yellow-crowned night-heron (*Nyctanassa violacea*), green heron (*Butorides virescens*), great blue heron (*Ardea herodias*), great egret (*Casmerodius albus*), and white ibis (*Eudocimus albus*) usually nest in colonies, and nest sites may be immediately adjacent to the water or a considerable distance from foraging sites (Palmer 1961). Wood storks (*Mycteria americana*) nest in cypress stands and once occurred from South Carolina to Texas. Their range is now largely restricted to Florida with some rookeries in Georgia and South Carolina (Ernst and Brown 1989). Permanently flooded sites are excellent foraging areas for anhingas (*Anhinga anhinga*). Hooded mergansers (*Lophodytes cucullatus*) always nest in tree cavities over or immediately adjacent to water (Morse et al. 1969). Wood ducks (*Aix sponsa*) and mallards (*Anas platyrhynchos*) are common wintering waterfowl. Wood ducks commonly nest in cavities over water (Sharitz and Mitsch 1993). The red-shouldered hawk (*Buteo lineatus*) is a characteristic raptor in this forest. American swallow-tailed kites (*Elanoides forficatus*) feed and nest in these forests (Wharton et al. 1982). Terwilliger (1987) reported that prairie warblers (*Dendroica discolor*), prothonotary warblers, hooded warblers (*Wilsonia citrina*), worm-eating warblers (*Helmitheros vermivorus*), and common yellowthroats (*Geothlypis trichas*) account for a majority of birds found in a study of Atlantic white-cedar stands in the Great Dismal Swamp. Numerous other migratory birds use these forests seasonally and temporarily, taking advantage of enhanced foraging opportunities because of the fluctuating water levels.

Fish

Sloughs and backwater swamps serve as spawning and feeding sites for fish and shellfish during the flooding season (Lambou 1963, 1990, Patrick et al. 1967, Bryan et al. 1976, Wharton et al. 1982). Deepwater swamps also serve as a reservoir for fish when floodwaters recede, even though conditions are less than optimal for aquatic life because of fluctuating water levels and low oxygen conditions (Mitsch and Gosselink 1993). Fish adapted to low oxygen conditions include bowfin (*Amia* sp.), gar (*Lepisosteus* sp.), and certain top minnows (e.g., *Fundulus* spp. and *Gambusia affinis*).

Invertebrates

Macroinvertebrates dominate deepwater swamp invertebrate communities. A wide diversity and high number of invertebrates have been reported in permanently flooded areas (Mitsch and Gosselink 1993, Sharitz and Mitsch 1993). The types of invertebrates found in these forests depend on water depth, duration of flooding, current, substrate, food availability, and oxygen level (Sklar 1983). Characteristic species include crayfish, clams, oligochaete worms, snails, freshwater shrimp, midges, amphipods, and various immature insects (Mitsch and Gosselink 1993). In the Atchafalaya swamp of Louisiana, cypress-tupelo forests contain more invertebrates than do bayous, lakes, canals, and rivers, likely due to the abundance of detritus in the forest (Beck 1977). High densities of invertebrates in natural cypress-tupelo forests in Louisiana (Sklar and Conner 1979) and stream floodplains of Virginia (Gladden and Smock 1990) indicate that periodic flooding also contributes to the density and diversity of invertebrate communities. The Hessel's hairstreak butterfly (*Mitoura hesseli*), the larva of which feed exclusively on Atlantic white-cedar, has been reported in the Great Dismal Swamp and in Dare County, North Carolina (Beck and Garnett 1983, Laderman 1989).

MANAGEMENT ISSUES

Past Practices

During the 1700s, French settlers along the lower Mississippi River paid for imported goods mainly with shipments of lumber. Although oak and pine were exported in small quantities, cypress was the staple commodity of the colonial lumber industry in Louisiana and the principal cash product for most colonists of the lower Mississippi Valley until the 1790s, when sugar products became profitable (Moore 1967).

Early loggers coming from the drier pine forests of the North had to devise new harvesting methods for the wet swamplands where cypress grew. Axemen preferred working in the swamps during low river stages in order to have comparatively firm ground. Log planks were cut on the spot with simple two-man handsaws because of the difficulty of moving the heavy green logs. During periods of high water, the axemen cut the trees while standing in boats, dropping the trees as close to shore as possible. Trees were dragged onto dry ground where handsaws could be used. When large timbers were required, green logs were lashed to rafts constructed of buoyant woods. Unfortunately, many logs broke free during transportation and sank (Moore 1967).

By 1725, loggers realized that by simply girdling the trees during the late summer and winter, the trees dried sufficiently enough to float out of the swamp during spring high water (Moore 1967, Burns 1980). Loggers working from boats or scaffolding were able to fell the trees, trim the branches from the floating logs, cut the boles to log lengths, and bind the logs into rafts without setting foot on dry land (Moore 1967). The May Brothers Company of Garden City, Louisiana, erected a levee approximately 1.8 m (6 ft) in height around sections of swamp 400 ha (1,000 acre) or more in size and flooded it to a depth of 1 m (3.3 ft) after girdling the trees. Later, they returned and cut the dried logs and floated them out of the constructed pond (Anonymous 1959, Davis 1975, Prophit 1982).

Many of the early sawmills along the lower Mississippi River were powered by water. Settlers dug ditches from the swamp through their land and the river levee into the river. Swamp water flowing through the ditch carried the logs from the swamp to the mill and supplied power to turn the water wheel attached to the sawmill (Moore 1967, Eisterhold 1972). Because of the relatively short time between the river cresting and the emptying of the swamps, the mills could operate no more than five months of the year. Thus, operations were generally small, and the owners were planters first and lumbermen second (Prophit 1982).

Large-scale commercial logging of cypress did not begin until the Homestead Act of 1866 was repealed by the Timber Act of 1876. The Homestead Act declared swamp lands unfit for cultivation and unavailable to private individuals. When the act was repealed, large tracts of swamp lands were sold for 60 cents to $1.25 per hectare (25 to 50 cents/acre) (Davis 1975). During the 1890s, the pullboat, and later the overhead-cableway skidder, increased the range of the logger and the amount of timber that could be brought out of the forest. By the close of the 19th century, 7.08 million m^3 (3 billion board ft) of baldcypress had been logged in Louisiana (Kerr 1981). Nationwide, the production of cypress sawtimber rose from 68.4 thousand m^3 (29 million board ft) in 1869 to slightly more than 2.36 million m^3 (1 billion board ft) in 1913, with the majority of the timber coming from Louisiana (Mattoon 1915).

Many of the logging operations maintained their own dredges to prevent delays in digging access canals (Davis 1975). The average size of the canals was 3 to 12 m (10 to 40 ft) wide and 2.4 to 3 m (8 to 10 ft) deep, resulting in partial drainage of many swamps (Mancil 1969, 1980). In other areas, railway lines were constructed. The mileage of railroads in Louisiana between 1880 and 1910 increased from 1,050 km (650 mi) to 8,942 km (5,557 mi). By 1920, however, the mileage began to decrease because of the abandonment of the logging operations (Mancil 1969). With the use of pullboat barges, trees could be pulled in from as far as 1,524 m (5,000 ft) from the canal through runs spaced about 46 m (150 ft) apart in a fan-shaped pattern. The runs were cleared of all trees and stumps and the logs pulled to the canal. This skidding of timber across the swamp floor damaged and destroyed much young growth, and the continual use of a run resulted in a mud-and-water-filled ditch 1.8 to 2.4 m (6 to 8 ft) deep for the length of the run (Mancil 1980). This operation left distinctive wagon wheel-shaped patterns that can still be seen on current aerial photographs.

The earliest settlements in North and South Carolina occurred in 1655 and 1670, respectively, and Atlantic white-cedar was used for cabins, shingles, and boats. Population levels were low and extraction methods primitive, so little impact was made on the resource prior to 1732. The introduction of the water-powered sawmill in 1732 hastened the harvest of most readily accessible timber. The introduction of steam dredging and logging railroad technology in the 1850s foreshadowed the harvest of essentially every known stand of Atlantic white-cedar in the Carolinas. Many stands regenerated following harvesting, however, the area regenerating to white-cedar forests was apparently smaller than the area of cedar harvested initially (Frost 1987). Drainage occurred on only a minor scale by the early 1900s, and it is speculated that the reduction in area regenerating to Atlantic white-cedar was due primarily to the fact that logging created a different, and less favorable, regeneration environment than did natural fire disturbances (Little 1950, Frost 1987).

Unfortunately, the early exploitation of these swamp lands occurred with little regard for sustainability. According to one logger, "We just use the old method of going in and cutting down the swamp and tearing it up and bringing the cypress out. When a man's in here with all the heavy equipment, he might as well cut everything he can make a board foot out of; we're not ever coming back in here again" (Van Holmes 1954). Nearly all of the virgin swamp lands were logged of cypress and Atlantic white-cedar. In some cases, landowners were encouraged to drain their cut-over lands and convert them to agriculture or to plant fast-growing black willow and tupelo trees (Norgress 1947).

Present Management Practices

Little silvicultural information is available for deepwater swamp forests, and management of these areas has been largely limited to clearcutting and highgrading (Johnson 1979, Williston et al. 1980). Only recently have studies begun to investigate the response and recovery of these forests to harvesting practices (Aust 1989, Mader et al. 1989, Mader 1990, Aust et al. 1989, 1991, 1997, Aust and Lea 1991, 1992). Most stands today are second-growth, are fairly dense, and support high basal areas (Table 11.3). Timber volumes can exceed 170 m^3/ha (2,429 ft^3/acre) (McGarity 1977). Deepwater stands should be managed on an even-aged basis because of the species' silvical characteristics, the nature of the existing stands, and the sites they inhabit (Korstian and Brush 1931, Putnam et al. 1960, Stubbs 1973, Smith and Linnartz 1980). Baldcypress and water tupelo regenerate well in swamps where the seedbed is moist and competitors are unable to cope with flooding, but extended dry periods are necessary for the seedlings to grow tall enough to survive future flooding. Naturally seeded baldcypress seedlings often reach heights of 20 to 36 cm (8 to 14 in) the first growing season and 40 to 60 cm (16 to 24 in) the second season (Mattoon 1915). Early height growth is important because seedlings can be killed by four to five weeks of total submergence during the growing season (Mattoon 1916, Johnson and Shropshire 1983). Baldcypress seedlings can endure partial shading but require overhead light for normal growth (Williston et al. 1980). Coppice regeneration is also a possibility in cut-over areas. Mattoon (1915) reported that stumps of vigorous stock up to 60 years old can generally be counted on to send

TABLE 11.3
Density and basal area (BA) of deepwater swamps of the southern United States.

Forest Type	Density (stems/ha)	BA (m^2/ha)	Reference
water tupelo	2730	69.0	Brinson et al. 1980
water tupelo	916	52.4	Applequist 1959
water tupelo/ogeechee gum	2210	32.8	Leitman et al. 1983
water tupelo/swamp tupelo	2050	66.1	ibid
swamp tupelo	746	251.4	Hall and Penfound 1939b
swamp tupelo	988	51.7	Applequist 1959
tupelo/cypress	703–1484	77.4–77.6	Good and Whipple 1982
tupelo/cypress	830–1423	77.1–93.9	Hall and Penfound 1943
tupelo/cypress	1588	55.0	White 1983
tupelo/cypress	3558	46.6	Hall and Penfound 1939a
tupelo/cypress	1120	59.2	Leitman et al. 1983
tupelo/cypress	900	46.0	Megonigal et al. 1997
cypress	1644	32.5	Brown 1981
cypress	856	80.2	Duever et al. 1984a
cypress	1560	59.3	Dabel and Day 1977
cypress	—	138.1	Marks and Harcombe 1981
cypress	2186	80.4	Schlesinger 1976
cypress	272	52.3	Schmelz and Lindsey 1965
cypress	530	26.6	Megonigal et al. 1997
cypress/tupelo	372–535	33.7–41.9	Robertson et al. 1978
cypress/tupelo	325–449	55.5–62.7	Anderson and White 1970
cypress/tupelo	1235	56.2	Conner and Day 1976
cypress/tupelo	930	54.5	Megonigal et al. 1997
cypress/tupelo	560	46.7	ibid

up healthy sprouts. Although many stumps sprout during the first growing season after logging, few of these sprouts survive in either baldcypress (Prenger 1985, Conner et al. 1986) or water tupelo (DeBell 1971, Kennedy 1982), although results from a study in the Mobile-Tensaw delta disagree (Goelz et al. 1993).

Because of the exacting requirements for germination and establishment (Stubbs 1973, Brandt and Ewel 1989) and the variable success of stump sprouting (Hook et al. 1967, Kennedy 1982, Conner 1988) and natural regeneration (Hamilton 1984, Gunderson 1984, Conner et al. 1986), planting of seedlings in these flooded environments may be necessary to ensure regeneration success (Bull 1949, Conner et al. 1986). While there has been little success in planting tupelo (Silker 1948, DeBell et al. 1982), much better results have been obtained with baldcypress. Rathborne Lumber Company planted nearly 1 million baldcypress seedlings on cutover land in Louisiana. Ninety percent of the seedlings planted in 1949 and 1950 survived into 1951 and grew 30 to 46 cm (12 to 18 in) in height by the end of the 1950 growing season. An additional 141,000 seedlings were planted in early 1951, with

80 to 95 percent survival (Rathborne 1951). In another Louisiana project, 8,500 seedlings were planted during January to March 1951 in water 15 to 50 cm (6 to 20 in) deep. In April 1951, nearly 95% of them were growing vigorously and had increased in height by an average of 7.5 cm (3 in) (Peters and Holcombe 1951). Unfortunately, both projects were abandoned and no further records maintained.

Planting of one-year-old baldcypress seedlings at least 1 m (3.3 ft) tall and larger than 1.25 cm (0.5 in) at the root collar improves early survival and growth (Faulkner et al. 1985). Planting is recommended in the late fall and winter so that seedlings become established during low water periods (Mattoon 1915). A 2.4 × 2.4 m (8 × 8 ft) spacing is generally recommended, although regular spacing may not be possible unless the area was clearcut (Mattoon 1915, Williston et al. 1980). Even when planted in permanent standing water, height growth averages 20–30 cm (8–12 in) per year for baldcypress when there are no herbivory problems (Conner 1988, Conner and Flynn 1989). A simple planting technique has been successfully tested for planting seedlings in standing water areas (Conner and Flynn 1989, Conner 1995, Funderburk 1995, McLeod et al. 1996). Root pruning, or trimming off the lateral roots and cutting the taproot to approximately 20 cm (8 in), allows the planter to grasp the seedling at the root collar and push it into the sediment until his hand hits the sediment. This method has worked well in trials with baldcypress and water tupelo, but not as well with green ash and swamp tupelo.

While data are limited, it appears that plantation-grown baldcypress grow better than natural stands and may even grow better than hardwood species (Krinard and Johnson 1987). Planted baldcypress grew more than 2 m in height in 5 years in a Louisiana crayfish pond (Conner et al. 1993). In Mississippi, a plantation established on an abandoned agricultural field had baldcypress trees up to 21 m (69 ft) tall at age 41 years (Williston et al. 1980). Another Mississippi baldcypress plantation contained trees 21.6 m (71 ft) tall and 36 cm (14 in) in diameter after 31 years (Krinard and Johnson 1987). In comparison, Mattoon (1915) reported height growth of 13–16 m (43–52 ft) by age 40 years for naturally established second-growth baldcypress in Maryland and Louisiana.

Cypress tends to grow well at high densities (Wilhite and Toliver 1990), but there is some evidence that thinning may enhance diameter growth in baldcypress. Data for pondcypress are conflicting (Terwilliger and Ewel 1986, Ewel and Davis 1992), but crown thinning in baldcypress forests to 50% of original basal area increases diameter growth 2.5 to 2.75 times that of unthinned stands (McGarity 1977, Toliver et al. 1987, Dicke and Toliver 1988). Thinning to that level, however, may produce an abundance of epicormic branches (increase from <1% of trees in unthinned stand to 28% in thinned stand) which may lower future timber value. Dicke and Toliver (1988) recommended removing approximately 40% of the original basal area as the best alternative since this level produced good growth with fewer epicormic branches.

The results of thinning in tupelo stands are mixed. While McGarity (1977) also reported that thinning increased growth of residual tupelo trees, Kennedy (1983) found that thinning intensity had no significant effect on diameter and height growth. Defoliation of trees in the latter study by the forest tent caterpillar (*Malacosoma disstria*) may explain the difference in response. Many tupelo forests along the Gulf of Mexico are defoliated annually and, while the trees do not usually die, their growth is retarded (Morris 1975, Conner et al. 1981).

The most commonly used regeneration method in deepwater swamps is usually clearcutting (Stubbs 1973, McKnight and Johnson 1975). In shallow swamps less than 1 m (3 ft) deep, bombadiers and wide-tracked tractors can be used. In deeper swamps, pullboats or some type of floatation logging may be required. Logging with helicopters has had some acceptance although it can be very costly (Jackson and Morris 1986, Willingham 1989, DeCosmo et al. 1990). One study in Louisiana investigated the use of hot air balloons to extract timber from the Atchafalaya Basin, but this method has not gained acceptance (Trewolla and McDermid 1969).

In an intensive study in Alabama, changes to a water tupelo/baldcypress site caused by helicopter vs. rubber-tired skidder clearcut logging operations were investigated. During the first two years following logging, vegetative growth was best in the helicopter logged area (Mader 1990). Seven years following treatment, average densities, total heights, diameters, and aboveground biomass of the skidder treatment was equal to or greater than in the helicopter treatment area (Aust et al. 1997). An important factor affecting results is that both treatments were conducted in areas with rapid natural reproduction and where no major changes had occurred in site conditions. If natural hydrologic conditions have been changed, natural regeneration may be hampered and recovery rates may be much slower or nonexistent (Sharitz and Lee 1985b, Conner et al. 1986).

Where Atlantic white-cedar is managed using natural regeneration, the guidelines developed by Little (1950) are still appropriate; i.e., manage in even-aged stands harvested by clearcutting, reduce slash, and control competing species and deer browse. The successful application of these simple guidelines is dependent upon many factors including hydrology and the availability of a suitable seed source. Yield tables for natural stands of Atlantic white-cedar were produced by Korstian and Brush (1931), with no significant improvements made to date. Attempts to develop nursery methods for Atlantic white-cedar seedlings have met with limited success, due in part to the large variability in seedling size. Techniques for rooting Atlantic white-cedar cuttings are becoming more widely used to produce a consistently uniform and healthy plant for regeneration purposes (Crutchfield, pers. comm., Hughes, pers. comm., and C. G. Williams, pers. comm.). Rooted cuttings are being used to examine the relationships between tree and stand development and density, and the growth and dynamics of Atlantic white-cedar stands created and maintained with intensive silviculture methods similar to those used for loblolly pine (Buford et al. 1991, Phillips et al. 1993). Conclusive results are not yet available from these studies.

The first consideration in choosing from the numerous silvicultural and management options available is determining clearly: 1) the objective to be achieved and 2) the existing land and forest condition. The objective may be as simple as maximizing one particular output (e.g., timber) or as complex as optimizing for a suite of objectives (e.g., songbird habitat, timber production, and wastewater application).

The desired objective, or condition, can often be described in terms of species composition and tree size distribution. These are determined by many factors such as monetary goals, habitat requirements for specific faunal species, nutrient uptake capability, and owner preference, singly or in combination.

As stated earlier, the silvical characteristics of the deepwater swamp species indicate that they should be managed in an even-aged system. If an uneven-aged condition is desired, it must be maintained by clearcutting groups of at least 1.2–2 ha (3–5 acres) in size to ensure appropriate light conditions for the regenerating stand (Williston et al. 1980, Toliver and Jackson 1989). The desire to maintain the uneven-aged forest condition must be weighed against potential impacts of the multiple entries required to maintain the condition (Toliver and Jackson 1989). Clearcutting with subsequent control of competing vegetation, coupled with planting or reliance on natural regeneration, is the preferred reproduction method for deepwater swamp species. Where a reliable seed source is not available, planting is the preferred method for both baldcypress and Atlantic white-cedar (Williston et al. 1980, Phillips et al. 1993). Results with coppice and planting water tupelo have been conflicting, and the optimum method for consistently regenerating this species without reliance on natural seed or direct seeding is currently unclear (Hook et al. 1967, Kennedy 1982). The control of competing vegetation that is crucial to regeneration success of deepwater species can be achieved through mechanical, chemical, or hydrologic means.

The hydrologic regime necessary to regenerate and maintain the stand must exist on the site. If the hydrologic regime has been significantly altered through impoundment, dredging or soil loss, for example, then it must be controllable to create conditions fostering regeneration, establishment, and growth of the desired species. This is crucial when restoration or creation of a deepwater habitat is necessary. Drayton and Hook (1989) gave a detailed description of a project designed to create a water management system to restore the hydroperiods of a baldcypress-water tupelo swamp. The objective of the project was to favor regeneration and growth of the deepwater species, thereby enhancing the habitat values associated with the forest type and to control the water level to facilitate access at harvest. Restoring or constructing the necessary infrastructure to control the hydrologic regime of a site can involve obtaining the appropriate permits under Section 404 of the Clean Water Act. Any project that involves construction or alteration of the site hydrologic regime may require federal, state, and/or local permits.

Each project will require access, usually a road system used for tending and extraction and for owner and tenant access. Road construction for silvicultural purposes in jurisdictional wetlands does not require a permit. However, to qualify, the road system must comply with the Best Management Practices (BMPs) outlined in Section 404 of the Clean Water Act. The primary factors governing road construction are protecting water quality, wildlife habitat, and the hydrologic regime of areas traversed. All southern states have established BMP guidelines which should be followed in developing access systems for deepwater swamps

RESEARCH NEEDS

Even though data exist on the biota and productivity of deepwater swamps, many aspects of their ecology and management are still poorly understood. Some basic management principles have been devised for these wetlands, but we must improve our long-term predictive capabilities and knowledge of the fine structure of these systems in order to produce a realistic, flexible, and permanent framework for planning and management (Livingston and Loucks 1979). Major areas of research identified in the late 1970s and 1980s, including hydrology, biogeochemistry, effects of perturbations, biological productivity, spatial patterns, and coupling with other ecosystems (Clark and Benforado 1981, Gosselink et al. 1990), require further attention to allow us to accurately forecast long-term management consequences. In addition, new needs have recently arisen as researchers and managers have attempted to determine the functions and values of these wetlands, the factors contributing to their loss, and appropriate techniques for creating and restoring these wetlands (Mitsch and Gosselink 1993).

Hydrology is one of the most important driving forces in forested wetlands, and the length, depth, and timing of flooding determines the diversity and productivity of these systems. Changes in normal hydrology patterns due to stream channelization or construction of roads, canals, levees, or dams affect the establishment and growth of forest species (Conner et al. 1981, Sharitz et al. 1990). Even species adapted to flooding are severely impacted by increased flooding (Harms et al. 1980, Megonigal et al. 1997). The long-term impacts of changes in hydrology should be a major area of research.

Another aspect of hydrology that needs consideration, especially in coastal areas, is eustatic sea level rise (Gornitz et al. 1982) and subsidence (Gosselink 1984). Recent projections by the Environmental Protection Agency suggest that there will be a rise of 30 cm (12 in) by the year 2100. Penland and Ramsey (1990) have shown that there is already a significant increase in water level along the entire Louisiana coast primarily due to subsidence. Flood control levees along the Mississippi and Atchafalaya Rivers prevent the flooding of these wetlands by sediment-laden waters, and subsidence generally exceeds sedimentation in many areas. Most of coastal Louisiana is presently experiencing an apparent water level rise of about 1 m (3 ft)/century (Salinas et al. 1986), impacting forested wetland species composition and growth (Conner and Day 1988, Conner and Brody 1989). More data are needed on stand level responses to refine forest growth models to better predict the impacts of changes in these systems.

Biogeochemical cycling in deepwater swamps is difficult to study because of the complex hydrologic linkages with associated streams and adjacent uplands (see Chapter 7). Additional quantitative work on the mass flow of nutrients and carbon within these wetlands is needed (Livingston and Loucks 1979, Laderman 1989). In addition, it is important to understand the role of these wetlands in mediating the form and timing of nutrient export to downstream ecosystems and what role these exports play in maintaining secondary production levels.

Deepwater swamps are sometimes considered to be "subclimax" forests. Tree species composition remains fairly constant because few species other than baldcypress and the tupelos can tolerate extended flooding. Wetland management goals based on the concepts of this stability, however, may be highly disruptive to the productivity of wetlands. There is growing evidence that periodic physical disruption (by man or nature) is necessary to maintain continued high productivity and species composition in these wetlands and associated systems (Livingston and Loucks 1979), and many of these coastal forests have developed through time as a result of this disturbance (Little 1950, Frost 1987, Conner et al. 1989). In the case of Atlantic white-cedar, we must understand the intricate balance among fire, hydrologic regime, inter-species competition, and stand establishment and growth.

Nutria herbivory is a growing problem in Gulf coastal deepwater swamps. Although nutria are a recent import from South America, they have made a distinct impression on Louisiana wetlands. During the 1950s, the Soil Conservation Service recommended that planting of baldcypress be suspended until some means of nutria control were developed (Blair and Langlinais 1960). The problem has not been solved (Conner 1988, Brantley and Platt 1992), and nutria have been reported to damage even mature trees (Hesse et al. 1996). Animal control or seedling protection must be developed if successful regeneration of forest species is expected in cutover swamps.

Long-term, multidisciplinary studies must be conducted on representative plants and animals as well as whole communities if development of reliable management regimes is desired. These long-term studies are desirable and necessary, but often are difficult to maintain because of funding uncertainty, shifts in personnel, complexity of data analysis, and the pressure to produce frequent publications. However, only long-term research can consider both the short- and long-term fluctuations of key driving forces. In conjunction with long-term studies, we should examine processes across the entire southern United States to determine if all wetlands function similarly. Although deepwater swamps can be fairly similar in species composition across their range (Sharitz and Mitsch 1993), their response to hydrology and nutrients may vary (Mitsch et al. 1991, Keeland 1994, Megonigal et al. 1997).

There has been relatively little research on the creation and restoration of deepwater swamp forests, and techniques for evaluating the success or failure of these projects have not been developed (Sharitz and Mitsch 1993). Regeneration successes have occurred, but failures are common. Success can be improved by paying attention to hydrology and matching species to site conditions (Hodges 1997).

Deepwater swamps have been used as sites for treatment of secondarily treated wastewater in order to avoid the construction of expensive tertiary treatment plants (Brandt and Ewel 1989, Breaux and Day 1994). While the application of wastewater to swamp forests has enhanced tree growth in some areas (Nessel and Bayley 1984, Hesse 1994), this is not always the case (Kuenzler 1987). The reasons for these different responses needs to be elucidated. Other unknowns to be considered include how long these forests can assimilate additional nutrients, the impact of wastewater on animal populations, the effects on wood quality, and the impacts of timber management activities on the water quality function of the system.

Because of the nearly constant to permanent flooding of these forests, traditional timber harvesting methods do not work well. As a result, research needs to be conducted on wood extraction activities and the impacts of these activities on wetland functions and values. Because of the expansive clays found in swamps and continued sediment deposition, these sites are generally not as sensitive to harvesting perturbations as upland sites (Hodges 1997). When ground-based operations are not feasible, alternatives such as helicopters or balloons may prove beneficial, although more research is needed to determine how effective these methods are and whether or not they can be cost-effective (Jackson and Stokes 1991). Another method that may have application in deepwater swamps involves the use of water level management. By controlling water levels, harvesting can theoretically be done without significantly altering the character of the wetland and allows for natural regeneration to occur before flooding is reintroduced (Drayton and Hook 1989). The long-term impacts of this type of management needs study.

SUMMARY

Southern deepwater swamps are freshwater systems characterized by standing water for most or all of the year. Stands of baldcypress, pondcypress, water tupelo, swamp tupelo, and Atlantic white-cedar (either in pure stands or intermixed) occur throughout the southern United States in a wide variety of geomorphic situations ranging from broad, flat floodplains to isolated basins. Slight changes in elevation produce quite different hydrologic conditions, soils, and plant communities. Since many deepwater swamps are found along the floodplains of rivers, their soils generally have ample nutrients, and these forests are highly productive. Primary productivity in these forests is closely tied to hydrologic conditions, with highest productivity occurring in forests that receive high inputs of water, sediments, and nutrients. These swamps once provided large amounts of timber and have the potential to do so again. Once thought to be useless areas, recent research has demonstrated that these wetlands perform many functions beneficial to man. Little silvicultural information is available for deepwater swamp forests, and management of these areas is generally restricted to clearcutting and natural regeneration. In order to produce a realistic, flexible, and permanent framework for planning and management of these lands, we need to better understand the impacts of logging activities, hydrologic modifications, and wastewater additions on these forests.

12 Major Alluvial Floodplains

Robert C. Kellison, Michael J. Young, Richard R. Braham, and Edwin J. Jones

CONTENTS

1-56670-228-3/97/$0.00+$.50
© 1998 by CRC Press LLC

INTRODUCTION

Major alluvial floodplains are inextricably connected to major rivers. Unimpeded major rivers are distinguished from minor rivers and large creeks by bank caving, formation of "point bars," and meander cutoffs. Point bars, through sedimentation, eventually evolve to new land and then to front land (Putnam et al. 1960). Front land joins a suite of land types that includes levees, sloughs, and low imperfectly drained ridges that are separated by varying expanses from first and second bottoms.

Major floodplains owe their existence to upstream soil erosion and subsequent deposition of soil particles downstream as the waters lose turbidity. The land building associated with soil deposition causes the river to meander, always seeking an outlet with least expenditure of energy by spreading the gradient over a longer distance. Thus, major alluvial floodplains are relatively flat. Seen from above, abandoned river channels and oxbow lakes nestle against the still-active river, first on one side and then on the other. This landscape extends as far as the eye can see for the largest of all southern rivers, the Mississippi. The landscape is relatively narrowly confined on rivers like the Red in Arkansas and Louisiana and the Roanoke in North Carolina where the distance from bluff to bluff is reduced to as little as 0.8 km (0.5 mi).

Unless altered by humans, major alluvial floodplains are invariably forested. The predominant forest cover is angiosperms although a few gymnosperms occur, most commonly cypress (*Taxodium* spp.). Despite the dense tree cover and the difficulty in land clearing, this ecosystem was the first in the southern United States to be converted to agricultural crops. The villages of native Americans, including agricultural fields, were almost always located along major streams, which the villagers also used for transportation. They were mimicked by the early colonists with their rice culture, first in the vicinity of Charleston, South Carolina, and then along the coast to the boundaries of rice production. On the fringes of the rice paddies and beyond, corn, wheat, and cotton were the major crops supplanting the hardwood forests. William Bartram, in his travels of the 1790s, told of extensive fields of corn being grown in the floodplains of St. Marys River, Florida (Van Doren 1955), and cotton was a major crop in the Mississippi-Yazoo Delta by the Civil War (Dickson 1937). Accounts also exist of Governor Williams of South Carolina diking about 8,100 ha (20,000 acres) of land along the Peedee River in the early 1800s in Darlington and Marlboro counties for the production of corn, wheat, and cotton (Cook 1916).

Priority given to clearing the bottomland forest was due to the inherent fertility of the soils. Governor Williams' prescription for cropping the bottomlands was, "harvest the timber and clear the land for row crops. Continue row cropping until the soil loses its fertility and weeds make farming impractical. Abandon the land in favor of new land that is brought into production" (Cook 1916). At that time,

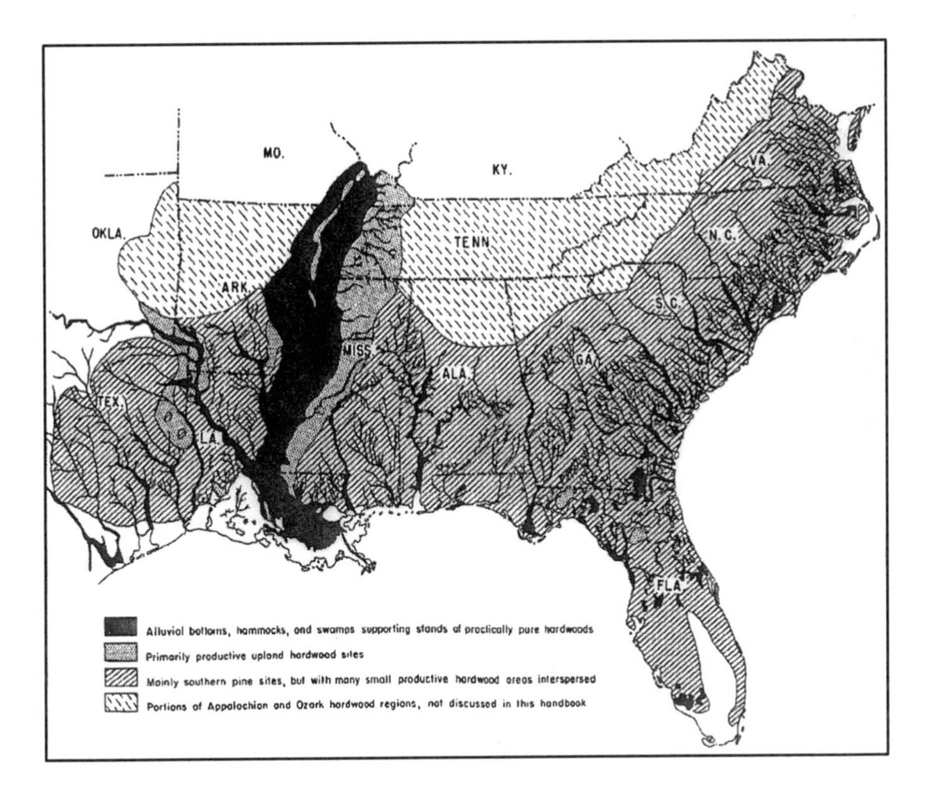

FIGURE 12.1 Major forest associations of the southern United States, with major alluvial floodplains occupying the heavily shaded areas (Putnam et al. 1960).

sustainable farming of uplands was secondary to bottomlands because of the absence of commercial fertilizers.

The timber in major alluvial floodplains has been in demand for fruit boxes, shoe heels, paneling for automobiles and houses, flooring, furniture, and decorative items (Heavrin 1981). With the advent of computers and copy machines, the resource is in great demand for high quality paper.

Following attempts to control water flow in the major alluvial floodplains, first by private enterprise (Cook 1916, Dickson 1937) and then by public agencies, especially the U.S. Army Corps of Engineers, the forests were increasingly cleared for agricultural crops. Compared to the original forests (Figure 12.1), only about half remained by the 1930s. During the following 50 years, conversion continued at an accelerated pace, reducing the forest area from 4.8 million ha (11.8 million acres) to 1.7 million ha (4.3 million acres) (Allen and Kennedy 1989). Conversion was especially rapid during the 1960s and 1970s when the price for farm crops, especially soybeans, reached unprecedented levels. Much of the converted land has proven to be nonproductive for agriculture crops because of poorly drained soils, leading to a modest program to artificially reestablish some of the bottomland hardwood forests (Allen and Kennedy 1989).

Major alluvial floodplains occur in conjunction with two basic types of rivers: red rivers and black rivers (Kellison et al. 1988, Nutter and Brinson 1994). Along the Atlantic Coast, red rivers have their origin in the mountains or Upper Piedmont; along the Gulf Coast, the origin of red rivers is inland from the Upper Coastal Plain. In either place, the waters at flood stage are reddish or brownish as a result of organic sediments being transported from inland sources. Sediment loads are deposited downstream when the waters lose turbidity. This phenomenon is especially noticeable along the Atlantic Coast below the fall line that separates the Piedmont from the Coastal Plain. Above the fall line, the water channel is narrowly defined, equating to minor alluvial bottoms. Below the fall line the river basin becomes measurably broader. Red rivers of the Western Gulf, without the fall line, grade imperceptibly from minor to major floodplain. Examples of redwater rivers are the Pee Dee (South Carolina), Alabama (Alabama), Red (Oklahoma, Arkansas, Louisiana), and Mississippi. Growth rates of merchantable timber in these and companion rivers are 6 to 7 m^3/ha/yr (80 to 100 ft^3/ac/yr) (Smith et al. 1975).

Black rivers have their origin in the Coastal Plain or, in a few instances, in the Lower Piedmont. They get their name from the color of the water which has the appearance of tea, a result of the release of fulvic acids from decaying organic matter. Waters of black rivers are significantly less turbid than red rivers, largely because of the relatively low difference in elevation between head and mouth, and the smaller watershed. Examples of black rivers are the Alligator (North Carolina), Ogeechee (Georgia), Escambia (Florida), Homochitto (Mississippi), and Calcasieu (Louisiana). Even then, major differences exist in black rivers. Those found in southeastern Georgia and eastern Florida (St. Marys, St. Johns) are commonly dominated by swamp tupelo (*Nyssa sylvatica* var. *biflora*); those to the north and west are comprised of many species (Figure 12.2). Growth rates of merchantable timber in the most-southern black river floodplains are about 3 m^3/ha/yr (40 ft^3/ac/yr), compared to twice that for the other blackwater rivers (Smith et al. 1975).

Red and black rivers become indistinguishable near the coast. Tidal influence and delta formation are the major reasons for this phenomenon, forming extensive swamps, lakes, and bayous.

PHYSICAL ENVIRONMENT

CLIMATE

Precipitation in the major alluvial floodplains of the southern United States ranges from about 1200 mm (48 in) to 1600 mm (64 in). Precipitation is generally greater during the months of June, July, and August, ranging from 200 mm (8 in) in east Texas to 600 mm (24 in) along the Gulf Coast. However, the amount of precipitation received is not a reliable indicator of the magnitude or duration of flood waters; upstream precipitation in large watersheds, like that of the Santee River, which is about 2.8 million ha (7 million acres), has a greater impact on downstream flooding than local precipitation.

Average July temperatures for the major alluvial floodplains, regardless of their location in the southern United States, are about 27°C (80°F); January temperatures

FIGURE 12.2 Well-developed stand of cypress along a black river (Salkehatchie River, South Carolina).

average about 15°C (60°F) at the most-southern locations and about 5°C (40°F) at the northern (inland) extensions. Mean annual potential evapotransporation, an expression of a plant's need for water, ranges from about 500 mm (20 in) in the Piedmont and mountain regions of Virginia and North Carolina to more than 1300 mm (50 in) in southern Florida. Virtually all regions of the southeast incur moderate to severe moisture deficits during the growing season.

Geomorphology of Soils

Weathering of residual parent material (parent material in its original position above bedrock) is not a dominant soil formation process of major alluvial floodplains. Rather, soils develop by weathering of materials deposited by water. We generally recognize three classes of alluvial deposits: (a) floodplains, (b) alluvial fans (point bars), and (c) deltas.

In floodplains, streamflow energy results in an orderly pattern of deposition of transported material which gives rise to a unique landscape and diverse soil types. As flooding occurs, coarse-textured materials drop from suspension nearest the main channel. As distance increases from the main channel and floodwater energy decreases, progressively finer-textured materials are deposited.

The most expansive forested floodplains in the southern United States occur along the Mississippi River, where floodplain deposits measure up to 121 km (75 mi) in width over a 800-km (500-mi) range. The alluvium can come from as far north as Canada and as far west and east as Montana and Pennsylvania. Many streams are well channelized and move very little over time; however, as waterflow intensity and energy fluctuates, streambeds shift and deposit alluvial material on the inside, and remove material from the outside of curves in a continual and dynamic balance. The resulting floodplain is made of alluvial deposits that can range in depth from 5 to 80 m (16 to 260 ft) (Mitsch and Gosselink 1986). When meandering becomes severe, the stream can make an abrupt traverse across a point, leaving an abandoned horseshoe- or crescent-shaped channel, which becomes an oxbow and lagoon.

A unique physiographic feature of these systems is the formation of first and second bottoms and terraces on one or both sides of the stream. The first bottom is commonly known as the Halocene formation, the second one the Deweyville. In southern floodplains, including the Altamaha (Georgia), Pee Dee (South Carolina), Cape Fear (North Carolina), Pearl and Pascagoula (Mississippi), and the Sabine, Trinity, and Brazos (Texas), second bottoms are easily distinguishable. Third bottoms (terraces), although present, are often not distinguished from upland. Where unprotected, second bottoms flood periodically by overland flow and backwater flooding, but third bottoms flood only during the most severe events (Hodges 1994a). Present-day flood control measures have virtually eliminated flooding of some second bottoms.

Deltas, formed from alluvial deposits, are found at the mouths of major rivers. They are a continuation of the floodplain, or its "front." Deltas consist of fine-textured deposits (silts and clays) making these areas largely poorly drained (Brady 1990). They are covered by marsh and are rarely forested. Conversely, the floodplain of the Mississippi River south of Memphis, Tennessee, is commonly referred to as "The Delta" region. Because that nomenclature is widely accepted, we will continue its use in this chapter.

The accretion and removal of deposits associated with subsidence of flood waters create not only bars, flats, natural levees, ridges, sloughs, and swamps (Hodges and Switzer 1979), but also topographic features, which can occur over elevational changes of only a few centimeters, and which have unique hydrologic conditions that lead to differences in soil texture, organic matter content, composition, aeration, water holding capacity, color, and available nutrients (Patrick 1981). A zonal classification of the soils of alluvial rivers provides a convenient framework for depicting this process of deposition and development (Clark and Benforado 1981, Wharton et al. 1982) and is summarized in Figure 12.3.

Geochemistry of Soils

Flooding alters the chemical properties of alluvial floodplain soils by (1) depositing and replenishing mineral nutrients, (2) producing anaerobic conditions, and (3) importing and removing organic matter (Wharton et al. 1982). In fact, alluvial floodplains vary greatly in their physiochemical properties depending on the source and age of the alluvium and the physiographic position within the floodplain. Physiochemical properties in alluvial floodplains arise from two major classes of weathering. Red rivers, being rock-dominated, derive inorganic chemical inputs are derived from weathering and leaching of the parent rocks and soils. In contrast, black rivers are precipitation-dominated, and rainfall contributes most of the inorganic chemical inputs. In general, red rivers usually have higher nitrogen (N), phosphorus (P), and

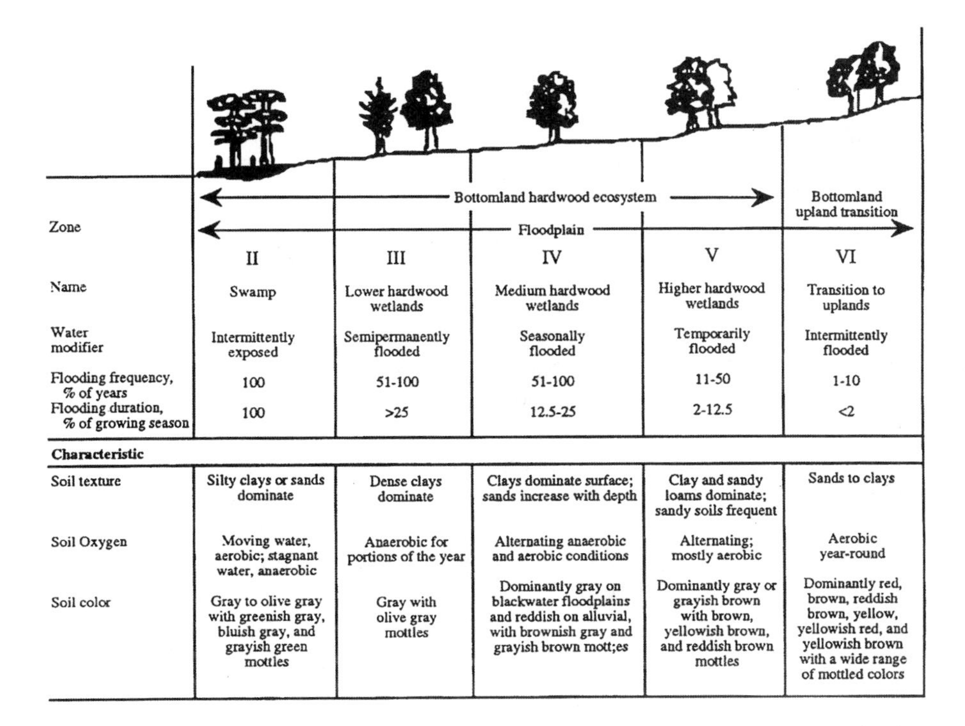

	← Bottomland hardwood ecosystem →				Bottomland upland transition
	← Floodplain →				
Zone	II	III	IV	V	VI
Name	Swamp	Lower hardwood wetlands	Medium hardwood wetlands	Higher hardwood wetlands	Transition to uplands
Water modifier	Intermittently exposed	Semipermanently flooded	Seasonally flooded	Temporarily flooded	Intermittently flooded
Flooding frequency, % of years	100	51-100	51-100	11-50	1-10
Flooding duration, % of growing season	100	>25	12.5-25	2-12.5	<2
Characteristic					
Soil texture	Silty clays or sands dominate	Dense clays dominate	Clays dominate surface; sands increase with depth	Clay and sandy loams dominate; sandy soils frequent	Sands to clays
Soil Oxygen	Moving water, aerobic; stagnant water, anaerobic	Anaerobic for portions of the year	Alternating anaerobic and aerobic conditions	Alternating; mostly aerobic	Aerobic year-round
Soil color	Gray to olive gray with greenish gray, bluish gray, and grayish green mottles	Gray with olive gray mottles	Dominantly gray on blackwater floodplains and reddish on alluvial, with brownish gray and grayish brown mott;es	Dominantly gray or grayish brown with brown, yellowish brown, and reddish brown mottles	Dominantly red, brown, reddish brown, yellow, yellowish red, and yellowish brown with a wide range of mottled colors

FIGURE 12.3 Zonal classification and soil properties of bottomland hardwood wetlands (from Mitsch and Gosselink 1993; with permission).

content than black river systems. Black rivers generally are more acidic and have higher concentrations of total organic carbon, but lower concentrations of dissolved inorganic ions, making them less fertile for agricultural and forest production.

Soils of both river systems experience anaerobic conditions during flooding; in the presence of organic matter, soils become reduced, and mobilization of P, N, magnesium (Mg), sulfur (S), iron (Fe), manganese (Mn), boron (B), copper (Cu), and zinc (Zn) occurs, leading to greater availability of these nutrients (Mitsch and Gosselink 1986). The associated wetting and drying of these soils is not only important for building soil structure but also in the further release of nutrients from litter decomposition. The generally high concentration of essential nutrients and relatively good soil structure suggest that the major limiting factor in floodplains is the stress from limited root oxygen (Sharitz and Mitsch 1993).

Floodplain systems are "open" nutrient cycles, having large inputs and outputs. Floodplain forests are also "nutrient transformers" because the inputs and outputs are in proportion to each other, but the nutrients are different. Studies by Elder and Mattraw (1982) and Elder (1985) found net increases in particulate organic N, dissolved organic N, particulate P and dissolved P, concomitant with net decreases in dissolved inorganic P, and soluble reactive P. Thus floodplain forests are net importers of inorganic forms and net exporters of organic forms of nutrients (Sharitz and Mitsch 1993).

In comparison, deepwater alluvial (tidal) swamps are nutrient "sinks" for mineral inputs. Kitchens et al. (1975) reported a 50 percent reduction in P outflow from a backswamp (stream head) of the Santee River (South Carolina). Day et al. (1977) found similar removals of P and N from a lake-swamp complex in Louisiana. Sequestering these nutrients has been attributed to adsorption of P to clays in the sediment and to denitrification.

Hydrology

Without man-made obstructions, red rivers periodically flood. In extreme years, the first bottom may flood most of the year. More typically, first bottoms flood in summer only in response to unusually heavy precipitation, and flood duration is only several days. In winter, however, first bottoms generally flood for one to three months, often to depths more than 1.2 m (4 ft). Even though precipitation is greater in summer, flooding is more common in winter because of the reduction in evapotranspiration, caused largely by the deciduousness of floodplain tree species.

Because of human influence within the last two centuries, the hydrology of almost all major alluvial floodplains has been inextricably altered. The alteration has resulted from the construction of dams, causeways, and channelization. The earliest dams on the major rivers were constructed for transportation and then for hydroelectric power, such as for the burgeoning textile industry in North Carolina, South Carolina, and Georgia. On the Santee River system alone, 49 powerhouses and 47 dams were constructed by the mid 1950s (Savage 1956). However, justification for additional dams during the last 50 years has been less for power generation than for preventing or ameliorating downstream flooding and for recreation. An added justification for dam construction, and especially for maintaining even flows of water from the dams, has been the Endangered Species Act of 1973 (Kusler 1992). The limited population of Roanoke bass (*Ambloplites cavifrons*) in the Roanoke River serves as a good example of a body of water managed for species conservation.

Recreation has superseded all other reasons for the construction and maintenance of dams on the major rivers of the southern United States. Recreational areas, retirement villages, and permanent and seasonal housing occupy the periphery of every large reservoir. Many property owners have built boat houses and docks to complement their lifestyle. These owners exert sufficient influence with the political process to ensure that their property is usable throughout the year. Thus, while the lake levels are held relatively constant, the alluvial floodplains downstream are drier during periods of low rainfall and wetter during periods of high rainfall compared to pre-dam construction. This alteration can seriously impact the timber resources in the affected areas.

The impact is especially noticeable during late spring and early summer. The released water keeps the water table higher than it would be in unaltered streams. The accumulated surface water becomes deoxygenated from increased summer temperatures. Deoxygenation is especially troublesome to seedling and sprout regeneration, and given sufficient duration, the surface water can adversely affect trees of even the healthiest forest. Experience with greentree reservoirs where the water is held in place during the dormant season and into late spring is showing tree decline similar to that experienced downstream from constructed dams (King 1995). Subtle impacts to other parts of the ecosystem also occur in conjunction with the decline of forest trees.

Channelization, both in the construction of levees to keep water confined within river banks and ditching to drain wetlands, may change species composition of bottomland forests. For example, loblolly pine (*Pinus taeda*) and red maple (*Acer rubrum*) are succeeding swamp tupelo in the lower reaches of the Santee River as a result of diverting water from the Santee River to the Cooper River. This change in species is only now becoming evident following the diversion in the early 1930s. It will be interesting to note the effect of the recent rediversion of water from Lake Moultrie on the Cooper River to the Santee River from whence it came 160 km (100 mi) upstream. Perhaps another 50 years will elapse before the outcome will be evident.

DAMAGING INFLUENCES

Timber in the major alluvial floodplains is host to a range of insects and diseases, none of which are catastrophic but many of which lessen the value of the resource. The greatest insect threat to defoliation is the forest tent caterpillar (*Malacosoma disstria*). This pest commonly defoliates large areas of water tupelo (*Nyssa aquatica*) and swamp tupelo. Great fluctuations occur in the population from year to year, and since the trees commonly suffer only one defoliation each year, usually in May, the trees refoliate within the next month. Even though growth impacts occur from defoliation, tree mortality is limited to those with low vigor. Because of the close proximity of these stands to open bodies of water, no attempts are advocated to control the larvae with insecticides or other means.

Great loss in timber value, especially to high-value oaks and ashes, occurs from borers and beetles. Notable among these pests are the roundhead borers (*Romaleum* spp., *Xyleborus* spp., *Stenodontes* spp.) (Solomon 1995).

Heart rots are the major cause of degrade in southern hardwoods. Entrance of the fungi, primarily *Polyporus hispidus*, is often gained at the base of a tree through a wound caused by logging, fire, or other damage (USFS 1989). Movement of the fungus up the tree varies by species, from 0.6 m (2 ft) per year for overcup oak (*Quercus lyrata*) and sugarberry (*Celtis laevigata*) to less than 0.3 m (1 ft) per year for sweetgum (*Liquidambar styraciflua*) and elms (*Ulmus* spp.) (Putnam et al. 1960).

Another damaging fungus of significance is *Poria spiculosa* which enters through dead branch stubs. This pathogen progresses more rapidly than *Polyporus hispidus*, and it is most common in the upper boles.

Even though the timber in major alluvial floodplains is somewhat sheltered from environmental extremes and fires, windthrow is a common occurrence. In addition to damage from hurricanes (Pittman et al. 1993), localized tornadoes and downdrafts are especially destructive to the shallow-rooted hardwoods common to the alluvial floodplains.

American beavers (*Castor canadensis*) are locally damaging to floodplain forests. Some timber is lost to the rodents from girdling and felling of trees for food and for dam and lodge construction, but most damage is wrought from water impoundment, killing woody plants by deoxygenation of the surface and subsurface water at temperatures above about 21°C (70°F). Even though the cumulative effects from beavers to the floodplain forest is great, the value is secondary to the damage wrought by man through construction of dams and inappropriate placement of culverts, causeways, and bridges in road construction.

VEGETATION COMMUNITIES

TYPES

Several classification systems for alluvial floodplains have been developed. The simplest classification system used since the time of the earliest explorers lumps all communities together, calling them "Bottomland Hardwoods." This name, entirely appropriate when recognizing individual communities, is somewhat of a misnomer, because at least six conifers and three palms may occur. Loblolly pine, spruce pine (*Pinus glabra*), baldcypress (*Taxodium distichum*), pondcypress (*Taxodium distichum* var. *nutans*), Atlantic white-cedar (*Chamaecyparis thyoides*), and sometimes eastern redcedar (*Juniperus virginiana*) are conifers that may occur in alluvial floodplains, although typically not in the same stand. Cabbage palmetto (*Sabal palmetto*), saw-palmetto (*Serenoa repens*), and dwarf palmetto (*Sabal minor*) are palms that may occur in black river floodplains.

Many more-definitive classifications, based upon species composition or topography, are available: Pinchot and Ashe (1897), Wells (1928), Westveld (1949), Braun (1950), Fredrickson (1979), and Eyre (1980). For regional surveys, the U.S. Forest Service (USFS) combines all floodplain communities into a single type, oak-gum-cypress (e.g., Nelson and Zillgitt 1969). Although some state surveys still recognize only one type, bottomland hardwoods (e.g., Rosson 1993a, 1993b), most surveys recognize two types, oak-gum-cypress and elm-ash-cottonwood (e.g., Sheffield and Knight 1986, Thompson 1989, Bechtold et al. 1990).

Recently, classification systems intended to organize data (Cowardin et al. 1979) or conceptualize trends, not to characterize specific sites, have become important. A classification developed by Huffman and Forsythe (1981), and adopted by the National Wetlands Technical Council, divides floodplains into five zones.

Although useful for research or the development of idealized transects across floodplains, some recent classifications have limited utility for field application because they either (1) require detailed measurements not easily taken in the field, (2) make distinctions that are difficult to see in floodplain stands, or (3) require that floodplains be divided into mosaics of very small-sized communities, some containing only several trees. Gaddy and Smathers (1980), for example, recognized 29 communities, including a few in upland communities, in the Congaree Swamp National Monument of South Carolina.

Classifications covering limited geographic areas, a few counties, or one state, may work well because the limited area reduces variation of many factors, including site types, species distribution, genetics of individual species, and climate. For example, classifications for bottomlands in North Carolina (Schafale and Weakley 1990) and Florida (Harper 1914, Ewel 1990b) are useful, but they have limited applicability to the entire South.

Species Composition

Despite development of various classification systems of woody species, few are universally applicable, because (1) few floodplain species have geographic distributions extending throughout the South (Table 12.1), (2) the extreme microsite complexity of bottomlands invariably provides exceptions, and (3) variation within and among communities is clinal, not discrete, often making demarcations arbitrary. Except for the relatively species-poor vegetation of sloughs and point bars, vegetation among floodplains differs largely in the relative abundances of the same species, not in the kind of species. Thus a complete list of woody species from two bottomlands with different site conditions would look somewhat similar.

Although exceptions occur, seven species may be found in pure stands on sloughs, levees, first bottoms, and terraces: swamp tupelo, black willow (*Salix nigra*), sweetgum, sycamore (*Platanus occidentalis*), eastern cottonwood (*Populus deltoides*), river birch (*Betula nigra*), and water tupelo. They generally form early successional communities, developing either on bare ground after catastrophic disturbance, after agricultural land has been abandoned, or on point bars or other nonvegetated land (Figure 12.4). Swamp tupelo stands are most common in southern Georgia. Black willow, sycamore, and eastern cottonwood stands are most common along the lower Mississippi River, southward to about Baton Rouge, Louisiana. Sweetgum and river birch stands are equally common throughout southern alluvial floodplains, but upriver from tidally influenced landscapes. Water tupelo forms pure stands on very low first bottoms that in topography and function are essentially very wide sloughs. Sloughs often support pure stands because few species tolerate the long hydroperiod. Pondcypress forms pure stands in blackwater floodplains, and baldcypress does the same in redwater floodplains.

Four basic geographic distribution patterns of floodplain species are found: (1) distributed throughout, (2) locally absent, (3) restricted distribution, and (4) rare. Distribution differences are often ignored. For example, pumpkin ash (*Fraxinus profunda*), waterlocust (*Gleditsia aquatica*), Atlantic white-cedar, and swamp cottonwood (*Populus heterophylla*) have been described as "common" to southern alluvial floodplains (Smith and Linnartz 1980, Christensen 1988), and yet all have restricted geographic distributions.

Only about 13 tree species are widely distributed and found in most southern floodplains (Table 12.1). In addition, 29 species occur in over one-half of southern floodplains, 21 occur in less than one-half of the floodplains, and 10 species are so infrequently encountered that, by definition, they are exceptions to overall trends (Table 12.1). Some of these species (e.g., franklinia [*Franklinia alatamaha*]) are truly rare, and others (e.g., swamp white oak [*Quercus bicolor*]) are rare in the lower South, but have large distributions at the more northerly latitudes. Because the vegetation of red and black rivers differs somewhat by species, and principally by relative species abundance, the two systems will be treated separately in this chapter.

TABLE 12.1
Tree species of alluvial floodplains of southern forests by distribution pattern. Tree names and geographic distributions follow Little (1971, 1977).

Tree species occurring throughout southern bottomlands	
Acer rubrum	red maple
Carya aquatica	water hickory
Celtis laevigata	sugarberry
Cephalanthus occidentalis	buttonbush
Diospyros virginiana	persimmon
Fraxinus pennsylvanica	green ash
Ilex opaca	American holly
Liquidambar styraciflua	sweetgum
Morus rubra	red mulberry
Quercus michauxii	swamp chestnut oak
Quercus nigra	water oak
Ulmus americana	American elm
Taxodium distichum	baldcypress
Tree species found largely throughout, but locally absent	
Aesculus pavia	red buckeye
Alnus serrulata	hazel alder
Betula nigra	river birch
Carpinus caroliniana	American hornbeam
Carya cordiformis	bitternut hickory
Cercis canadensis	eastern redbud
Fraxinus caroliniana	Carolina ash
Hamamelis virginiana	witch-hazel
Ilex decidua	possumhaw
Magnolia virginiana	sweetbay
Myrica cerifera	southern bayberry
Nyssa aquatica	water tupelo
Nyssa sylvatica var. *biflora*	swamp tupelo
Persea palustris	swampbay
Platanus occidentalis	sycamore
Populus deltoides	eastern cottonwood
Quercus falcata var. *pagodaefolia*	cherrybark oak
Quercus laurifolia	swamp laurel oak
Quercus lyrata	overcup oak
Quercus phellos	willow oak
Quercus shumardii	Shumard oak
Sabal minor	dwarf palmetto
Salix caroliniana	coastal plain willow
Salix nigra	black willow
Styrax americana	storax
Symplocos tinctoria	sweetleaf

TABLE 12.1 (continued)
Tree species of alluvial floodplains of southern forests by distribution pattern. Tree names and geographic distributions follow Little (1971, 1977).

Taxodium distichum var. *nutans*	pondcypress
Ulmus alata	winged elm
Viburnum nudum	possumhaw viburnum
Tree species with restricted distributions in southern alluvial floodplains	
Acer negundo	boxelder
Acer saccharinum	silver maple
Asimina triloba	pawpaw
Bumelia lycioides	buckthorn bumelia
Carya illinoensis	pecan
Catalpa bignonioides	southern catalpa
Cornus drummondii	roughleaf dogwood
Foresteria acuminata	swamp-privet
Fraxinus profunda	pumpkin ash
Gleditsia aquatica	waterlocust
Gleditsia triacanthos	honeylocust
Halesia diptera	two-wing silverbell
Ilex verticillata	winterberry
Nyssa ogeche	Ogeechee gum
Planera aquatica	water-elm
Populus heterophylla	swamp cottonwood
Quercus nuttallii	Nuttall oak
Sabal palmetto	cabbage palm
Staphylea trifolia	bladdernut
Ulmus crassifolia	cedar elm
Viburnum obovatum	Walter viburnum
Rare tree species of southern alluvial floodplains	
Carya myristicaeformis	nutmeg hickory
Celtis occidentalis	hackberry
Franklinia alatamaha	franklinia
Ilex amelanchier	sarvis holly
Leitneria floridana	corkwood
Quercus bicolor	swamp white oak
Quercus oglethorpensis	Oglethorpe oak
Quercus palustris	pin oak
Salix floridana	Florida willow
Salix sericea	silky willow

FIGURE 12.4 Black willow regeneration on recently exposed alluvium (Mississippi River, Kentucky).

Vegetation of Black Rivers

Levees and First Bottoms

Most black rivers have a narrow zone of highly flood-tolerant vegetation that parallels the channel. The zone extends from the channel through the transition zone, forming areas permanently to seasonally inundated. Pondcypress, swamp tupelo, red maple, and sometimes water tupelo are the major tree species found in this zone. The understory is sparse, owing to the long hydroperiod, but lizard's tail (*Saururus cernuus*) and knotweed (*Polygonum* spp.) may occur.

Black river levees are absent or poorly developed, owing to low sediment loads. When present, the levee varies from 1 to 20 m (3 to 65 ft) in width and from 10 to 20 cm (4 to 8 in) in height (Lollis 1981). The levee is rarely continuous, interrupted frequently by openings that drain the first bottom.

In area, first bottoms dominate most floodplains. First bottoms are usually flooded in winter and early spring and remain moist or flood episodically after heavy rains in summer and fall. On a local scale, representing one black river stand, Carolina ash (*Fraxinus caroliniana*), red maple, and swamp tupelo comprised more than 75 percent of the trees with diameter breast height (dbh) equal to or greater than 10 cm (4 in) (Table 12.2). Over a larger area, representing 10 black rivers in North Carolina, Flinchum (1977) found a combination of eight species in the upper canopy: swamp tupelo, red maple, cypress, swamp laurel oak (*Quercus laurifolia*), sweetgum, green ash (*Fraxinus pennsylvanica*), American elm (*Ulmus americana*), and water tupelo (Figure 12.5). These species comprise roughly 60 percent of the

TABLE 12.2
Percent of total dominance for trees more than 10 cm (4 in) dbh for six 0.08 ha (0.2 acre) fixed plots along the White Oak River, Jones County, North Carolina. Total dominance equals 31.2 m²/ha (136.1 ft²/acre) growing on Muckalee soil. Site index for loblolly pine at 50 years is 32 m (105 ft).

Species	Plot Number 1	2	3	4	5	6	Mean
Sweetgum	0.0	7.4	0.0	2.7	0.8	0.0	1.8
Carolina ash	47.0	19.3	32.3	36.7	39.1	18.1	32.1
Swamp tupelo	11.7	0.0	24.2	4.9	20.0	58.6	19.9
American elm	7.8	3.5	15.8	9.7	11.9	0.0	8.1
Red maple	31.6	33.4	26.9	36.7	15.9	16.5	26.8
Pondcypress	1.8	17.4	0.0	6.8	0.0	0.0	4.3
Laurel oak	0.0	19.0	0.0	0.0	1.5	5.7	4.4
American hornbeam	0.0	0.0	0.7	2.5	0.0	0.0	0.5
Loblolly pine	0.1	0.0	0.0	0.0	2.4	0.0	0.4
Eastern redcedar	0.0	0.0	0.0	0.0	3.4	0.0	0.6
Swamp redbay	0.0	0.0	0.0	0.0	5.0	0.0	0.8
Possumhaw*	0.0	0.0	0.0	0.0	0.0	1.1	0.2

* *Ilex decidua*

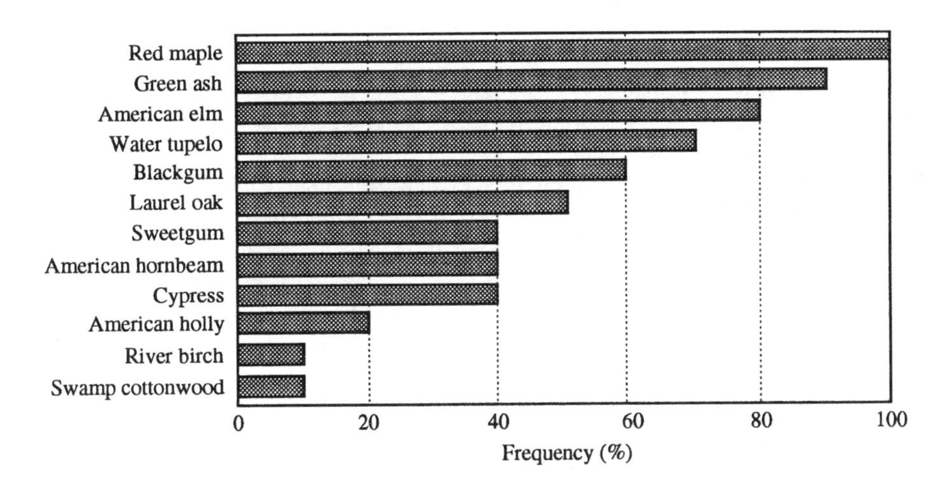

FIGURE 12.5 Absolute frequency of 12 species along 10 black rivers in Eastern North Carolina (Flinchum 1977).

total basal area that varies considerably from about 23 m²/ha (100 ft²/acre) to 72 m²/ha (300 ft²/acre) (Applequist 1959). In Mississippi, Louisiana, Arkansas, Texas, and parts of Alabama, baldcypress replaces pondcypress on both levees and first bottoms. The only species found on black river levees that might not be found on first bottoms are river birch and swamp chestnut oak (*Quercus michauxii*), but these species are not always present.

About 10 to 20 percent of upper-canopy basal area is composed of associate species: water oak (*Quercus nigra*), water tupelo, river birch, and loblolly pine. In North and South Carolina, swamp cottonwood and Atlantic white-cedar may occur, and in most areas except for southeastern Georgia and peninsular Florida, willow oak (*Quercus phellos*) occurs. In Florida, cabbage palm is common. Mistletoe (*Phoradendron serotinum*) generally occurs on canopy trees, especially on red maple, oaks, and ashes. Spanish moss (*Tillandsia usneoides*) is generally present, but is most abundant in Florida and in stands on the lower terraces of the Atlantic and Gulf Coastal Plain.

Except in stands severely degraded by logging where the lower canopy assumes greater importance, the lower canopy generally occupies about 10 to 30 percent of the total basal area. It contains saplings of red maple, swamp tupelo, willow oak, water oak, American elm, and mature trees of Carolina ash, American hornbeam (*Carpinus caroliniana*), American holly (*Ilex opaca*), sweetbay (*Magnolia virginiana*), southern bayberry (*Myrica cerifera*), and swampbay (*Persea palustris*). In many stands Carolina ash dominates the lower canopy. Except in North Carolina, water-elm (*Planera aquatica*) also occurs, sometimes sharing lower-canopy dominance with Carolina ash. In Florida, cabbage palms may also occur.

Understory density varies from dense to sparse, owing to differences in flooding and stand history. Stands that flood for long periods may lack understory species; stands that seldom flood or have been disturbed may have a thick understory that grows to 3 m (10 ft) in height and are nearly impenetrable to humans. The cover of this layer may range from 5 to 100 percent. It contains mostly Virginia willow (*Itea virginica*), fetterbush (*Lyonia lucida*), large gallberry (*Ilex coriacea*), sweetleaf (*Symplocos tinctoria*), sweet pepperbush (*Clethra alnifolia*), and giant cane (*Arundinaria gigantea*), but blueberry (*Vaccinium elliottii*), dwarf palmetto, and dog-hobble (*Leucothoe axillaris*) may occur also.

Within the last 20 years, Chinese privet (*Ligustrum sinense*), Chinese tallowtree (*Sapium sebiferum*), and shining privet (*Ligustrum lucium*) have invaded the shrub layer of many stands. They generally occur in better drained areas: on causeways built across the bottomland, on well-developed levees, on drier microsites within the first bottom, or throughout the terrace. The long-term effects of these exotics on species composition are not known, but they probably reduce the size and number of native species.

The vine layer contains poison-ivy (*Toxicodendron radicans*), climbing hydrangea (*Decumaria barbara*), and crossvine (*Anisostichus capreolata*), but greenbrier (*Smilax walteri*) and trumpet creeper (*Campsis radicans*) may also occur. Stand edges often contain thick drapings of greenbrier, and upper surfaces of fallen trees and stumps that extend above the high-water level are invariably covered with poison-ivy.

The herb layer is sparse for either of two reasons: flooding or a thick shrub layer. Flinchum (1977) documented the inverse relationship between frequency of flooding and understory biomass. In stands long-flooded, foliar cover varies from about 5 to 15 percent, and lizard's tail, knotweed, pennywort (*Hydrocotyle* spp.), and netted chain fern (*Woodwardia areolata*) are the only common species. Drier areas may have higher foliar cover, varying from about 10 to 25 percent, and containing sedges (*Carex* spp.), false-nettle (*Boehmeria cylindrica*), beardtongue (*Penstemon* spp.), seedbox (*Ludwigia* spp.), and violets (*Viola* spp.). Moist microsites that rarely flood — crevices in the bole of buttressed trees, tip-up mounds, tops of rotting stumps, and fallen trees — provide important micro-habitat for false nettle and partridgeberry (*Mitchella repens*). Mosses, liverworts, and lichens frequently grow on tree bases to a height of 3 m (10 ft).

Terraces

With some exceptions, terraces (second bottoms) of black rivers are similar to first bottoms in species composition, but the relative abundance of species may differ. The commonality of species and the typically gradual transition from first bottoms to terraces make separating these communities difficult. Pondcypress, water tupelo, swamp tupelo, Carolina ash, swamp cottonwood, Atlantic white-cedar, and water-elm occur less frequently on terraces, owing to improved drainage. Sweetgum, swamp laurel oak, swamp chestnut oak, water oak, willow oak, river birch, and loblolly pine are more abundant than on first bottoms.

In the understory, lizard's tail and Virginia willow become less common, and Japanese honeysuckle (*Lonicera japonica*), poison-ivy, supplejack (*Berchemia scandens*), and Virginia creeper (*Parthenocissus quinquefolia*) are more common than on first bottoms. Among understory species, Flinchum (1977) found that an increase in the relative frequency of Virginia creeper best indicates a decrease in average yearly water table level.

In the last few decades, Japanese grass (*Microstegium virmineum*) has become well established throughout the South on well-developed levees, well-drained first bottoms, and terraces. It is very common along roads built on causeways across bottomlands. In some stands, foliar cover of this noxious weed from Asia exceeds 80 percent, overwhelmingly dominating the herbaceous layer. Unlike some exotics, dominance by Japanese grass is not limited to disturbed stands. It may invade stable bottomland communities. The effect of Japanese grass on native vegetation is not clear, but like Chinese privet, it probably impacts all native species with which it comes in contact.

Sloughs

Blackwater sloughs contain only two principal species (swamp tupelo and pondcypress), although scattered individuals of red maple, Carolina ash, and American elm may also occur. The abundance of pondcypress has been reduced in many sloughs by high-grade logging. Owing to the long hydroperiod, the shrub layer is generally lacking, and the herb layer contains only scattered lizard's tail.

Vegetation of Red Rivers

The greater sediment loads and faster flood-water currents of red rivers make levees higher, first bottoms and terraces wider, and sloughs larger and deeper than black rivers. These differences are particularly pronounced for the Mississippi River watershed, much of which developed during Pleistocene glacial times when ice sheets to the north blocked rivers, diverting them south- and eastward. These movements significantly enlarged the Mississippi River watershed (Flint 1971). The enlarged watershed carried especially large amounts of melt-water and suspended sediment during the various retreats of the Pleistocene ice sheets (McKnight et al. 1981), and these forces contributed to the development of a very large floodplain.

With few exceptions, red rivers contain all woody species found in black rivers, but the reverse is not true. Swampbay and sweetbay are not typical of red rivers, and eastern cottonwood, bitternut hickory (*Carya cordiformis*), sycamore, cherrybark oak (*Quercus falcata* var. *pagodifolia*), Shumard oak (*Quercus shumardii*), Nuttall oak (*Quercus nuttallii*), silver maple (*Acer saccharinum*), pecan (*Carya illinoensis*), catalpa (*Catalpa bignonioides*), black walnut (*Juglans nigra*), and honeylocust (*Gleditsia triacanthos*) are not typical of black rivers.

Levees

The upper canopy of levees predominantly contains about 15 species in various mixtures: sweetgum, sycamore, sugarberry, bitternut hickory, red maple, river birch, willow oak, water oak, swamp laurel oak, green ash, loblolly pine, winged elm *(Ulmus alata),* American elm, pecan, eastern cottonwood, and persimmon (*Diospyros virginiana*). Except for portions of the Atlantic Coast, boxelder (*Acer negundo*) is also common. In the western Gulf states and along the Mississippi River, honeylocust occurs sporadically. Along the Atlantic Coast, honeylocust has occasionally spread from old home sites (Braham 1992) to levees. Silver maple may also occur in portions of the Mississippi River and Gulf States.

When compared to levees of the lower Coastal Plain, levees near the fall line are generally larger and topographically higher because the sudden reduction in current velocity causes large-sized sediments to settle quickly. As a result, flooding of levees near the fall line is rare and temporary, and species more typical of upland, nonalluvial rivers and major creeks may occur. White oak (*Quercus alba*), Florida maple (*Acer barbatum*), black walnut, American beech (*Fagus grandifolia*), flowering dogwood (*Cornus florida*), witch-hazel (*Hamamelis virginiana*), and yellow-poplar (*Liriodendron tulipifera*) may be found on well-drained levees. Except for Virginia and northeastern North Carolina, southern magnolia (*Magnolia grandifolia*) may also occur.

Levees support many species common to other bottomland communities, but the presence of bitternut hickory, winged elm, persimmon, and, in selected geographic areas, boxelder, honeylocust, and silver maple help differentiate the levee community. In addition, location on a topographic high point, occurrence on light-textured soils, and sometimes the presence of upland and branch bottom species further identify levee communities.

The lower canopy of levees almost always contains American hornbeam, American holly, and possumhaw, but hawthorns (*Crataegus viridis* and *C. marshallii*), red mulberry (*Morus rubra*), eastern redbud (*Cercis canadensis*), bladdernut (*Staphylea trifolia*), and pawpaw (*Asimina triloba*) may also occur. Except for Virginia and northeastern North Carolina, red buckeye (*Aesculus pavia*) occurs sporadically.

The shrub layer contains mostly Chinese privet, southern bayberry, strawberry-bush (*Euonymus americanus*), and spicebush. In addition, silky dogwood (*Cornus amomum*) and viburnum (*Viburnum nudum*) also occur sporadically. Sebastian bush (*Sebastiania ligustrina*) and snowbell (*Styrax* spp.) occur commonly in the Lower Mississippi River and Western Gulf communities.

The vine layer is especially well developed, particularly along stand edges, river banks and canopy openings. Frequent scouring of levees from flood waters may favor these disturbance-tolerant species. Japanese honeysuckle is invariably present. Greenbriers (*Smilax rotundifolia*, *S. bona-nox*, and *S. glauca*), trumpet creeper, and supplejack develop thick tangles to a height of about 3 m (10 ft), making human travel difficult. Muscadine (*Vitis rotundifolia*) and poison-ivy often climb high into the crowns of overstory trees, although they also grow along the soil surface or in tangled masses on the ground, the result of broken tree branches that once supported them. Virginia creeper and crossvine are also present, but generally in lower numbers.

Species composition of the herb layer is not well documented, owing to their ephemeral growth habit, the emphasis on documenting woody species, and the difficulty in making precise identifications. Among the taxa present are: rattlesnake fern, sedge, violet, Christmas fern (*Polystichum acrostichoides*), giant cane, uniola grass (*Chasmanthium laxum*), and may-apple (*Podophyllum peltatum*). Foliar cover of the herb layer is generally less than 35 percent, with the remaining areas covered with either leaf litter or bare soil. Large, well-drained levees may also support some of the wildflowers typical of lower mesic slopes, especially trout lily (*Erythronium americanum*), may-apple, and bloodroot (*Sanguinaria canadensis*).

Bottoms

Except for sloughs, first bottoms have the longest hydroperiod of alluvial floodplains. As a result, more flood-tolerant species occur there than are found on either levees or terraces.

The upper canopy of first bottoms is dominated by eight species: sweetgum, swamp laurel oak, red maple, water hickory, overcup oak, green ash, swamp tupelo, and American elm, with water hickory and overcup oak occupying the wettest soils. These eight species occupy roughly 70 percent of the total basal area that varies from 27 to 48 m^2/ha (117 to 208 ft^2/acre). In Mississippi, Arkansas, Louisiana, and adjacent states, Nuttall oak and cedar elm (*Ulmus crassifolia*) also occur. Baldcypress and water tupelo may occur in sloughs within first bottoms or scattered among the eight dominant species in the ecotone surrounding sloughs. In North and South Carolina, swamp cottonwood may occur with baldcypress and water tupelo. About 15 percent of the basal area is composed of associate species, especially willow oak, water oak, swamp chestnut oak, sycamore, and river birch.

With large crowns and clear boles, upper-canopy species of well-developed stands clearly dominate the community, forming "cathedral-like" canopies that commonly reach heights more than 30 m (100 ft). Seen from outside the stand, the understory of many first bottoms appears thick with vegetation and impenetrable. This impression results from seeing high-density understories and vines along stand edges; it is not typical of stand interiors. Inside well-developed stands, the vegetation

FIGURE 12.6 High quality, mixed hardwood stand along a red river (Tombigbee River, Alabama).

layers below the upper canopy are sparse, owing to flooding, site capture by the upper canopy, and lack of understory tolerant tree species (Figure 12.6).

Basal area of all lower layers is about 15 percent of the total, with individual plants widely scattered. The lower canopy invariably contains American hornbeam, possumhaw, American holly, hawthorn, and Virginia willow, in addition to some of the shrubs found on levees.

The vine layer contains poison-ivy and crossvine. The herb layer contains false-nettle, violets, giant cane, sedges and uniola grass. Most of the ground surface is covered with a thin layer of leaf litter.

Terraces

Terraces, inclusive of second bottoms, of red rivers are typically well drained, becoming saturated only after heavy precipitation. The upper canopy of well-developed stands is similar in basal area and form to first bottoms. It contains about a dozen species: sweetgum, red maple, green ash, sugarberry, loblolly pine, American elm, swamp chestnut oak, swamp laurel oak, water oak, and willow oak. In some stands, cherrybark oak is also a dominant. Associate species of the dominants are white oak, persimmon, shagbark hickory, black walnut, river birch, yellow-poplar, American beech, and Shumard oak.

The lower canopy and understory of terraces are similar in species composition to levees. Terraces are less subject to flooding than bottoms, and, consequently, foliar cover of the understory layers is up to 50 percent greater. In addition to the species listed above, spicebush (*Lindera benzoin*) and pepperbush may occur, and along stand edges or in localized canopy openings, common blackberry (*Rubus argutus*) occurs.

PLANT ADAPTATION TO WATER-SATURATED OR FLOODED SITES

Plant adaptations to water-saturated soils or flooding may be divided into three broad types: morphological, physiological, and reproductive. Not all alluvial floodplain species have all types, but all species have at least one, especially physiological adaptation. At least five morphological mechanisms occur in alluvial species: shallow rooting, hypertrophied lenticels, adventitious roots, buttressing, and knees are typical of trees in areas with long hydroperiods.

ROOT MODIFICATIONS

Most, if not all, alluvial floodplain species develop shallow roots that allow trees to utilize the upper-most soil regions, where anaerobic conditions are minimized. Seedlings and saplings of some species, notably waterlocust, red maple, silver maple, and black willow, form especially large lenticels when flooded (USACE 1987). Because lenticels are specialized epidermal openings that facilitate gas exchange of the phloem, cambium, and sap wood, hypertrophied lenticels probably improve aeration.

Adventitious roots in alluvial floodplain species develop above ground on the lower bole (on stem tissue). They are thought to facilitate air and nutrient absorption because of their location near the water line where oxygen is less limiting. Adventitious roots have been observed in boxelder, sycamore, eastern cottonwood, pin oak (*Quercus palustris*) (USACE 1987), and in other species.

Swollen and possibly convoluted lower boles are called buttresses. In the temperate zone they develop on trees growing on sites with long hydroperiods, and buttress size is positively correlated with hydroperiod. On sites with short hydroperiods, swelling is localized, only extending vertically for a distance of 1 to 2 m (3 to 7 ft) above ground level and horizontally for a distance only one-half times the diameter above the buttress. On sites with long hydroperiods, swelling can be pronounced, extending 3 to 4 m (10 to 15 ft) vertically and 2 to 3 times the width of the bole above the buttress. Large buttresses present special challenges in measuring stem diameter, because procedures require measuring 45 cm (18 in) above the buttress (Avery and Burkhart 1994). Buttressing has been observed in green ash, water tupelo, Ogeechee gum, swamp tupelo, the cypresses (USACE 1987), and other species.

Knees are modified roots that grow upward above the soil surface and often above the high water line. They have specialized low-density tissue, called aerenchyma (Fahn 1990), containing large intercellular spaces, and they have been observed on cypress and water tupelo. Traditionally thought to help aerate the root system through the large intercellular spaces, the exact function of knees is not understood (Harlow et al. 1996).

Physiological Mechanisms

At least three physiological mechanisms to survive water-saturated soil or flooding exist: oxidation of the rhizosphere, low soil oxygen requirements, and specialized chemical reactions. Green ash, swamp tupelo, water tupelo, and crack willow (*Salix fragilis*) are able to increase gaseous oxygen content of soil pores surrounding feeder roots (USACE 1987) by moving gaseous oxygen outward from roots to the surrounding soil. Increased oxygen levels maintain water and nutrient absorption of roots.

Some willows and some monocots are able to continue metabolic processes in soils with low oxygen concentrations (USACE 1987) that would stop many processes of species lacking this adaptation. Some monocots and swamp tupelo have specialized or alterable chemical processes that allow continued metabolism in water-saturated or flooded soils. These species may accummulate malate instead of ethanol, produce high levels of nitrate reductase, or reduce ethanol production by reducing alcohol dehydrogenase activity. See Chapter 8 for further discussion of physiological adaptations to flooding.

Reproductive Mechanisms

At least four reproductive mechanisms of bottomland species occur: diaspore release in spring, diaspores that float, flood-tolerant seedlings, and fragmentation.

Some alluvial floodplain species release diaspores (seeds or fruits) in the spring when water levels are high and dispersal distance by water is maximized. Red maple, silver maple, American elm, river birch, cottonwoods, and many willows complete their entire flowering and fruiting processes within a few weeks in the spring. Other species like sycamore mature diaspores in the fall, but release them the following spring.

Floating diaspores are also common among bottomland species, and water buoyancy is accomplished in many ways. Sycamore, cottonwoods, and willows have diaspores with hairs that allow for extended transport by wind and water, while American elm, river birch and maples have winged diaspores. Cypresses, water tupelo, swamp tupelo, and overcup oak have very low-density diaspores, mechanisms which enhance seed dispersal along water bodies. Bladdernut produces a capsule fruit that develops into papery "air bladders."

Seedlings of some bottomland species, green ash for example, tolerate total submersion, an obvious advantage in flood-prone bottomlands (USACE 1987). Fragmentation (reproduction from abscissed twigs) occurs in cottonwoods and willows, where preformed adventitious root buds occur at many nodes (Fahn 1990). This feature is extremely useful to bottomland species because branches can break, float downstream and root in moist soil. Fragmentation allows eastern cottonwood plantations to be established from cuttings stuck in the soil. Both the common and scientific names of crack willow refer to the ease with which branches break at preformed abscission zones.

SUCCESSIONAL PATTERNS

Severe disturbance of alluvial floodplains or colonization of nonvegetated sites such as old-fields and point bars reduces species-richness of all vegetation layers, when compared to mature stands. A result of such disturbance is pure stands. With time, however, the pure stand is succeeded by species with greater shade tolerance. On well-drained front land of the Mississippi River, for example, the species succession is eastern cottonwood, sycamore, silver maple, cherrybark oak, green ash, sugarberry and boxelder. On low ridges, the sequence is white oak, swamp chestnut oak, swamp tupelo, green ash, hickory and water oak, willow oak, or pin oak (Putnam et al. 1960).

Disturbance of an alluvial floodplain stand by harvesting will yield a young stand with equal or greater diversity than that of an old-growth forest. The major reason for the diversity is that regeneration is from both sprouts and seeds. This phenomenon occurs regardless of the site type, either upland or bottomland forest. Such stands go through a bottleneck in species diversity, at 15 to 20 years after regeneration, when many colonizing species are being severely impacted by successional species (Elliott and Swank 1994). Bowling and Kellison (1983) showed that swamp tupelo and American hornbeam declined in frequency relative to sweetgum and water oak at about 20 years following a regeneration harvest on a creek bottom in Mississippi.

Species succession following a harvest is different in sloughs, bayous, oxbow lakes, and deepwater swamps than it is for the alluvial floodplain. There, the predominant species of baldcypress, water tupelo, swamp tupelo, green ash, and Carolina ash will succeed themselves, largely from stump sprouts, without the presence of successional species (Mader 1990). In the wettest and poorest drained areas, and in areas where the harvested timber is so old that it does not readily sprout, herbaceous species such as bulrush (*Scirpus* spp.), cattail (*Typha* spp.), and alligator-weed (*Alternanthera philoxeroides*) will first occupy the site. The first woody species to invade are red maple and black willow. If the hydrology has not been significantly altered by changes in topography, the pioneering species will form elevated seed beds in which water tupelo and its cohorts will become established to form the succeeding stand. In extreme cases, shallow ponds may develop, and re-establishment of forest communities delayed for many decades.

WILDLIFE

Major alluvial floodplains are one of the most important wildlife habitats in the southern forest. The high fertility that supports high forest productivity produces more high-quality food and browse than is found on poorer sites.

Although few species of wildlife are restricted to alluvial floodplains, it is a preferred habitat of many. The extensiveness and denseness of unbroken floodplain forests are as important as the habitat itself. It is said that prior to logging and clearing, a squirrel could travel from the loess bluffs to the Mississippi River and never touch the ground. These broad expanses were important for species like the ivory-billed woodpecker (*Campephilus principalis*) and the panther (*Felis concolor*), species either extinct or extirpated from these habitats.

Game species that are present in alluvial floodplains, prized by early settlers and today's hunters, include white-tailed deer (*Odocoileus virginianus*) and wild turkey (*Meleagris gallopavo*). The presence and often substantial populations of these two game animals make hardwood floodplain forests popular areas for hunting. Other game and furbearing animals existing in these areas include common raccoon (*Procyon lotor*), bobcat (*Lynx rufus*), black bear (*Ursus americanus*), American beaver, squirrel (*Sciurus* spp.), common muskrat (*Ondatra zibethicus*), northern river otter (*Lutra canadensis*), mink (*Mustela vison*), and rabbit (*Sylvilagus* spp.).

Additional mammals found in these habitats include opossum (*Didelphis virginiana*), nutria (*Myocastor coypus*), short-tailed shrew (*Blarina brevicauda*), southeastern shrew (*Sorex longirostris*), marsh rice rat (*Oryzomys palustris*), and white-footed mouse (*Peromyscus leucopus*).

Numerous amphibians and reptiles occur in alluvial floodplains. Wharton et al. (1982) and others have described the species found in these forests to include southern toad (*Bufo terrestris*), southern leopard frog (*Rana utricularia*) and treefrogs (*Hyla* spp.). Numerous species of snakes (*Agkistrodon* spp., *Nerodia* spp., *Storecia* spp. and others), eastern mud turtle (*Kinosternon subrubrum subrubrum*), eastern box turtle (*Terrapene carolina carolina*), and common musk turtle (*Sternotherus odoratus*) can also be found in these habitats for at least part of the year.

Birds are perhaps the most studied vertebrates inhabiting major alluvial floodplains. Birds use these forests as breeding, wintering, or migrating stop-over habitat. Smith et al. (1993) reported that of the 236 landbirds in eastern North America, 200 (85 percent) can be found in the Mississippi Alluvial Valley forest. When comparing studies of avifauna in bottomland hardwoods to pine and pine-hardwood forests, Dickson (1978) reported that the former had among the highest diversity of bird species, the greatest number of species, and the greatest density. Hamel (1992) has identified the species with restricted populations in bottomland hardwood forests as Mississippi (*Ictinia mississippiensis*) and swallow-tailed (*Elanoides forficatus*) kites. Characteristic species include the pileated woodpecker (*Drycopus pileatus*), red-bellied woodpecker (*Melanerpes carolinus*), red-headed woodpecker (*Melanerpes erythrocephalus*), Acadian flycatcher (*Empidonax virescens*), great crested flycatcher (*Myiarchus crinitis*), blue-gray gnatcatcher (*Polioptila caerulea*), Prothonotary warbler (*Protonotaria citrea*), northern parula (*Parula americana*), yellow-rumped warbler (*Dendroica coronata*), yellow-throated warbler (*D. dominica*), Kentucky warbler (*Oporornis formosus*), hooded warbler (*Wilsonia citrina*), American redstart (*Setophaga ruticilla*), white-throated sparrow (*Zonotrichia albicollis*), Carolina wren (*Thryothorus ludovicianus*), American robin (*Turdus migratorius*), yellow-billed cuckoo (*Coccyzus americanus*), tufted titmouse (*Parus bicolor*), rusty blackbird (*Euphagus carolinus*), hermit thrush (*Catharus guttatus*), northern cardinal (*Cardinalis cardinalis*), blue jay (*Cyanocitta cristata*), barred owl (*Strix varia*), red-shouldered hawk (*Buteo lineatus*), American woodcock (*Scolopax minor*), and wild turkey. Additionally, waterfowl such as wood duck (*Aix sponsa*), mallard (*Anas platyrhynchos*), and teal (*Anas* spp.) are frequent inhabitants and users of this habitat.

FOREST MANAGEMENT

Forests of the major alluvial floodplains will be managed for timber production, and as a result, wildlife habitat, watershed management, aesthetics, and recreation will be accompanying benefits. In some cases, the primary emphasis will be on other values, such as wildlife management (Langdon et al. 1981). In our opinion, however, "preservation" of the forest is not an accepted alternative. Alluvial floodplain forests exist because of disturbance. In the absence of disturbance, the upper-canopy forest, composed of shade-intolerant species (e.g., eastern cottonwood) to medium shade-tolerant species (e.g., sweetgum and oaks), will reach physiological maturity and die. They will be succeeded by shade-tolerant species such as hawthorn, American hornbeam, American holly, dogwood, eastern redbud, and American beech. The mistaken assumption that a mature forest of today, with its pristine beauty, was always there and will always exist is like taking a snapshot in time (Shear et al. in press). Failure to recognize the dynamics of a forest, and to restrict activity, is an act of selfishness that will rob future generations of the same beauty. Without forest intervention, they will be looking at a decadent forest with recruits that lack value or beauty.

To successfully manage the alluvial floodplain forest, the difference between a regeneration harvest, partial harvest, and intermediate stand management must be made. Because so much of the alluvial forest has been degraded by selective logging, the major emphasis of today's professional forester is often on harvesting to obtain desired reproduction. In stands of good quality, however, recommended practices can include cleaning to favor desired species and thinning to favor individual trees in the stand. Thinning can be of many sorts: from below, from above, or a combination of the two, accomplished at any time after crown closure (at 8 to 10 years of age), to 10 to 15 years before harvest of the crop trees. A basic tenet of silviculture is that thinning is done to benefit the crop trees and only secondarily to recover forest products (Nyland 1996).

Even though the value of intermediate stand management is readily recognized for increasing timber quality in alluvial floodplains (Putnam et al. 1960), the practice is difficult to properly implement because of flooding during some portions of the year, some depressions remain wet and boggy during even the driest times of the year (thereby restricting use of wheeled and crawler logging equipment), and the use of logging machinery causes damage to tree boles and root systems. For example, Meadows (1993) showed that 62 percent of the residual trees received some bole damage (debarking) when thinning an alluvial floodplain stand to 60 percent of its original basal area, and that 35 percent of those trees were damaged enough to adversely affect log quality at time of harvest. For those reasons, thinning alluvial floodplain stands is often discouraged unless loggers use meticulous care in the process. Tree-length logging, except for helicopter logging which is usually prohibitively expensive, is especially discouraged.

Although exceptions occur, mixed hardwoods in the major alluvial floodplains have been logged one to several times since the first sawmill was built in the United States in 1633 by Dutch settlers (Heavrin 1981). The method of logging has generally been degenerative (removal of only the best and the largest trees while leaving the smallest and worst trees to form the new stand). This form of timber harvesting, commonly known as selective harvesting is, in reality, highgrading, which is condemned by foresters. This degenerative practice is not to be confused with the silviculturally sound selection system where the desired tree-species mix of all size classes is maintained.

Diameter-limit cutting, improperly applied, is another form of high-grading. The principle is to harvest only those trees above a certain size, such as 40 cm (14 in) dbh, and leave the remainder to develop into the succeeding stand. The assumption is that the small trees will grow into large trees of good quality in perpetuity. The fallacy is that natural stands of timber do not perpetuate themselves by like-producing-like. The openings created by removal of the larger trees will be occupied by the expanding crowns of the edge trees or by shade-tolerant understory trees that are already in place. The succeeding trees decrease the value of the stand for timber production and wildlife habitat with each successive partial harvest. In alluvial floodplains, cherrybark oak would likely be succeeded by green ash, green ash would be replaced by sugarberry, and sugarberry would finally succumb to boxelder and American hornbeam. Generations of selective, incomplete harvests have reduced many of our bottomland hardwoods to poorly stocked, low value stands.

Proper management of major alluvial floodplain forests is to control the undesirable trees at the same time desired ones are harvested, and to maintain natural water-flow patterns and cycles (Kellison et al. 1988). Fortunately, the practices best suited for accomplishing these goals are also best suited for timber production, wildlife management, and maintenance of the flora and fauna associated with the alluvial forest.

Stands that have been repeatedly harvested often contain two, but rarely more than three, age groups, with each age group dating to a previous harvest. Even though the species composition of the older age classes is usually desirable, a high component of the trees have cull-timber value. Conversely, a high component of the youngest age class of timber is usually of poor species composition, resulting from the development of shade-tolerant trees in the understory of the residual crown classes. However, many of these stands, especially those occupying sites of high soil quality, are worthy of timber stand improvement (TSI), in which the undesirable trees are controlled to release the desirable trees in the intermediate crown class.

A rule of thumb to assess stands suited for reclamation was developed by Kellison et al. (1988). Stands with more than 5, 7, and 10 m^2/ha (20, 30, and 40 ft^2/acre) of well-distributed, desirable species at 20, 30, and 40 years, respectively, are candidates for reclamation; those not achieving these goals are candidates for regeneration harvests.

Silvicultural Systems and Regeneration Methods for Natural Stands

When a harvest is planned, an assessment is made to determine how the stand will be regenerated. The information needed includes: (1) condition and size-class distribution of overstory trees by species, (2) quantity and condition of understory trees (desired initial and advanced reproduction), (3) kind and amount of competing vegetation, and (4) regeneration method (from seeds and seedlings or from stump and root sprouts). Results for regenerating bottomland hardwood forests, from many years of research involving many scientists, have been compiled by McKelvin (1992) and Kellison and Young (1997).

Even-Aged Systems

Experience has shown that stands occupying major alluvial floodplains will regenerate themselves following clean harvest of the timber from a single entry (clearcutting) or from two entries (shelterwood cutting). The regeneration from unaltered timber stands of trees less than about 100 years old will be largely from stump and root sprouts (Mader 1990). Stands of an older age class, and those with altered hydrology, will largely regenerate from seed, either in place at time of harvest or transported to the site by wind, water, and fauna. The types of even-aged regeneration systems having application to major alluvial floodplains are: clearcutting, patch clearcutting, shelterwood cutting, and seed-tree cutting.

Clearcutting

Clearcutting hardwood forests that have the propensity to regenerate themselves from stump and root sprouts reduces species succession almost to the pioneering sere. It is only one stage short of a catastrophic event such as a hurricane, where stump and root sprouting of merchantable timber is severely limited because of windthrow, and perhaps two stages short of a cleared bottomland field where all initial regeneration must be from seeds or planting.

Despite protests from people other than foresters, clearcutting is often the best way to regenerate hardwoods, especially degraded stands. The regeneration will largely be from advanced reproduction and sprouts, but seedling reproduction will form a part of the succeeding stand in patches where sprout or advanced reproduction is absent. Seedling reproduction has little chance of developing into the succeeding stand if it occurs three or more years after sprout development. Species succession of advanced regeneration and sprouts proceeds much as it does with seedling reproduction, with shade-intolerant species showing fastest initial growth.

Opposition to clearcutting often results from the visual impact of the treatment. To maintain the biological and ecological benefits of this often-abused system, we recommend that 1) the size not exceed 8 ha (20 acres), 2) the area be configured to the landscape with scalloped edges, 3) declining, over-mature trees be left standing for wildlife purposes (approximately 5/ha [2/acre]), and 4) dead and downed trees be left on site for associated flora and fauna. It is axiomatic that the harvest must conform to regulations and Best Management Practices.

Patch Clearcutting

This system is a variation of clearcutting, with the size of the affected area being the major difference. The configuration implied is noncontiguous patches or strips. Areas of about 2 ha (5 acres) are usually considered optimum. Smaller areas are adversely affected by edge trees, the influence of which extends into the opening about the distance of the height of the dominant trees. The edge trees limit the growth of shade intolerant species at the expense of shade tolerant ones.

The limitation of patch clearcutting is frequent stand entry, which eventually results in many small patches. The small patches create innumerable problems in stand management and inventory, and they are poorly suited for forest interior-dwelling birds and certain other fauna (Sietz and Segers 1993).

Shelterwood Cutting

Shelterwood cutting removes a portion of the overstory canopy in combination with the removal or control of the understory stratum. Best results are obtained when the overstory canopy is reduced to about 50 percent of its original cover. This level of reduction allows sufficient sunlight to reach the ground to promote seedling and sprout reproduction.

Experience has shown that the regeneration from clearcuts and shelterwood cuts are initially the same, but that the intolerant species under a shelterwood will start to decline if the overstory trees are not removed within five to ten years. Shelterwood cuts can buffer against rising water tables in areas where the soil water table has risen due to altered hydrology. In some situations, the system is advocated for the regeneration of oaks, especially cherrybark oak. The necessity of this practice in alluvial floodplains is not universal because species such as water oak and willow oak can regenerate equally well with or without a partial overstory stand (Leach and Ryan 1987). In the deepest water systems such as muck swamps, shelterwood systems appear to be no more effective in obtaining desired reproduction than clearcut systems.

Seed Tree Cutting

The prescription for seed tree cutting is to leave 10 to 20 trees/ha (four to eight/acre) following removal of the remaining overstory and understory trees. The theory is that seeds will be disseminated over the area, helping to ensure success in regeneration. However, seed trees are usually a wasted effort in alluvial floodplains because most heavily harvested hardwood stands regenerate from sprouts, from seeds buried in the duff, and from seeds disseminated by water, wind, and fauna. The primary reasons for leaving such trees is for wildlife, ecologic, and aesthetic values.

Uneven-Aged Systems

Stands of trees of widely different ages can be maintained by the selection system where harvesting, regeneration, and intermediate stand treatments are applied at the same time. Stands are entered at intervals ranging from every year to perhaps every 10 years. Each cutting removes financially mature and high-risk trees, adjusts stand density to create room for the best trees to grow, and makes space for new regeneration. In practice, a specific stand structure is created by leaving desired basal area levels in several diameter classes.

Single-Tree Selection

This system applies to the individual crop tree in which the best tree within a given ground space is favored in each diameter class. In theory, it creates an all-aged stand with a negative exponential diameter distribution (Nyland 1996). In practice, however, it is impractical to achieve this goal because of frequent stand entry and because the smallest diameter classes do not develop in the shade of trees of the larger diameter classes.

This is the system so often advocated by the adversaries of even-aged management in which clearcutting or shelterwood cutting is practiced. The principles of the system are sound, but the application is so difficult that the exercise approximates a selective or diameter-limit cut. Because of these limitations, the system is rarely recommended by professional foresters.

Group Selection

This variant of single-tree selection involves removal of groups of trees of similar age, size, or species on areas usually not exceeding 0.1 ha (0.25 acre). For this regeneration system to work, care must be given to removal of undesirable as well as desirable trees. Group selection differs from patch clearcutting in that it maintains the stand in different ages, but there is no attempt to record the separate ages. The limitation of the system is the frequent stand entry required to implement the practice.

Two-Aged System

Leave-tree (deferment) cutting is receiving increasing attention for regenerating southern hardwoods. Implementation of the practice includes leaving 5 to 7 m^2/ha (20 to 30 ft^2/acre) of basal area until the end of the following rotation in combination with the regeneration that develops in the openings created from partial harvest of the parent stand. As opposed to the shelterwood system, where the residual overstory trees are removed to allow the regeneration to develop, leave-tree cutting maintains the overstory trees until the end of the rotation. At that time, the residual trees will be removed in combination with about 75 percent of the basal area of the regenerated stand. The cycle is repeated with the next rotation, which perpetuates the presence of an overstory at all stages of stand development.

An added value of this system is that a mixture of crop trees can be retained for the next rotation. Some of the trees might be selected for their timber value and some for wildlife and other values. This system is equivalent to "high forests with reserves" of European forestry (Matthews 1989).

Plantation Management

Procedures have been reasonably well developed for the establishment of plantation hardwoods in alluvial floodplains (Malac and Herren 1979). The greatest effort by industrial foresters has been establishment of eastern cottonwood and sycamore plantations. Eastern cottonwood has been the most promising species in the Mississippi River Delta, whereas sycamore and to some extent, sweetgum, green ash, water oak, and willow oak have proven more adaptable to the other alluvial floodplains of the South. About 50,000 ha (125,000 acres) of commercial plantations currently exist. Despite successes, with growth rates of 15 to 18 m^3/ha/yr (3 to 4 cords/ac/yr), the trend is to establish hardwood plantations outside of the alluvial floodplains. The causes for this shift in site location include environmental concerns and the difficulty in managing and harvesting the resource in areas with episodic flooding. Even though debate about wetlands continues, enough uncertainty exists about complying with regulations of the Clean Water Act that few industrial forestry organizations are willing to invest in plantation forestry in alluvial floodplains.

The greatest effort today in alluvial floodplain plantation forestry is in land reclamation, primarily in the Mississippi River Delta. About 20,000 ha (50,000 acres) of a 200,000 ha (500,000 acres) landbase have been seeded or planted. The poorly drained land, cleared for agricultural crops in the 1960s and 1970s, has since been abandoned because of substandard yields and limited accessibility. The primary objective of the reclamation project is for wildlife, with the primary species being oaks, especially Nuttall oak (Allen and Kennedy 1989).

Recommendations

The recommended procedure for perpetuating viable hardwood forests is by management. Managed by an even-aged system, development of the harvested forest occurs in five recognizable stages: (1) impenetrable thicket — weeds, briars, vines, and tree seedlings and sprouts, which lasts to about age 10 years, (2) sapling stage — seedlings and sprouts begin to exert dominance and show formation of the new forest at ages of 10 to 15 years, (3) pole stage — competition among the seedlings and sprouts causes the strong to prosper at the expense of the weak from 15 to 25 years, (4) small sawlog stage — development of the stand from fiber to sawtimber at 25 to 40 years, and (5) large sawlog or mature forest stage — suitable for management in perpetuity or harvested at anytime beyond age 50 years.

As with plant succession, changes in fauna occur with each of these developmental stages. The impenetrable thicket is impenetrable only to humans. It is prime habitat for northern bobwhite (*Colinus virginianus*), rabbits, snakes, white-tailed deer, and low-nesting birds. The sapling stage of stand development retains the allure for white-tailed deer, and it becomes the home for an increasing array of neotropical birds. The pole-size stand is the least attractive of any of the stages of forest development because the vegetation required for browse is largely lacking, and the trees are too young for hard-mast production. By the small sawlog stage, enough hard and soft mast are being produced to attract avian populations and game animals. The large sawlog stage is equivalent to the impenetrable thicket in many aspects because it is home to a diverse array of fauna. In addition, the trees have developed so that cavity builders are increasingly at home, and hard mast production is at its ultimate. White-tailed deer, turkeys, squirrels, raccoons, and a host of other mammals find refuge there until the forest starts to decline, at which time the cycle is ready for renewal.

RESEARCH NEEDS

During the last decade, major research emphasis has been directed toward environmental concerns associated with management of alluvial floodplains. Of special significance is the work done on the effects of timber harvesting impacts and land restoration. Harvesting impact studies include those on hydrology, water quality, nutrient cycling, decomposition of organic matter, sedimentation, neotropical birds, herpetofauna, and recolonization of herbaceous and woody species following site disturbance (Shepherd 1992). On land restoration, the emphasis has been on performance of a suite of planted or seeded woody species on lands that have been cleared of timber and drained from 10 to 20 years ago. The reclamation treatments include plugging drainage ditches and planting (seeding) the land with and without soil tillage and competition control (Smith and Tome 1992).

While research of environmental concerns was in progress, little silvicultural research on hardwood ecosystems was conducted, especially in alluvial floodplains. With the pending shortage of hardwood timber (Cubbage et al. 1996), more effort is needed on improving timber productivity on private lands. The emphasis needs to be on private lands because society has largely determined that public lands have greater value for aesthetic and recreational values than they do for timber production. This trend is sobering because it dictates that the United States, already a net importer, will be importing additional wood and wood products from developing countries. Almost without exception, those countries have less stringent environmental regulations than we do. As a consuming nation, we will be making a greater negative impact on the world's environment than if we were self sufficient, or were net exporters of wood and wood products. Among the research needs to help offset the timber shortfall are:

(1) Evaluate regeneration systems best suited for major alluvial floodplains,
(2) Match optimum regeneration systems with practices that are environmentally and socially acceptable,
(3) Determine the effect of thinning and fertilizing, and the interaction thereof, of natural stands of hardwoods at an age when the stand can be effectively manipulated:

Age	Manipulative age
2–3	Seedling and sprout regeneration has developed
6–10	Crown closure
10–15	Sapling stage — Natural mortality of least adapted species and poorest trees within a species
15–25	Pole-sized timber
25–40	Small sawtimber
50+	Large sawtimber

(4) Evaluate mechanical means for controlling hardwood stocking, with emphasis on small machinery to minimize damage to residual trees,
(5) Evaluate herbicides to control unwanted vegetation, with emphasis on minimal environmental impacts, at optimum costs,
(6) Evaluate ways to successfully regenerate premium-grade hardwoods, such as cherrybark oak, at the expense of fast-growing pioneering species,
(7) Develop improved growth and yield models for bottomland hardwoods,
(8) Develop heuristic models to allow the practicing forester to prescribe optimum hardwood management regimes,
(9) Monitor site impact studies to determine how the presence and habits of neotropical birds, herpetofauna, and other fauna change with increasing age of a forest, and
(10) Research and develop markets for use of low-grade hardwoods.

SUMMARY

Major alluvial floodplains are differentiated from minor alluvial floodplains by bank erosion and formation of point bars which evolve into front land and by meanders that occasionally cut through to create oxbow lakes and bayous. The floodplain includes levees, first and second bottoms, low ridges, terraces, and sloughs. Red rivers have their origin in the mountains or upper Piedmont along the Atlantic and Gulf Coasts and are greatly extended into the Upper Coastal Plain in the area west of the Mississippi River. Black rivers have their origin in the Coastal Plain or occasionally in the Lower Piedmont. By their origin, red rivers are rock-dominated whereas black rivers are precipitation-dominated. As a result, some fauna and flora are unique to one river system compared to another, but other species can be found in both.

Four basic geographic distribution patterns of alluvial floodplain tree species are found: (1) distributed throughout, (2) locally absent, (3) restricted distribution, and (4) rare. Thirteen tree species occur throughout alluvial floodplains, while 29 species have wide distribution but are locally absent. In addition, 21 species have restricted distribution and 10 are rare. These species have one or more adaptations to the water-saturated soils at some stage in their life: morphological, physiological and reproductive. The morphological features include shallow rooting, hypertrophied lenticels, adventitious roots, buttressing, and knees.

By about 1930, roughly half of the alluvial floodplain forest had been cleared, reducing the area to 4.7 million ha (11.8 million acres). During the last 50 years, the area had been further reduced to 1.7 million ha (4.3 million acres). That small reserve is now increasing, due to restrictions on conversion of wetlands by the Clean Water Act of 1977, and to reclamation of about 200,000 ha (500,000 acres) of underperforming farm land in the basin of the Mississippi River and its southern tributaries. In contrast, industrial plantation forestry in the alluvial floodplain is declining because of environmental restrictions and episodic flooding which hamper plantation management and timber harvesting.

Good timber management in the alluvial floodplains has been more by accident than by design. Many of the stands have been repeatedly harvested, leaving cull overstory trees and trees of high shade tolerance (with low timber and wildlife value) in the understory. Prescriptions to optimize timber management are to do nothing or to thin periodically in stands that are adequately stocked with desirable species. Care has to be exercised in thinning and partial harvesting because of damage to the boles and root systems of the residual trees.

For stands that are inadequately stocked, the options are to implement TSI systems or to schedule regeneration harvests. Both uneven-aged and even-aged management systems are suitable, but uneven-aged systems are difficult to successfully implement. Examples of uneven-aged systems are single-tree selection and group selection. Even-aged systems include biological clearcuts, patch clearcuts, shelterwood cuts, and seed-tree cuts. A two-aged system, which combines attributes of both even-aged and uneven-aged systems, has considerable potential for successfully regenerating alluvial floodplain species while meeting the expectations of society.

Foremost research needs are studies to determine how best to manage naturally regenerated forests, from reproduction to rotation age. Cleaning and thinning by mechanical means and herbicides will be priority items, and the use of fertilizers will be evaluated with and without thinning. Modeling of forest stands for growth and yield projections and for management options will be essential. The lack of market for the low-grade material is one of the greatest limitations to successful management of hardwoods.

13 Minor Alluvial Floodplains

John D. Hodges

CONTENTS

INTRODUCTION

Processes leading to stream formation in the South have produced a variety of floodplain types that differ in hydrology, geomorphology, and soils. Because the composition and productivity of forest communities reflect these differences, the floodplains have been given different names. The terms "major bottom," "minor bottom," and "upland drainage" (or "branch bottom") are frequently used in the South to describe floodplains and stream systems, but distinctions among the floodplain systems have not been defined.

Table 13.1 compares several characteristics of the stream systems of the South. The most obvious difference between major and minor floodplains is that the streams and associated floodplains are generally much larger in major bottoms. However, differences in characteristics like hydrology, geomorphology, and soils are more important in determining the composition and productivity of forest communities. In major bottoms, flooding is usually seasonal, occurring primarily in late winter and early spring. It is primarily over-bank flooding with long duration. In minor bottoms, flooding is more frequent and associated more with local rainfall. Flood waters can come from riparian sources (Brinson 1993a) as well as over-bank flooding and are of fairly short duration. Major bottoms have broad fronts, ridges, flats, and sloughs, while these features are relatively narrow in minor bottoms. Soils of major bottoms can be transported from great distances and will vary greatly in texture and mineralogy. Soils of minor bottoms are of local origin with less variation in texture and mineralogy, and the sites are generally, but not always, less productive than those of major bottoms (Hodges and Switzer 1979).

In terms of vegetation, minor alluvial floodplains are generally considered to be part of Küchler's (1964) type 113 (southern floodplain forest). However, as noted above, there are major differences among floodplains. One important difference is

1-56670-228-3/97/$0.00+$.50
© 1998 by CRC Press LLC

TABLE 13.1
Comparison of floodplain types.

Characteristic	Major Floodplain	Minor Floodplain	Upland Drainage
Stream size	Large	Moderate	Small
Flow	Permanent	Permanent	Intermittent
Floodplain width	Few km to 160+ km	<0.6 km to to several km	Narrow, few hundred feet at most
Flooding			
Type	Overbank	Overbank and riparian transport	Riparian transport
Frequency	Infrequent-seasonal Winter-Spring	Frequent-late Fall–Spring	Frequent-Any season
Source	Regional	Local	Small watershed
Duration	Weeks-months	Few days	Short, 1 to 2 days
Geomorphology	Topographic features (fronts, ridges, flats, and sloughs) are broad	Topographic features are narrow	Topographic features do not occur
Soils — source of	Hundreds of km	Local origin	Very local — surrounding uplands

that in minor floodplains, water tupelo (*Nyssa aquatica*) and baldcypress *(Taxodium distichum)* are usually not as dominant as in major floodplains.

Most bottomland hardwood sites (floodplains) occur in the Atlantic and Gulf Coastal Plains, but a significant acreage also occurs in the Piedmont of Alabama, Georgia, and the Carolinas (Kundell and Woolf 1986, Brinson et al. 1996). Tansey and Cost (1990) estimate that 86% of floodplain wetlands, which are a subset of bottomland hardwood sites in the Southeast, occur in the Coastal Plain. No estimates are available for the extent of minor alluvial floodplains, but recent inventories (1987 to 1992) by the U.S. Forest Service (USFS) report about 12.1 million ha (29.8 million acres) of bottomland hardwoods in the South. Tansey and Cost (1990) report that 67% of forested wetlands in the Southeast are found along smaller drainages, and that value for wetlands is assumed to be the same for all bottomland hardwood sites. If correct, the area in minor floodplain forest is about 8.1 million ha (20 million acres).

Information is not available on the original extent of minor floodplain forests. Clearing for agriculture was much easier than in major bottoms, so it is quite likely that an even larger percentage of the area was cleared than in major floodplains. If so, based on values for clearing of all floodplains, there may have been 12.1 to 14.2 million ha (30 to 35 million acres) in minor floodplain forests at one time.

Minor bottomland forests are of tremendous economic benefit to private landowners in the South. While specific ownership patterns are not known for minor bottoms, private nonindustrial owners, mostly farmers, control about 66% of all bottomland forests, and forest industries own about 25% (Saucier and Cost 1988, McWilliams and Faulkner 1991).

Recent USFS inventories (Saucier and Cost 1988, McWilliams and Faulkner 1991) show that bottomland forests contain about 1.3 billion m^3 (45 billion ft^3) of growing stock and that sawtimber volume is approximately 10.7 million m^3 (160 billion bd ft). Net annual growth averages less than 1.4 m^3/ha (20 ft^3/acre). Despite this slow growth rate, growth has exceeded removal until very recently, but there is a decrease in volume of top-quality hardwoods (Beltz et al. 1990).

In addition to the value of timber harvests, mixed bottomland forests often provide a direct economic return from the wildlife resource (Haygood 1970, Glasgow and Noble 1971). Hunting leases of $25/ha/yr ($10/ac/yr) are common, and some exceed $31/ha/yr ($12.50/ac/yr). Not only do the leases provide a direct economic return, but the leasees often assist the landowner with security and protection, as well as upkeep of the forest roads.

PHYSICAL ENVIRONMENT

Climate of the Atlantic and Gulf Coastal Plains is discussed in detail in Chapter 4. The major implications for the ecology and management of minor floodplain forests are that ample rainfall is usually well distributed throughout the year, and moisture is usually not a severe limitation to growth. However, moderate drought occurs irregularly every few years, and severe prolonged drought may occur once every two or three decades. Catastrophic climatic events such as ice storms, hurricanes, prolonged drought, and tornadoes are infrequent but significantly affect the condition and composition of floodplain forests on a local scale. For example, the blight (dieback and eventual mortality) in the 1950s of sweetgum (*Liquidambar styraciflua*) growing on heavy clay soils has been attributed to drought over several consecutive years (Toole and Broadfoot 1959). Also, there is ample evidence that blowdown by hurricanes and tornadoes has had a major influence on composition and successional patterns of hardwood stands.

Alluvial floodplains can occur along most streams of the United States, but they are most common and are most extensive in the Atlantic Coastal Plain, East Gulf Coastal Plain, Mississippi Alluvial Plain, and West Gulf Coastal Plain geographic sections of the Coastal Plain Province for two reasons. First, the soils are derived from poorly consolidated, sedimentary material that permits easy erosion and formation of the broad valleys. Second, there is a marked decrease in stream gradients on the Coastal Plain as opposed to the Piedmont and Appalachian Mountains, which in turn results in high rates of deposition and floodplain formation.

The Coastal Plain landscape was formed during the Quaternary period (Fenneman 1938). A sequence of marine deposition of sand, silts, and clays associated with intermittent rises in sea level and subsequent uplifting of the land surface resulted in development of a series of marine terraces that make up the Coastal Plain of the southeastern United States. While the land surface was being raised, an undulating topography was formed, and seaward drainage patterns developed. Continual erosion of the uplands and deposition of eroded materials along the streams and at the sea outlets have resulted in the formation of the alluvial bottomlands.

Although the stream valleys may be geologically very old, most floodplain sediments, and perhaps the youngest terrace deposits, occurred in recent geologic times (Wharton et al. 1982) and are less than 18,000 years old. There are significant differences between floodplains of the Coastal Plain and the Mississippi Alluvial Valley and between major and minor floodplains, but the processes involved in formation are identical (Hodges 1997). Floodplain formation on the Atlantic Coastal Plain is predominately marine influenced and comprises a succession of marine and continental (alluvial) deposits (Colquhoun 1974). During periods of high rainfall, the marine deposits were degraded by heavy runoff from the Appalachian Mountains, thus shaping the valleys and floodplains. Strong evidence of the influence of climate changes on stream valleys and floodplains is shown by the common occurrence of "underfitted" streams in the Coastal Plain (Dury 1977). In such valleys, the flow and meander dimensions of present streams are far too small to account for the wide floodplain. The most logical explanation for formation of these floodplains is that discharges into the valley were much higher in prehistoric times (Dury 1977, Froehlich et al. 1977). In fact, Dury (1977) estimated that discharges in the past were as much as 18 times greater than at present. Similar processes have occurred on the Gulf Coastal Plain. Larger stream valleys of the South and Southeast typically have an active floodplain and terrace system with three or more terraces being common (Wharton et al. 1982). These terraces (abandoned floodplains) generally are not as productive for growth of bottomland hardwoods as the current floodplains, but they are extremely important for their hydrologic control over the modern floodplain (Wharton et al. 1982). Except for the youngest, most recent terrace, they are seldom flooded.

Current floodplains are young surfaces on which erosion and deposition are occurring, but deposition is the primary process on most floodplains. However, exceptions occur, as in the Piedmont, where cultural operations, mostly farming in the late 19th and early 20th centuries, resulted in massive amounts of deposition in the floodplains (Trimble 1970, Dowd et al. 1993). With the establishment of forests on upland areas, sediment deposition has been dramatically reduced, many streams are becoming incised (eroded downward), and the sediment is exported downstream (Richter et al. 1995, Burke 1996). Similar but less dramatic processes have occurred in some stream systems of the Gulf Coastal Plain. Most minor floodplains vary topographically (Figure 13.1), which results from stream meanders and differences in depositional patterns. The influence of stream meander on floodplain topography and deposition is illustrated in Figure 13.2. The pattern and types of deposition which occur in flood events is illustrated in Figure 13.3. When streams overflow their banks, the velocity of the water slows and rapid deposition occurs, forming natural levees or "fronts" that often represent the highest elevations in the floodplain. Slower rates of deposition of finer textured material occur on the flats or slackwater areas. When stream meander occurs (Figure 13.2), new fronts are formed and the old fronts are then recognized as levee ridges within the floodplain. The abandoned stream channel at first becomes a lake, but may eventually fill with sediment. Stream meanders can also result in formation of "ridge and swale" topography (Figure 13.4). These ridges are formed as bars, usually are sandy, and are normally much narrower than levee

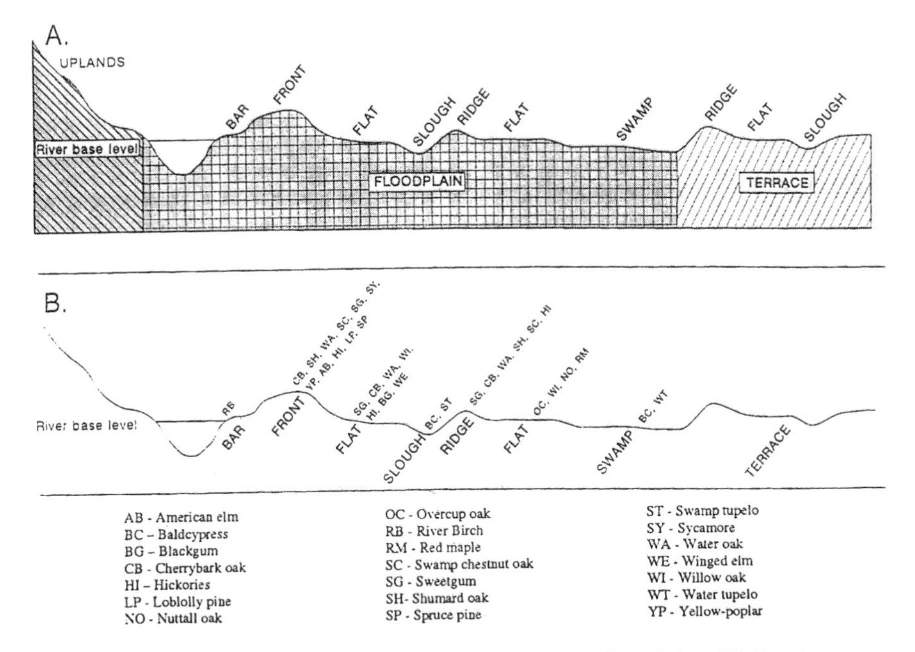

FIGURE 13.1 (A) Topographic site positions within minor floodplains. (B) Species occurrence as related to topographic site positions in minor floodplains.

ridges. Ridge and swale topography is more typical of streams whose headwaters originate in sandy material, and floodplains are composed mostly of sandy or sandy-loam material.

The nature or character of the soils in minor bottoms is determined by materials in the stream's drainage area, which serves as the erosional source of the deposits of mineral soil. Indistinct profile development is a feature common to all bottomland soils even though the soils are very complex and differ markedly in age and in physical and chemical properties. Texture ranges from newly deposited sands to older deposits of clay. Soil structure may vary from the single grains of sands to the compact blocky structure of clays. Soils of most minor bottoms, especially those in the lower Coastal Plain, are acidic. An exception occurs in minor bottoms of the Blackland Prairies of Alabama and Mississippi where some soils can be above pH 7 because of the chalk from which they are formed. Expansive clays (smectites) are common for floodplain sites of the Prairie and also occur in other minor bottoms, especially in the Gulf Coastal Plain. Soils usually contain ample available water and nutrients and moderate amounts of organic matter (Lytle 1960). However, nutrient status varies greatly among floodplains and within floodplains depending on differences in soil colloids. Soil aeration and drainage also vary considerably, and variations in species associations most often reflect aeration and drainage.

Soils of the fronts (Figure 13.1) along stream channels usually consist of silt loams, very fine sandy loams, or silty clay loams. Soil series will vary between regions, but in central Mississippi, soil series typical of front sites are Iuka and

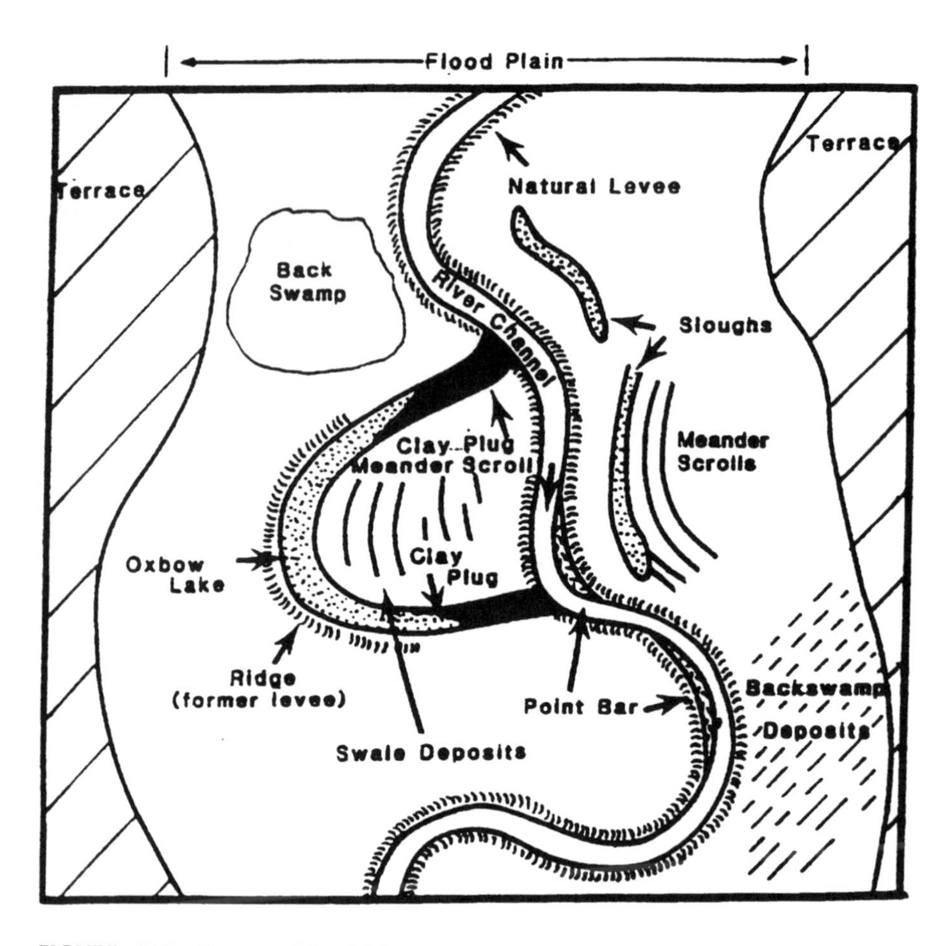

FIGURE 13.2 Features of floodplains as influenced by stream meanders (after Mitsch and Gosselink 1986).

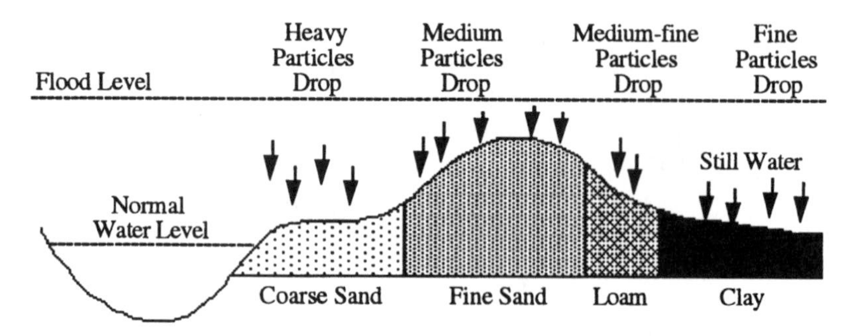

FIGURE 13.3 Depositional patterns in minor floodplains and the relationship to topographic features of the floodplain.

FIGURE 13.4 Formation of ridge and swale topography in a minor floodplain. (A) Formation of swale (old stream channel) and (B) new ridge being formed by sandy deposits from the current stream which is to the right in the photograph.

Ochlockonee. These are most often the highest and best drained areas in the floodplain with good external and internal drainage. Soils vary from well-drained to somewhat-poorly drained. Soils on the ridges (old fronts) may have similar texture, but are less well drained, and some leaching of nutrients may have occurred. Mantachee and Mathiston are soil series typical of ridge sites. On the flats, soils are generally clay loams or clays, and they vary from somewhat-poorly drained to poorly drained. Typical soil series on the flats are Rosebloom and Urbo. Soils in the sloughs typically are very clayey and very-poorly drained. Water may pond into the growing season. Within the floodplain, site productivity and plant species diversity tend to decrease in the order: front > ridge > flat > slough (Hodges, unpublished).

The topographic features of relief described above may not be as prominent in floodplains of younger and smaller streams where meander movement is less frequent or has not occurred, particularly near coastal areas. For the "blackwater" streams of the lower Gulf Coastal Plain, where stream gradient and velocity are low, natural levees (fronts) of streams are usually lower than for the upper Coastal Plain. Stream meanders occur, but the residual ridges and sloughs are less apparent than for stream systems of the upper Coastal Plain. Soils on the fronts are generally sands or sandy loams with finer textured sands and silts on the flats and sloughs. Soil texture and drainage class generally vary less across the floodplain than on floodplains of the upper Coastal Plain.

Disturbance is a natural and integral part of floodplain ecosystems. Major forms of disturbance include flooding, drought, fire, wind, and harvesting. Each of these forms will influence various components of the ecosystem in different ways. This chapter will emphasize their influence on the vegetation.

Flooding is most frequent during late winter and early spring (February to April), but can occur from late fall to late spring (November to June). The number of flood events is most often two or three per year, with as many as five in some years. Duration of a flood event is often only three or four days, but can last for a week or more.

Effects of flooding on the ecosystem vary depending on the time and duration. For primary producers, especially the trees, dormant season flooding causes little direct harm and may improve subsequent growth as a result of deposition and soil moisture recharge. However, spring flooding after foliation that results in complete inundation may kill many herbaceous plants and seedlings and saplings of most tree species (Broadfoot and Williston 1973). Large trees of most species can tolerate short periods of flooding even during the growing season, but few species can tolerate continuous flooding throughout the growing season (Broadfoot and Williston 1973). Interference with existing drainage patterns resulting from beaver impoundment, dam and road construction, and use of greentree reservoirs (GTRs) (Karr et al. 1990, Young et al. 1995a) can result in loss of vigor and growth, tree mortality, and composition changes in hardwood stands.

Little information is available on soil moisture relations of minor bottoms. Drought and severe soil moisture deficits do occur, but are not common. Soil moisture is usually adequate to maintain good plant growth (Schoenholtz, unpublished). Winter and spring flooding is usually sufficient for complete recharge of soil moisture, and the floodplain is efficient at retaining water from summer storms.

Fires seldom occur in bottomland hardwood stands because the sites are too wet and/or there is not enough litter to support a fire. Those using prescribed fires in upland areas often do not construct fire lines around minor bottoms because the fires do not spread there. Under severe drought, fires do occur in floodplain forests and are very detrimental, especially to seedlings and saplings. Fire damage to older trees usually results in rot, stain, and entrance of insects in the damaged areas of the trees.

Although very hot wildfires are detrimental to established stands, disturbances such as cool fires, grazing, and mowing have played an important role in the establishment of many stands (Aust et al. 1985). The role of such disturbances is most common after a complete clearcut or on abandoned agricultural lands. Many stands with a high percentage of oaks apparently originated following such disturbances (Aust et al. 1985). Although the tops of oaks may be removed or killed by the disturbances, large root systems are allowed to develop. Once the disturbances are eliminated, the vigorous sprouts produced by the oaks can outgrow competing woody vegetation.

Wind damage in minor floodplains occurs from hurricanes, tornadoes, and storms. Damage by hurricane winds occurs in coastal areas, but storm winds from hurricanes can occur hundreds of miles inland. Although hurricanes and tornadoes cause blow-down, significant damage may be due to breakage. Overall, most wind damage in hardwood stands is blow-down as the result of storms that occur mostly during the spring when soil moisture is excessively high and root firmness is low.

VEGETATION COMMUNITIES

Six forest types, as described by the Society of American Foresters (Eyre 1980), frequently occur on minor floodplains of the South. These types are: willow oak-water oak-diamondleaf (swamp laurel*) oak (88), swamp chestnut oak-cherrybark oak (91), sweetgum-willow oak (92), sycamore-sweetgum-American elm (94), baldcypress (101), and baldcypress-tupelo (102). Other types which occur less frequently are sweetgum-yellow-poplar (87), sugarberry-American elm-green ash (93), black willow (95), overcup oak-water hickory (96), and water tupelo-swamp tupelo (103). Tree and shrub species occurring in bottomland hardwood forests and their national wetland status are given by Sharitz and Mitsch (1993).

Species distribution within the floodplain is determined primarily by differences in hydrology, which in turn is related to the geomorphic site variations discussed above. Differences in relief of only a few centimeters have a marked influence on distribution and growth of species (Figure 13.1).

Species-site relationships (adaptations) within floodplains of minor stream valleys have been presented by Hodges and Switzer (1979) and Hodges (1994a). River birch (*Betula nigra*) is most often the pioneer species on bars and new deposits inside and outside the natural levees or fronts. The greatest variety of species within the floodplain occurs on the stream fronts and may include numerous oaks such as cherrybark (*Quercus falcata* var. *pagodaefolia*), water (*Quercus nigra*), Shumard (*Quercus shumardii*), and swamp chestnut (*Quercus michauxii*), sweetgum, sycamore (*Platanus occidentalis*), several hickory (*Carya*) species, green ash (*Fraxinus pennsylvanica*), and eastern cottonwood (*Populus deltoides*). These same species may also occur on high ridges (old fronts) of the floodplains. Species such as American beech (*Fagus grandifolia*), black walnut (*Juglans nigra*), yellow-poplar (*Liriodendron tulipifera*), white oak (*Quercus alba*), laurel oak (*Quercus hemisphaerica*), spruce pine (*Pinus glabra*), and loblolly pine (*Pinus taeda*) are less common in the floodplains, but when they do occur, they are usually on the front sites and occasionally on high ridges.

Species occurrence on flat sites within minor bottoms depends primarily on drainage. Overcup oak (*Quercus lyrata*), swamp laurel oak (*Quercus laurifolia*), willow oak (*Quercus phellos*), Nuttall oak (*Quercus nuttallii*), green ash, pumpkin ash (*Fraxinus profunda*), sugarberry (*Celtis laevigata*), hackberry (*Celtis occidentalis*), common persimmon (*Diospyros virginiana*), red maple (*Acer rubrum*), and Drummond red maple (*Acer rubrum* var. *drummondii*) are common on the less-well-drained flats. Less common species include black willow (*Salix nigra*) and water-locust (*Gleditsia aquatica*). The term "pin oak flat" is used locally for almost pure stands of willow oak that sometimes occur on these sites. True pin oak (*Quercus palustris*) occurs on similar sites in the northern latitudes of the southern hardwood region. Sweetgum, cherrybark oak, water oak, swamp chestnut oak, blackgum (*Nyssa sylvatica*), American elm (*Ulmus americana*), winged elm (*Ulmus alata*), boxelder (*Acer negundo*), green ash, sugarberry, hackberry, and hickories are common on the

---

* Some sources (Little 1979) do not distinguish between laurel and swamp laurel oak, classifying both as *Quercus laurifolia*. This author, however, feels like they are two different species occupying two different ecological niches.

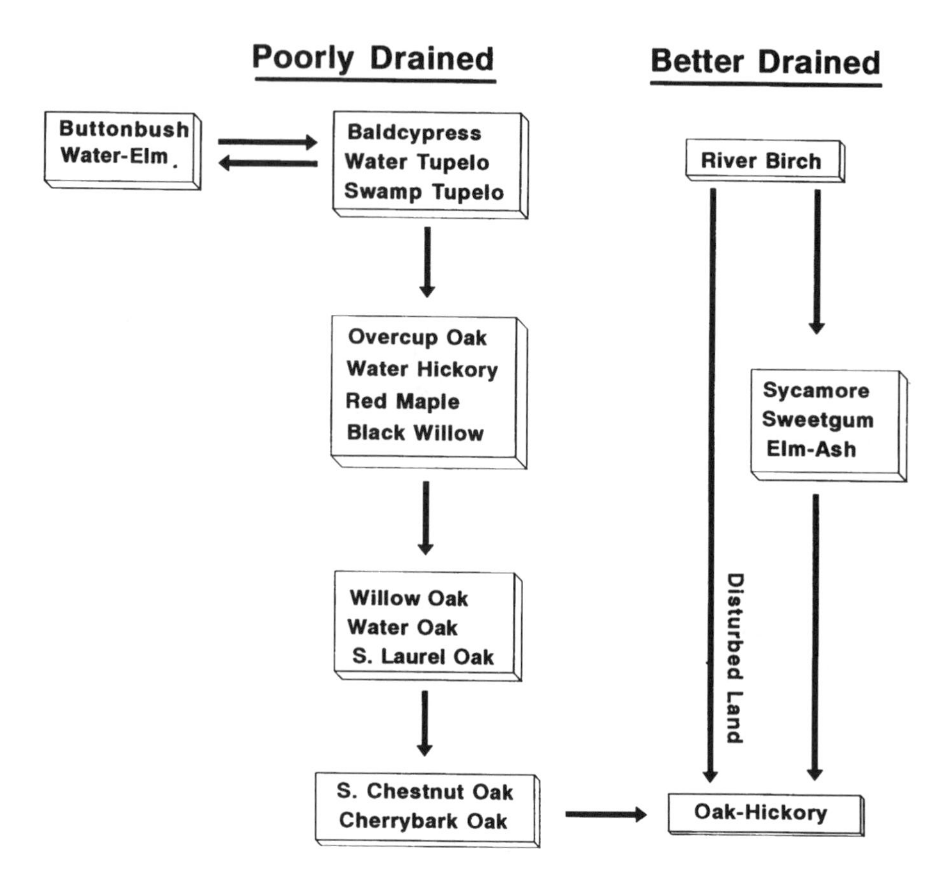

FIGURE 13.5 Successional patterns for tree species occurring on minor floodplains.

better drained flats and low ridges. Less common associates include sweet pecan (*Carya illinoensis*) and honeylocust (*Gleditsia triacanthos*).

Slough sites most often have stands composed of baldcypress, water tupelo and/or swamp tupelo (*Nyssa sylvatica* var. *biflora*), common persimmon, and water-elm (*Planera aquatica*). In the Gulf Coastal Plain, water hickory (*Carya aquatica*) is sometimes a component also.

Patterns of succession on minor floodplains are shown in Figure 13.5 and discussed in detail by Hodges (1994b, 1996). Type, pattern, and rate of succession are strongly influenced by geomorphic position and types and rates of deposition. Primary succession occurs on bars and mud flats formed by the stream where river birch is most often the pioneer tree species. As the site gains elevation by deposition, river birch may be replaced by sycamore, sweetgum, winged elm, and green ash (Figure 13.5). As deposition continues and the site matures, the regional oak-hickory "climax" community will appear.

Following disturbances such as clearcutting, secondary succession may be initiated on flats and ridges by dense stands of river birch, with a rapid juvenile growth rate. As growth slows, river birch is quickly replaced by a mixture of bottomland hardwoods consisting primarily of sweetgum, oaks, and hickories. Without further disturbance, the oak-hickory association will eventually dominate.

On very wet sites with permanent or near permanent flooding, the major tree species are baldcypress, water tupelo, and/or swamp tupelo. On these sites, succession is in an "arrested" state, and progression depends on site change. However, with disturbance, retrogression to a buttonbush (*Cephalanthus occidentalis*)-water-elm type is rapid and common, if advance regeneration of baldcypress and tupelo is not present or the site does not dry sufficiently to allow reproduction from seed. Progression from the baldcypress- tupelo type to the oak- hickory type (Figure 13.5) is driven by deposition, which in turn alters the hydrology of the site. It should be noted that there will be some overlap between the types shown in Figure 13.5 and that species such as Nuttall oak, green ash, pumpkin ash, and Drummond red maple may be common in the stages following baldcypress-tupelo. Sweetgum and sugarberry may be common associates in the later stages of succession.

Productivity of minor bottoms is extremely variable, depending primarily on the alluvial deposits that are of local origin. For example, timber production generally is not as good on sandy floodplains of the lower Gulf Coastal Plain as on those of the upper Coastal Plain where deposits are more variable, but generally are a mixture of sand, silt, and clays. Within the floodplain, the most productive sites are usually on the fronts and high ridges. Site indices (height of dominant and co-dominant trees at a base age) of 27 to 30 m (89 to 98 ft; 50-year base) are common on these sites, and indices of 35+ m (115+ ft) have been recorded on floodplains in Central Mississippi. Fully stocked stands of mixed bottomland hardwoods can yield more than 1.46 m^3/ha/yr (250 bd ft/ac/yr) on a 60- or 80-year rotation (Putnam et al. 1960). Volume growth of cherrybark oak on good sites has been estimated to range from 5.32 to 7.79 m^3/ha/yr (76 to 111 ft^3/ac/yr) (Krinard 1990). Mixed bottomland hardwoods seldom exceed 34.5 m^2/ha (150 ft^2/acre) of basal area/hectare (Johnson and Shropshire 1983).

A guide developed by Baker and Broadfoot (1979), or the interactive computer version (SITEQUAL) developed by Harrington and Casson (1986), can be used to evaluate site quality (site index) for bottomland hardwoods. The evaluation is based on easily observable site factors and can be used in degraded stands or where no trees are present.

Actual productivity on many minor floodplain sites is far less than potential because of past harvesting practices. Stocking in many stands is far less than desirable (Saucier and Cost 1988, McWilliams and Faulkner 1991), and net annual growth averages less than 1.4 m^3/ha/yr (20 ft^3/ac/yr) for all bottomland stands. What may be more important for timber managers is that, even with an overall increase in inventory, there is a decrease in volume of top-quality hardwoods (Beltz et al. 1990).

ANIMAL COMMUNITIES

Some of the most productive and valuable fish and wildlife habitats in the United States are found in bottomland hardwood forests including those on minor bottoms. Fertile alluvial soils, abundant water, high-quality food of great variety, relative freedom from fire, usually accessible agricultural fields, and good escape cover all contribute to provide this favorable habitat (Glascow and Noble 1971).

Wildlife habitat and wildlife species occurring in forested wetlands are described in Chapter 9. In this chapter, the intent is to provide more detailed information on habitat and relate wildlife habitat to management practices, especially those involving timber harvesting.

MANAGEMENT

Values and functions of forested wetlands are discussed in Chapter 2. The values desired from floodplain forests vary by ownership, but most owners generally have multiple objectives, including timber and wildlife. In considering appropriate management practices for floodplain forests, an important concept to keep in mind is that they must be secured or maintained by some form of silvicultural practice. These practices vary from doing nothing in the stand to intensive regeneration practices. The important point is that silvicultural practices (stand manipulations) are simply tools designed to manipulate the environment to provide maximum resources for the product desired and are equally as beneficial for securing values such as wildlife habitat as for timber production.

Initially, extensive areas of minor bottoms throughout the South were cleared for agriculture. Logging in the remaining forests generally took only the largest trees of the best species, but with time, species utilization standards changed and quality requirements were lowered each time the area was cut. Commercial clearcutting was not prevalent until after about 1900, and when used, there was usually no consideration for regeneration. These practices have generally continued into the present and have resulted in many stands in which stocking, composition, and quality are generally below the desired levels.

An even-aged silvicultural system with natural stand management is now used for most mixed hardwood stands on minor bottoms. That system is preferred because most of the desirable species such as oaks and sweetgum are intolerant or only moderately tolerant of shade. The most successful and most commonly used method of regeneration for mixed hardwoods employs a complete clearcut. However, the method is not a clearcut in the classical sense. Some seedlings do originate from seed that germinate after harvest, but the most important source of reproduction is usually advance regeneration established beneath the older stand. In addition, coppice from stumps and roots can be very important as a source of regeneration. Thus, the regeneration method is actually a combination of even-aged methods (shelterwood, coppice, and clearcut).

Stocking, composition, and quality of hardwood stands are generally less than desirable because of past harvesting practices. The first decision to be made for such stands is whether to manage what is present or to initiate regeneration procedures. In the past that decision had to be made with few available guidelines, but recently an expert system (Manuel et al. 1993) was developed to aid in that decision.

When management goals indicate regeneration of a stand, a procedure similar to that given by Clatterbuck and Meadows (1993) for oaks can be used (Figure 13.6).

The first step should be an assessment of the regeneration potential. The technique developed by Johnson (1980a), and modified by Hart et al. (1995), can be used for mixed bottomland hardwoods. It is based on an assessment of number and size of advance regeneration, sprout potential of severed stumps, and presence of light-seeded species in the overstory. When regeneration potential is adequate, a complete clearcut is recommended. Full release is recommended, but the size of the opening is not especially critical for obtaining regeneration. However, development of the more desirable, less tolerant species is much slower in openings of 0.2 ha (0.5 acre) or less than in larger openings (Johnson and Shropshire 1983).

If regeneration potential is inadequate, a modified shelterwood method can be used to obtain advance regeneration (Hodges 1987, Loftis 1990). Light at the level of the forest floor is most often the limiting factor for establishment and growth of advance regeneration. Adequate light may in some cases be provided by normal shelterwood cuts, but most often it will require control of less desirable species in the midstory and understory (Janzen and Hodges 1985, Hodges 1987, Loftis 1990). After advance regeneration is well established, usually three to five years, the overwood is removed in one operation.

Although even-aged methods are easier to use and are more reliable for regenerating hardwood stands on minor bottoms, the group selection method can be used. Single-tree selection is not appropriate because it will favor more shade tolerant and less desirable species (Johnson and Shropshire 1983). With group selection, regeneration potential should be evaluated in a manner similar to that for even-aged regeneration, as previously discussed. Opening size is determined by the sprouting ability of trees to be cut and advance regeneration present. Volume regulation can be adhered to, but success of the method depends on creation of an environment favorable to the desired species. Thus, opening size is often larger than that used with classical group selection cuts. Control of competing vegetation by cutting and/or use of herbicides may be necessary for some stands.

Artificial regeneration is an alternative for restoration efforts on cleared lands and for regeneration of existing stands where a desirable seed source is not present. Large-scale restoration or commercial plantings have been made for a number of species including cottonwood, American sycamore, sweetgum, green ash, cherrybark oak, Nuttall oak, water oak, willow oak and yellow-poplar. Experimental plantings have been successfully established for other species including pecan, water tupelo, and baldcypress (Johnson 1978, Johnson and Shropshire 1983).

Extensive plantings of cottonwood have been made in the Mississippi River floodplain with only limited plantings in minor bottoms. Genetic improvement has progressed further with this species than any other southern hardwood, and techniques for establishment and tending of plantations are well documented (McKnight and Biesterfeldt 1968, McKnight 1970, Mohn et al. 1970, Zsuffa 1976, Krinard and Johnson 1980). An interesting concept involving planting of a mixture of cottonwood and oaks is currently receiving considerable attention. The practice entails planting cottonwood cuttings on a 3.7 m ×3.7 m (12 ft ×12 ft) spacing, use of herbicides, and intensive cultivation for two years to control weedy competition, then interplanting of oaks, usually Nuttall, on a 7.32 m ×7.32 m (24 ft ×24 ft) spacing. The

Pre-treatment Regeneration Evaluation

Oak Prospects Poor

Oak Prospects Good

1. Take steps to promote advance reproduction by increasing light to the forest floor through understory removals and/or partial overstory cuttings
2. Supplemental planting
3. Artificial regeneration

1. Treat and harvest during dormant season
2. Control residual stems prior to next growing season

Evaluate at age 3

Less than 150 free-to-grow oaks per acre

More than 150 free-to-grow oaks per acre

1. Go with less than adequate oak stocking or
2. Convert to plantation

1. Leave alone or
2. Clean, weed or thin if needed

FIGURE 13.6 Regeneration procedures for bottomland oaks (after Clatterbuck and Meadows 1983)

anticipation is that the cottonwood can be harvested at about age 10, allowed to coppice, and then harvested again about 10 years later. The oaks would then be large enough to occupy the site and form the final stand (Portwood, pers. comm.).

A major problem with hardwood planting is the high cost. Cottonwood, sycamore, and sweetgum usually require intensive site preparation and intensive vegetation control, usually by cultivation, for one to five years (Johnson 1978). The oaks and green ash will usually survive and form an acceptable stand without cultivation, but growth is greatly improved by cultivation. Traditionally, most hardwood planting has been done by pulp companies to insure a source of a given species to meet mill requirements, especially during wet seasons when logging in natural stands is impossible. Species most often planted were the fast growing intolerants such as cottonwood, sycamore, and sweetgum. In the last 10 years, thousands of hectares of agricultural land across the South have been planted to hardwoods under the Conservation Reserve Program and the Wetland Reserve Program, as well as on lands purchased for mediation purposes and wildlife refuges. Oaks have been the major species planted.

For oaks, direct seeding is an alternative to planting for restoration or for regeneration of recently harvested sites (Allen and Kennedy 1989). Pecan has also been seeded successfully. Probability of successful regeneration is higher for planting than for direct seeding (Allen 1990), but the cost of direct seeding is considerably less (Bullard et al. 1992). Restoration of hardwood for wildlife habitat has emphasized the oaks, and much of it has been done by direct seeding (Allen 1990). Oaks are generally seeded, one acorn per spot, at a rate of about 3,705/ha (1500/acres) and a spacing of 0.8 m ×3.7 m (2.5 ft ×12 ft). For regeneration of recently harvested sites, openings should be at least 0.8 ha (2 acres) in size to reduce depredation of the acorns by rodents (Johnson 1981).

There is, justifiably, a concern for the influence of timber management practices on ecosystems of minor bottoms, particularly on the site itself. Disturbances are a natural feature of floodplain ecosystems, and silvicultural practices are simply one form of stand disturbance. However, the potential for damage exists, primarily due to harvesting practices, but it is important to understand that these sites and soils are in a constant state of change and this in turn is reflected in longer-term changes in the vegetation. Because of the relatively flat topography, the annual deposition, and the nature of the expansive clays (smectites) on many floodplain sites, they are not as sensitive as most upland sites (i.e., they are fairly resistant to perturbations such as harvesting). Most site damage occurs as a result of equipment use on wet soils and is preventable by restricting operations to dry seasons.

As discussed previously, bottomland forests provide very productive fish and wildlife habitat. Silvicultural practices normally used for natural stands are highly compatible with maintenance of that good habitat for most wildlife species (Johnson and Shropshire 1983). Changes in management practices to improve wildlife habitat, if needed at all, involve timing, size, and distribution of cuts.

Numerous studies have examined the influence of silvicultural practices on wildlife and wildlife habitat. Swamp rabbit (*Sylvilagus aquaticus*) numbers are higher in selectively logged bottomland hardwood forests than in uncut forests (Terrell 1972), and cuttings (improvement cuts, thinnings, harvest cuts of various types) in bottomland hardwood stands improve swamp rabbit habitat (Garner 1969, Mullin 1982, Hurst and Smith 1985). Weights of swamp rabbits found in clearcuts do not differ from those in mature stands (Palmer et al. 1991). Hurst and Bourland (1996) examined the influence of clearcut regeneration methods on breeding birds and found that number of breeding bird species was similar in mature forests and in clearcuts where residuals, either injected or alive, were left standing. Highest breeding bird densities occurred in the clearcut areas with live residuals (20 m^2/ha; 87 ft^2/acre). Wehrle et al. (1995) showed that cutting practices in minor floodplains can influence waterfowl habitat. Invertebrate biomass in an older clearcut (kept open by mowing) in a minor floodplain was similar to GTRs on the same floodplain, but less than in naturally flooded forests. Invertebrate resources were found to be low in fresh clearcuts but recover rapidly with time.

Where wildlife habitat is a primary objective, and where it is desirable to provide habitat for a variety of species, a range of stand conditions involving variations in composition, age, and structure must be represented. Management decisions to accomplish these habitat objectives must be made on a forest or landscape basis. The first requirement is a detailed description and mapping of existing stand conditions. The maps and descriptions should then be used to prepare an ecosystem management plan that makes recommendations for wildlife stand improvements to optimize diversity, dispersion, and juxtaposition of wildlife habitat types across the landscape.

A practice that has become important for minor floodplain sites is the establishment of GTRs for waterfowl habitat. These reservoirs are designed for dormant season flooding, and early indications were that they had little influence on the ecology of the area and might even stimulate tree growth (Broadfoot 1967) and acorn production (Minkler and McDermott 1960, Minkler and Jones 1965). However, more recent studies have shown that prolonged use of GTRs on some sites will result in reduced tree growth, poor regeneration of some oak species, and a change in composition to more water-tolerant species (Newling 1981, Wigley and Filer 1989, King 1995, Young et al. 1995a). In some reservoirs the major problem may not be dormant season flooding but the inability to remove growing season flood waters fast enough (Young et al. 1995a). There is much to be learned about GTR management. Present indications are that the key to their use, without affecting the ecology of the area, is the ability to remove floodwaters quickly, especially during the growing season and some alternation of years with and without artificial flooding.

Cultural practices, mainly farming and forest management on adjacent uplands, can have a major impact on minor floodplains. Perhaps the best example is that cited previously for Piedmont streams. Cotton farming on the uplands during the late 19th and early 20th centuries resulted in severe erosion and subsequent deposition onto the floodplains. Trimble (1970) reported depths of deposition of 5.5 m (18 ft) for some floodplains, and this decreased stream gradients and changed flooding frequencies and duration (Dowd et al. 1993), creating wetter conditions in the floodplain. Forests have been re-established on much of the farmland, and sediment deposition onto the floodplains has been greatly reduced (Brinson et al. 1996). Many of the streams have eroded downward and exported sediment downstream (Burke 1996). This incising process has influenced flooding regimes and will no doubt lead to drier conditions on the floodplain (Richter et al. 1995) and eventually to changes in the vegetation. Similar but less dramatic changes have occurred on some floodplains in the Coastal Plain.

SUMMARY

Minor alluvial floodplains are young sites in which the dominant geologic process generally is deposition. Two of the most important characteristics of these floodplain ecosystems are change and diversity. Because of flooding, deposition, and stream movement, floodplain sites are constantly changing, and this change has a marked influence on natural succession. Primary succession occurs on new land formed by deposition (e.g., sand bars and mud flats), and secondary succession on other sites is often driven or influenced by depositional patterns. Site diversity occurs as a result of minor differences in relief which are the result of stream meander and depositional patterns. Species occurrence and diversity within the floodplains are strongly related to these site differences, primarily because of differences in hydrology (drainage). Management practices within these floodplains should consider the topographic site differences and attempt to match the species to the site. This consideration is especially important for restoration projects and artificial regeneration of harvested areas.

An even-aged system of silviculture appears to be most appropriate for these floodplain forests. Clearcutting with natural regeneration is the method most often used, but adequate advance regeneration of the desired species, especially oaks, must be present at time of harvest. Size of the clearcuts can be adjusted to meet other management objectives such as improvement of wildlife habitat. Modifications of the shelterwood system have been used successfully and may be especially appropriate where advance regeneration is not present.

Minor alluvial floodplains provide very productive fish and wildlife habitat, and maintenance of that habitat is very compatible with silvicultural practices designed to produce and enhance timber quality. Management decisions to maintain or enhance wildlife habitat need to be made on a forest or landscape basis. Cutting practices, which are also needed for timber production, can then be used to optimize diversity, dispersion, and juxtaposition of wildlife habitat types across the landscape.

14 Pocosins and Carolina Bays

Rebecca R. Sharitz and Charles A. Gresham

CONTENTS

1-56670-228-3/97/$0.00+$.50
© 1998 by CRC Press LLC

INTRODUCTION

Pocosins and Carolina bays share several hydrologic and physical characteristics that distinguish them from other forested wetland types; they are nutrient-poor, typically have peat or organic substrates, and are non-alluvial. Their geographic distributions overlap, and they often have similar vegetation types. Pocosins and bays receive all or most of their water from precipitation, in contrast to deepwater swamps and riverine floodplains. Like pocosins, pine and hardwood flatwoods occur on large interstream areas of the outer Coastal Plain, but the flatwoods are on relatively fertile, mineral soil with little to no peat accumulation. Although some southeastern mountain bogs have peat substrates, their vegetation and location clearly separate them from pocosins. Finally, some Carolina bays have stands of pondcypress (*Taxodium distichum* var. *nutans*), but their geology and geographical distribution distinguish them from cypress domes.

Definition and Classification

Pocosin wetlands of the southeastern Atlantic Coastal Plain are characterized by non-alluvial hydrology (fed by rainwater or highly oligotrophic groundwater), strongly acid soils (peat or wet mineral soils with an organic layer), and a generally dense shrub layer of species that are generally characterized by evergreen, sclerophyllous leaves (Weakley and Schafale 1991b). Carolina bays are isolated ovate depressional wetlands, also of the southeastern Coastal Plain, that have a unique and characteristic geomorphic structure (shape, alignment, and surrounding rim), hydrology dominated by precipitation inputs and evapotranspiration losses, thus

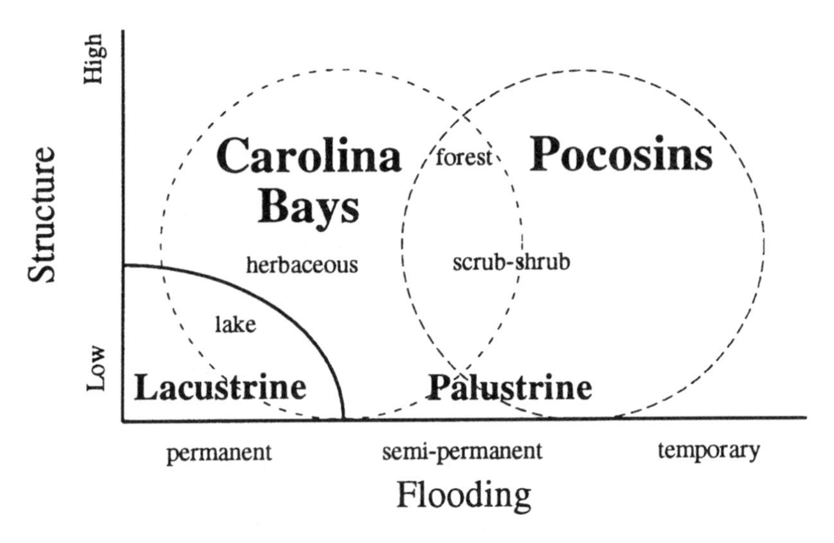

FIGURE 14.1 The relationship between pocosins and Carolina bays. Both are palustrine wetlands, although some bays may be lacustrine. Pocosins range from scrub–shrub to forested; bays may contain pocosin vegetation but may support other cover types as well.

highly variable, and plant communities that may include aquatic macrophyte beds, emergent grass and sedge marshes, evergreen shrub thickets, and wetland forests (often pondcypress or blackgum [*Nyssa sylvatica*]). Both pocosins and Carolina bays are nutrient-poor, usually without permanent standing water, but are strongly influenced by the hydrologic regime. They may have similar shrub bog vegetation and fauna, but they differ in geological origin.

All pocosins are classified as palustrine wetland ecosystems (Figure 14.1) since they are non-tidal, and trees, shrubs, persistent emergents, emergent mosses, or lichens cover 30% or more of the area (Cowardin et al. 1979). Most Carolina bays are also palustrine wetlands, with some species overlap with pocosins; however, many bays are dominated by other non-pocosin vegetation. A few Carolina bays have a large proportion of open water with little plant cover and are classified as lacustrine wetlands (Figure 14.1). This chapter focuses on palustrine pocosin and Carolina bay ecosystems, especially on the shrub bog and associated vegetation.

Pocosins

The name is derived from the Algonquin Indian word "poquosin," meaning "swamp-on-a-hill," and there are numerous early spellings of the name and descriptions of pocosins by Native Americans and early European explorers (Sharitz and Gibbons 1982, Richardson and Gibbons 1993). They were defined geologically (Kerr 1875) as "flatwoods with no natural drainage ways, different from swamps in that they are not alluvial, occurring on divides between rivers and sounds and frequently elevated above streams of which they were the source." More holistic definitions (e.g., Wells 1928, Woodwell 1958, Kologiski 1977) described pocosins as occurring in broad shallow basins, drainage basin heads, and on broad flat uplands of the southeastern Coastal Plain. These areas have long hydroperiods, temporary surface water, soils of sandy humus, muck, or peat; and they burn periodically.

Carolina Bays

Carolina bays are isolated elliptical depressions (Figure 14.2) that occur across the southeastern Coastal Plain, most abundantly in southern North Carolina and in South Carolina. They range in size from a few hectares up to several thousand hectares, and their long axis is usually, but not always, oriented in a northwest-southeast direction, with a sand rim to the southeast (Prouty 1952, Thom 1970). They were called "bays" by early European explorers who observed the evergreen shrubs and bay trees (loblolly-bay [*Gordonia lasianthus*], redbay [*Persea borbonia*], or swamp-bay [*P. palustris*]) growing either on the margins or in the centers of the depressions. Carolina bay soils range from highly organic to predominantly mineral, but most are underlain with sand alternating with impervious clay layers which retard water movement and may result in a perched water table. A distinction is sometimes made between "clay-based" bays that have mineral soil overlying the clay layers and "peat-based" bays that have thick deposits of peat within their basins (Bennett and Nelson 1991). Carolina bays characteristically have no natural drainages or artesian water sources, thus their water inputs are primarily direct precipitation and run-off from surrounding areas. Evapotranspiration, especially during the warm growing season, lowers the water table and can result in the complete drying of shallow bays. Thus, many smaller Carolina bays contain temporary aquatic habitats that may dry up seasonally under periods of low precipitation.

Geographical Extent

The distributions of pocosins and Carolina bays overlap, but pocosins typically occur in the lower and middle coastal plain terraces, whereas bays extend further inward into the upper terraces. Pocosins range from southeastern Virginia between Norfolk and the Dismal Swamp (the only occurrence in Virginia [Ludwig, pers. comm.]) to the Georgia-Florida border. They are most extensive in North Carolina (Richardson et al. 1981) where they are found from the Albemarle-Pamlico Sound and the Atlantic Ocean to approximately 150 km (93 mi) inland. In South Carolina, they occur from within 1 km (0.6 mi) of the ocean (Porcher 1995) to about 75 km (47 mi) inland, and in Georgia they extend inward from the coastline about 270 km (169 mi) (Ambrose, pers. comm.). The westernmost pocosin community is reported to be in Turner County, Georgia (Wharton 1978). Florida does not contain pocosins similar to those in the Carolinas (Hermann, pers. comm.) but does have shrub bogs with pocosin species as well as slash pine (*Pinus elliottii*), loblolly pine (*Pinus taeda*), and baldcypress (*Taxodium distichum* var. *distichum*) (Myers and Ewel 1990).

The northern limit of Carolina bays is north of the Chesapeake & Delaware Canal on the upper Delmarva Peninsula in Maryland and south of Newark, Delaware (Sneddon, pers. comm., Clancy, pers. comm.). Like pocosins they also extend southward through Georgia but not into Florida (Hermann, pers. comm.). In Virginia,

FIGURE 14.2 Cluster of peat-based Carolina bays in Horry County, SC, showing typical NW–SE orientation and well-developed sand rims. Undisturbed bays support pocosin vegetation. Note that two bays in the northern part of the image have been ditched and partially converted to agriculture.

Carolina bays are found throughout the Delmarva Peninsula (Tyndall et al. 1990, Ludwig, pers. comm.) and east of the fall line south of Richmond (Ludwig, pers. comm.). They are most abundant in the Carolinas, where they occur in a band from approximately 30 km (19 mi) west of the North Carolina coastline inward to 150 km (94 mi) (Frey 1950), and from within 3 km (2 mi) of the South Carolina coastline in Horry County (Dudley 1986) inward to 230 km (144 mi) (Bennett and Nelson 1991). In Georgia, they extend from 40 km (25 mi) west of the coastline to 190 km (119 mi) inland, as far as Brooks County (Ambrose, pers. comm.).

Physical alterations such as draining and conversions to forestry or agriculture have greatly reduced the extent and number of both pocosin and Carolina bay ecosystems. In the early 1950s, the total area of pocosin vegetation in North Carolina was estimated to be 907,993 ha (2,243,550 acres); by 1979 only 281,000 ha (695,000 acres) still existed in their natural state (Richardson et al. 1981, Richardson 1983). Drainage and conversion of pocosins have been especially extensive in Washington and Tyrrell Counties, North Carolina (Hefner and Moorhead 1991). Prouty (1952) suggested that as many as 500,000 Carolina bays may have once occurred, but more recent surveys report far smaller numbers (Frey 1950, Bennett and Nelson 1991).

It is likely that the total number of Carolina bays exceeds 10,000 and may be as high as 20,000 (Richardson and Gibbons 1993). Like pocosins, many Carolina bays have been converted to agricultural use. In South Carolina, Bennett and Nelson (1991) estimated that only 15 to 18% of the clay-based bays have less than 10% of their surface area disturbed, and they reported drainage ditches or ditch scars in virtually every bay surveyed.

Associated Ecosystems

Pocosin vegetation is found across a range of topographic, hydrologic, geographic, and pedologic conditions. A classification system for pocosins of North Carolina has been developed by Weakley and Schafale (1991b), integrating vegetation, site factors, and ecological dynamics. Of the eight pocosin community types that they identified, three (low pocosin, high pocosin, and small depression pocosin) will be emphasized in this chapter. Two others, bay forest and pond pine (*Pinus serotina*) woodland, are often associated with domed peatlands and will be discussed as part of these systems. The other three types (streamhead pocosin, streamhead Atlantic white-cedar [*Chamaecyparis thyoides*] forest, and peatland Atlantic white-cedar forest) will be discussed in other chapters. In addition, non-riverine wet hardwood forests and non-riverine swamp forests may occur on the margins of large peatlands and may grade into pond pine woodlands or other pocosin communities.

Depressional wetlands other than Carolina bays occur in the southeastern Coastal Plain (Lide in press), including cypress domes (Chapter 16) and limestone sinks. The vegetation of limestone sinks is often similar to that of depression meadow Carolina bays (Wharton 1978, Sutter and Kral 1994, Kirkman 1995). Terrestrial or ecotonal systems associated with bays include transitional shrub borders, xeric pine and scrub oak sandhill communities on the sand rims, and occasional oak-hickory forests on moderately steep slopes (Bennett and Nelson 1991).

Noteworthy Areas

The best examples of intact interfluvial pocosins occur in North Carolina: in The Nature Conservancy's holding in Green Swamp (Brunswick County), Croatan National Forest (Carteret and Craven Counties), Holly Shelter Swamp managed by the North Carolina Wildlife Commission (Pender County), Hofmann Forest managed by the Forestry Department at North Carolina State University (Jones and Onslow Counties), Bushy Lake State Natural Area (Cumberland County), and in the Alligator River National Wildlife Refuge (Dare and Hyde Counties) (Schafale and Weakly 1990). Carolina bays in North Carolina owned by The Nature Conservancy include Antioch Church Bay (Holk County), Dunahoe, Goose Pond, Pretty Pond, and Oak Savannah Bays (Robeson County), and the McIntire bay complex (four bays) and State Line Prairie Bay (Scotland County). Bladen Lakes State Forest (Bladen County) contains Carolina bays preserved by the North Carolina Division of Forest Resources. The North Carolina Division of Parks and Recreation preserves bays in Bushy Lake State Natural Area and in Jones Lake State Park. In South Carolina, Woods Bay in Woods Bay State Park (near Olanta) is fed by artesian water and contains marsh grass and a cypress-tupelo forest. Cathedral Bay Heritage Preserve near Olar has a pure stand of pondcypress. Juniper Bay near Fork (between Mullins and Dillon) has a stand of the relatively uncommon Atlantic white-cedar. Bennett's Bay Heritage Preserve near Foreston is one of the few large intact Carolina bays and has pocosin cover. Lewis Ocean Bay Heritage Preserve, between Conway and Myrtle Beach, protects more than 20 Carolina bays with pocosin vegetation.

Functional Value of Pocosins and Carolina Bays

In contrast to the commercial value of pocosins and bays for timber, row crop, and peat production, their functional value is more difficult to recognize and to quantify. Lugo et al. (1990c) listed the following categories of general wetland values: water (storage, conservation, desalination), organic productivity (net primary productivity and organic export), biogeochemical (nutrient recycling, carbon storage, pollutant sequestering), geomorphic (erosion control, buffer storm damage), biotic (nurseries and gene banks), and other (recreation, research, waste disposal). Undisturbed pocosins and bays have many of these functional values. Pocosins buffer adjacent estuaries from freshwater run-off, and the peat layer sequesters carbon (Bridgham et al. 1991). Both pocosins and bays provide habitats for a variety of wildlife and endangered plants.

PHYSICAL ENVIRONMENT

Climate

The climate throughout the geographic range of both pocosins and Carolina bays is characterized by mild temperatures and adequate rainfall (Table 14.1). Average daily January minimum air temperatures range from slightly below freezing at the northern limit to well above freezing in the southern part of their distribution. July average daily maximum temperatures range from around 31°C (about 88°F) in the north to around 33°C (about 91°F) in the south. The average annual precipitation increases slightly from north to south (Table 14.1); it is lowest in Virginia (1092 mm or 43 in), increasing in South Carolina and then decreasing somewhat in Georgia. The highest rainfall (1335 mm or 52.5 in) occurs in the southern coastal extreme of the range of pocosins. The growing season extends from <210 days in the north to 250–270 days in the southern extent of Carolina bays and pocosins.

Rainfall is typically greater in the summer months (July through September) with relatively dry periods before (April and May) and after (October through December) and moderate levels from January through March. In response to seasonal rainfall and temperature patterns, Carolina bays are often wetter in the winter and spring when the precipitation is moderate and evaporative water loss is low. They may gradually dry during the summer when evapotranspiration is high but be temporarily refilled by late summer rainfall events that are often associated with thunderstorms and hurricanes. The resulting unpredictability of the hydrologic regime greatly influences the plant and animal communities in these wetlands.

TABLE 14.1
Climatic conditions throughout the ranges of pocosins and Carolina bays.

		Average January daily minimum air temperature[1] °C (°F)	Average July daily maximum air temperature[1] °C (°F)	Average annual precipitation[1] mm (in)	Average growing season length[2] days
Northern limit of range	Pocosins	–1.4° (29.5°)	31.2° (88.2°)	1188 (46.84)	<210
	Carolina bays	–3.2° (26.2°)	31.0° (87.8°)	1092 (42.98)	<210
Southern limit of range	Pocosins	5.2° (41.3°)	33.3° (91.1°)	1335 (52.57)	270
	Carolina bays	3.7° (38.6°)	33.6° (92.5°)	1279 (50.36)	250
Western edge of range[3]	Pocosins	–1.4° (29.5°) VA	31.2° (88.2°) VA	1188 (46.84) VA	<210 VA
		4.3° (39.7°) GA	33.6° (92.5°) GA	1282 (50.49) SC	250 GA
				1153 (45.40) GA	
	Carolina bays	–3.5° (25.7°) VA	32.4° (90.3°) VA	1095 (43.11) VA	<210 VA
		2.3° (36.1°) GA	34.0° (93.2°) GA	1287 (50.68) SC	250 GA
				1173 (46.17) GA	

[1] National Climatic Data Center's Validated Historical Daily Data project (EarthInfo Inc. 1995)
[2] Nelson and Zillgitt (1969)
[3] States from which data were obtained are indicated.

Geology, Geomorphology, and Soils

Pocosins

Tectonic and other geologic processes have greatly influenced the physiography and hydrology of the South Atlantic Coastal Plain (Horton and Zullo 1991, Murphy 1995). The vast pocosin soils of North Carolina developed over an eroded plain in response to decreased drainage brought about by rising sea level and climatic changes (Daniel 1981). During the Wisconsin Ice Age, sea level was approximately 122 m (400 ft) below present levels, exposing a large area of continental shelf. Sea level started to rise about 10,000 years before present (YBP) when glacial ice had retreated north, and the rate of rise has been estimated to be 100 cm (39 in) per century (Peltier 1988). This rising sea level slowed stream flows on the eroding plain, causing alluvium deposition in stream channels and on interstream areas as well. Also, the milder climate promoted growth of aquatic vegetation in these shallow water areas, which eventually resulted in peat accumulations (Dolman and Buol 1967, Daniels et al. 1977). Carbon-14 dates from Dismal Swamp and Chesapeake Bay peats place the deposition of the earliest organic clays and overlying fibrous peat between 10,340 (±130) and 8,135 (±160) YBP (Daniel 1981). The growth of peat deposits in North Carolina's lower Coastal Plain probably began about the same time (Dolman and Buol 1967, Oats and Coch 1973). Carbon-14 dates from buried Atlantic white-cedar logs in the Hofmann Forest indicated that they were 7,930 to 5,685 (±185) years old (Daniels et al. 1977).

Soils of pocosins are classified as Histosols (Medisaprists), which must contain at least 20% organic matter if no clay is present, or at least 30% if there is up to 60% clay. The surface organic horizon can be greater than 130 cm (51 in) thick (Lilly 1981a). Major soil series include Pungo, Dare, and Pamlico which have a thick (at least 130 cm [51 in]) organic surface horizon over a sand or clay base, Mattamuskeet, with a 40 to 130 cm (16 to 51 in) organic horizon over a sandy base of mixed mineralogy, and Ponzer, Belhaven, and Scuppernong which have shallow organic horizons (40 to 130 cm [16 to 51 in]) over a loamy base of mixed mineralology (Daniels et al. 1984).

Peatland soils have been shown to be nutrient deficient (Bridgham and Richardson 1993), especially in phosphorus (P) (Richardson 1985). Walbridge (1991) demonstrated that the gradient from short pocosin to medium/tall pocosin to bay forest followed a gradient of increasing P availability and a decrease in the nitrogen (N):phosphorus ratio of plant leaves and surface soils. Other common characteristics of peatland soils include high water table, low pH (3.8 to 4.2), low cation exchange capacity, and low iron (Fe) and aluminum (Al) (Gilliam 1991).

Carolina Bays

The geologic origin of Carolina bays is their least understood feature. Technical and lay literature discuss theories of their origin and evidence supporting them (Johnson 1942, Bliley and Pettry 1979, Savage 1982, Sharitz and Gibbons 1982, Ross 1987, Richardson and Gibbons 1993, and Murphy 1995). An early theory involved solute removal and slow slumping of the soil (Johnson 1944), but Wells and Boyce (1953) provided strong paleoecological evidence that the bays formed quickly, supporting the meteorite theory. According to this explanation, a small meteorite shower occurred about 250,000 YBP, hitting the Atlantic Coastal Plain at a low angle. This theory explains the orientation, uniform shape, sand rim, and regional distribution of Carolina bays. However, seismic and magnetic measurements, drilling, and ground penetrating radar surveys of several bays have found no evidence to support this theory (Kaczorowski 1992). According to Kaczorowski, shallow, circular, open bodies of water formed, and unidirectional winds caused erosion on unconsolidated lake shorelines where the wave angle was high and deposition where the wave angle was low. Creating an elliptical lake from a circular one with this mechanism requires balanced opposing prevailing winds. Kaczorowski (1977) also estimated the age of Carolina bays to be 40,000 to 15,000 YBP, based on carbon-14 dating of basal peats, although other investigators believe the peats to be only 5,000 years old (Cohen, pers. comm.). Carbon-14 dates of organic horizons buried by rims of bays found on the northern Delmarva Peninsula indicated that these rims began forming between 20,000 and 15,000 YBP (Stolt and Rabenhorst 1987b). Recent geological and archaeological research (Brooks et al. 1996) has shown that basin morphology was still changing during the Holocene. To date, there is no agreement on the timing and mechanisms of bay formation.

Carolina bays undisturbed by burning or land management practices commonly have organic surface horizons overlying sand and clay layers (Figure 14.3). More than 20% of the undisturbed bays of the northern Delmarva Peninsula examined by Stolt and Rabenhorst (1987a, 1987b) had basin soils that were Histosols (Medisaprists) or had an organic horizon that was up to 50 cm (20 in) deep. The underlying mineral soils formed from either a sandy textured sediment or silty loess basin fill which originated from the Chesapeake and Delaware bays during glacial melting; they are classified as Histic, Cumulic or Fluvaquentic Humaquepts, Terric Medisaprists, Typic Umbraqualfs, or Typic Umbraquults. This contrasts to bays on the Virginia Eastern Shore where the interior soils are loam or sandy loam over sandy clay loam (Bliley and Pettry 1979).

Carolina bay basin soils in North Carolina are primarily Histosols (Typic and Terric Medisaprists) or wet, fine to coarse textured mineral soils such as Coxville, Rains, or Murville series, classified as Typic Paleaquults or Haplaquods (Daniels et al. 1984). Some of the larger bays in eastern North Carolina may have peat layers up to 4.5 m (15 ft) thick (Ingram and Otte 1981). In South Carolina, typical bay interior soil series are Pocomoke fine sandy loam, Rutlege loamy sand, Johnston loam, and Hobcaw loam for the northern coastal counties of Horry (Dudley 1986) and Georgetown (Stuckey 1982). Bays in the northern interior counties of Marion (Pitts 1980) and Williamsburg (Ward 1989) typically have Pantego loam, Coxville fine sandy loam, Rutlege loamy sand, and Byars sandy loam. Bays in southern interior counties of Aiken (Rogers 1985) and Barnwell (Rogers 1977) have interior soils in the Rembert loam and McColl loam series. The Rutlege and Johnston series are Aquepts, and the others are Aquults. Bay soils from coastal counties are generally characterized by 28 cm (11 in) of loam over loamy sand or sand, whereas bay soils from interior counties have loam over sandy clay loam, clay loam, or clay.

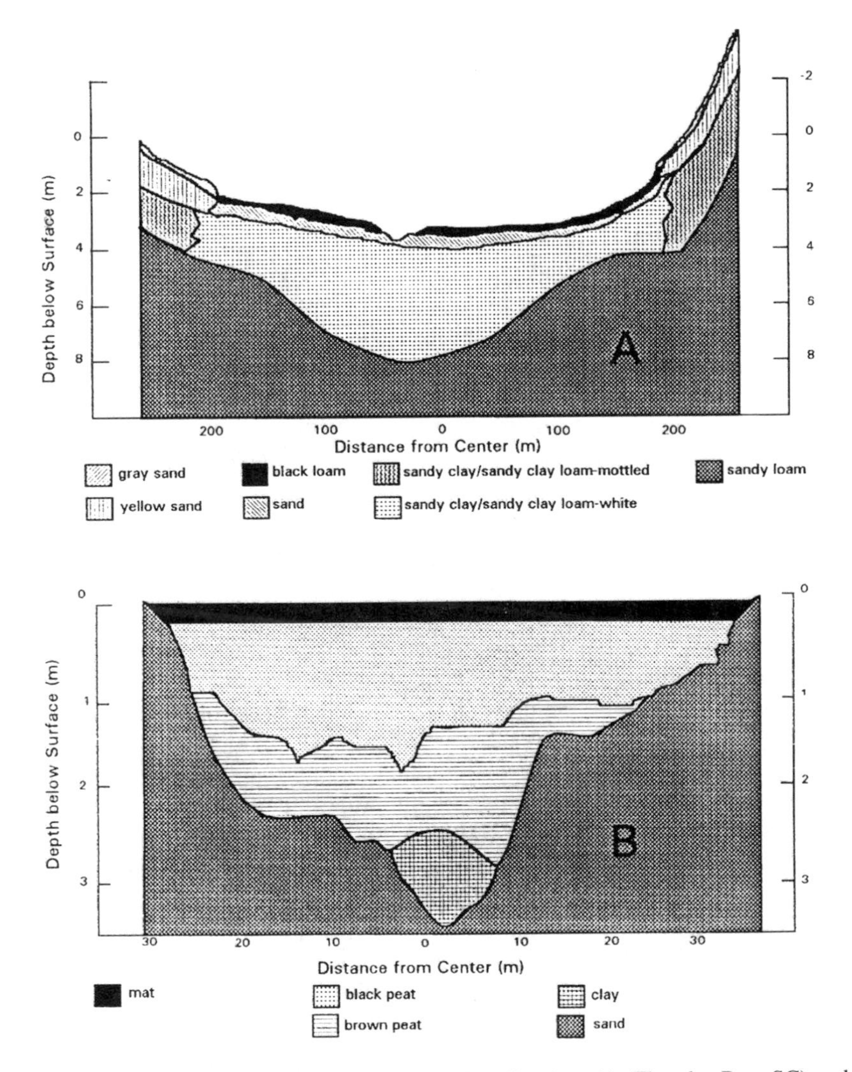

FIGURE 14.3 Cross-sections of a clay-based Carolina bay (A: Thunder Bay, SC) and a peat-based Carolina bay (B: Jerome Bog, NC), showing a comparison of substrates.

The stratigraphy and soils of Thunder Bay, a 7-ha (17.3-acre) bay in South Carolina (Figure 14.3) have been studied in detail (Lide et al. 1995). The central, and frequently ponded, portion of the bay contains somewhat mucky black loam (to a depth of 60 cm [24 in]) overlying a layer of highly permeable white sand (up to 30 cm [12 in] thick). This soil is mapped as a Williman sand, a hydric soil that is a loamy, siliceous, thermic Arenic Ochraquult. Beneath these sediments is a sandy clay or sandy clay loam hardpan which is at least 4 m (13 ft) thick in the center of the bay and thins to 1 to 2 m (3 to 6 ft) at the margins of the ponded area. The estimated hydraulic conductivity of the upper surface of the hardpan ranged from

1.1×10^{-6} to 3.6×10^{-9} m/sec (3.6 to 11.8 ft/sec), suggesting that it is moderately to slowly permeable (Lide et al. 1995).

Hydrology

Pocosins

The domed, interfluvial peatlands are generally ombrotrophic (Daniel 1981), receiving an annual average of 1200 to 1370 mm (47 to 54 in) of precipitation (Skaggs et al. 1991, Bridgham and Richardson 1993). From 10 to 30% of the precipitation is intercepted (Skaggs et al. 1991), and the remainder enters the soil to raise the water table. The water table fluctuates from just above the surface when the soil is saturated to 1.2 to 2.0 m (4 to 6.5 ft) below the surface (Daniel 1981, Skaggs et al. 1991). The water balance is maintained by surface runoff in stream channels or sheet flow at the soil surface just under the accumulated litter (Heath 1975, Gilliam and Skaggs 1981). Actual evapotranspiration ranges from 40 to 70% of rainfall (Skaggs 1996), and seepage loss is small (Daniel 1981). Thus, during the summer, when the water table is low, the water delivered by a heavy rainfall is captured in the soil and transpired by the vegetation, with no runoff (Ash et al. 1983). Heath (1975) gave an example of a water budget with 1300 mm (51 in) of rainfall balanced by 918 mm (36 in) of evapotranspiration, 369 mm (14 in) of runoff, and 13 mm (0.5 in) of seepage.

Carolina Bays

Most bays are hydrologically isolated ombrotrophic ponds with rainfall inputs, evapotranspiration losses, and water table dynamics determined by the difference between input and loss. The characteristic sandy clay hardpan is often assumed to limit surface water-groundwater interactions and result in a perched water table. Rainfall dominates water input to most bays (Sharitz and Gibbons 1982, Lide et al. 1995). Other sources of water may include artesian wells (Wells and Boyce 1953, Wharton 1978), groundwater (Newman and Schalles 1990, Lide et al. 1995), and inlet channels (Knight et al. 1989). Water loss is by evapotranspiration, seepage into the groundwater (Lide et al. 1995), and outlet channels. Although a few Carolina bays contain permanent open water, many have semi-permanent ponds that dry during prolonged drought, while others dry almost every year. Timing and duration of ponding can differ greatly among bays, even those in close proximity and receiving similar precipitation (Figure 14.4).

The net balance between inputs and losses gives Carolina bays a water table that fluctuates from 1 to 2 m (3 to 6 ft) above the surface (Schalles and Shure 1989, Lide et al. 1995) to 1 m (3 ft) or more below ground surface (Knight et al. 1989). This produces the seasonally flooded hydroperiod characteristic of most bays. Chemically, bay surface water is acidic (pH 4.5), has variable conductivity (32–320 μmhos/cm), and has moderate dissolved organic carbon (15 mg/l [0.9 grains/gal]) (CH2M Hill 1986, Schalles 1989). Groundwater is less acidic (pH 6.2) and has higher conductivity (367 μmhos/cm), half the total N (6.4 mg/l [0.4

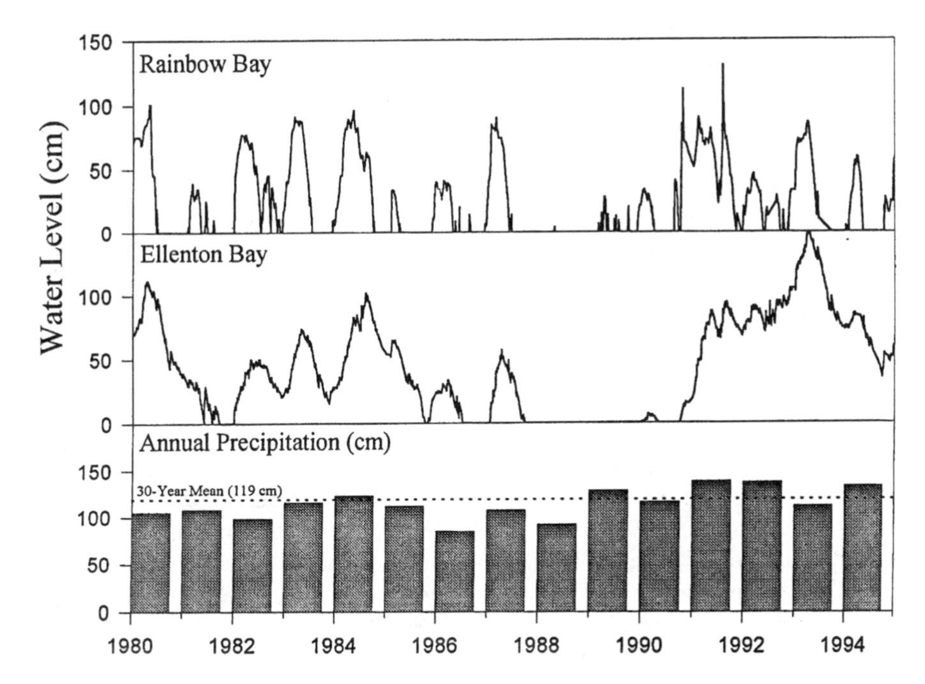

FIGURE 14.4 Hydrographs of a 1-ha (2.5-acre) Carolina bay (top) and a 10-ha (25-acre) Carolina bay (bottom) in Aiken County, SC. Although the bays are within 12 km (7.5 mi) of each other and receive similar precipitation inputs, their hydrologies (especially period of ponding) differ greatly.

grains/gal]), and far more total P (1.4 mg/l [0.08 grains/gal]) than does surface water (CH2M Hill 1986).

Biogeochemistry

Pocosin biogeochemistry and hydrology are inseparably linked. These are very nutrient-deficient systems with low levels of total and extractable P, N, and basic cations (Bridgham and Richardson 1992), and low soil pH (less than 4). Short pocosins, in the ombrotrophic center of the raised bog complex, have a soil organic matter content of about 95%. Tall pocosins have an average soil organic matter content from 76 to 93% in the top 30 cm (12 in) (Bridgham and Richardson 1992). In such nutrient-deficient systems, the composition of the vegetation is sensitive to increases in nutrient availability, especially P (Bridgham et al. 1991, Walbridge 1991), but it is not clear that vegetation height is controlled by soil nutrients (Bridgham and Richardson 1993). Pocosins have tight nutrient cycles, and runoff water is also nutrient poor (Ash et al. 1983). Bridgham et al. (1991) report that litter nutrient concentration, litterfall mass, and litter decomposition rates all increased along the gradient from short pocosin to tall pocosin to a more nutrient-rich gum swamp.

Few generalizations can be made about carbon and macronutrient cycling in Carolina bays because the hydrology and vegetative cover range from open water to forest. The biogeochemistry of peat-based Carolina bays is, presumably, similar to that of pocosins. The clay-based bays of the upper Coastal Plain have been described as dystrophic (low in pH and high in DOC) and dilute in nutrients (Schalles and Shure 1989). The surface soils of these clay-based bays are most commonly acidic sandy loams with pH ranging from 3.5 to 5.5. The soils tend to be quite low in organic matter (2 to 7%) relative to pocosins and other wetlands. There are relatively few data available on the nutrient status of these wetlands, particularly available N. Available P, ranging from 3 to 15 ppm in depression meadow bays (Miller, pers. comm.), is quite low compared to other freshwater marshes which range from 26 to 203 ppm (Mitsch and Gosselink 1993).

Fire

Fire in pocosins is a low frequency (13 to 50 years for short pocosin, 25 to 50 years for tall pocosin), widespread, high impact event (Frost 1995). Every three to four years, single or multiple fires burn >4,000 ha (>10,000 acres) of North Carolina pocosins (McNab et al. 1979), and in 1976, a wildfire burned more than 12,000 ha (30,000 acres) of pocosin land in Horry County, South Carolina. These fires are often intense, burning (though not always consuming) the aboveground vegetation and occasionally the peat (Christensen 1981). Natural factors contributing to pocosin fires include a heavy and highly combustible fuel load (27 to 34 mt/ha [12 to 15 tons/acre]) and the susceptibility of pocosin vegetation to late spring frosts that kill the normally evergreen foliage and succulent spring growth. Brief droughts expose the combustible organic soil in early spring also, increasing fuel availability.

Fire has apparently been a recurring phenomenon in Carolina bays as well. These depression wetlands are positioned within landscapes that were once, if not presently, dominated by fire-maintained pine forests. Presettlement fire frequency may have been as high as every one to three years (Frost 1995). During droughts, the probability of fire spreading into bay depressions from surrounding uplands would have been high. Fire periodically destroys peat accumulation (Wharton 1978, Schafale and Weakley 1990), and charcoal fragments have been found throughout pollen and peat profiles in several Carolina bays (Buell 1946, Wells and Boyce 1953). Fire in intermittently flooded Carolina bays may temporarily increase species richness but probably does not have a long-term effect on the vegetative composition (Kirkman and Sharitz 1994).

VEGETATIONAL COMMUNITIES

Types

A diversity of vegetation types occurs in pocosins and Carolina bays, occupying a range of topographic, hydrologic, geographic, and pedologic situations. For example, Schafale and Weakley (1990) identified 12 distinct and recurring assemblages of plant populations associated with pocosins or with bay depressions in North Carolina. Of these, some are found in both the domed peatlands of the outer Coastal Plain and in Carolina bays with extensive peat deposits, whereas others are largely found in only one system. Although, in general, the plant communities of these systems have not been well studied from an ecological standpoint, several phytosociological surveys and/or classifications include those by Kologiski (1977) for the Green Swamp, Snyder (1980) for the Croatan National Forest, Otte (1981) for North Carolina pocosins, and Bennett and Nelson (1991) for Carolina bays in South Carolina. The following pocosin and Carolina bay vegetation type descriptions are adapted from Schafale and Weakley (1990), Weakley and Schafale (1991b), and Bennett and Nelson (1991), which should be consulted for greater detail.

Pocosins

The vegetation structure of pocosins is characterized primarily by a dense shrub layer, with or without canopy or subcanopy strata. Herbs are generally sparse. The best indicator species of pocosins is pond pine, which is an important component of nearly all pocosins, typically forming a very open to closed canopy over the dense shrub layer. Pond pine is rarely a significant component of non-pocosin communities in the Carolinas; thus, it has a high fidelity and constancy in pocosin vegetation. Another strong indicator is zenobia (*Zenobia pulverulenta*), which is essentially endemic to pocosins of the Carolinas. Weakley and Schafale (1991b) recognized eight pocosin types, based on a combination of correlated vegetational, hydrologic, edaphic, and landscape factors. Of these types, five are described below; the other three are considered to be related plant communities and are described elsewhere (see Associated Ecosystems).

Low pocosin vegetation occurs in the central parts of the domed peatlands on poorly-drained interstream flats and in peat-filled Carolina bays, usually on peat deposits 1 to 5 m (3 to 16 ft) deep, although rarely on shallower peats or over very oligotrophic wet sands. Where the peat is deep and water saturated, plant roots never reach the underlying mineral soil. Thus, low pocosins are distinguished by the persistent low stature of the shrubby vegetation (less than 1.5 m [5 ft] tall) and the sparse distribution and low stature of the trees. These communities are extremely nutrient-poor, with normal nutrient input only from rainfall. The centers of domed peatlands are higher than the surrounding lands; thus, no surface or groundwater drains into them, and they are ombrotrophic. Wilbur and Christensen (1983) found P to be strongly limiting to plant production in low pocosins. Severe fires occur periodically associated with droughts and may burn into the peat as well as killing the aboveground vegetation. The low, dense shrub layer is usually dominated by fetterbush (*Lyonia lucida*), swamp cyrilla or titi (*Cyrilla racemiflora*), or zenobia, frequently intertwined with greenbrier (*Smilax* spp.). Widely scattered, stunted pond pine usually occur, as well as occasional swamp redbay, loblolly-bay, and sweetbay (*Magnolia virginiana*). Herbs are rare except in pools or openings. Stature of the vegetation increases with decreasing peat depth, and low pocosin grades into high pocosin in slightly drier areas with shallower peat (Figure 14.5).

High pocosin vegetation occurs in central-to-intermediate parts of domed peatlands on poorly drained interstream flats and in peat-filled Carolina bays, usually on peat deposits 1.5 m (5 ft) or less in depth or on oligotrophic wet sands. The peat is deep and saturated enough that plant roots can reach the underlying mineral soil

FIGURE 14.5 Low (left) and high (right) pocosin vegetation.

only during droughts. High pocosin is distinguished by shrubby vegetation 1.5 to 3.0 m (5 to 10 ft) tall and by the sparse distribution and relatively low stature of the few trees. Little surface or groundwater drains into high pocosins on domed peatlands, other than from adjacent low pocosin areas. Thus, they are largely ombrotrophic, but nutrient limitation is less severe than in low pocosins. The vegetation composition is similar to that of low pocosins, but the shrub layer is taller except when recovering from fire. Fetterbush, swamp cyrilla, and gallberry (*Ilex glabra*) usually dominate, with abundant greenbrier and switch cane (*Arundinaria gigantea*). Scattered pond pine, swamp redbay, loblolly-bay, and sweetbay usually occur (Figure 14.5). Herbs are essentially absent beneath the dense shrub layer.

Pond pine woodlands is a broad category covering a range of environments and vegetation (Society of American Foresters cover type 98, Eyre 1980). They occupy the outer parts of the domed peatlands on shallow organic deposits, and they may occur in shallow peat-filled Carolina bays but are uncommon (Bennett and Nelson 1991). In these areas, the water table drops to the underlying mineral soil during the dry season, allowing plants to root there. They may also receive some influx of water with nutrients from adjacent areas. Thus, these woodlands are wet and nutrient poor, but less so than low or high pocosins. They may be susceptible to fire during dry periods and the large amount of fuel may result in intense burns, but fire is still infrequent. Pond pine forms an open to nearly closed canopy; loblolly-bay, sweetbay, red maple (*Acer rubrum*), loblolly pine, and occasionally Atlantic white-cedar may also occur in the canopy or understory. The shrub layer is tall (greater than 5 m [16 ft]) and very dense unless recently burned; in addition to the high pocosin species, it may include large gallberry (*Ilex coriacea*), dangleberry (*Gaylussacia frondosa*), sweet pepperbush (*Clethra alnifolia*), maleberry (*Lyonia lingustrina*), and switch cane. Greenbrier is usually common. Herbs are generally absent under the dense woody cover.

Bay forests often occur on shallow organic deposits along drainages at the edges of peatlands and rarely in Carolina bays (Bennett and Nelson 1991). Like the other peatland communities, they are wet and nutrient poor, although less so than the low and high pocosins. Fires are very infrequent but may be severe. The distinguishing feature is a canopy dominated by loblolly-bay, sweetbay and swamp redbay. Pond pine, swamp tupelo *(Nyssa sylvatica* var. *biflora*), red maple, loblolly pine, and Atlantic white-cedar may be significant components. The shrub layer is dense to somewhat open, consisting of the same species as in the pond pine woodlands.

Small depression pocosins occur in small Carolina bays, lime sinks, and other small depressions. The soils are usually peaty sands or occasionally thicker organic accumulations. These depressions are seasonally flooded or intermittently exposed, and there may be some drainage received from surrounding upland sandy areas. Because of their small size and intermittent flooding, they may be susceptible to fire sweeping through the adjacent upland landscape. The dense to fairly dense shrub layer contains species typical of other pocosin communities and may also include myrtle dahoon (*Ilex myrtifolia*), maleberry, highbush blueberry (*Vaccinium corymbosum*), and Carolina sheepkill (*Kalmia angustifolia* var. *caroliniana*). The wettest areas may have zenobia and leatherleaf (*Cassandra calyculata*). A sparse to fairly dense canopy of bay forest species may occur, and pondcypress may also be present. The combination of more frequent disturbance and greater nutrient input from surrounding areas may account for the higher woody plant diversity in these small depression pocosins. The floors of these pocosins are highly shaded, with a depauperate herbaceous flora.

Carolina Bays

In addition to the pocosin communities described above, Carolina bay depressions support a variety of vegetation types that are strongly influenced by the hydroperiod of the bay. The following additional plant communities are classified by Schafale and Weakley (1990).

Vernal pools are found in small, seasonally flooded depressions with gently sloping sides. Because of their small size, they often are not distinguished on soil maps. Although such vernal pools may occur in depressions that are not Carolina bays (e.g., lime sinks), it may be difficult to determine the nature of the depression. The variation in water level, both among seasons and among years, is the primary environmental factor influencing the plant communities. When dry, vernal pools are subject to fires spreading from adjacent uplands. The vegetation typically is dominated by a dense to sparse herbaceous layer, often with strong zonation of dominant species, and composition may vary greatly. Grasses such as maidencane (*Panicum hemitomum*), andropogon (*Andropogon scoparius*), and leersia (*Leersia hexandra*) are common dry season dominants; during the wet season, aquatic plants such as bladderwort (*Utricularia* spp.) may become important. A shrub zone may surround the herbaceous marsh.

Cypress savannas occur in clay-based Carolina bays that are seasonally to temporarily flooded. These communities are apparently dependent on a combination of flooding and fire to maintain their open savanna structure. During droughts, they become invaded by pines and hardwoods which are usually eliminated during subsequent flooding. It is likely that they burn periodically, especially during the summer when they are often dry. These cypress savanna bays have an open to sparse canopy of pondcypress, with or without swamp tupelo, loblolly pine, pond pine, sweetgum (*Liquidambar styraciflua*), and other broad-leaved wetland trees. Shrubs include fetterbush, swamp cyrilla, swamp leucothoe (*Leucothoe racemosa*), and sarvis holly (*Ilex amelanchier*) . Major herbs are maidencane and other panic grasses and an often rich assemblage of sedges, rushes, and forbs. Many plant species of concern (rare) are associated with these cypress savannas.

Bennett and Nelson (1991) described two related communities (non-alluvial swamps and pondcypress ponds) in bays in South Carolina. The non-alluvial swamps experience cycles of flooding and are dominated by combinations of broad-leaved tree species and pondcypress in varying proportions. The cypress ponds have a very long hydroperiod and are nearly permanently flooded; thus they have a greater abundance of aquatic plants such as water lily (*Nymphaea odorata*), bladderwort, and pickerelweed (*Pontederia cordata*).

Depression meadows (Bennett and Nelson 1991) were considered by Schafale and Weakley (1990) to be a variant of the cypress savanna community type. These open, wet, grassy flats have at most a few scattered pondcypress, but they also occur in clay-based Carolina bays and usually have an herbaceous flora similar to the cypress savannas. Reasons for the absence of pondcypress are not clear. There is often a distinct zonal pattern of dominant species, with aquatic macrophytes in the center, followed by emergent grasses and sedges, and ringed by wetland shrubs and trees. Droughts cause shifts in these species zones, and fires and other disturbances result in changes in herbaceous species composition and dominance (Kirkman and Sharitz 1994).

Small depression ponds are distinguished from the related depressional wetlands by having permanent, or nearly permanent, water in the deeper parts. This is often accompanied by the presence of obligate wetland plants. The pond vegetation is a complex of zones, often concentric but sometimes irregular. The deepest parts may be open water or may support aquatic plants such as water lily, watershield (*Brasenia schreberi*), cowlily (*Nuphar luteum*), and bladderworts. Shallow water areas may contain a variety of emergent and wetland plants, including maidencane and other panic grasses, spikerushes (*Juncus* spp.), and beakrushes (*Rhynchospor*a spp.). Scattered pondcypress and swamp tupelo sometimes occur in both deep and shallow water zones. Most ponds are surrounded by a dense zone of shrubs that may be pocosin-like, with species such as swamp cyrilla, fetterbush, and gallberry, or may contain shoreline species such as buttonbush (*Cephalanthus occidentalis*). Overall, the plant communities in these small depression ponds are extremely variable, depending on water level fluctuations, substrate, slope of the depression, water chemistry, vegetation of the surrounding landscape, geographic location, and accidents of dispersal. It should be noted that these small depression ponds are arbitrarily distinguished from lake aquatic communities by an open water acreage of less than 8 ha (20 acres) (Schafale and Weakley 1990). Vegetated shorelines of the large Carolina bay lakes often support many of the same wetland plants.

Natural lake shorelines of the few large Carolina bay lakes often support a complex of vegetation zones that may include emergent graminoids and other herbs, shrub thickets, swamps of pondcypress or blackgum, or various other bottomland hardwood tree species. Shallow open water areas commonly have emergent and floating vegetation, including such species as water lily, yellow lotus (*Nelumbo lutea*), watershield, and heartleaf (*Nymphoides* spp.), along with many grasses and sedges.

Dominant Plant Species

Few studies provide detailed data on the woody species composition of either pocosins or Carolina bays. The apparent lack of interest is no doubt due in part to their perceived low commercial value and impenetrable vegetation. Short and tall pocosin sites in the Croatan National Forest, North Carolina, and a tall pocosin in the Francis Marion Forest, South Carolina, show considerable similarity in composition (Table 14.2). The majority of the species are considered to be wetland plants (obligate or facultative wet). Similarly, three wooded bays with peat substrates in Horry County, South Carolina, are dominated by shrub bog species (Table 14.3).

Adaptations

Many of the characteristic pocosin plant species are not only fire tolerant, but they actually depend upon fire for completing their life cycles. Pond pine trees are fire resistant and sprout from the roots and from epicormic buds (along the trunk) following fire. In addition, heat stimulates opening of the serotinous cones, some of which may remain closed after seed maturation for up to 10 years. Such opening of the cones following fire ensures seed dispersal at a time when open areas are available for pine seedling establishment (Bramlett 1990). Other species, such as loblolly-bay, sprout from the stump and roots following fire (Gresham 1983, Gresham and Lipscomb 1990).

Many pocosin shrubs also have evergreen sclerophyllous leaves (reinforced with lignin and having thickened cuticles). This feature is usually considered an adaptation to drought rather than to high moisture conditions, however, sclerophyllous leaves may conserve nutrients in these nutrient-deficient habitats (Monk 1966b). Schlesinger and Chabot (1977) have further suggested that some evergreen pocosin species (e.g., fetterbush) may have a higher nutrient use efficiency (especially of N) than do their deciduous counterparts. The leaves of many pocosin species also contain large quantities of secondary chemicals that make them more resistant to herbivory but also increase their flammability (Christensen 1981).

Several of the trees that occur in the more deeply and continuously inundated areas of Carolina bays exhibit characteristic morphological, anatomical, and physiological adaptations to waterlogging stress (described in Chapter 8). In particular,

TABLE 14.2
A comparison of tree and shrub layer composition in short and tall pocosins. Numbers for trees and shrubs are stems/hectare; shrub layer includes some tree seedlings.

Species Name	Short Pocosin[1] Tree	Short Pocosin[1] Shrub	Tall Pocosin[1] Tree	Tall Pocosin[1] Shrub	Tall Pocosin[2] Tree	Tall Pocosin[2] Shrub	National Status [3]
Pond pine	(28) 400	(14) 1800	(11) 175		538		FACW+, OBL
Loblolly-bay	(9) 462	(19) 1070	(5) 50	(5) 3500	125	255	FACW
Atlantic white-cedar			(7) 150	(7) 1250			OBL
Pondcypress					103	255	
Swamp tupelo					77	42	FACW+, OBL
Redbay		(5) 500		(9) 1250	18	1231	FACW
Sweetbay					7	594	FACW+, OBL
Red maple			(3) 25	(2) 500	7	446	FAC
Dahoon					4	64	FACW
Southern bayberry				(3) 4000	11		FAC, FAC+
Fetterbush		(20) 1575		(15) 1330		10144	FACW
Gallberry		(4) 4500		(8) 1000		2016	FACW-, FACW
Blueberry		(5) 31000				1401	FACW-, FACW
Red chokecherry		(5) 1000		(2) 1000		191	FACW
Swamp cyrilla	(2) 50	(29) 8500		(14) 8670			FACW, FACW+
Greenbrier		(13) 1000		(13) 2500		3480	FACW
Large gallberry						403	FACW
Huckleberry						446	
Dwarf azalea		(1) 2500		(1) 1000			
Zenobia		(2) 2500		(5) 500			FACW, OBL
Maleberry						21	FACW
Poison oak						85	
Total	**912**	**55945**	**400**	**27000**	**890**	**21074**	

[1] Croatan National Forest (data recalculated from Christensen et al. 1988). Parentheses indicate number of plots in which a species occurred (of 35 plots in short pocosin and 20 plots in tall pocosin).
[2] Francis Marion National Forest (Jones 1981)
[3] National Status (Reed 1988): OBL, obligate wetland; FACW, facultative wetland; FAC, facultative; FACU, facultative upland. A + sign indicates a frequency toward the higher end of the category (more frequently found in wetlands); a – sign indicates a frequency toward the lower end of the category.

pondcypress and swamp tupelo commonly have swollen buttresses and produce knees, both of which may be an adaptation for anchoring the trees.

At the community level, Carolina bays with a pronounced hydrologic cycle of flooding and drying may have a rich and persistent seed bank remaining in the soil (Kirkman and Sharitz 1994, Sutter and Kral 1994, Poiani and Dixon 1995). In four bays in South Carolina, Kirkman and Sharitz discovered the seed bank to be the most diverse of any reported in wetland habitats. Many of the species were not

TABLE 14.3
Vegetation of three wooded Carolina bays in Horry County, SC. Data are relative importance values on a scale of 1–300 for canopy species and 1–200 for other species[1].

Species Name	Bay1	Bay2	Bay3	National Status[2]
	Canopy Species — IV(300)[3]			
Pond pine	211.73	226.47	220.19	FACW+, OBL
Loblolly-bay	65.29	53.56	79.81	FACW
Redbay	16.92	15.29		FACW
Swamp cyrilla	6.06	4.68		FACW, FACW+
	Subcanopy, Shrub and Herbaceous Species — IV (200)[4]			
Red maple		0.97		FAC
Red chokecherry	3.19	3.22		FACW
Leatherleaf		0.97		
Sweet pepperbush		5.55		FAC+, FACW
Swamp cyrilla	27.66	20.73	28.09	FACW, FACW+
Loblolly-bay	4.97	3.62	14.40	FACW
Large gallberry	24.16	30.37	25.39	FACW
Gallberry	13.86	8.53	3.57	FACW-, FACW
Fetterbush	51.18	49.79	67.55	FACW
Sweetbay	1.20	6.19		FACW+, OBL
Evergreen bayberry	2.13	5.15		FAC, FACW
Southern bayberry	1.54			FAC, FAC+
Swamp tupelo	1.20	4.02		FACW+, OBL
Swampbay	5.46	4.02	8.41	
Pond pine	5.46	1.93	3.43	FACW+, OBL
Catbriar	15.42	14.88	12.36	FACU, FAC
Zenobia	14.45	10.06	8.81	FACW, OBL
Dahoon		1.13		FACW
Myrtle dahoon		1.13		
Virginia willow	1.20			
Dwarf sumac	1.20			
Highbush blueberry			2.55	
Chain-fern	10.84	13.91	9.29	OBL
Sphagnum	5.32	4.02	5.85	
Other mosses	8.52	8.86	10.30	

[1] Data from CH2M Hill (1994)
[2] National Status (Reed 1988); abbreviations as in Table 2
[3] Importance value (sum to 300) = relative dominance (basal area) + relative density + relative frequency
[4] Importance value (sum to 200) = relative cover + relative frequency

present in the aboveground vegetation when the bays were ponded, but germinated during periods of drying and drought-related soil disturbances. Thus, the seed bank may be considered an adaptation by species to persist under variable and unpredictable hydrologic conditions, and it serves to maintain a diverse community structure in an ecosystem that may undergo extreme environmental changes in short time periods.

Successional Patterns

Pocosins

There are two contrasting theories of pocosin succession. The first assumes that the frequency and intensity of fire control successional development (Wells 1928). Short pocosin is considered a pioneer stage, succeeding to high pocosin and subsequently to bay forest climax within a few hundred years if disturbance is prevented. The second theory assumes that nutrient levels are the controlling factor (Otte 1981), and the successional sequence is marsh to swamp forest to bay forest to tall pocosin to short pocosin. Pollen analysis supports Otte's theory and suggests that succession has taken place over the past 5,000 years as paludification has resulted in an extensive peat-covered landscape on part of the southeastern outer Coastal Plain.

A delicate balance of interrelated environmental conditions, particularly thickness of the peat, length of the hydroperiod, and severity of fire, controls the distribution and structure of vegetation in pocosins. Since the highly organic soils burn readily when dry, the potential severity of a fire is related to the water table depth. Pocosin vegetation recovers quickly from shallow peat fires that destroy surface vegetation and the surface layer of peat but do not damage the root and rhizome mat. A somewhat deeper burn that destroys the rhizome and root mat will be followed by recolonization from seeds and vegetation growing inward from the edge of the burned area. A fire that burns the peat so deeply that the roots of the recolonizing vegetation can more readily reach the mineral soil may result in recovery of the same species as originally present, however, because of increased nutrient availability, their growth may be enhanced (Otte 1981). A severe burn over shallow peat that removes the organic material above the mineral sediments will probably lead to the development of a non-pocosin community. On the other hand, a severe burn in deep peat during a dry period could open up an area that would become a lake upon return to a high water table (Otte 1981).

Carolina Bays

Some of the early studies of Carolina bay vegetation viewed the zonal distribution of plant communities as suggestive of a successional process in which ponded bays would gradually become filled and dominated by more upland plant species (e.g., Kelley and Batson 1955). Although more recent studies of bay vegetation no longer support this concept, stratigraphic studies of bays on the upper Coastal Plain show that they have been gradually filling (i.e., sediment depth has been increasing) during the last 4,000 to 4,500 years (Brooks et al. 1996).

On a shorter time scale, fire influences the successional development of vegetation in pocosin vegetated bays. Severe fires may kill the shrub vegetation and burn holes in the peat layer. Recovery is fast, because of root sprouting and the availability of nutrients, and species diversity may increase (Schafale and Weakley 1990). Canebrake communities may dominate at a high fire frequency (Frost 1995); pond pine and other pocosin species replace canebrakes as fire frequency decreases. Similarly, Atlantic white-cedar stands may become established after severe crown fires remove the previous canopy. Schafale and Weakley (1990) considered bay forests to be a late successional community replacing pond pine and Atlantic white-cedar in the absence of fire.

The fluctuating hydrologic regimes and episodic disturbances may account for changes in plant species dominance in depression meadow Carolina bays. Kirkman (1995) described a cyclic pattern in which droughts (occurring perhaps every 10 to 30 years) shift the species dominance from aquatic macrophytes to wetland emergent species and, ultimately, to invasive upland plants and fugitive species that germinate from the seed bank. Fire and other disturbances to the soil are likely to occur under such dry conditions. When rainfall increases and the bays again become ponded, wetland and aquatic species again dominate.

Productivity

There have been few studies of the natural productivity or standing crop biomass of either pocosin or Carolina bay communities. Christensen et al. (1981) cited data indicating that in very nutrient-poor short pocosins, aboveground biomass, excluding tree trunks, may be 1,200 to 1,800 g/m^2 (10,684 to 16,026 lbs/acre). In tall pocosins, aboveground biomass may be 3,300 to 4,700 g/m^2 (29,381 to 41,845 lbs/acre). In a depression meadow bay, root and rhizome materials were found to dominate the biotic structure of the ponded area, with an average dry weight standing crop of 780 g/m^2 (6,945 lbs/acre). Root/shoot ratios were high, ranging between 4 and 13.5 (average 8).

ANIMAL COMMUNITIES

Major Species

Pocosins

Ecological studies of animals in pocosin habitats have been so limited that generalizations are difficult. Pocosins are considered critical habitats on a regional basis because they are often the only extensive natural areas remaining for wildlife (Sharitz and Gibbons 1982). No species of plants or animals are endemic to pocosins, although they may serve as key habitat for certain faunal and floral assemblages. Common vertebrate species include white-tailed deer (*Odocoileus virginianus*), common gray fox (*Urocyon cineroargenteus*), common raccoon (*Procyon lotor*), opossum (*Didelphis virginiana*), northern river otter (*Lutra canadensis*), mink (*Mustela vison*), bobcat (*Lynx rufus*), marsh rabbit (*Sylvilagus palustris*), and gray squirrel

(*Sciurus carolinensis*) (Monschein 1981, Clark et al. 1985). Characteristic small mammals are short-tailed shrew (*Blarina brevicauda*), southeastern shrew (*Sorex longirostris*), Eastern harvest mouse (*Reithrodontomys humilis*), cotton mouse (*Peromyscus gossypinus*), golden mouse (*Ochrotomys nuttalli*), and meadow vole (*Microtus pennsylvanicus*) (Clark et al. 1985, Mitchell et al. 1995). The southern bog lemming (*Synaptomys cooperi*) has recently been reported as far south as central North Carolina, but only in undisturbed pocosins (Mitchell et al. 1995). Large tracts of pocosin may serve as refuge habitats for black bears (*Ursus americanus*).

Many bird species in the region feed or nest in high pocosin as readily as in other related vegetation types, but short pocosin typically has low diversity of breeding birds. Karriker (1993) reported that three species, common yellowthroats (*Geothlypis trichas*), eastern wood-pewees (*Contopus virens*), and rufous-sided towhees (*Pipilo erythrophthalmus*), were abundant in short pocosins. In tall pocosins, pileated woodpeckers (*Dryocopus pileatus*), downy woodpeckers (*Picoides pubescens*), Acadian flycatchers (*Empidonax virescens*), yellow-throated vireos (*Vireo flavifrons*), red-eyed vireos (*Vireo olivaceus*), prothonotary warblers (*Protonotaria citrea*), worm-eating warblers (*Helmitheros vermivorus*), and Kentucky warblers (*Oporornis formosus*) all reached peak abundance. Potter (1982) reported 83 species of wintering birds in pocosins but concluded that the species were those expected to be found during winter elsewhere in the lower Coastal Plain. Pocosins likely provide habitat for fragmentation-sensitive species. Waterfowl frequent the lakes associated with pocosins, and permanent aquatic sites have species of fish typical of the region.

Game fish species include the redfin pickerel (*Esox americanus americanus*), chain pickerel (*Esox niger*), yellow perch (*Perca flavescens*), warmouth (*Lepomis gulosus*), flier (*Centrarchus macropterus*), black crappie (*Pomoxis nigromaculatus*), bluegill (*Lepomis macrochirus*), and largemouth bass (*Micropterus salmoides*) (Monschein 1981).

Presumably, most reptile or amphibian species whose geographic ranges encompass pocosins could be found in these habitats. For example, the spotted turtle (*Clemmys guttata*), the eastern diamondback rattlesnake (*Crotalus adamanteus*), and the American alligator (*Alligator mississippiensis*) are known to occur in pocosins.

Carolina Bays

Certain components of the Carolina bay fauna (e.g., zooplankton, invertebrates, amphibians, and reptiles) have been extensively studied in a few bays, but thorough surveys throughout the range and across the various types of bay habitats have not been conducted. Carolina bays in some regions are critical focal points for breeding and feeding of a large variety of nonaquatic vertebrate species.

Temporarily ponded Carolina bays studied on the upper Coastal Plain in South Carolina support exceptionally rich zooplankton communities. From a survey of 23 bays on the Savannah River Site, South Carolina, Mahoney et al. (1990) reported 44 species of cladocerans and seven species of calanoid copepods (including a new species endemic to bays) (DeBaise, pers. comm.). In general, larger bays with longer periods of ponding supported more taxa. Certain orders of aquatic insects (e.g., dipterans) are also common, whereas others (e.g., odonates and ephemeropterans) are abundant in some bays and virtually absent from others (Schalles and Shure 1989, Schalles et al. 1989).

Although some Carolina bays have fish (Frey 1951), the majority probably do not support permanent fish populations because of their ephemeral aquatic status. Of 63 isolated bays on the Savannah River Site, only 13 contained fish assemblages (1 to 6 species/bay, 12 species total); presence or absence of fish was related to frequency of drying and proximity of the bays to permanent aquatic habitats that could serve as sources of colonists (Snodgrass et al. 1996). Lake Waccamaw, a 3,200-ha (7,900-acre) bay lake in North Carolina, has 25 fish species (Frey 1951), of which three may be endemic. This lake also has three endemic species of mollusks.

The absence of predatory fish allows semiaquatic vertebrates with aquatic larval stages to become extremely abundant in some bays. For example, in a one-year study of two 1-ha (2.5-acre) clay-based bays on the Savannah River Site, Gibbons and Semlitsch (1981) captured more than 72,000 amphibians moving to or from the water, including nine species of salamanders and 16 species of frogs. A 16-yr study of amphibian species in Rainbow Bay found hydroperiod to be the primary source of variation in community structure (Semlitsch et al. 1996). Other vertebrate species are also common users of the bays; for example, Gibbons and Semlitsch (1981) also captured six species of turtles, nine species of lizards, 19 species of snakes, and 13 species of small mammals. Many other terrestrial vertebrates and birds that occur in pocosins may use bays for water or breeding sites. Wood ducks (*Aix sponsa*) often nest around bays with standing water if mature trees with cavities are available, and heronries are found in some bays.

Secondary Productivity and Trophic Relationships

There is little information on secondary productivity in Carolina bays and virtually none for pocosins. Congdon and Gibbons (1989) estimated the total biomass production of six freshwater turtles in a 10-ha (24.7-acre) bay as 9.7 kg/ha/yr (8.65 lb/ac/yr), a value that is less than $5 \times 10^{-2}\%$ of the estimated total primary productivity. Such figures do not appear to be available for other vertebrates, although amphibians are likely to be the major contributors to secondary productivity in many bays. Likewise, there is little quantitative information on food web dynamics or the interactions among trophic levels in pocosins or bays.

Endangered or Rare Species

Several uncommon species, including the swallowtail butterfly (*Papilio palamedes*) and the Hessel's hairstreak butterfly (*Mitoura hesseli*), are found in pocosins. The once federally endangered pine barrens tree frog (*Hyla andersoni*) is always associated with pocosin vegetation, but it is not known to occur in the large peatlands of the outer Coastal Plain (Wilbur 1981). The federally endangered red-cockaded woodpecker (*Picoides borealis*) inhabits mature pond pines in pocosins, but the species is more abundant in mature longleaf and loblolly pines elsewhere. The most northern and inland nesting colony of the federally endangered wood stork (*Mycteria americana*) is at Big Dukes Pond in Jenkins County, Georgia (Hodgson et al. 1988), and wood storks forage extensively in Carolina bays and other shallow wetlands of the region.

The primary habitat value of both pocosins and Carolina bays may be as refuges for fragmentation-sensitive species. For example, the black bear, which once roamed freely throughout the Coastal Plain, is now mostly found in pocosins because they tend to be the last remaining unexploited areas (Hellgren et al. 1991). Likewise, Carolina bays are sources of water and wetland habitat in an otherwise upland and drier landscape. In urban or intensely farmed areas, many species of amphibians, reptiles, and small mammals unquestionably have clumped distribution patterns, with the areas of highest densities being associated with Carolina bays.

MANAGEMENT

TIMBER

Pocosins

The clearest and most definitive example of forest management in pocosin wetlands is in eastern North Carolina where forestry has been practiced for at least 100 years. In the late 1800s and early 1900s, much of North Carolina's coastal pocosin area was railroad-logged for Atlantic white-cedar, prized for roofing shingles (Lilly 1981a, 1981b). Several logging companies were formed and obtained tracts of 1,000 ha to more than 400,000 ha (200,000 to more than 988,000 acres). After harvesting the pocosins, these companies did not regenerate them; instead, they sold them to other timber companies or to potential developers. In the early 1900s, several government projects established drainage districts to support area-wide development. The development process included burning the cut-over land to remove unwanted vegetation and logging debris. Wildfires were also common and often intense since drainage had allowed the debris to dry. Because of this regular burning, these pocosin areas did not naturally regenerate to pine.

Foresters recognized that timber production in harvested pocosins was low, not only because of poor drainage but also because of frequent, severe wildfire (Miller and Maki 1957). In the 1940s, the forest industry pocosin owners and the North Carolina Forest Service established a wildfire control program that greatly reduced fire frequency and extent (Campbell and Hughes 1991). This allowed loblolly and pond pine to become established in cut-over pocosin lands.

In the 1950s and 1960s, there was large-scale clearcutting of these naturally regenerated pine stands by a few major pulp and paper companies (Campbell and Hughes 1981). Because these companies were committed to managing for a sustained yield of pine, they tried to regenerate clearcut land shortly after harvest. Most of the early efforts to replant clearcut pocosins failed because they used upland site preparation techniques on wetlands. These regeneration failures led to an extensive and, in some cases, continuing series of research projects to determine how to manage pocosin wetlands primarily for loblolly pine pulpwood (Miller and Maki 1957, Terry and Hughes 1975).

Three major problems were identified as the cause of regeneration failures: poor (at best) drainage, improper site preparation, and low soil fertility (Campbell and Hughes 1991). These problems were recognized as early as the middle 1950s, when at least two forestry research centers were established in eastern North Carolina by paper companies (Crutchfield and Trew 1957, Campbell and Hughes 1981), and North Carolina State University established the Hofmann Research Forest (Ash et al. 1983).

A drainage system was required to remove surface water and to lower the water table enough for seedlings to survive and grow to such a size that their transpiration could maintain a low water table (Campbell and Hughes 1981). Local drainage and site access were accomplished by a combination of dragline ditching, mounding the resulting material, and crowning it into a road. Early research on the Hofmann Forest showed that near the road canal (within 52 m [170 ft]), loblolly pines at age 17 were 15 cm (5.9 in) diameter at breast height (dbh) and 12 m (41 ft) tall; at 131 m (430 ft) from the canal, they were 13 cm (5.0 in) dbh and 9 m (29 ft) tall; and in an undrained area they were 7.6 cm (3.0 in) dbh and 5 m (17 ft) tall (Miller and Maki 1957).

These results indicated that drainage could make pocosins productive and led to numerous and extensive projects dealing with ditch design, density, and spacing (Campbell 1977). Current practices use a grid system of primary road canals at 0.8 km (0.5 mi) spacing and perpendicular secondary ditches 90 to 180 m (300 to 600 ft) apart. This intensity of roads and ditches not only allows fairly rapid movement of rainfall, but also allows rapid access for wildfire control (Campbell and Hughes 1981). Drainage outlet riser dams and ponds allow management of water flows to conserve water for increased tree growth and to minimize surges of freshwater downstream (Hughes 1996). After a drainage system is installed, cut-over pocosin lands must be site prepared to ensure seedling survival and growth. This site preparation usually consists of shearing off residual trees with a shear blade, raking this debris into windrows which are then burned, and creating mounds about 1 m (3 ft) high and 2 m (6 ft) wide at the base (bedding) (Gresham 1989). Bedding raises planted seedlings above the water table to ensure high survival and plows under other vegetation to temporarily reduce competition. This process of burning residual debris and bedding has greatly increased the productivity of pocosins for loblolly pine.

The final problem limiting loblolly pine growth is low available P (MacCarthy and Davey 1976). Ralston and Richter (1980) surveyed 30 pocosin sites in eastern North Carolina and found that 75% of them had less than 5 ppm available P (average of 3.6 ppm). Fertilization guidelines recommend at least 4 ppm for loblolly pine in poorly drained soils, and this deficiency can be solved by either aerial or ground fertilization. One company uses a fertilizer hopper attached to the front of the crawler pulling the bedding plow so that a measured amount of diammonium phosphate is placed on the bed prior to planting. Liming pocosin soils for loblolly pine establishment and growth is not an operational practice.

Large-scale drainage and intensive forest management in pocosin areas has caused concern about downstream effects, including altering the timing and volume of surface runoff and increasing nutrient export to nearby streams and estuaries (Kirby-Smith and Barber 1979, Street and McClees 1981, Walbridge and Richardson 1991). Hydrologic water management models (DRAINMOD and DRAINLOB) have successfully simulated on-site and off-site hydrology in drained pocosin areas (McCarthy and Skaggs 1992, Amatya et al. 1994, Richardson and McCarthy 1994). These models enable forest managers to test water flow management strategies (by adjusting flow control structures) to both increase tree growth through drainage and minimize downstream effects.

Nutrient losses from pocosins intensively managed as loblolly pine plantations are likely to be less than losses from unmanaged pocosins (Hughes 1996) or from pocosins converted to cropland (Walbridge and Richardson 1991). Pine plantations are harvested on 20- to 30-year rotations, whereas agricultural crops are harvested once or twice a year; thus the plantation nutrient retention mechanisms are disturbed less frequently. In pine plantations, there is a large amount of fast-growing root biomass to take up applied nutrients and a heavy canopy to shelter the soil from intense rainfall that would cause sediment loss through erosion. Finally, converted pocosin soils are nutrient deficient (thus a nutrient sink) and are fertilized only once or twice every 30 years for silvicultural purposes and at rates lower than for agriculture. Freshwater outputs are low from managed pine forests (Hughes 1996) because the high leaf area intercepts rainfall, thus reducing the amount of water reaching the soil; 70 to 80% of the rainfall input is lost as evapotranspiration (not as drainage); and deep-growing roots remove moisture from a large soil volume that can thus hold more rainfall. Amatya et al. (1996) monitored the water quality from three ditched and drained watersheds that supported loblolly pine plantations on an Umbraquult soil. Their results after two years of sampling indicated that controlled drainage decreased total N and P loss, and only losses of nitrate and nitrite exceeded those from unmanaged forests; these higher losses were attributed to recent fertilization.

This is not to say that there is little concern about intensive forest management on pocosin lands. Forest industries and forestry associations developed Best Management Practices (BMPs) to provide guidelines for self-regulation. Field surveys of voluntary compliance to BMPs in South Carolina indicated that compliance was 95% by forest industry and 78 to 86% by private landowners (Hook et al. 1991b). Compliance by all users in North Carolina was 88% (White 1992).

During the early 1990s, several environmental protection organizations (Environmental Defense Fund, North Carolina Wildlife Federation, and Sierra Club) brought legal action against a major paper company in eastern North Carolina, claiming that its forest management practices did not qualify for the silvicultural exemption of the Clean Water Act Section 404 and thus required a permit from the U.S. Environmental Protection Agency. After several years of litigation, the intensive forest management practices were determined to be in compliance with existing regulations (USEPA 1994). However, the EPA's ruling caused a demand for clarification of what could and could not be done in pocosin wetlands without a permit. The EPA determined (USEPA 1995) that mechanical site preparation requires a permit in the following sites: cypress-gum swamps, muck and peat swamps, cypress domes, riverine bottomland hardwood wetlands, Atlantic white-cedar swamps, Carolina bay wetlands, non-riverine wet hardwood flats, non-riverine swamp forests (dominated by baldcypress, swamp tupelo, or Atlantic white-cedar), low pocosins, wet marl forests, freshwater marshes, and maritime forests of the barrier islands. Permits are not required for mechanical site preparation in pine flatwoods, pond pine woodlands, and wet flats. This EPA memorandum also considered timber harvesting to be consistent with Section 404(f) of the Clean Water Act, thus harvesting and regenerating pine plantations on pocosin land does not require a permit.

Carolina Bays

Prior to legislation protecting wetlands, Carolina bays were often treated in the same manner as adjacent flatwoods and were harvested and site prepared when dry enough to support equipment. Productivity was probably low because soil fertility problems were not well understood at that time. For example, in a Carolina bay planted with loblolly pine in the early 1970s, the average dbh in 1985 was 8 cm (3 in), compared to 20 cm (8 in) dbh on a productive wet flat soil. Stand density was 230 stems/ha (93 stems/acre) in 1985 compared to a probable planting density of 1345 stems/ha (545 stems/acre), indicating a 17% survival (CH2M Hill 1988).

Managing Carolina bay wetlands for timber production requires clearing the existing vegetation, installing drainage ditches within the bay and through the rim, bedding the bay soil, and planting the desired crop tree species. Any of these activities would greatly alter the structure and/or function of the bay ecosystem.

Farming

Pocosins

The use of pocosin wetlands for agriculture is best documented by the numerous publications describing the clearing, intensive culture, and off-site concerns of vast tracts in eastern North Carolina (Lilly 1981a). Organic soil swamps were cleared for agriculture as early as the 1780s with the building of the Dismal Swamp Canal and Lake Phelps Canal, and drainage projects continued to become more widespread until the Civil War. After the war, logging became much more intensive, especially south of Albemarle Sound, and interest in farming these blacklands increased. Drainage districts were established, and by 1911 more than 280,000 ha (700,000 acres) were in drainage districts. As drainage projects failed, landowners sold lands to state and Federal agencies, which lead to the creation of the Hofmann Forest, Lake Mattamuskeet, and the Croatan National Forest.

A few early successes in draining pocosins were recognized, and several large ownerships (e.g., First Colony Farms was 152,000 ha [376,000 acres] at its peak) cleared large tracts, installed intensive drainage systems, and farmed the blacklands for corn and soybeans. Pocosin soils that have been drained, fertilized, and limed have become some of the most productive soils in North Carolina (Barnes 1981, Gilliam 1991). Liming a drained Belhaven soil produced up to 2,284 kg/ha/yr (34 bu/ac/yr) of soybeans and 7,776 kg/ha/yr (124 bu/ac/yr) of corn (Lilly 1981a). Modern herbicides have allowed cotton lint yields of 880 to 1,232 kg/ha/yr (786 to

1100 lb/ac/yr) on organic soils (Hudson 1996). Clearing pocosins for agriculture is no longer practiced, however, because recent Farm Bills direct that all subsidies be terminated if a landowner destroys a wetland for agricultural purposes. Today many of the cleared pocosins are in pasture.

Water quality research of the late 1970s identified three main impacts of converting pocosin and pine forests into intensive agriculture (Kirby-Smith and Barber 1979, Skaggs et al. 1994). First, salinity in adjacent estuaries decreased, especially during heavy rainfall, as a result of increased runoff from fields. Drainage altered the runoff hydrograph by increasing the peak flow rate to three to four times that of undrained areas and decreasing the period of flow. Second, turbidity increased. For example, natural streams associated with pocosins typically have turbidities of four to five Jackson Turbidity Units (JTU); ditches that drained recently developed lands had a maximum turbidity of 200 JTU. After development, the turbidity declined to 20 JTU, in some cases 10 times predevelopment levels. Such changes in salinity and turbidity in the adjacent estuaries may greatly reduce the quality of habitat in primary nursery areas for fish and shellfish. Third, concentrations of phosphate, nitrate, and ammonia in streams and estuaries increased. Nitrate concentrations increased dramatically in the warmer months, dropping back to "normal" during winter. Ammonia concentrations were more variable than nitrate but still increased. Phosphate concentrations were highest near the drainage ditches and during the summer. These impacts were correlated with runoff volumes from the developed land.

Although public sensitivity to these problems is still high (Burkholder 1996, Kidwell 1996, Spruill 1996), the problems can be minimized with proper water management (Evans et al. 1991, Skaggs et al. 1994). Drainage to downstream areas can be controlled by setting flashboard risers in the ditches to maintain the water table at desired levels. This not only provides moisture for crops but also results in the slow discharge of surface water. Subsurface drainage has also been shown to reduce nutrient exports (Xu 1996).

Carolina Bays

It is estimated that 97% of the Carolina bays in South Carolina have been disturbed by agriculture (71%), logging (34%), or both (Bennett and Nelson 1991). Agriculture is the predominant and probably the oldest use of bays. Machinery to drain bays became widely available in the late 1940s (Hardy, pers. comm.), and farmers probably cleared and cultivated smaller bays surrounded by fields even earlier. In the late 1950s, the Agricultural Conservation Program (ACP) of the Soil Conservation Service (now the National Resource Conservation Service) provided up to a 50% federal funds match for improved drainage of croplands (Hardy, pers. comm.). This program ultimately resulted in greater agricultural use of bays. The ACP ended in the 1970s, and the 1985 and 1990 Farm Bills now require farmers to protect wetlands to remain eligible for USDA farm program benefits (NRCS 1995b).

Soils of Carolina bays are highly organic (10% or greater) and have a high capacity to hold nutrients if the pH is raised to 5.0 or higher (Clemson University 1982). There are four problems in farming bay soils that can be readily resolved.

First, drainage is necessary and can be accomplished with underground tile lines emptying into ditches dug through the middle and around the perimeter of the bay, or by directly draining the tile field with submersible pumps (Doty 1973). Second, the low pH must be increased to 5.0–5.5 by liming. Third, the high organic matter ties up minor nutrients, especially Al, manganese, and copper, at low pHs. These minor elements can be supplied by soil or foliar sprays. Finally, weeds, especially grasses and sedges, are a problem on organic soils and must be controlled with herbicides. It should be noted that these activities greatly alter the structure and/or function of the Carolina bay ecosystem.

Converted bays in South Carolina are 10 to 15% more productive for corn and soybeans than upland soils (Palmer, pers. comm.). A comparison of Carolina bay soils with adjacent upland soils in Horry (Dudley 1986), Marion (Pitts 1980), Williamsburg (Ward 1989), and Georgetown (Stuckey 1982) Counties in South Carolina illustrates their productivity. In Horry County, typical bay soils are Rutlege and Johnston, which are capable of producing 4,480 kg/ha/yr (80 bu/ac/yr) of corn and 2,016 to 2,688 kg/ha/yr (30 to 40 bu/ac/yr) of soybeans. Adjacent upland soils are Leon, Centenary, and Echaw, which can produce 2,800 to 3,920 kg/ha/yr (50 to 70 bu/ac/yr) of corn and 1,343 to 2,016 kg/ha/yr (20 to 30 bu/ac/yr) of soybeans. Likewise, bay soils in Marion County are capable of 4,480 to 6,440 kg/ha/yr (80 to 115 bu/ac/yr) of corn and 2,016 to 3,023 kg/ha/yr (30 to 45 bu/ha/yr) of soybeans compared to 3,080 to 5,040 kg/ha/yr (55 to 90 bu/ac/yr) of corn and 1,344 to 2,688 kg/ha/yr (20 to 40 bu/ha/yr) of soybeans for adjacent upland soils. Bay soils in Williamsburg County (Byars series: 6,160 kg corn/ha [110 bu/acre], 2,688 kg soybeans/ha [40 bu/acre]) and Georgetown County (Hobcaw: 6,160 kg corn/ha [110 bu/acre], 2,688 kg soybeans/ha [40 bu/acre]) are as productive as adjacent upland soils. These data indicate why Carolina bays were drained and cropped before other wetland values were realized and legislation passed to protect them.

The Food Security Act of 1985 as amended (the 1985 and 1990 Farm Bills) established the Wetland Reserve Program which enables the USDA to purchase 30-year or permanent easements on up to 404,000 ha (1,000,000 acres) of eligible agricultural lands for wetland restoration (NACD 1995, NRCS 1995b). Eligible agricultural lands are those with restorable wetlands, former or degraded wetlands, riparian areas along streams, land adjacent to wetlands that contribute significantly to wetland functions and values, and previously restored wetlands. USDA will share the costs (up to 75%) of restoring wetlands and implementing associated conservation practices on a permanent easement area. This program does allow haying, grazing, and some timber harvesting on wetland acreage if the functional value of the wetland is not diminished.

Peat Mining

Interest in mining peat was increased by the oil shortage of the early 1970s (Campbell 1981). Ingram and Otte (1981) estimated that 560 million metric tons (617 million English tons) of moisture-free peat exist on 243,000 ha (600,000 acres) of pocosin land. The peat is mined by removing the undecomposed wood from the upper soil surface and extruding the upper layer into cylinders which are left in place to dry (Ash et al. 1983). After air drying, the cylinders are collected and stored. To date, no large-scale peat mining operations have been undertaken.

Although there are no peat mining operations in Carolina bays, many bays contain up to 1.5 m (5 ft) of peat (Kaczorowski 1992), and some have deposits up to 4.6 m (15 ft) deep (Ingram and Otte 1981). The quality of the peat for energy production is reflected in high bulk densities (0.144 to 0.171 oven dry mt/m^3 [9.0 to 10.7 oven dry lb/ft ^{3}]), high Btu values (23,244 J/gm to 27,893 J/gm [10,000 to 12,000 Btu/lb]), low ash content (2 to 5%), and relatively high Von Post values (h7 to h9) (Cohen and Strack 1992). The sand rims that surround Carolina bays may shield the interior from the influx of foreign material such as mineral soil and ash, which allows the production of high organic matter, low ash peats. Also, these peats are well decomposed, as indicated by the high Von Post numbers, reflecting the long time of unburned accumulation.

Ecosystem Preservation and Restoration

Pocosins

Managing pocosins to preserve the habitat is primarily a matter of managing fire (Pearsall, pers. comm.). A very narrow window of opportunity exists between being too wet to burn and too dry to burn safely. Even when proper fuel and weather conditions are present, burning is still difficult because of the heavy thick smoke load, the tendency of pocosin fires to spot ahead, and the impossibility of putting them out when they go underground and smolder in a thick peat layer. Fires have occurred naturally in peatlands throughout their history (Frost 1995), and the interaction of fire and peat depth strongly influences the vegetation (Weakley and Schafale 1991b).

Restoring wetland functions to pocosins that were converted to croplands requires restoring wetland hydrology. Evans (1996) is monitoring soil and soil water properties of croplands that have two water table management treatments (drainage weir set at either 15 cm [6 in] above or below soil surface), two microtopographic treatments (smooth and random disking), and two plant establishment treatments (planting wetland tree seedlings on a 6 $\times$6 m [20 $\times$20 ft] grid or regeneration from natural seed sources). Preliminary results indicate that disking is needed for seedling establishment.

Carolina Bays

Few unaltered Carolina bays remain (Bennett and Nelson 1991, Kirkman et al. 1996). Most have been ditched and at least partly drained at some time in the past; many have been converted to agriculture. Preservation of unaltered bays primarily involves maintaining a fire regime that simulates naturally-occurring fires and protecting the bays from human alteration. Fire is needed to maintain longleaf or other pine communities on the bay rims (Regier, pers. comm.). Shallow or intermittently ponded bays may burn readily during droughts, and burns may be necessary to maintain populations of some plants such as Iva (*Iva microcephala*) (Kirkman and Sharitz 1994), Venus' fly trap (*Dionaea muscipula*) (Stowe, pers. comm.), and the rare white wikey (*Kalmia cuneata*). Bays that are ponded for longer periods may rarely burn. Protecting bays from human alteration involves controlling spread of invasive species from adjacent fields and limiting destructive trespassing. It should be noted that the value of Carolina bays as habitat for certain animal species is greatly increased by maintaining buffer zones around the wetlands. For example, several species of aquatic turtles common to bays nest and hibernate on land. Burke and Gibbons (1995) reported that protecting 90% of the nest and hibernation sites requires a 73-m (240-ft) buffer zone.

Restoring altered Carolina bays generally begins with restoring wetland hydrology, usually by filling drainage ditches and restoring the rim to its natural topography if necessary. Return to a more natural hydrologic regime will encourage the recovery of wetland vegetation which may germinate from the seed bank (Kirkman and Sharitz 1994) or be dispersed into the bay. Logging and removal of non-wetland trees, with concomitant disturbance of the soil and increased light, enhanced recovery of herbaceous wetland vegetation in a restored Carolina bay in South Carolina (Singer, pers. comm.).

Other Management Activities

Honey Bee Grazing

Bays in the Carolinas have a flora that both supports the honey bee industry and is detrimental to it (Lord 1979). In the spring, nectar and pollen are provided by red maple, followed by American holly (*Ilex opaca*), gallberry, and blueberry. Yellow jasmine (*Gelsemium sempervirens*) also flowers at this time and can poison bee colonies. Sourwood (*Oxydendron arboreum*) blooms during summer and is a good nectar source, however, swamp cyrilla also flowers then and may be responsible for killing much of a brood. In the late summer, redbay, loblolly-bay, sweetbay, and sweet pepperbush bloom. Cleared bays in row crops often support fall blooming goldenrod (*Solidago* spp.) and asters (*Aster* spp.). Grazing honey bees in Carolina bay communities requires no alteration of the soil, vegetation, hydrology, or fire frequency, and thus has no impact on the structure and function of the bay ecosystem.

Tertiary Sewage Effluent Treatment

Four Carolina bays in Horry County (Myrtle Beach, South Carolina) are being used in a pilot project to test the feasibility of using such systems to provide tertiary (also called advanced) treatment of domestic wastewater (CH2M Hill 1988, 1994; Kadlec and Knight 1996). The major wastewater constituents of interest to the South Carolina Department of Health and Environmental Control (DHEC) are those contributing to the ultimate oxygen demand, calculated as 4.5 times the total ammonia N plus 1.5 times the biochemical oxygen demand. DHEC was also concerned about the ability of the bays to assimilate N, P, metals, and water of relatively high pH (7 to 8).

Starting in January 1987, treated effluent was pumped into a forested bay (without a peat layer) at the rate of 600,000 l/day (0.16 million gal/day), which kept the water table at or above the surface near the discharge and within 0.6 m (2 ft) of the surface at the opposite end of the bay. For the first five years, basal area of the canopy increased each year, and total plant diversity increased. Surface water quality in the bay decreased, primarily as a result of increased chloride concentrations. By 1992, canopy basal area and density began to decline, and marsh grasses appeared in open areas.

In 1993, effluent was pumped into a peat-based bay with low pocosin cover, maintaining the water table at 15 cm (6 in) above the surface. After 18 to 24 months, the fetterbush died, possibly due to growing season flooding stress and increased sodium (from 297 to 397 ppm) in the surface water (CH2M Hill 1994). As of 1996, data are being reviewed to determine if further discharge into bays is acceptable (Schwartz, pers. comm.). If permitting agencies are convinced that biological criteria can be met, effluent discharges will be rotated among three of the bays, and the fourth will be held as a control.

RESEARCH NEEDS

Future research on Carolina bays should continue to examine their development and history. Use of new techniques, such as ground penetrating radar to examine subsurface topography and the distribution of siliceous microfossils (phytoliths, diatoms, and sponge spicules) in sediment cores, may reveal much about the physical development of the bays as well as provide information on historic regional climate. An examination of oxygen isotope ratios in precipitation, surface water, and groundwater may be useful in characterizing bay hydrology. There is also a need for greater understanding of the relationship between hydrology and species composition, especially the expression of the plant seed bank in clay-based bays. One of the most important values of Carolina bays is as wildlife habitat, but the sizes and types of buffers between the wetland and upland areas needed to support populations of amphibians and other vertebrates is poorly understood. While bays are naturally isolated habitats, they have been further fragmented by conversion and isolated by an agricultural landscape. Carolina bays may be excellent systems with which to explore effects of fragmentation and isolation and to study methods of wetland restoration.

Pocosins cover large areas and immediately border sensitive ecosystems such as estuaries, thus more intensive research must address ways to reduce off-site effects of forestry and agricultural management. Recent advances in controlled drainage are encouraging, and ongoing projects examining these effects on an ecosystem basis are essential to understand the dynamics of peatlands. The role of fire in preserving pocosins is slowly being understood, but this deserves far more research. Work in the Alligator River National Wildlife Refuge has indicated that fire is the single most important management tool, therefore, we need a far more complete understanding of its short- and long-term impacts.

SUMMARY

Pocosins and Carolina bays are palustrine, oligotrophic, and usually ombrotrophic wetlands confined to the southeastern Atlantic Coastal Plain. Pocosins are located on broad, flat, interfluvial areas and are most extensive in the outer coastal region of North Carolina. Carolina bays are isolated elliptical depressions of constant orientation, occurring in both the lower and upper regions of the Atlantic Coastal Plain. Both pocosins and bays support a variety of cover types, with several in common. Pocosins are characterized by a dense shrub layer and may have a sparse to more dense canopy dominated by pond pine, loblolly-bay, and redbay; they can also support a pure pond pine canopy or a bay forest (pond pine, swamp tupelo, red maple, and loblolly pine). The Carolina bay topography can support pocosin communities, vernal pools, herbaceous meadows, cypress savannas, or bottomland hardwood forest species.

Most Carolina bays have been disturbed by drainage prior to row crop agriculture or loblolly pine silviculture. The land use history of pocosins is well documented from the initial logging through widespread conversion to large farms and pine plantations. Converting and managing pocosins for loblolly pine has been very controversial and resulted in the clarification of EPA regulations defining when a permit is needed for silvicultural operations in all wetland areas. Clearly, further research is needed to define the functional values of pocosin and Carolina bay wetlands and to determine the extent that farming and forestry activities affect these functions.

ACKNOWLEDGMENTS

We thank Robert D. Sutter and an unknown reviewer for valuable comments on an early draft. Preparation of this chapter was partly supported by Financial Assistance Award Number DE-FC09-96SR18546 between the U.S. Department of Energy and the University of Georgia.

15 Southern Mountain Fens

Kevin K. Moorhead and Irene M. Rossell

CONTENTS

INTRODUCTION

GEOGRAPHICAL EXTENT, PAST AND PRESENT

Mountain fens are unique and uncommon wetlands located in the Appalachian and Interior Highlands of the southern United States. Many of these wetlands are referred

1-56670-228-3/97/$0.00+$.50
© 1998 by CRC Press LLC

TABLE 15.1
Estimated number and area of fens in the Appalachian Highlands[1].

State	Estimated number	Estimated area (ha)
West Virginia	100+	2,000
Virginia	20	85
North Carolina	115	200
Tennessee	8	15
South Carolina	1	3
Georgia	30	NA
Kentucky	60	50
Total	334	2,369

[1] Weakley and Moorhead (1991) and Moorhead (unpublished data). Data compiled from state inventory documents or from conversations with state inventory personnel.

to colloquially as bogs or glades (e.g., Big Run Bog or the Cranberry Glades in West Virginia), but they lack the principal characteristics that distinguish bogs of more northern latitudes. There are probably fewer than 500 sites and less than 2,500 ha (6,175 acres) remaining in the Appalachian Highlands (Table 15.1). Most sites are <1 ha (2.5 acres) in size and many have not been described due to their remote location and inaccessibility.

The Appalachian Highlands encompass the Appalachian Plateau, Ridge and Valley, and Blue Ridge provinces (Figure 15.1). Although fens may also occur in the Piedmont province, our discussion will focus primarily on fens located at higher elevations. Elevations in the Appalachian Highlands range from 300–2,000 m (985–6560 ft). In West Virginia, most fens are located between 900 and 1,300 m (2,950 to 4,265 ft). In contrast, most North Carolina fens occur at lower elevations, between 700 and 1,000 m (2,295 to 3,280 ft). Elevations in the Interior Highlands, which include the Ozark Plateaus and the mountains of the Ouachita, range from 200 to 800 m (655 to 2,625 ft).

At one time, fens may have occurred in every mountain county in the South (Richardson and Gibbons 1993). Their current distribution is more limited, as many wetlands have been degraded or converted to other uses. In West Virginia, most fens are concentrated in Tucker, Randolph, Preston, and Pocahontas Counties. In North Carolina, fens are concentrated in two areas: a band between Henderson and Clay Counties in the southern mountains, and in Avery, Watauga, Ashe, and Alleghany Counties in the northern mountains (Early 1989). Weakley and Schafale (1991a) classify fens in these two areas of North Carolina as southern and northern sub-types, based primarily on floristic composition. Northern sub-type fens are characterized by northern disjunct species, while southern sub-type fens are characterized by coastal plain disjunct species. Fens also occur in Kentucky, Virginia, Tennessee, Georgia, and South Carolina. In the Ozark Plateaus of the Interior Highlands, fens are located in seven counties in northern Arkansas and several counties in southern

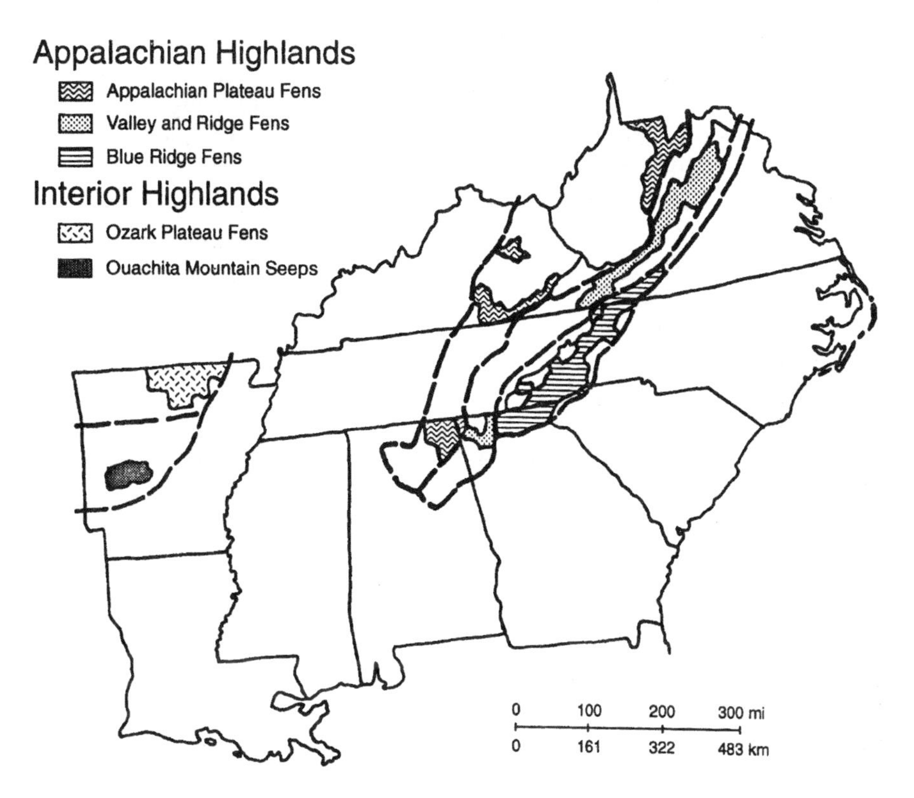

FIGURE 15.1 The physiographic provinces of the Appalachian and Interior Highlands. Distribution of fens is based primarily on county records (adapted from Richardson and Gibbons 1993).

Missouri. In the Ouachita Province, acid seeps are located in two counties in Arkansas.

There has been considerable debate over whether southern mountain fens should be classified as fens or bogs. The criteria used to distinguish fens and bogs in northern regions (Sjors 1950, Heinselmann 1970, Moore and Bellamy 1974) do not fit the southern ecosystems as well. Generally, northern bogs are described as ombrotrophic, acidic (pH <4.1) wetlands with a surface layer of sphagnum (*Spaghnum* spp.) and other mosses, and with hydrology and ion input influenced predominantly by precipitation. True ombrotrophic bogs develop peat layers higher than their surroundings and receive nutrients and other minerals exclusively from precipitation (Mitsch and Gosselink 1993). In contrast, northern fens are peatlands with hydrology derived from groundwater that has passed through mineral soil (minerotrophic), and often are dominated by sedges.

Many southern mountain fens have characteristics of both fens and bogs. Since they receive groundwater inputs from surrounding mineral soils, they should be classified as fens, with the degree of minerotrophy varying from rich to poor (Stewart and Nilsen 1993). Soils of rich fens have higher calcium (Ca) and magnesium (Mg) concentrations and lower nitrogen (N) concentrations than soils of poor fens (Moore and Bellamy 1974, Wieder 1985). Wieder (1985) describes Big Run Bog in West Virginia as a minerotrophic fen that is floristically and chemically similar to northern ombrotrophic bogs. Walbridge (1994) concluded that most West Virginia peatlands are fens that receive nutrients from groundwater flow. In the southern Appalachians, Weakley and Schafale (1994) apply the term "bog" to sites dominated by mosses (especially sphagnum), herbs, and shrubs. The term "glade" has been applied to poorly drained, ombrotrophic sites lacking trees and dominated by bryophytes (Walbridge and Lang 1982). Although most glades have been reported in West Virginia, some have been described in Virginia and North Carolina (Ogle 1982). Bridgham et al. (1996) suggest that the terms "bog" and "fen" be used colloquially to describe sites based on vegetation, alkalinity, and acidity, irrespective of hydrology. They recommend using the general term "peatland" for sites lacking hydrologic data.

We will break from historical and traditional nomenclature and refer to these unique wetlands collectively as southern mountain fens. It is important to note that peat-forming fens occur primarily in West Virginia, while similar "fen-like" wetlands found in mountains farther to the south lack peat deposits. Vegetation in many of these sites is similar to that occurring in both bogs and fens.

Southern mountain fens occur in four principal landscape positions: headwater regions of mountain streams, slopes (especially toe slopes) intercepting the water table and subject to constant seepage from groundwater, stream valleys no longer subject to flooding, and isolated systems over resistant rock strata (Walbridge 1994, Weakley and Schafale 1994). In West Virginia, most large fen complexes developed adjacent to streams, while the smaller but more numerous seepage wetlands developed where springs are frequent and drainage is impeded by underlying resistant rock (Gibson 1982). The seep-fens of Arkansas and Missouri are located along the upper reaches of second and third order streams within deeply dissected, relatively small drainage basins (Orzell and Pell 1985). Sedge-shrub fens in this area occur on toe slopes near upper and middle reaches of streams of moderately to deeply dissected valleys.

Classification

Southern mountain fens were not included in Küchler's 1964 classification. They would be classified by Cowardin et al. (1979) as palustrine systems; class: shrub-scrub, forested, or moss-lichen; subclass: moss, broadleaf deciduous or evergreen; water regime: saturated; water chemistry: poor fens-acid (pH <5.5), rich fens-circumneutral (pH 5.5 to 7.4). Many fens occupy such small areas that they are not included on National Wetland Inventory maps based on the Cowardin et al. (1979) system.

A comprehensive classification of non-alluvial mountain wetlands in the southern Blue Ridge province was developed by Weakley and Schafale (1994). Their classification is based on floristic composition and structure, landscape position, elevation, and degree of ombrotrophy/minerotrophy. The following community types are recognized in this classification: southern Appalachian bog, southern Appalachian fen, French Broad Valley bog, low mountain seepage bog, high elevation seep, and swamp forest-bog complex (two subtypes). The major distinguishing characteristics of each are listed in Table 15.2.

TABLE 15.2
Characteristics of fen and bog community types in the southern Blue Ridge province (after Weakley and Schafale 1994).

Community type	pH and soils	characteristic minerotrophy	vegetation species	patterns
Southern Appalachian Fen	organic-rich mineral soils	circumneutral, minerotrophic	some of northern origin, some calciphiles, mostly herbaceous	zonation complex
Southern Appalachian Bog	organic, or organic-rich mineral soils	very acidic to circumneutral	Sphagnum moss, many northern species	zonation concentric or patchy
French Broad Valley Bog	organic-rich mineral soils	acidic	some Coastal Plain species, many shrubs	zonation poorly developed
Low Mountain Seepage Bog	unspecified	acidic	numerous Coastal Plain species, some fire-adapted	unknown, due to disturbance
Swamp Forest-Bog Complex (typic subtype)	alluvial origin	varies	red maple, eastern hemlock, scattered sphagnum moss	forested thickets with small boggy depressions
Swamp Forest-Bog Complex (spruce subtype)	Toxaway series	varies	red spruce, scattered sphagnum moss	forested thickets with small boggy depressions
High Elevation Seep	rocky, gravelly, mucky	varies	dominated by forbs	zonation weak

The mountain wetlands of the Interior Highlands occurring on the Ozark Plateau are classified as streamside seep-fens and sedge-shrub fens (Orzell and Pell 1985); those occurring in the Ouachita Mountains are classified as forest acid seeps.

Associated Ecosystems

Many mountain fens located in headwater or floodplain areas are associated with bottomland hardwood or mixed hardwood/conifer forests. Many fens represent gaps in forest bottomlands where groundwater flow is restricted, where groundwater input has increased, or where beavers (*Castor canadensis*) have been active. At higher elevations, isolated fens may be surrounded by spruce and fir forests.

Economic Importance

Most southern mountain fens do not offer tangible economic benefits, although many are used for grazing cattle, and some have been logged (Gibson 1982). Cranberries are harvested in a few West Virginia and Virginia sites. The Cranberry Glades in West Virginia (Figure 15.2) and the Pink Beds in North Carolina are part of larger recreational areas within the Monongahela and Nantahala National Forests, respectively. Intangible benefits associated with fens include water storage, nutrient storage

FIGURE 15.2 The Cranberry Glades in West Virginia. Photograph by Kevin Colburn.

and transformation, food chain support, habitat for rare plants and animals, and educational and research opportunities.

Examples of Noteworthy Areas

Many southern mountain fens have been impacted severely by human intrusion, and few pristine areas remain in either public and private ownership. Canaan Valley, Cranesville Swamp, and the Cranberry Glades in West Virginia are three of the largest fen areas remaining in the Appalachian Highlands. A portion of the Canaan Valley system was dedicated recently as a National Wildlife Refuge. The Cranberry Glades are managed by the U.S. Forest Service, and Cranesville Swamp is owned by the Nature Conservancy. The Pink Beds and the Long Hope Valley fens in North Carolina represent excellent examples of headwater fens. Bluff Mountain Fen in North Carolina is protected by the Nature Conservancy. Noteworthy sites that have been used extensively for research include Big Run Bog and Tub Run Bog in West Virginia, and Flat Laurel Gap Bog and Tulula Bog in North Carolina.

PHYSICAL ENVIRONMENT

Climate

In West Virginia, fens occur in regions of high mean annual precipitation (139 to 154 cm [55 to 61 in]) and low mean annual temperatures (8 to 9°C [46 to 48°F]) (Walbridge 1994). In North Carolina, fens occur in areas receiving similar precipitation (135 cm

[53 in]) but with higher mean annual temperatures (12°C [54°F]) (NOAA 1994). The extreme variability of the southern mountain landscape has created a wide range of elevations, latitudes, and regional climatic conditions. For example, "frost pockets" created by local topography receive cold air draining from surrounding uplands and are impacted frequently by fog (Wieder et al. 1981, Walbridge 1994). The Interior Highlands have a more continental climate with mean annual precipitation ranging from 99 cm (39 in) in the Ozark Plateaus to 137 cm (54 in) in the Ouachita Mountains and mean annual temperatures ranging from 13 to 17°C (55 to 63°F).

Geology and Geomorphology

The Blue Ridge province is dominated by a series of large westward-vergent thrust faults and underlain by a complex mixture of metamorphic, igneous, and sedimentary rocks of Precambrian and Paleozoic age (Mills and Delcourt 1991). Common metamorphosed sediments include quartzite, slate, schist, and marble. The Ridge and Valley province is characterized by a series of prominent ridges and intervening valleys underlain by Paleozoic sedimentary rocks that have been folded and thrusted from the southeast (Mills and Delcourt 1991). Common sedimentary rocks in this region include sandstone, siltstone, limestone, and shale. The Appalachian Plateau is underlain chiefly by sandstones and shales of late Paleozoic age.

The Ozark Plateaus are a cratonic dome with a Precambrian core, surrounded by Paleozoic strata dipping gently away from the center of the dome (Madole et al. 1991). Dolostone and limestone are common sedimentary rocks in this region. The Ouachita Mountains are an east-west-trending belt of intensely folded and thrust-faulted Paleozoic sedimentary rocks (Madole et al. 1991). Common rocks include chert, shale, and sandstone.

Carbon dating of fen organic sediments has established that peat accumulations began 3,340 years before present (YBP) at Flat Laurel Gap Bog in North Carolina (Shafer 1988) and 13,080 YBP at Big Run Bog in West Virginia (Wieder 1985). Some fens may be of more recent origin due to logging, fire, and beaver activity (Weakley and Schafale 1994).

The relationship between fens and geomorphology has received little attention. Diehl and Behling (1982) described several geologic factors that may have affected the formation and presence of wetlands in the Appalachian Plateau. They include alluvial-filled valleys with low stream gradients and many meanders, a dipping resistant stratum that intersects stream beds at an acute angle, a flat resistant stratum capping a highland area that has been dissected by major streams, and an anticline in which the resistant stratum is breached so that a valley develops as it gradually widens and deepens. In West Virginia, Canaan Valley developed from a breached anticline, while Big Run and Tub Run Bogs developed in a dissected plateau with flat-lying resistant strata (Diehl and Behling 1982).

Weakley and Schafale (1994) reported that fens in North Carolina are found at high elevations over mafic rocks, and that bogs occur generally at low to moderate elevations over felsic rocks. Presumably, seepage from mafic rocks has a higher pH, conductivity, and cation concentration than seepage from felsic rocks (Weakley and Schafale 1994). The influence of underlying geology on ion input suggests differences in nutrient concentrations consistent with a poor fen/rich fen classification.

Mountain fens can be influenced by both alluvial and colluvial deposits. Shafer (1988) recognized three geomorphic surfaces with distinctive sedimentary deposits in Flat Laurel Gap in North Carolina. The three surfaces included a high colluvial fan, a complex of fluvial and colluvial terraces, and a modern floodplain and stream channels. Flat Laurel Gap Bog is located in the modern alluvial deposits.

Soils

From a broad perspective, fen soils can be classified as either organic or mineral. West Virginia fens, which are referred to frequently as peatlands, contain organic soils (Histosols) (Wieder 1985, Yavitt 1994). However, soils in many North Carolina fens do not fit the criteria used to distinguish organic soils (Soil Survey Staff 1987), and are classified as mineral soils.

Recent soil surveys have not been conducted in many mountain counties in the Appalachian Highlands. In any case, many fens in these counties are too small to be included on soil survey maps. Older surveys classified soils in mountain fens as muck and peat soils (Canaan Valley, Tucker County, WV) or medihemists (Cranberry Glades, Pocahontas County, WV). Two soil series have been proposed for mountain fens in Virginia; both are Histic Humaquepts with an organic surface horizon of 33 to 36 cm (13 to 14 in). Both are formed in colluvium and located in nearly level or gently sloping soils in upland depressions and drainageways.

By default, many fen soils have been classified as Entisols (Typic Fluvaquents) or Inceptisols (Cumulic Humaquepts or Fluventic Dystrochrepts) because they are located in headwater areas or adjacent to stream channels, therefore, they are classified frequently as floodplain soils. The soils of many fens adjacent to streams show an alluvial origin, although they no longer flood regularly (Weakley and Schafale 1994).

Research on soil properties has been conducted in very few fens. The pH of most mountain fens ranges from 3.6 to 4.9, and organic matter content ranges widely, from 7 to 95% (Table 15.3). Fens in West Virginia have a higher organic matter content and lower pH than fens located farther south. The depth of the surface peat horizon (O-horizon) or mineral A-horizon is typically <1 m (39 in) and in most cases <50 cm (20 in) (Wieder et al. 1994, Moorhead, unpublished data). The shallow surface peat or mineral horizon contrasts with northern peatlands where peat accumulations may exceed several meters. Fens with high organic matter content have low bulk densities (Stewart and Nilsen 1993), ranging between 0.05 and 0.80 g/cm^3 (0.03 and 0.46 oz/in^3), which increase with depth as organic matter decreases (Wieder 1985, Mowbray and Schlesinger 1988, Stewart and Nilsen 1993).

Cation exchange capacity (CEC) ranges from 15 to 150 $cmol_c$/kg and increases as organic matter increases (Table 15.3). In comparison, northern bogs have a CEC of >100 $cmol_c$/kg (Richardson et al. 1978). Wieder (1985) found significant differences in CEC at Big Run Bog that were based on vegetation community type and soil depth. Base saturation of the CEC was <10% in Big Run Bog (Wieder 1985) and Tulula Bog (Moorhead et al. 1995), but more than 20% at Bluff Mountain Fen (Mowbray and Schlesinger 1988).

Fens with mineral soils have a wide range of soil texture. In the surface horizons of three North Carolina fens, sand ranged from 52 to 74%, silt ranged from 16–37%,

TABLE 15.3
A comparison of pH, organic matter, and cation exchange capacity (CEC) of fens in the Appalachian Highlands. Data reported for the surface 0 to 20 cm (0 to 8 in) soil layer unless otherwise noted.

Site	pH	Organic Matter (%)	CEC ($cmol_c$/kg)	Reference
West Virginia				
Droop Mountain	4.0	91	—	Stewart and Nilsen 1993
Droop Mountain	3.7	94	—	Stewart and Nilsen 1993
Flag Glade	3.6	94	—	Stewart and Nilsen 1993
Round Glade	3.9	95	—	Stewart and Nilsen 1993
Big Run[1]	—	80	126	Wieder 1985
Tennessee				
Cross Mountain	4.7	42	—	Stewart and Nilsen 1993
North Carolina				
Bluff Mountain[2]	4.9	30	91	Mowbray and Schlesinger 1988
Tulula	4.7	15	46	Moorhead et al. 1995
McClures[3]	4.5	11	17	Moorhead, unpublished data
Deep Gap[3]	4.8	7	16	Moorhead, unpublished data
Dulaney[4]	4.7	14	—	Moorhead, unpublished data

[1]. Data averaged for four vegetation communities.
[2] 0 to 5 cm (0 to 2 in) depth.
[3] A1 horizon.
[4] 0 to 10 cm (0 to 4 in) depth.

and clay ranged from 7 to 14% (Moorhead, unpublished data). At Tulula Bog, there is a clayey horizon (~35% clay) 40 to 60 cm (16 to 24 in) below the surface, while at Deep Gap Bog, the soil horizon directly beneath the A horizon is approximately 80% sand. The clayey subsurface at Tulula Bog may restrict water flow, while the sandy subsurface at Deep Gap Bog may promote the rapid inflow of subsurface waters from surrounding slopes.

Hydrology

Few studies have examined the hydrology of southern mountain fens. Many fens are sustained by precipitation, groundwater inputs, and poor internal drainage. Groundwater sources include seepage from unconfined or confined aquifers. Fen hydrology also is influenced by overland surface water inputs from adjacent uplands (Novak and Wieder 1992). Internal fen hydrology is influenced by surface or groundwater outflow and by sphagnum mosses that absorb tremendous volumes of water. The fen areas of Cranesville Swamp in West Virginia contribute little to base streamflow of Muddy Creek and do not appear to dampen high storm streamflows (Eli and Rauch 1982).

The water table at Big Run Bog has been shown to fluctuate by 20 cm (8 in) annually, with frequent pools of standing surface water (Yavitt et al. 1990). The hydrology of McClures Bog (North Carolina) is maintained by discharge from two aquifers separated by a 60-cm (24-in) clay layer (Reberg-Horton 1994). The lower aquifer discharges to a localized area of the fen, while the upper aquifer, which originates from an adjacent slope, has a broader discharge zone of surficial groundwater (Reberg-Horton 1994).

In contrast to northern bogs, where precipitation is considered to drive hydrology, southern mountain fens receive substantial groundwater inputs (Weakley and Schafale 1994). However, input from groundwater may vary due to landform complexity. Data from Tulula Bog (Moorhead, unpublished data) support data from McClures Bog and Eller Seep (Reberg-Horton 1994), demonstrating the rapid response of the water table to precipitation events. This response is due primarily to surficial groundwater discharge from adjacent slopes.

Biogeochemistry

Carbon (C) storage and cycling have been investigated extensively in West Virginia fens (Yavitt et al. 1987, Wieder et al. 1990, Wieder et al. 1994, Yavitt 1994). Appalachian fens are characterized by shallow peat deposits despite high rates of primary production and up to 13,000 years of peat accumulation (Maxwell and Davis 1972, Wieder 1985, Wieder et al. 1990). Although shallow peat deposits are common in West Virginia fens, accumulations are limited in fens occurring farther south.

Peat accumulations result when the input of C from organic matter production exceeds the amount of C mineralized through organic matter decomposition (Yavitt et al. 1987). The higher latitudes and colder temperatures of West Virginia may explain the greater peat accumulation. Wieder et al. (1994) compared three West Virginia fens with two northern bogs and found similar rates of organic matter accumulation over the past 200 years, although they reported a general trend of older, thinner, and more highly decomposed peat deposits with decreasing latitudes. With decreasing latitude, decomposition apparently increases proportionately more than net annual production (Damman 1979).

Annual anaerobic carbon mineralization to methane (CH_4) or carbon dioxide (CO_2) at Big Run Bog was 52.8 mol/m^2, although CH_4 production accounted for only 12% of the C mineralized (Wieder et al. 1990). Most of the CO_2 produced was due to sulfate reduction, despite low dissolved sulfate concentrations. Temperature controls the mineralization of surface peat (Yavitt et al. 1987), with most CO_2 and CH_4 produced during the warmer summer months (Wieder et al. 1990). Yavitt et al. (1987) concluded that lower temperatures have less impact on CO_2 production than on CH_4 production during mineralization of peat. Peat decomposes throughout the year, resulting in CO_2 escaping to the atmosphere and partial decomposition products accumulating as dissolved organic C (Yavitt 1994).

Carbon mineralization of fresh sphagnum moss litter in surface peat proceeds rapidly, leaving recalcitrant residue in peat. The subsequent low rate of C mineralization of these recalicitrant residues results from the high polyuronic acid content of the sphagnum litter (Yavitt et al. 1987).

A large proportion (76 to 98%) of the total sulfur (S) in southern mountain fens is organic S (Wieder and Lang 1988, Novak and Wieder 1992). However, S cycling is dominated by fluxes through the inorganic pools. Total annual sulfate(SO_4) reduction in the top 35 cm (14 in) of peat was about 17 mol/m^2 in Big Run Bog; 66% of this occurred during the warmer months (Wieder et al. 1990). Despite the high rates of SO_4 reduction, the reduced inorganic S end products did not accumulate in the peat. Rates of SO_4 oxidation have been shown to be comparable to rates of SO_4 reduction (Wieder and Lang 1988). Vertical S profiles were similar for several fen peat deposits (Novak and Wieder 1992).

The biogeochemistry of the peat or mineral soils associated with southern mountain fens has received little attention. The understanding of N and P storage and cycling in these systems is remarkably limited. Wieder (1985) attributed spatial variation in peat chemistry across the surface of Big Run Bog to differences in vegetation communities. Four plant communities in this fen were distinguished by total calcium (Ca), exchangeable Ca, and base saturation in peat. Walbridge (1994) also found differences in community composition related to water chemistry in four West Virginia fens.

The surface water chemistries of West Virginia fen peatlands are comparable with those recorded in North American bogs (Wieder 1985, Walbridge 1994). The peat chemistry data of Wieder (1985) and water chemistry data of Walbridge (1994) suggest very low levels of exchangeable and soil solution Ca, potassium (K), Mg, and sodium (Na). Data from three fens in North Carolina (Moorhead, unpublished data) support these findings.

PLANT COMMUNITIES

SPECIES COMPOSITION

The floristic composition and successional patterns characterizing southern mountain fens differ from those characterizing northern bogs (Weakley and Schafale 1994). While vegetation in northern bogs frequently occurs in well-defined concentric zones (Mitsch and Gosselink 1993), the patterns in southern mountain fens are less distinct. In southern Blue Ridge fens, vegetation patterns have been described as patches or as poorly-developed concentric zones (Weakley and Schafale 1994), and floristic patterns in West Virginia fens have been described as an irregular mosaic (Wieder et al. 1981).

Typical plant species in southern mountain fens include bryophytes such as liverworts, sphagnum, and haircap (*Polytrichum* spp.) mosses; dwarf shrubs such as cranberry (*Vaccinium macrocarpon*) and swamp dewberry (*Rubus hispidus*); graminoids such as grasses, sedges (*Carex* spp.), rushes (*Juncus* spp.), beakrushes (*Rhynchospora* spp.), and cottongrass (*Eriophorum virginicum*); ferns such as cinnamon (*Osmunda cinnamomea*) and royal (*Osmunda regalis* var. *spectabilis*) ferns; forbs such as asters (*Aster* spp.), bedstraw (*Galium* spp.), tearthumb (*Polygonum sagittatum*), and skunk cabbage (*Symplocarpus foetidus*); tall shrubs such as silky willow (*Salix sericea*), tag alder (*Alnus serrulata*), red chokeberry (*Aronia arbutifolia*), and black chokeberry (*Aronia melanocarpa*); and trees such as red maple (*Acer rubrum*) and white pine (*Pinus strobus*). Common species occurring in southern mountain fens are listed in Table 15.4.

TABLE 15. 4
Plants commonly associated with southern mountain fens.

Scientific name	Common name
Bryophytes	
Aulacomnium palustre	Moss
Bazzania trilobata	Leafy liverwort
Polytrichum commune	Haircap moss
Sphagnum affine	Peatmoss
Sphagnum magellanicum	Peatmoss
Sphagnum palustre	Peatmoss
Sphagnum recurvum	Peatmoss
Graminoids	
Carex canescens	Hoary sedge
Carex folliculata	Sedge
Carex gynandra	Sedge
Carex leptalea	Bristly-stalk sedge
Carex stricta	Tussock sedge
Dulichium arundinaceum	Three-way sedge
Eriophorum virginicum	Cottongrass
Juncus brevicaudatus	Narrow-panicle rush
Juncus effusus	Soft rush
Leersia oryzoides	Rice cutgrass
Rhynchospora capitellata	Brownish beakrush
Scirpus cyperinus	Woolgrass
Herbs	
Aster novae-angliae	New England aster
Drosera rotundifolia	Round-leaved sundew
Orontium aquaticum	Golden club
Polygonum sagittatum	Arrow-leaf tearthumb
Sagittaria latifolia var. *pubescens*	Broad-leaf arrowhead
Sarracenia purpurea	Northern pitcher plant
Solidago patula	Rough-leaf goldenrod
Symplocarpus foetidus	Skunk cabbage
Pteridophytes	
Onoclea sensibilis	Sensitive fern
Osmunda cinnamomea	Cinnamon fern
Osmunda regalis var. *spectabilis*	Royal fern
Shrubs	
Alnus serrulata	Tag alder
Aronia arbutifolia	Red chokeberry
Aronia melanocarpa	Black chokeberry
Hypericum densiflorum	Bushy St. Johnswort
Ilex verticillata	Winterberry
Kalmia latifolia	Mountain laurel

TABLE 15. 4 (continued)
Plants commonly associated with southern mountain fens.

Scientific name	Common name
Leucothoe fontanesiana	Doghobble
Lyonia ligustrina var. *ligustrina*	Maleberry
Rhododendron maximum	Rosebay
Rosa palustris	Swamp rose
Rubus hispidus	Swamp dewberry
Salix sericea	Silky willow
Viburnum cassinoides	Witherod
Trees	
Acer rubrum	Red maple
Betula allegheniensis	Yellow birch
Betula lenta	Sweet birch
Nyssa sylvatica	Blackgum
Picea rubens	Red spruce
Pinus rigida	Pitch pine
Pinus strobus	Eastern white pine
Tsuga canadensis	Eastern hemlock

Factors such as elevation, soil and water pH, nutrient input, microtopography, geomorphology, and hydrology collectively determine the assemblage of species occurring in individual fens. Walbridge (1994) reported that plant communities in West Virginia were distributed according to pH and degree of minerotrophy. Wetlands with a pH between 4.0 and 4.4 were dominated by dwarf shrubs and bryophytes, while those with more base cations and a pH between 4.6 and 5.0 were dominated by taller shrubs and trees. Forty percent of the species in his study were unique to individual sites.

In the southern Blue Ridge province, 31% of plant species occurring in nonalluvial wetlands are of northern origin (e.g., cottongrass, cranberry) (Weakley and Schafale 1994). Many of these northern species, such as eastern hemlock (*Tsuga canadensis*), are at or near the southern limits of their range (Wieder et al. 1984, Weakley and Schafale 1994). Others, such as bog rose (*Arethusa bulbosa*) and bog buckbean (*Menyanthes trifoliata*), are relict populations of boreal species that migrated south during the last glaciation and now occur only in cold, moist sites at high elevations (Wieder et al. 1981, Delcourt and Delcourt 1985, Pittillo 1994). Several southern Appalachian fens are known to harbor a rich pollen record of boreal species (Delcourt 1985).

In addition to northern species, southern Blue Ridge wetlands also are characterized by plants that are more widely distributed. Thirty-six percent of the flora, including species such as red maple, tag alder, and swamp rose (*Rosa palustris*), occurs throughout eastern North America (Weakley and Schafale 1994). Another 20% of the flora, including netted chainfern (*Woodwardia areolata*) and swamp pink (*Helonias bullata*), is classified as disjunct Coastal Plain species, some of which originated in more northern areas. The remaining 13% of the flora consists of southern and central Appalachian species. A few, such as Gray's lily (*Lilium grayi*) (Pittillo 1994), mountain sweet pitcher plant (*Sarracenia rubra* ssp. *jonesii*), and bunched arrowhead (*Sagittaria fasciculata*) (Murdock 1994, Weakley and Schafale 1994), are endemic to southern Appalachian fens.

Some of the unique floristic patterns characterizing southern fens may be due to microtopographical relief. Some larger complexes have been described as small boggy depressions interspersed among low ridges in floodplain forests (Weakley and Schafale 1994). Within individual fens, moss hummocks (including sphagnum and haircap mosses) may support specific assemblages of species (Gibson 1970, Wieder et al. 1981, Gibson 1982, Walbridge and Lang 1982). At Alder Run Bog in West Virginia, Gibson (1982) noted that the driest hummocks were composed of haircap moss and supported upright shrubs, while the wettest hummocks were composed of sphagnum moss and supported graminoids, sundews (*Drosera* spp.), and cranberry.

Seed banks, which represent past, present, and future plant communities, have been poorly studied in both northern bogs and southern fens. We found only one published study examining the seed bank of a southern mountain fen. McGraw (1987) reported large seed reserves in both sedge- and moss-dominated communities at Big Run Bog; some occurred in numbers as high as those reported for prairie marshes. Soft rush (*Juncus effusus*) contributed the greatest number of seeds at this site. The seed bank of Tulula Bog is also dominated by rushes, although sedge seeds are almost as numerous (Wells and Rossell, unpublished data).

Rare and Unique Plants

Small, isolated fens in the southern Appalachians are important habitats for rare endemic and disjunct populations of plants. In North Carolina alone, mountain fens support 77 species of rare, threatened, and endangered plants (Murdock 1994). Examples include the federally-endangered bunched arrowhead, the federally-threatened swamp pink, and the state-threatened Gray's lily. Some of the rare plants found in southern mountain fens are listed in Table 15.5. More complete lists can be found in Richardson and Gibbons (1993) and Murdock (1994).

Carnivorous plants, including sundews, pitcher plants, and bladderworts (*Utricularia* spp.), may be found in southern mountain fens. In some cases, as with the northern pitcher plant (*Sarracenia purpurea*) at Big Run Bog, they were introduced to the site (Wieder et al. 1981). Federally endangered carnivorous plants found in southern mountain fens include the mountain sweet pitcher plant and the green pitcher plant (*Sarracenia oreophila*) (Murdock 1994).

Productivity

Primary productivity in southern fens generally exceeds that recorded for northern bogs (Wieder and Lang 1983, Wieder et al. 1989). Wieder et al. (1989) estimated the annual aboveground net primary production (NPP) at Big Run Bog to be 1045 g/m^2 (3.42 oz/ft^2). Of this, 43% was attributed to bryophytes, 27% to upright shrubs,

TABLE 15.5
Some rare plants associated with southern mountain fens (after Murdock 1994).

Scientific name	Common name
Arethusa bulbosa	Bog rose
Caltha palustris	Marsh marigold
Carex oligosperma	Few-seeded sedge
Cladium mariscoides	Twig-rush
Eriocaulon lineare	Narrow pipewort
Helonias bullata	Swamp pink
Juncus caesariensis	Rough rush
Lilium canadense ssp. *editorum*	Red Canada lily
Lilium grayi	Gray's lily
Menyanthes trifoliata	Bog buckbean
Myrica gale	Sweet gale
Parnassia asarifolia	Grass of Parnassus
Poa paludigena	Bog bluegrass
Rhynchospora alba	White beakrush
Sagittaria fasciculata	Bunched arrowhead
Sarracenia rubra ssp. *jonesii*	Mountain sweet pitcher plant
Sarracenia oreophila	Green pitcher plant
Saxifraga pensylvanica	Swamp saxifrage
Solidago uliginosa	Bog goldenrod
Sphagnum fallax	Peatmoss
Vaccinium macrocarpon	Cranberry

and nearly 20% to graminoids. In northern bogs, NPP estimates are lower, ranging from 129 to 857 g/m^2 (0.42 to 2.81 oz/ft^2) (Wieder et al. 1989). Similarly, Wieder and Lang (1983) reported that annual NPP of sphagnum decreased as latitude increased. They attributed the higher productivity in southern wetlands to the warmer climate and longer growing season, and to the fact that southern fens tend to occur low in their respective watersheds, where they may receive nutrient inputs from upland forests containing deciduous trees. However, since these processes slow the accumulation of peat, deposits in southern fens tend to be thinner and more decomposed than those in bogs located farther north (Wieder and Lang 1983).

Succession

Beaver have played an important role in the successional patterns in many southern mountain fens (Weakley and Schafale 1994). Flooding from beaver dams causes the water table to rise rapidly, which can kill woody vegetation and create open water. The open water is colonized first by submerged and emergent macrophytes and may be encroached on by sphagnum mats. Secondary succession eventually leads to wet meadows, shrub fens, and forest-fen complexes (Gibson 1982, Weakley and Schafale

1994). In West Virginia, Walbridge (1994) reported that the diverse hydrologic conditions created by beaver increased overall species diversity in peatlands.

Paludification by the spread of sphagnum has also maintained and possibly increased the area covered by mountain fens. When trees are windthrown or removed by logging, overall transpiration and water losses decrease, enhancing the spread of sphagnum (Gaddy 1981). Conversely, when trees and shrubs establish in fens, large amounts of water may transpire, contributing to terrestrialization (Murdock 1994).

Successional patterns also may be influenced by other factors. For example, successional trends at Flat Laurel Gap Bog in North Carolina may have resulted in part from the allelopathic influence of ericaceous shrubs (Shafer 1986). Logging activities in the late 1800s and early 1900s changed the species composition and successional patterns in many southern Appalachian forested fens, particularly where logging was followed by intense fires (Gibson 1982).

ANIMALS

INVERTEBRATES

Little or no research has investigated invertebrates in northern bogs or southern mountain fens (Murdock 1994). Research in the Midwest has documented several butterflies that are restricted to bogs and fens; some are associated with sedges and cranberries (Shuey 1985). The Baltimore butterfly (*Euphydryas phaeton*), a checkerspot butterfly, which is significantly rare in North Carolina, inhabits mountain fens, marshes, and wet meadows, and feeds primarily on turtlehead (*Chelone* spp.) (Murdock 1994). A preliminary survey of the insects inhabiting Tulula Bog has shown that members of the orders Hymenoptera and Diptera are particularly well-represented (Humphries and Rossell 1996).

Wieder et al. (1984) studied the flowering phenology at Big Run Bog. Of the plant species encountered in their transects, half were insect-pollinated. However, many wetland plants depend on insects from surrounding terrestrial areas for pollination (Pearson 1994). For example, the green pitcher plant is pollinated by bumblebees (*Bombus* spp.) (Murdock 1994), which probably are not restricted to wetland habitats.

The assemblage of invertebrates inhabiting the leaves of the northern pitcher plant is well known. The liquid contained in the leaves harbors nematodes, mites, copepods, and the larvae of three Dipterans (Fish and Hall 1978). In each leaf pitcher, one sarcophagid (flesh fly) larva feeds on floating insect prey, several mosquito larvae feed on suspended particulates, and several chironomid (midge) larvae feed on settled remains (Fish and Hall 1978). In addition, several species of moths are known to feed on the leaves, rhizomes, flowers, and seeds of pitcher plants (Murdock 1994).

VERTEBRATES

Most mountain fens in the southern Blue Ridge are above the reach of regular flooding by streams and small rivers (Weakley and Schafale 1994). This ensures the availability of fish-free vernal pools that are required for breeding by a variety of amphibians whose eggs are vulnerable to predation by fish. Four-toed salamanders (*Hemidactylium scutatum*), which are vernal pool breeders, lay and guard their eggs in sphagnum moss along the edges of pools; other vernal pool breeders include spotted salamanders (*Ambystoma maculatum*) and wood frogs (*Rana sylvatica*) (Martof et al. 1980, Murdock 1994).

Terrestrial forests surrounding mountain fens often are as critical to fen inhabitants as the wetland itself (Pearson 1994). Spotted salamanders inhabit burrows in upland forests during most of the year, but seek wetland pools for breeding in the spring (Martof et al. 1980). Many other animals neither nest nor breed in mountain fens, but use them regularly. For example, black bear (*Ursus americanus*), bobcat (*Lynx rufus*), white-tailed deer (*Odocoileus virginianus*), wild turkey (*Meleagris gallaparo*), and raccoon (*Procyon lotor*) travel through or forage in wetlands, and upland birds may feed, roost, or perch in wetlands. Rossell et al. (1997) documented 16 amphibian, 10 reptile, 19 mammal, and 74 bird species at Tulula Bog. Animals that commonly occur in or use southern mountain fens are listed in Table 15.6.

Small mountain fens in North Carolina harbor five species of rare, threatened, or endangered animals (Table 15.7) (Murdock 1994). In addition, the meadow jumping mouse (*Zapus hudsonius*), golden-winged warbler (*Vermivora chrysoptera*), and Swainson's warbler (*Limnothlypis swainsonii*) have been documented at Tulula Bog (Rossell et al. 1997); all three are on the Watch List of the North Carolina Natural Heritage Program (LeGrand 1993).

Boynton (1994) reviewed the species assemblages that might occur in different successional stages of mountain wetlands. In summary, beaver ponds with open water harbor great blue herons (*Ardea herodias*), wood ducks (*Aix sponsa*), belted kingfishers (*Ceryle alcyon*), bullfrogs (*Rana catesbeiana*), and fish such as bluegills (*Lepomis macrochirus*). Snags attract cavity-nesting species such as screech owls (*Otus asio*), tree swallows (*Tachycineta bicolor*), and pileated woodpeckers (*Dryocopus pileatus*). Early successional sedge meadows and shrub fens support bog turtles (*Clemmys muhlenbergii*), white-tailed deer, meadow jumping mice, golden-winged warblers, American woodcock (*Scolopax minor*), and song sparrows (*Melospiza melodia*). Forested areas support woodland jumping mice (*Napaeozapus insignis*), golden mice (*Ochrotomys nuttalli*), ruffed grouse (*Bonasa umbellus*), and Swainson's warblers, as well as species requiring fallen logs and woody debris, such as spotted salamanders, black rat snakes (*Elaphe obsoleta*), masked shrews (*Sorex cinereus*), and southern bog lemmings (*Synaptomys cooperi*).

One of the most-studied animals of southern fens is the bog turtle (Figure 15.3), a boreal species with a disjunct southern range. It is listed as a threatened species in North Carolina and is a candidate for federal listing. In the South, the greatest number of bog turtles has been recorded in North Carolina. However, less than half of the estimated 200 ha (500 acres) of mountain fens in this state support bog turtles (Herman 1994). The next largest population has been found in Virginia, and very few turtles have been found in Georgia, South Carolina, or Tennessee (Tryon 1990). In North Carolina, nearly 60% of bog turtles inhabit small fens between 600 to 900 m (1,970 to 2,950 ft) in elevation (Herman 1994) and are found primarily in wet meadows with scattered shrubs that lack dense forested canopies. Bog turtles may use small mammal runs and muskrat tunnels while foraging in wet meadows (Herman 1989).

TABLE 15.6
Animals commonly associated with southern mountain fens.

Scientific name	Common name
Amphibians	
Psuedacris crucifer	Spring peeper
Hyla chrysoscelis	Gray treefrog
Rana sylvatica	Wood frog
Bufo americanus	American toad
Rana catesbeiana	Bullfrog
Rana clamitans melanota	Green frog
Ambystoma maculatum	Spotted salamander
Desmognathus ochrophaeus	Mountain dusky salamander
Notophthalmus viridescens	Red-spotted newt
Reptiles	
Agkistrodon contortrix	Southern copperhead
Crotalus horridu	Timber rattlesnake
Elaphe obsoleta obsoleta	Black rat snake
Thamnophis sirtalis sirtalis	Eastern garter snake
Chelydra serpentina	Snapping turtle
Terrapene carolina carolina	Eastern box turtle
Eumeces fasciatus	Five-lined skink
Birds	
Aix sponsa	Wood duck
Ardea herodias	Great blue heron
Scolopax minor	American woodcock
Meleagris gallaparo	Wild turkey
Bonasa umbellus	Ruffed grouse
Otus asio	Eastern screech owl
Ceryle alcyon	Belted kingfisher
Tachycineta bicolor	Tree swallow
Dendroica pensylvanica	Chestnut-sided warbler
Limnothlypis swainsonii	Swainson's warbler
Vermivora chrysoptera	Golden-winged warbler
Wilsonia citrina	Hooded warbler
Passerina cyanea	Indigo bunting
Mammals	
Sorex cinereus	Masked shrew
Blarina brevicauda	Northern short-tailed shrew
Peromyscus leucopus	White-footed mouse
Ochrotomys nuttalli	Golden mouse
Zapus hudsonius	Meadow jumping mouse
Napaeozapus insignis	Woodland jumping mouse
Procyon lotor	Common raccoon
Odocoileus virginianus	White-tailed deer
Lynx rufus	Bobcat
Ursus americanus	Black bear
Castor canadensis	American beaver

TABLE 15.7
Rare animals associated with southern mountain fens (after Murdock 1994).

Scientific name	Common name
Ambystoma talpoideum	Mole salamander
Clemmys muhlenbergii	Bog turtle
Condylura cristata	Star-nosed mole
Euphydryas phaeton	Baltimore butterfly
Hemidactylium scutatum	Four-toed salamander
Synaptomys cooperi	Southern bog lemming

FIGURE 15.3 The bog turtle (*Clemmys muhlenbergii*). Photograph by Dennis Herman.

MANAGEMENT

Past Land Use Practices

Many fens have been destroyed by grazing, logging, mining, inundation for artificial reservoirs, and drainage for agriculture or development (Richardson and Gibbons 1993). Others have been impacted by adjacent land activities that lowered the water table or reduced surface inputs. In eastern Tennessee, the Army Corps of Engineers drained more than 4,000 ha (10,000 acres) of fens for farmland in the 1920s (Tryon 1990). Draining lowers the water table and encourages invasion by woody plants (Mohan et al. 1993), which lowers the water table even further, due to increased transpiration. Figure 15.4 shows clearly the effects of woody plants on the water table of Tulula Bog. This site has an open canopy area dominated by herbaceous vegetation that is adjacent to a closed canopy area dominated by mature red maple. In the open canopy area, the water table remained at 20 cm (8 in) or less from the surface for most of the two-year period. In the closed canopy area, the water table often fluctuated between depths of >20 cm (8 in) and >35 cm (14 in) between July and early October 1994. Less precipitation during 1995 resulted in lower water tables in both areas, but the effects were more dramatic in the closed canopy area, where the water table was frequently >50 cm (20 in) from the surface between July and October.

Although grazing has been nearly universal in fens, cattle can trample sensitive forbs and byrophytes (Weakley and Schafale 1994). Consumption and trampling by cattle have been implicated in the decline of the green pitcher plant in a North Carolina fen (Govus 1987). However, limited grazing can help control woody plants and prevent the development of a thick canopy. Excluding cattle from a fen in North Carolina resulted in the invasion of red maple and shrubs, which negatively affected the bog turtle population (Herman 1994). Other disturbances that remove or prevent the growth of shrubs can benefit herbaceous plants by reducing competition for light. Govus (1987) noted that a site in North Carolina that had been cleared by bulldozing lacked a shrub layer, which benefitted the green pitcher plant and a species of bladderwort.

Many fens are in close proximity to agricultural fields, pastures, orchards, Christmas tree farms, and nurseries (Murdock 1994), where they are subject to pesticide drift and nutrient loading from fertilizer and manure runoff. Stewart and Nilsen (1993) predict that eutrophication could cause a decline in evergreen species, such as bog ericads, and an increase in species such as swamp dewberry. In addition, the growth of sphagnum is decreased when NH_4–N increases (Rudolph and Voigt 1986).

Some disturbances, such as beaver activity, continue to play an integral role in maintaining fens in the landscape (Walbridge 1994). Although beaver were once abundant in this region, they were extirpated from the southern Appalachians by the late 1800s (Boynton 1994). In the absence of regular flooding, woody vegetation has become increasingly dominant in some fens. Beaver have now recolonized some regions of the southern mountains and are actively altering the landscape once more.

Another problematic disturbance in southern Appalachian fens is the unauthorized and uninformed collection of unique and endangered plants and animals (Murdock 1994). For example, an uninformed person collected the last remaining green pitcher plant from a fen in Tennessee (Dennis 1980). In North Carolina, there has been at least one unauthorized collection of a large number of seedpods of the northern pitcher plant from a privately-owned fen. The bog turtle, which brings exorbitant prices in international pet markets, has been illegally collected in North Carolina, particularly during seasons when the turtles are mobile (Herman 1994).

Although fire has been implicated as an important factor in maintaining southern mountain fens (Kiviat 1978), some authors believe that most fens in this region were not influenced by regular fires, despite the presence of some fire-adapted species (Weakley and Schafale 1994).

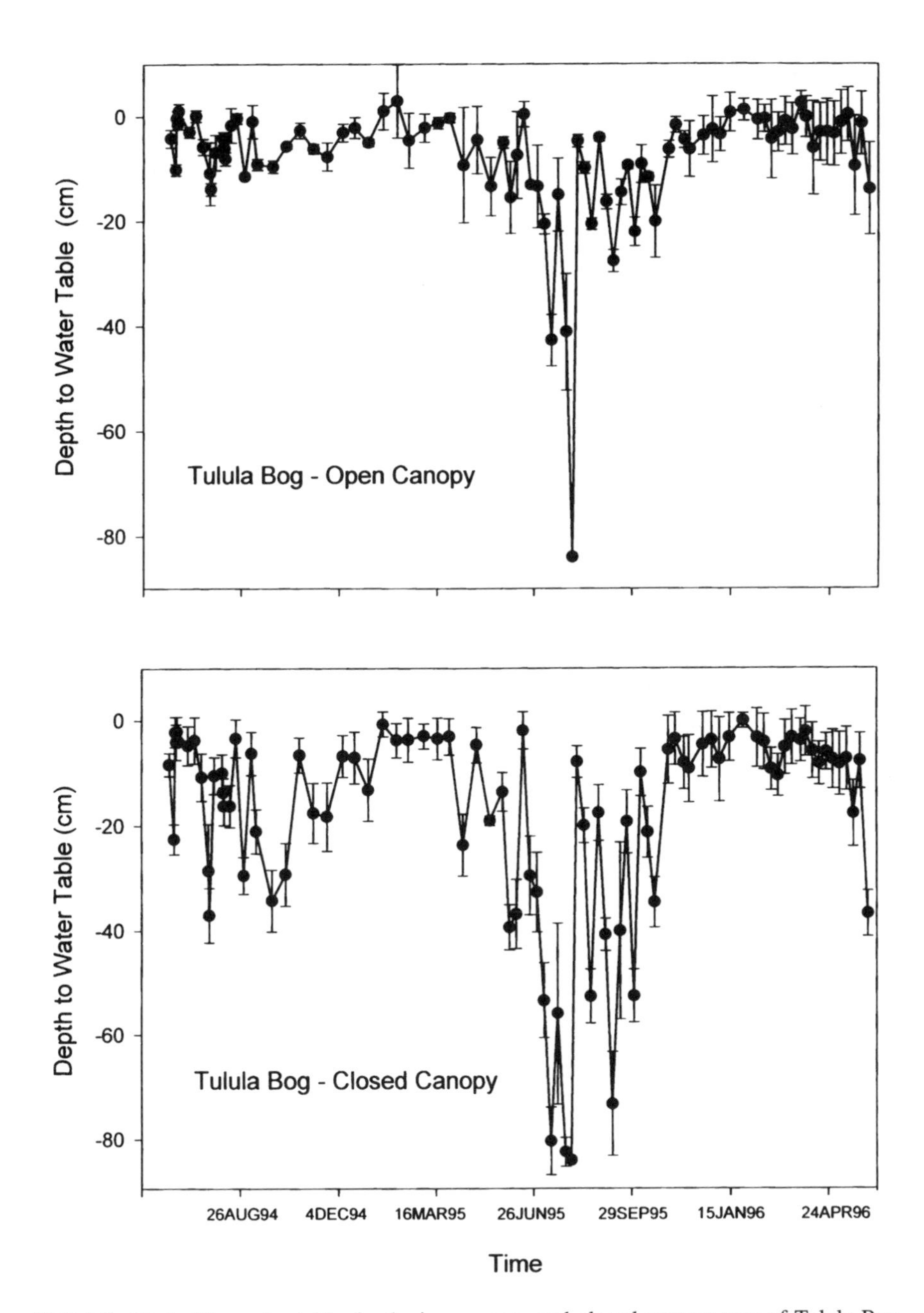

FIGURE 15.4 The water table depths in an open- and closed-canopy area of Tulula Bog from June 1994 until May 1996.

Effects from Management of Adjacent Ecosystems

The hydrology of many mountain fens has been altered by development in adjacent areas. Changes in land cover can divert precipitation as runoff and impose a cycle of inundation and drought on fens deprived of the slower, more constant release of groundwater (Pearson 1994). Excessive well-drilling and de-watering of shallow aquifers has threatened the integrity of some Arkansas fens (Orzell and Pell 1985). Channelizing adjacent streams can effectively lower the water table of a fen and accelerate shrub succession (Murdock 1994). In coal-producing regions, acid mine drainage can impact mountain fens. Tub Run Bog effectively minimizes the impact of acid mine drainage on adjacent streams by removing H^+, Ca^{+2}, Mg^{+2}, Fe^{+2} and SO_4^{-2} (Wieder and Lang 1984). Additional impacts of mine drainage on the fen have not been determined.

Fens may be highly vulnerable to acidification by acid deposition. Mowbray and Schlesinger (1988) concluded that the surface soil at Bluff Mountain Fen has a high capacity to buffer changes in soil pH, but that deeper soils are more vulnerable because buffer capacity decreases with decreasing organic matter content. Austin and Wieder (1987) determined that the growth of sphagnum moss was inhibited at a pH of 3 or less. However, they also concluded that elevated concentrations of H^+, SO_4^{-2}, NO_3^- and NH_4^+ did not have severe negative short-term effects on three species of sphagnum.

Current Management Strategies

In North Carolina, the scarcity of mountain fens and the concentration of rare species in these wetlands have led the Natural Heritage Program to focus on identifying and protecting high-quality sites (Weakley and Schafale 1994). However, complete surveys of mountain fens are lacking in the Appalachian Highlands, and there is no uniform classification system for this region. Minimal survey work and inconsistent classification have led to confusion over the number and exact locations of remaining fens.

Few southern mountain fens are actively managed. Those that receive direct management are either in relatively good condition or are managed to correct past land use practices. Common management activities include fencing to exclude grazing, filling ditches to improve water retention, and removing woody vegetation to improve conditions for herbs, byrophytes, and bog turtles. In a southwestern Virginia fen, Ogle (1982) reported that removing tag alder increased the diversity of herbaceous flora. Herman (1994) suggested that limited grazing, controlling invasive plants, selective cutting of woody vegetation, and controlled burning would prevent canopy closure and enhance bog turtle habitat.

Although fens are not prime areas for timber resources, those located in bottomland hardwood forests are vulnerable to the impacts of logging. Several states have developed voluntary Best Management Practices (BMPs) for forestry in wetlands. Since fens are sensitive areas that often contain rare plants and animals, discretionary buffer zones should be established around them prior to initiating logging on adjacent sites.

Active management of some southern mountain fens has focused on protecting or enhancing habitat for bog turtles or rare plants. The North Carolina Herpetological Society has initiated Project Bog Turtle to help conduct bog turtle surveys, purchase and protect habitat, and restore damaged habitats (Somers, pers. comm.). However, conservation programs for rare species require adequate data on life history and habitat use; these are lacking for many plants and animals (Herman 1994).

Since many mountain fens are privately owned and not actively managed or protected (Weakley and Moorhead 1991), educating landowners on the values and significance of fens is an important first step for protecting them. Options available for protecting mountain fens include conservation easements, management agreements, leases, and bargain or market-value sales (Weakley and Schafale 1994). Many fens that are located on public lands (national forests and parks, the Blue Ridge Parkway, state parks) also are not actively managed. Although public ownership of large tracts of land gives the impression that fens on these lands are at least passively protected, long-term protection of these sites is not guaranteed.

Restoration

A few restoration projects have been initiated in mountain fens. Several federal and state agencies have become involved with these projects, including the U.S. Fish and Wildlife Service (USFWS), U.S. Army Corps of Engineers, U.S. Natural Resources Conservation Service, North Carolina Wildlife Resources Commission, North Carolina Department of Transportation (NCDOT), and the Nature Conservancy (TNC).

The USFWS is working with TNC and two private landowners to restore the hydrology in four damaged fens in North Carolina (Murdock 1994). All four sites harbor rare plants or animals. One site in the Shady Valley fen complex in Tennessee is being restored for bog turtles.

The NCDOT has purchased Tulula Bog, a 90-ha (225-acre) site in western North Carolina, to develop a mitigation bank to mitigate wetland impacts associated with highway projects. The site was degraded in the mid 1980s during construction of a golf course, and will be restored to its original condition. The largest fen is still intact, and degraded fen remnants are scattered throughout the site. Restoration strategies will include recreating the original stream channel, removing spoil, filling ditches, constructing vernal pools, and revegetating portions of the site. The Tennessee DOT has expressed interest in restoring a portion of the Shady Valley fen complex as a wetland mitigation bank for transportation projects in eastern Tennessee.

Legal Aspects

Only two federal laws influence activities that may occur in mountain fens. Section 404 of the Clean Water Act regulates dredging and filling wetlands such as mountain fens. Historically, the implementation of Section 404 regulations in the Appalachian Highlands has been poorly enforced, partly due to the small size and remote locations of many fens.

The Endangered Species Act is an important tool for protecting rare, threatened, or endangered species in mountain fens. However, the isolated nature of many fens makes it difficult to document the presence of species that may be eligible for listing under this act and to enforce the act, since rare plant and animal species can command high prices on the commercial market.

RESEARCH NEEDS

Southern mountain fens provide countless opportunities for research. Very little information is available on their community structure, successional patterns, hydrology, soils, geology, nutrient cycling, water quality, wildlife, rare species, and responses to disturbance. The gradient of fen soil and peat properties between West Virginia and Georgia has not been examined. Other areas in need of scientific investigation include the contributions of surface water and groundwater to fen hydrology, cycling of N and P, habitat use and requirements of plant and animal species, and the role of underlying geology in fen formation and nutrient dynamics. The lack of information on virtually all ecological aspects of southern mountain fens has hindered the development and implementation of effective management and restoration strategies.

SUMMARY

Southern mountain fens represent a diversity of unique habitats in the Appalachian and Interior Highlands of the United States. Although most of these wetlands are referred to as bogs by resource agencies and laypeople, they should be considered fens due to the influence of groundwater from surrounding mineral soils. Our knowledge of southern mountain fens is extremely limited. Since these ecosystems span a range of six degrees latitude, and are found with elevational differences of up to 1,800 m (5,900 ft), each has distinct geological, chemical, and physical characteristics. Rare plant and animal species have become a focal point for managing fens, although many sites are not actively managed on either public or private land. The degradation of many fens has provided an immediate need for increased preservation and restoration activities. The preservation and management of southern mountain fens should consider the origin, maintenance, and future needs of these unique wetlands. Fens of recent origin, such as those resulting from renewed beaver activity, may require different management strategies than older fens containing developed peat soils. The unusual assemblage of plant and animal species, the overall rarity and diversity of fen habitat, and our considerable lack of knowledge of such wetlands suggest that resource agencies and the scientific community should devote more effort toward understanding these wetlands.

ACKNOWLEDGMENTS

Information on fens has been provided by state natural heritage programs and other similar programs, as well as by federal agencies such as the U.S. Fish and Wildlife

Service and the Natural Resources Conservation Service. Our work on North Carolina fens has been supported by the Center for Transportation and the Environment, the North Carolina Nature Conservancy, and the Undergraduate Research Program at the University of North Carolina at Asheville. We thank L. L. Moorhead, C. R. Rossell, Jr., and two anonymous reviewers for helpful comments.

16 Pondcypress Swamps

Katherine C. Ewel

CONTENTS

INTRODUCTION

Swamps dominated by pondcypress (*Taxodium distichum* var. *nutans*) are a characteristic feature of the southeastern Coastal Plain of the United States. They occur as isolated cypress domes, cypress ponds, sinkhole ponds (sometimes called limesinks), and other depressional ponds. They also occur as linear cypress strands, which drain into lakes or rivers, or as broad savannas with dwarf cypress trees. Pondcypress swamps often occur in high densities across the landscape, but until recently were widely considered, along with most wetlands, as wastelands. An appreciation of the many services that these wetlands can provide to society is finally emerging, but a complete understanding of their ecological functions and economic utility is yet to be achieved. The purpose of this chapter is to summarize our current knowledge about the ecological characteristics of pondcypress swamps, the ways in which these swamps are valued, and the effects of management practices.

1-56670-228-3/97/$0.00+$.50
© 1998 by CRC Press LLC

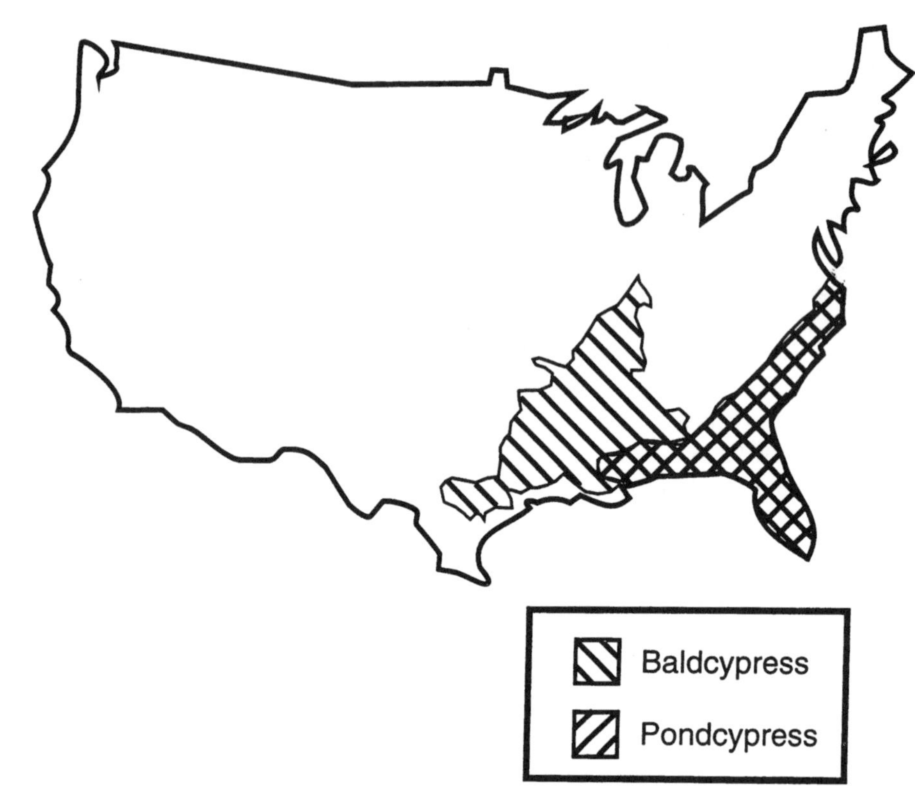

FIGURE 16.1 Distribution of baldcypress and pondcypress (after Little 1971).

THE TREE: DIFFERENCES BETWEEN PONDCYPRESS AND BALDCYPRESS

Pondcypress ranges from eastern Louisiana to southern Virginia and is considered either a subspecies of or a separate species from baldcypress (*T. distichum*), which ranges further north and west (Figure 16.1). The two trees have several characteristics in common, and the most recent taxonomic treatment considers the differences not worthy of recognition at the species level (Watson 1985). Both are deciduous, have seeds that will not germinate underwater, are resistant to fire, and will sprout from the root collar or stump after being burned or cut. Both taxa can develop outgrowths from their roots called knees, and both can develop swollen bases or buttresses. They both withstand hurricanes well, perhaps because of these structures (Putz and Sharitz 1991).

Differences between pondcypress and baldcypress are subtle (Table 16.1). Extremes in the degree of spreading of the needles and the color and shagginess of the bark can be easy to detect. Differences in branchlet orientation and in leaflet morphology are less marked under low light conditions, when pondcypress seedlings are particularly likely to resemble baldcypress seedlings (Gunderson 1977, Neufeld 1983). Baldcypress seedlings tolerate low light conditions better than pondcypress seedlings (Mattoon 1915, Neufeld 1983).

TABLE 16.1
Differences between baldcypress and pondcypress trees.

Characteristic	Baldcypress	Pondcypress
Leaf morphology	Splayed out from the twig	Appressed to the twig
Leaf shape	Elliptical	Triangular
Branchlet orientation	Drooping	Vertical
Bark morphology	Reddish; tightly bound to the stem	Gray; shaggy
Growth rate of young (<100 yrs) trees	Rapid	Slow

Gradations between the two morphological extremes are so fine that few biologists feel confident enough to identify unambiguously any individual in the field. In fact, the two are often distinguished according to the site in which they grow (Table 16.2) as well as by morphological features. Baldcypress trees are more common in floodplain forests and other bottomland hardwood swamps and on the margins of lakes. These swamps tend to have a richer nutrient supply. Plant communities in swamps with a mixture of baldcypress and hardwoods are generally more productive and diverse (Ewel 1990b). Swamps where baldcypress is common are also less likely to burn. The difference in exposure to fire may be the environmental factor that has generated the genetic differences between pondcypress and baldcypress (Harper 1927, Ewel 1995).

Characteristics and Types of Pondcypress Swamps

Pondcypress swamps comprise a spectrum from very low to moderately high productivity (Table 16.3). Occupying the center of this spectrum are cypress domes, or ponds, which are small (<10 ha [25 acres]) and occupy seasonally flooded depressions, often in high density, in pine forests or freshwater marshes (Figure 16.2). These swamps are fed mainly by rainfall, groundwater flow, and occasionally runoff. Evapotranspiration (ET) is the largest water loss (Heimburg 1984).

TABLE 16.2
Selected ecological differences between pondcypress and baldcypress swamps.

Characteristic	Baldcypress Swamp	Pondcypress Swamp
Exposure to fire	Seldom	Frequent
Gross primary productivity	High	Low-Moderate
Aboveground net primary productivity	High	Low-Moderate
Nutrient limitations	None	N, P
Hydroperiod	Long	Moderately long
Other common trees	Many	Swamp tupelo

FIGURE 16.2 Cypress dome in a managed slash pine plantation in north Florida. A drainage channel leads to a nearby ditch.

Cypress strands are linear swamps in which water flows intermittently during the year. They may form from cypress domes that gradually coalesce, especially when water levels are high. Soil profiles suggest that in many cases adjacent domes may have once been part of a strand that broke up into smaller units, perhaps because of a drying trend or groundwater diversion.

One of the largest swamps dominated by pondcypress is the Okefenokee Swamp in south Georgia. Covering approximately 1740 km^2 (672 mi^2), it is bounded on the east by an old beach ridge and is underlain by impermeable sediments that perch the swamp waters above the limestone aquifer (Rykiel 1984). It is actually a mosaic of pondcypress swamp, bay swamp, marsh, and open water, with islands interspersed. Rainfall dominates its water inflow, and its major outflows are ET and surface water flows; the Okefenokee gives rise to the Suwannee and St. Mary's Rivers (Rykiel 1984).

The least productive pondcypress swamps are savannas in which dwarf cypress is common. These swamps form over shallow, nutrient-poor soils, such as the marls of the Florida Everglades or the clay soils of the Florida panhandle. In south Florida, the Big Cypress National Preserve, so called because of the extensiveness of the swamp it protects rather than the size of the trees, consists of a matrix of dwarf cypress savanna interspersed with cypress domes and strands. When seen from the air, the tops of trees rise and fall in "undulating waves," marking where depressions have accumulated sufficient organic matter and nutrients to generate taller trees (Wade et al. 1980).

Similar to cypress domes are gum ponds, which are small swamps dominated by swamp tupelo (*Nyssa sylvatica* var. *biflora*), and are common on clay soils in north Florida and Georgia. They are more likely to be underlain by clay and appear to have longer hydroperiods than cypress ponds (Wharton 1978), but no studies have been conducted to determine why these wetlands differ from each other.

PHYSICAL ENVIRONMENT

Pondcypress swamps occupy shallow depressions in sandy soils underlain by clay and eventually dolomite and limestone. In Florida, some of these wetlands appear to have formed from slumping in the substrate, perhaps a result of long-term percolation of acid water; others appear to have formed where clay has caused water to pond (Spangler 1984). In Georgia, most are found in the Coastal Plain in depressions that originated from underground solution of limestone; a few may have resulted from wind erosion in former marine dune systems (Wharton 1978). Pondcypress also dominates many Carolina bays, which are discussed in Chapter 14.

HYDROLOGIC CYCLE

The combination of seasonally abundant precipitation, well-drained, sandy soils overlying deeper clays, and flat topography causes these swamps to fill as the water table rises throughout the entire region. Water level fluctuations can be dramatic in these wetlands. In south Florida, most of the rainfall occurs in the summer. In north Florida and south Georgia, two rainy seasons (summer and winter) are often separated by long dry periods in the fall and spring. Water level fluctuation is important to all cypress swamps because seeds will not germinate underwater (Demaree 1932), and high water seems to be necessary to keep fast-growing, more mesic vegetation from outcompeting the young cypress trees (Marois and Ewel 1983).

Pondcypress swamps are flooded for about nine months in a normal rainfall year in north Florida (Marois and Ewel 1983); large swamps may remain flooded for longer periods, but most dry out nearly every year (Wharton 1978). Rainfall is probably the major source of water for most swamps, although the magnitude of the groundwater fluxes in a pond's water budget has not yet been measured (Heimburg 1984). Groundwater flows into and out of a swamp depending on the differences in hydraulic head between the swamp itself and the water table in the uplands. Because of low relief in the landscape, rates of lateral groundwater movement are slow, ranging from <1 to 56 cm/day (0.4 to 22.0 in/day) (Riekerk 1992, Crownover et al. 1995). ET is a major outflow, representing 42% (Heimburg 1984) to 81% (Ewel and Smith 1992) of rainfall, not including interception. Direct measurements with a gas analyzer (Brown 1981), and indirect estimates derived from hydrographs (Ewel and Smith 1992), suggest that ET is low relative to baldcypress swamps, hardwood swamps, and nearby upland forests. Recent studies indicate that transpiration from pondcypress may be no different than ET from slash pine *(Pinus elliottii),* however (Sun 1995, Liu, pers. comm.).

TABLE 16.3
Productivity of pondcypress swamps and related ecosystems (after Ewel 1990b).

Swamp	Aboveground Net Primary Productivity (Mg dry wt/ha/yr)	Source
Pondcypress swamps		
Dwarf cypress savanna (south Florida)	2.7	Flohrschutz 1978, Brown 1981
Cypress domes (north and central Florida)	8.1–10.3	Deghi et al. 1980, Brown 1981
Cypress strands		
(mixed bald- and pondcypress; south Florida)	9.0–15.4	Burns 1984, Duever et al. 1984b
Other ecosystems		
Hardwood-baldcypress swamps (north Florida, Louisiana)	11.2–16.1	Conner and Day 1976, Brown 1981
Slash pine plantation (north Florida)	5.8	Ewel and Gholz 1991

Nutrient Supplies

In a typical pondcypress swamp, water quality reflects the low nutrient content of its major inflow, rainfall, and the sandy, nutrient-poor soils that surround the swamp. Surface water in one cypress dome in north Florida that was monitored for a year was low in specific conductance ($\bar{x}$ = 52 μmho/cm) and pH ($\bar{x}$ = 4.51) and high in dissolved organic carbon ($\bar{x}$ = 39.9 mg/l) due to humic materials that impart a deep reddish-brown color to the water; dissolved oxygen was consistently less than saturation ($\bar{x}$ = 2.03 mg/l) (Dierberg and Brezonik 1984b). Increased growth response to nutrient enrichment (Brown 1981, Nessel et al. 1982, Lemlich and Ewel 1984, Brown and van Peer 1989) is evidence of nutrient limitation, although the growth rates of trees in many cypress domes are moderate (Table 16.3). In many wetlands, water level fluctuation promotes denitrification, but the low pH in these swamps slows decomposition and limits nitrification (Dierberg and Brezonik 1982). Low levels of orthophosphate in the water column may be offset by low redox potential in the soil, which may make inorganic phosphorus and ammonium more available in the rhizosphere.

Primary productivity rates in cypress strands are higher than in cypress domes, presumably because of the greater supply of nutrients and oxygen brought in by flowing water. Large strands, like Corkscrew Swamp in south Florida, are dominated in the center by baldcypress, which grades into pondcypress along the edges. When water levels are high enough to flood both baldcypress and pondcypress, there is little flow throughout the strand. Instead, flow rates are fastest when water levels are rising and falling, probably affecting baldcypress more than pondcypress (Duever, pers. comm.).

Where pondcypress trees grow in particularly nutrient-poor conditions, they are stunted and gnarled, and they are often called dwarf cypress, scrub cypress, or hatrack cypress. Gross productivity in such a stand in south Florida was less than half the rate measured in a pondcypress swamp in north Florida (Brown et al. 1984).

Fire

The distribution of pondcypress swamps and other fire-maintained ecosystems in the southeastern United States corresponds closely with high incidence of lightning strikes and wildfires (Ewel 1995). Thunderstorms are common in the summer in the southeastern Coastal Plain, whereas winter precipitation is associated with frontal storms. In south Florida, where pondcypress swamps are common, the dry season is long, and lightning-caused fires are frequent during the rainy season. The frequency with which pondcypress swamps burn is probably lower than in surrounding ecosystems, which are likely to be marshes in south Florida and pine flatwoods throughout the rest of the Coastal Plain.

PLANT COMMUNITIES

Pondcypress swamps can be nearly monospecific, especially in south Florida, or, particularly further north, co-dominant with swamp tupelo. Slash pine, loblolly-bay (*Gordonia lasianthus*), swampbay (*Persea palustris*), red maple (*Acer rubrum*), and sweetbay (*Magnolia virginiana*) may also be found in these north Florida swamps, especially around the edges. Plant communities in the Okefenokee Swamp comprise a mosaic of monospecific pondcypress stands along with these same hardwoods. In addition to red maple and swampbay, hardwoods such as Carolina ash (*Fraxinus caroliniana*) and coastal plain willow (*Salix caroliniana*) are common in cypress strands in south Florida, particularly in the center.

Phytoplankton are seldom abundant in the water column of pondcypress swamps because of the darkly stained water and long dry periods. Ferns (especially Virginia chain fern [*Woodwardia virginica*] in north Florida and swamp fern [*Blechnum serrulatum*] in south Florida) and a variety of vascular macrophytes grow on hummocks and in shallow water. There is often a ground cover of *Sphagnum* spp. Pondcypress swamps often have a conspicuous shrub layer dominated by southern bayberry or wax-myrtle (*Myrica cerifera*), fetterbush (*Lyonia lucida*), and Virginia willow (*Itea virginica*).

In cypress domes and strands, trees are characteristically shorter on the edges and taller in the middle, giving these swamps a dome-like profile. They tend to grow faster in the centers of these swamps (Duever et al. 1984a, Ewel and Wickenheiser 1988), but age differences may also be important, particularly in large swamps where the center seldom burns.

Aboveground net primary productivity is moderate in pondcypress swamps in comparison with other swamps (Ewel 1990b). In Florida and Louisiana, hardwood and baldcypress swamps are more productive, whereas mature slash pine plantations are less productive (Table 16.3). Growth rates of pondcypress trees are variable

TABLE 16.4
Density, basal area, and basal area increment of selected cypress ponds in central Florida (Ewel and Davis 1992).

	Annual basal area increment		Basal area		Density	
Swamp	**(mm²)**	**(in²)**	**(m²/ha)**	**(ft²/ac)**	**(stems/ha)**	**(stems/ac)**
Clay Sink Road	916	1.42	53	230	1,149	465
North Carter Pond Road	594	0.92	52	226	865	350
Revel Road	316	0.49	28	122	1,149	465
Range (n = 8)	316–916	0.49–1.42	16–70	69–303	865–1,421	350–575

The three swamps listed represent the fastest, slowest, and closest to average in annual basal area increment of eight pondcypress swamps

(Table 16.4) but appear to be equivalent to stands of slash pine trees (Brandt and Ewel 1989) and slower than baldcypress trees (Ewel and Davis 1992). In a study involving nine pondcypress swamps in central Florida, basal area of all trees and density of pondcypress trees proved to be significant predictors of growth rates; most rapid growth rates were correlated with low to moderate stocking levels (Ewel and Wickenheiser 1988).

Fire is believed to play a major role in succession in cypress swamps (Gunderson 1984), not only because some tree species are more susceptible, but also because loss of peat can change hydrologic relationships dramatically. Periodic fire can maintain dominance of cypress in a community, and, in the absence of fire, cypress swamps are believed to develop into hardwood swamps in south Florida and bay swamps further north (Monk 1966a, Clewell 1971, Duever et al. 1984b). This is probably not a simple relationship, particularly for bay swamps, which have surface water outflows (usually missing from cypress ponds). Moreover, many bay swamps are actually dominated in some areas by cypress (Casey, 1977). Fire and logging can generate a variety of habitats that change both spatially and temporally as succession in pondcypress swamps proceeds after each disturbance (Hamilton 1984).

When pondcypress swamps burn, epicormic branches sprout from damaged pondcypress stems, and coppice sprouts from the root crown if the stem is destroyed. Destruction of the surface layers of dead litter kills only seedlings, but more intensive fires, such as during a drought, will kill pines and hardwoods before cypress (Ewel and Mitsch 1978). Mature pondcypress trees are more likely to be killed in such fires if they are in the center of a swamp, where organic matter is deeper and the fire can smolder longer, than around the edges, where the organic matter is thinner and roots are less likely to be affected (Ewel and Mitsch 1978).

The fact that baldcypress is more resistant to fire than many other tree species in the Southeast has been documented (Hare 1965), but pondcypress has not been tested. Pondcypress bark is notably thicker and shaggier than baldcypress bark and may in fact offer more protection.

ANIMAL COMMUNITIES

Detritus on the forest floor probably forms the basis of the food chain in pondcypress swamps. Because the needles of pondcypress (and most conifers) are unpalatable and slow to decompose, a thick organic layer lining the bottom of the swamps is common; fire may be important in limiting its development. The benthic invertebrate community in pondcypress swamps in north-central Florida is diverse, and species composition varies seasonally and spatially, both within a swamp (e.g., from edge to center) and from one swamp to another (Leslie 1996). Generalists that feed on detritus dominate; there are very few filter-feeders, and the presence of *Sphagnum* spp. may influence species composition.

Fish may live in pondcypress swamps when they are flooded, particularly in larger swamps that seldom dry out completely. Thirty-six species have been collected in the Okefenokee Swamp (Laerm et al. 1984). Reptiles and amphibians are the most common vertebrates in these wetlands in north Florida, particularly in the summer (Harris and Vickers 1984). The abundance of each species varies seasonally and with the degree of flooding. The southern leopard frog (*Rana utricularia*) and the ground skink (*Scincella lateralis*) dominate the amphibian and reptile communities, respectively, in north Florida (Domingue O'Neill 1995).

Upland birds are much more common than water birds in many mature pondcypress swamps in north Florida, feeding primarily on insects such as ants in the canopy and shrubs; water birds seldom use these swamps either to roost or feed, unless open water is available nearby to provide feeding habitat (Workman 1996). Bird densities in north Florida pondcypress swamps are highest in the winter when year-round residents are augmented by northern migrants (Harris and Vickers 1984). Wading bird rookeries are more common in pondcypress swamps in south Florida, particularly when the swamps are flooded (J. Ogden, cited in Duever et al. 1986). Wood storks and bald eagles, both endangered species, nest in pondcypress trees.

Large mammals such as white-tailed deer (*Odocoileus virginianus*) and black bear (*Ursus americanus*), as well as moderate-sized mammals such as raccoons (*Procyon lotor*), use pondcypress swamps, but there are probably no mammals that use them exclusively or are endemic. The few mammals that live in these wetlands (e.g., southern flying squirrel [*Glaucomys volans*], golden mouse [*Ochrotomys nuttalli*], and eastern gray squirrel [*Sciurus carolinensis*]) are arboreal and are concentrated around the edges (Harris and Vickers 1984).

IMPORTANCE OF PONDCYPRESS SWAMPS

Pondcypress swamps are valuable in several ways (Table 16.5). Some of the goods and services they provide can be appreciated with little direct intervention by people, but others require active management. Such management may in turn invoke actions that prevent other values from being realized. Decisions involving changes in abiotic influences, plants, animals, and organic matter storages all can affect the structure and function of an ecosystem, thereby changing its ability to deliver services (Ewel 1990a).

TABLE 16.5
Services performed by pondcypress swamps.

Services
Services provided without direct intervention
Animal habitat
Groundwater recharge
Flood control
Environmental education and ecotourism
Services that require active management
Fiber production
Water quality improvement

Implicit Services

With no overt management, pondcypress swamps provide several services to society, including habitat for a variety of animals, regional flood control and groundwater recharge, and opportunities for environmental education and ecotourism. None of these services is easily quantified, but it is likely that the value of each of them increases with the density of swamps in a landscape.

Pondcypress swamps are important in maintaining regional biodiversity. Although they are not centers of endemism nor of particularly high diversity, these wetlands provide open water, nesting cavities, cover, and food chain support for a variety of animals (Ewel 1990b). Managed pine flatwoods have greater diversity and density of birds where pondcypress swamps and edge habitats are available, and these wetlands are important refugia for birds after pines are harvested (Marion and O'Meara 1982). The cypress savannas and other wetlands in the Big Cypress National Preserve in south Florida are commonly used by many vertebrates because of their great spatial extent, but they are not considered critical to any (Duever et al. 1986).

Pondcypress swamps are also important in establishing regional hydrology patterns. They occupy depressions that fill when rainfall exceeds ET, delaying overland flow until the upland soils are saturated and the ponds are full. Draining these ponds therefore increases the rate at which downstream floodwaters rise during large storms or extended rainfall. Maintaining the natural hydrologic cycle of these ponds can therefore be important to managing flooding in downstream areas.

The role that pondcypress swamps play in groundwater recharge is unclear and depends on the rate of transpiration from pondcypress trees as well as regional geomorphology. If ET rates are indeed low, then pondcypress swamps can provide groundwater recharge. Low ET rates in a flooded swamp would keep the hydraulic head high in the surrounding surface aquifer, increasing the rate at which water infiltrates to the deeper aquifer, particularly through sinkholes. However, the ability of water to percolate from shallow to deep water tables would be affected by clay lenses that perch the water table as well as by the density of sinkholes or other transport pathways.

Many communities throughout the Southeast capitalize on the novelty of the distinct boundaries of pondcypress swamps and the knees and impressive butt swell of cypress trees by constructing boardwalks through these swamps and scheduling nature walks as a regular part of environmental education programs. Children and adults can take advantage of these easily visible contrasts and characteristics to learn about ecosystems, hydrology, and adaptations of vegetation to flooding (and occasionally about the role of fire in wetlands).

Active Management in Pondcypress Swamps

Fiber production and water quality improvement are two services that pondcypress swamps can provide, but sustaining these uses requires careful management and long-term monitoring. These activities may conflict with some of the intrinsic services listed above, mandating that management plans should therefore be made on a landscape basis, with attention given to the many functions and values of wetlands.

Fiber Production

Timber products derived from both baldcypress and pondcypress have been important in southern economies for at least two centuries. Baldcypress in particular achieves a very large size over several centuries and develops a rot-resistant heartwood that is especially prized and increasingly rare. Pondcypress probably does not reach the same size or develop the same amount of heartwood as baldcypress. Unfortunately, the reputation of cypress as a rot-resistant tree prevails, and lumber cut from second-growth pondcypress and baldcypress trees is frequently sold without sufficient notice to buyers that protection from weathering is needed if the wood is exposed. Until recently, pondcypress was used primarily for minor wood products such as fence posts, ladders, and crab traps, and most pondcypress swamps show signs of having been harvested some time in the past (Terwilliger and Ewel 1986).

Pondcypress is now harvested primarily for mulch, a practice that began to increase in the early 1980s after Atlantic white-cedar (*Chamaecyparis thyoides*) was no longer available. In Florida, a small industry based primarily on pondcypress expanded rapidly, and approximately 70% of the cypress now harvested in Florida is pondcypress (Irvin, pers. comm.). Florida leads the region in the amount of cypress harvested (Table 16.6). In addition to 21 full-time sawmills in Florida, there is an equal number of full-time mulching operations and another 15 to 20 operations that process other woods besides cypress (Irvin, pers. comm.). In other states, baldcypress is the predominant target. Baldcypress and pondcypress together represent approximately 30% of the timber harvested in Florida but less than 1% in the other states.

Mensuration

Estimating volume of pondcypress trees is difficult because the butt swell often requires that measurements be taken much higher up the tree than at standard breast height (1.37 m [4.5 ft]). Several of the volume tables available depend on diameter at head height (2 m [9.8 ft]) as well as diameter at breast height (dbh) (Swinford 1948). Biomass equations have also been compiled (Brown 1978).

Determining the age of both baldcypress and pondcypress trees is also difficult, because of their propensity to form false growth rings or no growth rings at all. Examination of the cell structure of a tree core under magnification is necessary to

TABLE 16.6
Pattern of cypress harvesting in southern states.

State	Annual live timber removal		Reference	Mills
	m^3	ft^3		
Florida	1,181,622	41,724,000	Brown 1994	21
Georgia	341,737	12,067,000	Thompson 1989	16
Louisiana	252,435	8,913,680	Rosson 1995	5
North Carolina	231,997	8,192,000	Johnson 1991	0
South Carolina	186,034	6,569,000	Conner 1993	0
Mississippi	44,604	1,574,995	Hartsell and London 1995	6
Alabama	8,569	302,564	Vissage and Miller 1991	2

No distinction is made between baldcypress and pondcypress. Number of mills in each state dealing primarily with cypress is also identified (Irvin and Johnson, pers. comm.).

distinguish between false and true rings. Rates of increase in basal area can be determined with some accuracy by averaging ring widths over 6- to 10-year periods (Ewel and Parendes 1984).

Silvicultural Systems and Regeneration

Pondcypress swamps are usually harvested by clearcutting, because anything less than full extraction from small wetlands is uneconomical. The increase in growth rate (1.2%) of pondcypress trees after thinning (Figure 16.3) may be statistically significant, but the increments are nevertheless small and less than twice the growth rate of trees in unthinned ponds (Ewel and Davis 1992).

A rotation age of 50 years is based on the time required to achieve the basal area measured in the only (apparently) unharvested pondcypress swamp in north Florida (Terwilliger and Ewel 1986). Another rotation age of 100 to 120 years was suggested for Avon Park Air Force Range in central Florida, based on the time it took to achieve a dbh of 30 cm (12 in) (Duever, pers. comm.). The rotation time needed for mulching operations may be much shorter than these estimates, but there has been no analysis of the economic or ecological implications of shorter rotations.

Pondcypress appears to regenerate well in clearcuts, where it depends on both seed production and sprouting (Ewel et al. 1989). Leaving some mature trees to produce seeds is often advocated because of the uncertainty of both sprouting (see below) and year-to-year variation in seed production (Mattoon 1915, Schlesinger 1978). Consistency and amount of pondcypress seed produced increase after nutrient enrichment from secondary wastewater (Brown 1978) and after harvesting (Ewel, unpublished data). Fire and other forms of physical damage may also stimulate cone production (Owens and Blake 1985). Sprouting is important to cypress regeneration, not only because harvested (or burned) stems are replaced rapidly, but also because cones are produced by the coppice within one or two years (Figure 16.4). A high-graded swamp may therefore retain its genetic diversity, provided the seeds produced by these cones are viable.

FIGURE 16.3 Cypress dome in central Florida thinned to 25 m^2/ha (109 ft^2/acre).

FIGURE 16.4 Cypress cones being produced on recent coppice.

The sprouting response is not completely reliable, however. Among several swamps in which 0.2-ha (0.5-acre) plots were thinned or larger patches or strips were clearcut, 17% of the stumps produced sprouts that survived at least two years (Ewel 1996). Sprouts were more likely to survive on stumps left in large openings and cut in the winter. They were also more successful on stumps with smaller diameters (<60 cm [24 in]) and those cut closer to the ground (<70 cm [28 in]). A more recent study suggests that stumps larger than 10 cm (4 in) in diameter and cut less than 20 cm (8 in) from the ground may not produce sprouts (Vince, unpublished data). Additional observations on the controls over sprouting should make it possible to influence (to a limited degree) the quality of restocking by cutting stumps at different heights.

Pondcypress is seldom planted for regeneration. Most of the cypress seedlings available commercially are baldcypress.

Use of Fire

The degree to which fire should be used to manage pondcypress swamps is not clear, in spite of the obvious dependence on fire for maintenance of a characteristic species complement. Ponds can be control-burned along with a surrounding pine plantation, but if organic matter ignites, it can smolder for several days or weeks, killing trees and causing respiratory distress to people in nearby communities. Swamps in drained landscapes are especially susceptible. Burning a swamp lightly, before water levels have fallen too far, is a sensible strategy (Wade et al. 1980).

Cypress coppice is susceptible to fire, and burning after logging should be avoided (Ewel 1995). One clearcut cypress swamp that had burned intensively after harvesting was revegetated by swamp tupelo (Ewel et al. 1989). The coastal plain willow is a common dominant species in burned cypress swamps in south Florida (Gunderson 1984).

Effects on Animals

Although it appears that pondcypress swamps can regenerate with minimum intervention, clearcutting can affect availability of animal habitat. Soil structure changes associated with heavy machinery may have been responsible for distinct changes in benthic invertebrate species composition that began to appear three months after harvesting in three pondcypress swamps in north Florida (Leslie 1996). Removal of trees may also have changed the benthic environment of these wetlands by reducing soil redox potential slightly but significantly (Casey, unpublished data). More amphibians were trapped in clearcut swamps compared to control swamps at the same site following post-harvest water level increases (Domingue O'Neill 1995). Wading bird densities also increased in clearcut ponds, along with species more adapted to open areas (Workman, 1996). Mammal use is also likely to change, with fewer nesting and den sites but more vertebrates as prey.

With fewer natural areas remaining in the Southeast, it may be necessary to consider the impacts of overall change on wildlife habitat before large-scale harvesting operations are conducted. Within a given landscape, for instance, harvest schedules for pondcypress swamps could be staggered to provide a variety of ponds at different stages in succession in order to accommodate different habitat needs.

Water Quality Improvement

Pondcypress swamps proved useful as a model in the development of low-energy strategies for recycling secondarily treated wastewater (Ewel and Odum 1984, Ewel

1997). With no surface outflow, this kind of wetland protects downstream water bodies from eutrophication. Instead, wastewater discharged into pondcypress swamps percolates downward through organic matter, then moves slowly laterally through sands and clays. Most of the introduced nutrients, such as phosphorus, are retained in the soil during this filtration process; concentrations of microbes are reduced as well (Ewel 1997). The clear, tannin-stained water becomes covered by a carpet of duckweed (*Lemna* spp. and *Spirodella oligorrhiza*) and water fern (*Azolla caroliniana*), which turns over rapidly (Dierberg and Ewel 1984), altering habitat for both benthic-dwelling and surface-feeding animals and attracting a different species complement of birds (Harris and Vickers 1984). Trees grow faster for many years with this nutrient enrichment, even after wastewater inflow ceases (Nessel et al. 1982, Lemlich and Ewel 1984, Brown and van Peer 1989), but they do not take up a very large proportion of the introduced nutrients (Dierberg and Brezonik 1984a). Trees often respond to nutrient enrichment with increased disease incidence (Foster 1968), however. Although there have been no such reports yet in the literature, declines in growth rates after even decades of sustained or increased growth rates have followed other kinds of anthropogenic changes in growing conditions, as described below. Most wetland-based wastewater-treatment facilities now utilize constructed wetlands, but pondcypress swamps still attract attention for treatment of other high-nutrient wastewaters such as stormwater runoff.

This use of a pondcypress swamp for water quality improvement may compromise the other values listed above. Wildlife habitat is changed significantly. Flood control benefits during high-rainfall periods may be lost because storage capacity is already being used. Groundwater recharge benefits are maintained, however, because of the greater hydraulic head. Fiber production may increase because of the faster tree growth rate (but with the caution noted above). Many communities around the country have incorporated wastewater-treatment wetlands of various kinds into environmental education programs (USEPA 1993a).

Inadvertent Effects of Management

Pondcypress swamps can be dramatically affected by management practices that may not have been intended to affect them. Increasing the hydroperiod of a swamp that receives more runoff when surrounding lands are developed may affect tree growth rates gradually, without becoming apparent for several years. For instance, decreases of growth rates of southeastern bottomland hardwoods that were impounded to increase waterfowl use in the winter (green-tree reservoirs) have only become apparent after decades (Guntenspergen et al. 1993); similar decreases in growth rates of impounded baldcypress trees have taken 30 years to be manifested (G. L. Young et al. 1995b). Such changes in growth rates may eventually lead to mortality and changes in tree species composition. Pondcypress swamps are likely to respond much more rapidly to decreases in hydroperiod, such as might be caused by drainage of a pine flatwood during conversion to a plantation, or withdrawal of groundwater for municipal use. Shortening a swamp's hydroperiod even slightly opens it to invasion from other species that are less flood-tolerant. Invading shrubs and increased tree productivity rates can increase fuel load, organic content in the soil, the potential for fire during the longer dry periods, and fire severity as well (Marois and Ewel 1983). More severe drainage can lead to subsidence caused first by shrinkage of the soil and then by compaction, oxidation, and erosion (Bacchus 1995). Even without fire, such a change can destroy a wetland.

In theory, creation or restoration of pondcypress swamps could be carried out as mitigation for loss of existing wetlands in a development project. This practice is still in its infancy (Clewell and Lea 1990). One potential pitfall was mentioned earlier in this paper; most of the cypress seedlings available in nurseries are baldcypress, and revegetating pondcypress habitat with baldcypress seedlings may create an environment that superficially resembles a pondcypress swamp but differs from it in important respects. Differences in susceptibility to fire alone may compromise the wetland's ability to support appropriate organisms and to provide goods and services to people.

CONCLUSIONS

Pondcypress swamps are very important components of the southeastern Coastal Plain landscape. They provide habitat for the diversity of plants and animals that live there, they play important roles in maintaining a characteristic hydrologic regime, and with careful management they can provide water quality improvement and economically important wood products. Their apparent resiliency is impressive, but there are still substantial gaps in our understanding of how they function. A keener appreciation of the physiological differences between pondcypress and baldcypress, of the role that pondcypress swamps play in regional biodiversity, and of the role of fire in these ecosystems will be important contributions for ensuring that pondcypress swamps continue to define the landscape of the southeastern United States.

ACKNOWLEDGMENTS

I thank Susan Vince, Katherine Kirkman, Michael Duever, Paul Scowcroft, and Sydney Bacchus for reviewing the manuscript, and Leon Irvin, Florida Division of Forestry, and Tony G. Johnson and James F. Rosson, Jr., USDA Forest Service, Southern Research Station, for providing information on cypress harvesting. Support for some of the research reported was provided by USDA Competitive Grants Program, No. 93-371019102 to K. C. Ewel and S. W. Vince at the University of Florida, Gainesville.

17 Wet Flatwoods

William. R. Harms, W. Michael Aust, and James A. Burger

CONTENTS

1-56670-228-3/97/$0.00+$.50
© 1998 by CRC Press LLC

INTRODUCTION

The forested wetlands described in this chapter are a diverse mixture of non-riverine hardwood and pine dominated site-types that occur within the flatwoods section of the Atlantic and Gulf Coastal Plains from Maryland and southeastern Virginia to southeastern Texas (Buol 1973, Allen and Campbell 1988, Christensen 1988, Stout and Marion 1993). Flatwoods are lowland areas of generally poorly drained soils formed in the lower coastal plain marine terraces of Pleistocene age. They range in elevation from sea level to about 15 m (49 ft) and include higher elevation interior areas of flatwoods in Alabama, Mississippi, Louisiana, and east Texas. Forested wetlands occupy the broad, interstream areas of the flatwoods and occur on poorly drained flats, in small to large basins or depressions, and in drainage ways. These wetlands are not connected to stream or river systems except during periods of high rainfall. Sources of water are rainfall, shallow ground water, seepage, and overland flow from adjacent uplands, thus they tend to be nutrient-limited systems (Ewel 1990b).

Vegetation community types vary tremendously across a hydrology-soils-fire continuum. On the drier, sandier, and more frequently burned end of this continuum, fire-adapted species such as longleaf pine (*Pinus palustris*) occur. This portion of the continuum has relatively few jurisdictional wetlands. On the wetter, more organic, and less frequently burned end of the continuum, water-tolerant species such as swamp tupelo (*Nyssa sylvatica* var. *biflora*) and pondcypress (*Taxodium distichum* var. *nutans*) dominate, and jurisdictional wetlands are common. Many of the hardwood tree species are the same as those that occur in alluvial swamps and bottoms (Chapters 12 and 13), but there may be differences in species associations (Hodges and Switzer 1979). The dominance of deciduous hardwood species distinguishes hardwood site-types from the pine flatwoods, pocosins, and bay forests with which they are associated (see Chapter 14). Pine wet flats are commonly dominated by facultative vegetation species that occur with equal frequency in both upland and wetland environments.

Flatwoods wetlands include some of the more controversial categories of forested wetlands from a jurisdictional and management perspective. Their mosaic of spatial patterns on the landscape often complicates determination of wetland boundaries, and because many wet pine forests are managed intensively for the production of wood and fiber (Allen and Campbell 1988), concerns about environmental and ecosystem degradation are sometimes raised. Pre-European settlement successional patterns and the geographic ranges of these site-types are difficult to identify because natural disturbances such as fire, and human interventions such as fire control, road and drainage projects, agricultural clearing, and intensive forest management have occurred repeatedly on most flatwoods sites, obliterating many of the distinguishing features.

Classifications for Wet Flatwoods Forests

Wet flatwoods site-types are referenced by a wide variety of descriptive names, including bay swamps (Sharitz and Gibbons 1982), bays (Allen and Campbell 1988), boggy swamps (Hodges 1980), coastal slash pine flatwoods (Hodges 1980), flatland hardwoods (Marks and Harcombe 1981), flatwoods (Christensen 1988, Stout and Marion 1993), shallow freshwater swamps (Penfound 1952), basin swamp forests (Allard 1990), hydric hammocks (Simons et al. 1989, Vince et al. 1989), non-alluvial swamp forest (Applequist 1959,1960, Nelson 1986), non-riverine wet hardwoods (Schafale and Weakley 1990), pine barrens (Sharitz and Gibbons 1982), pine hammocks (Kellison et al. 1982), pine flatwoods (Monk 1966b, Stout and Marion 1993), wet pine savannas (Sharitz and Gibbons 1982, Allen and Campbell 1988, Stout and Marion 1993), piney woods (Sharitz and Gibbons 1982), mixed swamps (Monk 1966a), pitcher plant flats (Allen and Campbell 1988), pocosins (Kellison et al. 1982), wet flats (Kellison et al. 1982), wet flat hardwoods (Stubbs 1962, 1966), and wet marl forests (Schafale and Weakley 1990). Examples of confusion over various local names for the wet flatwood sites abound. For example, the poorly drained mineral flats of the Virginia coastal plain are known variously as wet flats, flatwoods, and pocosins. These "pocosins" do not have the histic soil characteristics, vegetation, or water chemistry commonly associated with the pocosins of North and South Carolina. Also, in South Carolina, the term bay includes the true Carolina bays as well as the true pocosins and the wet, mineral flatwood sites. In the Gulf Coast states, the term flatwood includes an extremely wide range of site types, ranging from the xeric upland sites dominated by upland species to the wet, mineral flats that may have jurisdictional wetland status. Several classification systems have attempted to solve the dilemma caused by regional names.

Kellison et al. (1982) provided one of the more descriptive definitions for wet flats, describing them as areas "topographically similar to peat swamps and pocosins, they all lie in broad interstream divides where drainage systems are poorly developed. However, wet flats are better drained than their associates because of higher elevation. The non-alluvial soils may possess some accumulation of organic matter, but fertility is superior to peat swamps and pocosins because of superior parent material. Abandoned rice fields of the southern lower coastal plain often fall in this category. Examples are: Jordan and J & W pocosin in North Carolina. Species generally encountered: (1) on wetter portion — sweetgum, red maple, laurel and willow oaks, ashes, loblolly pine, elms and other species; (2) on islands with better drainage — cherrybark oak, Shumard and swamp chestnut oaks, yellow-poplar, hickories, and occasionally beech."

Küchler (1964) classified wet flats within the Broadleaf and Needleleaf Forests group Type 101, the oak-hickory-pine forest (*Quercus-Carya-Pinus*), and Type 102, the southern mixed forest (beech-sweetgum-magnolia-pine-oak [*Fagus-Liquidambar-Magnolia-Pinus-Quercus*]).

The National Wetland Inventory (NWI) classification system separates hardwood flats and pines at the subclass level (Cowardin et al. 1979), primarily based upon the upper canopy vegetation. The NWI hierarchical classification of wet flats is:

System: Palustrine wetlands
Class: Forested wetlands
Subclasses: broad-leaved deciduous (hardwood flats), needle-leaved evergreen (wet pine flats), and broad-leaved deciduous (transitional areas)
Water regime: saturated, semi-permanently, intermittently, or seasonally flooded
Water chemistry: fresh, acid
Soils: mineral or organic

The Hydrogeomorphic (HGM) model for wetland characterization (Brinson 1993b) identifies flatwood wetlands as occurring on mineral flats or depressions, having precipitation as the primary source of water, and having vertical and unidirectional horizontal water movement. Pocosins and Carolina bays would have a very similar rating, except organic soils would be a more common feature.

The Society of American Foresters (Eyre 1980) identifies the coastal plain region as the Southern Forest Region. Within this region, several specific forest cover types are identified that commonly exist on wet flatwoods sites. Wet pine-flat forest cover types (and numbers) are: longleaf pine (70), longleaf-slash pine (83), loblolly-shortleaf pine (80), loblolly pine (81), slash pine (84), south Florida slash pine (111), loblolly pine-hardwood (82), slash pine-hardwood (85), and pond pine (98). Hardwood wet flats and depression sites correspond in whole or in part to southern redcedar (73), willow oak-water oak-diamondleaf (laurel) oak (88), sweetgum-American elm-green ash (92), overcup oak-water hickory (96), water tupelo-swamp tupelo (103), and sweetbay-swamp tupelo-redbay (104) (Eyre 1980).

For purposes of discussion in this chapter, we have distinguished between flatwoods wetlands that are dominated by pine species and those dominated by hardwood species. Pine site-types occur on wet flats and include longleaf pine savannas, loblolly pine, slash pine, and pond pine flats. The dominant species is determined by flooding regime, fire frequency, and/or management goals. We classify three hardwood site-types: hardwood flats, non-alluvial swamps, and hydric hammocks. These site-types are differentiated by topographic setting, vegetation composition, edaphic conditions, and hydrologic characteristics.

Geographical Extent

Pine Wet Flats

The past and current range of pine wet flats is difficult to determine because of their occurrence between and within other types of wetlands and because of patterns of human-induced disturbances. Sargent (1884) estimated that the pine flatwoods ecosystem accounted for over one-half of the forestland acreage in the coastal plain prior to European settlement. Cubbage and Flather (1993) determined forested wetland areas in 14 southern states and found a total of 2.6 million ha (6.4 million acres) dominated by loblolly pine-hardwood, loblolly pine, slash pine-hardwood, slash pine, pond pine, other oak-pine, and shortleaf pine-oak forests in 1982. Brown (1996) estimated that pine flats accounted for 41 percent of the total forested wetland acreage in South Carolina in 1996. If these statistics hold true for the southern states of the Atlantic and Gulf Coastal Plains, then pine flats include approximately 1 million ha (2.47 million acres), although this area is considerably larger if non-jurisdictional, but similar, areas are included.

One of the difficulties in determining the extent of pine flats is that they may grade into mixed pine-hardwood flats as the soil saturation regime increases, or they may be intermingled with more organic pocosins and Carolina bays. Stout and Marion (1993) correlated the dominant pine species of wet flats to the hydrologic regime and the periodicity of fire. In the Atlantic Coastal Plain, the more xeric and frequently burned flatwoods are dominated by the longleaf pine community, while more mesic sites are dominated by slash pine (*Pinus elliottii*) or loblolly pine (*Pinus taeda*). As the hydroperiod increases, pond pine (*Pinus serotina*) becomes more common. Noteworthy examples of wet flats include many of those located on divides between streams that flow into the Great Dismal Swamp in southeastern Virginia and northeastern North Carolina, the wet flats surrounding the Okefenokee Swamp in Georgia, the Jordan Pocosin in North Carolina, the Hellhole Swamp in South Carolina, and wetter portions of the Interior Flatwoods of Mississippi and Alabama. In the Gulf Coastal Plain flatwoods, longleaf pine and/or slash pine dominate (Christensen 1988), and in East Texas shortleaf pine (*Pinus echinata*) enters the community (Marks and Harcombe 1981).

Hardwood Types

The extent and geographical importance of hardwood-dominated wetlands occurring in the flatwoods are poorly documented. Because of their often small individual size, wide dispersion on the landscape, and a relative lack of interest in them as an economic resource, they have attracted little scientific interest. As a group, they probably occupy relatively less area than at the time of European settlement because of drainage and conversion to pine plantations, exploitative timber harvest, agriculture, and other disturbances. However, there may have been gains in some wet-flat hardwoods because the virtual elimination of wildfire in the South has allowed the expansion of the fire-sensitive species into mesic pine wet flat sites. Hydric hammocks, the most restricted of these site-types, are estimated to have occupied about 200,000 ha (494,000 acres) in Florida at the time of European settlement. At present they cover 80,000 to 100,000 ha (197,600 to 247,000 acres), the decrease in area due to type conversion to pine plantations, real estate development, and clearing for agriculture (Simons et al. 1989, Vince et al. 1989).

Economic Importance

Pine-dominated wet flats have been and still are some of the most economically important types of forested wetlands in the flatwoods. During the late 1800s and early 1900s, a majority of the wet flats were harvested for timber. The naval stores industry utilized many of the longleaf pine and slash pine stands for production of turpentine and rosin, and a free range policy for grazing existed in many areas well into the 1960s. Since the 1950s, the pine flats have become areas of intensive forest management for the production of loblolly and slash pine wood and fiber. In many rural counties of the lower coastal plain, it is common for forest industry-related jobs that rely on wood and fiber production from pine flats to support more than 75 percent of the local economies.

PHYSICAL ENVIRONMENT

CLIMATE

The climate of the flatwoods is varied, but most of the region is characterized by relatively long growing seasons (six or more months) and mean annual air temperatures of 10° to 24°C (50 to 75°F). Annual precipitation ranges from 1625 mm (64 in) near New Orleans, Louisiana to 1200 mm (47 in) in southeastern Virginia. Localized thunderstorms often cause wet periods within the "normal" dry periods, and moisture deficits are also common in the growing season, typically in the summer months when evapotranspiration rates are greatest (Daniels et al. 1973).

SOILS AND SITES

Soils of the flatwoods were formed from marine and alluvial sediments deposited in the Tertiary and Cretaceous periods. In general, the wet flats of the Atlantic Coastal Plain are located on seaward-facing marine terraces, whereas in the Gulf Coastal Plain they are located on "belted" coastal plains that often face inland (Daniels et al. 1973). Soils of the area are diverse, and common soil orders include Ultisols, Alfisols, Entisols, Inceptisols, and Spodosols (Buol 1973, Allen and Campbell 1988). Organic soils (Histosols) are more commonly restricted to the pocosin wetlands.

Buol (1973) provided broad mapping units for a 13-state region in the southeastern United States. The sites in the lower coastal plain of Virginia, North Carolina, South Carolina, and Georgia that support pine flats are predominantly Ochraquults, Umbraquults, and Haplaquods. North Carolina also has large areas of Medisaprists, but much of this area would be classified as true pocosin. The lower coastal plain of South Carolina and Georgia includes Ochraqualfs and Albaqualfs. Many of the wet flatwoods soils in Florida are Haplaquods, Ochraqualfs, Paleaquults, and Ochraquults. Paleaquults are common throughout the lower coastal plain of Mississippi and Alabama. In Louisiana and east Texas, Glossaqualfs, Albaqualfs, and Ochraqualfs are typical of wet flatwoods.

Wet Flat Pine Sites

Allen and Campbell (1988) interpreted results from Fisher and Garbett (1980) and Fisher (1981) and synthesized the major soils and diagnostic features of pine flats (Table 17.1). The argillic horizons have profound effects on the hydrology of wet flats. Typically, coarser textured soils of the surface horizons have hydraulic conductivity values that are several orders of magnitude greater than that of the finer textured argillic horizons. Lateral water movement is predominantly through the coarser, more porous A and E horizons. Therefore, water movement in the lower argillic horizons tends to be slow, with vertical movement controlled by precipitation,

TABLE 17.1
Common soils of pine flats. Adapted from Allen and Campbell (1988), as interpreted from Fisher and Garbett (1980) and Fisher (1981).

General Description	Diagnostic criteria	Representative subgroups
Fine textured savanna	No spodic horizon argillic horizon within 50 cm	Typic aquult Albic aquult Plinthic aquult Umbric aquult
Coarse textured savanna	No spodic horizon argillic horizon below 50 cm	Aerenic aquult Aerenic aquept Aerenic aquent
Ultic Flatwood	Spodic and argillic horizons present	Ultic aquods Ultic humods
Typic Flatwood	Spodic, but no argillic, horizon present	Typic aquod Aeric aquod Aerenic aquod

evaporation, and transpiration. Several studies have evaluated the impacts of harvesting-related traffic on such sites and concluded that compaction and rutting by heavy equipment on wet flats can restrict lateral water movement (Wimme 1987, Aust et al. 1993, 1995, Scheerer et al. 1995).

Both pine and hardwood wet flats may experience saturated soil conditions at any season of the year, but drier conditions usually occur during the growing season (Figure 17.1) when evapotranspiration rates are higher. The depth of the water table above or below the soil surface is affected by numerous factors, including microtopography, the presence or absence of roadside ditches, minor drainage, and age and type of the vegetation.

Wet Flat Hardwood Sites

Hardwood wet flats lie at slightly higher elevations than adjacent natural drainage channels and are intermingled with bays, ponds, swamps, and pocosins (Stubbs 1966, Schafale and Weakley 1990). Excessive soil moisture is common, and these flats are seasonally saturated or briefly flooded during periods of high rainfall because of high water tables, poor drainage, and runoff from adjacent uplands. Surface and internal drainage of hardwood flats is slow, and soil profiles usually have thin surface horizons over gleyed subsoil clays. The soils are extremely diverse, principally loamy or clayey, and typically Ochraquults, Umbraqualfs, and Paleudults (Ultisols) (Buol 1973, Schafale and Weakley 1990).

Non-Alluvial Hardwood Swamp Sites

Non-alluvial hardwood swamps occur as inclusions in the flatwoods. They are found in interstream areas as low-lying headwater collection basins of sluggish wet-flat

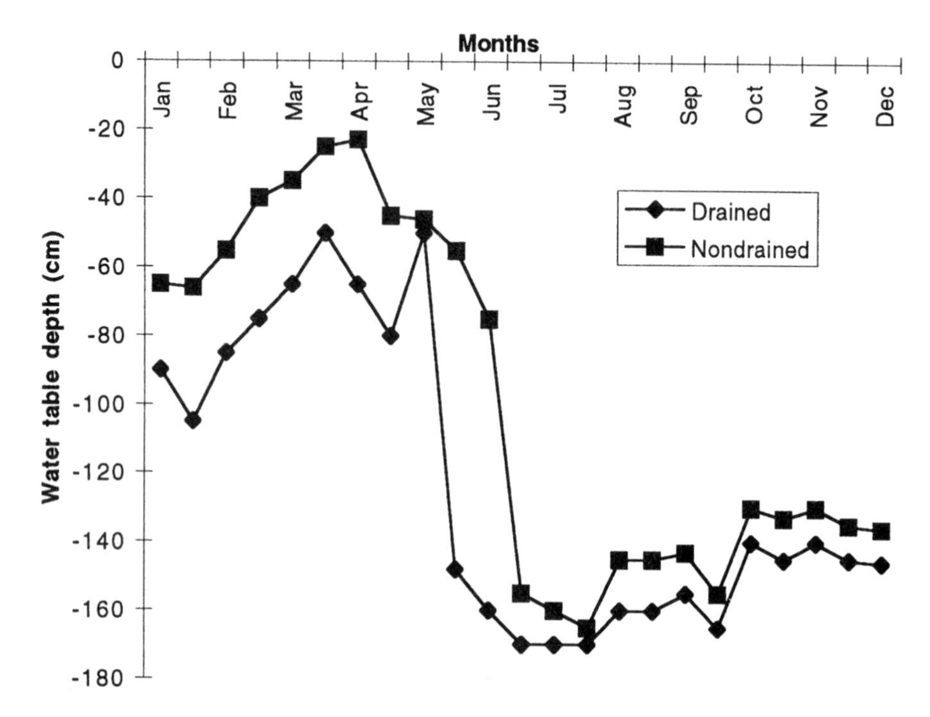

FIGURE 17.1 Average water table fluctuations for 1991–1992 for drained and non-drained 23-year-old loblolly pine plantations located on wet flats in the Virginia coastal plain (modified from Andrews 1993).

stream systems. They receive water from rainfall and runoff from surrounding uplands and discharge slowly into creeks and rivers. The type also occurs in depressions and along creeks and in drainage ways. These swamps are the lowest, wettest areas of the flatwoods, and although water movement occurs, it is so slow that surface flooding prevails except during drought periods. Soils vary from thin organic loams at swamp margins, to mucks, to deep woody peats where drainage is poorest, and are generally underlain by clay (Applequist 1959, Stubbs 1966, Schafale and Weakley 1990). Typical soils include Typic or Terric Medisaprists (Histosols), Umbraquults, and Umbric Paleaquults (Ultisols).

Hydric Hammock Sites

The term hammock was used by early European settlers to distinguish and name the islands of dense hardwood forests that occurred in the extensive pine wet flats in Florida (Simons et al. 1989). Hammocks that are situated on poorly drained soils, and soils with high water tables that flood often enough to support vegetation adapted to anaerobic soil conditions, are called hydric hammocks. Hydric hammocks are largely restricted to the Florida peninsula north of Lake Okeechobee, principally at low elevations along the Gulf Coast from Aripeka to St. Marks and at various inland sites (Vince et al. 1989). Although Wharton (1978) indicated that similar vegetation occurs in coastal Georgia near Savannah, vegetation surveys for other states suggest that hammocks outside of Florida are rare and primarily mesic and xeric types (Vince et al. 1989). Soils of hydric hammocks formed mainly in sandy and loamy marine sediments over limestone and other alkaline materials such as marl, and are characteristically slightly acid to mildly alkaline. Organic matter content of surface soils is generally low, often less that five percent (Vince et al. 1989). Typical soils have sandy surface layers and sandy loam subsoils, are nearly level, and are somewhat poorly to poorly drained. Representative soils include Ochraqualfs (Alfisols), and Ochraquults and Umbraquults (Ultisols) (Buol 1973). The sources of water for hydric hammocks are rainfall supplemented by seepage from adjacent uplands, overland flow, and, depending on proximity to rivers, streams, and springs, periodic inundation. A high water table, seasonally or year-round, is characteristic, and most are subject to occasional flooding (Vince et al. 1989). The essential site feature that sets hydric hammocks apart from other hardwood wetlands is the presence of a limestone substrate.

PLANT COMMUNITIES

Flatwoods wetlands support a characteristically rich and diverse mixture of woody plant species. The most common species that occur on one or more of the site-types include eight conifers, three palms, 31 hardwoods, and more than 20 shrubs and vines (Table 17.2). Only two of the species, loblolly pine and sweetgum(*Liquidambar styraciflua*), occur on all of the site-types. Other species that are common on the majority of the site-types include slash pine, red maple (*Acer rubrum*), and sweetbay (*Magnolia virginiana*).

Pine-Dominated Communities

Pine flats historically were dominated by pine species due to the frequent occurrence of fire, but the species of dominant pine varied by hydrologic regime and latitude. In Florida and Georgia, slash pine dominated the wetter sites, while pond pine was more common on the wetter sites of South Carolina, North Carolina, and Virginia (Harms 1996). In the Gulf Coastal Plain, loblolly pine was the most common pine species. On the drier end of the spectrum, longleaf pine stands were more common. However, due to past agricultural manipulations, exclusion of fire, hydrologic modifications associated with roads, and deliberate silvicultural manipulations, loblolly pine dominates the pine flats today (Table 17.2).

Hardwood-Dominated Communities

Wet Flat Hardwoods

These forests are distinguished by dominance of various water-tolerant and nutrient-demanding bottomland oaks or mixed hardwoods. Canopy codominants include swamp laurel oak (*Quercus laurifolia*), water oak (*Q. nigra*), willow oak (*Q. phellos*), overcup oak (*Q. lyrata*), with associates such as sweetgum, red maple, swamp tupelo, green ash (*Fraxinus pennsylvanica*), American elm (*Ulmus americana*), and other

TABLE 17.2
Characteristic woody plants on wet-flat site types.

Common Name	Pine Types				Hardwood Types		
	Longleaf Pine Savanna	Loblolly Pine Flat	Pond Pine Flat	Slash Pine Flat	Hardwood Flat	Hardwood Swamp	Hydric Hammock
Conifers							
Baldcypress	•		•			•	
Loblolly pine	•	•	•	•	•	•	•
Longleaf pine	•			•			
Pondcypress	•		•	•		•	
Pond pine	•	•	•	•		•	
Slash pine	•	•	•	•	•	•	
Southern redcedar				•			•
Spruce pine		•					
Palms							
Cabbage palm				•		•	•
Dwarf palmetto					•		•
Saw-palmetto	•			•			
Hardwoods							
American beech					•		
American elm					•		
American holly					•		
Blackgum	•	•					
Carolina ash						•	
Cherrybark oak					•		
Drummond red maple						•	
Florida elm						•	•
Green ash					•	•	

Laurel oak	•			•	•		•
American hornbeam					•	•	•
Live oak							•
Loblolly-bay				•			•
Magnolia		•					
Overcup oak					•	•	
Redbay		•				•	
Red maple	•	•		•	•	•	•
Sassafras		•					
Serviceberry		•					
Sugarberry					•		•
Swamp chestnut oak					•	•	
Swamp dogwood						•	
Swamp tupelo		•		•	•	•	
Sweetbay	•	•	•	•		•	•
Sweetgum	•	•	•	•	•	•	•
Water-elm						•	
Water oak	•	•		•	•		•
Water hickory						•	
Water tupelo	•					•	
Willow oak		•			•		
Yellow-poplar					•		
Shrubs							
Alabama supplejack					•		
Blueberry	•	•	•				
Buckwheat tree				•			
Buttonbush						•	
Deciduous holly	•	•		•			
Dwarf huckleberry				•			
Dahoon				•		•	

TABLE 17.2 (continued)
Characteristic woody plants on wet-flat site types.

Common Name	Pine Types				Hardwood Types		
	Longleaf Pine Savanna	Loblolly Pine Flat	Pond Pine Flat	Slash Pine Flat	Hardwood Flat	Hardwood Swamp	Hydric Hammock
Fetterbush		•	•	•		•	
Gallberry	•	•	•	•			
Rhododendron		•					
Swamp cyrilla	•		•	•		•	
Sweet pepperbush	•	•				•	
Virginia willow		•				•	
Waxmyrtle	•	•	•	•		•	
Yaupon			•	•			

Adapted from Applequist (1959), Monk (1966a), Stubbs (1966), Grelen (1980), Hodges (1980), J. W. Johnson (1980), R. L. Johnson (1980b), Shropshire (1980), Switzer (1980), Marks and Harcombe (1981), Sharitz and Gibbons (1982), Wharton et al. (1982), Nelson (1986), Vince et al. (1989), Schafale and Weakley (1990), and Hauser et al. (1993).

species typical of bottomlands. On better drained areas, swamp chestnut oak (*Q. michauxii*), cherrybark oak (*Q. falcata* var. *pagodaefolia*), hickories (*Carya* spp.), beech (*Fagus americana*), and yellow-poplar (*Liriodendron tulipifera*) may enter the stand. Loblolly pine is a common associate in the Atlantic coast wet flats, and it is joined by slash pine in Georgia and Florida, and the Gulf wet flats. Pine often predominates to form pine-hardwood stands. The understory includes species such as American hornbeam (*Carpinus caroliniana*), red maple, and American holly (*Ilex opaca*). The shrub layer is generally sparse to moderate and may include dwarf palmetto (*Sabal minor*), greenbriers (*Smilax* spp.), poison-ivy (*Toxicodendron radicans*), Alabama supplejack (*Berchemia scandens*), and peppervine (*Ampelopsis arborea*), among others (Table 17.2) (Stubbs 1966, Shropshire 1980, R. L. Johnson 1980b, Marks and Harcombe 1981, Schafale and Weakley 1990).

Non-Alluvial Swamps

This site type usually is dominated by even-age swamp tupelo, which may occur in pure stands or mixed with baldcypress (*Taxodium distichum*). The overstory vegetation in the Atlantic Coastal Plain reflects the gradation in moisture regimes in a gradual transition of species from the moist pine-hardwood margins, where loblolly pine and pond pine, in association with sweetgum, swamp chestnut oak, willow oak, and swamp laurel oak, grade to red maple and swamp tupelo in areas of nominal flooding, and to water tupelo (*Nyssa aquatica*), baldcypress, or pondcypress in the deeply flooded reaches of the swamp. Understory development also is controlled by the degree of flooding, ranging from very dense near swamp margins to virtual absence in the interior regions, except on hummocks that form around swamp tupelo trunks. Shrubs include fetterbush (*Lyonia lucida*), redbay (*Persea borbonia*), swamp cyrilla (*Cyrilla racemiflora*), and Virginia willow (*Itea virginica*) (Wharton et al. 1982, Nelson 1986, Schafale and Weakley 1990). For mixed swamps in Florida, Monk (1966a) adds Carolina ash (*Fraxinus caroliniana*), cabbage palm (*Sabal palmetto*), Florida elm (*Ulmus americana* var. *floridiana*), water hickory (*Carya aquatica*), and sweetbay. Important understory species are buttonbush (*Cephalanthus occidentalis*), dahoon (*Ilex cassine*), southern bayberry (*Myrica cerifera*), and American hornbeam. Applequist (1959), in a study of swamp and water tupelo growth in the coastal plain swamps of Georgia, recorded the following additional species: slash pine, green ash, snowbell (*Styrax americana*), overcup oak, sweet pepperbush (*Clethra alnifolia*), water-elm (*Planera aquatica*), myrtle dahoon (*Ilex myrtifolia*), hazel alder (*Alnus serrulata*), and swamp dogwood (*Cornus stricta*) (Table 17.2).

Hydric Hammocks

The dominant vegetation of hydric hammocks is a dense cover of forest approximately 20 m (65 ft) tall and commonly includes cabbage palm, live oak (*Quercus virginiana*), laurel oak, sweetgum, red maple, southern redcedar (*Juniperus silicicola*), and loblolly pine. Florida elm is also common. Vines such as trumpet creeper (*Campsis radicans*), pepper vine, poison ivy, and wild grape (*Vitis* spp.) are also common. Understory and shrub layers are variable and may be sparse or absent.

Where an understory is present, cabbage palm and sweetgum are frequent, and American hornbeam dominates on some sites (Vince et al. 1989). Dwarf palmetto and greenbriar are often common in the shrub layer. The ground layer (when present) may be composed of various ferns, sedges, and grasses (Vince et al. 1989) (Table 17.2).

Successional Patterns

Pine Cover Types

Pine flats have a long history of both natural and human-induced disturbances. Natural disturbances include both fire and hurricanes. For example, the importance of hurricanes as a natural disturbance was reemphasized in 1989 by the destruction of older stands of pine and pine-hardwood by Hurricane Hugo on flatwoods sites in the Francis Marion National Forest in South Carolina (Hook et al. 1991a). Such hurricanes may not totally remove the overstory, but the fuel loads from the debris are increased to levels that allow catastrophic fires in areas where pre-settlement fire was common. When disturbances such as fire and hurricanes are excluded from pine flats, the hardwood component generally increases, which results in a mixed pine and hardwood stand. Examples of this conversion toward more shade-tolerant species abound in areas where fire control has been imposed (Smith 1986). Such fire exclusions have also converted previous longleaf stands to loblolly and slash pine stands (McCulley 1950). Forest managers have capitalized on the natural fire ecology of pine flats by using prescribed fire as a silvicultural tool to retard the growth of the hardwood component and protect the areas from more severe fires. Controlled burns are also used for the management of certain game species like white-tailed deer (*Odocoileus virginianus*), wild turkey (*Meleagris gallopavo*), and northern bobwhite (*Colinus virginianus*). Overall, the relationships among hydroperiod, soils, and fire modify the successional pattern of wet flats (Figure 17.2). Areas that have longer hydroperiods tend to accumulate more organic matter and have less frequent fires. Wetter flats that seldom burn become hardwood stands or mixed pine-hardwood stands. On the other end of the spectrum, where hydroperiods are short and fire occurrence is frequent, longleaf pine stands develop.

Wet Flat Hardwoods

Successional dynamics of these communities are poorly understood, but they are probably stable associations, provided that the hydrology is not altered (Schafale and Weakley 1990). Because of wet soils and a lack of a substantial litter layer, fire is infrequent in these communities. Fires are possible during times of drought, and may result in conversion to pine if the canopy is opened up, as a pine seed source is usually present. On sites where water oaks are predominant, heavy cutting or other disturbance that opens up the stand may allow species such as sugarberry (*Celtis laevigata*), green ash, American elm, and red maple to capture the site (Shropshire 1980). In some situations, the elimination of frequent fire from neighboring pine

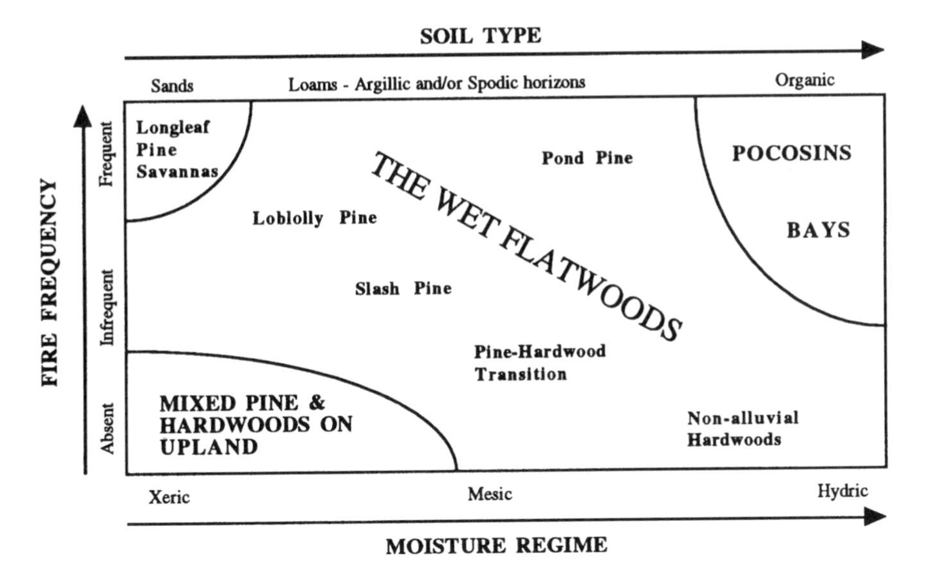

FIGURE 17.2 Proposed relationships among moisture regimes, soil type, and fire in flatwood wetlands environments. Adapted from Allen and Campbell (1988), Sharitz and Gibbons (1982), and Stout and Marion (1993).

flatwoods and savannas favors the encroachment of fire-sensitive hardwoods from wet flats onto these pine sites. This may result in a shift in dominance from pine to pine-hardwood or pure hardwoods (Christensen 1981).

Non-Alluvial Swamps

The dynamics of hardwood swamps are not well known, but the natural processes that control growth and development of the vegetation are primarily hydrological and determine the successional patterns following disturbance. These communities are normally flooded for most of the year and, therefore, tend to be stable. Because of wet or flooded soil, fire is not normally a factor. When destructive fires do occur, or when a large-scale disturbance such as a hurricane occurs, the recovery of the vegetation is largely a function of the severity of the disturbance and the state of the hydrology. Water-tolerant species such as cypress and tupelo will regenerate if a seed source is present and a suitably exposed mineral soil is available for a sufficiently long period during the growing season to allow germination and height growth. Seed of these species will not germinate in flooded soil nor will submerged seedlings survive (DeBell and Hook 1969). Species, such as the tupelos that sprout readily from stumps, will provide additional regeneration (DeBell 1971), as will existing advanced reproduction of the canopy species. Given a stable hydrology and a source of reproduction, these communities will recover and return to a pre-disturbance state (DeBell and Hook 1969, Burns and Honkala 1990, Hook et al. 1993).

Hydric Hammocks

Four types of hydric hammocks have been distinguished based on relative abundance of species, hydrologic regime, and physiographic setting (Simons et al. 1989, Vince et al. 1989). The most common is the coastal type, composed principally of live oak, cabbage palm, and southern red-cedar. It occurs extensively along the Atlantic and Gulf coasts on sandy loam or sandy clay soils with limestone or shell near the surface. It is usually bordered on the downhill side by marsh or prairie and on the uphill side by pine wet flats or mesic hammock. The second type is the inland hydric hammock, which may be dominated by pure live oak or mixed with water oak, swamp laurel oak, sweetgum, loblolly pine, cabbage palm, Florida elm, sugarberry, and red maple. American hornbeam is common in the understory. The type occurs on sandy loam or sandy clay soils with limestone near the surface and is bordered by swamp rather that prairie or marsh. The third type is dominated by loblolly pine in association with cabbage palm, live oak, water oak, and sweetgum. Stands of loblolly pine may be extensive. This type occurs on sandy clay soils and grades into pine wet flats or the interior hydric hammock type. The fourth type occurs along spring runs and is dominated by cabbage palm, red maple, sweetbay, and swamp laurel oak. Loblolly-bay (*Gordonia lasianthus*) may be present, while live oak and southern redcedar are absent; needle palm (*Raphidophyllum hystrix*) is often in the shrub layer. This type receives a constant supply of high-calcium water, and the soil is often a deep, organic muck (Simons et al. 1989). Succession in hydric hammocks has not been studied, but it is thought that fire, wind, and grazing probably all play a role in development following disturbance (Vince et al. 1989).

ANIMAL COMMUNITIES

With the exception of hydric hammocks, little has been published about the animal communities of these wetland site types, but they probably support a diverse fauna similar to that reported for other coastal plain hardwood forests. Wharton et al. (1982), Echternacht and Harris (1993), and Sharitz and Mitsch (1993) provide extensive treatments of animal communities in the bottomland hardwood swamps of the Southeast, and Hamel et al. (1982) have summarized data on bird-habitat relationships of the major Southeastern forest types. These data, together with the studies of Vince et al. (1989) and Simons et al. (1989) in hydric hammocks, delineate the range of species likely to be characteristic of non-alluvial forest wetlands. Rather than list all animal species reported by these authors, an attempt will be made to note only the most common or characteristic species. The original sources should be consulted for additional data and a complete listing. There is insufficient site-specific information on fish and invertebrate species to permit discussion.

Reptiles and Amphibians

The herpetofauna of hydric hammocks includes at least 64 species (Simons et al. 1989). Common reptiles are the southern black racer (*Coluber constrictor priapus*), eastern indigo snake (*Drymarchon corais couperi*), yellow rat snake (*Elaphe obsoleta quadrivittata*), eastern coral snake (*Micrurus fulvius fulvius*), eastern diamondback rattlesnake (*Crotalus adamanteus*), Florida box turtle (*Terrapene carolina bauri*), green anole (*Anolis carolinensis*), ground skink (*Scincella alterale*), and broadhead skink (*Eumeces laticeps*). Common amphibians include the southern toad (*Bufo terrestris*), green treefrog (*Hyla cinerea*), gray treefrog (*Hyla chrysoscelis*), squirrel treefrog (*Hyla squirella*), spring peeper (*Pseudacris crucifer*), and eastern narrow-mouth frog (*Gastrophyrne carolinensis*). Some of the species listed by Sharitz and Mitsch (1993), such as the dusky salamander (*Desmognathus* spp.), the many-lined salamander (*Stereochilus marginatus*), the dwarf salamander (*Eurycea quadridigitata*), and the red-backed salamander (*Plethodon cinereus*), may also be common in these site types. Species inhabiting flooded sites include the green frog (*Rana clamitans melanota*) and leopard frogs (*Rana* spp.), mud turtles (*Kinosternon subrubrum* and *K. baurii*), glossy crayfish snake (*Regina rigida*), mud snake (*Farancia abacura*), plainbelly water snake (*Nerodia erythrogaster*), and eastern cottonmouth moccasin (*Agkistrodon piscivorus piscivorus*).

Birds

Hamel et al. (1982) provide bird census data for the oak-gum-cypress forest type, which they define as bottomland forests and swamps in which water tupelo, swamp tupelo, sweetgum, oaks, or cypress, singly or in combination, make up a plurality of the stand. They consider that the most important environmental variable that affects bird communities within this type is the extent and duration of flooding. Birds that commonly utilize understory vegetation and the ground are not found on the wetter sites, whereas prothonotary warblers (*Protonotaria citrea*) are confined to the wetter sites (Hamel et al. 1982). Species listed as characteristic of this type include red-shouldered hawk (*Buteo lineatus*), yellow-billed cuckoo (*Coccyzus americanus*), barred owl (*Strix varia*), pileated woodpecker (*Dryocopus pileatus*), red-bellied woodpecker (*Melanerpes carolinus*), red-headed woodpecker (*M. erythrocephalus*), Acadian flycatcher (*Empidonax virescens*), tufted titmice (*Parus bicolor*), blue-gray gnatcatcher (*Polioptila caerulea*), prothonotary warbler, northern parula warbler (*Parula americana*), yellow-rumped warbler (*Dendroica coronata*), yellow-throated warbler (*D. dominica*), and rusty blackbird (*Euphagus carolinus*). Birds restricted to this type include American swallow-tailed kite (*Elanoides forficatus*), Mississippi kite (*Ictinia mississippiensis*), and Bachman's warbler (*Vermivora bachmanii*). Birds resident year-round in hydric hammocks include the red-shouldered hawk, barred owl, red-bellied woodpecker, pileated woodpecker, northern flicker (*Colaptes auratus*), American crow (*Corvus brachyrhynchos*), fish crow (*C. ossifragus*), blue jay (*Cyanocitta cristata*), Carolina wren (*Thryothorus ludovicianus*), tufted titmouse, Carolina chickadee (*Parus carolinensis*), and the northern cardinal (*Cardinalis cardinalis*). The most common summer residents are the great crested flycatcher (*Myiarchus crinitus*), northern parula warbler, and the summer tanager (*Piranga rubra*). Common winter residents include the eastern phoebe (*Sayornis phoebe*), American robin (*Turdus migratorius*), house wren (*Troglodytes aedon*), ruby-crowned kinglet (*Regulus calendula*), yellow-rumped warbler, American goldfinch (*Carduelis tristis*), and the white-throated sparrow (*Zonotrichia albicollis*)

(Vince et al. 1989). Flocks of migrating robins have been observed in the fall of the year in headwater swamps in South Carolina feeding on swamp tupelo fruits (DeBell and Auld 1968). Study showed that the birds ate the fruit whole and passed cleaned seed. This feeding habit resulted in a uniform distribution of seed over the forest floor and may play an important role in establishing new reproduction (DeBell and Hook 1969). Other birds observed feeding on tupelo fruits included cedar waxwing (*Bombycilla cedrorum*), eastern bluebird (*Sialia sialis*), and blackbird (*Euphagus* spp.).

Mammals

Mammals important in these site types include mink (*Mustela vison*), common raccoon (*Procyon lotor*), white-tailed deer, bobcat (*Lynx rufus*), and swamp rabbit (*Sylvilagus palustris*) (Sharitz and Mitsch 1993). Common on hydric hammocks are the Virginia opossum (*Didelphis virginiana*), southeastern shrew (*Sorex longirostris*), southern short-tailed shrew (*Blarina carolinensis*), nine-banded armadillo (*Dasypus novemcinctus*), eastern gray squirrel (*Sciurus carolinensis*), southern flying squirrel (*Glaucomys volans*), cotton mouse (*Peromyscus gossypinus*), common raccoon, feral hog, and white-tailed deer (Vince et al. 1989).

MANAGEMENT OF FLATWOODS WETLAND FORESTS

Wet Flat Pine Management Considerations

Pine management on wet flats can be characterized into two broad categories: extensive forest management activities and intensive management activities. Extensive management activities would usually be directed toward protection of the resource from catastrophic events (e.g., crown fires), and direct economic returns from the resource would be less of a consideration. Examples of this type of management would include management of wildlife sanctuaries, wilderness areas, multiple-use areas, and old-growth (Harms 1996). Intensive management would usually be directed toward the production of wood fiber, although other amenities would also be available. Lands belonging to non-industrial private landowners could be managed in many different ways, ranging from very intensive to extensive, depending upon the landowner's objectives, financial resources, and perspectives (Table 17.3).

Pine Flats — Intensive Management

Site manipulations on pine flats are specific attempts to increase the productivity of pine species (usually loblolly or slash pine) by modifying the genotype of the new stands, site hydrology, soil nutritional status, and competing vegetation. All of the major pine species that might be considered for intensive management are relatively shade-intolerant species and require full sunlight in order to survive and grow well. If a landowner desires to manage and maintain pine on wet flats, regeneration strategies are usually limited to some type of even-aged regeneration system. Examples of such systems include seedtree, shelterwood, clearcut with natural regeneration, or

TABLE 17.3
Examples of alternative management strategies for wet flats.

Stand Type	Landowner/ Objectives	Management Strategy	Typical Operations
Loblolly Pine on wet flats	Industrial/ maximize fiber production	Intensive Plantations	Clearcut harvest Drainage or water management via road ditches Bedding Plant bare root seedlings Phosphorus fertilization Herbaceous weed control Prescribed burning Thinning Harvest at stand age 20 to 40 years
Longleaf Pine savannas	Federal agency/ multiple use, special emphasis on red cockaded woodpecker	Extensive, natural regeneration via seed tree	Longer rotations (100+ years) Prescribe burn, followed by seed tree cut Protection from wildfire, periodic prescribed burns Protection of cavity trees Optional thinnings
Loblolly Pine-Hardwoods on wet flats	Private landowner/ source of periodic income recreational use, equity for heirs	Extensive to Intensive	Extremely varied. Ranges from selective harvests (highgrading) to establishment of plantations
Non-alluvial hardwoods	Public Lands/ natural areas for recreation, wildlife	Extensive	Protection from catastrophic wildfire, sanitation and improvement cuts, provision for regeneration cuts as needed

clearcut with artificial regeneration, all of which mimic the natural disturbances of fires and hurricanes and allow the shade-intolerant pines to become established.

The seedtree regeneration system has been successfully used to regenerate loblolly pine stands on many types of sites throughout the southeastern United States. It has the advantage of being relatively inexpensive, but success of the regeneration is not ensured, and the eventual removal of the seed trees may result in damage to the new seedlings. The use of this system requires additional planning and preparation to ensure that the operations are conducted during good seed crop years, and the seedbed is properly prepared. Because this system does not allow for the use of mechanical site preparation techniques, this method usually requires the use of controlled burns prior to the harvest to prepare the soil surface for pine seed establishment.

The shelterwood system is classically used for species that need a sheltering or partially shaded environment in order to regenerate. Loblolly and slash pine do not require such conditions, but the shelterwood system is sometimes used for maintenance of stand aesthetics or wildlife habitat in areas that have unique wildlife species or in areas that are visually sensitive. This system has the same disadvantages and advantages of the seedtree system, but damage to the new stand may be more of a problem because there are repeated harvest removals.

Clearcutting is the most common technique, followed by artificial regeneration methods consisting of broadcast seeding or tree planting. Planting tree seedlings is the more favored approach because it allows control of initial spacing, although it is more expensive. Clearcutting and planting allow intensive plantation culture of pines, which is commercially more productive than the less intensive regeneration systems. Plantation culture may include site clearing, hydrologic site modification, fertilization, vegetation control by tillage and herbicides, and the use of genetically improved seedlings.

Sites are converted from the original vegetation community to pine-dominated plantation communities by removing the existing canopy vegetation by first harvesting merchantable trees, and then by using techniques such as shearing, burning, and piling the remaining slash debris to remove the remaining vegetation. Site clearing opens the area for other management operations that can be used to modify the soil water regime if necessary. Numerous studies have indicated that establishment and early growth of loblolly and slash pine are limited by poorly to very poorly drained conditions commonly found on pine flats (Ralston 1965, Terry and Hughes 1975, Langdon and Hatchell 1977, Kozlowski 1986, McKee and Hatchell 1987, Allen and Campbell 1988, Olszewski 1988, Scheerer et al. 1995). The hydrology of some flats has been altered by a complex system of primary, secondary, and tertiary drainage ditches that lower the water tables in regenerating stands. Such ditch networks may be open ditches (free drainage), or they may contain water control structures (flashboard risers) that are used to maintain water levels at optimum soil depths (Hughes 1996). Some areas have also been drained by less deliberate attempts with ditch systems along road networks. Research has shown that the quality of water coming from such areas is little affected by such management (Amatya et al. 1996, Hughes 1996).

When wet sites are cleared of transpiring vegetation, the water table commonly rises. Plantations on pine flats are commonly "bedded" whereby continuous rows of mounds are created with disc harrows. Trees are planted on the beds, and the elevated position keeps the roots of planted seedlings above the water table as it rises to the surface during wet periods. Bedding greatly enhances tree survival and early growth by keeping the soil aerated and by reducing competition from shrub and ground layer vegetation. Bedding also repairs soils compacted or puddled during wet-site timber harvests.

Productivity of many wet flat soils, particularly several subgroups of Ultisols and Spodosols described above, is limited by phosphorus (P). Total soil P levels can be quite low, and plant-available P is inadequate and complicated by unique soil chemistry. The productivity of many wet pine flats has been increased dramatically with relatively small additions (30 kg/ha; 27 lbs/acre) of P fertilizer prior to tree planting. Phosphorus fertilizer is usually incorporated into the soil during bedding operations.

Prepared plantation areas are commonly planted with one-year-old nursery-grown seedlings that are grown from seed that has been genetically selected for fast growth, good tree form, and disease resistance. After trees are planted on prepared sites, herbicides are used to selectively control the growth of herbaceous plants competing with the pines. After closure of the pine-dominated canopy, nitrogen (N) fertilizers are applied to further speed the growth toward shortened rotations. This combination of silvicultural practices greatly increases pine volume, shortens rotation lengths, and increases the value of a unit volume of wood by improving site quality, by reallocating site resources to plantation trees, and by improving the genotypes of the planted trees. While the commercial value of the forest is greatly increased, studies show that plant biodiversity and wildlife habitat are maintained at levels nearly equivalent to non-plantation sites (Hauser 1992), particularly when plantations are managed as landscape mosaics of various ages with interconnections of non-plantation corridors.

Hardwood Management Considerations

Management of non-alluvial wetland hardwood forests is complicated by the extreme diversity of plant and animal species and the wide range of sites on which they occur. Most natural stands are mixtures of tree species ranging in age from seedlings to mature trees. Of the 28 named canopy trees found on these site types, 24 are commercially important species (Putnam et al. 1960). The highest economic value species include baldcypress, cherrybark and swamp chestnut oak, green ash, swamp and water tupelo, sweetgum, red maple, yellow-poplar, and loblolly and slash pine. Medium-value species are water and willow oak, American elm, pondcypress, water hickory, sweetbay, pond pine, and southern redcedar. The species of lowest economic value are American holly, beech, overcup and laurel oak, and the hickories. All of these species can occur together in various mixtures and proportions on wet-flat hardwood sites, non-alluvial swamps, and hydric hammocks, and each has particular soil and site requirements for best growth (Burns and Honkala 1990). On many of the stands of these site-types, especially wet-flat sites, natural disturbances and past cutting practices have left behind degraded ecosystems. Biodiversity has been reduced by removal of trees of highest economic value, leaving a high proportion of low-grade, over-mature, overcrowded, damaged, or cull trees. In these stands, it is necessary to restore health and diversity by applying sanitation and improvement cuts to enhance species composition and to provide an environment suitable for reestablishment and growth of the depleted species. In some cases, stands may be in such an altered or disturbed state that it is necessary to clearcut and establish new stands of the desired species composition (Putnam et al. 1960, Stubbs 1973, McKnight and Johnson 1980, Johnson and Shropshire 1983).

It is biologically feasible to employ any one of five silvicultural systems to govern tree harvesting on these site-types to ensure the natural regeneration of the desired species and the subsequent development of the desired forest structure (McKnight and Johnson 1980). The systems are single-tree selection, group selection, shelterwood, seed-tree, and clearcut (Burns 1983). Selection of a silvicultural system depends on the site-type, the ecological characteristics and requirements of the species, and the management objectives of the landowner. The principal ecological characteristics that must be considered are species tolerance to flooding and the environment necessary for seedling establishment and growth. Flooding and hydroperiod have the most important effects on species composition when the stand is being regenerated. Seedlings of most of the tree species found on these sites can withstand flooding when dormant or if some of the foliage remains above water, but few can withstand extended periods of complete inundation during the growing season; repeated flooding during the growing season can effectively prevent establishment (Hodges and Switzer 1979, Hook 1984b). Broad guidelines for selecting a silvicultural system for managing coastal plain hardwood forests are given by Stubbs (1973), Johnson and Shropshire (1983), and Toliver and Jackson (1989). More specific silvicultural guidelines are given for wet flat hardwoods by Putnam et al. (1960), hydric hammocks by Simons et al. (1989), and non-alluvial swamps by Hook et al. (1993) and DeBell and Hook (1969).

Silvicultural operations in wetlands must be carefully planned and executed to avoid physical damage to the soils and to the forest stand. Tree harvest almost always involves the operation of heavy machinery on wet soils, and therefore, special precautions must be taken to minimize damage. Logging operations can seriously damage site productivity by destroying soil structure, blocking drainage, damaging seedlings, and scarring or breaking residual trees. Best Management Practices (BMPs) have been drafted by many of the states in the southeastern United States (e.g., South Carolina Forestry Commission 1994) to provide guidelines for forestry operations that will minimize on-site and off-site damage to the environment due to road construction, skid trails, soil compaction, soil erosion, minor drainage, and nonpoint source water pollution. McKee et al. (1985) and Jackson and Stokes (1991) discussed problems of wet-site logging and provide guidelines and methods for minimizing damage.

LEGAL CONSIDERATIONS

Management of wet flats is related to several important legal considerations, including the Federal Clean Water Act of 1972 and its amendments, the Endangered Species Act of 1973, and operational BMPs developed for forested lands and/or forested wetlands by the various states. In general, roads and associated ditches are some of the areas in which landowners may not comply with provisions of the Clean Water Act. Forest road construction is allowed under the silvicultural exemption clauses of the Clean Water Act, but these roads must follow several specific federal BMPs and must have the primary purpose of supporting forest operations. Roads to hunting camps, residences, etc., may require a permit. Recently, the Environmental Protection Agency (EPA) and Corps of Engineers developed a memorandum that specifies the site types that are subject to the permitting process for site preparation activities. This memorandum also specifies site preparation guidelines that should be considered on wet flats.

Endangered species are also an important consideration in wet flats. One of the more well-known endangered species, the red-cockaded woodpecker (*Picoides borealis*), is generally limited to areas and stand conditions common to longer rotation stands on pine flats. As the habitats and requirements of endangered or threatened species become better known, management strategies may require revisions in order to protect these species.

RESEARCH NEEDS

Pine Wet Flats

The hydrology of pine flats is poorly understood because these areas do not lend themselves to watershed projects of a size that can be efficiently monitored. Most projects to date have monitored areas within larger watersheds for relatively short periods of time. Forest operations such as wet weather harvesting have the potential to interrupt the subsurface flow of water and perhaps alter site productivity. The entire issue of harvest impacts and long-term site productivity has only begun to be addressed by various researchers. Also, although the relatively flat nature of the terrain indicates that off-site water quality issues may be minimal, there are few studies that have actually monitored the quality of water leaving intensively managed pine flats. The mosaic arrangement of wet flats is known, but at present, the effects of human-induced disturbances within these wet flats and the subsequent effects on landscape level diversity are poorly understood.

Hardwood Types

There remains much to be learned about wetland hardwood site types before their varied structure and functions are well enough understood to ensure their health, productivity, and continued use. In a regional context, the location, extent, and physical condition of these site-types are largely unknown. Current regional forest inventories need to be refined to describe and quantify these forest types as distinct categories in future inventories. This information will enable us to monitor their health, productivity, and land use and will provide a data base for study of ecological functions and linkages to other ecosystems in the regional landscape.

The wet pine flat silvicultural knowledge base is relatively large, but the ecological and silvicultural knowledge base for the hardwood site types is small. Comprehensive ecological studies are needed to describe the species composition and structure of the plant and animal communities, describe and quantify biotic community development and successional dynamics, determine the major ecological and hydrological functions of each of the site types, establish the role of site hydrology in community development and stability, and quantify biomass productivity relationships for each site-type.

Silvicultural research is needed to quantify the effects of individual silvicultural practices on biodiversity, species composition, age structure, spatial heterogeneity, and forest stand structure, of each site-type, determine which combination of practices will provide for tree harvesting, stand regeneration, and intermediate stand treatments, with least impact on the ecosystem, define BMPs for each site-type, and develop a predictive forest growth and yield relationship for each site type that will enable the landowner to evaluate biological and economic productivity.

SUMMARY

Flatwoods forested wetlands are typically located on the broad, flat interstream divides located between larger drainages. Hydrologic inputs are primarily restricted to rainfall, yet these sites often experience soil saturation or ponding due to the poor subsurface drainage and low topographic gradients. In some areas of the lower coastal plain, wet flat forests are the single largest type of forested wetland, yet their importance is sometimes overlooked because they do not fit the stereotypical idea of a swamp. Flatwoods site types are typically dominated by either pine or hardwood species, depending on the hydrologic regime and the resultant soil formation processes and fire disturbance frequency. As hydroperiods increase, soils tend to become more organic and fire frequency decreases. Such sites are typically dominated by hardwood species similar to those found in major and minor river bottoms. These hardwood site-types have, in general, received minimal silvicultural treatments or have been converted to other land uses. On sites having shorter periods of inundation, mineral soils are more common and fire frequency increases, so that pine species predominate. The species of pine vary naturally on a soil mineralogy and fire frequency continuum, but loblolly pine has become the most common due to forest plantation management. Loblolly pine has been favored over longleaf pine where the natural fire patterns have been eliminated because loblolly pine has more favorable growth characteristics for commercial timber production. Many of these commercial stands are intensively managed, which usually involves clearcut harvests, some modification of site hydrology via drainage and bedding, control of competing vegetation, and fertilization. These plantations are very efficient at producing wood and fiber and have a negligible effect on water quality. However, it is not certain at present how such management may affect diversity at the landscape level and whether such intensive management is sustainable over multiple rotations. Research projects have been established by university, industry, and government organizations in the past decade to address these issues.

18 Mangrove Wetlands

Robert R. Twilley

CONTENTS

INTRODUCTION

The purpose of this chapter is to characterize the ecology and management of mangrove wetlands in the Gulf of Mexico and Caribbean Sea relative to human impacts and resource utilization. There are many excellent reviews of mangrove ecology, including those specific to the Gulf of Mexico and Caribbean regions (Davis 1940, Lugo and Snedaker 1974, Walsh 1974, Odum et al. 1982, Yáñez-Arancibia and Day 1982, Britton and Morton 1989, Lugo et al. 1990b, Odum and McIvor 1990, Gilmore and Snedaker 1993, Twilley 1995, and Rützler and Feller 1996). In addition, there are numerous reviews and books that describe the ecology and

1-56670-228-3/97/$0.00+$.50
© 1998 by CRC Press LLC

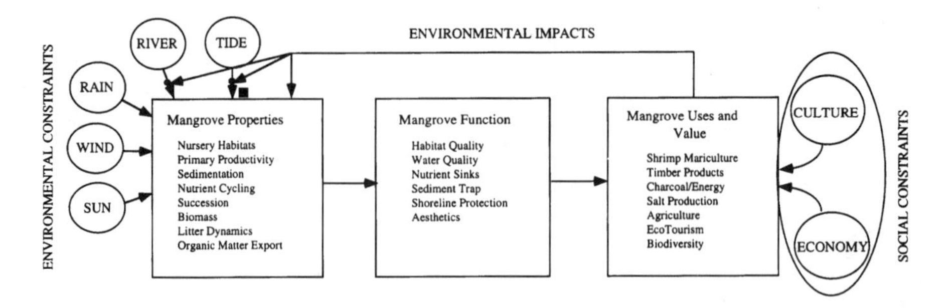

FIGURE 18.1 Linkages among environmental setting, ecosystem properties, ecological functions, and uses of mangrove ecosystems, including feedback effects of human utilization on ecosystem properties (modified from Twilley 1995).

management of mangroves around the world (Chapman 1976, Clough 1982, Tomlinson 1986, Yáñez-Arancibia and Day 1988, Robertson and Alongi 1992, Twilley et al. 1996), including a reference that describes techniques to study the ecology of mangrove wetlands (Snedaker and Snedaker 1984, Field 1996). The management of mangroves has also received considerable attention, because these coastal forested wetlands have value and use by the forestry and fishery industries; and in many circumstances, these two commercial enterprises provide very complex management issues, particularly where there are urban and residential reclamations of these coastal areas. I will describe an approach to classifying mangrove wetlands and provide a conceptual framework to integrate both the ecological processes and human utilization of mangrove ecosystems. The Gulf of Mexico and Caribbean represent a diverse group of mangrove wetlands to develop these ideas with a unique set of ecological and management issues. It is very important to keep in mind, though, that this review is not a global perspective of mangrove ecology and management, but an overview of very regional characteristics and literature of the new world subtropics of the Northern Hemisphere.

The ecological functions of mangroves are linked to the processes of mangrove wetlands that are constrained by the environmental setting of a particular coastal region, which are indicated by the forcing functions in Figure 18.1 (Twilley 1995). This has been well developed by geologists and plant geographers that have associated specific patterns of forest structure to different types of geomorphology (e.g., Thom 1982, Lugo et al. 1990b). However, links between the geophysical energies of an environmental setting (which include tides, winds, waves, and river discharge) with the ecological function of mangrove wetlands are less clear and will be the first topic in this review. This is an important concept, since the function of mangrove wetlands may be summarized by the physical processes of a region, helping managers quickly understand the value of mangroves in a regional setting. For example, ecological properties of mangroves including productivity, nutrient cycling, litter dynamics, succession, and sedimentation contribute to the habitat and water quality of estuarine and coastal waters (Figure 18.1). It is the combination of these ecological properties in mangrove wetlands that makes these coastal margin ecosystems some of the most productive regions of the world (Twilley 1988, 1995).

Mangroves refer to a unique group of forested wetlands that dominate the intertidal zone of tropical and subtropical coastal landscapes, generally between 25° N and 25° S latitude (Lugo and Snedaker 1974, Tomlinson 1986). Mangroves are trees considered as a group of halophytes with species from 12 genera in eight families (Waisel 1972, Lugo and Snedaker 1974). A total of 36 species has been described from the Indo-West-Pacific area, but fewer than 10 species are found in the new world (Macnae 1968). Species richness is even lower in the Gulf of Mexico and Caribbean region with only four mangroves commonly found, including red mangrove (*Rhizophora mangle*), black-mangrove (*Avicennia germinans*), white-mangrove (*Laguncularia racemosa*), and button-mangrove or buttonwood (*Conocarpus erectus*) (Britton and Morton 1989, Carlton 1975). The term mangroves may best define a specific type of tree, whereas mangrove wetlands refers to whole plant associations with other community assemblages in the intertidal zone, similar to the term 'mangal' introduced by Macnae (1968) to refer to swamp ecosystems. These are important distinctions since the biodiversity of mangroves may be considered species poor in some regions if considering only tree species richness, but the biodiversity of mangrove wetlands including all the flora and fauna may equal that of other tropical ecosystems (Rützler and Feller 1996, Twilley et al. 1996). In addition, the habitats of tropical estuaries are diverse, consisting of a variety of primary producers and secondary consumers distributed in bays and lagoons that have the intertidal zone dominated by mangrove wetlands. These may be referred to as mangrove-dominated estuaries. Thus, there is a hierarchical use of the term mangrove extending from individual trees, to forested wetland ecosystems, to tropical estuarine landscapes.

MANGROVE ECOSYSTEM STRUCTURE AND FUNCTION

Ecogeomorphology of Mangroves

An ecogeomorphic classification (Twilley 1995, Twilley et al. 1996) of mangroves (Figure 18.2) describes the diversity of mangrove communities as a combination of geophysical processes of coastal environments along with ecological processes within a local habitat (Hedgpeth 1957, Thom 1984, Woodroffe 1992). The landform characteristics of a coastal region, together with geophysical processes, control the basic patterns in forest structure and growth (Thom 1982, 1984). The two basic types of geomorphological settings are those with terrigenous inputs compared to those located in carbonate environments. Within each of these two categories, there are different types of settings depending on the nature of circulation, sediment transport, regional topography, and physical processes. Carbonate systems include continental margins and islands that differ depending on the soil type and amount of freshwater flow from upland watersheds. Continental regions include those with peat and marl sediments, whereas islands can have either large or small catchments (high and low islands). The hypothesis of this classification scheme is that a gradient in the forcing functions (geophysical processes) of a coastal region will result in variation in energy flow and material cycling of mangrove wetlands (Twilley 1995).

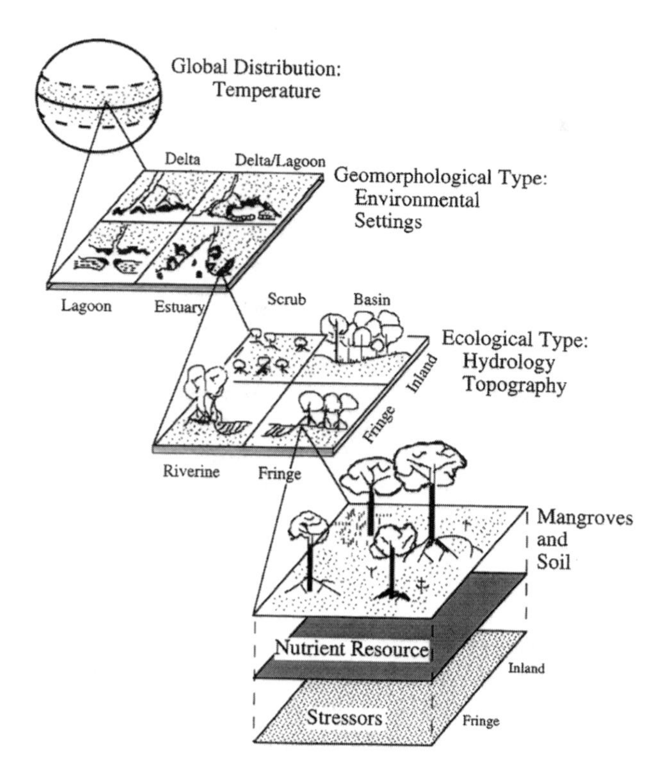

FIGURE 18.2 Hierarchical classification system to describe diverse patterns of mangrove structure and function based on global, geomorphological (regional), and ecological (local) factors that control the concentration of nutrient resources and stressors in soil along gradients from fringe to more inland locations from shore (modified from Twilley et al. 1996).

This will be reflected in different patterns of forest structure, biomass, productivity, biogeochemistry, and exchange of nutrients and organic matter with coastal waters. Thus, both the structure and function of mangrove ecosystems are linked to the geomorphology and geophysical processes of coastal environments.

A combination of the physiographic and structural attributes of mangrove wetlands, together with local conditions of topography and hydrology, were used to formulate the classic ecological classification system by Lugo and Snedaker (1974) (Figure 18.2, midsection). This ecological scheme uses the dominant environmental factors of the Caribbean, such as soil resources and stressors, to classify mangrove forests into four types including riverine, fringe, basin, and scrub forests (Figure 18.2, midsection). The formation and physiognomy of these forest types appear to be controlled strongly by local patterns of tides, terrestrial surface drainage, soil characteristics, and biological interactions. Along an inland transect, tidal frequency changes, together with changes in slope, may cause an increase in salinity and other edaphic factors that limit the development of mangrove forests (mechanisms of mangrove zonation). Other ecological factors are associated with the biology of these systems such as crab consumption of propagules and leaf litter (Robertson and Daniel 1989, Smith 1987b, 1992). The effect of these ecological factors on forest structure are twofold. They can decrease the maximum development of mangrove forests allowed by the existing environmental constraints of the geomorphological subclass, and they can control the relative zonation of mangrove species.

A combination of ecological types of mangroves can occur within any one of the geomorphological settings described above, depending on the distribution of soil resources and stressors (Figure 18.2, lower section). These ecological types are based on microtopographic and biological factors that can influence the structure and function of mangroves from shoreline to more inland locations. Local patterns in tidal inundation have long been related to the structure and species colonization of mangrove forests (Watson 1928, Chapman 1944, Walsh 1974, Chapman 1976). The juxtaposition of mangroves in a lagoon to river input will influence the fertility and hydrology of these wetlands. I will refer to this hierarchical classification as the ecogeomorphic typology of mangrove wetlands that includes two important descriptors in classifying mangrove wetlands: (1) location of the wetland within a specific type of coastal geomorphology (environmental settings); and (2) location of the wetland zone along a transect perpendicular to shore (topographic settings). Thus, a fringe, basin, or scrub mangrove within a specific type of geomorphological setting may have somewhat different ecological characteristics than similar types in a different environmental setting. In addition, the response of these ecogeomorphic types of mangroves to regional land use management plans is specific to these two landscape parameters due to changes in both regional and habitat factors that control the ecology of forested wetlands (Lugo et al. 1990b).

Mangrove Distribution and Zonation

The frequency, duration, and severity of freezing temperature are prime factors governing the distribution and abundance of mangroves in the northern Gulf of Mexico (Sherrod et al. 1986). Although red mangrove seedlings can survive temperatures of 2 to 4°C (36 to 39°F) for 24 hours, this species is more severely impacted by cold than other Caribbean mangrove species due to its inability to resprout following foliar damage from freezing (West 1977, Sherrod and McMillan 1985, Sherrod et al. 1986, Olmsted et al. 1993). The greater resprouting ability of black- and white-mangroves results in greater recovery from freeze damage (Lugo and Patterson-Zucca 1977, Sherrod and McMillan 1985, Olmsted et al. 1993). Black- and white-mangroves have extensive secondary meristematic tissues and can quickly recover by developing sprouts following defoliation of the canopy (Snedaker et al. 1992). Evidence of black-mangrove tolerance to freezing is the distribution of this species into Texas, Louisiana, and Mississippi (Sherrod and McMillan 1981, Patterson and Mendelssohn 1991, Patterson et al. 1993). Black-mangroves can survive recurrent exposure to chilling temperature of 2 to 4°C (36 to 39°F) (McMillan 1975, McMillan and Sherrod 1986). Comparison of the three mangroves under winter conditions on the Texas coast at Port Aransas showed that black-mangroves have the greatest tolerance to winter conditions (Markley et al. 1982).

Regional patterns of mangrove zonation following the global constraints of temperature depend on moisture content of intertidal soils, which is controlled by the balance of rainfall and evapotranspiration, together with the inundation by rivers and tides (Figure 18.3). Climate and regional hydrology (to a large degree) control the distribution of nutrient resources and stressors in mangrove soils that determine forest structure along the intertidal zone. Evapotranspiration is linked to seasonal patterns of temperature and, together with rainfall, determines dry or humid coastal areas. The structure of mangrove wetlands in several locations of the Caribbean is limited in coastal regions, with dry life zones based on relative amounts of precipitation and evapotranspiration (Holdridge 1967, Pool et al. 1977, Cintrón et al. 1978, Blasco 1984, Lugo 1990a). The location of dry coastal climates in the Gulf of Mexico and Caribbean is described using climate diagrams (Figure 18.3) that plot average monthly precipitation relative to monthly temperature, such that deficits or excesses can be indicated. Dry coastal climates include the Texas coast of the United States (Port Arthur, Figure 18.3) and the upper coast of the Yucatan Peninsula (Progresso, Figure 18.3). Moist coastal regions are represented by the southwestern Yucatan Peninsula (Coatzacoalcos).

In dry coastal tropical regions where precipitation deficits occur, limits to mangrove zonation and forest structure can be canceled by the presence of freshwater inputs to the coast from upland watersheds. Thus, the amount of moisture in mangrove soils is a complex interaction of regional climate and hydrology, including upland freshwater input. The major coastal regions of the Gulf of Mexico that receive a significant freshwater subsidy from upland regions include the Mississippi River region, the Everglades region of south Florida (Duever et al. 1994, Light and Dineen 1994), and the Usamacinta delta near the Yucatan Peninsula of Mexico (Yáñez-Arancibia and Day 1988). Even though mangroves may be located in a dry climatic zone, they may have characteristic zonation of mangroves in humid coastal environmental settings due to the presence of freshwater inputs. Freshwater subsidies from upland watersheds increase the moisture of mangrove soils and influence the functional ecology of mangroves. The environmental subsidy of upland runoff will be an important consideration in the management of mangrove resources, as described in a later section of this chapter.

Local zonation in mangroves is the result of differential responses of trees to environmental gradients caused by the effects of topography on the distribution of nutrient resources and stressors in mangrove soils (see Tomlinson [1986] and Ball [1988] for reviews of mangrove ecophysiology). As described above, these environmental gradients are a complex interaction among the frequency and duration of tides along the intertidal zone, degree of upland freshwater input (surface or groundwater), and regional climate (balance of precipitation and evapotranspiration). Mangroves are considered forested wetlands that inhabit stressed environments (Lugo 1980), but each tree species has adaptations to specific tolerances within the intertidal zone. Lack of nutrient resources can be a form of stress and limit mangrove development (constraining forest development), as can the lack of light resources in mature forests (McKee 1995a). Stressors in the intertidal zone include hypersaline conditions as well as the possible effect caused by the absence of salt that may limit the competition of mangroves with freshwater wetlands. Another stressor in the intertidal zone that has recently received attention is hydrogen sulfide (McKee et al. 1988, McKee 1993, McKee 1995b). The responses of mangroves to specific inundation classes of mangroves and impoundments indicate that hydroperiod is also

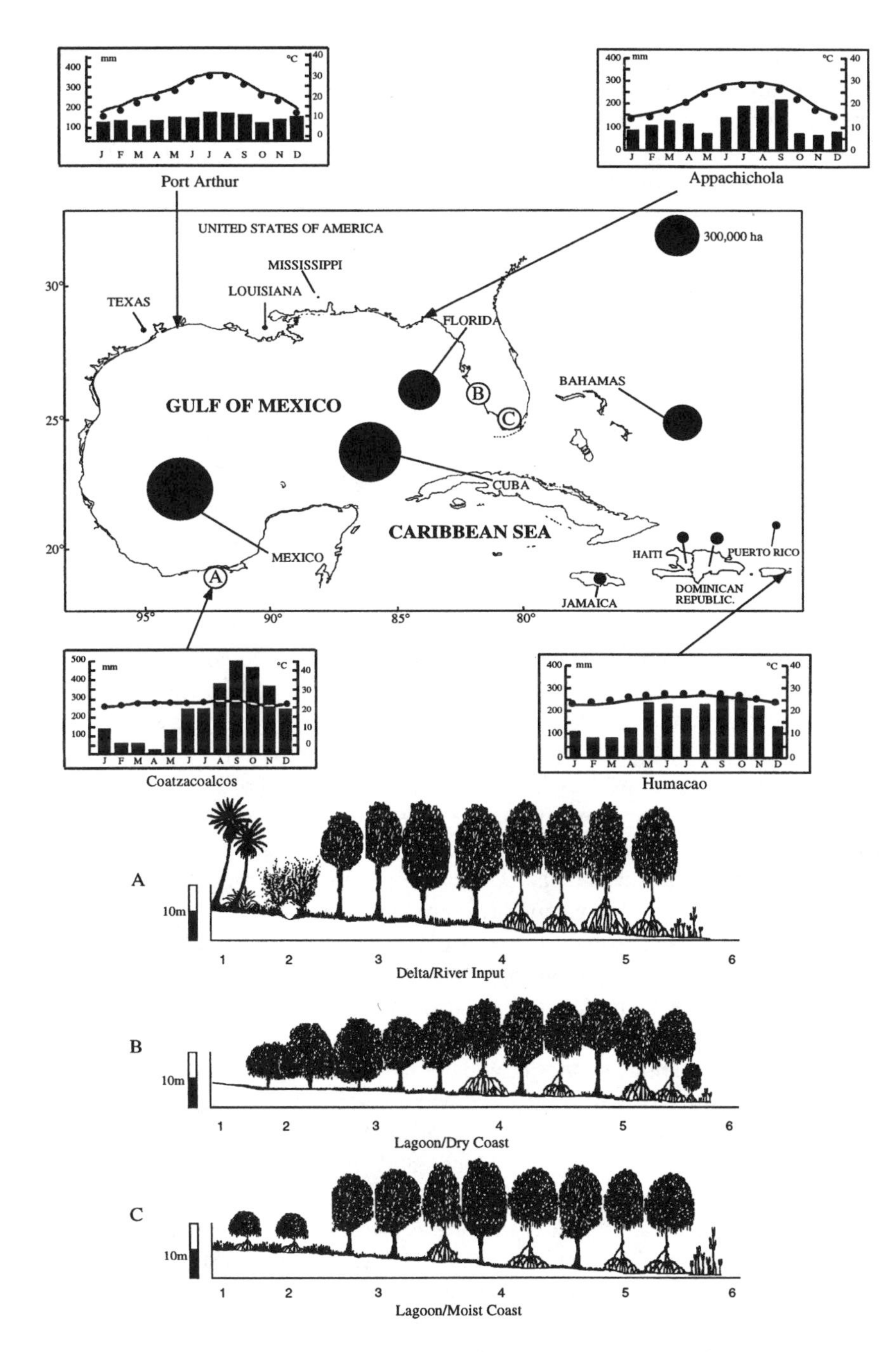

FIGURE 18.3 Map of the Gulf of Mexico and Caribbean region showing the distribution of mangroves (after Gilmore and Snedaker 1993) and climatic conditions of specific coastal environments (temperature shown with points and precipitation with bars). The lower panel shows three examples of mangrove zonations perpendicular to shore in three types of coastal environments (A — delta, B — lagoon with minor upland runoff, C — lagoon with upland freshwater input).

important to patterns of zonation (Watson 1928, Chapman 1944, McKee and Mendelssohn 1987). There are indications of some interesting interactions of hydroperiod with nutrient resource availability and susceptibility of mangroves to stressors. Thus, even though mangroves are adapted to the conditions of the intertidal zone, optimum growth is limited by the distribution of nutrient resources, stressors, and possibly increased duration of flooding (hydroperiod) (Figure 18.2).

In mangrove wetlands there is often a distinct elevation gradient from the water's edge (lower intertidal zone) to the upland transition zone (upper intertidal zone), and often the lower intertidal zone consists of a streamside levee or berm that is frequently inundated by coastal waters. The physiognomy of mangroves is more robust in this lower intertidal zone, referred to as fringe mangroves, which are identified by the presence of red mangrove with taller trees compared to more inland mangrove vegetation (Figure 18.3a-c) (Lugo and Snedaker 1974, Lugo 1990a). This fringe mangrove zone is very similar among diverse geomorphological settings, differing in tree height depending on the fertility of the site, which is usually linked to the presence of river sediment (riverine vs. fringe mangroves) (Twilley 1995). In higher latitudes, the fringe zone is restricted to black-mangroves and is usually shorter in stature (temperature effect described above) (Patterson et al. 1993).

The higher elevations of the upper intertidal zone can have different mangrove tree composition and forest stature depending on the climate and geomorphology of the region (Figure 18.3a-c) . In drier climates with little upland freshwater input, the increased elevation results in more monospecific black-mangrove forests with shorter tree heights, changing from basin to scrub mangrove forests (Figure 18.3b). More inland areas that are less frequently inundated by tides and have no upland discharge of fresh water have hypersaline soils with salinity >100 g/kg (1.6 oz/lb). This constraint leads to formation of salt pannes or salinas that prevent colonization by all but the most salt adapted plants, even though no mangroves grow there. Zones that are void of mangrove vegetation (referred to as die back zones) may also form in the center of low islands in the Caribbean where dead tree trunks delimit historical wetland distributions. In addition to extreme hypersaline conditions (Cintrón et al. 1978), some of these die back zones are also noted to have toxic levels of hydrogen sulfide in mangrove soils (McKee et al. 1988).

The higher elevations of the intertidal zone in more moist coastal regions are also represented by mangrove vegetation with reduced stature, changing from basin to scrub mangrove zones, but there are some distinct differences compared to the zonation described above in dry coastal climates (Figure 18.3a, c). In moist zones, mangrove vegetation gradually mixes with marsh vegetation such as saw grass (*Cladium jamaicense*) and black rush (*Juncus roemarianus*), forming a continuous ecotone of vegetation with upland environments (Koch 1996). In addition to the lack of salinas, the scrub mangroves in the upper intertidal zone are dominated by red mangroves in soils that are not hypersaline (Figure 18.3c). In the case of the Everglades, this scrub ecotone is thought to include the extreme tidal range of storm events, causing patches of scrub red mangroves far from coastal waters. Low nutrient resources are argued to be the key limiting factor to plant stature, replacing stressors (hypersalinity and hydrogen sulfide) in dry coasts as the constraint of mangrove development in the upper intertidal zone (Koch 1996). Mangrove scrub communities on the interior of carbonate islands along the coast of Belize have also been shown to be limited in stature by levels of phosphorus (P) concentrations in soils (Feller 1995). These scrub mangrove communities are distinct from dwarf red mangroves that include distinct stunted trees that are not limited by environmental conditions. The presence of scrub red mangroves in the upper intertidal zone of coastal environments may indicate low salinity and low nutrient resource soil conditions, depending on the climate and hydrology of the region.

The environmental settings of mangroves are a complex behavior of regional climate, tides, river discharge, wind, and oceanographic currents (Figure 18.1). The next section will describe some of the ecological processes of mangroves related to these same environmental settings to see if ecosystem properties can also be linked to regional patterns in climate, geomorphology, and geophysical processes.

Ecological Processes

Succession

Succession in mangroves has often been equated with zonation (Davis 1940), wherein "pioneer species" would be found in the fringe zones, and zones of vegetation more landward would "recapitulate" the successional sequence towards terrestrial communities. Zonation in mangrove communities has variously been accounted for by a number of biological factors including salinity tolerance of individual species (e.g., Snedaker 1982), seedling dispersal patterns resulting from different sizes of mangrove propagules (Rabinowitz 1978), differential consumption by grapsid crabs (Smith 1987b), and interspecific competition (Ball 1980). Snedaker (1982) proposed the establishment of stable monospecific zones wherein each species is best adapted to flourish due to the interaction of physiological tolerances of species with environmental conditions.

Zones of mixed species composition have been thought of primarily as transition zones or ecotones between monospecific zones, and as such have been interpreted as being temporary responses to disturbance (Lugo and Snedaker 1974, Lugo 1980). However, analyses of long-term transects established by Lugo and Snedaker in Rookery Bay have shown that mixed associations of mangroves may be stable communities of the coastal landscape (Warner 1990). Remote sensing of land-use changes in southwest Florida showed that mixed species zones occupied similar percentages of mangrove wetlands from 1952 to 1984 (Patterson 1986). Geological surveys of the intertidal zone of Tabasco, Mexico, demonstrated that the zonation and structure of mangrove wetlands are responsive to eustatic changes in sea level, and that mangrove zones can be viewed as steady-state zones migrating toward or away from the sea, depending on its level (Thom 1967). Thus, both monospecific and mixed vegetation zones of mangrove wetlands represent steady state adjustments rather than successional stages (Lugo 1980, Snedaker 1982).

The structure of mangrove forests is influenced by a combination of geomorphological, climatic, and ecological factors that determine the patterns of zonation along shorelines. The trajectory of vegetation dynamics is constrained by the geomorphological and climatic characteristics of coastal environments and modified by the ecological interactions within a mangrove forest. Thus, general patterns may be observed within geographic regions, but there are diverse patterns globally based on different land forms that occur in tropical coastlines. In addition, changes in environmental settings caused by drought, subsidence, or tropical cyclones can interrupt patterns in vegetation development. Since we seldom have sufficient long-term records to distinguish these patterns in mangrove wetlands, there are no models of mangrove succession that can be applied outside specific geographical boundaries. A simulation model of mangroves in the Shark River estuary of south Florida by Chen (1996) showed that at least 200 years of forest development without disturbance was necessary to express the steady state dominance of mangrove species, although the maximum basal area and biomass may be obtained in nearly 30 years following hurricane disturbance, similar to the projections of Lugo et al. (1976).

Biomass and Productivity

Tree height and above-ground biomass of mangrove wetlands throughout the tropics decrease at higher latitudes, indicating the constraint of climate on forest development in the upper regions of the Gulf of Mexico and Caribbean (Cintrón and Schaeffer-Novelli 1984, Saenger and Snedaker 1992, Twilley et al. 1992). In addition, mangrove biomass can vary dramatically within any given latitude, an indication that local effects may significantly limit the potential for forest development at all latitudes. These local effects include topography and hydrology, including the effects of rivers and tides on soil characteristics as described above (Figure 18.1). Cintrón et al. (1978) found that with increasing salinity, the values of a number of structural and functional parameters of mangrove forests decreased. These included litterfall, tree density, basal area (total cross-sectional area of trunks), and tree height. For example, tree height (Y, m) of mangroves in Puerto Rico was inversely related to soil salinity by the equation $Y = -0.20(X) + 16.58$ ($r^2 = 0.72$), where X is soil salinity in g/kg (Cintrón et al. 1978). Studies of mangroves on this Caribbean Island summarized that salinities above 50 g/kg (0.8 oz/lb) have a negative effect on mangrove forest structure.

Another important constraint on the development of mangrove biomass in the Caribbean is the frequency of hurricanes. Mangrove study sites in Terminos Lagoon (Mexico) and Puerto Rico are both along the 25°N latitude; those with the highest biomass are located on the west coast of the Yucatan peninsula where hurricanes seldom occur (Ruffner 1978). Flores-Verdugo et al. (1986) attributed the poorly developed forest structure in El Verde Lagoon on the Pacific coast of Mexico to frequent hurricanes. Although hurricanes may stimulate certain ecological processes in coastal ecosystems (Conner et al. 1989), forested wetlands such as mangroves in the Gulf of Mexico region are susceptible to damage from high winds and hurricanes, and the normal frequencies of this region can result in a dominance of young successional forests (Lugo et al. 1976). However, in contrast to the projections of Smith et al. (1994), most of these forested wetlands have historically recovered, particularly in the south Florida region (see section on Climate Change).

The primary productivity of mangroves is most often evaluated by measuring the rate of litterfall (see Brown and Lugo 1982, Odum et al. 1982, Twilley et al.

1986, and Odum and McIvor 1990 for reviews of mangrove productivity). Regional rates in litter production in mangroves are a function of water turnover within the forest, and the rank among ecological types is as follows: riverine > fringe > basin > dwarf (Pool et al. 1975, Twilley et al. 1986). Fewer tides and lower freshwater input cause higher soil salinities (Cintrón et al. 1978) and/or the accumulation of toxic substances such as hydrogen sulfide (Carlson et al. 1983, Nickerson and Thibodeau 1985, McKee et al. 1988), which can result in increased stress on these inland mangrove forests (Hicks and Burns 1975, Lugo 1978). The mechanisms by which higher water turnover stimulates litter productivity are complex and may include a combination of factors including increased fertility, control of stressors in pore waters, aeration of soil matrix, and supply of silts and clays (Wharton and Brinson 1979).

Nitrogen (N) and P have been implicated as the nutrients most likely limiting primary productivity of mangrove ecosystems (Boto and Wellington 1983, 1984). Onuf et al. (1977) found greater leaf production in guano enriched overwash mangrove islands, as well as significantly higher foliar N concentrations in oligotrophic waters of the Florida coast. Although enhanced productivity of these mangrove islands was not associated with a particular nutrient, leaf tissue analyses indicated that guano was enriched with more P than N. Growth of scrub mangroves in a carbonate island off the coast of Belize was stimulated when fertilized with mixtures of N and P (Twilley 1995). Highest growth rates in leaf production, increased stem length, and tree height occurred in plots fertilized with tree spikes annually from 1986 to 1988. More detailed studies of mangroves in this research site operated by the Smithsonian Institution showed that P is more important than N in limiting mangrove growth (Feller 1995). Both studies showed that mangroves in these sites are also influenced by a combination of fertility and hydroperiod. Fertilization studies of scrub mangroves in south Florida has also shown that P rather than N can stimulate growth in red mangroves (Koch 1996). Forest biomass along the estuarine gradient of the Shark River estuary and at sites in the Caribbean demonstrates a relation with soil nutrients follows a Monod model (Bridgham et al. 1995) such that the optimum soil P content is 36.3 g/m^2 (0.12 oz/ft^2) to a depth of 40 cm (15.7 in) (Chen 1996). The half-saturation constant of soil P, based on the nonlinear model, was 29.3 g/m^2 (0.1 oz/ft^2). These types of specific mathematical functions of nutrient resource concentration and forest development have been used to develop ecological models of mangrove wetlands (Chen 1996).

Litter Dynamics

Leaf litter produced in the canopy of mangrove forests represents a potential source of organic matter and nutrients that may be transported to adjacent coastal waters and can support commercially important fisheries (Odum and Heald 1972, Macnae 1974, Thayer et al. 1987, Yáñez-Arancibia et al. 1988, Robertson and Duke 1990). The dynamics of mangrove litter, including productivity, decomposition, and export, can determine the coupling of mangroves to the secondary productivity and biogeochemistry of coastal ecosystems (Twilley 1988, 1995, Alongi et al. 1992, Robertson et al. 1992). Patterns of leaf litter turnover may be specific among the ecogeomorphological types

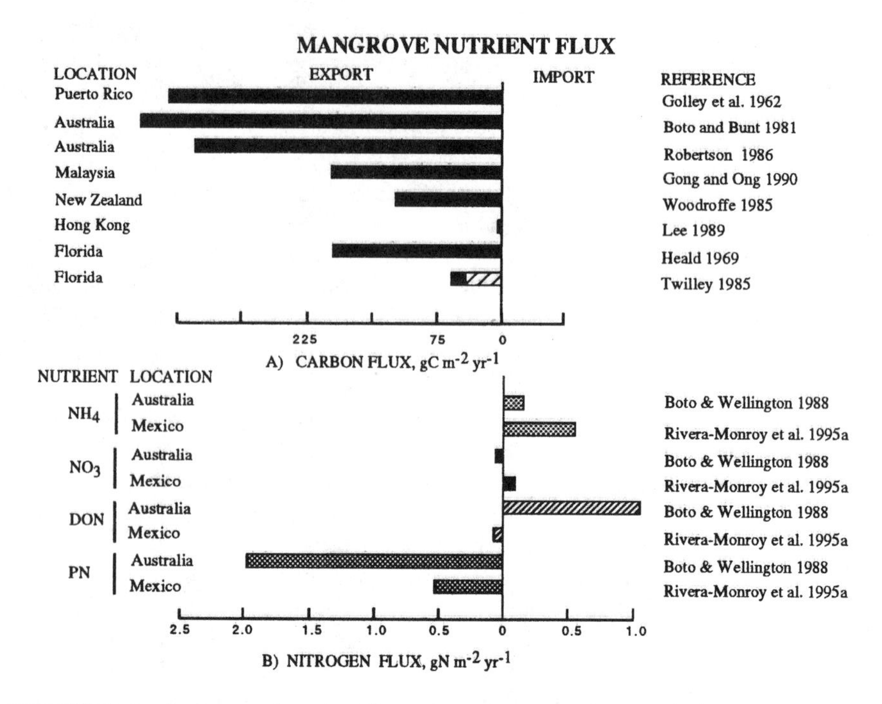

FIGURE 18.4 Summary of nutrient fluxes in mangrove wetlands including carbon (upper) and nitrogen (lower) exchange.

of mangroves depending on the ecological type and the environmental setting (Lugo and Snedaker 1974, Twilley et al. 1986, Twilley 1995). Accordingly, the rank of leaf litter turnover is usually riverine > fringe > basin, with greater litter export in sites with increasing tidal inundation. Therefore, regional scale environmental settings that depend on geophysical processes may be important to describing the variation in litter dynamics among mangrove wetlands (Twilley 1995). Based on this hypothesis, mangrove sites with frequent tides and river inundations are more productive, and higher proportions of leaf fall are exported to coastal waters.

Carbon (C) export from mangrove ecosystems ranges from 1.86 to 401 g C/m^2 yr (0.006 to 1.31 oz/ft^2/yr) (Twilley et al. 1992), with an average rate of about 210 g C/m^2/yr (0.69 oz/ft^2/yr) (Figure 18.4a). Carbon export from mangrove wetlands is nearly double the rate of average C export from salt marshes (Nixon 1980), which may be associated with the more buoyant mangrove leaf litter, higher precipitation in tropical wetlands, and greater tidal amplitude in mangrove systems studied (Twilley 1988). Rates of organic C export from basin mangroves are dependent on the volume of tidal water inundating the forest each month, and accordingly, export rates are seasonal in response to the seasonal fluctuation in sea level. As tidal amplitude increased, the magnitude of organic material exchanged at the boundary of the forests increased (Twilley 1985). Trends for litter productivity and export suggest that as geophysical energies increase, the exchange of organic matter between mangroves and adjacent estuarine waters also increases.

Several studies in the old world tropics in higher energy coastal environments of Australia and Malaysia have emphasized the influence of crabs on the fate of mangrove leaf litter (Malley 1978, Leh and Sasekumar 1985, Robertson 1986, Robertson and Daniel 1989, Lee 1989, Camilleri 1992). In these coastal environments, a strictly geophysical model of litter dynamics is limited for many mangrove sites where crabs consume 28 to 79% of the annual leaf fall (Robertson 1986, 1988, Robertson and Daniel 1989). These studies suggest that biological factors may influence leaf litter turnover more than the geophysical processes as suggested in Twilley et al. (1986) for new world mangrove wetlands. A study in the Guayas River estuary in Ecuador found that the fate of leaf litter in new world mangroves was influenced by the crab *Ucides occidentalis*, similar to the studies of crab processing of leaf litter in the old world tropics (Twilley et al. 1997). Rates of litter turnover by crabs in mangrove wetlands in both hemispheres are similar and much higher than in wetlands with minor influence of crab consumption on leaf litter dynamics. Differences in litter turnover rates among mangrove wetlands are a combination of species specific degradation rates, hydrology (tidal frequency), soil fertility, and biological factors such as crabs that influence the rates of litter dynamics. Mangrove crabs are considered less abundant in mangroves of the Gulf of Mexico and Caribbean (Jones 1984), suggesting that they have less influence on ecological processes in mangrove wetlands of this region. However, where they do occur in significant numbers, mangrove crabs can have a significant influence on litter dynamics.

Nutrient Biogeochemistry of Mangroves

The biogeochemistry of mangrove wetlands includes the fate and effects of sediment, nutrients, organic matter, dissolved gases, trace elements, and toxic substances on ecological properties as a function of diverse coastal landscapes. Each of these constituents of biogeochemistry are related to management issues, such as eutrophication, contaminants, global and land-use environmental changes, and sustainability. I will focus on nutrient biogeochemistry of mangrove wetlands and will discuss the management issues relative to the eutrophication of coastal waters (Figure 18.5). The function of mangrove wetlands as either a nutrient source or sink depends on the process of material exchange at the interface between mangrove wetlands and the estuary, which is largely controlled by tides (TE in Figure 18.5). The specific nature of material flux at the estuarine interface is controlled by the balance of many ecological processes within the mangrove wetland, and therefore, exchange is not strictly dependent upon the geophysical properties of an estuary (Figure 18.5). Defining the relative influence of ecological and geophysical processes associated with the biogeochemical functions of mangrove wetlands as sink, source or transformer of nutrients is clearly one of the most misunderstood properties of these ecosystems.

Nutrient exchanges may occur with either coastal waters (TE) or with the atmosphere (AE), depending on whether the nutrient has a gas phase (Figure 18.5). For instance, P dynamics in mangrove wetlands is representative of a sedimentary cycle and is usually not influenced by atmospheric exchange (except for minor precipitation inputs). However, substantial amounts of C and N can exchange with the atmosphere depending on the balance of certain biochemical processes. Thus

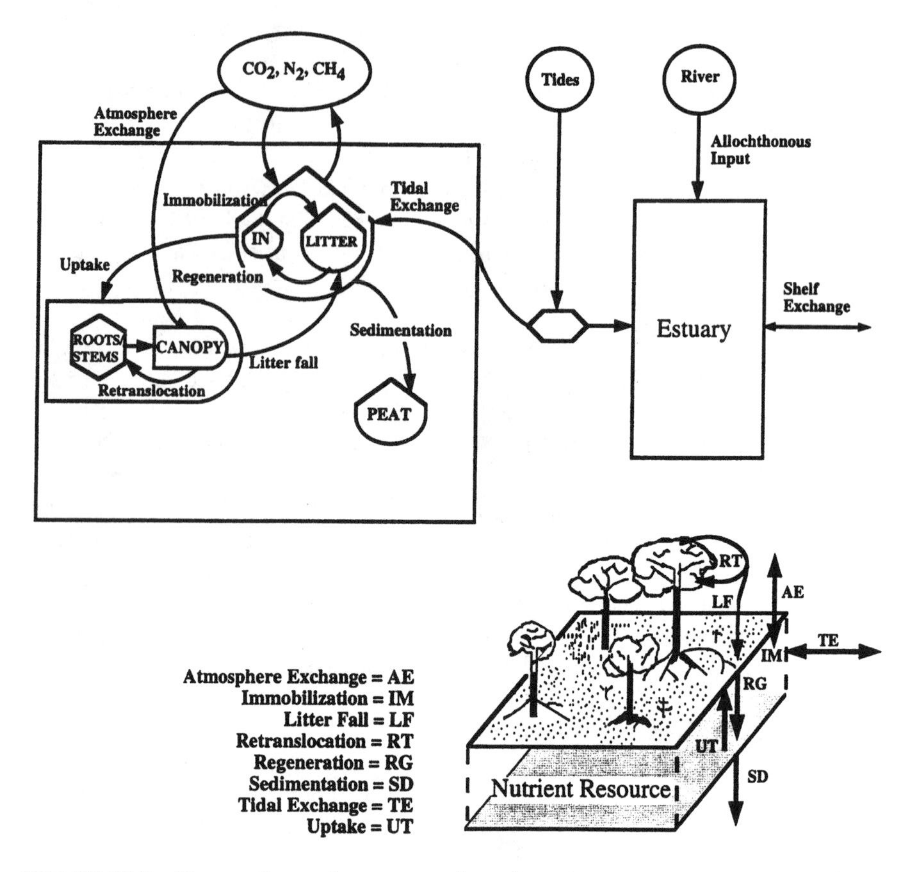

FIGURE 18.5 Fluxes of organic matter and nutrients in a mangrove ecosystem (upper panel), including exchange with the estuary (IN = inorganic nutrients). Diagram of a mangrove wetland with soil nutrient resources is presented in the lower panel to describe the spatial linkages in these ecological processes (modified from Twilley et al. 1996).

the N cycle of mangroves is very complex, with several mechanisms both at the interface with coastal waters and with the atmosphere that influence the mass balance of this nutrient. In addition, there are processes of nutrient uptake, recycling and remineralization that cycle nutrients among the canopy and soils of mangrove wetlands. These nutrient cycles are similar to those described for other forested wetlands including root uptake (UT), retranslocation in the canopy (RT), litterfall (LF), decomposition (RG), immobilization (IM), and sedimentation (SD) (Figure 18.5). One of the global hypotheses in mangrove nutrient biogeochemistry, as with other wetlands, is that the balance of these nutrient flows will determine the exchanges across the wetland boundary. I will focus on just a couple of these ecological processes to demonstrate the link between nutrient cycling and exchange within mangrove wetlands.

Surveys of N exchange demonstrate some of the principles of determining the function of mangrove wetlands as a nutrient sink. Export and import flux of N, including two inorganic species (NH_4 and NO_3) and two organic forms (PON and PN) are shown for a site in Mexico (Rivera-Monroy et al. 1995a) and Australia (Boto and Wellington 1988) in Figure 18.4b. For both sites, there is an import of NH_4 compared to a slight release of NO_3 in Australia and import in Mexico. One of the main differences between the two sites is the flux of DON, which is imported into the wetlands in Australia and exported from the site in Mexico. The largest N flux at both sites is export of PN, which coincides with net flux of organic C from most mangroves (Figure 18.4a). Compared to other flux studies of mangroves, there seems to be a pattern of net inorganic fluxes into the wetlands and corresponding flux of organic nutrients out (Rivera-Monroy et al. 1995a). However, the net balance of these fluxes is unclear given the large statistical error associated with these flux calculations. The best summary may be that mangrove wetlands transform the tidal import of inorganic nutrients into organic nutrients that are then exported to coastal waters.

Flux studies in mangrove sediments indicate that dissolved inorganic N (DIN) in tidal waters are "trapped" in the first cm of the sediment surface (Boto et al. 1985, Kristensen et al. 1988, Boto and Wellington 1988, Rivera-Monroy et al. 1995a). Rates of net NO_3 exchange by mangrove sediments in Mexico was about 179 μmol/m^2 hr (16.6 μmol/ft^2 hr) (Rivera-Monroy et al. 1995a), similar to the medium range of rates measured in mangrove sediments described by Nedwell (1975). Nitrate uptake in mangrove sediments is considered denitrification and results in a loss of this nutrient from the wetland. Estimates of denitrification based on NO_3 uptake range from a low of 0.53 μmol/m^2 h (0.05 μmol/ft^2 h) at Hinchinbrook Island, Australia (Iizumi 1986), where concentrations are <10 μM, compared to ranges of 9.7 to 261 μmol N/m^2 h (0.9 to 24.3 μmol/ft^2 h) in mangrove forests receiving effluents from sewage treatment plants with concentrations >1000 μM (Nedwell 1975, Corredor and Morell 1994). However, studies of denitrification based on NO_3 uptake are indirect measures of N gas production. Measures of denitrification using small amendments of $^{15}NO_3$ followed by direct measures of $^{15}N_2$ production have shown that denitrification accounts for <10% of the applied isotope, suggesting that NO_3 uptake is not a sink of N (Rivera- Monroy et al. 1995b). Rates of denitrification linked to nitrification in mangrove sediments are also low, and thus, little exchange of N gas can be associated with DIN uptake (Rivera-Monroy and Twilley 1996). The lack of $^{15}N_2$ gas production indicates that much of the net DIN exchange at the boundary of mangroves described above may not be loss to the atmosphere via denitrification, but accumulated in the litter on the forest floor.

The net exchange of N gas in mangrove ecosystems depends on the inputs of N via fixation relative to the loss via denitrification. As discussed above, the loss of N gas in most natural mangrove wetlands is 2.5 g N/m^2 yr (0.23 oz/ft^2 yr, increasing to 24.5 g N/m^2 yr (2.28 oz/ft^2 yr) with nutrient enrichment. Nitrogenase activity has been observed in decomposing leaves, root surfaces (prop roots and pneumatophores) and sediment, but few of these studies have interpreted these rates relative to the N budget of mangrove forests (Kimball and Teas 1975, Gotto and Taylor 1976, Zuberer and Silver 1978, Potts 1979, Gotto et al. 1981). Results from mangrove sediments in south Florida indicate that N fixation rates range from 0.4 to

3.2 g N/m^2 yr (0.001 to 0.01 oz/ft^2 yr) (Kimball and Teas 1975, Zuberer and Silver 1978), similar for the natural rates of denitrification. These studies have shown that decaying mangrove leaves are sites of particularly high rates of fixation, and thus may account for some of the N immobilization in leaf litter on the forest floor (Gotto et al. 1981, van der Valk and Attiwill 1984). However, the spatial analysis of N fixation is still inadequate to provide a clear estimate of this contribution to the N budget of mangrove wetlands.

The other nutrient sink in mangrove wetlands is the burial of N and P associated with sedimentation. Sediments suspended in the water column are deposited in mangroves during flooding, and this material enriches mangrove soils. A survey of sedimentation and nutrient accumulation among five sites in south Florida and Mexico indicates patterns associated with the ecological types of mangroves (Lynch 1989, Lynch et al. 1989). The contribution of inorganic material ranged from 133 to 1,404 g/m^2 yr (0.44 to 4.6 oz/ft^2 yr). The higher values occurred in riverine mangrove forest in Mexico's Terminos Lagoon where the Candelaria, Chumpan, and Palizada rivers discharge more than 190 m^3/s (6,709 ft^3/s) of freshwater. Riverine mangroves have higher sedimentation rates due to the higher input of inorganic materials to these ecosystems. The accumulation of N among the riverine and basin forests was similar at about 5.5 g/m^2 yr (0.02 oz/ft^2 yr), which is higher than the range of denitrification and N fixation described above. Thus this is an important fate of N in mangrove ecosystems.

Intrasystem nutrient recycling mechanisms may reduce the loss of nutrients to export and reduce the demand for new nutrients to support primary productivity (Switzer and Nelson 1972, Turner 1977, Ryan and Bormann 1982). The canopy may be a site of N conservation in mangroves and, together with leaf longevity, could influence the N demand of these ecosystems. It is unclear if the relative amounts of N conserved in the canopy via retranslocation responds to amount of soil fertility. Another site of nutrient conservation is the forest floor, where the concentration of some nutrients in leaf litter during decomposition usually increases (Heald 1969, Rice and Tenore 1981, Twilley et al. 1986, Day et al. 1987). If this increase of nutrients, particularly N, is proportionately greater than the loss of leaf mass during decomposition, then there will be a net input of N to the forest floor. The source of this N may be absorption and adsorption processes by bacterial and fungal communities (Fell and Master 1973, Rice and Tenore 1981, Rice 1982) and N fixation (see discussion above). Twilley et al. (1986) found that this process of N immobilization occurs in decomposing red mangrove leaf litter in a basin mangrove forest. However, the significance of this ecological process to the nutrient budget of different mangrove wetlands has not been determined.

The nutrient biogeochemistry of mangrove wetlands consists of complex ecological processes which are still poorly understood (Figure 18.5). The role of intrasystem recycling mechanisms relative to the flux of nutrients at the boundary of coastal waters is important to understanding patterns of productivity and forest structure of these forested wetlands. The fate of nutrients is important to the management of mangrove wetlands for maintaining good water quality in coastal ecosystems (see Water Quality section).

Mangrove Food Webs

The function of mangrove wetlands as a source of habitat and food to estuarine-dependent fisheries is one of the most celebrated values of forested wetlands (Figure 18.6). These ecological processes will not be described in detail, given there are several excellent reviews that describe the secondary productivity of tropical mangrove ecosystems (Odum and Heald 1972, Yáñez-Arancibia 1985, Yáñez-Arancibia et al. 1988, Robertson and Duke 1990, Odum and McIvor 1990). The "outwelling hypothesis" of mangroves has been revised from the original paradigms described by Odum and Heald (1972) for the estuaries in south Florida (see Figure 18.6b). These modifications have been developed based on comparisons of food webs among different mangrove estuaries (Thayer et al. 1987, Yáñez-Arancibia et al. 1988, Robertson et al. 1992) and the use of natural isotope abundance to trace mangrove organic matter through estuarine food chains (Peterson and Fry 1984, Rodelli et al. 1984). There are seasonal and spatial differences in the amount of mangrove detritus that can be measured in shrimp and fish that inhabit mangrove estuaries. As distance from the source of mangrove detritus increases, the proportion of C in the tissue of shrimp from mangrove detritus decreases as the signal of C phytoplankton increases (Rodelli et al. 1984). Also, in tropical estuaries where mangrove habitat may be minimal compared to other primary producers such as seagrasses and phytoplankton, mangrove C in tissues of resident fauna is low. The seasonal timing of mangrove export of detritus relative to the migration of estuarine-dependent fisheries may also dilute the contribution of mangrove detritus from the food webs among diverse sites. In addition, mangrove detritus low in N relative to C may be modified by the microbial community and then utilized by higher trophic levels, masking the direct utilization of this organic matter as an energy source.

The secondary productivity of tropical estuarine fishes in mangrove wetlands have three contributions to the energy and nutrient dynamics of estuarine ecosystems; they can be stores of nutrients and energy, control rates and magnitude of energy flow through grazing of food sources, and move energy and nutrients across ecosystem boundaries (Yáñez-Arancibia 1985). The migratory nature of many of the nekton communities and the seasonal pulsing of both organic detritus input and primary productivity result in very complex linkages of forested wetlands with estuarine-dependent fisheries (Figure 18.6). The sequential pulsing of primary production by planktonic and macrophyte communities, coupled with seasonal export of mangrove detritus, suggests that the delivery of organic matter sustains high estuarine secondary production and species diversity of estuarine-dependent consumers (Figure 18.6a) (Rojas et al. 1992). The seasonal coupling of primary and secondary production in mangrove ecosystems, along with variation in environmental conditions, shows how the functional assemblages of fish use tropical lagoon habitats in time and space to reduce the effect of competition and predation. Assemblages of macroconsumers within functional groups play an important ecological role by coupling life-history strategies with the environmental gradients within the estuary (Figure 18.6). In Figure 18.6a, we can see the two main mangroves habitats (fringe and riverine), where the same fish population assemblages utilize the habitats in a sequential manner from one season to the next. This diversity in behavior

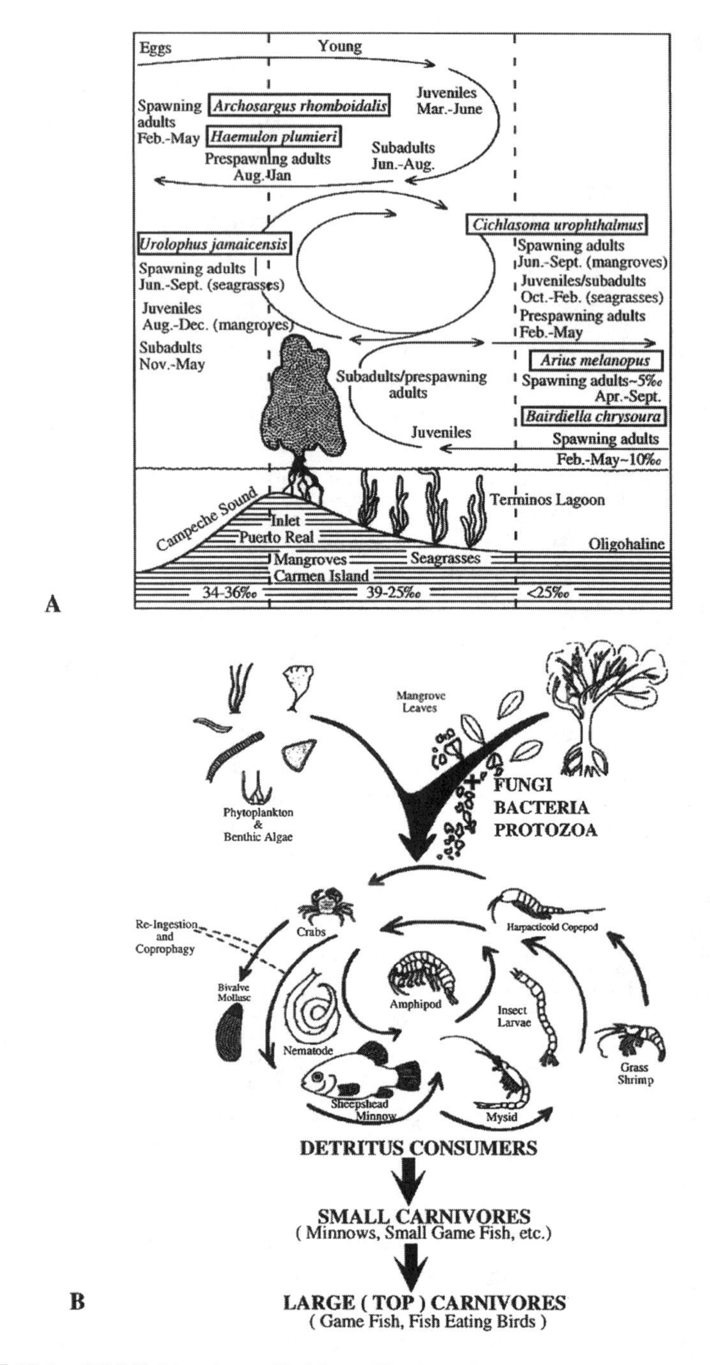

FIGURE 18.6 (A) Life histories and habitat utilization of six selected dominant fish species, including marine-estuarine spawners, estuarine spawners, and freshwater-estuarine spawners in Terminos Lagoon (from Yáñez-Arancibia et al. 1988). (B) Conceptual diagram describing the fate of mangrove leaf litter in the food chain of an estuary in south Florida (from Odum and Heald 1972).

suggests that the lag in fluctuations of total biomass through the year, as is common in high-latitude estuaries, is a consequence of the sequential use of the different mangrove habitats by different species (seasonal programming). Although common species appear in fringe and riverine mangroves, there are different peaks of abundance regulated by climatic changes which control the biological and physical characteristics of the two habitats (Yáñez-Arancibia et al. 1988, 1993). Thus the dominant species (or keystone species) act as controllers of the structure and function of the whole macroconsumer community, while the physical variability and mangrove productivity modulate their species diversity. The seasonal nature of these processes secure the recruitment and functional species diversity of estuarine nekton communities (Yáñez-Arancibia et al. 1993).

ECOSYSTEM FEEDBACKS AND DISTURBANCE

USE AND VALUE OF MANGROVES

The ecogeomorphic classification of mangroves (Figure 18.2) can be linked in a few instances to the structure and ecological processes of mangrove wetlands (Figure 18.1). Collectively these ecological properties of mangrove wetlands determine the function of these ecosystems in mangrove-dominated estuaries such as controlling water quality and providing habitat value. Mangroves produce a variety of forest products, support the productivity of economically important estuarine-dependent fisheries, and modify the water quality in warm-temperate and tropical estuarine ecosystems. These goods and services lead to increased human utilization of mangrove resources that vary throughout the tropics depending on economic and cultural constraints (Figure 18.1). Economic constraints are usually in the form of available capital to fund land-use changes in coastal regions, as well as river basin development. Cultural constraints are complex and determine the degree of environmental management and natural resource utilization. However, the sustainable utilization of coastal resources are to a large degree controlled by these two social conditions of a region. Human use and value of mangrove wetlands are therefore a combination of both the ecological properties of these coastal ecosystems together with patterns of social exploitation. Any Best Management Plan designed to provide for the sustainable utilization of mangrove wetlands has to consider both the ecological and social constraints of the region. Humans are part of all ecosystems, and management of natural resources is a combination of policies that seek to regulate the actions of societies within limitations that are imposed by the environment.

The utilization of mangrove resources by humans can have negative feedback effects on the ecological processes of mangrove ecosystems (Figure 18.1). These feedbacks can be indirect impacts, such as diversion of freshwater, which can cause changes in the productivity, litter export, and biogeochemistry of mangroves, among other processes. Direct impacts may include inputs of excess nutrients or toxic substances such as oil that limit the ecological processes of mangrove wetlands. If these impacts limit the ecological processes and, thus, the function of mangroves, then the impact connects all the way back to the use and value of these natural resources. Thus the properties of mangrove wetlands are influenced both by the environmental constraints (forcing functions in Figure 18.1) and social constraints (economics and culture in Figure 18.1) of a coastal region. Environmental and social constraints of mangrove wetlands are inextricably linked by the nature of environmental feedbacks from human exploitation. Thus the ecology of mangroves has to include the impacts of humans and nature. Sustainable development of mangrove wetlands may be viewed as the proper balance between environmental and social constraints that allow for the continued use and value of these coastal resources.

Land-Use Disturbance of Mangroves

Direct Exploitation and Reclamation

Deforestation of mangrove wetlands is associated with many uses of coastal environments including urban, agriculture, and aquaculture reclamation, as well as the use of forest timber for furniture, energy, chip wood, and construction materials (Watson 1928, Saenger et al. 1983). In the Gulf of Mexico and Caribbean, there are very few examples of mangrove use in silviculture as throughout most of the Neotropics, in comparison to Indonesia where mangrove plantations have been in existence for several decades (Snedaker 1986). One example of timber utilization in the Caribbean occurred in Puerto Rico where white-mangroves were used for fence posts, nearly resulting in the entire loss of this forest type from the island (Wadsworth 1959). Throughout the coast of Mexico and on many islands in the Caribbean, there is evidence of wood extraction from mangrove forests for many personal uses, but there is no commercial exploitation of this timber resource. The only plantations in this region have apparently been developed in Cuba over the last 20 years, although there is little published information on this operation. It is unclear why mangrove wetlands have not historically been used for timber in the Caribbean, although there are examples in South America where mangrove wood is used for production of charcoal (Snedaker 1986, Twilley et al. 1992). The site specific use of mangroves in silviculture is one example of how the cultural constraints on mangrove use and exploitation are important in the development of regional management plans (Figure 18.1).

Two reclamation activities that have contributed to examples of massive mangrove deforestation are agriculture and aquaculture enterprises. Agriculture impacts to mangroves are most noted in West Africa and parts of Indonesia (Ponnamperuma 1984b), but there are few direct reclamations of mangrove intertidal areas for rice production in the Neotropics. This is associated again to cultural and economic utilization of coastal landscapes in the Gulf and Caribbean, but also to environmental constraints of dry coastal environments. Many of the large agricultural uses are found in humid coastal areas or deltas where freshwater is abundant and intertidal lands are seasonally available for crop production. Mariculture use of the tropical intertidal zones has become one of the fastest growing aquaculture industries in the past decade. One of the largest and most publicized is the expansion of shrimp farming in Ecuador, where in the late 1980s, there were more shrimp produced in the ponds of this Pacific coastal zone than anywhere else in the world (Twilley et al. 1992).

The environmental impact of shrimp farming on mangrove wetlands and water quality of tropical estuaries may threaten the sustainability of this economic activity (Twilley et al. 1996). However, mariculture has not shown the same explosive impact in the intertidal zone of other Gulf and Caribbean regions, although there are isolated attempts that have prompted regional concerns of environmental quality. The Yucatan coastline of Mexico, Belize, and some of the Caribbean islands are sites of mariculture activities, but in several instances there have been attempts to minimize impacts to mangrove wetlands. In the United States, strict coastal zone management regulations limit the development of shrimp ponds in areas of mangrove wetlands.

One of the most obvious reclamation activities of mangroves in the Gulf of Mexico and more recently throughout the Caribbean region is deforestation associated with urban real estate including residential and recreational development. The impact of large-scale residential development on mangrove forests has been best documented in southwest Florida at Marco Island (Patterson 1986) and the region of Big Cypress Swamp (Carter et al. 1973). In the Marco Island area, the landscape cover of mangrove wetlands decreased from 4567 ha (11,285 acres) in 1952 to 3471 ha (8,577 acres) in 1984, largely due to the construction of residential development; this represents a 24% loss of these forested wetlands (Patterson 1986). The continued expansion of tourism throughout this region has increased the conflict with coastal zone management and particularly the loss of mangrove wetlands. The south Florida region has played a major role in establishing the ecological significance of mangroves that has influenced other efforts to protect this coastal resource in the Gulf of Mexico and Caribbean region.

Today, there is still much controversy over the continued permitting of marginal mangrove habitats for the construction of golf courses and residential housing in south Florida. In addition, there are preliminary plans to continue expanding recreational facilities along the Belizean reef system and the Quantan Roo region of the Yucatan Peninsula. These urban projects are threatening mangrove wetlands throughout the Gulf and Caribbean intertidal zones. Not only do these projects threaten deforestation of wetland areas by dredge and fill activities, but perimeter mangroves are trimmed to allow for residents to view the coastal scenery. In south Florida, one of the most controversial impacts of mangroves is the relaxation of tree trimming laws in coastal counties, resulting in massive removal of mangrove canopies (Beever 1989, Snedaker et al. 1992). The cumulative impacts of mangrove tree trimming, along with the direct impacts of residential development activities, are uncertain and provide for one of the major environmental conflicts in the south Florida region.

Freshwater Diversion

Human alterations of upland watersheds causing diversion of freshwater (dams and canals) can have severe impacts on coastal forested wetlands, particularly in dry coastal environments. Regional scale changes in freshwater surface inflow into coastal areas can alter the landscape distribution of mangrove wetlands and reduce the productivity and habitat quality of these forested wetlands. Together, these disturbances can alter the secondary productivity of tropical estuarine ecosystems due to degradation of habitat and environmental quality and alter the terrestrial and aquatic food webs that mangrove ecosystems support. The most dramatic impact of freshwater diversion in the Neotropics has been documented in Colombia, where the construction of highways and levees restricted the flow of the Magdelena River to the Cienaga Grande de Santa Marta, resulting in the death of nearly 25,000 ha (61,775 acres) of mangrove wetlands over the last decade (Botero 1990). The sensitivity of mangrove wetlands to slight changes in freshwater input in dry coastal regions that dominate the Caribbean and most of the Gulf Coast has been well documented (Cintrón et al. 1978). Hypersaline conditions in the intertidal zone of lagoons in Texas has increased with the diversion of freshwater from the coast, limiting the distribution of mangroves due to hypersaline soils (Britton and Morton 1989).

The distribution of freshwater is also a major issue in the Everglades, one of the largest mangrove areas of the Gulf of Mexico, where South Florida Water Management District and Corps of Engineers have developed a series of canals and water control structures to accommodate the freshwater supply and flood control from Lake Okeechobee to Florida Bay (Light and Dineen 1994). The water management practices of this south Florida region have resulted in modification of flow patterns from attenuated to more pulsed conditions, unnatural pooling and overdrainage, and unnatural and reduced flow of freshwater to Florida Bay (Light and Dineen 1994). The implications of these freshwater diversions to the natural resources of the region are evident in loss of wetland habitat and reduction in wading bird populations (see section on Fragmentation). The potential impact of these changes on the ecology of mangrove wetlands in this region are presently unknown. However, it has been proposed that the increased mangrove dieback on overwash islands in Florida Bay may be related to salinization associated with reduced freshwater discharge from the Everglades (McIvor et al. 1994).

The control of mosquitos in south Florida and other Gulf regions uses a water management strategy of keeping wetlands either continuously wet or dry to disrupt the life cycle of this insect pest. Impoundments have been constructed throughout the coast of Florida for the eradication of mosquitos (e.g., the Indian lagoon region). These impoundments consist of permanent dikes that prevent the exchange of surface tides and accumulate freshwater contribution by rainfall. The elevated water levels usually have an immediate increase in tree mortality, depending on the scheduled increase in water levels. Increased hydroperiod relative to the ability of tree roots to exchange gases with the atmosphere determines their susceptibility to stress. Impoundments have also been caused by highway construction (Zucca 1978, Patterson 1986); this leads to a decrease in the productivity and organic matter export of mangrove wetlands. The feedback effects of freshwater diversions and impoundments on mangrove ecosystems are depicted in Figure 18.1 by the interaction of environmental impacts with rivers and tides as a result of human disturbance. The sensitivity of ecological processes such as productivity and litter dynamics will depend on the humidity province of the coast; alterations to mangroves in drier coastal environments will have more impact than those in more humid regions.

Water Quality

Other human impacts on mangrove wetlands, such as the introduction of contaminants that can disrupt natural ecological processes, can be more direct (Figure 18.1). Deterioration of water quality includes inputs of excessive nutrients to coastal waters (eutrophication) and toxic materials (heavy metals, oil spills, pesticides). The increased attraction of coastal environments by humans has made coastal regions more vulnerable to these water quality problems. In addition, coastal waters are important channels of navigation that provide access for petroleum and chemical industries. The urbanization of coastal regions, specifically mangrove habitats, results in unique management conflicts that in some instances could be minimized by using the free services of mangroves as a form of mitigating these impacts to coastal waters.

The increase in residential and recreational real estate in coastal regions of the Gulf and Caribbean not only causes direct impacts such as deforestation, but this land use also increases the amount of nutrient discharge to coastal waters. Mangrove wetlands have the potential to be used in a landscape design to remove excess nutrients from waste waters before they are discharged to coastal systems. This potential depends on the nutrient biogeochemistry described above, such that the nutrient sink capacities of mangroves are optimized relative to the nature of the input. There have been some descriptions of how to use mangrove wetlands for nutrient abatement in waste waters (Sell 1977, Clough et al. 1983, Day in press). As described above, mangrove wetlands have a high capacity of N removal via denitrification (Nedwell 1975, Corredor and Morell 1994), and nutrient burial in mangrove soils may also be an important nutrient removal process. However, the detailed designs of hydrology and environmental engineering to optimize for these losses need better specifications, since there are few pilot studies of these operations. This management approach using the nutrient biogeochemistry of mangrove wetlands represents a very important point in the management of these coastal forests (Figure 18.1); the natural free services of these forested wetlands as a nutrient sink can minimize environmental impacts. Engineering mangrove wetlands for the use of wastewater treatment, rather than removing these forested wetlands, can be a positive feedback to sustain human activities in the coastal zone.

Oil spills represent contaminants to mangroves that can alter succession, productivity, and nutrient cycling. These impacts have been well documented in ecological studies in Puerto Rico (Cintrón et al. 1981), Panama (Duke and Pinzon 1993, Garrity et al. 1994, and citations within), and Gulf of Mexico (Getter et al. 1981). Negative feedbacks of contaminants such as oil in tropical coastal waters are the products of both oil exploration and transportation. These inputs become constraints on the ecological processes of mangroves by limiting their ability to support ecological functions (e.g., fisheries and forestry) that provide goods and services in tropical coastal regions. Efforts to minimize the damage of oil spills and enhance the recovery of mangroves are important to minimize the complex loss of ecological functions to the coastal zone.

An oil slick in a mangrove wetland will cause tree mortality, depending on the concentration of hydrocarbons and species of trees, as well as the edaphic stress levels already existing at the site (Cintrón et al. 1981, Duke and Pinzon 1993). Thus, those mangroves in dry coastal environments of the Gulf of Mexico and Caribbean may be more vulnerable to oil spills than those in more humid environments. Also, Duke and Pinzon (1993) found that the exposure of the shore to wind and wave energy was an important factor in determining the degree of tree mortality and reductions in growth. Mangrove disturbance was patchy, depending on the transport of oil into fringe zones of this microtidal environment, and few inland mangrove zones were perturbed. Oil impacts can also affect the recruitment of mangrove seedlings into young tree cohorts, depending on the effects of oil type and concentration in the plot (Duke and Pinzon 1993). The leaf area index of a mangrove canopy decreases as a response of oil toxicity, and this additional leaf litter is processed in the ecosystem. The reduced canopy also increases light resources in forests, and this will influence the relative competition of subdominants and seedlings within the gap. However, this increase in light resources has to be evaluated simultaneously with negative effects of soil hydrocarbons on plant growth and recruitment. Thus an increase in light resources may have minimal positive effect on forest recovery and patterns of zonation following oil disturbance.

Climate Disturbances

Tropical Storms and Hurricanes

Warmer temperatures are expected to increase the intensity and/or the frequency of storms (DeSylva 1986, Emanuel 1987, Hobgood and Cerveny 1988). Since mangroves of the Gulf of Mexico and Caribbean are distributed in latitudes with frequent hurricanes, it is important to understand how tropical storms and hurricanes affect forest development (i.e., forest structure, species composition, etc.) and community dynamics in mangrove wetlands. For example, studies in Florida (Davis 1940, Egler 1952, Craighead and Gilbert 1962, Alexander 1967), Puerto Rico (Wadsworth 1959, Glynn et al. 1964), Mauritius (Sauer 1962), and British Honduras (Vermeer 1963) reported qualitative information on the effects of hurricanes on defoliation and tree mortality. Recent studies in Nicaragua (Roth 1992) and Florida (Smith et al. 1994) have provided a more quantitative evaluation of the effects of hurricanes on mangrove forests. Damage by Hurricane Joan was more severe on large trees and favored the establishment of abundant regeneration in all mangrove species in the area (Roth 1992). Tree mortality was 36% in the sample area, representing 68% of the basal area of a pre-existing stand. Seventeen months after the hurricane, trees of short stature and small diameter recovered best. This fast recovery of short trees has been reported in other areas (Wadsworth 1959, Craighead 1971). Roth (1992) stressed that species attributes and availability of propagules are important factors, along with the severity of storm and sediment disturbance, in projecting recovery patterns.

Frequent storm disturbance tends to favor species capable of constant or timely flowering, abundant seedling or sprouting, fast growth in open conditions, and early reproductive maturity (Roth 1992). Although mangrove trees show these "traits," competition with other species introduced by humans and with similar characteristics can change the plant diversity and affect community metabolism of mangrove ecosystems (e.g., Ball 1980). Thus, in order to evaluate the effect of hurricanes on biodiversity in mangrove ecosystems, it is important to consider the cumulative impact of human activities on these ecosystems in conjunction with the complex natural cycle of regeneration and growth of mangrove forests (Smith et al. 1994).

Sea Level Rise

The intertidal nature of mangrove wetlands has lead to recent speculation regarding the continued existence of these ecosystems in relation to various scenarios of accelerated rates in sea level rise (SLR) (Woodroffe 1990, Ellison and Stoddard 1991, Ellison 1993, Parkinson et al. 1994, Semeniuk 1994, Wanless et al. 1994). The continued existence of mangroves within the intertidal zone depends on numerous processes that contribute to vertical accretion at a rate that balances the increase in regional sea level. Critical rates in SLR have been estimated above which there is a projected collapse of mangrove ecosystems (Woodroffe 1990). In the absence of terrigenous sediments, mangroves can tolerate SLR of 0.8 to 0.9 mm/yr (0.031 to 0.035 in/yr), are under stress at SLR ranging from 0.9 to 1.2 mm/yr (0.035 to 0.047 in/yr), but cannot sustain existence at SLR >1.2 mm/yr (0.047 in/yr) (Ellison and Stoddart 1991). Given projected rates of accelerated SLR of 10 to 20 mm/yr (0.394 to 0.787 in/yr), most mangrove wetlands will collapse as viable coastal ecosystems. For example, a SLR of 1.43 mm/yr (0.056 in/yr) is responsible for die-back areas of mangroves in Bermuda (Ellison 1993), and geologic records of coastal south Florida suggest that any prolonged rate at or faster than 2.3 mm/yr (0.091 in/yr) will lead to rapid and complete coastal erosion (Wanless et al. 1994).

There is evidence that mangroves sustained periods of accelerated rates of SLR above these critical rates and maintained an existence in the intertidal zone of estuaries over the last several thousand years. The mean rate of SLR for Gulf coast and Caribbean stations of 2.3 mm/yr (0.091 in/yr) (Gornitz et al. 1982) is very similar to the mean accretion rates (2.2 mm/yr [0.087 in/yr]) of several mangrove wetlands in this region (Lynch et al. 1989). An integrated rate of SLR over the last 100 yrs, based on tidal records from Key West, is 2.02 mm/yr (0.080 in/yr), which is within one standard error of the recent accretion rate (1.8 mm/yr ± 0.2 mm/yr [0.071 ± 0.008 in/yr]) at Rookery Bay. Mangroves in Australia can keep pace with changes in rates of SLR, with fluctuations ranging from 0.2 to 6 mm/yr (0.008 to 0.236 in/yr) in the south Alligator tidal river, and mangrove forests in many estuaries in northern Australia have tolerated SLR of 8 to 10 mm/yr (0.315 to 0.394 in/yr) in the early Holocene (Woodroffe et al. 1986, Woodroffe 1990). These rates are higher than many projections of accelerated rates of SLR associated with global climate change. Many of these mangroves receive terrigenous sediments and exist in macrotidal environments, with critical rates that are much different than for mangroves in microtidal and oligotrophic environments. The continuous existence of mangroves on carbonate islands in both the Caribbean and Pacific Oceans, in response to shifts in SLR from mid to late Holocene periods, are further evidence that sedimentation rates of mangroves in microtidal environments can keep pace with changes in SLR from 0.4 to 5 mm/yr (0.016 to 0.197 in/yr) (Woodroffe 1990, Bacon 1994, Snedaker et al. 1994).

CONSERVATION AND MANAGEMENT

BIODIVERSITY

The low species richness of mangrove trees in the Gulf of Mexico and Caribbean supports a disproportionately rich diversity of animals, the dimensions of which are only now being documented. Mangrove ecosystems support a variety of marine and estuarine food webs involving an extraordinarily large number of animal species (Odum and Heald 1972, Yáñez-Arancibia et al. 1988, Thayer et al. 1987, Gilmore and Snedaker 1993) and a complex heterotrophic microorganism food web (Odum 1971, Fell and Master 1973, Snedaker 1989, Robertson et al. 1992). In the new world tropics, extensive surveys of the composition and ecology of mangrove nekton have found 26 to 114 species of fish (Robertson and Blaber 1992). In addition to the marine and estuarine food webs and associated species, there are a relatively large number and variety of animals that range from terrestrial insects to birds that live in and/or that feed directly on mangrove vegetation. These include sessile organisms such as oysters and tunicates, arboreal feeders such as foliovores and frugivores, and ground-level seed predators. A highly diverse variety of sponges, tunicates, and other forms of epibionts live on prop roots of mangroves (Sutherland 1980, Rützler and Feller 1988, Ellison and Farnsworth 1992), especially along mangrove shorelines with little terrigenous input. There are four distinct spatial guilds of residents and casual faunal populations in south Florida mangroves that may have well over an estimated 200 species, many of which are as yet non-cataloged (Gilmore and Snedaker 1993). In addition, over 200 species of insects have been documented in mangroves in the Florida Keys (Simberloff and Wilson 1969). This same richness of insects and faunal biota has been observed in other parts of the Caribbean (Rützler and Feller 1988, Bacon 1990, Feller 1993, Farnsworth and Ellison 1991, 1993).

One of the most published links between mangrove biodiversity and ecosystem function may be the presence of crabs in mangrove wetlands (Jones 1984, Smith 1987a, Twilley et al. 1996). In general, the mangrove crab fauna is dominated by representatives of two families, the Ocypodidae and Grapsidae, and each family by one genus, *Uca* and *Sesarma,* respectively. Furthermore, within the Grapsidae the genus *Sesarma* accounts for more than 60 species of crabs predominantly associated with mangroves (Jones 1984), of which the tropical Americas have only three to five species, compared to more than 30 in other tropical continents. Crabs can influence forest structure (Smith 1987a, 1987b, Smith et al. 1989, Robertson et al. 1990), litter dynamics (Robertson and Daniel 1989), and nutrient cycling (Smith et al. 1991) of mangrove wetlands, suggesting that they are a keystone guild in these forested ecosystems. The selective consumption by crabs on mangrove propagules influences rates of sapling recruitment by removing some species from zones of crab habitation, although this process seems to be less significant in the Gulf of Mexico wetlands than in Australia (Smith et al. 1989). In the Neotropics, crab consumption is less intense, and recruitment generally follows the distribution patterns of propagules and the constraints of edaphic conditions.

FRAGMENTATION

The fragmentation of mangrove-dominated landscapes is believed to create the same types of problems for migratory organisms that are associated with the fragmentation of upland forests, yet there has been little, if any, research on this topic. There is no documentation concerning diel or seasonal migration patterns of resident species within mangroves, or how such species might be affected by the impact of fragmentation. For instance, seagrass and adjacent mangrove habitats are used by many species of nekton and are generally characterized by high fish abundance and diversity. From Yáñez-Arancibia et al. (1993), it is clear that the utilization of the two interacting habitats by fishes is spatially distinct but linked by the life cycles of organisms (Figure 18.6a). They found that there is a strong correlation between the life history patterns of migratory fish and the pattern of primary production, using the two habitats sequentially in a time period. The fragmentation of mangrove-seagrass landscapes probably will reduce ecosystem complexity and nekton diversity. It has been speculated that one of the consequences of reduced mangrove area and/or increased fragmentation will be reduction in population numbers (and this may be important for commercial species) or outright local extinctions of certain species. However, as reviewed by Robertson and Blaber (1992), there is no empirical evidence that such a consequence will occur.

The historical use of wetland resources by large flocks of wading birds in the Everglades may also represent the impact of landscape fragmentation on sustaining wildlife populations (Ogden 1994). The peak number of nesting wading birds declined from 180,000 to 245,000 birds in 1933–34 to 50,000 birds in 1976. Four of the five major nesting colonies in the Everglades are associated with mangrove-dominated habitats including the lower Ten Thousand Islands between the mouth of Lostmans River and Everglades City, a cluster of mangrove islands in northwestern Florida Bay, the southern mainland estuary from Cape Sable to Taylor Slough, and the mangrove-marsh ecotone at the headwaters of Whitewater Bay (Ogden 1994). The reduction in wading birds in these regions is associated with changes in the timing, location, and strength of the foraging opportunities. One hypothesis of explaining the historical change in populations is the link between alterations in upland hydrology (freshwater delivery) causing a compartmentalized wetland system creating barriers that have limited emigration and seasonal recolonization (see citations in Ogden 1994 and Frederick and Spalding 1994). One of the major factors in the reduction of birds is the reduction of prey and feeding refugia, particularly in the mangrove-marsh ecotone in the upper estuarine zone. The fragmentation of this estuarine zone linked to changes in hydrological conditions has been recommended as a focus of the restoration effort of the Everglades ecosystem.

Mangrove Restoration

Restoration of ecosystems is considered the ultimate test of our scientific understanding of the ecological processes that control the structure and function of the landscape (Ewel 1987). There have been several reviews of mangrove restoration (Carlton 1974, Teas 1981, Lewis 1982, Lewis 1990a, 1990b, Cintrón 1990, Field 1996) that have collectively alluded to the concept that since these forested wetlands are adapted to stressed environments, they are relatively amenable to restoration efforts. The success of mangrove restoration is the establishment of the proper environmental settings (described in this chapter) that control the characteristic structure and function of mangrove wetlands. Degraded or disturbed mangroves can be described as a divergence in mangrove structure and function away from the natural or reference condition (reference site). The degree of this divergence in some selected properties of the degraded site away from the natural site depends on the magnitude of the impact. The goal of ecological restoration is to return the degraded site back to either the natural condition (restoration) or to some other new condition (rehabilitation). In most cases, the damage to mangrove environments is so severe that restoration efforts are no longer available. The rates of change in the ecological characteristics of mangrove wetlands between natural, degraded, and some rehabilitated condition are known as trajectories. The nature of these trajectories will depend on the type of environmental impact (feedback from uses and values in Figure 18.1), the magnitude of the impact, and the ecogeomorphic type of mangrove wetland that is impacted. Alterations in the environmental setting have to be restored or maintained to obtain a restoration trajectory whereby a degraded site returns to a natural condition; deviations in the environmental settings will more likely result in some altered or rehabilitated mangrove wetland. Trajectories are a very important concept in the ecological restoration of any ecosystem, since they include both the structural and functional characteristics of the sites relative to natural conditions and the time required to obtain a rehabilitated state or condition.

The success of any mangrove restoration project depends on the establishment of proper site conditions, including currents from waves and winds, tidal flooding frequency, temperature, soil nutrients, and stressors (salinity and hydrogen sulfide), along with ecological processes of the site such as the availability of propagules and the recruitment of these individuals to the sapling stage of development. Thus, the seasonal nature of planting and propagule density, along with increased selective mortality of propagules from physiological stress and increased mortality by crab consumption and disease, will establish the initial trajectory of any restoration project (Lewis 1982, Cintrón 1990). Some of the key parameters of a restoration project include the elevation of the landscape to provide the proper hydrology of the site, recognizing the significance of natural processes to sustain the restored condition, and proper planting techniques to enhance recruitment. There are several examples of mangrove restoration projects in Florida and Puerto Rico that include some recommendations concerning site preparation and mangrove planting techniques (see references in Lewis 1990a, 1990b, Cintrón 1990, and Snedaker and Biber 1996). The lack of post-project monitoring has limited the ability to determine what specific techniques deliver the optimum results. Recent studies on factors that affect tolerance of seedlings to environmental and ecological processes discussed above provide more specific information on the nature of site preparations needed to establish mangrove forests.

RESEARCH DIRECTIONS

Mangrove wetlands have been of tremendous botanical interest since they are the only trees that can tolerate saline waters, along with other interesting characteristics in reproductive biology (Tomlinson 1986). In the last two decades, since the classic paper by Lugo and Snedaker (1974), there has been particular interest in the ecology of mangroves. This focus was initially on the function of mangroves in coastal landscapes as a source of nutrition for economically important estuarine fisheries, as demonstrated by the work of Odum and Heald (1972). Studies of organic matter production, decomposition, and export dominated the literature for several years, with limited contribution to understanding the biogeochemistry of mangroves. One of the most controversial ecological properties of mangroves is succession, which has been debated since Davis (1940) followed the general paradigm that zonation recapitulates mangrove development. There have been renewed interest in patterns of mangrove zonation and forest development testing concepts of tropical forest gap dynamics in mangrove wetlands. There has also been recent discoveries in the physiological ecology of mangroves, trying to link patterns of tree distribution to tolerance of edaphic conditions such as salinity and sulfide. Finally, the stress of nutrient limitation and more comprehensive work on nutrient cycling in mangrove wetlands has added much to our knowledge of how these unique ecosystems work. This is an exciting time in the study of mangrove ecosystems, with anticipation that these hierarchical studies will uncover some of the mysteries of these complex tropical forests that live in the sea. The development of mangrove research has occupied two distinct geographical hemisphere with competing ideas from the old world and new world tropics. The landmark paper by Lugo and Snedaker (1974) that captured the ideas of mangrove ecology in the Gulf of Mexico and the Caribbean has been recently balanced with a summary of mangroves in Australia by Robertson and Alongi (1992). The increased interaction among mangrove scientists of the two hemispheres has produced evidence that there may be distinct differences in the properties of forested wetlands of these two continental regions.

Appendix Common and Scientific Names of Plants and Animals Mentioned in the Text

PLANTS[1]

Common name	Scientific name
Alabama supplejack, supple-jack, rattan-vine	*Berchemia scandens* (Hill) K. Koch
Alder	
European alder	*Alnus glutinosa* (L.) Gaertn.
hazel alder, tag alder	*Alnus serrulata* (Ait.) Willd.
Japanese alder	*Alnus japonica* (Thunb.) Steud.
red alder	*Alnus rubra* Bong.
	Alnus hirsuta Turcz.
alligator-weed	*Alternathera philoxeroides* (Mart.) Griseb.
American beech	*Fagus grandifolia* Ehrh.
American hornbeam, ironwood, blue beech	*Carpinus caroliniana* Walt.
andropogon	*Andropogon scoparius* Michx.
arrowhead	
broad-leaf arrowhead	*Sagittaria latifolia* (Willd.) var. *pubescens* (Muhl.) J.G. Sm.
bunched arrowhead	*Sagittaria fasciculata* E. O. Beal
arrow-leaf tearthumb	*Polygonum sagittatum* L.
ash	
Carolina ash, water ash, pop ash	*Fraxinus caroliniana* Mill.
green ash, red ash	*Fraxinus pennsylvanica* Marsh.
pumpkin ash	*Fraxinus profunda* (Bush) Bush
white ash	*Fraxinus americana* L.
aster	*Aster* spp.
Atlantic white-cedar	*Chamaecyparis thyoides* (L.) B.S.P.
bay	
loblolly-bay	*Gordonia lasianthus* (L.) Ellis
redbay	*Persea borbonia* (L.) Spreng.
swampbay	*Persea palustris* (Raf.) Sarg.
sweetbay	*Magnolia virginiana* L.
beakrush	*Rhynchospora* spp.

bean	*Phaseolus vulgaris* L.
beardtongue	*Penstemon* sp.
beech, American	*Fagus grandifolia* Ehrhart
birch	
European white birch	*Betula pendula* Roth
river birch	*Betula nigra* L.
sweet birch	*Betula lenta* L.
white birch	*Betula pubescens* J. F. Ehrh.
yellow birch	*Betula alleghaniensis* Britt.
blackberry	*Rubus argutus* Link
black cherry	*Prunus serotina* Ehrh.
black chokeberry	*Aronia melanocarpa* (Michx.) Ell.
black walnut	*Juglans nigra* L.
bladdernut	*Staphylea trifolia* L.
bladderwort	*Utricularia* spp.
bloodroot	*Sanguinaria canadensis* L.
blueberry	*Vaccinium elliottii* Chapm.
blueberry, highbush	*Vaccinium corymbosum* L.
bluestem, turkeyfoot	*Andropogon gerardii* Vitman
bog bluegrass	*Poa paludigena* Fern. & Wieg.
bog buckbean	*Menyanthes trifoliata* L.
bog goldenrod	*Solidago uliginosa* Nutt.
bog-rose	*Arethusa bulbosa* L.
boxelder	*Acer negundo* L.
broom sedge	*Andropogon virginicus* L.
brownish beakrush	*Rhynchospora capitellata* (Michx.) Vahl
buckthorn bumelia	*Bumelia lycioides* (L.) Pers.
buckwheat-tree	*Cliftonia monophylla* (Lam.) Britt.
bulrush	*Scripus lacustris* L.
bulrush	*Scirpus* sp.
buttonbush, common	*Cephalanthus occidentalis* L.
bushy St. Johnswort	*Hypericum densiflorum* Pursh.
Carolina sheep-kill	*Kalmia angustifolia* var. *caroliniana* (Small) Fern.
cattail	*Typha* sp.
cedar	
eastern redcedar	*Juniperus virginiana* L.
southern redcedar	*Juniperus silicicola* (Small) Bailey
climbing hempweed	*Mikania scandens* (L.) Willd.
climbing hydrangea	*Decumaria barbara* L.
common reed	*Phragmites* spp.
corkwood	*Leitneria floridana* Chapm
cottongrass	*Eriophorum virginicum* L.
cottonwood	
black cottonwood	*Populus trichocarpa* Torr. & Gray
eastern cottonwood	*Populus deltoides* (Bartr.) ex Marsh.
swamp cottonwood	*Populus heterophylla* L.

cowlily	*Nuphar luteum* (L.) Sobthorp & Smith
cranberry	*Vaccinium macrocarpon* Ait.
cross vine	*Anisostichus capreolata* (L.) Bureau
cypress	
baldcypress	*Taxodium distichum* (L.) Rich.
pondcypress	*Taxodium distichum* var. *nutans* (Ait.) Sweet
dangleberry	*Gaylussacia frondosa* (L.) T. & G.
dock	*Rumex acetosa* L.
	Rumex maritimus L.
	Rumex palustris Sm.
dog-hobble	*Leucothoe axillaris* (Lam.) D. Don
	Leucothoe fontanesiana (Steud.) Sleumer
swamp leucothoe	*Leucothoe racemosa* (L.) Gray
dogwood	
flowering dogwood	*Cornus florida* L.
roughleaf dogwood	*Cornus drummondii* C.A. Meyer
silky dogwood	*Cornus amomum* Miller
duckweed	*Lemna* spp., *Spirodella oligorrhiza* (Kurz) Hegelm.
dwarf huckleberry	*Gaylussacia dumosa* (Andrz.) T. & G.
eastern hemlock	*Tsuga canadensis* (L.) Carr.
eastern redbud	*Cercis canadensis* L.
elm	
American elm	*Ulmus americana* L.
cedar elm	*Ulmus crassifolia* Nutt.
Florida elm	*Ulmus americana* var. *floridana* (Chapm.) Little
winged elm	*Ulmus alata* Michx.
eucalyptus	*Eucalyptus marginata* Donn ex Smith
false nettle	*Boehmeria cylindrica* (L.) Swartz
fern	
Christmas fern	*Polystichum acrostichoides* (Michx.) Schott
cinnamon fern	*Osmunda cinnamomea* L.
netted chain fern	*Woodwardia areolata* (L.) Smith
royal fern	*Osmunda regalis* L. var. *spectabilis* (Willd.) A.Gray
sensitive fern	*Onoclea sensibilis* L.
fetterbush	*Lyonia lucida* (Lam.) Koch
franklinia	*Fanklinia alatamaha* Bartr.
giant cane	*Arundinaria gigantea* (Walt.) Muhl.
golden club	*Orontium aquaticum* L.
goldenrod	*Solidago* spp.
grass-of-Parnassus	*Parnassia asarifolia* Vent.
greenbriar	
catbriar	*Smilax bona-nox* L.
coral greenbriar	*Smilax walteri* Pursh.
greenbriar	*Smilax rotundifolia* L.
sawbriar	*Smilax glauca* Walt.

hackberry	
hackberry	*Celtis occidentalis* L.
sugarberry	*Celtis laevigata* Willd.
haircap moss	*Polytrichum commune* Hedw.
hawthorns	
green hawthorn	*Crataegus viridis* L.
parsley hawthorn	*Crataegus marshalli* Eggl.
heartleaf	*Nymphoides* spp.
hickory	
bitternut hickory	*Carya cordiformis* (Wang.) Koch.
nutmeg hickory	*Carya myristicaeformis* (Michx. f.) Nutt.
pecan, sweet	*Carya illinoensis* (Wang.) Koch
water hickory, bitter pecan	*Carya aquatica* (Michx. f.) Nutt.
holly	
American holly	*Ilex opaca* Ait.
dahoon	*Ilex cassine* L.
gallberry	*Ilex glabra* (L.) Gray
large gallberry	*Ilex coriacea* (Pursh) Chapm.
myrtle dahoon	*Ilex myrtifolia* Walt.
possumhaw, deciduous holly	*Ilex decidua* Walt.
sarvis holly	*Ilex amelanchier* M. A. Curtis
yaupon	*Ilex vomitoria* Ait.
huckleberry	*Gaylussacia* spp.
Iva	*Iva microcephala* Nutt.
Japanese grass	*Microstegium virmineum* (Trin.) A. Camus
Japanese honeysuckle	*Lonicera japonica* Thunberg
knotweed	*Polygonum* spp.
leafy liverwort	*Bazzania trilobata* (L.) S. Gray
leatherleaf	*Cassandra calyculata* (L.) D. Don
leersia	*Leersia hexandra* Swartz.
lily	
roan lily	*Lilium grayi* Watson
red Canada lily	*Lilium canadense* var. *editorum* Fernald
lizard's tail	*Saururus cernus* L.
locust	
black locust	*Robinia pseudoacacia* L.
honeylocust	*Gleditsia triacanthos* L.
waterlocust	*Gleditsia aquatica* Marsh.
maidencane	*Panicum hemitomum* Schultes
magnolia, southern	*Magnolia grandiflora* L.
maleberry	*Lyonia ligustrina* (L.) DC var. *ligustrina*
mango	*Mangifera indica* L.
mangrove	
black-mangrove	*Avicennia germinans* (L.) L.
button-mangrove, buttonwood	*Conocarpus erectus* L.

red mangrove	*Rhizophora mangle* L.
white-mangrove	*Laguncularia racemosa* (L.) Gaertn.
maple	
Drummond red maple	*Acer rubrum* var. *drummondii* (Hook & Arn.)Sarg.
Florida maple	*Acer barbatum* Michx.
red maple	*Acer rubrum* L.
silver maple	*Acer saccharinum* L.
marshhay cordgrass	*Spartina patens* (Ait.) Muhl
marsh-marigold	*Caltha palustris* L.
mayapple	*Podophyllum peltatum* L.
mistletoe	*Phoradendron serotinum* (Raf.) M. C. Johnston
moss	*Aulacomnium palustre* (Hedw.) Schwaegr.
mountain laurel	*Kalmia latifolia* L.
muscadine grape, scuppernong	*Vitus rotundifolia* L.
narrow pipewort	*Eriocaulon lineare* Small
needle palm	*Rhaphidophyllum hystrix* (Pursh.) Wendl.
New England aster	*Aster novae-angliae* L.
oak	
black oak	*Quercus velutina* Lam.
bur oak	*Quercus macrocarpa* Michx.
cherrybark oak	*Quercus falcata* var. *pagodaefolia* Ell.
laurel oak, Darlington oak	*Quercus hemisphaerica* Bartr.
live oak	*Quercus virginiana* Mill.
northern red oak	*Quercus rubra* L.
Nuttall oak	*Quercus nuttalli* Palmer
Oglethorpe oak	*Quercus oglethorpensis* Duncan
overcup oak	*Quercus lyrata* Walt.
pin oak	*Quercus palustris* Muenchh.
scarlet oak	*Quercus coccinea* Muenchh.
Shumard oak	*Quercus shumardii* Buckl.
swamp chestnut oak	*Quercus michauxii* Nutt.
swamp laurel oak, diamondleaf oak	*Quercus laurifolia* Michx.
swamp white oak	*Quercus bicolor* Willd.
water oak	*Quercus nigra* L.
white oak	*Quercus alba* L.
willow oak	*Quercus phellos* L.
palm, Jauari	*Astrocaryum jauari* Mart.
palmetto	
cabbage palmetto	*Sabal palmetto* (Walt.) Lodd
dwarf palmetto	*Sabal minor* (Jacq.) Pers.
saw-palmetto	*Serenoa repens* (Bartr.) Small
partridge berry	*Mitchella repens* L.
pawpaw	*Asimina triloba* (L.) Dunal
pea	*Pisum sativum* L.
peatmoss	*Sphagnum fallax* (Klinggr.) Klingrr.
	Sphagnum affine Ren. & Card.

	Sphagnum magellanicum Brid.
	Sphagnum palustre L.
	Sphagnum recurvum P. Beauv.
pennywort	*Hydrocotyle* sp.
pepper-vine	*Ampelopsis arborea* (L.) Koehne
persimmon, common	*Diospyros virginiana* L.
pickerelweed	*Pontederia cordata* L.
pine	
eastern white pine	*Pinus strobus* L.
Honduras Caribbean pine	*Pinus caribaea* var. *hondurensis* (Sénécl) Barr.et Golf.
loblolly pine	*Pinus taeda* L.
lodgepole pine	*Pinus contorta* Dougl. ex Loud.
longleaf pine	*Pinus palustris* Mill.
Mediterranean pine	*Pinus halepensis* Mill.
pitch pine	*Pinus rigida* Mill.
pond pine	*Pinus serotina* Michx.
sand pine	*Pinus clausa* (Chapm. ex Engelm.) Vasey exSarg.
Scotch pine	*Pinus sylvestris* L.
shortleaf pine	*Pinus echinata* Mill.
slash pine	*Pinus elliottii* Engelm.
spruce pine	*Pinus glabra* Walt.
Virginia pine	*Pinus virginiana* Mill
pitcher-plant	
green pitcher plant	*Sarracenia oreophila* Kearney ex Wherry
mountain sweet pitcher plant	*Sarracenia rubra* Walt. ssp. *jonesii* (Wherry) Wherry
northern pitcher plant	*Sarracenia purpurea* L.
poison-ivy	*Toxicodendron radicans* (L.) Kuntze
poison sumac	*Rhus vernix* L.
privet	
Chinese privet	*Ligustrum sinense* Lour.
shining privet	*Ligustrum japonicum* Thunberg
quaking aspen	*Populus tremuloides* Michx.
red buckeye	*Aesculus pavia* L.
red chokeberry	*Aronia arbutifolia* (L.) Ell.
red mulberry	*Morus rubra* L.
red spruce	*Picea rubens* Sarg.
rhododendron	*Rhododendron* spp.
rice	*Oryza sativa* L.
rice cutgrass	*Leersia oryzoides* (L.) Sw.
rosebay rhododendron	*Rhododendron maximum* L.
rough-leaf goldenrod	*Solidago patula* Muhl.
round-leaved sundew	*Drosera rotundifolia* L.
rubber tree, Para rubber	*Hevea brasiliensis* Müll. Arg.
rush	
black rush	*Juncus roemerianus* Scheele

narrow-panicle rush	*Juncus brevicaudatus* (Engelm.) Fern.
rough rush	*Juncus caesariensis* Coville
soft rush	*Juncus effusus* L.
sassafras	*Sassafras albidum* (Nutt.) Nees
saw grass	*Cladium jamaicense* Crantz
Sebastian bush	*Sebastiania ligustrina* (Michx.) Muell-Arg.
sedge	
bristly-stalk sedge	*Carex leptalea* Wahl.
few-seeded sedge	*Carex oligosperma* Michx.
hoary sedge	*Carex canescens* L.
sedge	*Carex folliculata* L.
sedge	*Carex gynandra* Schw.
tussock sedge	*Carex stricta* Lam.
seedbox	*Ludwigia* sp.
serviceberry	*Amelanchier* spp.
skunk cabbage	*Symplocarpus foetidus* (L.) Nutt.
smooth cordgrass	*Spartina alterniflora* Loisel.
snowbell, storax	*Styrax americana* Lam.
sourwood	*Oxydendron arboreum* (L.) DC
South American elodea	*Egeria densa* Planchon
southern bayberry, wax-myrtle	*Myrica cerifera* L.
southern catalpa	*Catalpa bignonioides* Walt.
Spanish moss	*Tillandsia usneoides* L.
spicebush	*Lindera benzoin* (L.) Blume
spikerush	*Juncus* spp.
stiffcornel dogwood, swamp dogwood	*Cornus stricta* Lam.
strawberry bush	*Eunonymus americanus* L.
sunflower	*Helianthus annuus* L.
supplejack	*Berchemia scandens* (Hill) K. Koch
swamp cyrilla, titi, he-huckleberry	*Cyrilla racemiflora* L.
swamp dewberry	*Rubus hispidus* L.
swamp fern	*Blechnum serrulatum* Rich.
swamp leucothoe, fetter-bush	*Leucothoe racemosa* (L.) Gray
swamp-pink	*Helonias bullata* L.
swamp-privet	*Forestiera acuminata* (Michx.) Poir.
swamp rose	*Rosa palustris* Marsh.
swamp saxifrage	*Saxifraga pensylvanica* L.
sweetgum	*Liquidambar styraciflua* L.
sweet gale	*Myrica gale* L.
sweetleaf	*Symplocos tinctoria* (L.) L'Hér.
sweet pepperbush	*Clethra alnifolia* L.
sweet potato	*Ipomoea batatas* (L.) Lam.
switch cane	*Arundinaria gigantea* (Walt.) Muhl.
sycamore, American	*Platanus occidentalis* L.
tallowtree, Chinese tallow	*Sapium sebiferum* (L.) Roxb.

three-way sedge	*Dulichium arundinaceum* (L.) Britt.
tomato	*Lycopersicon esculentum* Miller
tupelo gum	
blackgum	*Nyssa sylvatica* Marsh.
Ogeechee tupelo	*Nyssa ogeche* Bartr.
swamp tupelo	*Nyssa sylvatica* var. *biflora* (Walt.) Sarg.
water tupelo	*Nyssa aquatica* L.
trout lily, dog-tooth violet	*Erythronium americanum* Ker
trumpet creeper, cow-itch-vine	*Campsis radicans* (L.) Seem.
twig-rush	*Cladium mariscoides* (Muhl.) Torrey
two-wing silverbell	*Halesia diptera* Ellis
Venus' fly trap	*Dionaea muscipula* Ellis
Viburnum	
arrowwood	*Viburnum* spp.
possomhaw viburnum	*Viburnum nudum* L.
Walter viburnum	*Viburnum obovatum* Walt.
witherod	*Viburnum cassinoides* L.
violet	*Viola* sp.
Virginia chain fern	*Woodwardia virginica* (L.) Smith
Virginia creeper	*Parthenocissus quinquefolia* (L.) Planch.
Virginia willow, tassel-white	*Itea virginica* L.
water-elm, planertree	*Planera aquatica* Walt. ex. J. F. Gmel.
water fern, mosquito fern	*Azolla caroliniana* Willd.
water lily	
fragrant water lily, sweet water lily	*Nymphaea odorata* Aiton
white water lily	*Nymphaea alba* L.
yellow floating-heart water lily	*Nymphoides peltata* (S.G. Gmel) Kuntze
watershield	*Brasenia schreberi* Gmelin
white beakrush	*Rhynchospora alba* (L.) Vahl
white wikey	*Kalmia cuneata* Michx.
willow	
black willow	*Salix nigra* Marsh.
coastal plain willow	*Salix caroliniana* Michx.
crack willow	*Salix fragilis* L.
Florida willow	*Salix floridana* Chapm.
silky willow	*Salix sericea* Marsh.
winterberry, common	*Ilex verticillata* (L.) Gray
wire grass	*Aristida stricta* Michx.
witch-hazel	*Hamamelis virginiana* L.
woolgrass	*Scirpus cyperinus* (L.) Kunth
yellow jessamine, poor-man's-rope	*Gelsemium sempervirens* (L.) Jaume St. Hil.
yellow lotus	*Nelumbo lutea* (Willd.) Pers.
yellow-poplar	*Liriodendron tulipifera* L.
zenobia	*Zenobia pulverulenta* (Bartram) Pollard

REPTILES AND AMPHIBIANS[2]

American alligator	*Alligator mississippiensis*
American crocodile	*Crocodylus acutus*
frogs	
barking treefrog	*Hyla gratiosa*
bird-voiced treefrog	*Hyla avivoca*
boreal chorus frog	*Pseudacris triseriata*
bronze frog	*Rana clamitans clamitans*
bullfrog	*Rana catesbeiana*
carpenter frog	*Rana virgatipes*
crawfish frog	*Rana areolata*
Cuban treefrog	*Osteopilus septentrionalis*
eastern narrowmouth frog	*Gastrophryne carolinensis*
eastern spadefoot	*Scaphiopus holbrooki*
gopher frog	*Rana capito*
gray treefrog	*Hyla chysoscelis* and *Hyla versicolor*
green frog	*Rana clamitans melanota*
green treefrog	*Hyla cinerea*
greenhouse frogs	*Eleutherodactylus planirostris*
leopard frog	*Rana* spp.
little grass frog	*Limnaoedus ocularis*
mountain chorus frog	*Pseudacris brachyphona*
ornate chorus frog	*Pseudacris ornata*
northern cricket frogs	*Acris crepitans*
pig frog	*Rana grylio*
pine barrens treefrog	*Hyla andersonii*
pine woods treefrog	*Hyla femoralis*
river frog	*Rana heckscheri*
southern chorus frog	*Pseudacris nigrita nigrita*
southern cricket frog	*Acris gryllus gryllus*
southern leopard frog	*Rana utricularia*
spring peeper	*Pseudacris crucifer*
squirrel treefrog	*Hyla squirella*
wood frog	*Rana sylvatica*
lizards	
bark anole	*Anolis distichus*
broadhead skink	*Eumeces laticeps*
brown anole	*Anolis sagrei*
coal skink	*Eumeces anthracinus*
eastern glass lizard	*Ophisaurus ventralis*
fence lizard	*Sceloporus undulatus*
five-lined skink	*Eumeces fasciatus*
green anole	*Anolis carolinensis*
ground skink	*Scincella lateralis*
six-lined racerunner	*Cnemidophorus sexlineatus sexlineatus*

slender glass lizard	*Ophisaurus attenuatus*
southeastern five-lined skink	*Eumeces inexpectatus*
salamanders	
dwarf salamander	*Eurycea quadridigitata*
dwarf siren	*Pseudobranchus striatus*
eastern tiger salamander	*Ambystoma tigrinum tigrinum*
flatwoods salamander	*Ambystoma cingulatum*
four-toed salamander	*Hemidactylium scutatum*
greater siren	*Siren lacertina*
lesser siren	*Siren intermedia*
longtail salamander	*Eurycea longicauda longicauda*
many-lined salamander	*Stereochilus marginatus*
marbled salamander	*Ambystoma opacum*
mole salamander	*Ambystoma talpoideum*
mountain dusky salamander	*Desmognathus ochrophaeus*
northern dusky salamander	*Desmognathus fuscus fuscus*
northern slimy salamander	*Plethodon glutinosus*
northern two-lined salamander	*Eurycea bislineata*
redback salamander	*Plethodon cinereus*
red-spotted newt, central newt	*Notophthalmus viridescens viridescens*
rusty mud salamander	*Pseudotriton montanus floridanus*
smallmouth salamander	*Ambystoma texanum*
southern dusky salamander	*Desmognathus auriculatis*
southern red salamander	*Pseudotriton ruber vioscai*
spotted salamander	*Ambystoma maculatum*
spring salamander	*Gyrinophilus porphyriticus*
two-toed amphiuma	*Amphiuma means*
snakes	
black swamp snake	*Seminatrix pygaea*
brown snake	*Storeria* spp.
brown water snake	*Nerodia taxispilota*
Chicago garter snake	*Thamnophis sirtalis semifasclatus*
corn snake	*Elaphe guttata guttata*
diamondback water snake	*Nerodia rhombifer rhombifer*
eastern coachwhip	*Masticophis flagellum flagellum*
eastern coral snake	*Micrurus fulvius fulvius*
eastern cottonmouth, moccasin	*Agkistrodon piscivorus piscivorus*
eastern diamondback rattlesnake	*Crotalus adamanteus*
eastern garter snake	*Thamnophis sirtalis sirtalis*
eastern hognose snake	*Heterodon platirhinos*
eastern indigo snake	*Drymarchon corais couperi*
eastern kingsnake	*Lampropeltis getula getula*
eastern milk snake	*Lampropeltis triangulum triangulum*
eastern ribbon snake	*Thamnophis sauritus sauritus*
eastern worm snake	*Carphophis amoenus amoenus*
glossy crayfish snake	*Regina rigida*

Mississippi green water snake	*Nerodia cyclopion*
mud snake	*Farancia abacura*
northern brown snake	*Storeria dekayi dekayi*
northern pine snake	*Pituophis melanoleucus melanoleucus*
pine woods snake	*Rhadinaea flavilata*
plainbelly water snake	*Nerodia erythrogaster*
queen snake	*Regina septemvittata*
rat snake	*Elaphe obsoleta*
redbelly snake	*Storeria occipitomaculata*
rough earth snake	*Virginia striatula*
rough green snake	*Opheodrys aestivus*
scarlet kingsnakes	*Lampropeltis triangulum elapsoides*
scarlet snake	*Cemophora coccinea*
smooth earth snake	*Virginia valeriae*
southeastern crowned snake	*Tantilla coronata*
southern black racer	*Coluber constrictor priapus*
southern copperhead	*Agkistrodon contortrix contortrix*
southern hognose snake	*Heterodon simus*
southern ringneck snake	*Diadophis punctatus punctatus*
southern water snake	*Nerodia fasciata*
timber rattlesnake	*Crotalus horridus*
western ribbon snake	*Thamnophis proximus proximus*
yellowbelly water snake	*Nerodia erythrogaster flavigaster*
yellow rat snake	*Elaphe obsoleta quadrivittata*
toads	
American toad	*Bufo americanus*
eastern narrowmouth toad	*Gastrophyrne carolinensis*
giant toad	*Bufo marinus*
Gulf Coast toad	*Bufo valliceps valliceps*
oak toad	*Bufo quercicus*
southern toad	*Bufo terrestris*
Woodhouse's toad	*Bufo woodhousii*
turtles	
bog turtle	*Clemmys muhlenbergii*
chicken turtle	*Deirochelys reticularia*
common musk turtle, stinkpot	*Sternotherus odoratus*
eastern box turtle	*Terrapene carolina carolina*
eastern mud turtle	*Kinosternon subrubrum subrubrum*
Florida box turtle	*Terrapene carolina bauri*
Florida redbelly turtle	*Pseudemys nelsoni*
Florida softshell	*Apalone ferox*
gopher tortoises	*Gopherus polyphemus*
green turtle	*Chelonia mydas*
loggerhead musk turtle	*Sternotherus minor minor*
northern diamondback terrapin	*Malaclemys terrapin terrapin*
red-eared slider	*Trachemys scripta elegans*

snapping turtle	*Chelydra serpentina*
spotted turtle	*Clemmys guttata*
striped mud turtle	*Kinosternon baurii*

BIRDS[3]

Acadian flycatcher	*Empidonax virescens*
American coot	*Fulica americana*
American crow, common crow	*Corvus brachyrhynchos*
American goldfinch	*Carduelis tristis*
American redstart	*Setophaga ruticilla*
American robin	*Turdus migratorius*
American swallow-tailed kite	*Elanoides forficatus*
American woodcock	*Scolopax minor*
anhinga	*Anhinga anhinga*
Bachman's warbler	*Vermivora bachmanii*
Bachman's sparrow	*Aimophila aestivalus*
bald eagle	*Haliaeetus leucocephalus*
barred owl	*Strix varia*
belted kingfisher	*Ceryle alcyon*
black-and-white warbler	*Mniotilta varia*
black-throated green warbler	*Dendroica virens*
black vulture	*Coragyps atratus*
blue-gray gnatcatcher	*Polioptila caerulea*
blue grosbeak	*Guiraca caerulea*
blue jay	*Cyanocitta cristata*
brown-headed cowbird	*Molothrus ater*
brown-headed nuthatch	*Sitta pusilla*
brown pelican	*Pelicanus occidentalis*
brown thrasher	*Toxostoma rufum*
Canada warbler	*Wilsonia canadensis*
Carolina chickadee	*Parus carolinensis*
Carolina wren	*Thryothorus ludovicianus*
cattle egret	*Bubulcus ibis*
cedar waxwing	*Bombycilla cedrorum*
cerulean warbler	*Dendroica cerulea*
chestnut-sided warbler	*Dendroica pensylvanica*
chimney swift	*Chaetura pelagica*
chipping sparrow	*Spizella passerina*
chuck-will's-widow	*Caprimulgus carolinensis*
common grackle	*Quiscalus quiscula*
common nighthawk	*Chordeiles minor*
common yellowthroat	*Geothlypis trichas*
Cooper's hawk	*Accipiter cooperii*
downy woodpecker	*Picoides pubescens*
eastern bluebird	*Sialia sialis*

eastern kingbird	*Tyrannus tyrannus*
eastern meadowlark	*Sturnella magna*
eastern phoebe	*Sayornis phoebe*
eastern screech owl	*Otus asio*
eastern wood-pewee	*Contopus virens*
field sparrow	*Spizella pusilla*
fish crow	*Corvus ossifragus*
golden-winged warbler	*Vermivora chrysoptera*
gray catbird	*Dumetella carolinensis*
green heron	*Butorides virescens*
great blue heron	*Ardea herodias*
greatcrested flycatcher	*Myiarchus crinitus*
great egret	*Casmerodius albus*
hairy woodpecker	*Picoides villosus*
hermit thrush	*Catharus guttatus*
hooded merganser	*Lophodytes cucullatus*
hooded warbler	*Wilsonia citrina*
house wren	*Troglodytes aedon*
indigo bunting	*Passerina cyanea*
ivory-billed woodpeckers	*Campephilus principalis*
Kentucky warbler	*Oporornis formosus*
little blue heron	*Egretta caerulea*
Louisiana waterthrush	*Seiurus motacilla*
magnificent frigatebird	*Fregata magnificens*
mallard	*Anas platyrhynchos*
mangrove cuckoo	*Coccyzus minor*
Mississippi kite	*Ictinia mississippiensis*
mourning dove	*Zenaida macroura*
northern cardinal	*Cardinalis cardinalis*
northern bobwhite	*Colinus virginianus*
northern flicker	*Colaptes auratus*
northern mockingbird	*Mimus polyglottos*
northern oriole	*Icterus galbula*
northern parula	*Parula americana*
olive-sided flycatcher	*Contopus borealis*
osprey	*Pandion haliaetus*
ovenbird	*Seiurus aurocapillus*
painted buntings	*Passerina ciris*
pileated woodpecker	*Dryocopus pileatus*
pine warbler	*Dendroica pinus*
prairie warbler	*Dendroica discolor*
prothonotary warbler	*Protonotaria citrea*
purple finch	*Carpodacus purpureus*
red-bellied woodpecker	*Melanerpes carolinus*
red-cockaded woodpecker	*Picoides borealis*
red-eyed vireo	*Vireo olivaceus*

red-headed woodpecker	*Melanerpes erythrocephalus*
red-shouldered hawk	*Buteo lineatus*
red-winged blackbird	*Agelaius phoeniceus*
roseate spoonbill	*Ajaia ajaja*
rose-breasted grosbeak	*Pheucticus ludovicianus*
ruby-crowned kinglet	*Regulus calendula*
ruby-throated hummingbird	*Archilochus colubris*
ruffed grouse	*Bonasa umbellus*
rufous-sided towhee	*Pipilo erythrophthalmus*
rusty blackbird	*Euphagus carolinus*
scarlet tanager	*Piranga olivacea*
short-tailed hawk	*Buteo brachyurus*
snowy egret	*Egretta thula*
summer tanager	*Piranga rubra*
Swainson's warbler	*Limnothlypis swainsonii*
teal	*Anas* spp.
tree swallow	*Tachycineta bicolor*
tufted titmouse	*Parus bicolor*
turkey vulture	*Cathartes aura*
veery	*Catharus fuscescens*
white-breasted nuthatch	*Sitta carolinensis*
white-eyed vireo	*Vireo griseus*
white ibis	*Eudocimus albus*
white-throated sparrow	*Zonotrichia albicollis*
wild turkey	*Meleagris gallopavo*
willow flycatcher	*Empidonax traillii*
wood duck	*Aix sponsa*
wood stork	*Mycteria americana*
wood thrush	*Hylocichla mustelina*
worm-eating warbler	*Helmitheros vermivorus*
yellow-billed cuckoo	*Coccyzus americanus*
yellow-breasted chat	*Icteria virens*
yellow-crowned night-heron	*Nyctanassa violacea*
yellow-rumped warbler	*Dendroica coronata*
yellow-throated vireo	*Vireo flavifrons*
yellow-throated warbler	*Dendroica dominica*

MAMMALS[4]

American beaver	*Castor canadensis*
big brown bat	*Eptesicus fuscus*
black bear	*Ursus americanus*
black rat	*Rattus rattus*
bobcat	*Lynx rufus*
common gray fox	*Urocyon cinereoargenteus*
common muskrat	*Ondatra zibethicus*

common raccoon	*Procyon lotor*
cotton mouse	*Peromyscus gossypinus*
eastern cottontail	*Sylvilagus floridanus*
eastern fox squirrel	*Sciurus niger*
eastern gray squirrel	*Sciurus carolinensis*
eastern harvest mouse	*Reithrodontomys humulis*
eastern mole	*Scalopus aquaticus*
eastern pipistrelle	*Pipistrellus subflavus*
eastern red bat	*Lasiurus borealis*
evening bat	*Nycticeius humeralis*
Florida panther	*Felis concolor coryi*
Florida woodrat	*Neotoma floridana*
fulvous harvest mouse	*Reithrodontomys fulvescens*
golden mouse	*Ochrotomys nuttalli*
hispid cotton rat	*Sigmodon hispidus*
house mouse	*Mus musculus*
Key rice rat	*Oryzomys argentatus*
least shrew	*Cryptotis parva*
long-tailed weasel	*Mustela frenata*
Louisiana black bear	*Ursus americanus luteolus*
manatee	*Trichechus manatus*
marsh rabbit	*Sylvilagus palustris*
marsh rice rat	*Oryzomys palustris*
masked shrew	*Sorex cinereus*
meadow jumping mouse	*Zapus hudsonius*
meadow vole	*Microtus pennsylvanicus*
mink	*Mustela vison*
nine-banded armadillo	*Dasypus novemcinctus*
northern river otter	*Lutra canadensis*
northern short-tailed shrew	*Blarina brevicauda*
nutria	*Myocastor coypus*
oldfield mouse	*Peromyscus polionotus*
opossum, Virginia	*Didelphis virginiana*
panther	*Felis concolor*
pygmy shrew	*Microsorex hoyi*
Rafinesque's big-eared bat	*Plecotus rafinesquii*
red squirrel	*Tamiasciurus hudsonicus*
rock vole	*Microtus chrotorrhinus*
smoky shrew	*Sorex fumeus*
southeastern shrew	*Sorex longirostris*
southern bog lemming	*Synaptomys cooperi*
southern flying squirrel	*Glaucomys volans*
southern red-backed vole	*Clethrionomys gapperi*
southern short-tailed shrew	*Blarina carolinensis*
star-nosed mole	*Condylura cristata*
swamp rabbit	*Sylvilagus aquaticus*

water shrew	*Sorex palustris*
white-footed mouse	*Peromyscus leucopus*
white-tailed deer	*Odocoileus virginianus*
wild pig	*Sus scrofa*
woodland jumping mouse	*Napaeozapus insignis*
woodland vole	*Microtus pinetorum*

FISH[5]

American eel	*Anguilla rostrata*
bowfin	*Amia calva*
bullhead catfishes	
black bullhead	*Ameiurus melas*
yellow bullhead	*Ameiurus natalis*
brown bullhead	*Ameiurus nebulosus*
tadpole madtom	*Noturus gyrinus*
catfish	
white catfish	*Ictalurus catus*
common carp	*Cyprinus carpio*
drums	
freshwater drum	*Aplodinotus grunniens*
spot	*Leiostomus xanthurus*
gars	
spotted gar	*Lepisosteus oculatus*
longnose gar	*Lepisosteus osseus*
shortnose gar	*Lepisosteus platostomus*
Florida gar	*Lepisosteus platyrhincus*
killifishes	
sheepshead Minnow	*Cyprinodon variegatus*
goldspotted killifish	*Floridichthys carpio*
golden topminnow	*Fundulus chrysotus*
marsh killifish	*Fundulus confluentus*
starhead topminnow	*Fundulus dispar*
gulf killifish	*Fundulus grandis*
mummichog	*Fundulus heteroclitus*
striped killifish	*Fundulus majalis*
blackspotted topminnow	*Fundulus olivaceus*
blackstripe topminnow	*Fundulus notatus*
bayou topminnow	*Fundulus notti*
longnose killifish	*Fundulus similis*
Seminole killfish	*Fundulus seminolis*
Waccamaw killifish	*Fundulus waccamensis*
flagfish	*Jordanella floridae*
pygmy killfish	*Leptolucania ommata*
bluefin killifish	*Lucania goodei*
rainwater killifish	*Lucania parva*

livebearers	
western mosquitofish	*Gambusia affinis*
eastern mosquitofish	*Gambusia holbrooki*
mangrove gambusia	*Gambusia rhizophorae*
least killifish	*Heterandria formosa*
sailfin molly	*Poecilia latipinna*
mangrove rivulus	*Rivulus marmoratus*
minnows	
blacktail shiner	*Cyprinella venusta*
bluehead shiner	*Notropis hubbsi*
coastal shiner	*Notropis petersoni*
taillight shiner	*Notropis maculatus*
blackmouth shiner	*Notropis melanostomus*
weed shiner	*Notropis texanus*
pugnose minnow	*Opsopoeodus emiliae*
fathead minnow	*Pimephales promelas*
bluenose shiner	*Pteronotropis welaka*
mojarras	
spotfin mojarra	*Eucinostomus argenteus*
silver jenny	*Eucinostomus gula*
mudminnows	
central mudminnow	*Umbra limi*
eastern mudminnow	*Umbra pygmaea*
mullets	
striped mullet	*Mugil cephalus*
needlefishes	
Atlantic needlefish	*Strongylura marina*
redfin needlefish	*Strongylura notata*
paddlefish	*Polyodon spathula*
perches	
bluntnose darter	*Etheostoma chlorosomum*
swamp darter	*Etheostoma fusiforme*
Waccamaw darter	*Etheostoma perlongum*
cypress darter	*Etheostoma proeliare*
Gulf darter	*Etheostoma swaini*
slough darter	*Etheostoma gracile*
yellow perch	*Perca flavescens*
pikes	
redfin pickerel	*Esox americanus americanus*
grass pickerel	*Esox americanus vermiculatus*
chain pickerel	*Esox niger*
pirate perch	*Aphredoderus sayanus*
Roanoke bass	*Ambloplites cavifrons*
shad	*Dorosoma* spp.
silversides	
hardhead silverside	*Atherinomorus stipes*

brook silverside	*Labidesthes sicculus*
inland silverside	*Menidia beryllina*
Waccamaw silverside	*Menidia extensa*
tidewater silverside	*Menidia peninsulae*
snooks	
common snook	*Centropomus undecimalis*
suckers	
smallmouth buffalo	*Ictiobus bubalus*
black buffalo	*Ictiobus niger*
creek chubsucker	*Erimyzon oblongus*
lake chubsucker	*Erimyzon succetta*
spotted sucker	*Minytrema melanops*
sunfishes	
mud sunfish	*Acantharchus pomotis*
flier	*Centrarchus macropterus*
blackbanded sunfish	*Enneacanthus chaetodon*
bluespotted sunfish	*Enneacanthus gloriosus*
banded sunfish	*Enneacanthus obesus*
warmouth	*Lepomis gulosus*
bluegill	*Lepomis macrochirus*
dollar sunfish	*Lepomis marginatus*
redspotted sunfish	*Lepomis miniatus*
blackspotted sunfish	*Lepomis punctatus*
bantam sunfish	*Lepomis symmetricus*
black crappie	*Pomoxis nigromaculatus*
largemouth bass	*Micropterus salmoides*
sunfishes, pygmy	
Carolina pygmy sunfish	*Elassoma boehlkei*
Everglades pygmy sunfish	*Elassoma evergladei*
bluebarred pygmy sunfish	*Elassoma okatie*
Okefenokee pygmy sunfish	*Elassoma okefenokee*
banded pygmy sunfish	*Elassoma zonatum*
swampfish	*Chologaster cornuta*
tarpon	*Megalops atlanticus*
white bass	*Morone americana*

[1] Tree and shrub nomenclature follows Little (1979) except in the case of laurel oak and swamp laurel oak which follows Harlow et al. 1996. Supplementary sources include Radford et al. (1968) and Godfrey and Wooten (1981).

[2] Nomenclature follows Conant and Collins (1991).

[3] Nomenclature follows Bull and Farrand (1995).

[4] Nomenclature follows Whittaker (1996).

[5] Nomenclature follows Mayden et al. (1992), Lee et al. (1980), and Boschung et al. (1983).

References

A

Aarssen, L. W. 1984. On the distinction between niche and competitive ability: implications for the coexistence theory. Acta Biotheoretica 33: 67–83.

Abdul, A. S. and R. W. Gillham. 1984. Laboratory studies of the effects of the capillary fringe on streamflow generation. Water Resources Research 10: 691–698.

Abdul, A. S. and R. W. Gillham. 1989. Field studies on the effects of the capillary fringe on streamflow generation. Journal of Hydrology 112: 1–18.

Abrahamson, W. G. and D. C. Hartnett. 1990. Pine flatwoods and dry prairies. p. 103–149. *In* R. L. Myers and J. J. Ewel (eds.) Ecosystems of Florida. University of Central Florida Press, Orlando, FL.

Adams, D. A. 1993. Renewable Resource Policy: The Legal–Institutional Foundations. Island Press, Washington, DC.

Adamus, P. R., E. J. Clairain, Jr., R. D. Smith, and R. E. Young. 1987. Wetland evaluation technique (WET). Waterways Experiment Station, U.S. Army Corps of Engineers, Vicksburg, MS.

Adis, J., K. Furch, and U. Irmler. 1979. Litter production of a central-Amazonian black water inundation forest. Tropical Ecology 29: 236–245.

AF&PA. 1997. The Tulloch Rule litigation (4th update). American Forest and Paper Association, Washington, DC. Resources Legal Issues February 25.

Akerman, A. 1923. The white cedar of the Dismal Swamp. Virginia Forestry Publication 30. Virginia Geological Commission, Charlottesville, VA.

Alabama Forestry Planning Committee. 1993. Considerations for forest management on Alabama soils. Montgomery, AL.

Albrecht, V. S. and B. N. Goode. 1994. Wetland Regulation in the Real World. Beveridge and Diamond, Washington, DC.

Alexander, T. R. 1967. Effect of Hurricane Betsy on the southeastern Everglades. The Quarterly Journal of the Florida Academy of Sciences 30: 10–24.

Allard, D. J. 1990. Southeastern United States Ecological Community Classification, Interim Report, Version 1.2. The Nature Conservancy, Southeast Regional Office, Chapel Hill, NC.

Allen, C. E. 1980. Feeding habits of ducks in a green-tree reservoir in eastern Texas. Journal of Wildlife Management 44: 232–236.

Allen, D. H. and J. R. Sweeney. 1990. Home range and habitat use of eastern bluebirds in South Carolina. Proceedings of the Annual Conference of Southeastern Association of Fish and Wildlife Agencies 44: 295–303.

Allen, D. L. 1954. Our Wildlife Legacy. Funk and Wagnalls Co., New York, NY.

Allen, H. L. and R. G. Campbell. 1988. Wet site pine management in the southeastern United States. p. 173–184. *In* D. D. Hook et al. (eds.) The Ecology and Management of Wetlands. Vol. 2. Management, Use, and Value of Wetlands. Timber Press, Portland, OR.

Allen, J. A. 1990. Establishment of bottomland oak plantations on the Yazoo Wildlife Refuge complex. Southern Journal of Applied Forestry 14: 206–210.

Allen, J. A. and H. E. Kennedy, Jr. 1989. Bottomland Hardwood Reforestation in the Lower Mississippi Valley. U.S. Department of Agriculture Forest Service, Southern Forest Experiment Station, Stoneville, MS.

Alongi, D. M., P. Christoffersen, F. Tirendi, and A. I. Robertson. 1992. The influence of freshwater and material export on sedimentary facies and benthic processes within the Fly Delta and adjacent Gulf of Papua (Papua New Guinea). Continental Shelf Research 12: 287–326.

Amatya, D. M., R. W. Skaggs, and J. D. Gregory. 1994. Hydrologic modeling of drained forested watersheds. p. 27–55. *In* Water Management in Forested Wetlands. U.S. Department of Agriculture Forest Service, Atlanta, GA. Technical Publication R8-TP-20.

Amatya, D. M., R. W. Skaggs, and J. D. Gregory. 1996. Effects of controlled drainage on the hydrology of drained pine plantations in the North Carolina Coastal Plain. Journal of Hydrology 181: 211–232.

Amatya, D. M., J. W. Gilliam, R. W. Skaggs, and C. D. Blanton. 1996. Quality of water from drained managed plantations in North Carolina coastal plain. *In* Solutions: A Technical Conference on Water Quality. J. S. McKimmon Extension Education Center, North Carolina State University, Raleigh, NC.

Ambrose, J. 1996. Personal communication. Department of Natural Resources, Atlanta, GA.

Ambus, P. and R. Lowrance. 1991. Comparison of denitrification in two riparian soils. Soil Science Society of America Journal 55: 994–997.

Anderson, M. P. and W. W. Woessner. 1992. Applied Groundwater Modeling. Academic Press, New York, NY.

Anderson, R. C. and J. White. 1970. A cypress swamp outlier in southern Illinois. Illinois Academy of Science 63: 6–13.

Andrews, L. M. 1993. Loblolly pine response to drainage and fertilization of hydric soils. M.S. Thesis. College of Forestry and Wildlife Resources, Virginia Polytechnic Institute and State University, Blacksburg, VA.

Angelov, M. N., S. S. Sung, R. L. Doong, W. R. Harms, P. P. Kormanik, and C. C. Black, Jr. 1996. Long- and short-term flooding effects on survival and sink-source relationships of swamp-adapted tree species. Tree Physiology 16: 477–484.

Anger, K. 1995. The conquest of freshwater and land by marine crabs: adaptations in life history patterns and larval bioenergetics. Journal of Experimental Marine Biology and Ecology 193: 119–145.

Anonymous. 1959. Swamp loggers operate submarine show for cypress and tupelo in bayous of Louisiana. The Timberman 60: 52–55.

Applegate, H. 1996. Personal communication. Forest Management Chief, Tennessee Division of Forestry, Nashville, TN.

Applequist, M. B. 1957. Service tests of native Louisiana fence post materials. Louisiana State University, Baton Rouge, LA. Forestry Note 12.

Applequist, M. B. 1959. A study of soil and site factors affecting the growth and development of swamp blackgum and tupelogum stands in southeastern Georgia. D.F. Thesis. Duke University, Durham, NC.

Applequist, M. B. 1960. Soil site studies of southern hardwoods. p. 49–63. *In* P.Y. Burns (ed.) Southern Forest Soils, Eighth Annual Forestry Symposium. Louisiana State University Press, Baton Rouge, LA.

Armstrong, J. and W. Armstrong. 1988. *Phragmites australis* — a preliminary study of soil-oxidizing sites and internal gas transport pathways. New Phytologist 108: 373–382.

Armstrong, W. 1967. The oxidizing activity of roots in waterlogged soils. Physiologia Plantarum 20: 920–926.

Armstrong, W. 1968. Oxygen diffusion from the roots of woody species. Physiologia Plantarum 21: 539–543.

Armstrong, W. 1972. A re-examination of the functional significance of aerenchyma. Physiologia Plantarum 27:173–177.

Armstrong, W. 1979. Aeration in higher plants. p. 225–332. *In* H.W.W. Woolhouse (ed.) Advances in Botanical Research. Academic Press, London.

Armstrong, W. and D. J. Boatman. 1967. Some field observations relating the growth of bog plants to conditions of soil aeration. Journal of Ecology 55: 101–110.

Armstrong, W. and D. J. Read. 1972. Some observations on oxygen transport in conifer seedlings. New Phytologist 71:55–62.

Aronen, T. S. and H. M. Haggman. 1994. Occurrence of lenticels in roots of Scots pine seedlings in different growth conditions. Journal of Plant Physiology 143: 325–329.

Arshad, M. and W. T. Frankenberger, Jr. 1990. Production and stability of ethylene in soil. Biology and Fertility of Soils 10: 29–34.

ASCS. No date. Water Bank Program (WBP) description. U.S. Department of Agriculture, Agricultural Stabilization and Conservation Service, Washington, DC. Fact Sheet.

Ash, A. N., C. B. McDonald, E. S. Kane, and C. A. Pories. 1983. Natural and modified pocosins: literature synthesis and management options. U.S. Fish and Wildlife Service, Biological Services Program, Washington, DC. FWS/OBS-83-04.

Ashe, W. W. 1894. The forests, forest lands and forest products of eastern North Carolina. North Carolina Geological Survey, Raleigh, NC. Bulletin 5.

Atkins, P. W. 1987. Physikalische Chemie. VCH Verlagsgesel-schaft, Weinheim, FRG.

Atwell, B. J. and H. Greenway. 1987. Carbohydrate metabolism of rice seedlings grown in oxygen deficient solution. Journal of Experimental Botany 38(188): 466–478.

Aust, W. M. 1989. Abiotic functional changes of a water tupelo-baldcypress wetland following disturbance by harvesting. Ph.D. Dissertation. North Carolina State University, Raleigh, NC.

Aust, W. M. and R. Lea. 1991. Soil temperature and organic matter in a disturbed forested wetland. Soil Science Society of America Journal 55: 1741–1746.

Aust, W. M. and R. Lea. 1992. Comparative effects of aerial and ground logging on soil properties in a tupelo-cypress wetland. Forest Ecology and Management 50:57–73.

Aust, W. M., J. D. Hodges, and R. L. Johnson. 1985. The origin, growth and development of pure, even-aged stands of bottomland oak. p. 163–170. *In* E. Shoulders (ed.) Proceedings Third Biennial Southern Silvicultural Research Conference. U.S. Department of Agriculture Forest Service, New Orleans, LA. General Technical Report SO-54.

Aust, W. M., S. F. Mader, and R. Lea. 1989. Abiotic changes of a tupelo-cypress swamp following helicopter and rubber-tired skidder timber harvest. p. 545–551. *In* J. H. Miller (compiler) Proceedings of the Fifth Biennial Southern Silvicultural Conference. U.S. Department of Agriculture Forest Service, Southern Forest Experiment Station, New Orleans, LA. General Technical Report SO-74.

Aust, W. M., T. W. Reisinger, J. A. Burger, and B. J. Stokes. 1991. Site impacts associated with three timber harvesting systems operating on wet pine flats — preliminary results. p. 342–350. *In* S. S. Coleman and D. G. Neary (eds.) Proceedings of the Sixth Biennial Southern Silvicultural Conference. U.S. Department of Agriculture Forest Service, Southeastern Forest Experiment Station, Asheville, NC. General Technical Report SE-70.

Aust, W. M., T. W. Reisinger, J. A. Burger, and B. J. Stokes. 1993. Soil physical and hydrological changes associated with logging a wet pine flat with wide tired skidders. Southern Journal of Applied Forestry 17: 22–25.

Aust, W. M., S. H. Schoenholtz, T. W. Zaebst, and B. A. Szabo. 1997. Recovery status of a tupelo-cypress wetland seven years after disturbance: silvicultural implications. Forest Ecology and Management 90: 161–170.

Aust, W. M., M. D. Tippett, J. A. Burger, and W. H. McKee, Jr. 1995. Compaction and rutting during harvesting affect better drained soils more than poorly drained soils on wet flat pines. Southern Journal of Applied Forestry 19: 72–77.

Austin, K. A., and R. K. Wieder. 1987. Effects of elevated H^+, SO_4^{-2}, NO_3^-, and NH_4^+ in simulated acid precipitation on the growth and chlorophyll content of 3 North American Sphagnum species. The Bryologist 90: 221–229.

Avery, T. E. and H. E. Burkhart. 1994. Forest Measurements. McGraw–Hill Book Co., New York, NY.

Axt, J. and M. R. Walbridge. 1997. Phosphate removal capacity of selected freshwater wetlands in the Virginia Piedmont and Coastal Plain. Unpublished report to the Rappahannock Area Development Commission. Department of Biology, George Mason University, Fairfax, VA.

B

Bacchus, S. 1995. Groundwater levels are critical to the success of prescribed burns. p. 117–133. *In* S. I. Cerulean and R. T. Engstrom (eds.) Fire in Wetlands: A Management Perspective, Proceedings of the 19th Tall Timbers Fire Ecology Conference. Tall Timbers Research Station, Tallahassee, FL.

Bacon, P. R. 1990. The ecology and management of swamp forests in the Guianas and Caribbean region. p. 213–250. *In* A.E. Lugo, M. M. Brinson, and S. L. Brown (eds.) Forested Wetlands. Vol. 15. Ecosystems of the World. Elsevier Science Publishers B.V., Amsterdam, The Netherlands.

Bacon, P. R. 1994. Template for evaluation of impacts of sea level rise on Caribbean coastal wetlands. Ecological Engineering 3: 171–186.

Bailey, R. G., R. D. Pfister, and J. A. Henderson. 1978. Nature of land and resource classification — a review. Journal of Forestry 76: 650–655.

Baines, R. A. 1990. Prospects for white cedar: a North Carolina assessment. Forem 13: 8–11.

Baker, J. A., K. J. Killgore, and R. L. Kasul. 1991. Aquatic habitats and fish communities in the lower Mississippi River. Reviews in Aquatic Sciences 3: 313–356.

Baker, J. B. and W. M. Broadfoot. 1979. Site evaluation for commercially important southern hardwoods. U.S. Department of Agriculture Forest Service, Southern Forest Experiment Station, New Orleans, LA. General Technical Report S0-26.

Balassa, J. 1996. Personal communication. Kentucky Natural Resources and Environmental Protection Cabinet, Office of Information Services, Frankfort, KY.

Ball, M. C. 1980. Patterns of secondary succession in a mangrove forest of southern Florida. Oecologia 44: 226–235.

Ball, M. C. 1988. Ecophysiology of mangroves. Trees 2: 129–142.

Balling, R. C. 1992. The heated debate: greenhouse predictions vs. climate reality. Pacific Research Institute for Public Policy, San Francisco, CA.

Bandle, B. J. and F. P. Day. 1985. Influence of species, season and soil on foliar macronutrients in the Great Dismal Swamp. Bulletin of the Torrey Botanical Club 112: 146–157.

Barber, D. A., M. Ebert, and N. T. Evans. 1962. The movement of ^{15}O through barley and rice plants. Journal of Experimental Botany 13(39): 397–403.

Barclay, A. M. and R. M. M. Crawford. 1981. Temperature and anoxic injury in pea seedlings. Journal of Experimental Botany 32(130): 943–949.

Barnes, J. S. 1981. Agricultural adaptability of wet soils of the North Carolina Coastal Plain. p. 225–237. *In* C. J. Richardson (ed.) Pocosin Wetlands: An Integrated Analysis of Coastal Plain Freshwater Bogs in North Carolina. Hutchinson Ross Publication Co., Stroudsburg, PA.

Barnett, B. S. 1972. The freshwater fishes of the Hillsborough River drainage, Florida. M.A. Thesis. University of South Florida, Tampa, FL.

Barney, R. L. and B. J. Anson. 1920. Life history and ecology of the pigmy sunfish, *Elassoma zonatum*. Ecology 1: 241–256.

Barney, R.L. and B. J. Anson. 1921. Seasonal abundance of the mosquitofish destroying topminnow, *Gambusia affinis*, especially in relation to male frequency. Ecology 2: 53–69.

Barry, J. M. 1980. Natural Vegetation of South Carolina. University of South Carolina Press, Columbia, SC.

Bates, R. D. 1989. Biomass and primary productivity on an early successional floodplain forest site. p. 375–383. *In* R. R. Sharitz and J. W. Gibbons (eds.) Freshwater Wetlands and Wildlife Symposium: Perspectives on Natural, Managed and Degraded Ecosystems. U.S. Department of Energy, Office of Scientific and Technical Information, Oak Ridge, TN. CONF-8603101, DOE Symposium Series No. 61.

Beach, M. L. 1974. Food habits and reproduction of the taillight shiner, *Notropis maculatus* (Hay), in central Florida. Florida Scientist 37: 5–16.

Bear, J. and A. Verruijt. 1987. Modeling groundwater flow and pollution. D. Reidel Publishing Co., Dordrecht, The Netherlands.

Beasley, B. R., W. D. Marshall, A. H. Miglarese, J. D. Scurry, and C. E. Vanden Houten. 1996. Managing resources for a sustainable future: the Edisto River basin project report. Water Resources Division, South Carolina Department of Natural Resources, Columbia, SC. Report No. 12.

Bechtold, W. A., M. J. Brown, and R. M. Sheffield. 1990. Florida's forests, 1987. U.S. Department of Agriculture Forest Service, Southeastern Forest Experiment Station, Asheville, NC. Resource Bulletin SE-110.

Beck, A. F. and W. J. Garnett. 1983. Distribution and notes on the Great Dismal Swamp population of *Mitoura hesseli* Rawson and Ziegler (Lycaenidae). Journal of the Lepid. Society 37: 289–300.

Beck, L. T. 1977. Distribution and relative abundance of freshwater macroinvertebrates of the lower Atchafalaya River Basin, Louisiana. M.S. Thesis. Louisiana State University, Baton Rouge, LA.

Beck, R. E. 1994. The movement in the United States to restoration and creation of wetlands. Natural Resources Journal 34: 781–822.

Bedford, B. and D. R. Bouldin. 1994. Response to the paper "On the difficulties of measuring oxygen release by root systems of wetland plants" by B. K. Sorrell and W. Armstrong. Journal of Ecology 82: 185–186.

Bedford, B. L., D. R. Bouldin, and B. D. Beliveau. 1991. Net oxygen and carbon-dioxide balances in solutions bathing roots of wetland plants. Journal of Ecology 79: 943–959.

Bedinger, M. S. 1981. Hydrology of bottomland hardwood forests of the Mississippi Embayment. p. 161–176. *In* J. R. Clark and J. Benforado (eds.) Wetlands of Bottomland Hardwood Forests. Elsevier Scientific Publishing Company, Amsterdam, The Netherlands.

Beecher, H. A., W. C. Hixson, and T. S. Hopkins. 1977. Fishes of a Florida oxbow lake and its parent river. Florida Scientist 40: 140–148.

Beever, J. W. 1989. The effects of fringe red mangrove trimming for view in the Southwest Florida Aquatic Preserves. Southwest Florida Aquatic Preserves, Tallahassee, FL. Report No. 5.

Behler, J. L. and F. W. King. 1979. National Audubon Society Field Guide to North American Reptiles and Amphibians. Alfred A. Knopf, New York, NY.

Belk, M. C. and C. Lydeard. 1994. Effect of *Gambusia holbrooki* on a similar-sized, syntopic poeciliid, *Heterandria formosa*: competitor or predator. Copeia 1994: 296–302.

Bell, D. T. and S. K. Sipp. 1975. The litter stratum in the streamside forest ecosystem. Oikos 26: 391–397.

Bellis, V. J. and J. R. Keough. 1995. Ecology of maritime forests of the southern Atlantic coast: a community profile. U.S. Fish and Wildlife Service, National Biological Service, Washington, DC. Biological Report 30.

Belokon, W. 1996. Personal communication. Mississippi Automated Resource Information System, Jackson, MS.

Beltz, R. C., C. E. Burkhardt, and D. M. May. 1990. Interpreting 1987 forest survey results for Mississippi hardwood sawtimber. Southern Journal of Applied Forestry 14: 170–173.

Bennett, S. H. and J. B. Nelson. 1991. Distribution and status of Carolina bays in South Carolina. South Carolina Wildlife and Marine Resources Department, Columbia, SC. Nongame and Heritage Trust Publication No. 1.

Bent, A. C. 1953. Life histories of North American wood warblers, Vol. 1. Smithsonian Institution, U.S. National Museum, U.S. Government Printing Office, Washington, DC. Bulletin 203.

Berg, B. 1986. Nutrient release from litter and humus in coniferous forest soils — a mini review. Scandinavian Journal of Forest Research 1: 359–369.

Bertani, A., I. Brambilla, and R. Reggiani. 1982. Effect of exogenous nitrate on anaerobic root metabolism. Zeitochrift fur Pflanzenphysiologie Band 107: 255–264.

Betts, H. S. 1938. Southern cypress. American Woods. U.S. Department of Agriculture Forest Service, U.S Government Printing Office, Washington, DC.

Bevelander, G. 1934. The gills of *Amia calva* specialized for respiration in an oxygen deficient habitat. Copeia 1934: 123–127.

Birch, T. W. 1996. Private forest-land owners of the United States, 1994. U.S. Department of Agriculture Forest Service. Resource Bulletin NE-134.

Birkeland, P. W. 1984. Soils and Geomorphology. Oxford University Press, New York, NY.

Blair, R. M. and M. J. Langlinais. 1960. Nutria and swamp rabbits damage baldcypress plantings. Journal of Forestry 58: 388–389.

Blake, T. J. and D. M. Reid. 1981. Ethylene, water relations and tolerance to waterlogging of three *Eucalyptus* species. Australian Journal of Plant Physiology 8: 497–505.

Blasco, F. 1984. Climatic factors and the biology of mangrove plants. p. 18–35. *In* S. C. Snedaker and J. G. Snedaker (eds.) The Mangrove Ecosystem: Research Methods. UNESCO, Paris, France.

Bliley, D. J. and D. E. Pettry. 1979. Carolina bays on the eastern shore of Virginia. Soil Science Society of America Journal 43: 558–564.

Blom, C. W. P. M., G. M. Bogemann, P. Laan, A. J. M. Van der Sman, H. M. Van de Steeg, and L. A. C. J. Voesenek. 1990. Adaptations to flooding in plants from river areas. Aquatic Botany 38: 29–47.

Blom, C. W. P. M., L. A. C. J. Voesenek, M. Banga, W. M. H. G. Engelaar, J. H. G. M. Rijnders, H. M. Van de Steeg, and E. J. W. Visser. 1994. Physiological ecology of riverside species: adaptive responses of plants to submergence. Annals of Botany 74: 253–263.

Blumm, M. C. and D. B. Zaleha. 1989. Federal wetlands protection under the Clean Water Act: regulatory ambivalence, intergovernmental tension, and a call for reform. University of Colorado Law Review 60: 695–772.

BMP MOA. 1995. Memorandum of Agreement Between the U.S. Environmental Protection Agency and U.S. Department of the Army, Application of Best Management Practices to Mechanical Silvicultural Site Preparation Activities for the Establishment of Pine Plantations in the Southeast. November 28, 1995. EPA Office of Wetlands, Oceans and Watersheds, Washington, DC.

Boelter, D. H. 1974. The hydrologic characteristics of undrained organic soils in the Lake States. p. 33–46. *In* Histosols: Their Characteristics, Classification, and Use. Soil Science Society of America, Madison, WI. Special Publication No. 6.

Boelter, D. H. and E. S. Verry. 1977. Peatland and water in the northern Lake States. U.S. Department of Agriculture Forest Service, Northcentral Forest Experiment Station, St. Paul, MN. General Technical Report NC-31.

Bohlke, J. 1956. A new pigmy sunfish from southern Georgia. Notulae Naturae of the Academy of Natural Sciences of Philadelphia 294: 1–11.

Bohn, H. L., B. L. McNeal, and G. A. O'Conner. 1985. Soil Chemistry, 2nd Edition. John Wiley and Sons, NY.

Bolton, E. F. and A. E. Erickson. 1970. Ethanol concentration in tomato plants during soil flooding. Agronomy Journal 62: 220–224.

Bonnel, M. 1993. Progress in the understanding of runoff generation dynamics in forests. Journal of Hydrology 150: 217–275.

Bootlink, H. W. G., R. Hatano, and J. Bouma. 1993. Measurement and simulation of bypass flow in a structured clay soil: a physico-morphological approach. Journal of Hydrology 148: 149–168.

Borger, G. A. and T. T. Kozlowski. 1972. Effects of temperature on first periderm and xylem development in *Fraxinus pennsylvanica, Robinia pseudoacacia* and *Ailanthus altissima.* Canadian Journal of Forest Research 2: 198–205.

Bos, M. G. 1978. Discharge Measurement Structures, 2nd Edition. International Institute for Land Reclamation and Improvement, Wageningen, The Netherlands.

Bosch, J. M. and J. D. Hewlett. 1982. A review of catchment experiments to determine the effect of vegetation changes on water yield and evapotranspiration: influence of forests on water supplies. Journal of Hydrology 55: 3–23.

Boschung, H. T., Jr., J. D. Williams, D. W. Gotshall, D. K. Caldwell, and M. C. Caldwell. 1983. National Audubon Society Field Guide to North American Fishes, Whales, and Dolphins. Alfred A. Knopf, New York, NY.

Botero, L. 1990. Massive mangrove mortality on the Caribbean coast of Colombia. Vida Silvestre Neotropical 2: 77–78.

Boto, K. G. and J. S. Bunt. 1981. Tidal export of particulate organic matter from a Northern Australian mangrove system. Estuarine, Coastal and Shelf Science 13: 247–255.

Boto, K. G. and J. T. Wellington. 1983. Phosphorus and nitrogen nutritional status of a northern Australian mangrove forest. Marine Ecology Progress Series 11: 63–69.

Boto, K. G. and J. T. Wellington. 1984. Soil characteristics and nutrient status in a northern Australian mangrove forest. Estuaries 7: 61–69.

Boto, K. G. and J. T. Wellington. 1988. Seasonal variations in concentrations and fluxes of dissolved organic and inorganic materials in a tropical, tidally-dominated, mangrove waterway. Marine Ecology Progress Series 50: 151–160.

Boto, K., P. Saffingna, and B. Clough. 1985. Role of nitrate in nitrogen nutrition of the mangrove *Avicennia marina.* Marine Ecology Progress Series 21: 259–265.

Boulton, A. J. and P. I. Boon. 1991. A review of methodology used to measure leaf litter decomposition in lotic environments: time to turn over a new leaf? Australian Journal of Marine and Freshwater Research 42: 1–43.

Bouma, J. and L. W. Decker. 1978. A case study on infiltration into dry clay soil. I Morphological observations. Geoderma 20: 27–40.

Bouwer, H. H. and R. C. Rice. 1976. A slug test for determining hydraulic conductivity of unconfined aquifers with completely or partially penetrating wells. Water Resources Research 12: 423–428.

Bowden, W. B. 1987. The biogeochemistry of nitrogen in freshwater wetlands. Biogeochemistry 4: 313–348.

Bowling, D.R. and R. C. Kellison. 1983. Bottomland hardwood stand development following clearcutting. Southern Journal of Applied Forestry 7: 110–116.

Boyce, S. G. and N. D. Cost. 1974. Timber potentials in the wetland hardwoods. p. 131–151. *In* M. C. Blount (ed.) Water Resources: Utilization and Conservation in the Environment. Southeast Region Water Resources Symposium, Fort Valley, GA.

Boyer, W. D. 1979. Regenerating the natural longleaf pine forest. Journal of Forestry 77: 572–575.

Boynton, A. C. 1994. Wildlife use of southern Appalachian wetlands in North Carolina. Water, Air and Soil Pollution 77: 349–358.

Bradford, K. J. 1983a. Effects of soil flooding on leaf gas exchange of tomato plants. Plant Physiology 73: 475–479.

Bradford, K. J. 1983b. Involvement of plant growth substances in the alteration of leaf gas exchange on flooded tomato plants. Plant Physiology 73: 480–483.

Brady, N. C. 1974. The Nature and Properties of Soils, 8th Edition. Macmillan Publishing Co., New York, NY.

Brady, N. C. 1990. The Nature and Properties of Soils, 10th Edition. Macmillan Publishing Co., New York, NY.

Braendle, R. and R. M. M. Crawford. 1987. Rhizome anoxia tolerance and habitat specialization in wetland plants. p. 397–410. *In* R.M.M. Crawford (ed.) The Series Analytic: Plant Life in Aquatic and Amphibious Habitats. Blackwell Scientific Publications, Oxford. Special Publications of the British Ecological Society, No. 5.

Braham, R. R. 1992. Woody plants of Hill Forest. North Caolina State University, College of Forest Resources, Raleigh, NC. Technical Report #68.

Bramlett, D. L. 1990. *Pinus serotina* Michx. Pond pine. p. 470–475. *In* R. M. Burns and B. H. Honkala (tech. coords.) Silvics of North America: 1. Conifers. U.S. Department of Agriculture, Washington, DC. Agriculture Handbook 654.

Brandt, K. and K. C. Ewel. 1989. Ecology and management of cypress swamps: a review. Florida Cooperative Extension Service, Institute of Food and Agricultural Sciences, University of Florida, Gainesville, FL.

Brandt, L. A., K. L. Montgomery, A. W. Saunders, and F. J. Mazzotti. 1993. *Gopherus polyphemus* (gopher tortoise) burrows. Herpetological Review 24(4): 149.

Brantley, C. G. and S. G. Platt. 1992. Experimental evaluation of nutria herbivory on baldcypress. Proceedings of the Louisiana Academy of Science 55: 21–25.

Brasfield, J. F., V. W. Carlisle, and R. W. Johnson. 1983. Spodosols. p. 57–64. *In* S.W. Buol (ed.) Soils of the Southern States and Puerto Rico. Joint Publication of the Agricultural Experiment Stations of the Southern States and Puerto Rico Land Grant Universities. Southern Cooperative Series Bulletin No. 174.

Braun, E. L. 1950. Deciduous Forests of Eastern North America. Blakiston Company, Philadelphia, PA.

Braun, E. L. 1972. Deciduous Forests of Eastern North America. Hafner Publishing Company, New York, NY.

Breaux, A. M. and J. W. Day, Jr. 1994. Policy considerations for wetland wastewater treatment in the coastal zone: a case study for Louisiana. Coastal Management 22: 285–307.

Breder, C. M., Jr. 1936. The reproductive habits of the North American sunfishes (Family Centrarchidae). Zoologica 21: 1–11.

Breder, C. M. and D. E. Rosen. 1966. Modes of Reproduction in Fishes. Natural History Press, Garden City, NJ.

Brett, J. R. 1956. Some principles in the thermal requirements of fishes. Quarterly Review of Biology 31: 75–87.
Bridgham, S. and C. J. Richardson. 1992. Freshwater peatlands (pocosins) on the southeastern coastal plain of the USA: community description, nutrient dynamics and disturbance. p. 1–15. *In* D. N. Grubich and T. J. Malterer (eds.) Peat and Peatlands: The Resource and Its Utilization. Proceedings International Peat Symposium, Duluth, MN.
Bridgham, S. D. and C. J. Richardson. 1993. Hydrology and nutrient gradients in North Carolina peatlands. Wetlands 13: 207–218.
Bridgham, S. D., J. Pastor, C. A. McClaugherty, and C. J. Richardson. 1995. Nutrient-use efficiency: a litterfall index, a model, and a test along a nutrient-availability gradient in North Carolina peatlands. The American Naturalist 145: 1–21.
Bridgham, S. D., C. J. Richardson, E. Maltby, and S. P. Faulkner. 1991. Cellulose decay in natural and disturbed peatlands in North Carolina. Journal of Environmental Quality 20: 695–701.
Bridgham, S. D., J. Pastor, J. A. Janssens, C. Chapin, and T. J. Malterer. 1996. Multiple limiting gradients in peatlands: a call for a new paradigm. Wetlands 16: 45–65.
Briggs, S. V. and M. T. Maher. 1983. Litterfall and leaf decomposition in a Red River gum (*Eucalyptus camaldulensis*) swamp. Australian Journal of Botany 31: 307–316.
Brinson, M. M. 1977. Decomposition and nutrient exchange of litter in an alluvial swamp forest. Ecology 58: 601–609.
Brinson, M. M. 1990. Riverine forests. p. 87–141. *In* A. E. Lugo, M. M. Brinson, and S. L. Brown (eds.) Forested Wetlands. Vol. 15. Ecosystems of the World. Elsevier Science Publishers B.V., Amsterdam, The Netherlands.
Brinson, M. M. 1993a. Changes in the functioning of wetlands along environmental gradients. Wetlands 13: 65–74.
Brinson, M. M. 1993b. A hydrogeomorphic classification for wetlands. U.S. Army Corps of Engineers Waterways Experiment Station, Wetlands Research Program, Vicksburg, MS. Technical Report WRP-DE-4.
Brinson, M. 1995. The HGM approach explained. National Wetland Newsletter 17: 7–13.
Brinson, M. M. and R. Rheinhardt. 1996. The role of reference wetlands in functional assessment and mitigation. Ecological Applications 6: 69–76.
Brinson, M. M., H. D. Bradshaw, and E. S. Kane. 1984. Nutrient assimilative capacity of an alluvial floodplain swamp. Journal of Applied Ecology 21: 1041–1058.
Brinson, M. M., A. E. Lugo, and S. Brown. 1981a. Primary productivity, decomposition and consumer activity in freshwater wetlands. Annual Review of Ecology and Systematics 12: 123–161.
Brinson, M. M., B. L. Swift, R. C. Plantico, and J. S. Barclay. 1981b. Riparian ecosystems: their ecology and status. U.S. Fish and Wildlife Service, Biological Services Program, Washington, DC. FWS/OBS-81/17.
Brinson, M. M., H. D. Bradshaw, R. N. Holmes, and J. B. Elkins, Jr. 1980. Litterfall, stemflow, and throughfall nutrient fluxes in an alluvial swamp forest. Ecology 61: 827–835.
Brinson, M. M., W. L. Nutter, R. Rheinhardt, and B. Pruitt. 1996. Background and recommendations for establishing reference wetlands in the Piedmont of the Carolinas and Georgia. Final Report, Contract No. 3B0978TEX, U.S. Environmental Research Laboratory, National Health and Environmental Effects Laboratory, Western Division, Corvallis, OR.
Brinson, M. M., F. R. Hauer, L. C. Lee, W. L. Nutter, R. D. Rheinhardt, R. D. Smith, and D. Whigham. 1995. Guidebook for application of hydrogeomorphic assessments to riverine wetlands. U.S. Army Corps of Engineers Waterways Experiment Station, Vicksburg, MS. Technical Report TR-WRP-DE-11.

Brittingham, M. C. and S. A. Temple. 1983. Have cowbirds caused forest songbirds to decline? Bioscience 33: 31–35.

Britton, J. C. and B. Morton. 1989. Shore Ecology of the Gulf of Mexico. University of Texas Press, Austin, TX.

Broadfoot, W. M. 1964. Soil suitability for hardwoods in the midsouth. U.S. Department of Agriculture Forest Service, Southern Forest Experiment Station, New Orleans, LA. Research Paper SO-10.

Broadfoot, W. M. 1967. Shallow-water impoundment increases soil moisture and growth of hardwoods. Soil Science Society of America Proceedings 31: 562–564.

Broadfoot, W. M. 1976. Hardwood suitability for and properties of important Midsouth soils. U.S. Department of Agriculture Forest Service, Southern Forest Experiment Station, New Orleans, LA. Research Paper SO-127.

Broadfoot, W. M. and H. L. Williston. 1973. Flooding effects on southern forests. Journal of Forestry 71: 584–587.

Brooks, K. N. 1991. Hydrology and Management of Watersheds. Iowa State University Press, Ames, IA.

Brooks, M. J., B. E. Taylor, and J. A. Grant. 1996. Carolina bay geoarchaeology and Holocene landscape evolution on the Upper Coastal Plain of South Carolina. Savannah River Archeological Research Program, South Carolina Institute of Archeology and Anthropology, University of South Carolina, Columbia, SC.

Brown, C. A. 1972. Wildflowers of Louisiana and Adjoining States. Louisiana State University Press, Baton Rouge, LA.

Brown, C. A. 1984. Morphology and biology of cypress trees. p. 16–24. *In* K. C. Ewel and H. T. Odum (eds.) Cypress Swamps. University Presses of Florida, Gainesville, FL.

Brown, C. A. and G. N. Montz. 1986. Baldcypress the Tree Unique, the Wood Eternal. Claitor's Publishing Division, Baton Rouge, LA.

Brown, D. A., V. E. Nash, A. G. Caldwell, L. J. Bartelli, R. C. Carter, and O. R. Carter. No date. A monograph of the soils of the Southern Mississippi River Valley Alluvium. Arkansas Agricultural Experiment Station, Fayetteville, AR. Southern Cooperative Series Bulletin 178.

Brown, D. J. and T. J. Coon. 1994. Abundance and assemblage structure of fish larvae in lower Missouri River and its tributaries. Transactions of the American Fisheries Society 123: 718–732.

Brown, M. J. 1994. Forest statistics for Florida. U.S. Department of Agriculture Forest Service, Southern Research Station, Asheville, NC. Resource Bulletin SR-6.

Brown, M. J. 1996. Forest survey statistics for forested wetlands of the Carolinas. p. 138–146. *In* K. M. Flynn (ed.) Proceedings of the Southern Forested Wetlands Ecology and Management Conference. Clemson University, Clemson, SC.

Brown, M. T., H. T. Odum, R. C. Murphy, R. A. Christianson, S. J. Doherty, T. R. McClanahan, and S. E. Tennenbaum. 1995. Rediscovery of the world: developing an interface of ecology and economics. p. 216–250. *In* C. A. S. Hall (ed.) Maximum Power. University Press of Colorado, Niwot, CO.

Brown, N. C. 1934. Logging — Principals and Practices in the United States and Canada. John Wiley & Sons, Inc., New York, NY.

Brown, R. B., E. L. Stone, and V. W. Carlisle. 1990. Soils. p. 35–69. *In* R. L. Myers and J. J. Ewel (eds.) Ecosystems of Florida. University of Central Florida Press, Orlando, FL.

Brown, S. L. 1978. A comparison of cypress ecosystems in the landscape of Florida. Ph.D. Dissertation. University of Florida, Gainesville, FL.

Brown, S. L. 1981. A comparison of the structure, primary productivity, and transpiration of cypress ecosystems in Florida. Ecological Monographs 51: 403–427.

Brown, S. L. 1990. Structure and dynamics of basin forested wetlands in North America. p. 171–199. *In* A. E. Lugo, M. M. Brinson, and S. L. Brown (eds.) Forested Wetlands. Vol. 15. Ecosystems of the World. Elsevier Science Publishers B.V., Amsterdam, The Netherlands.

Brown, S. L. and A. E. Lugo. 1982. A comparison of structural and functional characteristics of saltwater and freshwater forested wetlands. p. 109–130. *In* B. Gopal, R. Turner, R. Wetzel, and D. Whigham (eds.) Wetlands Ecology and Management. National Institute of Ecology and International Scientific Publications, New Delhi, India.

Brown, S. L. and D. L. Peterson. 1983. Structural characteristics and biomass production of two Illinois bottomland forests. American Midland Naturalist 110: 107–117.

Brown, S. L. and R. van Peer. 1989. Response of pondcypress growth rates to sewage effluent application. Wetland Ecology and Management 1: 13–20.

Brown, S. L., M. M. Brinson, and A. E. Lugo. 1979. Structure and function of riparian wetlands. p. 17–31. *In* R. R. Johnson and J. F. McCormick (eds.) Strategies for Protection and Management of Floodplain Wetlands and Other Riparian Ecosystems. U.S. Department of Agriculture Forest Service, Washington, DC. General Technical Report WO-12.

Brown, S. L., E. W. Flohrschutz, and H. T. Odum. 1984. Structure, productivity, and phosphorus cycling of the scrub cypress ecosystems. p. 304–317. *In* K. C. Ewel and H. T. Odum (eds.) Cypress Swamps. University Presses of Florida, Gainesville, FL.

Bryan, C. F., D. J. DeMont, D. S. Sabins, and J. P. Newman, Jr. 1976. A limnological survey of the Atchafalaya Basin. Annual Report of the Louisiana Cooperative Fishery Research Unit, Louisiana State University, Baton Rouge, LA.

Buchel, H. B. and W. Grosse. 1990. Localization of the porous partition responsible for pressurized gas transport in *Alnus glutinosa* (L.) Gaertn. Tree Physiology 6: 247–256.

Buell, M. F. 1946. Jerome Bog, a peat-filled "Carolina bay." Bulletin of the Torrey Botanical Club 73: 24–33.

Buell, M. F. and R. L. Cain. 1943. The successional role of southern white cedar, *Chamaecyparis thyoides*, in southeastern North Carolina. Ecology 24: 85–93.

Buford, M. A., C. G. Williams, and J. H. Hughes. 1991. Growth and survival of Atlantic white-cedar on a South Carolina Coastal Plain site — first year results. p. 579–583. *In* S. S. Coleman and D. G. Neary (ed.) Proceedings of the Sixth Biennial Southern Silvicultural Research Conference. U.S. Department of Agriculture Forest Service, Southeastern Forest Experiment Station, Asheville, NC. General Technical Report SE-70.

Bull, H. 1949. Cypress planting in southern Louisiana. Southern Lumberman 179: 229–230.

Bull, J., and J. Farrand, Jr. 1995. National Audubon Society Field Guide to North American Birds. Alfred A. Knopf, New York, NY.

Bullard, S., J. D. Hodges, R. L. Johnson, and T. J. Straka. 1992. Economics of direct seeding and planting for establishing oak stands on old-field sites in the South. Southern Journal of Applied Forestry 16: 34–40.

Buol, S. W. (ed.). 1973. Soils of the Southern United States and Puerto Rico. Agricultural Experimental Stations of the Southern United States and Puerto Rico Land-Grant Universities, Southern Cooperative Series, Fort Worth, TX. Bulletin No. 174.

Burdick, D. M. and I. A. Mendelssohn. 1990. Relationship between anatomical and metabolic responses to soil waterlogging in the coastal grass *Spartina patens*. Journal of Experimental Botany 41(223): 223–228.

Burdick, D. M., D. Cushman, R. Hamilton, and J. G. Gosselink. 1989. Faunal changes and bottomland hardwood forest loss in the Tensas watershed, Louisiana. Conservation Biology 3: 282–292.

Burke, M. A. 1996. Historic evolution of channel morphology and riverine wetland hydrologic functions in the Piedmont of Georgia. M.S. Thesis. University of Georgia, Athens, GA.

Burke, V. J. and J. W. Gibbons. 1995. Terrestrial buffer zones and wetland conservation: a case study of freshwater turtles in a Carolina bay. Conservation Biology 9: 1365–1369.

Burkholder, J. 1996. Surface water quality issues in North Carolina. p. 11–15. *In* Solutions: A Technical Conference on Water Quality. North Carolina State University, Raleigh, NC.

Burns, A. C. 1980. Frank B. Williams: cypress lumber king. Journal of Forest History 24: 127–133.

Burns, L. A. 1978. Productivity, biomass, and water relations in a Florida cypress forest. Ph.D. Dissertation. University of North Carolina, Chapel Hill, NC.

Burns, L. A. 1984. Productivity and water relations in the Fakahatchee Strand of south Florida. p. 318–333. *In* K. C. Ewel and H. T. Odum (eds.) Cypress Swamps. University Presses of Florida, Gainesville, FL.

Burns, R. M. (ed.). 1983. Silvicultural Systems for the Major Forest Types of the United States. U.S. Department of Agriculture Forest Service, Washington, DC. Agricultural Handbook No. 445, Revised.

Burns, R. M. and B. H. Honkala. 1990. Silvics of North America: 2, Hardwoods. U.S. Department of Agriculture Forest Service, Washington, DC. Agricultural Handbook 654.

Burr, B. M. and L. M. Page. 1975. Distribution and life history notes on the tailliught shiner *Notropis maculatus* in Kentucky. Transactions of the Kentucky Academy of Science 36: 71–74.

C

Calnan, T. 1996. Personal communication. Texas General Land Office, Austin, TX.

Camilleri, J. C. 1992. Leaf-litter processing by invertebrates in a mangrove forest in Queensland. Marine Biology 114: 139–145.

Campbell, R. G. 1977. Drainage of lower coastal plain soils. p. 17–27. *In* W. E. Balmer (ed.) Sixth Southern Forest Soils Workshop. U.S. Department of Agriculture Forest Service. Asheville, NC.

Campbell, R. G. and J. H. Hughes. 1981. Forest management systems in North Carolina pocosins: Weyerhaeuser. p. 199–231. *In* C. J. Richardson (ed.) Pocosin Wetlands: An Integrated Analysis of Coastal Plain Freshwater Bogs in North Carolina. Hutchinson Ross Publication Co., Stroudsburg, PA.

Campbell, R. G. and J. H. Hughes. 1991. Impact of forestry operations on pocosins and associated wetlands. Wetlands 11: 467–480.

Campbell, R. N., Jr. 1981. Peat Energy Program: First Colony Farms, Inc. p. 214–224. *In* C. J. Richardson (ed.) Pocosin Wetlands: An Integrated Analysis of Coastal Plain Freshwater Bogs in North Carolina. Hutchinson Ross Publication Co., Stroudsburg, PA.

Campbell, R. N. and J. W. Clark. 1960. Decay resistance of baldcypress heartwood. Forest Products Journal 10:250–253.

Carlander, K. D. 1969. Handbook of Freshwater Fishery Biology, Volume One. Iowa State University Press, Ames, IA.

Carlson, J. E. and M. J. Duever. 1977. Seasonal fish population fluctuations in a south Florida swamp. Proceedings of the Annual Conference of Southeastern Association of Fish and Wildlife Agencies 31: 603–611.

Carlson, P. R., L. A. Yarbro, C. F. Zimmermann, and J. R. Montgomery. 1983. Pore water chemistry of an overwash mangrove island. Florida Scientist 46: 239–249.

Carlton, J. M. 1974. Land-building and stabilization by mangroves. Environmental Conservation 1: 285–294.

Carlton, J. M. 1975. A guide to common Florida salt marsh and mangrove vegetation. Florida Department of the Environment, St. Petersburg, FL. Research Publication No. 6.

Carter, M. R., L. A. Burns, T. R. Cavinder, K. R. Dugger, P. L. Fore, D. B. Hicks, H. L. Revells, and T. W. Schmidt. 1973. Ecosystem analysis of the Big Cypress Swamp and estuaries. Region IV, Atlanta, GA. U.S. EPA 904/9-74-002.

Cashner, R. C., B. M. Burr, and J. S. Rogers. 1989. Geographic variation of the mud sunfish, *Acantharcus pomotis* (Family Centrarchidae). Copeia 1989: 129–141.

Casey, W. 1997. Influences of environmental factors on successional relationships between cypress and bay swamps in north central Florida. MS thesis, University of Florida, Gainesville, FL.

Casterlin, M. E. and W. W. Reynolds. 1979. Thermoregulatory behavior of the bluespotted sunfish, *Enneacanthus gloriosus*. Hydrobiologia 64: 3–4.

Cerulean, S. I. and R. T. Engstrom (eds.). 1993. Fire in wetlands: a management perspective. Proceedings of the 19th Tall Timbers Fire Ecology Conference, Tall Timbers Research Station, Tallahassee, FL.

Chamie, J. P. M. and C. J. Richardson. 1978. Decomposition in northern wetlands. p. 115–130. *In* R. E. Good, D.F. Whigham, and R.L. Simpson (eds.) Freshwater Wetlands: Ecological Processes and Management Potential. Academic Press, New York, NY.

Chang, L. A., L. K. Hammett, and D. M. Pharr. 1983. Internal gas atmospheres, ethanol, and leakage of electrolytes from a flood-tolerant and a flood-susceptible sweet potato cultivar as influenced by anaerobiosis. Canadian Journal of Botany 61: 3399–3404.

Chapman, V. J. 1944. Cambridge University Expedition to Jamaica. I. A study of the botanical processes concerned in the development of the Jamaican shore-line. Journal of the Linnaean Society of London Botanical Series 52: 407–447.

Chapman, V. J. 1976. Mangrove Vegetation. J. Cramer, Vaduz, Germany.

Chapin, F. S., III and R. A. Kedrowski. 1983. Seasonal changes in nitrogen and phosphorus fractions and autumn retranslocation in evergreen and deciduous taiga trees. Ecology 64: 376–391.

Chen, R. 1996. Ecological analysis and simulation models of landscape patterns in mangrove forest development and soil characteristics along the Shark River estuary, Florida. Ph.D. Dissertation. University of Southwestern Louisiana, Lafayette, LA.

Chirkova, T. V. and S. V. Soldatenkov. 1965. Paths of movement of oxygen from leaves to roots kept under anaerobic conditions. Fiziologiya Rastenii 12: 216–225.

Choong, E. T., P. J. Fogg, and J. P. Jones. 1986. Natural decay resistance of baldcypress. Louisiana State University, Baton Rouge, LA. Wood Utilization Notes No. 38.

Chow, K. H. 1984. Alcohol dehydrogenase synthesis and waterlogging tolerance in maize. Tropical Agriculture 61: 302–304.

Chow, V. T. 1952. On the determination of transmissivity and storage coefficients from pumping test data. Transactions of the American Geophysical Union 33: 397–404.

Christensen, N. L. 1981. Fire regimes in southeastern ecosystems. p. 112–136. *In* H. A. Mooney, T. Bonnicksen, N. L. Christensen, W. A. Reiners, and J. Lotan (eds.) Fire Regimes and Ecosystem Properties. U.S. Department of Agriculture Forest Service, Washington, DC. General Technical Report WO-26.

Christensen, N. L. 1988. Vegetation of the southeastern coastal plain. p. 317–363. *In* M. G. Barbour and W. D. Billings (eds.) North American Terrestrial Vegetation. Cambridge University Press, New York, NY.

Christensen, N. L., R. B. Wilbur, and J. S. McLean. 1988. Soil-vegetation correlations in the pocosins of Croatan National Forest, North Carolina. U.S. Fish and Wildlife Service, Washington, DC. Biological Report 88(28).

Christensen, N. L., R. B. Burchell, A. Liggett, and E. L. Simms. 1981. The structure and development of pocosin vegetation. p. 43–61. *In* C. J. Richardson (ed.) Pocosin Wetlands: An Integrated Analysis of Coastal Plain Freshwater Bogs in North Carolina. Hutchinson Ross Publication Co., Stroudsburg, PA.

CH2M Hill. 1986. Carolina bay effluent discharge pilot program. Bear Bay baseline monitoring. July 1985–July 1986. CH2M Hill, Charleston, SC.

CH2M Hill. 1988. 201 Facilities Plan Amendment for the Grand Strand Region (Plant 'A' Service Area) Grand Strand Water and Sewer Authority. Vol. 3. Carolina Bay Effluent Discharge Pilot Program Final Project Report. CH2M Hill, Charleston, SC.

CH2M Hill. 1994. Carolina Bay Natural Land Treatment Program. Sixth Year Annual Report January–December 1993. CH2M Hill, Charleston, SC.

Chueng, N. and S. Brown. 1995. Decomposition of silver maple (*Acer saccharinum* L.) woody debris in a central Illinois bottomland forest. Wetlands 15: 232–241.

Cintrón, G. 1990. Restoration of mangrove systems. Symposium on Habitat Restoration, National Oceanic and Atmospheric Administration. Washington, DC.

Cintrón, G. and Y. Schaeffer-Novelli. 1984. Caracteristicas y desarrollo estructural de los manglares de Norte y Sur America. Programa Regional de Desarrollo Cientifico y Tecnologico 25: 4–15.

Cintrón, G., A. E. Lugo, D. J. Pool, and G. Morris. 1978. Mangroves of arid environments in Puerto Rico and adjacent islands. Biotropica 10: 110–121.

Cintrón, G., A.E. Lugo, R. Martinez, B.B. Cintrón, and L. Encarnacion. 1981. Impact of oil in the tropical marine environment. p. 18–27. *In* Technical Publication. Division of Marine Resources, Department of Natural Resources of Puerto Rico.

Clancy, K. E. 1996. Personal communication. Delaware Heritage Trust, Department of Natural Resources and Environmental Control, Dover, DE.

Clark, J. E. 1978. Fresh water wetlands: habitats for aquatic invertebrates, amphibians, reptiles, and fish. p. 330–343. *In* P. E. Greeson, J. R. Clark, and J. E. Clark (eds.) Wetland Functions and Values: The State of Our Understanding. American Water Resources Association, Minneapolis, MN.

Clark, J. R. and J. Benforado (eds.). 1981. Wetlands of Bottomland Hardwood Forests. Elsevier Scientific Publishing Company, Amsterdam, The Netherlands.

Clark, M. K., D. S. Lee, and J. B. Funderburg, Jr. 1985. The mammal fauna of Carolina Bays, pocosins, and associated communities in North Carolina: an overview. Brimleyana 11: 1–38.

Clatterbuck, W. K. and J. S. Meadows. 1993. Regenerating oaks in the bottomlands. p 184–195. *In* D. Loftis and C.E. McGee (eds.) Oak Regeneration: Serious Problems, Practical Recommendations. U.S. Department of Agriculture Forest Service. Geneneral Technical Report SE-84.

Clawson, R. G., B. G. Lockaby, and R. H. Jones. 1997. Amphibian responses to helicopter harvesting in forested floodplains of low order, blackwater streams. Forest Ecology and Management 90: 225–236.

Clemson University. 1982. Lime and fertilizer recommendations based on soil-test results. Cooperative Extension Service, Clemson University, Clemson, SC. Circular 476.

Clewell, A. F. 1971. The vegetation of the Apalachicola National Forest: an ecological perspective. Final Report to U.S. Department of Agriculture Forest Service, Atlanta, GA.

Clewell, A. F. and R. Lea. 1990. Creation and restoration of forested wetland vegetation in the southeastern United States. p. 195–231. *In* J. A. Kusler and M. E. Kentula (eds.) Wetland Creation and Restoration. Island Press, Washington, DC.

Clough, B. F. (ed.). 1982. Mangrove Ecosystems in Australia: Structure, Function and Management. Australian National University Press, Canberra, Australia.

Clough, B. F., K.G. Boto, and P.M. Attiwill. 1983. Mangroves and sewage: a re-evaluation. p. 151–162. *In* H.J. Teas (ed.) Biology and Ecology of Mangroves. Dr W. Junk Publishers, The Hague, The Netherlands.

Clymo, R. S. 1984. The limits to peat bog growth. Philosophical Transactions of the Royal Society of London B 303: 605–654.

Cohen, A. D. 1996. Personal communication. Department of Geological Sciences, University of South Carolina, Columbia, SC.

Cohen, A. D. and E. M. Stack. 1992. The peat resources of South Carolina. South Carolina Geology 34: 1–24.

Cohen, A. D., D. J. Casagrande, M. J. Andrejko, and G. R. Best. 1984. The Okefenokee Swamp: It's Natural History, Geology, and Geochemistry. Wetland Surveys, Los Alamos, NM.

Collette, B. B. 1962. The swamp darters of the subgenus Hololepis (Pisces, Percidae). Tulane Studies in Zoology 9: 115–211.

Collins, M. E. 1982. Definition and identification of natural soil drainage classes. p. 16–42. *In* N. B. Comerford and A. V. Mollitor (eds.) Guide to Classification and Management of Southeastern Coastal Plain Forest Soils. Cooperative Research in Forest Fertilization, University of Florida, Gainesville, FL.

Colquhoun, D. J. 1974. Cyclic surficial stratigraphic units of the middle and lower coastal plains, central South Carolina. p. 179–190. *In* R. Q. Oats and J. R. DuBar (eds.) Postmiocene Stratigraphy Central and Southern Atlantic Coastal Plain. Utah State University Press.

Conant, R. and J. T. Collins. 1991. A Field Guide to Reptiles and Amphibians of Eastern and Central North America, 3rd Edition, Houghton Mifflin Co., New York, NY.

Congdon, J. D. and J. W. Gibbons. 1989. Biomass productivity of turtles in freshwater wetlands: A geographic comparison. p. 583–592. *In* R. R. Sharitz and J. W. Gibbons (eds.) Freshwater Wetlands and Wildlife Symposium: Perspectives on Natural, Managed and Degraded Ecosystems. U.S. Department of Energy, Office of Scientific and Technical Information, Oak Ridge, TN. CONF-8603101, DOE Symposium Series No. 61.

Congressional Research Service. 1973. A Legislative History of the Water Pollution Control Act Amendments of 1972. U.S. Government Printing Office, Washington, DC. Serial No. 93–1.

Conner, R. C. 1993. Forest statistics for South Carolina. U.S. Department of Agriculture Forest Service, Southeastern Forest Experiment Station, Asheville, NC. Resource Bulletin SE-141.

Conner, W. H. 1988. Natural and artificial regeneration of baldcypress in the Barataria and Lake Verret Basins of Louisiana. Ph.D. Dissertation. Louisiana State University, Baton Rouge, LA.

Conner, W. H. 1994. Effect of forest management practices on southern forested wetland productivity. Wetlands 14: 27–40.

Conner, W. H. 1995. Preliminary study of the feasibility of root pruning baldcypress seedlings for planting in flooded sites. p. 430–434. *In* M. B. Edwards (compiler) Proceedings of the Eighth Biennial Southern Silvicultural Research Conference. U.S. Department of Agriculture Forest Service, Southern Research Station, Asheville, NC. General Technical Report SRS-1.

Conner, W. H. and M. Brody. 1989. Rising water levels and the future of southeastern Louisiana swamp forests. Estuaries 12: 318–323.

Conner, W. H. and J. W. Day, Jr. 1976. Productivity and composition of a baldcypress-water tupelo site and a bottomland hardwood site in a Louisiana swamp. American Journal of Botany 63: 1354–1364.

Conner, W. H. and J. W. Day, Jr. 1982. The ecology of forested wetlands in the southeastern United States. p. 69–87. *In* B. Gopal, R. E. Turner, R. G. Wetzel, and D. F. Whigham (eds.) Wetlands Ecology and Management. National Institute of Ecology and International Scientific Publications, Jaipur, India.

Conner, W. H. and J. W. Day, Jr. 1988. The impact of rising water levels on tree growth in Louisiana. p. 219–224. *In* D. D. Hook et al. (eds.) The Ecology and Management of Wetlands, Vol. 2: Management, Use, and Value of Wetlands. Croom Helm Ltd Publishers, England.

Conner, W. H. and J. W. Day, Jr. 1991. Leaf litter decomposition in three Louisiana freshwater forested wetland areas with different flooding regimes. Wetlands 11: 303–312.

Conner, W. H. and J. W. Day, Jr. 1992a. Diameter growth of *Taxodium distichum* (L.) Rich. and *Nyssa aquatica* L. from 1979–1985 in four Louisiana swamp stands. American Midland Naturalist 127: 290–299.

Conner, W. H. and J. W. Day. 1992b. Water level variability and litterfall productivity of forested freshwater wetlands in Louisiana. American Midland Naturalist 128: 237–245.

Conner, W. H. and K. Flynn. 1989. Growth and survival of baldcypress (*Taxodium distichum* (L.) Rich.) planted across a flooding gradient in a Louisiana bottomland forest. Wetlands 9(2): 207–217.

Conner, W. H. and R. A. Muller. 1986. Regeneration of baldcypress (*Taxodium distichum* (L.) Rich.) as related to drought in the LaBranche wetlands of southern Louisiana. p. 261–271. *In* G. C. Jacoby and J. W. Hornbeck (comps.) Proceedings of the International Symposium on Ecological Aspects of Tree-Ring Analysis. U.S. Department of Energy, Washington, DC.

Conner, W. H. and J. R. Toliver. 1990. Long-term trends in the baldcypress (*Taxodium distichum* (L.) Rich.) resource in Louisiana. Forest Ecology and Management 33/34: 543–557.

Conner, W. H., J. G. Gosselink, and R. T. Parrondo. 1981. Comparison of the vegetation of three Louisiana swamp sites with different flooding regimes. American Journal of Botany 68: 320–331.

Conner, W. H., R. T. Huffman, and W. M. Kitchens. 1990. Composition and productivity in bottomland hardwood forest ecosystems: the report of the vegetation workgroup. p. 455–479. *In* J. G. Gosselink, L. C. Lee, and T. A. Muir (eds.) Ecological Processes and Cumulative Impacts Illustrated by Bottomland Hardwood Wetland Ecosystems. Lewis Publishers, Chelsea, MI.

Conner, W. H., J. R. Toliver, and F. H. Sklar. 1986. Natural regeneration of cypress in a Louisiana swamp. Forest Ecology and Management 14: 305–317.

Conner, W. H., J. R. Toliver, and G. R. Askew. 1993. Artificial regeneration of baldcypress in a Louisiana crayfish pond. Southern Journal of Applied Forestry 17(1): 54–57.

Conner, W. H., J. W. Day, Jr., R. H. Baumann, and J. Randall. 1989. Influence of hurricanes on coastal ecosystems along the northern Gulf of Mexico. Wetlands Ecology and Management 1: 45–56.

Conservation Foundation, The. 1988. Protecting America's wetlands: an action agenda. Final Report of the National Wetlands Policy Forum. Washinton, DC.

Cook, H. T. 1916. The Life and Legacy of David Rogerson Williams. (publisher unlisted) New York.

Cooke, C. Q. 1936. Geology of the coastal plain of South Carolina. U.S. Geologic Survey Bulletin 867.

Cooper, H. H., Jr., J. D. Bredehoeft,and I. S. Papadopoulos. 1967. Response of a finite diameter well to an instantaneous charge of water. Journal of Hydrology 3: 263–269.

Copp, G. H. and B. Cellot. 1988. Drift of embryonic and larval fishes, especially *Lepomis gibbosus*, in the upper Rhone River. Journal of Freshwater Ecology 4: 419–424.

Corredor, J. E. and J. M. Morell. 1994. Nitrate depuration of secondary sewage effluents in mangrove sediments. Estuaries 17: 295–300.

Costanza, R., S. C. Farber, and J. Maxwell. 1989. The valuation and management of wetland ecosystems. Ecological Economics 1: 335–361.

Coultas, C. L., and M. J. Deuver. 1984. Soils of cypress swamps. p. 51–59. *In* K. C. Ewel and H. T. Odum (eds.) Cypress Swamps. University of Florida Press, Gainesville, FL.

Coulter, M. C. and A. L. Bryan, Jr. 1995. Factors affecting reproductive success of wood storks (*Mycteria americana*) in east-central Georgia. Auk 112: 237–243.

Coutant, C. C. 1977. Compilation of temperature preference data. Journal of the Fisheries Research Board of Canada 34: 739–745.

Coutts, M. P. 1981. Effects of waterlogging on water relations of actively-growing and dormant sitka spruce seedlings. Annals of Botany 47: 747–753.

Coutts, M. P. and W. Armstrong. 1976. Role of oxygen transport in the tolerance of trees to waterlogging. p. 361–385. *In* M. G. R. Cannell and F. T. Last (eds.) Tree Physiology and Yield Improvement. Academic Press, New York, NY.

Coutts, M. P. and J. J. Philipson. 1978a. Tolerance of tree roots to waterlogging. I. Survival of sitka spruce and lodgepole pine. New Phytologist 80: 63–69.

Coutts, M. P. and J. J. Philipson. 1978b. Tolerance of tree roots to waterlogging. II. Adaptation of sitka spruce and lodgepole pine to waterlogged soil. New Phytologist 80: 71–77.

Cowardin, L. M., V. Carter, F. C. Golet, and E. T. LaRoe. 1979. Classification of Wetlands and Deepwater Habitats of the United States. U.S. Fish and Wildlife Service, Washington, DC. FWS/OBS-79/31.

Cowell, B. C. and B. S. Barnett. 1974. Life history of the taillight shiner, *Notropis maculatus*, in central Florida. American Midland Naturalist 91: 282–292.

Cowles, S. 1975. Metabolism measurements in a cypress dome. M.S. Thesis. University of Florida, Gainesville, FL.

Craft, C., S. Faulkner, and C. Richardson. 1995. p. 2–32. *In* Lecture outline notes. Identifying and Delineation of Jurisdictional Wetlands Course. Duke University School of the Environment, Durham, NC.

Craighead, F. C. 1971. The Trees of South Florida, Vol. 1. University of Miami Press, Coral Gables, FL.

Craighead, F. C., and V. C. Gilbert. 1962. The effects of Hurricane Donna on the vegetation of southern Florida. Quarterly Journal of the Florida Academy of Science 25: 1–28.

Crawford, R. M. M. 1967. Alcohol dehydrogenase activity in relation to flooding tolerance in roots. Journal of Experimental Botany 18: 458–464.

Crawford, R. M. M. 1976. Tolerance of anoxia and the regulation of glycolysis in tree roots. p. 387–401. *In* M.G.R. Cannell and F. T. Last (eds.) Tree Physiology and Yield Improvement. Academic Press, Harcourt Brace Jovanovich, Publishers, London, England.

Crawford, R. M. M. 1982. Physiological responses to flooding. p. 453–477. *In* O. L Lange, P. S. Nobel, C. B. Osmond, and H. Ziegler (eds.) Physiological Plant Ecology II. Springer Verlag, Berlin.

Crawford, R. M. M. 1993. Plant survival without oxygen. Biologist 40:110–114.

Crawford, R. M. M. and M. A. Baines. 1977. Tolerance of anoxia and the metabolism of ethanol in tree roots. New Phytologist 79: 519–526.

Crawford, R. M. M. and D. M. Finegan. 1989. Removal of ethanol from lodgepole pine roots. Tree Physiology 5: 53–61.

Crawford, R. M. M. and M. McManmon. 1968. Inductive responses of alcohol and malic dehydrogenases in relation to flooding tolerance in roots. Journal of Experimental Botany 19: 435–441.

Crawford, R. M. M. and Z. M. Zochowski. 1984. Tolerance of anoxia and ethanol toxicity in chickpea seedlings (*Cicer arietinum* L.). Journal of Experimental Botany 35(159): 1472–1480.

Creaser, E. P. 1931. Some cohabitants of burrowing crayfish. Ecology 12: 243–244.

Crownover, S. H., N. B. Comerford, D. G. Neary and J. Montgomery. 1995. Horizontal groundwater flow patterns through a cypress swamp-pine flatwoods landscape. Soil Science Society of America Journal 59: 1199–1206.

Crutchfield, D. M. 1996. Personal communication. Westvaco (retired), Summerville, SC.

Crutchfield, D. M. and I. F. Trew. 1957. Report on Dare County drainage studies. West Virginia Pulp and Paper Co., Summerville, SC. North Carolina Research Project, Report NC-3.

Cubbage, F. W. and C. H. Flather. 1993. Forested wetland area and distribution — a detailed look at the South. Journal of Forestry 91: 35–40.

Cubbage, F. W. and T. Harris, Jr. 1988. Wetlands protection and regulation: Federal law and history. TOPS. Official Publication of the Georgia Forestry Association 22: 18–23, 26.

Cubbage, F. W. and W. C. Siegel. 1990. The impact of federal environmental law on forest resource management in the United States. Forestry Legislation: Report of the IUFRO Working Party S4.0803: 87–107. Zurich, Switzerland.

Cubbage, F. W, J. O'Laughlin, and C. S. Bullock III. 1993. p. 363–375. *In* Forest Resource Policy. John Wiley & Sons, Inc., New York, NY.

Cubbage, F. W., W. S. Dvorak, R. C. Abt, and G. Pacheco. 1996. World timber supply and prospects: models, projections, plantations and implications. North Carolina State University, CAMCORE Annual Meeting, Bali, Indonesia.

Cuffney, T. F. 1988. Input, movement and exchange of organic matter within a subtropical coastal blackwater river-floodplain system. Freshwater Biology 19: 305–320.

D

Dabel, C. V. and F. P. Day, Jr. 1977. Structural comparison of four plant communities in the Great Dismal Swamp, Virginia. Bulletin of the Torrey Botanical Club 104: 352–360.

Dacey, J. W. H. 1981. Pressurized ventilation in the yellow water-lily. Ecology 62: 1137–1147.

Dagan, G. 1987. Theory of solute transport by groundwater. Annual Review of Fluid Mechanics 19: 83–215.

Dahl, T. E. 1990. Wetlands Losses in the United States 1780's to 1980s. U.S. Fish and Wildlife Service, Washington, DC.

Dahl, T. E. 1996. Unpublished data. National Wetlands Inventory. U.S. Fish and Wildlife Service, Washington, DC.

Dahl, T. E. and C. E. Johnson. 1991. Wetlands status and trends in the conterminous United States mid 1970s to mid 1980s. U.S. Fish and Wildlife Service, Washington, DC.

Dahl, T. E., C. E. Johnson, and W. E. Frayer. 1991. Status and trends of wetlands in the conterminous United States, mid 1970s to mid 1980s. U.S. Fish and Wildlife Service, Washington, DC.

Damman, A. W. H. 1979. Geographic patterns in peatland development in eastern North America. p.42–57. *In* Proceedings of the International Symposium on Classification of Peat and Peatlands. International Peat Society, Hyytiala, Finland.

Dana, S. T. and S. K. Fairfax. 1980. Forest and Range Policy. McGraw–Hill Book Company, New York, NY.

Daniel, C. C. III. 1981. Hydrology, geology and soils of pocosins: a comparison of natural and altered systems. p. 69–108. *In* C. J. Richardson (ed.) Pocosin Wetlands: An Integrated Analysis of Coastal Plain Freshwater Bogs in North Carolina. Hutchinson Ross Publication Co., Stroudsburg, PA.

Daniels, R. B. and E. E. Gamble. 1978. Relations between stratigraphy, geomorphology and soils in Coastal Plain areas of southeastern U.S.A. Geoderma 21: 41–65.

Daniels, R. B., B. L. Allen, H. H. Bailey, and F. H. Beinroth. 1973. Physiography. p. 3–16. *In* S. W. Buol (ed.) Soils of the Southeastern United States and Puerto Rico. U.S. Department of Agriculture. Southern Cooperative Series Bulletin No. 174.

Daniels, R. B., E. E. Gamble, W. H. Wheeler, and C. S. Holzhey. 1977. The stratigraphy and geomorphology of the Hofmann Forest pocosin. Soil Science Society of America Journal 41: 1175–1180.

Daniels, R. B., H. J. Kleiss, S. B. Buol, H. J. Byrd, and J. A. Phillips. 1984. Soil systems in North Carolina. North Carolina Agricultural Research Service, North Carolina State University, Raleigh, NC. Bulletin 467.

Darke, A. K. 1997. The effects of artificial flooding and seasonal hydroperiod variation on Al and Fe crystallinity and P sorption capacity in a forested floodplain. Ph.D. Dissertation. George Mason University, Fairfax, VA.

Davis, D. W. 1975. Logging canals: a distinct pattern of the swamp in south Louisiana. Forests and People 25: 14–17, 33–35.

Davis, J. H. 1940. The ecology and geologic role of mangroves in Florida. Carnegie Institute, Washington, DC. Publication Number 517. Paper from the Tortugas Laboratory 32: 303–412.

Davison, E. M. and F. C. S. Tay. 1985. The effect of waterlogging on seedlings of *Eucalyptus marginata*. New Phytologist 101: 743–753.

Day, F. P., Jr. 1982. Litter decomposition rates in the seasonally flooded Great Dismal Swamp. Ecology 63: 670–678.

Day, F. P., Jr. 1984. Biomass and litter accumulation in the Great Dismal Swamp. p. 386–392. *In* K. C. Ewel and H. T. Odum (eds.) Cypress Swamps. University Presses of Florida, Gainesville, FL.

Day, F. P., Jr. 1987a. Effects of flooding and nutrient enrichment on biomass allocation in *Acer rubrum* seedlings. American Journal of Botany 74: 1541–1554.

Day, F. P., Jr. 1987b. Production and decay in a *Chamaecyparis thyoides* swamp in southern Virginia. p. 123–132. *In* A. D. Laderman (ed.) Atlantic White Cedar Wetlands. Westview Press, Boulder, CO.

Day, G. 1953. The Indian as an ecological factor in the northeastern forest. Ecology 34:329–346.

Day, J. W., Jr. In press. The design of mangroves for waste water treatment in Campeche. *In* A. Yáñez-Arancibia and A. L. Laura-Domínguez (eds.) Mangrove Ecosystems in Tropical America; Structure, Function and Management. University of Mexico Press, Mexico City.

Day, J. W., Jr., T. J. Butler, and W. H. Conner. 1977. Productivity and nutrient export studies in a cypress swamp and lake system in Louisiana. p. 255–269. *In* M. Wiley (ed.) Estuarine Processes, Vol. II. Academic Press, New York, NY.

Day, J. W., Jr., W. H. Conner, F. Ley-Lou, R. H. Day, and A. Machado. 1987. The productivity and composition of mangrove forests, Laguna de Términos, Mexico. Aquatic Botany 27: 267–284.

DeBaise, A. 1996. Personal communication. Savannah River Ecology Laboratory, Aiken, SC.

DeBell, D. S. 1971. Stump sprouting after harvest cutting in swamp tupelo. U.S. Department of Agriculture Forest Service, Southeastern Forest Experiment Station, Asheville, NC. Research Paper SE-83.

DeBell, D. S. and I. D. Auld. 1968. Robins and tupelo in low country swamp. South Carolina Wildlife 15: 16.

DeBell, D. S. and D. D. Hook. 1969. Seeding habits of swamp tupelo. U.S. Department of Agriculture Forest Service, Asheville, NC. Research Paper SE-47.

DeBell, D. S., D. D. Hook, W. H. McKee, Jr., and J. L. Askew. 1984. Growth and physiology of loblolly pine roots under various water table level and phosphorus treatments. Forest Science 30: 705–714.
DeBell, D. S., G. R. Askew, D. D. Hook, J. Stubbs, and E. G. Owens. 1982. Species suitability on a lowland site altered by drainage. Southern Journal of Applied Forestry 6: 2–9.
DeBusk, W. F. and K. R. Reddy. 1987. Removal of floodwater nitrogen in a cypress swamp receiving primary wastewater effluent. Hydrobiologia 153: 79–86.
DeCalesta, D. S. 1994. Effect of white-tailed deer on songbirds within managed forests in Pennsylvania. Journal of Wildlife Management 58: 711–718.
DeCosmo, J. M., T. V. Cambre, and D. H. Sims. 1990. Helicopter logging a low-impact harvesting method for the wetlands. U.S. Department of Agriculture Forest Service, Atlanta, GA. Cooperative Forestry Technology Update, Management Bulletin R8-MB 48.
Deghi, G. S., K. C. Ewel, and W. J. Mitsch. 1980. Effects of sewage effluent application on litter fall and litter decomposition in cypress swamps. Journal of Applied Ecology 17: 397–408.
Delcourt, H. R. 1985. Holocene vegetational changes in the southern Appalachian mountains, U.S.A. Ecologia mediterranea Tome XI (Fascicule 1): 9–15.
Delcourt, H. R. and P. A. Delcourt. 1985. Quaternary palynology and vegetational history of the southeastern United States. p.1–37. *In* V. M. Bryant, Jr. and R. G. Holloway (eds.) Pollen Records of Late-Quaternary North American Sediments. American Association of Stratigraphic Palynologists Foundation, Calgary, Alberta, Canada.
Delnicki, D. and K. J. Reinecke. 1986. Mid-winter food use and body weights of mallards and wood ducks in Mississippi. Journal of Wildlife Management 50: 43–51.
Demaree, D. 1932. Submerging experiments with *Taxodium*. Ecology 13: 258–262.
deMaynadier, P. G., and M. L. Hunter, Jr. 1995. The relationship between forest management and amphibian ecology: a review of the North American literature. Environmental Review 3: 230–261.
Dennis, J. V. 1988. The Great Cypress Swamps. Louisiana State University Press, Baton Rouge, LA.
Dennis, W. M. 1980. *Sarracenia oreophila* (Kearny) Wherry in the Blue Ridge Province of northeastern Georgia. Castanea 45: 101–103.
Dennison, M. S. and J. F. Barry. 1993. Wetlands: Guide to Science, Law, and Technology. Noyes Publications, Park Ridge, NJ.
DeSylva, D. P. 1986. Increased storms and estuarine salinity and other ecological impacts of the "greenhouse effect." p. 153–164. *In* J.G. Titus (ed.) Effects of Changes in Stratospheric Ozone and Global Climate. U.S. Environmental Protection Agency, Washington, DC.
Dicke, S. G. and J. R. Toliver. 1988. Effects of crown thinning on baldcypress height, diameter, and volume growth. Southern Journal of Applied Forestry 12: 252–256.
Dickson, H. 1937. The Story of King Cotton. Funk & Wagnalls Co., New York, NY.
Dickson, J. G. 1978. Forest bird communities of the bottomland hardwoods. p. 66–73. *In* R. M. DeGraaf (tech. coord.) Proceedings of Workshop on the Management of Southern Forests for Nongame Birds. USDA Forest Service General Technical Report SE-14.
Dickson, J. G. and M. L. Warren, Jr. 1994. Wildlife and fish communities of eastern riparian forests. p. 1–31. *In* Riparian Ecosystems in the Humid U.S.: Functions, Values and Management. National Association of Conservation Districts, Washington, DC.
Dickson, R. E., T. C. Broyer, and C. M. Johnson. 1972. Nutrient uptake by tupelogum and baldcypress from saturated or unsaturated soil. Plant Soil 37: 297–308.
Dickson, J. G., F. R. Thompson, III, R. N. Conner, and K. E. Franzreb. 1995. Silviculture in central and southeastern oak-pine forests. p. 245–266. *In* T. E. Martin and D. M. Finch (eds.) Ecology and Management of Neotropical Migratory Birds. Oxford University Press, New York, NY.

Diehl, J. W. and R. E. Behling. 1982. Geologic factors affecting formation and presence of wetlands in the North Central Section of the Appalachian Plateaus Province of West Virginia. p. 3–9. *In* B. R. McDonald (ed.) Proceedings of the Symposium on Wetlands of the Unglaciated Appalachian Region. West Virginia University, Morgantown, WV.

Dierberg, F. E. and P. L. Brezonik. 1982. Nitrifying population densities and inhibition of ammonium oxidation in natural and sewage-enriched cypress swamps. Water Research 16: 123–126.

Dierberg, F. E. and P. L. Brezonik. 1984a. Nitrogen and phosphorus mass balances in a cypress dome receiving wastewater. p. 112–118. *In* K. C. Ewel and H. T. Odum (eds.) Cypress Swamps. University Presses of Florida, Gainesville, FL.

Dierberg, F. E. and P. L. Brezonik. 1984b. Water chemistry of a Florida cypress dome. p. 34–50. *In* K. C. Ewel and H. T. Odum (eds.) Cypress Swamps. University Presses of Florida, Gainesville, FL.

Dierberg, F. E., and K. C. Ewel. 1984. The effects of wastewater on decomposition and organic matter accumulation in cypress domes. p. 164–170. *In* K. C. Ewel and H. T. Odum (eds.) Cypress Swamps. University Presses of Florida, Gainesville, FL.

Dollar, K. E., S. G. Pallardy, and H. G. Garret. 1992. Composition and environment of floodplain forests of northern Missouri. Canadian Journal of Forest Research 22: 1343–1350.

Dolman, J. D. and S. W. Buol. 1967. A study of organic soils (histosols) in the tidewater region of North Carolina. North Carolina Agricultural Experiment Station, Raleigh, NC. Technical Bulletin Number 181.

Domingue O'Neill, E. 1995. Amphibian and reptile communities of temporary ponds in a managed pine flatwoods. M.S. Thesis. University of Florida, Gainesville, FL.

Dorney, J. 1996. Personal communication. North Carolina Division of Environmental Management, Water Quality Section, Environmental Science Branch, Raleigh, NC.

Doty, C. W. 1973. Pump drainage in Carolina bays. Journal of the Irrigation and Drainage Division 99(IR4): 465–475.

Dowd, J., D. Wenner, and M. Carpenter. 1993. Piedmont geohydrology: implications for flow and transport. Proceedings of the 1993 Georgia Water Resources Conference, University of Georgia, Athens, GA.

Drayton, E. R. III and D. D. Hook. 1989. Water management of a baldcypress-tupelo wetland for timber and wildlife. p. 54–58. *In* D. D. Hook and R. Lea (eds.) Proceedings of the Symposium: The Forested Wetlands of the Southern United States. U.S. Department of Agriculture Forest Service, Southeastern Forest Experiment Station, Asheville, NC. General Technical Report SE-50.

Drew, M. C. and J. M. Lynch. 1980. Soil anaerobiosis, microorganisms, and root function. Annual Review of Phytopathology 18: 37–66.

Drew, M. C., P. H. Saglio, and A. Pradet. 1985. Larger adenylate energy charge and ATP/ADP ratios in aerenchymatous roots of *Zea mays* in anaerobic media as a consequence of improved internal oxygen transport. Planta 165: 51–58.

Drew, M. C., M. B. Jackson, S. C. Gifford, and R. Campbell. 1981. Inhibition by silver ions of gas space (aerenchyma) formation in adventitious roots of *Zea mays* L. subjected to exogenous ethylene or to oxygen deficiency. Planta 153: 217–224.

Dudal, R. 1992. Introductory Address: Wet Soils. p. 1–4. *In* J. M. Kimble (ed.) Proceedings of the 8th International Soil Correlation Meeting (VIII ISCOM): Characterization, Classification, and Utilization of Wet Soils. U.S. Department of Agriculture Soil Conservation Service, Soil Management Support Services, Washington, DC.

Dudley, T. A. 1986. Soil survey of Horry County South Carolina. U.S. Department of Agriculture, Soil Conservation Service, Conway, SC.

Duever, M. J. 1996. Personal communication. The Nature Conservancy, Walt Disney Wilderness Preserve, Kissimmee, FL.

Duever, M. J. and L. A. Riopelle. 1984. Tree-ring analysis in the Okefenokee Swamp. p. 180–188. *In* A. D. Cohen, D. J. Casagrande, M. J. Andrejko, and G. R. Best (eds.) The Okefenokee Swamp: Its Natural History, Geology, and Geochemistry. Wetland Surveys, Los Alamos, NM.

Duever, M. J., J. E. Carlson, and L. A. Riopelle. 1975. Ecosystem analyses at Corkscrew Swamp. p. 627–725. *In* H. T. Odum and K. C. Ewel (eds.) Cypress Wetlands for Water Management, Recycling and Management. Second Annual Report to National Science Foundation and Rockefeller Foundation, University Presses of Florida, Gainesville, FL.

Duever, M. J., J. E. Carlson, and L. A. Riopelle. 1984a. Corkscrew Swamp: a virgin cypress stand. p. 334–348. *In* K. C. Ewel and H. T. Odum (eds.) Cypress Swamps, University Presses of Florida, Gainesville, FL.

Duever, M. J., J. F. Meeder, and L. C. Duever. 1984b. Ecosystems of the Big Cypress Swamp. p. 294–303. *In* K. C. Ewel and H. T. Odum (eds.) Cypress Swamps. University Presses of Florida, Gainesville, FL.

Duever, M. J., J. F. Meeder, L. C. Meeder, and J. M. McCollom. 1994. The climate of south Florida and its role in shaping the Everglades ecosystem. p. 225–248. *In* S. M. Davis and J. C. Ogden (eds.) Everglades — The Ecosystem and Its Restoration. St. Lucie Press, Delray Beach, FL.

Duever, M. J., J. E. Carlson, J. F. Meeder, L. C. Duever, L. H. Gunderson, L. A. Riopelle, T. R. Alexander, R. F. Meyers, and D. P. Spangler. 1986. The Big Cypress National Preserve. National Audubon Society, New York, NY.

Dugan, D. 1993. Wetlands in Danger. Oxford University Press, New York, NY.

Duke, N. C. and Z. Pinzon. 1993. Mangrove forests. p. 447–553. *In* B. D. Keller and J. B. C. Jackson (eds.) Long-term assessment of the oil spill at Bahia las Minas, Panama. Synthesis Report, Volume II, Technical Report. U.S. Department of the Interior, Minerals Management Service, Gulf of Mexico OCS Regional Office, New Orleans, LA.

Dunne, T. 1983. Relation of field studies and modeling in the prediction of storm runoff. Journal of Hydrology 65: 25–48.

Dury, G. H. 1977. Underfit streams: retrospect, perspect, and prospect. p. 281–293. *In* K. J. Gregory (ed.) River Channel Changes. John Wiley and Sons, New York, NY.

E

Eager, D. 1996. Personal communication. Tennessee Division of Water Pollution Control, Nashville, TN.

Earley, L. S. 1987. Twilight for junipers. Wildlife in North Carolina 51: 9–15.

Earley, L. S. 1989. Wetlands in the highlands. Wildlife in North Carolina 53: 11–16.

EarthInfo Inc. 1995. Database guide for EarthInfo CD[2] NCDC summary of the day. EarthInfo Inc., Boulder, CO.

Echelle, A. A. and C. D. Riggs. 1972. Aspects of the early life history of gars (*Lepisosteus*) in Lake Texoma. Transactions of the American Fisheries Society 101: 106–112.

Echternacht, A. C. and L. D. Harris. 1993. The fauna and wildlife of the southeastern United States. p. 81–116. *In* W. H. Martin, S. G. Boyce, and A. C. Echternacht (eds.) Biodiversity of the Southeastern United States: Lowland Terrestrial Communities. John Wiley and Sons, Inc., New York, NY.

EDF. 1995. Weyerhaeuser, environmental groups reach agreement. Environmental Defense Fund, Press Release, December 5, 1995. Raleigh, NC.

Eggler, W. A. and W. G. Moore. 1961. The vegetation of Lake Chicot, Louisiana, after eighteen years of impoundment. Southwestern Naturalist 6: 175–183.

Egler, F. E. 1952. Southeast saline Everglades vegetation, Florida, and its management. Vegetatio 3: 213–265.

Egli, M. and P. Fitze. 1995. The influence of increased NH_4^+ deposition rates on aluminum chemistry in the percolate of acid soils. European Journal of Soil Science 46: 439–447.

Eisterhold, J. A. 1972. Lumber and trade in the lower Mississippi Valley and New Orleans, 1800–1860. Louisiana History 13: 71–91.

Elder, J. F. 1985. Nitrogen and phosphorus speciation and flux in a large Florida river-wetland system. Water Resources Research Bulletin 21: 724–732.

Elder, J. F. and D. J. Cairns. 1982. Production and decomposition of forest litter fall on the Apalachicola River flood plain, Florida. U.S. Department of Interior, Washington, DC. U.S. Geological Survey Water-Supply Paper 2196-B.

Elder, J. F. and H. C. Mattraw, Jr. 1982 Riverine transport of nutrients and detritus to the Apalachicola Bay estuary, Florida. Water Resources Research Bulletin 18: 849–856.

Eli, R. N. and H. W. Rauch. 1982. Fluvial hydrology of wetlands in Preston County, West Virginia. p. 11–28. *In* B. R. McDonald (ed.) Proceedings of the Symposium on Wetlands of the Unglaciated Appalachian Region. West Virginia University, Morgantown, WV.

Elliott, K. J. and W. T. Swank. 1994. Changes in tree species diversity after successive clearcuts in the Southern Appalachians. Vegetatio 114: 11–18.

Ellis, J. M., G. B. Farabee, and J. B. Reynolds. 1979. Fish communities in three successional stages of side channels in the upper Mississippi River. Transactions of the Missouri Academy of Science 13: 5–20.

Ellison, A. M. and E. J. Farnsworth. 1992. The ecology of Belizean mangrove-root fouling communities: patterns of epibiont distribution and abundance, and effects on root growth. Hydrobiologia 20: 1–12.

Ellison, J. C. 1993. Mangrove retreat with rising sea-level Bermuda. Estuarine, Coastal and Shelf Science 37: 75–87.

Ellison, J. C. and D.R . Stoddart. 1991. Mangrove ecosystem collapse during predicted sea-level rise: holocene analogues and implications. Journal of Coastal Research 7: 151–165.

Elwood, J. W., J. D. Newbold, A. F. Trimble, and R. W. Stark. 1981. The limiting role of phosphorus in a woodland stream ecosystem: effects of P enrichment on leaf decomposition and primary producers. Ecology 62: 146–158.

Emanuel, K. A. 1987. The dependence of hurricane intensity on climate. Nature 326: 483–485.

Enge, K. M. and W. R. Marion. 1986. Effects of clearcutting and site preparation on herpetofauna of a north Florida flatwoods. Forest Ecology and Management 14: 177–192.

Engler, R. M. and W. H. Patrick. 1974. Nitrate removal from floodwater overlying flooded soils and sediments. Journal of Environmental Quality 3: 409–413.

Ernst, J. P. and V. Brown. 1989. Conserving endangered species on southern forested wetlands. p. 135–145. *In* D. D. Hook and R. Lea (eds.) Proceedings of the Symposium: The Forested Wetlands of the Southern United States. U.S. Department of Agriculture Forest Service, Southeastern Forest Experiment Station, Asheville, NC. General Technical Report SE-50.

Ernst, W. H. O. 1990. Ecophysiology of plants in waterlogged and flooded environments. Aquatic Botany 38: 73–90.

Etherington, J. R. 1984. Comparative studies of plant growth and distribution in relation to waterlogging: X. Differential formation of adventitious roots and their experimental excision in *Epilobium hirsutum* and *Chamerion angustifolium*. Journal of Ecology 72: 389–404.

Evans, R. 1996. Techniques to restore wetland functions to prior converted croplands. p. 125–129. *In* SOLUTIONS: A Technical Conference on Water Quality, North Carolina State University, Raleigh, NC.

Evans, R. O., J. W. Gilliam, and R. W. Skaggs. 1991. Controlled drainage management guidelines for improving drainage water quality. North Carolina State University, Raleigh, NC. Cooperative Extension Service Publication Number AGRIC-443.

Everard, J. D. and M. C. Drew. 1989. Water relations of sunflower (*Helianthus annuus*) shoots during exposure of the root system to oxygen deficiency. Journal of Experimental Botany 40(220): 1255–1264.

Ewel, J. J. 1987. Restoration is the ultimate test of ecological theory. p. 31–33. *In* W. R. Jordan, III, M. E. Gilpin, and J. D. Aber (eds.) Restoration Ecology: A Synthetic Approach to Ecological Research. Cambridge University Press, Cambridge, England.

Ewel, K. C. 1976. Cypress ponds. p 14–49. *In* C. H. Wharton, H. T. Odum, K. C. Ewel, M. Duever, A. Lugo, R. Boyt, J. Bartholomew, E. DeBellevue, S. Brown, M. Brown, and L. Duever (eds.) Forested Wetlands of Florida — Their Management and Use. Center for Wetlands, University of Florida, Gainsville, FL.

Ewel, K. C. 1990a. Multiple demands on wetlands. BioScience 40: 660–666.

Ewel, K. C. 1990b. Swamps. p. 281–323. *In* R. L. Myers and J. J. Ewel (eds.) Ecosystems of Florida. University of Central Florida Press, Orlando, FL.

Ewel, K. C. 1995. Fire in cypress swamps in the southeastern United States. p. 111–116. *In* S. I. Cerulean and R. T. Engstrom (eds.) Fire in Wetlands: A Management Perspective, Proceedings of the 19th Tall Timbers Fire Ecology Conference, Tall Timbers Research Station, Tallahassee FL.

Ewel, K. C. 1996. Sprouting by pondcypress (*Taxodium distichum* var. *nutans*) after logging. Southern Journal of Applied Forestry 20: 209–213.

Ewel, K. C. 1997. Water quality improvement by wetlands. p. 329–344. *In* G. Daily (ed.) Nature's Services: Societal Dependence on Natural Ecosystems. Island Press, Washington, DC.

Ewel, K. C. and H. T. Davis. 1992. Response of pondcypress (*Taxodium distichum* var. *nutans*) to thinning. Southern Journal of Applied Forestry 16: 175–177.

Ewel, K. C., and H. L. Gholz. 1991. A simulation model of the role of belowground dynamics in a Florida pine plantation. Forest Science 37: 397–438.

Ewel, K. C. and W. J. Mitsch. 1978. The effects of fire on species composition in cypress dome ecosystems. Florida Scientist 4: 25–31.

Ewel, K. C. and H. T. Odum (eds.). 1989. Cypress Swamps. University Presses of Florida, Gainesville, FL.

Ewel, K. C. and L. A. Parendes. 1984. Usefulness of annual growth rings of cypress trees (*Taxodium distichum*) for impact analysis. Tree-Ring Bulletin 44: 39–43.

Ewel, K. C. and J. E. Smith. 1992. Evapotranspiration from Florida pondcypress swamps. Water Resources Bulletin 28: 299–304.

Ewel, K. C. and L. P. Wickenheiser. 1988. Effects of swamp size on growth rates of cypress (*Taxodium distichum*) trees. American Midland Naturalist 120: 362–370.

Ewel, K. C., H. T. Davis, and J. E. Smith. 1989. Recovery of Florida cypress swamps from clearcutting. Southern Journal of Applied Forestry 13: 123–126.

Ewing, M. S. 1991. Turbidity control and fisheries enhancement in a bottomland hardwood backwater system in Louisiana (U.S.A.). Regulated Rivers Research and Management 6: 87–99.

Eyre, F. H. 1980. Forest cover types of the United States and Canada. Society of American Foresters, Washington, DC.

F

Fahn, A. 1990. Plant Anatomy. Pergamon Press, Oxford, England.

Fan, Y., E. E. Wood, M. L. Baeck, and J. A. Smith. 1996. Fractional coverage of rainfall over a grid: analysis of NEXRAD data over the southern Plains. Water Resources Research 32: 2787–2803.

Farnsworth, E .J. and A. M. Ellison. 1991. Patterns of herbivory in Belizean mangrove swamps. Biotropica 23: 555–567.

Farnsworth, E. J. and A. M. Ellison. 1993. Dynamics of herbivory in Belizean mangal. Journal of Tropical Ecology 9: 435–453.

Faulkner, P. L., F. Zeringue, and J. R. Toliver. 1985. Genetic variation among open-pollinated families of baldcypress seedlings planted on two different sites. p. 267–272. *In* Proceedings of the 18th Southern Forest Tree Improvement Conference, Long Beach, MS. Sponsored Publication Number 40.

Federal Geographic Data Committee. 1992. Application of satellite data for mapping and monitoring wetlands — fact finding report. Wetlands Subcommittee, Federal Geographic Data Committee, Washington, DC. Technical Report 1.

Federal Geographic Data Committee. 1994. Strategic interagency approach to developing a national digital wetlands database (second approximation). Federal Geographic Data Committee, Washington, DC.

Federal Interagency Committee for Wetland Delineation. 1989. Federal manual for delineating jurisdictional wetlands. Cooperative publication, U.S. Army Corp of Engineers, U.S. Environmental Protection Agency, U.S. Fish and Wildlife Service, U.S. Soil Conservation Service, Washington, DC.

Federal Register. 1993. 58 Federal Register 45,008.

Federal Register. 1996a. 61 Federal Register 7242.

Federal Register. 1996b. 61 Federal Register 65,874.

Fell, J. W. and I. M. Master. 1973. Fungi associated with the degradation of mangrove (*Rhizophora mangle* L.) leaves in south Florida. p. 455–465. *In* L.H. Stevenson and R.R. Colwell (eds.) Estuarine Microbial Ecology. University of South Carolina Press, Columbia, SC.

Feller, I. C. 1993. Effects of nutrient enrichment on growth and herbivory of dwarf red mangrove. Ph.D. Dissertation. Georgetown University, Washington, DC.

Feller, I. C. 1995. Effects of nutrient enrichment on growth and herbivory of dwarf red mangrove (*Rhizophora mangle*). Ecological Monographs 65: 477–505.

Fenneman, N. M. 1938. Physiography of Eastern United States. McGraw Hill Book Co., New York, NY.

Field, C.D. (ed.). 1996. Restoration of Mangrove Ecosystems. International Society for Mangrove Ecosystems, Okinawa, Japan. South China Printing Co. Ltd., Hong Kong.

Findley, S., P. J. Smith, and J. L. Meyer. 1986. Effect of detritus addition on metabolism of river sediment. Hydrobiologia 137: 257–263.

Finger, T. R. and E. M. Stewart. 1987. Response of fishes to flooding regime in lowland hardwood wetlands. p. 86–92. *In* W. J. Matthews and D. C. Heins (eds.) Community and Evolutionary Ecology of North American Stream Fishes. University of Oklahoma Press, Norman, OK.

Fish, D. and D. W. Hall. 1978. Succession and stratification of aquatic insects inhabiting the leaves of the insectivorous pitcher plant, *Sarracenia purpurea*. American Midland Naturalist 99: 172–183.

Fisher, H. M. and E. L. Stone. 1990. Air-conducting porosity in slash pine roots from saturated soils. Forest Science 36: 18–33.

Fisher, R. F. 1981. Soil interpretations for silviculture in the southeastern coastal plain. p. 323–330. *In J. P.* Barnett (ed.) Proceedings of the First Biennial Southern Silvicultural Research Conference. U.S. Department of Agriculture Forest Service, Southern Forest Experiment Station, New Orleans, LA. General Technical Report SO-34.

Fisher, R. F. and W. S. Garbett. 1980. Response of semi-mature slash and loblolly pine plantations to fertilization with nitrogen and phosphorus. Soil Science Society of America Journal 44: 850–854.

Fleming, D. M., W. F. Wolff, and D. L. DeAngelis. 1994. Importance of landscape heterogeneity to wood storks in Florida Everglades. Environmental Management 18: 743–757.

Flinchum, D. M. 1977. Lesser vegetation as indicator of varying moisture regimes in bottomland and swamp forests of northeastern North Carolina. Ph.D. Dissertation. North Carolina State University, Raleigh, NC.

Flohrschutz, E. W. 1978. Dwarf cypress in the Big Cypress Swamp of southwestern Florida. M.S. Thesis. University of Florida, Gainesville, FL.

Flores-Verdugo, F., J. W. Day, Jr., and R. Briseño-Dueñas. 1986. Structure, litterfall, decomposition, and detritus dynamics of mangroves in a Mexican coastal lagoon with an ephemeral inlet. Marine Ecology Progress Series 35: 83–90.

Fog, K. 1988. The effect of added nitrogen on the rate of decomposition of organic matter. Biology Review 63: 433–462.

Foley, D. H., III. 1994. Short-term response of herpetofauna to timber harvesting in conjunction with streamside-management zones in seasonally-flooded bottomland-hardwood forests of southeast Texas. M.S. Thesis. Texas A&M University, College Station, TX.

Foster, A. A. 1968. Damage to forests by fungi and insects as affected by fertilizers. p. 42–46. *In* Forest Fertilization — Theory and Practice. Tennessee Valley Authority, Knoxville, TN.

Foster, J. R. 1992. Photosynthesis and water relations of the flood-plain tree, boxelder (*Acer negundo* L.). Tree Physiology 11: 133–149.

Foster, S. 1996. Personal communication. Alabama Department of Environmental Management, Montgomery, AL.

Foti, T. 1996. Personal communication. Arkansas Natural Heritage Commission, Little Rock, AR.

Foti, T., X. Li, M. Blaney, and K. G. Smith. 1994. A classification system for the natural vegetation of Arkansas. Proceedings of the Arkansas Academy of Science 48: 50–62.

Fowells, H. A. (compiler). 1965. Silvics of Forest Trees of the United States. U.S. Department of Agriculture Forest Service, Washington, DC. Agriculture Handbook 271.

Fowler, B. K. and C. Hershner. 1989. Primary production in Cohoke Swamp, a tidal freshwater wetland in Virginia. p. 365–374. *In* R. R. Sharitz and J. W. Gibbons (eds.) Freshwater Wetlands and Wildlife Symposium: Perspectives on Natural, Managed and Degraded Ecosystems. U.S. Department of Energy, Office of Scientific and Technical Information, Oak Ridge, TN. CONF-8603101, DOE Symposium Series No. 61.

Francis, J. K. 1985. Bottomland hardwood fertilization — the Stoneville experience. p. 346–350. *In* E. Shoulders (ed.) Proceedings Third Biennial Southern Silvicultural Research Conference. U.S. Department of Agriculture Forest Service, New Orleans, LA. General Technical Report SO-54.

Frayer, W. E., T. J. Monahan, D. C. Bowden, and F. A. Graybill. 1983. Status and trends of wetlands and deepwater habitats in the conterminous United States, 1950s to 1970s. Colorado State University, Ft. Collins, CO.

Frederick, P. C. and M. G. Spalding. 1994. Factors affecting reproductive success of wading birds (Ciconiiformes) in the Everglades ecosystem. p. 659–691. *In* S. M. Davis and J. C. Ogden (eds.) Everglades — The Ecosystem and Its Restoration. St. Lucie Press, Delray Beach, FL.

Fredrickson, L. H. 1979. Lowland hardwood wetlands: current status and habitat values for wildlife. p. 296–306. *In* P. E. Greeson, J. R. Clark, and J. E. Clark (eds.) Wetland Functions and Values: The State of Our Understanding. American Water Resources Association, Minneapolis, MN.

Fredrickson, L. H. 1991. Strategies for water level manipulations in moist-soil systems. U.S. Fish and Wildlife Service, Washington, DC. Fish and Wildlife Leaflet 13.4.6.

Freeman, B. J. and M. C. Freeman. 1985. Production of fishes in a subtropical blackwater ecosystem: the Okefenokee Swamp. Limnology and Oceanography 30: 686–692.

Frey, D. G. 1948. Lakes of the Carolina bays as biotic environments. Journal of the Elisha Mitchell Society 64: 179.

Frey, D. G. 1950. Carolina bays in relation to the North Carolina coastal plain. Journal of the Elisha Mitchell Scientific Society 66: 44–52.

Frey, D. G. 1951. The fishes of North Carolina's bay lakes and their intraspecific variation. Journal of the Elisha Mitchell Society 67: 1–44.

Froehlich, W. L., L. Kaszowski, and L. Starkel. 1977. Studies of present-day and past river activity in the Polish Carpathians. p. 411–428. *In* K. J. Gregory (ed.) River Channel Changes. John Wiley and Sons, New York, NY.

Frost, C. C. 1987. Historical overview of Atlantic white cedar in the Carolinas. p. 257–264. *In* A. D. Laderman (ed.) Atlantic White Cedar Wetlands. Westview Press, Boulder, CO.

Frost, C. C. 1995. Presettlement fire regimes in southeastern marshes, peatlands and swamps. p. 39–60. *In* S. I. Cerulean and R. T. Engstrom (eds.) Fire in Wetlands: A Management Perspective, Proceedings of the Tall Timbers Fire Ecology Conference No. 19. Tall Timbers Research Station, Tallahassee, FL.

Fuad, T. D. 1995. Phosphorus retention in non-tidal palustrine forested wetlands in the Virginia Coastal Plain. M.S. Thesis. George Mason University, Fairfax, VA.

Fulton, J. M. and A. E. Erickson. 1964. Relation between soil aeration and ethyl alcohol accumulation in xylem exudate of tomatoes. Soil Science Society of American Proceedings 28: 610–614.

Funderburk, E. L. 1995. Growth and survival of four forested wetland species planted on Pen Branch delta, Savannah River Site, S.C. M.S. Thesis. Clemson University, Clemson, SC.

G

Gaddy, L. L. 1981. The bogs of the southwestern mountains of North Carolina. Report to the North Carolina Natural Heritage Program, Raleigh, NC.

Gaddy, L. L. and G. A. Smathers. 1980. The vegetation of the Congaree Swamp National Monument. *In* H. Lieth and E. Landolt (eds.) Contribributions to the Knowledge of the Flora and Vegetation of the Carolinas, Vol. 2. Proceedings 16th International Phytogeographical Excursion, (IPE), 1978, Through the Southeastern United States.

Gafni, A., and K. N. Brooks. 1990. Hydraulic characteristics of four peatlands in Minnesota. Canadian Journal of Soil Science 70: 239–253.

Gambrell, R. P. and W. H. Patrick, Jr. 1978. Chemical and microbiological properties of anaerobic soils and sediments. p. 375–423. *In* D. D. Hook and R. M. M. Crawford (eds.) Plant Life in Anaerobic Environments. Ann Arbor Science Publishers Inc., Ann Arbor, MI.

Garabedian, S. P., D. R. LeBlanc, L. W. Gelhar, and M. A. Celia. 1991. Large-scale natural gradient tracer test in sand and gravel, Cape Cod, Massachusetts. 2. Analysis of spatial moments for a nonreactive tracer. Water Resources Research 19: 1387–1397.

Garcia-Novo, F. and R. M. M. Crawford. 1973. Soil aeration, nitrate reduction and flooding tolerance in higher plants. New Phytologist 72: 1031–1039.

Garner, G. W. 1969. Shortterm succession of vegetation following habitat manipulation of bottomland hardwoods for swamp rabbits in Louisiana. M.S. Thesis. Louisiana State University, Baton Rouge, LA.

Garrity, S. D., S. C. Levings, and K. A. Burns. 1994. The Galeta oil spill. I. Long-term effects on the physical structure of the mangrove fringe. Estuarine, Coastal and Shelf Science 38: 327–348.

Gee, J. H. 1980. Respiratory patterns and antipredator responses in the central mudminnow, *Umbra limi*, a continuous facultative, air-breathing fish. Canadian Journal of Zoology 58: 819–827.

Geiger, R., R. H. Aron, and P. Todhunter. 1995. The Climate Near the Ground, 5th Edition. F. Vieweg and Sohn, Weisbaden, Germany.

Gelhar, L. W. 1993. Stochastic Subsurface Hydrology. Prentice Hall, Englewood Cliffs, NJ.

General Accounting Office. 1988. Wetlands: The Corps of Engineers' Administration of the Section 404 program. U.S. Government Printing Office, Washington, DC. GAO/RCED-88–110.

Georgia Natural Heritage Program. 1996. State of Georgia landcover statistics by county. Georgia Department of Natural Resources, Atlanta, GA. Project Report 26.

Getter, C. D., G. I. Scott, and J. Michel. 1981. The effects of oil spills on mangrove forests: a comparison of five oil spill sites in the Gulf of Mexico and the Caribbean Sea. p. 65–111. *In* Proceedings of the 1981 Oil Spill Conference. API/EPA/USCG, Washington, DC.

Giaquinta, R. T., W. Lin, N. L. Sadler, and V. R. Franceschi. 1983. Pathway of phleom unloading of sucrose in corn roots. Plant Physiology 72: 362–367.

Gibbons, J. W. and R. D. Semlitsch. 1981. Terrestrial drift fences with pitfall traps: an effective technique for quantitative sampling of animal populations. Brimleyana 7: 1–16.

Gibson, J. R. 1970. The flora of Alder Run Bog, Tucker County, West Virginia. Castanea 35: 81–98.

Gibson, J. R. 1982. Alder Run Bog, Tucker County, West Virginia: its history and vegetation. p.101–105. *In* B. R. McDonald (ed.) Proceedings of the Symposium on Wetlands of the Unglaciated Appalachian Region. West Virginia University, Morgantown, WV.

Gilbert, C. R. 1992. Rare and Endangered Biota of Florida. University Press of Florida, Gainesville, FL.

Gilbert, F. S. 1980. The equilibrium theory of island biogeography: fact or fiction? Journal of Biogeography 7: 209–235.

Gilbert, K. M., J. D. Tobe, R. W. Cantrell, M. E. Sweeley, and J. R. Cooper. 1996. The Florida wetlands delineation manual. Florida Department of Environmental Protection, Tallahassee, FL.

Gill, C. J. 1970. The flooding tolerance of woody species-a review. Forestry Abstracts 31: 671–688.

Gilliam, J. W. 1991. Wet soils in the North Carolina lower coastal plain. Wetlands 11: 391–398.

Gilliam, J. W. and R. W. Skaggs. 1981. Drainage and agricultural development: effects on drainage waters. p. 109–124. *In* C. J. Richardson (ed.) Pocosin Wetlands: An Integrated Analysis of Coastal Plain Freshwater Bogs in North Carolina. Hutchinson Ross Publication Co., Stroudsburg, PA.

Gillham, R. W. 1984. The capillary fringe and its effect on water table response (groundwater flow, recharge). Journal of Hydrology 67: 307–324.

Gilmore, R. G. 1995. Environmental and biogeographic factors influencing ichthyofaunal diversity: Indian River Lagoon. Bulletin of Marine Science 57: 153–170.

Gilmore, R. G., Jr. and S. C. Snedaker. 1993. Mangrove forests. p. 165–198. *In* W. H. Martin, S. G. Boyce, and A. C. Esternacht (eds.) Biodiversity of the Southeastern United States: Lowland Terrestrial Communities. John Wiley and Sons, Inc., New York, NY.

Gindin, E., T. Tirosh, and S. Mayak. 1989. Effects of flooding on ultrastructure and ethylene production in chrysanthemums. ACTA Horticulturae 261: 171–183.
Gladden, J. E. and L. A. Smock. 1990. Macrophyte distribution and production on the floodplains of two lowland headwater streams. Freshwater Biology 24: 533–545.
Glascow, L. L. and R. E. Noble. 1971. The importance of bottomland hardwoods to wildlife. p. 30–43. *In* Proceedings Symposium on Southeastern Hardwoods. U.S. Department of Agriculture Forest Service, Southeastern Area S&PF, Atlanta, GA.
Glynn, P. W., L. R. Almodovar and J. G. Gonzalez. 1964. Effects of Hurricane Edith on marine life in La Parguera, Pueto Rico. Caribbean Journal of Science 4: 335–345.
Godfrey, R. K and J. W. Wooten. 1981. Aquatic and Wetland Plants of Southeastern United States, 2 vols. The University of Georgia Press, Athens, GA.
Goelz, J. C. G., J. S. Meadows, and P. W. Willingham. 1993. Precommercial thinning of water tupelo stands on the Mobile–Tensaw River delta. p. 105–108. *In* J. C. Brissette (ed.) Proceedings of the Seventh Biennial Silvicultural Conference. U.S. Department of Agriculture Forest Service, Southern Forest Experiment Station, New Orleans, LA. General Technical Report SO-93.
Golet, F. C. and D. J. Lowry. 1987. Water regimes and tree growth in Rhode Island Atlantic white cedar swamps. p. 91–110. *In* A. D. Laderman (ed.) Atlantic White Cedar Wetlands. Westview Press, Boulder, CO.
Golley, F., H. T. Odum, and R. Wilson. 1962. A synoptic study of the structure and metabolism of a red mangrove forest in southern Puerto Rico in May. Ecology 43: 9–18.
Gomez, M. and F. P. Day. 1982. Litter, nutrient content and production in the Great Dismal Swamp. American Journal of Botany 69: 1314–1321.
Gong, W. K. and J. E. Ong. 1990. Plant biomass and nutrient flux in a managed mangrove forest in Malaysia. Estuarine, Coastal and Shelf Science 31: 519–530.
Good, B. J. and W. H. Patrick, Jr. 1987. Gas composition and respiration of water oak (*Quercus nigra* L.) and green ash (*Fraxinus pennsylvanica* Marsh.) roots after prolonged flooding. Plant and Soil 97: 419–427.
Good, B. J. and S. A. Whipple. 1982. Tree spatial patterns: South Carolina bottomland and swamp forests. Bulletin of the Torrey Botanical Club 109: 529–536.
Gornitz, V., S. Lebedeff, and J. F. Hansen. 1982. Global sea level trend in the past century. Science 215: 1611–1614.
Gosselink, J. G. 1984. The ecology of delta marshes of coastal Louisiana: a community profile. U.S. Fish and Wildlife Service, Office of Biological Services, Washington, DC. FWS/OBS-84/09.
Gosselink, J. G. and L. C. Lee. 1989. Cumulative impact assessment in bottomland hardwood forests. Wetlands 9: 83–174.
Gosselink, J. G., L. C. Lee, and T. A. Muir (eds.). 1990. Ecological Processes and Cumulative Impacts: Illustrated by Bottomland Hardwood Wetland Ecosystems. Lewis Publishers, Chelsea, MI.
Gosselink, J. G., E. P. Odum, and R. M. Pope. 1974. The value of the tidal marsh. Center for Wetland Resources, Louisiana State University, Baton Rouge, LA. Publication LSU-SG-74-03.
Gosselink, J. G., S. E. Bayley, W. H. Conner, and R. E. Turner. 1981. Ecological factors in the determination of riparian wetland boundaries. p. 197–219. *In* J. R. Clark and J. Benforado (eds.) Wetlands of Bottomland Hardwood Forests. Elsevier Scientific Publishing Company, Amsterdam, The Netherlands.
Gotto, J. W. and B. F. Taylor. 1976. N_2 fixation associated with decaying leaves of the red mangrove (*Rhizophora mangle*). Applied and Environmental Microbiology 31: 781–783.

Gotto, J. W., F. R. Tabita, and C. V. Baalen. 1981. Nitrogen fixation in intertidal environments of the Texas gulf coast. Estuarine, Coastal and Shelf Science 12: 231–235.

Govus, T. E. 1987. The occurrence of *Sarracenia oreophila* (Kearney) Wherry in the Blue Ridge Province of southwestern North Carolina. Castanea 52: 310–311.

Graham, J. H. 1989. Foraging by sunfishes in a bog lake. p. 517–526. *In* R. R. Sharitz and J. W. Gibbons (eds.) Freshwater Wetlands and Wildlife Symposium: Perspectives on Natural, Managed and Degraded Ecosystems. U.S. Department of Energy, Office of Scientific and Technical Information, Oak Ridge, TN. CONF-8603101, DOE Symposium Series No. 61.

Graham, J. H. and R. W. Hastings. 1984. Distributional patterns of sunfishes on the New Jersey coastal plain. Environmental Biology of Fishes 10: 137–148.

Graves, M. R., M. R. Kress, and S. Bourne. 1995. Characterization of changes to bottomland hardwood forests and forested wetlands in the Cache River, Arkansas, watershed. Waterways Experiment Station, Wetlands Research Program, Vicksburg, MS. Technical Report WRP-SM-14.

Green, M. S. and J. R. Etherington. 1977. Oxidation of ferrous iron by rice (*Oryza sativa* L.) roots: a mechanism for waterlogging tolerance. Journal of Experimental Botany 28(104): 678–690.

Greenfield, R. R. 1985. Controlling nonpoint sources of pollution — the federal legal framework and the alternative of nonfederal action. *In* An Overview of the National Nonpoint Source Policy in Perspectives on Nonpoint Source Pollution: Proceedings of a National Conference, Kansas City, Missouri. U.S. Environmental Protection Agency. EPA 440/5-85-001.

Greeson, P. E., J. R. Clark, and J. E. Clark. 1979. Wetland Functions and Values: The State of Our Understanding. American Water Resources Association, Minneapolis, MN.

Grelen, H. E. 1980. Longleaf pine — slash pine. p. 52–53. *In* F. H. Eyre (ed.) Forest Cover Types of the United States and Canada. Society of American Foresters, Washington, DC.

Gresham, C. A. 1983. Selected ecological and silvical characteristics of *Gordonia lasianthus* (L.) Ellis in coastal South Carolina. p. 267–274. *In* R. R. Sharitz and J. W. Gibbons (eds.) Freshwater Wetlands and Wildlife Symposium: Perspectives on Natural, Managed and Degraded Ecosystems. U.S. Department of Energy, Office of Scientific and Technical Information, Oak Ridge, TN. CONF-8603101, DOE Symposium Series No. 61.

Gresham, C. A. 1989. A literature review of effects of developing pocosins. p. 44–50. *In* D. D. Hook and R. Lea (eds.) Proceedings of the Symposium: The Forested Wetlands of the Southern United States. U.S. Department of Agriculture Forest Service, Southeastern Forest Experiment Station, Asheville, NC. General Technical Report SE-50.

Gresham, C. A. and D. J. Lipscomb. 1990. *Gordonia lasianthus* (L.) Ellis. Loblolly bay. p. 365–369. In R. M. Burns and B. H. Honkala (tech. coords.) Silvics of North America: 2. Hardwoods. U.S. Department of Agriculture, Washington, DC. Agriculture Handbook 654.

Gresham, C. A., T. M. Williams, and D. J. Lipscomb. 1991. Hurricane Hugo wind damage to southeastern U.S. coastal forest tree species. Biotropica 23: 420–426.

Gries, C., L. Kappen, and R. Losch. 1990. Mechanism of flood tolerance in reed, *Phragmites australis* (Cav.) Trin. ex Steudel. New Phytologist 114: 589–593.

Grosse, W. 1989. Thermo-osmotic air transport in aquatic plants affecting growth activities and oxygen diffusion to wetland soils. p. 469–476. *In* D. A. Hammer (ed.) Constructed Wetlands for Wastewater Treatment. Lewis Publishers, Chelsea, MI.

Grosse, W. and J. Mevi-Schutz. 1987. A beneficial gas transport system in *Nymphoides peltata*. American Journal of Botany 74: 947–952.

Grosse, W. and P. Schroder. 1984. Oxygen supply of roots by gas transport in alder trees. Zeitschrift fuer Naturforschung 39C: 1186–1188.

Grosse, W. and P. Schroder. 1985. Aeration of roots and chloroplast free tissues of trees. Ber. Deutsch. Bot. Ges. 98: 311–318.

Grosse, W., H. B. Buchel, and H. Tiebel. 1991. Pressurized ventilation in wetland plants. Aquatic Botany 39: 89–98.

Grosse, W., J. Frye, and S. Lattermann. 1992. Root aeration in wetland trees by pressurized gas transport. Tree Physiology 10: 285–295.

Grosse, W., A. Schulte, and H. Fujita. 1993. Pressured gas transport in two Japanese alder species in relation to their natural habitats. Ecological Research 8: 151–158.

Grossnickle, S. C. 1987. Influence of flooding and soil temperature on the water relations and morphological development of cold-stored black spruce and white spruce seedlings. Canadian Journal of Forest Research 17: 821–828.

Guillory, V. 1979. Utilization of an inundated floodplain by Mississippi River fishes. Florida Scientist 42: 222–228.

Gunderson, L. H. 1977. Regeneration of cypress, *Taxodium distichum* and *Taxodium ascendens*, within logged and burned cypress strands at Corkscrew Swamp Sanctuary, Florida. M.S. Thesis. University of Florida, Gainesville, FL.

Gunderson, L. H. 1984. Regeneration of cypress in logged and burned strands at Corkscrew Swamp Sanctuary, Florida. p. 349–357. *In* K. C. Ewel and H. T. Odum (eds.) Cypress Swamps. University Presses of Florida, Gainesville, FL.

Gunning, G. E. and W. M. Lewis. 1955. The fish populations in a spring-fed swamp in the Mississippi Bottoms of southern Illinois. Ecology 36: 552–558.

Guntenspergen, G. R., J. R. Keough, and J. Allen. 1993. Wetland systems and their response to management. p. 383–390. *In* G. A. Moshiri (ed.) Constructed Wetlands for Water Quality Improvement. CRC Press, Inc., Boca Raton, FL.

Gunter, G., and L. N. Eleuterius. 1973. Some effects of hurricanes on the terrestrial biota, with special reference to Camille. Gulf Research Reports 4(2): 174–185.

H

Hagan, J. M., P. S. McKinley, A. L. Meehan, and S. L. Grove. 1995. Diversity and abundance of landbirds in a northeastern industrial forest landscape. National Council of the Paper Industry for Air and Stream Improvement, Inc., Research Triangle Park, NC. NCASI Technical Bulletin No. 705.

Haissig, B. E. 1974. Origins of adventitious roots. New Zealand Forest Science 4: 229–310.

Hall, T. F. and W. T. Penfound. 1939a. A phytosociological study of a cypress-gum swamp in southern Louisiana. American Midland Naturalist 21: 378–395.

Hall, T. F. and W. T. Penfound. 1939b. A phytosociological study of a *Nyssa biflora* consocies in southern Louisiana. American Midland Naturalist 22: 369–375.

Hall, T. F. and W. T. Penfound. 1943. Cypress-gum communities in the Blue Girth Swamp near Selma, Alabama. Ecology 24: 208–217.

Hallgren, S. W. 1989. Growth response of *Populus* hybrids to flooding. Annales des Sciences Forestieres 46: 361–372.

Hamel, P. B. 1989. Breeding bird populations on the Congaree Swamp National Monument, South Carolina. p. 617–628. *In* R. R. Sharitz and J. W. Gibbons (eds.) Freshwater Wetlands and Wildlife. U.S. Department of Energy, Office of Scientific and Technical Information, Oak Ridge, TN. DOE Symposium Series No. 61.

Hamel, P. B. 1992. Land manager's guide to the birds of the South. The Nature Conservancy, Southeastern Region, Chapel Hill, NC.

Hamel, P. B., W. P. Smith, and J. W. Wahl. 1993. Wintering bird populations of fragmented forest habitat in the Central Basin, Tennessee. Biological Conservation 66:107–115.

Hamel, P. W., H. E. LeGrand, Jr., M. R. Lennartz, and S. A. Gauthreaux, Jr. 1982. Bird-habitat relationships on southeastern forest lands. U.S. Department of Agriculture Forest Service, Asheville, NC. General Technical Report SE-22.

Hamilton, D. B. 1984. Plant succession and the influence of disturbance in Okefenokee Swamp. p. 86–111. *In* A. D. Cohen, D. J. Casagrande, M. J. Andrejko, and G. R. Best (eds.) The Okefenokee Swamp: Its Natural History, Geology, and Geochemistry. Wetland Surveys, Los Alamos, NM.

Hansen, K. L., E. G. Ruby, and R. L. Thompson. 1971. Trophic relationships in the water hyacinth community. Quarterly Journal of the Florida Academy of Science 34: 107–113.

Hanson, M. A. and M. R. Riggs. 1995. Potential effects of fish predation on wetland invertebrates: a comparison of wetlands with and without fathead minnows. Wetlands 15: 167–175.

Hanski, I. 1991. Single-species metapopulation dynamics: concepts, models and observations. Biological Journal of the Linnean Society 42: 17–38.

Hanski, I. and M. Gilpin. 1991. Metapopulation dynamics: brief history and conceptual domain. Biological Journal of the Linnean Society 42: 3–16.

Hantush, M. S. 1966. Analysis of data from pumping tests in anisotropc aquifers. Journal of Geophysical Research 71: 421–426.

Hardin, E. D. and W. A. Wistendahl. 1983. The effects of floodplain trees on herbaceous vegetation patterns, microtopography and litter. Bulletin of the Torrey Botanical Club 110: 23–30.

Hardy, G. 1996. Personal communication. Natural Resources Conservation Service, Columbia, SC.

Hare, R. C. 1965. Contribution of bark to fire resistance of southern trees. Journal of Forestry 63: 248–251.

Harlow, W. M. and E. S. Harrar. 1979. Textbook of Dendrology. McGraw–Hill, New York, NY.

Harlow, W. M., E. S. Harrar, J. W. Hardin, and F. M. White. 1996. Textbook of Dendrology. McGraw Hill, Inc., New York, NY.

Harmon, M. E., J. F. Franklin, F. J. Swanson, P. Sollins, S. V. Gregory, J. D. Lattin, N. H. Anderson, S. P. Cline, S. G. Aumen, J. R. Sedell, G. W. Lienkaemper, K. Cromack, and K. W. Cummins. 1986. Ecology of coarse woody debris in temperate ecosystems. Advances in Ecological Research 15: 133–276.

Harms, W. R. 1996. An old-growth definition for wet pine forests, woodlands, and savannas. U.S. Department of Agriculture Forest Service, Southern Research Station, Asheville, NC. General Technical Report SE-22.

Harms, W. R., H. T. Schreuder, D. D. Hook, and C. L. Brown. 1980. The effects of flooding on the swamp forest in Lake Ocklawah, Florida. Ecology 61: 1412–1421.

Harper, R. M. 1914. Geography and Vegetation of Northern Florida. Florida Geological Survey. Sixth Annual Report.

Harper, R. M. 1927. Natural resources of southern Florida. Florida Geological Survey. 18th Annual Report.

Harrington, C. A. 1987. Responses of red alder and black cottonwood seedlings to flooding. Physiologia Plantarum 69: 35–48.

Harrington, C. A. and B. M. Casson. 1986. SITEQUAL — A user's guide. Computerized site evaluation for 14 southern hardwood species. U.S. Department of Agriculture Forest Service, New Orleans, LA. General Technical Report SO-62.

Harrington, R. W. 1959. Delayed hatching in stranded eggs of marsh killifish, *Fundulus confluentus*. Ecology 40: 430–437.

Harrington, R. W. and E. S. Harrington. 1961. Food selection among fishes invading a high subtropical salt marsh: from onset of flooding through progress of a mosquito brood Ecology 42: 646–666.

Harris, L. D. and J. D. McElveen. 1981. Effect of forest edges on north Florida breeding birds. School Forest Resources and Conservation, University of Florida, Gainesville, FL. IMPAC Rep. 6(4): 1–24.

Harris, L. D. and T. E. O'Meara. 1989. Changes in southeastern bottomland forests and impacts on vertebrate fauna. p. 755–772. *In* R. R. Sharitz and J. W. Gibbons (eds.) Freshwater Wetlands and Wildlife Symposium: Perspectives on Natural, Managed and Degraded Ecosystems. U.S. Department of Energy, Office of Scientific and Technical Information, Oak Ridge, TN. CONF-8603101, DOE Symposium Series No. 61.

Harris, L. D. and G. Silva-Lopez. 1992. Forest fragmentation and the conservation of biological diversity. p. 197–237. *In* P. L. Fiedler and S. H. Jain (eds.) Conservation Biology: The Theory and Practice of Nature Conservation Preservation and Management. Chapman and Hall, New York/London.

Harris, L. D. and C. R. Vickers. 1984. Some faunal community characteristics of cypress ponds and the changes induced by perturbations. p. 171–185. *In* K. C. Ewel and H. T. Odum (eds.) Cypress Swamps. University of Florida Press, Gainesville, FL.

Harry, D. E. and T. W. Kimmerer. 1991. Molecular genetics and physiology of alcohol dehydrogenase in woody plants. Forest Ecology and Management 43: 251–272.

Harry, D. E., C. S. Kinlaw, and R. R. Sederoff. 1988. The anaerobic stress response and its use for studying gene expression in conifers. p. 275–290. *In* J.W. Hanover and D.E. Keathley (eds.) Genetic Manipulation of Woody Plants. Plenum Publishing Corp, New York, NY.

Hart, C. P., J. D. Hodges, and K. L. Belli. 1995. Testing and modification of a technique designed to evaluate regeneration potential of oaks and ash in minor bottoms of the Southeastern Coastal Plain. p. 434–442. *In* M. B. Edwards (compiler) Proceedings Eighth Biennial Southern Silvicultural Research Conference. U.S. Department of Agriculture Forest Service, Asheville, NC. General Technical Report SRS-1.

Hartsell, A. J. and J. D. London. 1995. Forest statistics for Mississippi counties — 1994. U.S. Department of Agriculture Forest Service, Southern Forest Experiment Station, New Orleans, LA. Resource Bulletin SO-190.

Harvey, B. C. 1987. Susceptibility of young of the year fishes to downstream displacement by flooding. Transactions of the American Fisheries Society 116: 851–855.

Hauser, J. W. 1992. Effects of hydrology-altering site preparation and fertilization on plant diversity and productivity in pine plantations in the coastal plain of Virginia. M.S. Thesis. Virginia Polytechnic Institute and State University, Blacksburg, VA.

Hauser, J. W., W. M. Aust, J. A. Burger, and S.M. Zedaker. 1993. Drainage effects on plant diversity and productivity in loblolly pine (*Pinus taeda*) plantations on wet flats. Forest Ecology and Management 61: 109–126.

Haygood, J. L. 1970. Wildlife recreation management in bottomland hardwoods. p. 122–134. *In* Silviculture and Management of Southern Hardwoods, 19th Annual Forestry Symposium, Louisiana State University, Baton Rouge, LA.

Heald, E. J. 1969. The production of organic detritus in a south Florida estuary. Ph.D. Dissertation. University of Miami, Coral Gables, FL.

Healy, W. M. and E. S. Nenno. 1983. Minimum maintenance vs. intensive management of clearings for wild turkeys. Wildlife Society Bulletin 11: 113–120.

Heath, R. C. 1975. Hydrology of the Albemarle-Pamlico region, North Carolina, preliminary report on the impact of agricultural developments. U.S. Geological Survey, Raleigh, NC. Water Resources Investigations 9–75.

Heavrin, C. A. 1981. Boxes, Baskets and Boards: A History of Anderson-Tully Company. Memphis State University Press, Memphis, TN.

Hedgpeth, J. W. 1957. Classification of marine environments. Geological Society of America, Memoir 67, 1:17–28.

Hedlund, A. 1969. Hurricane Camille's impact on Mississippi timber. Southern Lumberman 219(2728): 191–192.

Hefner, J. M. and K. K. Moorhead. 1991. Mapping pocosins and associated wetlands in North Carolina. Wetlands 11: 377–389.

Hefner, J. M., B. O. Wilen, T. E. Dahl, and W. E. Frayer. 1994. Southeast wetlands: status and trends mid 1970s to mid 1980s. U.S. Fish and Wildlife Service, Washington, DC.

Heimburg, K. 1984. Hydrology of north-central Florida cypress domes. p. 72–82. *In* K. C. Ewel and H. T. Odum (eds.) Cypress Swamps. University Presses of Florida, Gainesville, FL.

Heins, D. C. and D. R. Dorsett. 1986. Reproductive traits of the blacktail shiner, *Notropis venustus* (Girard), in southeastern Mississippi. Southwestern Naturalist 31: 185–189.

Heins, D. C. and F. G. Rabito, Jr. 1988. Reproductive traits in populations of the weed shiner, *Notropis texanus*, from the Gulf coastal plain. Southwestern Naturalist 33: 147–156.

Heinselman, M. L. 1970. Landscape evolution, peatland types, and the environment at the Lake Agassiz Peatlands Natural Area, Minnesota. Ecological Monographs 40: 235–261.

Heiskanen, J. 1995. Irrigation regime affects water and aeration conditions in peat growth medium and the growth of containerized Scots pine seedlings. New Forests 9:181–195.

Heitmeyer, M. E. 1985. Wintering strategies of female mallards related to dynamics of lowland hardwood wetlands in the upper Mississippi Delta. Ph.D. Dissertation. University of Missouri, Columbia, MO.

Hellgren, E. C., M. R. Vaughan, and D. F. Stauffer. 1991. Macrohabitat use by black bears in a southeastern wetland. Journal of Wildlife Management 55: 442–448.

Hellier, T. R., Jr. 1966. The fishes of the Santa Fe River system. Bulletin of the Florida State Museum 11: 1–46.

Hemond, H. 1983. The nitrogen budget of Thoreau's Bog, Massachusetts, U.S.A. Ecology 64: 99–109.

Henderson-Sellers, A. 1994. Numerical modelling of global climates. p. 99–124. *In* N. Roberts (ed.) The Changing Global Environment. Blackwell, Cambridge, MA.

Herman, D. W. 1989. Tracking the rare bog turtle. Wildlife in North Carolina 53: 17–19.

Herman, D. W. 1994. The bog turtle, *Clemmys muhlenbergii*, in North Carolina. An action plan for its conservation and management. Zoo Atlanta, Atlanta, GA.

Hermann, S. 1996. Personal communication. Tall Timbers Research Station, Tallahassee, FL.

Herting, G. E. and A. Witt. 1967. The role of physical fitness of forage fishes in relation to their vulnerability to predation by bowfin (*Amia calva*). Transactions of the American Fisheries Society 96: 427–430.

Hess, G. R. 1994. Conservation corridors and contagious disease: a cautionary note. Conservation Biology 8: 256–262.

Hesse, I. D. 1994. Dendroecological determination of municipal wastewater effects on *Taxodium distichum* (L.) Rich. productivity in a Louisiana swamp. M.S. Thesis. Louisiana State University, Baton Rouge, LA.

Hesse, I. D., W. H. Conner, and J. W. Day, Jr. 1996. Herbivory impacts on the regeneration of forested wetlands. p. 23–28. *In* K. M. Flynn (ed.) Proceedings of the Southern Forested Wetlands Ecology and Management Conference. Clemson University, Clemson, SC.

Hewlett, J. D. 1982. Principles of Forest Hydrology. University of Georgia Press, Athens, GA.

Hewlett, J. D. and A. R. Hibbert. 1967. Factors affecting the response of small watersheds to precipitation in humid areas. p. 275–290. *In* W. L. Sopper and W. B. Lull (eds.) International Symposium on Forest Hydrology. Pergamon Press, Oxford, England.

Hewlett, J. D., J. C. Fortson, and G. B. Cunningham. 1977. The effect of rainfall intensity on storm flow and peak discharge from forest land. Water Resources Research 13: 259–266.

Hibbert, A. R. and C. A. Troendle. 1988. Streamflow generation by variable source area. p. 111–127. *In* W. T. Swank and D. A. Crossley Jr. (eds.) Forest Hydrology and Ecology at Coweeta. Springer–Verlag, New York, NY.

Hicks, D. B. and L. A. Burns. 1975. Mangrove metabolic response to alterations of natural freshwater drainage to southwestern Florida estuaries. p 238–255. *In* G. Walsh, S. Snedaker, and H. Teas (eds.) Proceedings of the International Symposium on the Biology and Management of Mangroves. Institute of Food and Agricultural Sciences, University of Florida, Gainesville, FL.

Hill, L. G. 1972. Social aspects of aerial respiration of young gars (Lepisosteus). Southwestern Naturalist 16: 239–247.

Hill, L. G., G. D. Schnell, and A. A. Echelle. 1973. Effect of dissolved oxygen concentration on locomotory reactions of the spotted gar, *Lepisosteus oculatus* (Pisces: Lepisosteidae). Coepia 1973: 119–124.

Hillel, D. 1971. Soil and Water: Physical Principles and Processes. Academic Press, New York, NY.

Hinds, R. 1996. Personal communication. Louisiana Department of Natural Resources, Coastal Management Division, Baton Rouge, LA.

Hobgood, J. S. and R. S. Cerveny. 1988. Ice-aged hurricanes and tropical storms. Nature 333: 243–245.

Hocutt, C. H. and E. O. Wiley. 1986. The Zoogeography of North American Freshwater Fishes. John Wiley and Sons, New York, NY.

Hodges, J. D. 1980. Slash pine. p. 56–57. *In* F. H. Eyre (ed.) Forest Cover Types of the United States and Canada. Society of American Foresters, Washington, DC.

Hodges, J. D. 1987. Cutting mixed bottomland hardwoods for good growth and regeneration. p. 53–60. *In* Proceedings of the 15th Annual Hardwood Symposium of the Hardwood Research Council. Memphis, TN.

Hodges, J. D. 1994a. The southern bottomland hardwood region and brown loam bluffs subregion. p. 227–267. *In* J. W. Barrett (ed.) Regional Silviculture of the United States, 3rd ed. John Wiley and Sons, New York, NY.

Hodges, J. D. 1994b. Ecology of bottomland hardwoods. p. 5–11. *In* W. P. Smith and D. N. Pashley (eds.) A Workshop to Resolve Conflicts in the Conservation of Migratory Landbirds in Bottomland Hardwood Forests. U.S. Department of Agriculture Forest Service. General Technical Report SO-114.

Hodges, J. D. 1996. Personal communication. Anderson-Tully Company, Memphis, TN.

Hodges, J. D. 1997. Development and ecology of bottomland hardwood sites. Forest Ecology and Management 90: 117–126.

Hodges, J. D. and G. L. Switzer. 1979. Some aspects of the ecology of southern bottomland hardwoods. p. 360–365. *In* North America's Forests: Gateway to Opportunity, Proceedings of the 1978 Convention of the Society of American Foresters and the Canadian Institute of Forestry. Washington, DC.

Hodgson, M. E., J. R. Jensen, H. E. Mackey, Jr., and M. C. Coulter. 1988. Monitoring wood stork foraging habitat using remote sensing and geographic information systems. Photogrammetric Engineering and Remote Sensing 54: 1601–1607.

Hoese, H. D. 1985. Jumping mullet-the internal diving bell hypothesis. Environmental Biology of Fishes 13: 309–314.

Hohenstein, W. G. 1987. Forestry and the Water Quality Act of 1987. Journal of Forestry 85: 5–8.

Holdridge, L. R. 1967. Life Zone Ecology. Tropical Science Center, San Jose, Costa Rica.

Hollands, G. G. and D. W. Magee. 1986. A method for assessing the functions of wetlands. p. 108–121. *In* J. A. Kusler and P. Riexinger (eds.) Proceedings of the National Wetland Assessment Symposium. Portland, ME.

Holloway, A. D. 1954. Notes on the life history and management of the shortnose and longnose gars in Florida waters. Journal of Wildlife Management 18: 433–449.

Hontela, A. and N. E. Stacy. 1990. Cyprinidae. p. 53–77. *In* P. E. Munro, A. P. Scott, and T. J. Lam (eds.) Reproductive Seasonality in Teleosts: Environmental Influences. CRC Press, Boca Raton, FL.

Hook, D. D. 1984a. Adaptations to flooding with fresh water. p. 265–294. *In* T. T. Kozlowski (ed.) Flooding and Plant Growth. Academic Press, Inc., Harcourt Brace Jovanovich, Publishers, New York, NY.

Hook, D. D. 1984b. Waterlogging tolerance of lowland tree species of the South. Southern Journal of Applied Forestry 8: 136–149.

Hook, D. D. 1993. Wetlands: history, current status, and future. Environmental Toxicology and Chemistry 12: 2157–2166.

Hook, D. D. and C. L. Brown. 1967. Lactic acid production in the roots of *Nyssa sylvatica* var. *biflora* (Walt.) Sarg. p. 280–281. *In* Proceedings, Association of Southern Agricultural Workers, Inc. New Orleans, LA.

Hook, D.D. and C.L. Brown. 1970. Lenticel and water root development of swamp tupelo under various flooding conditions. Botanical Gazette 131:217–224.

Hook, D. D. and C. L. Brown. 1972. Permeability of the cambium to air in trees adapted to wet habitats. Botanical Gazette 133: 304–310.

Hook, D. D. and C. L. Brown. 1973. Root adaptations and relative flood tolerance of five hardwood species. Forest Science 19: 225–229.

Hook, D. D. and R. M. M. Crawford (eds.). 1978. Plant Life in Anaerobic Environments. Ann Arbor Science Publishers, Ann Arbor, MI.

Hook, D. D. and S. Denslow. 1987. Metabolic responses of four families of loblolly pine to two flood regimes. p. 281–292. *In* R.M.M. Crawford (ed.) The Series Analytic: Plant Life in Aquatic and Amphibious Habitats. Blackwell Scientific Publications, Oxford, England. Special Publications of the British Ecological Society, No. 5.

Hook, D. D. and M. R. McKevlin. 1988. Use of oxygen micro-electrodes to measure aeration in the roots of intact tree seedlings. p. 467–476. *In* D. D. Hook et al. (eds.) The Ecology and Management of Wetlands, Vol. 1: Ecology of Wetlands. Croom Helm, Long and Timber Press, Portland, OR.

Hook, D. D. and J. R. Scholtens. 1978. Adaptations and flood tolerance of tree species. p. 299–332. *In* D. D. Hook and R.M.M. Crawford (eds.) Plant Life in Anaerobic Environments. Ann Arbor Science Publishers, Ann Arbor, MI.

Hook, D. D., C. L. Brown, and P. P. Kormanik. 1970b. Lenticel and water root development of swamp tupelo under various flooding conditions. Botanical Gazette 131: 217–224.

Hook, D. D., C. L. Brown, and P. P. Kormanik. 1971. Inductive flood tolerance in swamp tupelo (*Nyssa sylvatica* var. *biflora* (Walt.) Sarg.). Journal of Experimental Botany 22: 78–89.

Hook, D. D., C. L. Brown, and R. H. Wetmore. 1972. Aeration in trees. Botanical Gazette 133: 443–454.

Hook, D. D., M. A. Buford, and W. R. Harms. 1993. Effect of residual trees on natural regeneration in a tupelo-cypress swamp after 24 years. p. 91–96. *In* J. C. Brissette (ed.) Proceedings of the Seventh Biennial Southern Silvicultural Research Conference. U.S. Department of Agriculture Forest Service, New Orleans, LA. General Technical Report SO-93.

Hook, D. D., M. A. Buford, and T. M. Williams. 1991a. Impact of Hurricane Hugo on the South Carolina coastal plain forest. Journal of Coastal Research SI-8: 291–300.

Hook, D. D., W. P. LeGrande, and O. G. Langdon. 1967. Stump sprouts on water tupelo. Southern Lumberman 215(2680): 111–112.

Hook, D. D., O. G. Langdon, and W. A. Hamilton. 1973. The swamp and its water nymph. American Forests 79: 40–42.

Hook, D. D., D. S. DeBell, W. H. McKee, Jr., and J. L. Askew. 1983. Responses of loblolly pine (mesophyte) and swamp tupelo (hydrophyte) seedlings to soil flooding and phosphorous. Plant and Soil 71:387–394.

Hook, D. D., O. G. Langdon, J. Stubbs, and C. L. Brown. 1970a. Effect of water regimes on the survival, growth, and morphology of tupelo seedlings. Forest Science 16: 304–311.

Hook, D. D., M. D. Murray, D. S. DeBell, and B. C. Wilson. 1987. Variation in growth of red alder families in relation to shallow water table levels. Forest Science 33: 224–229.

Hook, D. D., W. H. McKee, Jr., T. M. Williams, S. Jones, D.van Blariccom, and J. Parsons. 1994. Hydrologic and wetland characteristics of a piedmont bottom in South Carolina. Water Air and Soil Pollution 77: 293–320.

Hook, D. D., W. McKee, T. Williams, B. Baker, L. Lundquist, R. Martin, and J. Mills. 1991b. A survey of voluntary compliance of forestry BMPs. South Carolina Forestry Commission, Columbia, SC.

Hoover, J. J., K. J. Killgore, and M. A. Konikoff. 1995. Larval fish dynamics in a riverine wetland of the lower Mississippi Basin. Waterways Experiment Station, Wetland Research Program, Vicksburg, MS. Technical Report WRP-SM-10.

Hoover, J. P., M. C. Brittingham, and L. J. Goodrich. 1995. Effects of forest patch size on nesting success of wood thrushes. Auk 112: 146–155.

Horn, N. H. and C. D. Riggs. 1973. Effect of temperature and light on the rate of air breathing of the bowfin, *Amia calva*. Copeia 1973: 653–657.

Horton, J.W. and V. A. and Zullo (eds.). 1991. The Geology of the Carolinas. University of Tennessee Press, Knoxville, TN.

Horton, R. E. 1933. The role of infiltration in the hydrologic cycle. Transactions of the American Geophysical Union. 14: 446.

Howard, R. J. and J. A. Allen. 1989. Streamside habitats in southern forested wetlands: Their role and implications for management. p. 97–106. *In* D. D. Hook and R. Lea (eds.) Proceedings of the Symposium: The Forested Wetlands of the Southern U.S. U.S. Department of Agriculture Forest Service, Asheville, NC. General Technical Report SE-50.

Howarth, R. W. and S. G. Fisher. 1976. Carbon, nitrogen, and phosphorus dynamics during leaf decay in nutrient enriched stream microecosystems. Freshwater Biology 6: 221–228.

Hruby, T., W. E. Cesanek, and K. E. Miller. 1995. Estimating relative wetland values for regional planning. Wetlands 15: 93–107.

Hsu, P. A. 1989. Aluminum hydroxides and oxyhydroxides. p. 331–378. *In J. B.* Dixon and S. B. Weed (eds.) Minerals in Soil Environments, 2nd Edition. Soil Science Society of America, Madison, WI.

Hubbs, C. L. 1941. The relation of hydrological conditions to speciation in fishes. p. 182–195. *In* A Symposium on Hydrobiology. University of Wisconsin Press, LaCrosse, WI.

Hubbs, C. L. 1982. Life history dynamics of the *Menidia beryllina* from Lake Texoma. American Midland Naturlist 107: 1–12.

Hubbs, C. L., and E. R. Raney. 1946. Endemic fish fauna of Lake Waccamaw, North Carolina. Miscellaneous Publications of the Museum of Zoology University of Michigan 65: 5–29.

Hudson, C. 1976. The Southeastern Indians. University of Tennessee Press, Knoxville, TN.

Hudson, J. 1996. New technology boosts weed control. Southeast Farm Press, February 21, 1996: 25, 10.

Huffman, R. T. and S. W. Forsythe. 1981. Bottomland hardwood forest communities and their relation to anaerobic soil conditions. p. 187–196. *In* J. R. Clark and J. Benforado (eds.) Wetlands of Bottomland Hardwood Forests. Elsevier Scientific Publishing Company, Amsterdam, The Netherlands.

Hughes, J. H. 1996. Personnal communication. Weyerhaeuser Company, New Bern, NC.

Hughes, J. H. 1996. Managed pine forests in coastal North Carolina: a water quality solution? p. 56–60. *In* SOLUTIONS: A Technical Conference on Water Quality, North Carolina State University, Raleigh, NC.

Hulme, M. 1994. Historic records and recent climatic change. p. 69–98. *In* N. Roberts (ed.) The Changing Global Environment. Blackwell, Cambridge, MA.

Humphries, W. H. and I. M. Rossell. 1996. Unpublished data. University of North Carolina at Asheville, Asheville, NC.

Huner, J. V. 1978. Exploitation of freshwater crayfishes in North America. Fisheries 3: 2–5.

Hunt, B. P. 1953. Food relationships between Florida spotted gar and other organisms in the Tamiami Canal, Dade County, Florida. Transactions of the American Fisheries Society 82: 13–33.

Hunt, B. P. 1960. Digestion rate and food consumption of Florida gar, warmouth, and large-mouth bass. Transactions of the American Fisheries Society 89: 206–211.

Hunt, C. B. 1974. Natural Regions of the United States and Canada. Freeman, San Francisco, CA.

Hunter, M. L., Jr. 1990. Wildlife, Forests, and Forestry: Principles of Managing Forests for Biological Diversity. Prentice–Hall, Inc., Englewood Cliffs, NJ.

Hurst, G. A. and T. R. Bourland. 1996. Breeding birds on bottomland hardwood regeneration areas on the Delta National Forest. Journal of Field Ornithology 67: 181–187.

Hurst, G. A. and M. W. Smith. 1985. Effects of silvicultural practices in bottomland hardwoods on swamp rabbit habitat. p. 335–340. *In* E. Shoulders (ed.) Proceedings Third Biennial Southern Silvicultural Research Conference. U.S. Department of Agriculture Forest Service, New Orleans, LA. General Technical Report SO-54.

Hurst, G. H., and B. D. Stringer, Jr. 1975. Food habits of wild turkey poults in Mississippi. Proceedings of the National Wildlife Turkey Symposium 3: 76–85.

Hvorslev, M. J. 1951. Time lag and soil permeability in groundwater observations. U.S. Army Corps of Engineers, Waterways Experiment Station, Vicksburg, MS. Bulletin 36.

I

Iizumi, H. 1986. Soil nutrient dynamics. p. 171–180. *In* S. Cragg and N. Polunin (eds.) Workshop on Mangrove Ecosystem Dynamics. UNDP/UNESCO Regional Project (RAS/79/002), New Delhi, India.

Ingram, R. L. and L. J. Otte. 1981. Peat in North Carolina wetlands. p. 125–134. *In* C. J. Richardson (ed.) Pocosin Wetlands: An Integrated Analysis of Coastal Plain Freshwater Bogs in North Carolina. Hutchinson Ross Publication Co., Stroudsburg, PA.

Innes, W. T. 1948. Exotic Aquarium Fishes. Innes Publishing Co., Philadelphia, PA.

Irmler, F. and K. Furch. 1980. Weight, energy, and nutrient changes during the decomposition of leaves in the emersion phase of Central-Amazonian leaf litter. Pedobiologia 20: 118–130.

Irvin, L. 1996. Personal communication. Florida Division of Forestry, Tallahassee, FL.

J

Jackson, B. D. and R. A. Morris. 1986. Helicopter logging of baldcypress in southern swamps. Southern Journal of Applied Forestry 10: 20–23.

Jackson, B. D. and B. J. Stokes. 1991. Low-impact harvesting systems for wet sites. p. 701–709. *In* S. S. Coleman and D. G. Neary (compilers) Proceedings of the Sixth Southern Silvicultural Research Conference. U.S. Department of Agriculture Forest Service, Asheville, NC. General Technical Report SE-70.

Jackson, M. B. 1985. Ethylene and responses of plants to soil waterlogging and submergence. Annual Review of Plant Physiology 36: 145–174.

Jackson, M. B. 1990. Hormones and developmental change in plants subjected to submergence or soil waterlogging. Aquatic Botany 38: 49–72.

Jackson, M. B. and M. C. Drew. 1984. Effects of flooding on growth and metabolism of herbaceous plants. p. 47–128. *In* T. T. Kozlowski (ed.) Flooding and Plant Growth. Academic Press, Inc., Harcourt Brace Jovanovich, Publishers, Orlando, FL.

Jackson, M. B., D. D. Davies, and H. Lambers (eds.). 1991. Plant Life Under Oxygen Deprivation: Ecology, Physiology and Biochemistry. SPB Academic Publishing bv, The Hague, The Netherlands.

Jackson, M. B., K. Gales, and D.J. Campbell. 1978. Effect of waterlogged soil conditions on the production of ethylene and on the water relations of tomato plants. Journal of Experimental Botany 29: 183–193.

Jacobs, T. C. and J. W. Gilliam. 1985. Riparian losses of nitrate from agricultural drainage waters. Journal of Environmental Quality 14: 472–478.

Janzen, G. C. and J. D. Hodges. 1985. Influence of midstory and understory vegetation removal on the establishment and development of oak regeneration. p. 273–278. *In* E. Shoulders (ed.) Proceedings Third Biennial Southern Silvicultural Research Conference. U.S. Department of Agriculture Forest Service, New Orleans, LA. General Technical Report SO-54.

Jenkin, L. E. T. and T. ap Rees. 1986. Effects of lack of oxygen on the metabolism of shoots of *Typha angustifolia*. Phytochemistry 25: 823–827.

Jester, D. B., A. A. Echelle, W. J. Matthews, J. Pigg, C. M. Scott, and K. D. Collins. 1992. The fishes of Oklahoma, their gross habitats, and their tolerance of degradation in water quality and habitat. Proceedings of the Oklahoma Academy of Science 72: 7–19.

Johansson, M. B., B. Berg, and V. Meentemeyer. 1995. Litter mass-loss rates in late stages of decomposition in a climatic transect of pine forests. Long-term decomposition in a Scots pine forest. IX. Canadian Journal of Botany 73: 1509–1521.

John, C. D. and H. Greenway. 1976. Alcoholic fermentation and activity of some enzymes in rice roots under anaerobiosis. Australian Journal of Plant Physiology 1: 513–520.

Johnson, A. S. and L. L. Landers. 1982. Habitat relationships of summer resident birds in slash pine flatwoods. Journal of Wildlife Management 46: 416–428.

Johnson, D. and E. Odum. 1956. Breeding bird populations in relation to pant succession in the Piedmont of Georgia. Ecology 37: 50–62.

Johnson, D. W. 1942. The Origin of the Carolina Bays. Columbia University Press, New York, NY.

Johnson, D. W. 1944. Mysterious craters of the Carolina coast. American Journal of Science 5: 247–255.

Johnson, J. R., B. G. Cobb, and M. C. Drew. 1994. Hypoxic induction of anoxia tolerance in roots of ADH1 Null *Zea mays* L. Plant Physiology 105: 61–67.

Johnson, J. W. 1980. Pond pine. p. 58–59. *In* F. H. Eyre (ed.) Forest Cover Types of the United States and Canada. Society of American Foresters, Washington, DC.

Johnson, R. L. 1978. Hardwood culture in the eastern United States. p. 55–59. *In* Proceedings of Symposium on Utilization and Management of Alder. U.S. Department of Agriculture Forest Service, Portland, OR. General Technical Report PNW-70.

Johnson, R. L. 1979. Timber harvests from wetlands. p. 598–605. *In* P. E. Greeson, J. R. Clark, and J. E. Clark (eds.) Wetland Functions and Values: The State of Our Understanding. American Water Resources Association, Minneapolis, MN.

Johnson, R. L. 1980a. New ideas about regeneration of hardwoods. p. 17–19. *In* Proceedings of Hardwood Regeneration Symposium. Southeastern Lumberman Manufacturing Association, Atlanta, GA.

Johnson, R. L. 1980b. Sweetgum-willow oak. p. 64–65. *In* F. H. Eyre (ed.) Forest Cover Types of the United States and Canada. Society of American Foresters, Washington, DC.

Johnson, R. L. 1981. Oak seeding — it can work. Southern Journal of Applied Forestry 5: 28–30.

Johnson, R. L. 1990. *Nyssa aquatica* L. — water tupelo. p. 474–478. *In* R. M. Burns and B. H. Honkala (tech. coords.) Silvics of North America: 2. Hardwoods. U.S. Department of Agriculture, Washington, DC. Agriculture Handbook 654.

Johnson, R. L. and F. W. Shropshire. 1983. Bottomland hardwoods. p. 175–179. *In* R. M. Burns (ed.) Silviculture Systems for the Major Forest Types of the United States. U.S. Department of Agriculture Forest Service, Washington, DC. Agriculture Handbook 445, revised.

Johnson, T. G. 1991. Forest statistics for North Carolina. U.S. Department of Agriculture Forest Service, Southeastern Forest Experiment Station, Asheville, NC. Resource Bulletin SE-120.

Johnson, T. G. 1996. Personal communication. U.S. Department of Agriculture Forest Service, Southern Research Station, Asheville, NC.

Joly, C. A. 1994. Flooding tolerance: reinterpretation of Crawford's metabolic theory. Proceedings of the Royal Society of Edinburg Section B (Biological Sciences) 102: 343–354.

Joly, C. A. and R. M. M. Crawford. 1982. Variation in tolerance and metabolic responses to flooding in some tropical trees. Journal of Experimental Botany 33: 799–809.

Jones, D. A. 1984. Crabs of the mangal ecosystem. p. 89–109. *In* F.D. Por and I. Dor (eds.) Hydrobiology of the Mangal. Dr W. Junk Publishers, The Hague, The Netherlands.

Jones, H. E. 1971a. Comparative studies of plant growth and distribution in relation to waterlogging. II. An experimental study of the relationship between transpiration and the uptake of iron in *Erica cinerea* L. and *E. tetralix* L. Journal of Ecology 59: 167–178.

Jones, H. E. 1971b. Comparative studies of plant growth and distribution in relation to waterlogging. III. The response of *Erica cinerea* L. to waterlogging in peat soils of differing iron content. Journal of Ecology 59: 583–591.

Jones, H. E. and J. R. Etherington. 1970. Comparative studies of plant growth and distribution in relation to waterlogging. I. The survival of *Erica cinerea* L. and *E. Tetralix* and its apparent relationship to iron and manganese uptake in waterlogged soil. Journal of Ecology 58: 487–496.

Jones, R. 1972. Comparative studies of plant growth and distribution in relation to waterlogging. V. The uptake of iron and manganese by dune and dune slack plants. Journal of Ecology 60: 131–139.

Jones, R. H. 1981. A classification of lowland forests in the Northern Coastal Plain of South Carolina. M.S. Thesis. Clemson University, Clemson, SC.

Jones, R. H., B. G. Lockaby, and G. Somers. 1996. Effects of microtopography and disturbance on fine-root dynamics in wetland forests of low-order, stream floodplains. American Midland Naturalist 136: 57–71.

Junk, W. J., P. B. Bailey, and R. E. Sparks. 1989. The flood-pulse concept in river floodplain systems. Canadian Special Publication of Fisheries and Aquatic Sciences 106: 110–127.

Justin, S. H. F. W. and W. Armstrong. 1987. The anatomical characteristics of roots and plant response to soil flooding. New Phytologist 106: 465–495.

K

Kaczorowski, R. T. 1977. The Carolina bays: a comparison with modern oriented lakes. Ph.D. Dissertation. University of South Carolina, Columbia, SC.

Kaczorowski, R. T. 1992. Some unusually high quality peat deposits with a curious history. p. 97–108. *In* D. N. Grubich and T. J. Malterer (eds.) Peat and Peatlands: The Resource and Its Utilization. Proceedings of the International Peat Symposium, Duluth, MN.

Kadlec, R. H. and R. L. Knight. 1996. Treatment Wetlands. CRC Lewis Press, Boca Raton, FL.

Kalen, S. 1993. Commerce to conservation: the call for a national water policy and the evolution of federal jurisdiction over wetlands. North Dakota Law Review 69: 873–914.

Kangas, P. C. 1990. An energy theory of landscape for classifying wetlands. p. 15–23. *In* A. E. Lugo, M. M. Brinson, and S. L. Brown (eds.) Forested Wetlands. Vol. 15. Ecosystems of the World. Elsevier Science Publishers B.V., Amsterdam, The Netherlands.

Karr, B. L., G. L. Young, J. D. Hodges, B. D. Leopold, and R. M. Kaminski. 1990. Effect of flooding of greentree reservoirs. Water Resources Research Institute, Mississippi State University, MS. Technical Completion Report Project Number G1571-03.

Karriker, K. S. 1993. Effects of intensive silviculture on breeding and wintering birds in North Carolina pocosins. M.S. Thesis. North Carolina State University, Raleigh, NC.

Kawase, M. 1979. Role of cellulase in aerenchyma development in sunflower. American Journal of Botany 66: 183–190.

Kawase, M. 1981a. Effect of ethylene on aerenchyma development. American Journal of Botany 68: 651–658.

Kawase, M. 1981b. Anatomical and morphological adaptation of plants to waterlogging. HortScience 16: 30–34.

Kearney, T. H. 1901. Report on a botanical survey of the Dismal Swamp region. Contributions from the U.S. National Herbarium 5: 321–550.

Keeland, B. D. 1994. The effects of hydrologic regime on diameter growth of three wetland tree species of the southeastern U.S.A. Ph.D. Dissertation. University of Georgia, Athens, GA.

Keeland, B. D. and R. R. Sharitz. 1995. Seasonal growth patterns of *Nyssa sylvatica* var. *biflora*, *Nyssa aquatica*, and *Taxodium distichum* as affected by hydrologic regime. Canadian Journal of Forest Research 25: 1084–1096.

Keeland, B. D., J. A. Allen, and V. V. Burkett. 1995. Southern forested wetlands. p. 216–218. *In* E. T. LaRoe, G. S. Farris, C. E. Puckett, P. D. Doran, and M. J. Mac (eds.) Our Living Resources: A Report to the Nation on the Distribution, Abundance, and Health of U.S. Plants, Animals, and Ecosystems. U.S. Department of Interior, National Biological Survey, Washington, DC.

Keeley, J. E. 1979. Population differentiation along a flood frequency gradient: physiological adaptations to flooding in *Nyssa sylvatica*. Ecological Monographs 49: 89–108.

Kelley, W. R. and Batson, W. T. 1955. An ecological study of the land plants and cold-blooded vertebrates of the Savannah River project area. Part VI. Conspicuous vegetational zonation in a "Carolina bay." University of South Carolina, Columbia, SC. Publication Series III Biology 1:244–248.

Kellison, R.C. and M.J. Young. 1997. The bottomland hardwood forest of the southern United States. Forest Ecology and Management 90: 101–116.

Kellison, R. C., D. J. Frederick, and W. E. Gardner. 1982. A guide for regenerating and managing natural stands of southern hardwoods. North Carolina State University, Raleigh, NC. North Carolina Agricultural Research Service Bulletin 463.

Kellison, R. C., J. P. Martin, G. D. Hansen, and R. Lea. 1988. Regenerating and managing natural stands of bottomland hardwoods. American Pulpwood Association, Washington, DC. Technical Bulletin APA 88-A-6.

Kemp, G. P., W. H. Conner, and J. W. Day, Jr. 1985. Effects of flooding on decomposition and nutrient cycling in a Louisiana swamp forest. Wetlands 5: 35–51.

Kennedy, H. E., Jr. 1970. Growth of newly planted water tupelo seedlings after flooding and siltation. Forest Science 16: 250–256.

Kennedy, H. E., Jr. 1982. Growth and survival of water tupelo coppice regeneration after six growing seasons. Southern Journal of Applied Forestry 6: 133–135.

Kennedy, H. E., Jr. 1983. Water tupelo in the Atchafalaya Basin does not benefit from thinning. U.S. Department of Agriculture Forest Service, Southern Forest Experiment Station, New Orleans, LA. Research Note SO-298.

Kennedy, H. E., Jr. 1990. Personal communication. U.S. Department of Agriculture Forest Service, Southern Hardwoods Laboratory, Stoneville, MS (retired).

Kennedy, R. A., M. E. Rumpho and T. C. Fox. 1987. Germination physiology of rice seeds: Metabolic adaptations to anoxia. p. 193–203. *In* R.M.M. Crawford (ed.) The Series Analytic: Plant Life in Aquatic and Amphibious Habitats. Special Publications of the British Ecological Society, No. 5. Blackwell Scientific Publications, Oxford, England.

Kennedy, R. A., M. E. Rumpho, and T. C. Fox. 1992. Anaerobic metabolism in plants. Plant Physiology 100: 1–6.

Kerr, E. 1981. The history of forestry in Louisiana. Special publication of the Louisiana Forestry Association, Alexandria, LA.

Kerr, W. C. 1875. Physical geography, resume, economical geology, Vol. 1. Geological Survey of North Carolina, Raleigh, NC.

Kidwell, B. 1996. Farming in a fragile environment. Progressive Farmer January 1996: 30–31.

Killgore, K. J. and J. A. Baker. 1996. Patterns of larval fish abundance in a bottomland hardwood wetland. Wetlands 16: 288–295.

Kimball, M. C. and H. J. Teas. 1975. Nitrogen fixation in mangrove areas of southern Florida. p. 654–660. *In* G. Walsh, S. Snedaker, and H. Teas (eds.) Proceedings of the International Symposium on the Biology and Management of Mangroves. Institute of Food and Agricultural Sciences, University of Florida, Gainesville, FL.

Kimmerer, T. W. 1987. Alcohol dehydrogenase and pyruvate decarbohylase activity in leaves and roots of eastern cottonwood (*Populus deltoides* Bartr.) and soybean (*Glycine max* L.). Plant Physiology 84: 1210–1213.

Kimmerer, T. W. and M. A. Stringer. 1988. Alcohol dehydrogenase and ethanol in the stems of trees. Plant Physiology 87: 693–697.

Kimmins, J. P. 1987. Forest Ecology. McMillian, New York, NY.

King, S. L. 1995. Effects of flooding regime on two impounded bottomland hardwood stands. Wetlands 15: 272–284.

Kirby-Smith, W. W. and R. T. Barber. 1979. The water quality ramifications in estuaries of converting forest to intensive agriculture. Water Resources Research Institute, North Carolina State University, Raleigh, NC. Report No. 148.

Kirk, P. W., Jr. 1979. The Great Dismal Swamp. University Press of Virginia, Charlottesville, VA.

Kirkman, L. K. 1995. Impacts of fire and hydrological regimes on vegetation in depression wetlands of southeastern U.S.A. p. 10–20. *In* S. I. Cerulean and R. T. Engstrom (eds.) Fire in Wetlands: A Management Perspective, Proceedings of the Tall Timbers Fire Ecology Conference No. 19. Tall Timbers Research Station, Tallahassee, FL.

Kirkman, L. K. and R. R. Sharitz. 1994. Vegetation disturbance and maintenance of diversity in intermittently flooded Carolina bays in South Carolina. Ecological Applications 4: 177–188.

Kirkman, L. K, R. F. Lide, G. R. Wein, and R. R. Sharitz. 1996. Vegetation changes and land-use legacies of depression wetlands of the western coastal plain of South Carolina 1951–1992. Wetlands 16: 564–576.

Kitchens, W. M., Jr., J. M. Dean, L. H. Stevenson, and J. M . Cooper. 1975. The Santee Swamp as a nutrient sink. p. 349–366. *In* F. G. Howell, J. B. Gentry, and H. M. Smith (eds.) Mineral Cycling in Southeastern Ecosystems. U.S. Government Printing Office, Washington, DC. ERDA Symposium Series 740513.

Kiviat, E. 1978. Bog turtle habitat ecology. Bulletin of the Chicago Herpetological Society 13: 29–42.

Klee, A. J. 1968. A history of the aquarium hobby in America. Parts 2, 10, and 13. The Aquarium 1(3): 44–57; 1(11): 34–35, 50–57; 2(2): 32–33, 50–57.

Kludze, H .K., S. R. Pezeshki, and R. D. DeLaune. 1994. Evaluation of root oxygenation and growth in bald cypress in response to short-term hypoxia. Canadian Journal of Forest Research 24: 804–809.

Knight, R. L., J. S. Bays, and F. R. Richardson. 1989. Floral composition, soil relations and hydrology of a Carolina bay in South Carolina. p. 219–234. *In* R. R. Sharitz and J. W. Gibbons (eds.) Freshwater Wetlands and Wildlife Symposium: Perspectives on Natural, Managed and Degraded Ecosystems. U.S. Department of Energy, Office of Scientific and Technical Information, Oak Ridge, TN. CONF-8603101, DOE Symposium Series No. 61.

Koch, M. S. 1996. Resource availability and abiotic stress effects on *Rhizophora mangle* L. (red mangrove) development in south Florida. Ph.D. Dissertation. University of Miami, Coral Gables, FL.

Kologiski, R. L. 1977. The Phytosociology of the Green Swamp, NC. North Carolina Agricultural Experiment Station, Raleigh, NC. Technical Bulletin 250.

Korstian, C. F. and W. D. Brush. 1931. Southern white cedar. U.S. Department of Agriculture, Washington, DC. Technical Bulletin 251.

Kozlowski, T. T. 1982. Water supply and tree growth. Part II. Flooding. Forestry Abstracts 43: No. 3.

Kozlowski, T. T. (ed.). 1984. Flooding and Plant Growth. Academic Press, Inc. Harcourt Brace Jovanovich, Publishers, Orlando, FL.

Kozlowski, T. T. 1986. Soil aeration and growth of forest trees. Scandinavian Journal of Forest Research 1: 113–123.

Kraemer, P. J., W. S. Riley, and T. T. Bannister. 1952. Gas exchange of cypress knees. Ecology 33: 117–121.

Kramer, D. L. 1987. Dissolved oxygen and fish behavior. Environmental Biology of Fishes 18: 81–92.

Kress, M. R., M. R. Graves, and S. G. Bourne. 1996. Loss of bottomland hardwood forests and forested wetlands in the Cache River basin, Arkansas. Wetlands 16: 258–263.

Kreuzberg, K. 1984. Starch fermentation via a formate producing pathway in *Chlamydomonas reinhardii, Chlorogonium elongation,* and *Chlorella fusca.* Physiologia Plantarum 61: 87–94.

Krinard, R. M. 1990. Cherrybark oak. p. 644–649. *In* R. M. Burns and B. H. Honkala (tech. coords.) Silvics of North America: 2. Hardwoods. U.S. Department of Agriculture Forest Service, Washington, DC. Agriculture Handbook 654.

Krinard, R. M. and R. L. Johnson. 1980. Fifteen years of cottonwood plantation growth and yield. Southern Journal of Applied Forestry 4: 180–185.

Krinard, R. M. and R. L. Johnson. 1987. Growth of 31-year-old baldcypress plantation. U.S. Department of Agriculture Forest Service, Southern Forest Experiment Station, New Orleans, LA. Research Note SO-339.

Kristensen, E., F. Ø. Andersen, and L. H. Kofoed. 1988. Preliminary assessment of benthic community metabolism in a south-east Asian mangrove swamp. Marine Ecology Progress Series 48: 137–145.

Kruseman, G. P. and N. A. DeRidder. 1979. Analysis and evaluation of pump test data. International Institute for Land Reclamation and Improvement, Wageningen, The Netherlands.

Küchler, A. W. 1964. Potential Natural Vegetation of the Conterminous United States. American Geographical, New York, NY. Special Publication Number 36.

Kuenzler, E. J. 1987. Impacts of sewage effluent on tree survival, water quality and nutrient removal in coastal plain swamps. Water Resources Research Institute of the University of North Carolina, Chapel Hill, NC. Report No. 235.

Kuenzler, E. J. 1989. Value of forested wetlands as filters for sediments and nutrients. p. 85–96. *In* D.D. Hook and R. Lea (eds.) Proceedings of the Symposium: The Forested Wetlands of The Southern United States. U.S. Department of Agriculture Forest Service, Southeastern Forest Experiment Station, Asheville, NC. General Technical Report SE-50.

Kundell, J. E. and S. W. Woolf. 1986. Georgia wetlands: trends and policy options. Carl Vinson Institute of Government, University of Georgia, Athens, GA.

Kuo, S. and A. S. Baker. 1982. The effect of soil drainage on phosphorus status and availability to corn in long-term manure-amended soils. Soil Science Society of America Journal 46: 744–747.

Kuo, S. and D. S. Mikkelsen. 1979. Distribution of iron and phosphorus in flooded and nonflooded soil profiles and their relation to P absorption. Soil Science 127: 18–25.

Kurz, H. and D. Demaree. 1934. Cypress buttresses in relation to water and air. Ecology 15: 36–41.

Kushlan, J. A. 1973. Differential responses to drought in two species of Fundulus. Copeia 1974: 808–809.

Kushlan, J. A. 1974. Effects of a natural fish kill on the water quality, plankton, and fish population of a pond in the Big Cypress Swamp, Florida. Transactions of the American Fisheries Society 103: 235–243.

Kushlan, J. A. and M. S. Kushlan. 1980. Population fluctuations of the prawn, *Palemometes paludosus*, in the Everglades. American Midland Naturalist 103: 401–403.

Kusler, J. A. 1992. Wetlands delineation: an issue of science or politics? Environment 34: 7–37.

Kusler, J. A. and M. E. Kentula (eds.). 1990. Wetland Creation and Restoration: The Status of the Science. Environmental Protection Agency, Environmental Research Laboratory, Corvallis, OR. EPA 600/3-89/038.

L

Laan, P. and C. W. P. M. Blom. 1990. Growth and survival responses of *Rumex* species to flooded and submerged conditions: the importance of shoot elongation, underwater photosynthesis and reserve carbohydrates. Journal of Experimental Botany 41(228): 775–783.

Labisky, R. F. and J. A. Hovis. 1987. Comparison of vertebrate wildlife communities in longleaf pine and slash pine habitats in north Florida. p. 201–230. *In* H. A. Pearson, F. E. Smeins, and R. E. Thill (eds.) Ecological, Physical, and Socioeconomic Relationships Within Southern National Forests. U.S. Department of Agriculture Forest Service, New Orleans, LA.

Laderman, A. D. 1989. The ecology of the Atlantic white cedar wetlands: a community profile. U.S. Fish and Wildlife Service, Office of Biological Services, Washington, DC. Biological Report 85(7.21).

Laerm, J., B. J. Freeman, L. J. Vitt, and L. E. Logan. 1984. Appendix 4: Checklist of vertebrates of the Okefenokee Swamp. p. 682–701. *In* A. D. Cohen, D. J. Casagrande, M. J. Andrejko, and G. R. Best (eds.) The Okefenokee Swamp: Its Natural History, Geology, and Geochemistry. Wetland Surveys, Los Alamos, NM.

Laerm, J., B. J. Freeman, L. J. Vitt, J. M. Meyers, and L. E. Logan. 1980. Vertebrates of the Okefenokee Swamp. Brimleyana 4: 47–73.

Lambou, V. W. 1959. Fish populations in backwater lakes in Louisiana. Transactions of the American Fisheries Society 88: 7–15.

Lambou, V. W. 1961. Utilization of macrocrustaceans for food by freshwater fishes in Louisiana and its effects on the determination of predator-prey relations. The Progressive Fish-Culturist 23: 18–25.

Lambou, V. W. 1963. The commercial and sports fisheries of the Atchafalaya Basin floodway. Proceedings of the Annual Conference of Southeastern Association of Game and Fish Commissioners 17: 256–281.

Lambou, V. W. 1990. Importance of bottomland hardwood forest zones to fishes and fisheries: the Atchafalaya Basin, a case history. p. 125–193. *In* J. G. Gosselink, L. C. Lee, and T. A. Muir (eds.) Ecological Processes and Cumulative Impacts: Illustrated by Bottomland Hardwood Wetland Ecosystems. Lewis Publishers, Chelsea, MI.

Land Trust Alliance. 1993. Conservation Options: A Landowner's Guide. Washington, DC.

Langdon, O. G. and G. E. Hatchell. 1977. Performance of loblolly, slash, and pond pines on a poorly drained site with fertilization at ages 11 and 14. p. 121–126. *In* Proceedings of the Sixth Southern Forest Soils Workshops. U.S. Department of Agriculture Forest Service, Southeastern Area State and Private Forest, Atlanta, GA.

Langdon, O. G. and W. H. McKee, Jr. 1981. Can fertilization of loblolly pine on wet sites reduce the need for drainage? p. 212–218. *In* J. P. Barnett (ed.) Proceedings of the First Biennial Southern Silvicultural Research Conference. U.S. Department of Agriculture Forest Service, Southern Forest Experiment Station, New Orleans, LA. General Technical Report SO-34.

Langdon, O. G. and K. B. Trousdell. 1978. Stand manipulation: effects on soil moisture and tree growth in southern pine and pine-hardwood stands. p. 221–236. *In* W. E. Balmer (ed.) Proceedings: Soil Moisture — Site Productivity Symposium. U.S. Department of Agriculture Forest Service, State and Private Forestry, Atlanta, GA.

Langdon, O. G., J. P. McClure, D. D. Hook, J. M. Crockett, and R. Hunt. 1981. Extent, condition, management, and research needs of bottomland hardwood-cypress forests in the Southeastern United States. p. 71–85. *In* J. R. Clark and J. Benforado (eds.) Wetlands of Bottomland Hardwood Forests. Elsevier Scientific Publishing Company, Amsterdam, The Netherlands.

Lantz, K. E. 1970a. An ecological survey of factors affecting fish production in a Louisiana backwater area and river. Louisiana Wildlife and Fisheries Commission, Baton Rouge, LA. Bulletin Number 5.

Lantz, K. E. 1970b. An ecological survey of factors affecting fish production in a Louisiana natural lake and river. Louisiana Wildlife and Fisheries Commission, Baton Rouge, LA. Bulletin Number 6.

Lantz, K. E. 1974. Natural and controlled water level fluctuations in a backwater lake and three Louisiana impoundments. Louisiana Wildlife and Fisheries Commission, Baton Rouge, LA. Bulletin Number 11.

Larson, J. S. and D. B. Mazzarese. 1994. Rapid assessment of wetlands: history and application to management. p. 625–636. *In* W. J. Mitsch (ed.) Global Wetlands: Old World and New. Elsevier, Amsterdam, The Netherlands.

Larson, K. D., F. S. Davies, and B. Schaffer. 1991. Floodwater temperature and stem lenticel hypertrophy in *Mangifera indica* (Anacardiaceae). American Journal of Botany 78: 1397–1403.

Larson, K. D., B. Schaffer, and F. S. Davies. 1993. Floodwater oxygen content, ethylene production and lenticel hypertrophy in flooded mango (*Mangifera indica* L.) trees. Journal of Experimental Botany 44: 665–671.

Leach, G. N. and P. P. Ryan. 1987. Natural hardwood regeneration in the Escambia River bottom following logging. Champion International Corporation, Pensacola, FL. Research Note GS-87-03.

Leany, F. W., K. R. J. Smettem, and D. J. Chittleborough. 1993. Estimating contribution of preferential flow to subsurface runoff from a hillslope using deuterium and chloride. Journal of Hydrology 147: 83–101.

Leberg, P. L. and M. L. Kennedy. 1988. Demography and habitat relationships of raccoons in western Tennessee. Proceedings of the Annual Conference of the Southeastern Association of Fish and Wildlife Agencies 42: 272–282.

Lee, D. S. 1986. Pocosin breeding bird fauna. American Birds 40: 1263–1273.

Lee, D. S. 1987. Breeding birds of Carolina bays: succession-related density and diversification on ecological islands. Chat 51: 85–102.

Lee, D. S., C. R. Gilbert, C. H. Hocutt, R. E. Jenkins, D. E. McAllister, and J. R. Stauffer, Jr. 1980 et seg. Atlas of North American Freshwater Fishes. North Carolina State Museum of Natural History, Raleigh, NC.

Lee, S. Y. 1989. Litter production and turnover of the mangrove *Kandelia candel* (L.) Druce in a Hong Kong tidal shrimp pond . Estuarine, Coastal and Shelf Science 29: 75–87.

LeGrand, H. E. 1993. Natural Heritage Program List of the Rare Animal Species of North Carolina. North Carolina Natural Heritage Program, Raleigh, NC.

Leh, C. M. U. and A. Sasekumar. 1985. The food of sesarmid crabs in Malaysian mangrove forests. Malay Naturalist Journal 39: 135–145.

Leibowitz, S. G., B. Abbruzzese, P. R. Adamus, L. E. Hughes, and J. T. Irish. 1992. A synoptic approach to cumulative impact assessment: a proposed methodology. U.S. Environmental Protection Agency, Environmental Research Laboratory, Corvallis, OR. EPA/600/R-92/167.

Leighty, R. G., and S. W. Buol. 1983. Histosols. p. 92–94. *In* S. W. Buol (ed.) Soils of the Southern States and Puerto Rico. Joint Publication of the Agricultural Experiment Stations of the Southern States and Puerto Rico Land Grant Universities. Southern Cooperative Series Bulletin No. 174.

Leitman, H. M., M. R. Darst, and J. J. Nordhaus. 1991. Fishes in the forested flood plain of the Ochlockonee River, Florida, during flood and drought conditions. U.S. Geological Survey, Tallahassee, FL. Water Resource Investigations Report 90–4202.

Leitman, H. M., J. E. Solm, and M. A. Franklin. 1983. Wetland hydrology and tree distribution of the Apalachicola River floodplain, Florida. U.S. Geological Survey, Washington, DC. Water Supply Paper 2186-A.

Lemlich, S. K. and K. C. Ewel. 1984. Effects of wastewater disposal on growth rates of cypress trees. Journal of Environmental Quality 13: 602–604.

Leonard, D. L., Jr. 1994. Avifauna of forested wetlands adjacent to river systems in central Florida. Florida Field Naturalist 22: 97–105.

Leopold, L., M. G. Wolman, and J. Miller. 1964. Fluvial Processes in Geomorphology. W. H. Freeman and Co., San Francisco, CA.

Leslie, A. J. 1996. Structure of benthic macroinvertebrate communities in natural and clearcut cypress ponds of north Florida. M.S. Thesis. University of Florida, Gainesville, FL.

Lewis, R. R. 1982. Mangrove forests. p. 153–171. *In* R.R. Lewis (ed.) Creation and Restoration of Coastal Plant Communities. CRC Press, Boca Raton, FL.

Lewis, R. R. 1990a. Creation and restoration of coastal plain wetlands in Florida. p. 73–101. *In* J.A. Kusler and M.E. Kentula (eds.) Wetland Creation and Resotration. Island Press, Washington, DC.

Lewis, R. R. 1990b. Creation and restoration of coastal wetlands in Puerto Rico and the U.S. Virgin Islands. p. 103–123. *In* J.A. Kusler and M.E. Kentula (eds.). Wetland Creation and Resotration. Island Press, Washington, DC.

Lewis, W. M., Jr. 1970. Morphological adaptations of cyprinodontoids for inhabiting oxygen-deficient waters. Copeia 1970: 319–326.

Lewty, M. J. 1990. Effects of waterlogging on the growth and water relations of three *Pinus* taxa. Forest Ecology and Management 30: 189–201.

Lide, R. F. In press. When is a depression wetland a Carolina bay? Southeastern Geographer.

Lide, R. F., V. G. Meentemeyer, J. E. Pinder, and L. M. Beatty. 1995. Hydrology of a Carolina bay located on the upper coastal plain of western South Carolina. Wetlands 15: 47–57.

Light, H. M., M. R. Darst, and J. W. Grubbs. 1995. Hydrologic conditions, habitat characteristics, and occurrence of fishes in the Apalachicola River floodplain, Florida: second annual report of progress, October 1993–September 1994. U.S. Geological Survey, Tallahassee, FL. Open File Report 95-167.

Light, S. S., and J. W. Dineen. 1994. Water control in the Everglades: a historical perspective. p. 47–84. *In* S.M. Davis and J.C. Ogden (eds.) Everglades — The Ecosystem and Its Restoration. St. Lucie Press, Delray Beach, FL.

Lilly, J. P. 1981a. The blackland soils of North Carolina, their characteristics and management for agriculture. North Carolina Agricultural Research Service, Raleigh, NC. Technical Bulletin Number 270.

Lilly, J. P. 1981b. A history of swamp land development in North Carolina. p. 20–39. *In* C. J. Richardson (ed.) Pocosin Wetlands: An Integrated Analysis of Coastal Plain Freshwater Bogs in North Carolina. Hutchinson Ross Publication Co., Stroudsburg, PA.

Lin, J. and J. L. Beal. 1995. Effects of mangrove marsh management on fish and decapod communities. Bulletin of Marine Science 57: 193–201.

Lindenmayer, D. B. 1994. Wildlife corridors and the mitigation of logging impacts on fauna in wood-production forests in south-eastern Australia: a review. Wildlife Research 21: 323–340.

Lindsey, R. K., M. A. Kohler, and J. L. H. Paulus. 1958. Chapter 11. p. 270–273. *In* Hydrology for Engineers. McGraw Hill, New York, NY.

Little, E. L., Jr. 1971. Atlas of United States trees, Vol. 1, conifers and important hardwoods. U.S. Department of Agriculture, U.S. Government Printing Office, Washington, DC. Miscellaneous Publication 1146.

Little, E. L., Jr. 1977. Atlas of United States trees, Vol. 4, minor eastern hardwoods. U.S. Department of Agriculture, U.S. Government Printing Office, Washington, DC. Miscellaneous Publication 1342.

Little, E. L., Jr. 1979. Checklist of United States trees (native and naturalized). U.S. Department of Agriculture Forest Service, Washington, DC. Agriculture Handbook 541.

Little, S., Jr. 1950. Ecology and silviculture of whitecedar and associated hardwoods in southern New Jersey. School of Forestry, Yale University, New Haven, CT. Bulletin 56.

Little, S. and P. W. Garrett. 1990. *Chamaecyparis thyoides* (L.) B.S.P. Atlantic white cedar. p. 103–108. *In* R. M. Burns and B. M. Honkala (eds.) Silvics of North America: 1. Conifers. U.S. Department of Agriculture Forest Service, Washington, DC. Agriculture Handbook 654.

Liu, S. 1996. Personal communication. School of Forest Resource Conservation, University of Florida, Gainesville, FL.
Liu, Z. and D. I. Dickman. 1992. Abscisic acid accumulation in leaves of two contrasting hybrid poplar clones affected by nitrogen fertilization plus cyclic flooding and soil drying. Tree Physiology 11: 109–122.
Liu, Z. and D. I. Dickman. 1993. Responses of two hybrid *Populus* clones to flooding, drought, and nitrogen availability. II. Gas exchange and water relations. Canadian Journal of Botany 71: 927–938.
Livingston, R. J. and O. L. Loucks. 1979. Productivity, trophic interactions, and food-web relationships in wetlands and associated systems. p. 101–119. *In* P. E. Greeson, J. R. Clark, and J. E. Clark (eds.) Wetland Functions and Values: the State of Our Understanding. American Water Resources Association, Minneapolis, MN.
Lockaby, B. G., A. Murphy, and G. L. Somers. 1996. Hydroperiod effects on nutrient dynamics in decomposing litter. Soil Science Society of America Journal 60: 1267–1272.
Lockaby, B. G., J. Stanturf, and M. Messina. 1997a. Effects of silvicultural activity on ecological processes in floodplain forests of the southern United States: a review of existing reports. Forest Ecology and Management 90: 93–100.
Lockaby, B. G., F. C. Thornton, R. H. Jones, and R. G. Clawson. 1994. Ecological responses of an oligothrophic floodplain forest to harvesting. Journal of Environmental Quality 23: 901–906.
Lockaby, B. G., R. H. Jones, R. G. Clawson, J. S. Meadows, J. A. Stanturf, and F. C. Thornton. 1997b. Influences of harvesting on functions of floodplain forests associated with low-order, blackwater streams. Forest Ecology and Management 90: 217–224.
Lockaby, B. G., R. G. Clawson, K. Flynn, R. Rummer, S. Meadows, B. Stokes, and J. Stanturf. 1997c. Influence of harvesting on biogeochemical exchange in sheetflow and soil processes in an euthrophic floodplain forest. Forest Ecology and Management 90: 187–194.
Lockaby, B. G., R. S. Wheat, and R. G. Clawson. 1996. Influence of hydroperiod on conversion of litter to SOM in a floodplain forest. Soil Science Society of America Journal. 60(6): 1989–1993.
Loftis, D. L. 1990. A shelterwood method for regenerating red oak in the southern Appalachians. Forest Science 36: 917–929.
Lollis, J. P. 1981. The effects of stream channelization on wildlife habitat in eastern North America. M.S. Thesis. North Carolina State University, Raleigh, NC.
Lonard, R. I., E. J. Clairain, Jr., R. T. Huffman, J. W. Hardy, C. D. Brown, P. E. Ballard, and J. W. Watts. 1981. Analysis of methodologies used for the assessment of wetland values, Final Report. U.S. Army Corps of Engineers Waterways Experiment Station, U.S. Water Resources Council, Washington, DC.
Lord, W. G. 1979. Honey plants — Carolina bays: a unique feature of southeastern beekeeping. Gleanings in Bee Culture 107: 416–418.
Lorman, J. G. and J. J. Magnuson. 1978. The role of crayfishes in aquatic ecosystems. Fisheries 3: 8–10.
Lowrance, R., R. Todd, J. Fail, O. Hendrickson, R. Leonard, and L. Asmussen. 1984. Riparian forests as nutrient filters in agricultural watersheds. Bioscience 34: 374–377.
Ludwig, J. A. and J. F. Reynolds. 1988. Statistical Ecology. John Wiley and Sons, New York, NY.
Ludwig, J. C. 1996. Personal communication. Virginia Heritage Trust, Richmond, VA.
Lugo, A. E. 1978 . Stress and ecosystems. p. 62–101. *In* J. H. Thorp and J. W. Gibbons (eds.) Energy and Environmental Stress. Department of Energy, Washington, DC. DOE 771114.
Lugo, A. E. 1980. Mangrove ecosystems: successional or steady state? Biotropica 12: 65–72.

Lugo, A. E. 1990a. Fringe wetlands. p. 143–169. *In* A. E. Lugo, M. M. Brinson, and S. L. Brown (eds.) Forested Wetlands. Vol. 15. Ecosystems of the World. Elsevier Science Publishers B.V., Amsterdam, The Netherlands.

Lugo, A. E. 1990b. Introduction. p. 1–14. *In* A. E. Lugo, M. M. Brinson, and S. L. Brown (eds.) Forested Wetlands. Vol. 15. Ecosystems of the World. Elsevier Science Publishers B.V., Amsterdam, The Netherlands.

Lugo, A. E. and M. M. Brinson. 1979. Calculations of the value of salt water wetlands. p. 120–130. *In* P. E. Greeson, J. R. Clark, J. E. Clark (eds.) Wetland Functions and Values: The State of our Understanding. American Water Resources Association, Minneapolis, MN.

Lugo, A. E. and C. Patterson-Zucca. 1977. The impact of low temperature stress on mangrove structure and growth. Tropical Ecology 18: 149–161.

Lugo, A. E. and S. C. Snedaker. 1974. The ecology of mangroves. Annual Review of Ecology and Systematics 5: 39–64.

Lugo, A. E., M. M. Brinson, and S. L. Brown. 1990a. Synthesis and search for paradigms in wetland ecology. p. 447–460. *In* A. E. Lugo, M. M. Brinson, and S. L. Brown (eds.) Forested Wetlands. Vol. 15. Ecosystems of the World. Elsevier Science Publishers B.V., Amsterdam, The Netherlands.

Lugo, A. E., M. M. Brinson, and S. L. Brown (eds.). 1990b. Ecosystems of the World, Vol. 15. Forested Wetlands. Elsevier Science Publishers B.V., Amsterdam, The Netherlands.

Lugo, A. E., S. L. Brown, and M. M. Brinson. 1990c. Concepts in wetland ecology. p. 53–85. *In* A. E. Lugo, M. M. Brinson, and S. L. Brown (eds.) Forested Wetlands. Vol. 15. Ecosystems of the World. Elsevier Science Publishers B.V., Amsterdam, The Netherlands.

Lugo, A. E., M. Sell, and S. C. Snedaker. 1976. Mangrove ecosystem analysis. p. 113–145. *In* B. C. Patten (ed.) Systems and Simulation in Ecology. Academic Press, New York, NY.

Lutgens, F. K. and E. J. Tarbuck. 1994. The Atmosphere, 6th Edition. Prentice Hall, Englewood Cliffs, NJ.

Lynch, J. C. 1989. Sedimentation and nutrient accumulation in mangrove ecosystems of the Gulf of Mexico. M.S. thesis. University of Southwestern Louisiana, Lafayette, LA.

Lynch, J. C., J. R. Meriwether, B. A. McKee, F. Vera-Herrera, and R. R. Twilley. 1989. Recent accretion in mangrove ecosystems based on ^{137}Cs and ^{210}Pb. Estuaries 12: 284–299.

Lynch, J. M. 1982. Breeding birds of Hall Swamp Pocosin, Martin County, N.C. Chat 46: 93–101.

Lytle, S. A. 1960. Physiography and properties of southern forest soils. p. 1–8. *In* Proceedings of the Eighth Annual Forestry Symposium. Louisiana State University, Baton Rouge, LA.

Lytle, S. A. and M. B. Sturgis. 1962. General soil areas and associated soil series groups of Louisiana (map). Agricultural Experiment Station, Louisiana State University, Baton Rouge, LA.

M

MacArthur, R. H. and H. W. MacArthur. 1961. On bird species diversity. Ecology 42: 594–598.

MacCarthy, R. and C. B. Davey. 1976. Nutritional problems of *Pinus taeda* L. (loblolly pine) growing on pocosin soil. Soil Science Society of America Journal 40: 582–585.

MacDonald, R. C. and T. W. Kimmerer. 1991. Ethanol in the stems of trees. Physiologia Plantarum 82: 582–588.

Macnae, W. 1968. A general account of the fauna and flora of mangrove swamps and forests in the Indo-West-Pacific region. Advances in Marine Biology 6: 73–270.

Macnae, W. 1974. Mangrove forests and fisheries. FAO/UNDP Indian Ocean Programme, IOFC/DEV/7434.

Mader, S. F. 1990. Recovery of ecosystem functions and plant community structure by a tupelo-cypress wetland following timber harvest. Ph.D. Dissertation. North Carolina State University, Raleigh, NC.

Mader, S. F. 1991. Forested wetlands classification and mapping: a literature review. National Council of the Paper Industry for Air and Stream Improvement, New York, NY. Technical Bulletin 606.

Mader, S. F., W. M. Aust, and R. Lea. 1989. Changes in net primary productivity and cellulose decomposition rates in a water tupelo-bald cypress swamp following timber harvest. p. 539–543. *In* J. H. Miller (compiler) Proceedings of the Fifth Biennial Southern Silvicultural Conference. U.S. Department of Agriculture Forest Service, Southern Forest Experiment Station, New Orleans, LA. General Technical Report SO-74.

Madole, R. F., C. R. Ferring, M. J. Guccione, S. A. Hall, W. C. Johnson, and C. J. Sorenson. 1991. Quaternary geology of the Osage Plains and Interior Highlands. p. 503–546. *In* R. B. Morrison (ed.) Quaternary Nonglacial Geology: Conterminous U.S. Geological Society of America, Boulder, CO.

Mahoney, D. L., M. A. Mort, and B. E. Taylor. 1990. Species richness of calanoid copepods, cladocerans and other branchiopods in Carolina bay temporary ponds. American Midland Naturalist 123: 244–258.

Maki, T. E. 1974. Factors affecting forest production on organic soils. p. 119–136. *In* Histosols: Their Characteristics, Classification, and Use. Soil Science Society of America, Madison, WI. Special Publication No. 6.

Malac, B. F. and R. D. Herren. 1979. Hardwood plantation management. Southern Journal of Applied Forestry 3: 3–6.

Malley, D. F. 1978. Degradation of mangrove leaf litter by the tropical sesarmid crab *Chiromanthes onychophorum*. Marine Biology 49: 377–386.

Maltby, E., D. V. Hogan, C. P. Immirzi, J. H. Tellam, and M. J. van der Peijl. 1994. Building a new approach to the investigation and assessment of wetland ecosystem functioning. p. 637–658. *In* W. J. Mitsch (ed.) Global Wetlands: Old World and New. Elsevier, Amsterdam, The Netherlands.

Mancil, E. 1969. Some historical and geographical notes on the cypress lumbering industry in Louisiana. Louisiana Studies 8: 14–25.

Mancil, E. 1980. Pullboat logging. Journal of Forest History 24:135–141.

Manuel, T. M., K. L. Belli, and J. D. Hodges. 1993. A decision-making model to manage or regenerate southern bottomland hardwood stands. Southern Journal of Applied Forestry 17:75–79.

Marion, W. R. and T. E. O'Meara. 1982. Wildlife dynamics in managed flatwoods of north Florida. p. 63–67. *In* Proceedings of the 14th Annual Symposium on the Impacts of Intensive Forest Management Practices. University of Florida, Gainesville, FL.

Markewich, H. W., M. J. Pavich, and G. R. Buell. 1990. Contrasting soils and landscapes of the Piedmont and Coastal Plain, eastern United States. Geomorphology 3: 417–448.

Marks, P. L. and P. A. Harcombe. 1981. Forest vegetation of the Big Thicket, southeast Texas. Ecological Monographs 51: 287–305.

Markley, J. L., C. McMillan, and G. A. Thompson, Jr. 1982. Latitudinal differentation in reponse to chilling temperatures among populations of three mangroves, *Avicennia germinans, Laguncularia racemosa,* and *Rhizophora mangle* from the western tropical Atlantic and Pacific Panama. Canadian Journal of Botany 60: 2704–2715.

Marois, K. C. and K. C. Ewel. 1983. Natural and management-related variation in cypress domes. Forest Science 29: 627–640.

Marschner, H. 1995. Mineral Nutrition of Higher Plants, 2nd Edition. Academic Press, New York, NY.

Marshall, W. D. 1993. Assessing change in the Edisto River basin: an ecological characterization. South Carolina Water Resources Commission, Columbia, SC. Report Number 177.

Marsily, G. de. 1986. Quantitative Hydrogeology. Academic Press, New York, NY.

Marsinko, A. P. C., J. H. Syme, and R. A. Harris. 1991. Cypress: a species in transition. Forest Products Journal 41: 61–64.

Martin, A. C., F. M. Uhlr, and W. S. Bourn. 1953. Classification of wetlands of the United States. U.S. Fish and Wildlife Service, Washington, DC. Special Science Report 20.

Martin, C. 1973. Fire and forest structure in the aboriginal eastern forest. The Indian Historian 6: 38–42, 54.

Martin, R. G. 1977. Density dependent aggressive advantage in melanistic male *Gambusia affinis holbrooki* (Girard). Florida Scientist 40: 393–400.

Martin, W. H. and S. G. Boyce. 1993. Introduction: the southeastern setting. p. 1–46. *In* W. H. Martin, S. G. Boyce, and A. C. Esternacht (eds.) Biodiversity of the Southeastern United States: Lowland Terrestrial Communities. John Wiley and Sons, Inc., New York, NY.

Martof, B. S., Palmer, W. M., Bailey, J. R., and Harrison, J. R., III. 1980. Amphibians and Reptiles of the Carolinas and Virginia. University of North Carolina Press, Chapel Hill, NC.

Mather, J. R. 1974. Climatology: Fundamentals and Applications. McGraw–Hill, New York, NY.

Mather, J. R. 1978. The Climatic Water Budget in Environmental Analysis. Lexington Books, D.C. Heath & Co., Lexington, MA.

Matthews, J. D. 1989. Silvicultural Systems. Oxford University Press, Oxford.

Mattoon, W. R. 1915. The southern cypress. U.S. Department of Agriculture, Washington, DC. Agriculture Bulletin No. 272.

Mattoon, W. R. 1916. Water requirements and growth of young cypress. Society of American Foresters Proceedings 11: 192–197.

Mausbauch, M. J. and J. L. Richardson. 1994. Biogeochemical processes in hydric soil formation. Current Topics and Wetland Biogeochemistry 1: 68–128.

Maxwell, J. A. and M. B. Davis. 1972. Pollen evidence of Pleistocene and Holocene vegetation on the Allegheny Plateau, Maryland. Quaternary Research 2: 506–530.

Mayden, R. L., B. M. Burr, L. M. Page, and R. R. Miller. 1992. The native freshwater fishes of North America. p. 827–863. *In* R.L. Mayden (ed.) Systematics, Historical Ecology, and North American Freshwater Fishes. Stanford University Press, Stanford, CA.

Mayne, R. G. and H. Kende. 1986. Glucose metabolism in anaerobic rice seedlings. Plant Science 45: 31–36.

McCabe, G. J. 1989. A parabolic function to adjust Thornthwaite estimates of potential evapotranspiration in the Eastern United States. Physical Geography 10: 176–189.

McCarthy, E. J. and R. W. Skaggs. 1992. Simulation and evaluation of water management systems for a pine plantation watershed. Southern Journal of Applied Forestry 16: 48–56.

McComb, W. C. and R. C. Noble. 1980. Small mammal and bird use of some unmanaged and managed forest stands in the midsouth. Proceedings of the Annual Conference of the Southeastern Association of Fish and Wildlife Agencies 34: 482–491.

McCulley, R. D. 1950. Management of natural slash pine stands in the flatwoods of south Georgia and north Florida. U.S. Department of Agriculture, Washington, DC. Circular No. 845.

McDonald, M. G. and A. W. Harbaugh. 1988. A modular three-dimensional finite-difference ground-water flow model. U.S. Geological Survey, Reston, VA. U.S. Geologic Survey, Books and Open File Reports TWI 6-A1.

McDonnell, J. J. 1990. A rationale for old water discharge through macropores in a steep humid catchment. Water Resources Research 26: 2821–2832.

McGarigal, K. and W. C. McComb. 1994. Relationship between landscape structure and breeding birds in the Oregon Coastal Range. p. 35. *In* The Ecology and Management of Oregon Coast Range Forests. Speaker abstracts, Coastal Oregon Productivity Enhancement Symposium, March 29–31, Gleneden Beach, OR.

McGarity, R. W. 1977. Ten-year results of thinning and clearcutting in a muck swamp timber type. International Paper Company, Wood Products and Resources Group, Natchez Forest Research Center, Natchez, MS. Technical Note No. 38.

McGraw, J. B. 1987. Seed-bank properties of an Appalachian sphagnum bog and a model of the depth distribution of viable seeds. Canadian Journal of Botany 65: 2028–2035.

McIvor, C. C., J. J. Ley, and R. D. Bjork. 1994. Changes in freshwater inflow from the Everglades to Florida Bay including effects on biota and biotic processes: a review. p. 117–146. *In* S.M. Davis and J.C. Ogden (eds.) Everglades — The Ecosystem and Its Restoration. St. Lucie Press, Delray Beach, FL.

McKee, K. L. 1993. Soil physicochemical patterns and mangrove species distribution — reciprocal effects? Journal of Ecology 81: 477–487.

McKee, K. L. 1995a. Interspecific variation in growth, biomass partition, and defensive characteristics of neotropical mangrove seedlings: response to light and nutrient availability. American Journal of Botany 82: 299–307.

McKee, K. L. 1995b. Seedling recruitment patterns in a Belizean mangrove forest: effects of establishment ability and physico-chemical factors. Oecologia 101: 448–460.

McKee, K. L. and I. A. Mendelssohn. 1987. Root metabolism in the black mangrove [*Avicennia germinans* (L.) L]: response to hypoxia. Environmental and Experimental Botany 27: 147–156.

McKee, K. L., I. A. Mendelssohn, and D. M. Burdick. 1989. Effect of long-term flooding in root metabolic responses in five freshwater marsh plant species. Canadian Journal of Botany 67: 3446–3452.

McKee, K. L., I. A. Mendelssohn, and M. W. Hester. 1988. Reexamination of pore water sulfide concentrations and redox potentials near the aerial roots of *Rhizophora mangle* and *Avicennia germinans*. American Journal of Botany 75: 1352–1359.

McKee, W. H., Jr. 1996. Personal communication. U.S. Department of Agriculture Forest Service (retired), Charleston, SC.

McKee, W. H., Jr. and G. E. Hatchell. 1987. Pine growth improvement on logging sites with site preparation and fertilization. p. 411–418. *In D. R.* Phillips (ed.) Proceedings of the Fourth Biennial Southern Silvicultural Research Conference. U.S. Department of Agriculture Forest Service, Atlanta, GA. General Technical Report SE-42.

McKee, W. H., Jr. and M. R. McKevlin. 1993. Geochemical processes and nutrient uptake by plants in hydric soils. Environmental Toxicology and Chemistry 12: 2197–2207.

McKee, W. H., Jr. and L. P. Wilhite. 1986. Loblolly pine response to bedding and fertilization varies by drainage class on lower Atlantic coastal plain sites. Southern Journal of Applied Forestry 10: 16–21.

McKee, W. H., Jr., G. E. Hatchell, and A. E. Tiarks. 1985. Managing site damage from logging: a loblolly pine management guide. U.S. Department of Agriculture Forest Service, Southeastern Forest Experiment Station, Asheville, NC. General Technical Report SE-32.

McKee, W. H., Jr., D. D. Hook, D. S. DeBell, and J. L. Askew. 1984. Growth and nutrient status of loblolly pine seedlings in relation to flooding and phosphorus. Soil Science Society of America Journal 48: 1438–1442.

McKevlin, M. R. 1984. Nutritional and anatomical responses of *Pinus taeda* L. to root environment. M.S. Thesis. Clemson University, Clemson, SC.

McKevlin, M. R. 1992. Guide to regeneration of bottomland hardwoods. U.S. Department of Agriculture Forest Service, Southeastern Forest Experiment Station, Asheville, NC. General Technical Report SE-76.

McKevlin, M. R. and D. D. Hook. 1996. Morphological and physiological differences between rooted cuttings of Atlantic white cedar grown in continuously flooded or moist peat. p. 311–320. *In* M. B. Edwards (compiler) Proceedings of the Eighth Annual Southern Silvicultural Research Conference. U.S. Department of Agriculture Forest Service, Southern Research Station, Asheville, NC. General Technical Report SRS-1.

McKevlin, M. R., D. D. Hook, and W. H. McKee, Jr. 1995. Growth and nutrient use efficiency of water tupelo seedlings in flooded and well-drained soil. Tree Physiology 15: 753–758.

McKevlin, M. R., D. D. Hook, W. H. McKee, Jr., S. U. Wallace, and J. R. Woodruff. 1987a. Loblolly pine seedling root anatomy and iron accumulation as affected by soil waterlogging. Canadian Journal of Forest Research 17: 1257–1264.

McKevlin, M. R., D. D. Hook, W. H. McKee, Jr., S. U. Wallace, and J. R. Woodruff. 1987b. Phosphorus allocation in flooded loblolly pine seedlings in relation to iron uptake. Canadian Journal of Forest Research 17: 1572–1576.

McKnight, J. S. 1970. Planting cottonwood cuttings for timber production in the South. U.S. Department of Agriculture Forest Service, New Orleans, LA. Research Paper SO-60.

McKnight, J. S. and R. C. Biesterfeldt. 1968. Commercial cottonwood planting in the southern United States. Journal of Forestry 66: 670–675.

McKnight, J. S. and R. L. Johnson. 1975. Growing hardwoods in southern lowlands. Forest Farmer 34: 38–47.

McKnight, J. S. and R. L. Johnson. 1980. Hardwood management in southern bottomlands. Forest Farmer 39: 31–39.

McKnight, J. S., D. D. Hook, O. G. Langdon, and R. L. Johnson. 1981. Flood tolerance and related characteristics of trees of the bottomland forests of the southern United States. p. 26–69. *In* J. R. Clark and J. Benforado (eds.) Wetlands of Bottomland Hardwood Forests. Elsevier Scientific Publishing Company, Amsterdam, The Netherlands.

McKnight, T. R. 1996. Physical Geography: A Landscape Appreciation, 5th Edition. Prentice Hall, Upper Saddle River, NJ.

McLeod, K. W., M. R. Reed, and T. G. Ciravolo. 1996. Reforesting a damaged stream delta. Land and Water 40: 11–13.

McLeod, K. W., L. A. Donovan, N. J. Stumpff, and K. C. Sherrod. 1986. Biomass, photosynthesis, and water use efficiency of woody swamp species subjected to flooding and elevated water temperature. Tree Physiology 2: 341–346.

McManmon, M. and R. M. M. Crawford. 1971. A metabolic theory of flooding tolerance: the significance of enzyme distribution and behaviour. New Phytologist 70: 299–306.

McMillan, C. 1975. Adaptive differentiation to chilling in mangrove populations. p. 62–70. *In* G. E. Walsh, S. C. Snedaker, and H. J. Teas (eds.) Proceedings of the International Symposium on Biology and Management of Mangroves. Vol. I. Institute of Food and Agricultural Sciences, University of Florida, Gainesville, FL.

McMillan, C. and C. L. Sherrod. 1986. The chilling tolerance of black mangrove, *Avicennia germinans*, from the Gulf of Mexico coast of Texas, Louisiana and Florida. Contributions in Marine Science 29: 9–16.

McNab, W. H., R. W. Johansen, and W. B. Flanner. 1979. Cold winter and spring extend fire season in the pocosins. Fire Management Notes Fall 1979: 11–12.

McWilliams, W. H. and J. L. Faulkner. 1991. The bottomland hardwood timber resource of the coastal plain province in the South Central U.S.A. American Pulpwood Association Southcentral Technical Division, Washington, DC. Forest Management Commission 91-A-11.

McWilliams, W. H. and J. F. Rosson, Jr. 1990. Composition and vulnerability of bottomland hardwood forests of the Coastal Plain province in the south central United States. Forest Ecology and Management 33/34:485–501.

Meadows, J. S. 1993. Logging damage to residual trees following partial cutting in a green ash-sugarberry stand in the Mississippi Delta. p. 248–260. *In* Proceedings of the 9th Central Hardwood Forest Conference, Purdue University, West Lafayette, IN.

Meentemeyer, V. 1978. Macroclimate and lignin control of litter decomposition rates. Annals of the Association of American Geographers 74: 551–560.

Meentemeyer, V. and B. Berg. 1986. Regional variation in mass-loss of *Pinus sylvestris* needle litter in Swedish pine forests as influenced by climate and litter quality. Scandanavian Journal of Forest Research 1: 167–180.

Meffe, G. K. 1990. Offspring size variation in eastern mosquitofish (*Gambusia holbrooki*: Poeciliidae) from contrasting thermal environments. Copeia 1990: 10–18.

Meffe, G. K. and A. L. Sheldon. 1988. The influence of habitat structure on fish assemblage composition in southeastern blackwater streams. American Midland Naturalist 120: 225–240.

Megonigal, J. P. and F. P. Day. 1992. Effects of flooding on root and shoot production of bald cypress in large experimental enclosures. Ecology 73: 1182–1193.

Megonigal, J. P., W. H. Conner, S. Kroeger, and R. R. Sharitz. 1997. Aboveground production in southeastern floodplain forests: a test of the subsidy-stress hypothesis. Ecology 78: 370–384.

Melchiors, M. A. and C. Cicero. 1987. Streamside and adjacent upland forest habitats in the Ouachita Mountains. Proceedings of the Annual Conference of the Southeastern Association of Fish and Wildlife Agencies 41: 385–393.

Melick, D. R. 1990. Flood resistance of *Tristaniopsis laurina* and *Acmena smithii* from riparian warm temperate rainforest in Victoria. Australian Journal of Botany 38: 371–381.

Melillo, J. M., J. D. Aber, A. E. Linkins, A. Ricca, B. Fry, and K. J. Nadelhoffer. 1989. Carbon and nitrogen dynamics along the decay continuum: plant litter to soil organic matter. Plant and Soil 115: 189–198.

Menegus, L., F. Cattaruzza, A. Chersi, and G. Fronza. 1989. Differences in the anaerobic lactate-succinate productions and in the changes of cell sap pH for plants with high and low resistance to anoxia. Plant Physiology 90: 29–32.

Mendelssohn, I. A. and K. L. McKee. 1981. Determination of adenine nucleotide levels and adenylate energy charge ratio in two *Spartina* species. Aquatic Botany 11: 37–55.

Merendino, M. T. and L. M. Smith. 1991. Influence of drawdown date and reflood depth of wetland vegetation establishment. Wildlife Society Bulletin 19: 143–150.

Metraux, J. P. and H. Kende. 1983. The role of ethylene in the growth response of submerged deep water rice. Plant Physiology 72: 441–446.

Meyer, J. L. and R. T. Edwards. 1990. Ecosystem metabolism and turnover of organic carbon along a blackwater river continuum. Ecology 71: 668–677.

Middleton, B. A. 1994. Decomposition and litter production in a northern bald cypress swamp. Journal of Vegetation Science 5: 271–274.

Miller, K. V., R. L. Marchinton, and J. J. Ozoga. 1995. Deer sociobiology. p. 118–128. *In* K. V. Miller and R. L. Marchinton (eds.) Quality Whitetails: The Why and How of Quality Deer Management. Stackpole Books, Mechanicsburg, PA.

Miller, R. S. 1967. Pattern and process in competition. Advances in Ecological Research 4: 1–74.

Miller, S. P. 1996. Personal communication. Institute of Ecology, University of Georgia, Athens, GA.

Miller, W. D. and T. E. Maki. 1957. Planting pines in pocosins. Journal of Forestry 55: 659–663.

Mills, H. H. and P. A. Delcourt. 1991. Quaternary geology of the Appalachian Highlands and Interior Low Plateaus. p. 611–628. *In* R. B. Morrison (ed.) Quaternary Nonglacial Geology: Conterminous U.S. Geological Society of America, Boulder, CO.

Minkler, L. S. and R. E. McDermot. 1960. Pin oak acorn production and regeneration as affected by stand density, structure and flooding. Agricultural Experiment Station, University of Missouri. Research Bulletin 750.

Minkler, L. S. and D. Jones. 1965. Pin oak acorn production on normal and flooded areas. University of Missouri. Agricultural Experiment Station Research Bulletin 898.

Mitchell, L. J. 1989. Effects of clearcutting and reforestation on breeding bird communities of baldcypress-tupelo wetlands. M.S. Thesis. North Carolina State University, Raleigh, NC.

Mitchell, L. J. and R. A. Lancia. 1990. Breeding bird community changes in a bald cypress-tupelo wetland following timber harvesting. Proceedings of the Annual Conference of the Southeastern Association of Fish and Wildlife Agencies 44: 189–201.

Mitchell, L. J., R. A. Lancia, R. Lea, and S. A. Gauthreaux, Jr. 1991. Effects of clearcutting and natural regeneration on breeding bird communities of a baldcypress-tupelo wetland in South Carolina. p. 155–161. *In* J. Kusler and S. Daly (eds.) Proceedings of the Symposium on Wetlands and River Corridor Management. Association of Wetland Managers, Berne, NY.

Mitchell, M. S. 1994. Effects of intensive forest management on the mammal communities of selected North Carolina pocosin habitats. Management on the mammal communities of selected North Carolina pocosin habitats. National Council of the Paper Industry for Air and Stream Improvement, Inc., Research Triangle Park, NC. NCASI Technical Bulletin Number 665.

Mitchell, M. S., K. S. Karriker, E. L. Jones, and R. A. Lancia. 1995. Small mammal communities associated with pine plantation management of pocosins. Journal of Wildlife Management 59: 875–881.

Mitigation MOA. 1990. Memorandum of Agreement Between the U.S. Environmental Protection Agency and U.S. Department of the Army, Determination of Mitigation Under the Clean Water Act 404(b)(1) Guidelines. p. 331–336. *In* D. Wallenberg, D. and D. Adams (eds.) Wetlands Deskbook. 1993. Environmental Law Institute, Washington, DC.

Mitigation MOA. 1993. Memorandum of Agreement Between the U.S. Environmental Protection Agency and U.S. Department of the Army, Establishement and Use of Wetland Mitgation Banks in the Clean Water Section 404 Regulatory Program. Appen 9–73. In W. L. Want (ed.) 1996. Law of Wetlands Regulation. Clark, Broardman Callaghan, Deerfield, IL.

Mitsch, W. J. and K. C. Ewel. 1979. Comparative biomass and growth of cypress in Florida wetlands. American Midland Naturalist 101: 417–426.

Mitsch, W. J. and J. G. Gosselink. 1986. Wetlands. Van Nostrand Reinhold Co., New York, NY.

Mitsch, W. J. and J. G. Gosselink. 1993. Wetlands, 2nd Edition. Van Nostrand Reinhold Co., New York, NY.

Mitsch, W. J., C. L. Dorge, and J. R. Weimhoff. 1979a. Ecosystem dynamics and a phosphorus budget of an alluvial cypress swamp in southern Illinois. Ecology 60: 1116–1124.

Mitsch, W. J., M. D. Hutchinson, and G. A. Paulson. 1979b. The Momence Wetlands of the Kankakee River in Illinois — An Assessment of Their Value. Illinois Institute of Natural Resources, Chicago, IL. Doc 79/17.

Mitsch, W. J., J. R. Taylor, and K. B. Benson. 1991. Estimating primary productivity of forested wetland communities in different hydrologic landscapes. Landscape Ecology 5: 75–92.

Mladenoff, D. J., M. A. White, J. Pastor, and T. R. Crow. 1993. Comparing spatial pattern in unaltered old-growth and disturbed forest landscapes. Ecological Applications 3: 294–306.

Mocquot, B., C. Prat, C. Mouches, and A. Pradet. 1981. Effect of anoxia on energy charge and protein synthesis in rice embryo. Plant Physiology 68: 636–640.

Mohan, J., S. Benjamin, and R. Sutter. 1993. Site design for McClures Bog. The Nature Conservancy, Southeast Regional Office, Chapel Hill, NC.

Mohn, C. A., W. K. Randall, and J. S. McKnight. 1970. Fourteen cottonwood clones selected for midsouth timber production. U.S. Department of Agriculture Forest Service, New Orleans, LA. Research Paper SO-62.

Molz, F. J., R. H. Morin, A. E. Melville, and O. Güven. 1989. The impeller meter for measuring aquifer permeability variations: evaluation and comparison with other tests. Water Resources Research 25: 1677–1683.

Monk, C. D. 1966a. An ecological study of hardwood swamps in north-central Florida. Ecology 47: 649–654.

Monk, C. D. 1966b. The ecological significance of evergreenness. Ecology 47: 504–505.

Monk, L. S., R. Braendle, and R. M. M. Crawford. 1987. Catalase activity and post-anoxic injury in monocotyledonous species. Journal of Experimental Botany 38: 233–246.

Monscheim, T. D. 1981. Values of pocosins to game and fish species in North Carolina. p. 155–170. *In* C. J. Richardson (ed.) Pocosin Wetlands: An Integrated Analysis of Coastal Plain Freshwater Bogs in North Carolina. Hutchinson Ross Publication Co., Stroudsburg, PA.

Moon, M., M. R. Rattray, F. E. Putz, and G. Bowes. 1993. Acclimatization to flooding of the herbaceous vine, *Mikania scandens*. Functional Ecology 7: 610–615.

Moore, J. H. 1967. Andrew Brown and Cypress Lumbering in the Old Southwest. Louisiana State University Press, Baton Rouge, LA.

Moore, J. H. and J. H. Carter, III. 1987. Habitats of white cedar in North Carolina. p. 177–190. *In* A. D. Laderman (ed.) Atlantic White Cedar Wetlands. Westview Press, Boulder, CO.

Moore, P. A., Jr. and W. H. Patrick, Jr. 1988. Effect of zinc deficiency on alcohol dehydrogenase activity and nutrient uptake in rice. Agronomy Journal 80: 882–885.

Moore, P. D. and D. J. Bellamy. 1974. Peatlands. Springer–Verlag, New York, NY.

Moore, W. G. 1942. Field studies on the oxygen requirements of certain fresh-water fishes. Ecology 23: 319–329.

Moorhead, D. J., J. D. Hodges and K. J. Reinecke. 1991. Silvicultural options for waterfowl management in bottomland hardwood stands and greentree reservoirs. p. 710–721. *In* S. S. Coleman and D. G. Neary (eds.) Proceedings of the Sixth Biennial Southern Silvicultural Conference. U.S. Department of Agriculture Forest Service, Southeastern Forest Experiment Station, Asheville, NC. General Technical Report SE-70.

Moorhead, K. K. 1996. Unpublished data. Environmental Studies, University of North Carolina Asheville, Asheville, NC 28804.

Moorhead, K. K., I. M. Rossell, C. R. Rossell, Jr., and J. W. Petranka. 1995. Restoring wetlands for a mitigation bank for surface transportation projects in western North Carolina. Annual Progress Report to The Center for Transportation and the Environment, Raleigh, NC.

Morisset, C., P. Raymond, B. Mocquot, and A. Pradet. 1982. Plant adaptation to hypoxia and anoxia. Bulletin de la Societe Botanique de France 129: 73–91.

Morris, J. T. and J. W. H. Dacey. 1984. Effects of O_2 on ammonium uptake and root respiration by *Spartina alterniflora*. American Journal of Botany 71: 979–985.

Morris, L. A. and R. G. Campbell. 1991. Soil and site potential. p. 183–206. *In* M. L. Duryea, and P. M. Dougherty (eds.) Forest Regeneration Manual. Kluwer Academic Publishers, Amsterdam, The Netherlands.

Morris, R. C. 1975. Tree-eaters in the tupelo swamps. Forest and People 25: 22–24.

Morse, T. E., J. L. Jakabosky, V. P. McCrow. 1969. Some aspects of the breeding behavior of hooded mergansers. Journal of Wildlife Management 33: 596–604.

Mowbray, T. and W. H. Schlesinger. 1988. The buffer capacity of organic soils of the Bluff Mountain Fen, North Carolina. Soil Science 146: 73–79.

Mueller, T. L. 1995. Partners for Wildlife habitat conservation program. Land Trust Alliance Exchange Fall: 14–17.

Muench, D. G., O. W. Archibold, and A. G. Good. 1993. Hypoxic metabolism in wild rice (*Zizania palustris*): enzyme induction and metabolite production. Physiologia Plantarum 89: 165–171.

Mulholland, P. J. 1993. Hydrometric and stream chemistry evidence of three storm flowpaths in Walker Branch Watershed. Journal of Hydrology 151: 291–316.

Muller, R. A. 1975. A synoptic climatology for environmental baseline analysis: New Orleans. Journal of Applied Meteorology 16: 20–33.

Muller, R. A. and T. M. Oberlander. 1984. Physical Geography Today: A Portrait of a Planet, 3rd Edition. Random House, New York, NY.

Muller, R. A. and R. C. Thompson. 1987. Water budget analysis. p. 914–920. *In* J. E. Oliver and R. W. Fairbridge (eds.) The Encyclopedia of Climatology. Van Nostrand Reinhold Co., NY.

Muller, R. A. and N. L. Tucker. 1986. Climatic opportunities for the long-range migration of moths. p. 61–83. *In* A. N. Sparks (ed.) Long-Range Migration of Moths of Agronomic Importance to the United States and Canada: Specific Examples of Occurrence and Synoptic Weather Patterns Conducive to Migration. Agricultural Research Service — 43, National Technical Information Service, Springfield, VA.

Muller, R. A. and J. E. Willis. 1983. New Orleans weather 1961–1980: a climatology by means of synoptic weather types. School of Geoscience, Louisiana State University, Baton Rouge, LA. Miscellaneous Publication 83–1.

Mullin, K. D. 1982. Effects of selected silvicultural practices on swamp rabbit (*Sylvilagus aquaticus*) in disturbed bottomland hardwoods and associated pinewoods in west-central Louisiana. M.S. Thesis. Louisiana State University, Baton Rouge, LA.

Mullin, S. J. 1995. Estuarine fish populations among red mangrove prop roots of small overwash islands. Wetlands 15: 324–329.

Murdock, N. A. 1994. Rare and endangered plants and animals of southern Appalachian wetlands. Water, Air and Soil Pollution 77: 385–405.

Murdy, E. O. and J. W. E. Wortham, Jr. 1980. Contributions to the reproductive biology of the eastern pirateperch, *Aphredoderus sayanus*. Virginia Journal of Science 31: 20–27.

Murphy, C. H. 1995. Carolina Rocks! The Geology of South Carolina. Sandlapper Publishing Company, Orangeburg, SC.

Murray, N. L. 1992. Habitat use by nongame birds in central Appalachian riparian forests. M.S. Thesis. Virginia Polytechnic Institute and State University, Blacksburg, VA.

Murray, N. L. and D. F. Stauffer. 1995. Nongame bird use of habitat in central Appalachian riparian forests. Journal of Wildlife Management 59: 78–88.

Muzika, R. M., J. B. Gladden, and J. D. Haddock. 1987. Structural and functional aspects of succession in southeastern floodplain forests following a major disturbance. American Midland Naturalist 117: 1–9.

Myers, R. L. and J. J. Ewel. 1990. Ecosystems of Florida. University of Central Florida Press, Orlando, FL.

N

NACD. 1995. Wetlands Reserve Program, Restoring America's Wetlands Heritage. National Association of Conservation Districts. Pm-1482, revised March 1995.

National Climate Data Center. 1994. NCDC Summary of the Day. Earthinfo Inc., Boulder, CO. CD-ROM.

National Climatic Data Center. 1996. Billion dollar U.S. weather disasters 1980–1996. National Climatic Data Center Home Page, http://www.noaa.gov.

National Research Council. 1995. Wetlands: Characteristics and Boundaries. National Academy Press, Washington, DC.

Natural Resources Conservation Service. 1994. SCS Soils-5 Data Base. Earthinfo Inc., Boulder, CO. CD-ROM.

Navid, D. 1988. Developments under the Ramsar Convention. p. 21–27. *In* D.D. Hook et al. (eds.) The Ecology and Management of Wetlands. Vol. 2. Management, Use, and Value of Wetlands. Timber Press, Portland, OR.

NCASI. 1994. Southern regional review of state nonpoint source control programs and best management practices for forest management operations. National Council of the Paper Industry for Air and Stream Improvement, Inc., New York, NY. Technical Bulletin No. 686.

NCCES. 1995. Woodland owner notes: restoration of wetlands under the Wetlands Reserve Program. North Carolina Cooperative Extension Service, Raleigh, NC.

Neal, C., M. Neal, A. Warrington, A. Avila, J. Pinol, and F. Roda. 1991. Stable hydrogen and oxygen isotope studies of rainfall and streamwaters for two contrasting holm oak areas of Catalonia, northeastern Spain. Journal of Hydrology 140: 163–178.

Nedwell, D. B. 1975. Inorganic nitrogen metabolism in a eutrophicated tropical mangrove estuary. Water Research 9: 221–231.

Negus, M. T., J. M. Aho, and C. S. Anderson. 1987. Influences of acclimation temperature and developmental stage on behavioral responses of lake chubsuckers to temperature gradients. American Fisheries Society Symposium 2: 157–163.

Neill, W. T. 1950. An estivating bowfin. Copeia 1950: 240.

Neill, W. T. 1951. Notes on the role of crawfishes in the ecology of reptiles, amphibians, and fishes. Ecology 32: 764–766.

Neill, W. T. 1957. Historical biogeography of present-day Florida. Bulletin of the Florida State Museum 2: 175–220.

Nelson, J. B. 1986. The natural communities of South Carolina: initial classification and description. South Carolina Wildlife and Marine Resources Department, Division of Wildlife and Freshwater Fisheries, Columbia, SC.

Nelson, L. E., G. L. Switzer, and B. G. Lockaby. 1987. Nutrition of *Populus deltoides* plantations during maximum production. Forest Ecology and Management 20: 25–41.

Nelson, T. C. and W. M. Zillgitt. 1969. A forest atlas of the south. U.S. Department of Agriculture Forest Service, Southern Forest Experiment Station, New Orleans, LA.

Nessel, J. K. 1978a. Phosphorus cycling, productivity and community structure in the Waldo cypress strand. p. 750–801. *In* H. T. Odum and K. C. Ewel (eds.) Cypress Wetlands for Water Management, Recycling, and Conservation, Center for Wetlands, University of Florida, Gainesville, FL. 4th Annual Report.

Nessel, J. K. 1978b. Distribution and dynamics of organic matter and phosphorus in a sewage enriched cypress strand. M.S. Thesis. University of Florida, Gainesville, FL.

Nessel, J. K. and S. E. Bayley. 1984. Distribution and dynamics of organic matter and phosphorus in a sewage-enriched cypress swamp. p. 262–278. *In* K. C. Ewel and H. T. Odum (eds.) Cypress Swamps. University Presses of Florida, Gainesville, FL.

Nessel, J. K., K. C. Ewel, and M. S. Burnett. 1982. Wastewater enrichment increases mature pondcypress growth rates. Forest Science 28: 400–403.

Netsch, N. F. and A. Witt, Jr. 1961. Contributions to the life history of the longnose gar (*Lepisosteus osseus*) in Missouri. Transactions of the American Fisheries Society 91: 175–262.

Neubreck, W. L. 1939. American southern cypress. U.S. Department of Commerce, Washington, DC. Trade Promotion Series Number 194.

Neufeld, H. S. 1983. Effects of light on growth, morphology, and photosynthesis in baldcypress (*Taxodium distichum* (L.) Rich) and pondcypress (*T. ascendens* Brongn.) seedlings. Bulletin of the Torrey Botanical Club 110: 43–54.

Newling, C. J. 1981. Ecological investigation of a greentree reservoir in the Delta National Forest, Mississippi. U.S. Army Engineer Waterways Experiment Station, Vicksburg, MS. Miscellaneous Paper EL-81-5.

Newman, K. D. and T. T. Van Toai. 1991. Developmental regulation and organ-specific expression of soybean alcohol dehydrogenase. Crop Science 31: 1253–1257.

Newman, M. C. and J. F. Schalles. 1990. The water chemistry of Carolina bays: a regional study. Archiv fur Hydrobiologia 118: 147–168.

Nickerson, N. H. and F. R. Thibodeau. 1985. Association between pore water sulfide concentrations and the distribution of mangroves. Biogeochemistry 1: 183–192.

Nilsson, C. and G. Grelsson. 1990. The effects of litter displacement on riverbank vegetation. Canadian Journal of Botany 68: 735–741.

Nilsson, K. 1973. *Jordanella floridae*, the American flagfish. Tropical Fish Hobby 21: 5–15, 91–93.

Nixon, S. W. 1980. Between coastal marshes and coastal waters—a review of twenty years of speculation and research on the role of salt marshes in estuarine productivity and water chemistry. p. 437–525. *In* P. Hamilton and K. B. MacDonald (eds.) Estuarine and Wetland Processes With Emphasis on Modeling. Plenum Press, New York, NY.

NOAA. 1994. Climatological data annual summary, North Carolina. Volume 99, Number 13.

Noss, R. F. 1987. Corridors in real landscapes: a reply to Simberloff and Cox. Conservation Biology 1: 159–164.

Novak, M. and R. K. Wieder. 1992. Inorganic and organic sulfur profiles in nine *Sphagnum* peat bogs in the United States and Czechoslovakia. Water, Air, and Soil Pollution 65: 353–369.

Norgress, R. E. 1947. The history of the cypress lumber industry in Louisiana. The Louisiana Historical Quarterly 30(3): 46–61.

NRCS. 1995a. Field Indicators of Hydric Soils in the United States, Version 2.1. U.S. Department of Agriculture, Natural Resources Conservation Service, Washington, DC.

NRCS. 1995b. Wetlands and Agriculture: Section 404 of the Clean Water Act Swampbuster in the Food Security Act. Program Aid 1546. U.S. Department of Agriculture, Washington, DC.

Nurnberg, G. K. 1995. The anoxia factor, a quantitative measure of anoxia and fish species richness in central Ontario lakes. Transactions of the American Fisheries Society 124: 677–686.

Nusser, S. M. and J. J. Goebel. In press. The National Resources Inventory: A long-term multi-resource monitoring program. Environmental and Ecological Statistics.

Nutter, W. L. and M. M. Brinson. 1994. Application of hydrogeomorphic principles and function to assessment of drainage intensity in forested wetlands. p. 13–26. *In* Proceedings of the Workshop on Water Management in Forested Wetlands. U.S. Department of Agriculture Forest Service, Atlanta, GA. Technical Bulletin R8-TP 20.

Nyland, R. D. 1996. Silviculture Concepts and Applications. The McGraw-Hill Co., Inc., New York, NY.

O

Oats, R. Q., Jr. and N. K. Coch. 1973. Post-Miocene Stratigraphy and Morphology, Southeastern Virginia. Virginia Division of Mineral Resources, Richmond, VA. Bulletin 82.

O'Connel, P. E. 1982. Rainfall network design — a review. p. 18–50. *In* V. P. Singh (ed.) Modeling Components of the Hydrologic Cycle. Water Resources Publications, Littleton, CO.

Odum, E. P. 1978. The value of wetlands: a hierarchical approach. p. 16–25. *In* P. E. Greeson, J. R. Clark, and J. E. Clark (eds.) Wetland Functions and Values: the State of Our Understanding. American Water Resource Association, Minneapolis, MN.

Odum, E. P. 1989. Wetland values in retrospect. p. 1–9. *In* R. R. Sharitz and J. W. Gibbons (eds.) Freshwater Wetlands and Wildlife Symposium: Perspectives on Natural, Managed and Degraded Ecosystems. U.S. Department of Energy, Office of Scientific and Technical Information, Oak Ridge, TN. CONF-8603101, DOE Symposium Series No. 61.

Odum, H. T. 1971. Environment, Power, and Society. Wiley Inter-Science, New York, NY.

Odum, H. T. and D. K. Caldwell. 1955. Fish respiration in the natural oxygen gradient of an anaerobic spring in Florida. Copeia 1955: 104–106.

Odum, W. E. and E. J. Heald. 1972. Trophic analysis of an estuarine mangrove community. Bulletin Marine Science 22: 671–738.

Odum, W. E. and C. C. McIvor. 1990. Mangroves. p. 517–548. *In* R. L. Myers and J. J. Ewel (eds.) Ecosystems of Florida, University of Central Florida Press, Orlando, FL.

Odum, W. E., C. C. McIvor, and T. J. Smith, III. 1982. The ecology of the mangroves of south Florida: a community profile. U.S. Fish and Wildlife Service, Office of Biological Services, Washington, DC. FWS/OBS-81-24.

Ogden, J. C. 1994. A comparison of wading bird nesting colony dynamics (1931–1946 and 1974–1089) as an indication of ecosystem conditions in the southern Everglades. p. 533–570. *In* S. M. Davis and J. C. Ogden (eds.) Everglades — The Ecosystem and Its Restoration. St. Lucie Press, Delray Beach, FL.

Ogle, F. B. 1993. The ongoing struggle between private property rights and wetlands regulation: recent developments and proposed solutions. University of Colorado Law Review 64: 573–606.

Ogle, D. W. 1982. Glades of the Blue Ridge in Southwestern Virginia. p.143–147. *In* B. R. McDonald (ed.) Proceedings of the Symposium on Wetlands of the Unglaciated Appalachian Region. West Virginia University, Morganton, WV.

Oke, T. 1987. Boundary Layer Climates, 2nd Edition. Methuen, London and NY.

Olmsted, I., H. Dunevitz, and W. J. Platt. 1993. Effects of freezes on tropical trees in Everglades National Park, Florida, U.S.A. Tropical Ecology 34: 17–34.

Olszewski, R. J. 1988. Forestland drainage and regulation in the southeastern coastal plain. p. 59–62. *In D. D.* Hook and R. Lea (eds.) Proceedings of the Symposium: The Forested Wetlands of The Southern United States. U.S. Department of Agriculture Forest Service, Southeastern Forest Experiment Station, Asheville, NC. General Technical Report SE-50.

O'Meara, T. E. 1984. Habitat-island effects on the avian community in cypress ponds. Proceedings of the Annual Conference of the Southeastern Association of Fish and Wildlife Agencies 38: 97–110.

Onuf, C., J. Teal, and I. Valiela. 1977. The interactions of nutrients, plant growth, and herbivory in a mangrove ecosystem. Ecology 58: 514–526.

Orzell, S. and B. Pell. 1985. Notes on three palustrine natural community types in the Arkansas Ozarks. Arkansas Academy of Science Proceedings 39: 141–143.

Osonubi, O. and M. A. Osundina. 1989. Adventitious roots, leaf abscission and nutrient status of flooded *Gmelina* and *Tectona* seedlings. Tree Physiology 5: 473–483.

Osundina, M. A. and O. Osonubi. 1987. Comparison of the responses to flooding of seedlings and cuttings of *Gmelina*. Tree Physiology 3: 147–156.

Otte, L. J. 1981. Origin, development and maintenance of pocosin wetlands of North Carolina. North Carolina Natural Heritage Program, North Carolina Department of Natural Resources and Community Development, Raleigh, NC.

Owens, J. N. and M. D. Blake. 1985. Forest tree seed production. Petawawa National Forestry Institute, Chalk River, Ontario, Canada. Information Report PI-X-53.

P

Palik, B. J. and K. S. Pregitzer. 1992. A comparison of presettlement and present-day forests on two bigtooth aspen-dominated landscapes in northern lower Michigan. American Midland Naturalist 127: 327–338.

Paller, M. H. 1987. Distribution of larval fish between macrophyte beds and open channels in a southeastern floodplain swamp. Journal of Freshwater Ecology 4: 191–200.

Palmer, J. H. 1996. Personal communication. Department of Agronomy and Soils, Clemson University, Clemson, SC.

Palmer, R. S. 1961. Handbook of North American Birds, Volume 1. Yale University Press, New Haven, CT.

Palmer, W. E., G. A. Hurst, B. D. Leopold, and D. C. Cotton. 1991. Body weights and sex and age ratios for the swamp rabbit in Mississippi. Journal of Mammalogy 72: 620–622.

Pardue, G. B. 1993. Life history and ecology of the mud sunfish (*Acantharcus pomotis*). Copeia 1993: 533–540.

Parfitt, R. L. and R. St. C. Smart. 1978. The mechanism of sulfate adsorption on iron oxides. Soil Science Society of America Journal 42: 48–50.

Parker, N. C. and B. A. Simco. 1975. Activity patterns, feeding, and behavior of the pirate-perch, *Aphredoderus sayanus*. Copeia 1975: 372–374.

Parkinson, R. W., R. D. DeLaune, and J. R. White. 1994. Holocene sea-level rise and the fate of mangrove forests within the wider caribbean region. Journal of Coastal Research 10: 1077–1086.

Parsons, J. E., R. W. Skaggs, and C. W. Doty. 1991. Development and testing of a water management model (WATRCOM): Development. Transactions of the American Society of Agricultural Engineers 34: 120–128.

Paton, P. W. C. 1994. The effect of edge on avian nest success: how strong is the evidence. Conservation Biology 8: 17–26.

Patrick, R., J. Cairns, Jr., and S. S. Roback. 1967. An ecosystematic study of the fauna and flora of the Savannah River. Proceedings of the Academy of Natural Science of Philadelphia 118: 109–407.

Patrick, W. H., Jr. 1981. Bottomland soils. p. 177–185. *In* J. R. Clark and J. Benforado (eds.) Wetlands of Bottomland Hardwood Forests. Elsevier Scientific Publishing Company, Amsterdam, The Netherlands.

Patrick, W. H., Jr., G. Dissmeyer, D. D. Hook, V. W. Lambou, H. M. Leitman, and C. H. Wharton. 1981. Characteristics of wetlands ecosystems of southeastern bottomland hardwood forests. p. 275–300. *In* J. R. Clark and J. Benforado (eds.) Wetlands of Bottomland Hardwood Forests. Elsevier Scientific Publishing Company, Amsterdam, The Netherlands.

Patterson, C. S. and I. A. Mendelssohn. 1991. A comparison of physiochemical variables across plant zones in a mangal/salt marsh community in Louisiana. Wetlands 11: 139–161.

Patterson, C. S., I. A. Mendelssohn, and E. M. Swenson. 1993. Growth and survival of *Avicennia germinans* seedlings in a mangal/salt marsh community in Louisiana, U.S.A. Journal of Coastal Research 9: 801–810.

Patterson, G. G., G. K. Speiren, and B. H. Whetstone. 1985. Hydrology and its effects on distribution of vegetation in the Congaree Swamp National Monument, South Carolina. U.S. Geologic Survey, Columbia, SC. Water Resources Investigation Report 854256.

Patterson, S. G. 1986. Mangrove community boundary interpretation and detection of areal changes on Marco Island Florida: application of digital image processing and remote sensing techniques. U.S. Department of the Interior, National Wetlands Research Center, Washington, DC. Biological Report 86(10).

Pearsall, S. 1996. Personal communication. The Nature Conservancy, Wilmington, NC.

Pearson, J. and D. C. Havill. 1988. The effect of hypoxia and sulphide on culture-grown wetland and non-wetland plants. Journal of Experimental Botany 39(201): 431–439.

Pearson, S. M. 1994. Landscape-level processes and wetland conservation in the southern Appalachian mountains. Water, Air and Soil Pollution 77: 321–332.

Pedrazzini, F. R. and K. L. McKee. 1984. Effect of flooding on activities of soil dehydrogenases and alcohol dehydrogenase in rice roots. Soil Science and Plant Nutrition 30: 359–366.

Peltier, W. R. 1988. Global sea level and earth rotation. Science 240: 895–901.

Pelton, M. R. 1996. Personal communication. Department of Forestry, Wildlife and Fisheries, University of Tennessee, Knoxville, TN.

Penfound, W. T. 1949. Vegetation of Lake Chicot, Louisiana in relation to wildlife resources. Proceedings of the Louisiana Academy of Science 12: 47–56.

Penfound, W. T. 1952. Southern swamps and marshes. Botanical Review 18: 413–446.

Penland, S. and K. Ramsey. 1990. Relative sea-level rise in Louisiana and the Gulf of Mexico: 1908–1988. Journal of Coastal Research 6: 323–342.

Penn, G. H. 1950. Utilization of crawfishes by cold-blooded vertebrates in the eastern United States. American Midland Naturalist 44: 643–658.

Perata, P., A. Alpi, and F. LoSchiavo. 1986. Influence of ethanol on plant cells and tissues. Journal of Plant Physiology 126: 181–188.

Peschke, V. M. and M. M. Sachs. 1994. Characterization and expression of transcripts induced by oxygen deprivation in maize (*Zea mays* L.). Plant Physiology 104: 387–394.

Peterjohn, B. G., and J. R. Sauer. 1993. North American Breeding Bird Survey annual summary, 1990–1991. Bird Populations. 1: 1–15.

Peterjohn, W. T. and D. L. Correll. 1984. Nutrient dynamics in an agricultural watershed: observations on the role of a riparian forest. Ecology 65: 1466–1475.

Peters, M. A. and E. Holcombe. 1951. Bottomland cypress planting recommended for flooded areas by Soil Conservationists. Forest and People 1: 18, 32–33.

Peterson, B. J. and B. Fry. 1987. Stable isotopes in ecosystem studies. Annual Review of Ecology and Systematics 18: 293–320.

Peterson, D. L. and G. L. Rolfe. 1982. Nutrient dynamics and decomposition of litterfall in floodplain and upland forests of central Illinois. Forest Science 28: 667–681.

Peterson, M. S., R. E. Brockmeyer, and D. M. Scheidt. 1991. Hypoxia-induced changes in vertical position and activity in Juvenile snook, *Centropomus undecimalis*: its potential role in survival. Florida Scientist 54: 173–178.

Petranka, J. W., M. E. Eldridge, and K. E. Haley. 1993. Effects of timber harvesting on southern Appalachian salamanders. Conservation Biology 7: 363–370.

Pettry, D.E. 1996. Personal communication. Mississippi State University, Starkville, MS.

Pezeshki, S. R. 1991. Root responses of flood-tolerant and flood-sensitive tree species to soil redox conditions. Trees 5: 180–186.

Pezeshki, S. R. 1993. Differences in patterns of photosynthetic responses to hypoxia in flood-tolerant and flood-sensitive tree species. Photosynthetica 28: 423–430.

Pezeshki, S. R. and J. L. Chambers. 1985a. Responses of cherrybark oak seedlings to short-term flooding. Forest Science 31: 760–771.

Pezeshki, S. R. and J. L. Chambers. 1985b. Stomatal and photosynthetic response of sweet gum (*Liquidambar styraciflua*) to flooding. Canadian Journal of Forest Research 15: 371–375.

Pezeshki, S. R. and J. L. Chambers. 1986. Variation in flood-induced stomatal and photosynthetic responses of three bottomland tree species. Forest Science 32: 914–923.

Pezeshki, S. R., R. D. DeLaune, and W. H. Patrick, Jr. 1986 . Gas exchange characteristics of bald cypress (*Taxodium distichum* L.): evaluation responses to leaf aging, flooding, and salinity. Canadian Journal of Forest Research 16: 1394–1397.

Pezeshki, S. R., R. D. DeLaune, and W. H. Patrick, Jr. 1990. Differential response of selected mangroves to soil flooding and salinity: gas exchange and biomass partitioning. Canadian Journal of Forest Research 20: 869–874.

Pezeshki, S. R., J. H. Pardue and R. D. DeLaune. 1993. The influence of soil oxygen deficiency on alcohol dehydrogenase activity, root porosity, ethylene production, and photosynthesis in *Spartina patens*. Environmental and Experimental Botany 33: 565–573.

Pezeshki, S .R., J. H. Pardue, and R. D. DeLaune. 1996. Leaf gas exchange and growth of flood-tolerant and flood-sensitive tree species under low soil redox conditions. Tree Physiology 16: 453–458.

Pflieger, W. L. 1975. The fishes of Missouri. Missouri Department of Conservation, Jefferson City, MO.

Phelps, J. P. and R. A. Lancia. 1995. Effects of a clearcut on the herpetofauna of a South Carolina bottomland swamp. Brimleyana 22: 31–45.

Philipson, J. J. and M. P. Coutts. 1978. The tolerance of tree roots to waterlogging. III. Oxygen transport in lodgepole pine and Sitka spruce roots of primary structure. New Phytologist 80: 341–349.

Philipson, J. J. and M. P. Coutts. 1980. The tolerance of tree roots to waterlogging. IV. Oxygen transport in woody roots of Sitka spruce and lodgepole pine. New Phytologist 85: 489–494.

Phillips, J. D. 1992. The source of alluvium in large rivers of the Lower Coastal Plain of North Carolina. Catena 19: 59–75.

Phillips, R., W. E. Gardner, and K. O. Summerville. 1993. Plantability of Atlantic white-cedar rooted cuttings and bare-root seedlings. p. 97–104. *In* J. C. Brissette (ed.) Proceedings of the Seventh Biennial Southern Silvicultural Research Conference. U.S. Department of Agriculture Forest Service, Southern Forest Experiment Station, New Orleans, LA. General Technical Report SO-93.

Pinchot, G. and W. W. Ashe. 1897. Timber trees and forests of North Carolina. M.I. and J.C. Stewart, Public Printers, Winston, NC. North Carolina Geological Survey Bulletin Number 6.

Pittillo, J. D. 1994. Vegetation of three high-elevation southern Appalachian bogs and implications of their vegetational history. Water, Air and Soil Pollution 77: 333–348.

Pittman, K., R. C. Kellison, and R. Lea. 1993. Hurricane Hugo damage assessment of bottomland hadwoods in South Carolina. Hardwood Research Cooperative Project Report (Unpublished). North Carolina State University, Raleigh, NC.

Pitts, J. J. 1980. Soil Survey of Marion County South Carolina. U.S. Department of Agriculture, Soil Conservation Service, Marion, SC.

Poiani, K. A. and P. M. Dixon. 1995. Seed banks of Carolina bays: potential contributions from surrounding landscape vegetation. American Midland Naturalist 134: 140–154.

Ponnamperuma, F. N. 1972. The chemistry of submerged soils. Advances in Agronomy 24: 29–96.

Ponnamperuma, F. N. 1984a. Effects of flooding on soils. p. 9–45. *In* T. T. Kozlowski (ed.) Flooding and Plant Growth. Academic Press, Inc. Harcourt Brace Jovanovich, Publishers. New York, NY.

Ponnamperuma, F. N. 1984b. Mangrove swamps in south and southeast Asia as potential rice lands. p. 672–683. *In* E. Soepadmo, A.N. Rao, and D.J. McIntosh (eds.) Proceedings Asian Mangrove Symposium. University of Malaya, Kuala Lumpur.

Pool, D. J., A. E. Lugo, and S. C. Snedaker. 1975. Litter production in mangrove forests of southern Florida and Puerto Rico. p. 213–237. *In* G. Walsh, S. Snedaker, and H. Teas (eds.) Proceedings of the International Symposium on the Biology and Management of Mangroves. Institute of Food and Agricultural Sciences. University of Florida, Gainesville, FL.

Pool, D. J., S. C. Snedaker, and A. E. Lugo. 1977. Structure of mangrove forests in Florida, Puerto Rico, Mexico, and Costa Rica. Biotropica 9: 195–212.

Porcher, R. D. 1995. Wildflowers of the Lowcountry and Lower Pee Dee. University of South Carolina Press, Columbia, SC.

Portwood, J. 1996. Personal communication. Crown Vantage Corporation, Vicksburg, MS.

Post, H. A. and A. De La Cruz. 1977. Litterfall, litter decomposition, and flux of particulate organic material in a coastal plain stream. Hydrobiologia 55: 201–207.

Potter, E. 1982. A preliminary assessment of the wintering bird fauna of the First Colony Farm lands in Dare County, North Carolina. U.S. Fish and Wildlife Service, Division of Ecological Services, Raleigh, NC.

Potts, M. 1979. Nitrogen fixation (acetylene reduction) associated with communities of heterocystous and non-heterocystous blue-green algae on mangrove forests of Sinai. Oecologia 39: 359–373.

Pough, F. H., E. M. Smith, D. H. Rhodes, and A. Collazo. 1987. The abundance of salamanders in forest stands with different histories of disturbance. Forest Ecology and Management 20: 1–9.

Powell, S. W. and F. P. Day. 1991. Root production in four communities in the Great Dismal Swamp. American Journal of Botany 78: 288–297.

Power, M. E., A. Sun, G. Parker, W. E. Dietrich, and J. T. Wooton. 1995. Hydraulic food-chain models. BioScience 45: 159–168.

Pradet, A. and P. Raymond. 1983. Adenine nucleotide ratios. Annual Review of Plant Physiology 34: 199–224.

Prenger, R. S., Jr. 1985. Response of a second-growth natural stand of baldcypress trees [*Taxodium distichum* (L.) Rich.] to various intensities of thinning. M.S. Thesis. Louisiana State University, Baton Rouge, LA.

Preston, E. and B. L. Bedford. 1988. Evaluating cumulative effects on wetland functions: a conceptual overview and generic framework. Environmental Management 12: 565–583.

Priestley, C. H. B. and R. J. Taylor. 1972. On assessment of surface heat flux and evaporation using large-scale parameters. Monthly Weather Review 100: 81–92.

Pritchett, W. L. and N.B. Comerford. 1982. Long-term response to phosphorous fertilization on selected southeastern coastal plain soils. Soil Science Society of America Journal 46: 640– 644.

Pritchett, W. L. and R. F. Fisher. 1987. Properties and Management of Forest Soils, 2nd Edition. Wiley & Sons, New York, NY.

Prophit, W. 1982. The swamp's silent sentinel — a history of Louisiana cypress logging. Forests and People 33: 6–8, 32, 35.

Prouty, W. F. 1952. Carolina bays and their origin. Bulletin of the Geological Society of America 63: 167–224.
Pulliam, H. R. 1988. Sources, sinks, and population regulation. American Naturalist 132(5): 652–661.
Pulliam, W. M. 1993. Carbon dioxide and methane exports from a southeastern floodplain swamp. Ecological Monographs 63: 29–53.
Putnam, J. A. 1951. Management of bottomland hardwoods. U.S. Department of Agriculture Forest Service, Southern Forest Experiment Station, New Orleans, LA. Occasional Paper 116.
Putnam, J. A. and H. Bull. 1932. The trees of the bottomlands of the Mississippi River Delta region. U.S. Department of Agriculture Forest Service, Southern Forest Experiment Station, New Orleans, LA. Occasional Paper 27.
Putnam, J. A., G. M. Furnival, and J. S. McKnight. 1960. Management and inventory of southern hardwoods. U.S. Department of Agriculture Forest Service, Washington, DC. Agriculture Handbook 181.
Putz, F. E. and R. R. Sharitz. 1991. Hurricane damage to old-growth forest in Congaree Swamp National Monument, South Carolina, U.S.A. Canadian Journal of Forest Research 21: 1765–1770.
Pyron, M. and C. M. Taylor. 1993. Fish community structure of Oklahoma Gulf coastal plains. Hydrobiologia 257: 29–35.

Q

Qualls, R. G. 1984. The role of leaf litter nitrogen immobilization in the nitrogen budget of a swamp stream. Journal of Environmental Quality 13: 640–644.
Quinn, J. R. 1990. Our Native Fishes. Countryman Press, Woodstock, VT.

R

Rabinowitz, D. 1978. Early growth of mangrove seedlings in Panama, and an hypothesis concerning the relationship of dispersal and zonation. Journal of Biogeography 5: 113–133.
Radford, A. E., H. E. Ahles, and C. R. Bell. 1968. Manual of the Vascular Flora of the Carolinas. The University of North Carolina Press, Chapel Hill, NC.
Radulovich, R., P. Sollins, P. Baveye, and E. Solorzano. 1992. Bypass water flow through unsaturated microaggregated tropical soils. Soil Science Society of America Journal 56: 712–726.
Ralston, C. W. 1965. Forest drainage. p. 298–319. *In* A Guide to Loblolly and Slash Pine Plantation Management In Southeastern U.S.A. Georgia Forestry Research Council Report 14.
Ralston, C. W. and D. D. Richter. 1980. Identification of lower Coastal Plain sites of low soil fertility. Southern Journal of Applied Forestry 2: 84–88.
Raskin, I. and H. Kende. 1983. How does deep water rice solve its aeration problem. Plant Physiology 72: 447–454.
Raskin, I. and H. Kende. 1984. Effect of submergence on translocation, starch content, and amolytic activity in deep water rice. Planta 162: 556–559.
Rathborne, J. C. 1951. Cypress reforestation. Southern Lumberman 183(2297): 239–240.
Reberg-Horton, C. 1994. Mountain bog management. Honors Thesis. University of North Carolina, Chapel Hill, NC.

Reed, P. B., Jr. 1988. National list of plant species that occur in wetlands: 1988 national summary. U.S. Fish and Wildlife Service, Washington, DC. Biological Report 88(24).

Reed, T. M. 1983. The role of species-area relationships in reserve choice. Biological Conservation 25: 263–271.

Regier, S. 1996. Personal communication. North Carolina Division of Parks and Recreation, Raleigh, NC.

Rehfield, K. R., J. M. Boggs, and L. W. Gelhar. 1992. Field study of dispersion in a heterogeneous aquifer. 3. Geostatistical analysis of hydraulic conductivity. Water Resources Research 28: 3309–3324.

Renfro, J. L. and L. G. Hill. 1970. Factors influencing the aerial breathing and metabolism of gars (*Lepisosteus*). Southwestern Naturalist 15: 45–54.

Rey, J. R., J. Shaffer, D. Tremain, R. A. Crossman, and T. Kain. 1990. Effects of re-establishing tidal connections in two impounded subtropical marshes on fishes and physical conditions. Wetlands 10: 27–45.

Reynolds, W. W. and M. E. Casterlin. 1978. Ontogenetic change in preferred temperature and diel activity of the yellow bullhead, *Ictalurus natalis*. Comparative Biochemistry and Physiology 59A: 409–411.

Reynolds, W. W., M. E. Casterlin, and S. T. Millington. 1978. Circadian rhythm of preferred temperature in the bowfin *Amia calva*, a primitive holostean fish. Comparative Biochemistry and Physiology 60A: 107–109.

Rheinhardt, R. D. and C. Herschner. 1992. The relationship of below-ground hydrology to canopy composition in five tidal freshwater swamps. Wetlands 12: 208–216.

Rice, D. L. 1982. The detritus nitrogen problem: new observations and perspectives from organic geochemistry. Marine Ecology Progress Series 9: 153–162.

Rice, D. L. and K .R. Tenore. 1981. Dynamics of carbon and nitrogen during the decomposition of detritus derived from estuarine macrophytes. Estuarine, Coastal and Shelf Science 13: 681–690.

Rice, M., B. G. Lockaby, J. A. Stanturf, and B. J. Keeland. In press. Influence of hurricane disturbance on coarse woody debris decomposition in the Atchafalaya River Basin. Soil Science Society of America Journal.

Richardson, C. J. 1983. Pocosins: vanishing wastelands or valuable wetlands? BioScience 33: 626–633.

Richardson, C. J. 1985. Mechanisms controlling phosphorus retention capacity in freshwater wetlands. Science 228: 1424–1427.

Richardson, C. J. 1989. Freshwater wetlands: transformers, filters, or sinks? p. 25–46. *In R. R.* Sharitz and J. W. Gibbons (eds.) Freshwater Wetlands and Wildlife Symposium: Perspectives on Natural, Managed and Degraded Ecosystems. U.S. Department of Energy, Office of Scientific and Technical Information, Oak Ridge, TN. CONF-8603101, DOE Symposium Series No. 61.

Richardson, C. J. 1994. Ecological functions and human values in wetlands: a framework for assessing forestry impacts. Wetlands 14: 1–9.

Richardson, C. J., and J. W. Gibbons. 1993. Pocosins, Carolina bays, and mountain bogs. p. 257–310. *In* W. H. Martin, S. G. Boyce, and A. C. Esternacht (eds.) Biodiversity of the Southeastern United States: Lowland Terrestrial Communities. John Wiley and Sons, Inc., New York, NY.

Richardson, C. J. and E. J. McCarthy. 1994. Effect of land development and forest management on hydrologic response in Southeastern coastal wetlands. Wetlands 14: 56–71.

Richardson, C. J., R. Evans, and D. Carr. 1981. Pocosins: an ecosystem in transition. p. 3–19. *In* C. J. Richardson (ed.) Pocosin Wetlands: An Integrated Analysis of Coastal Plain Freshwater Bogs in North Carolina. Hutchinson Ross Publication Co., Stroudsburg, PA.

Richardson, C. J., M. R. Walbridge, and A. Burns. 1988. Soil chemistry and phosphorus retention capacity of North Carolina coastal plain swamps receiving sewage effluent. Water Resources Research Institute of the University of North Carolina, Raleigh, NC. UNC-WRRI-88-241.

Richardson, C. J., D. L. Tilton, J. A. Kadlec, J. P. M. Chamie, and W. A. Wentz. 1978. Nutrient dynamics of northern wetland ecosystems. p. 217–241. *In* R. E. Good, D. F. Whigham, and R. L. Simpson (eds.) Freshwater Wetlands: Ecological Processes and Management Potential. Academic Press, New York, NY.

Richter, D. D., K. Korfmacher, and R. Nau. 1995. Decreases in Yadkin River basin sedimentation: statistical and geographic time-trend analysis, 1951 to 1990. Water Resources Research Institute, University of North Carolina, Raleigh, NC. Report Number 297.

Riekerk, H. 1992. Groundwater movement between pine uplands and cypress wetlands. p. 644–654. *In* M. C. Landin (ed.) Wetlands, Proceedings of the 13th Annual Conference of the Society of Wetland Scientists. South Central Chapter, Society of Wetland Scientists, Utica, MS.

Rikard, M. 1988. Hydrologic and vegetation relationships of the Congaree Swamp National Monument. Ph. D Dissertation. Clemson University, Clemson, SC.

Rivera-Monroy, V. H. and R. R. Twilley. 1996. The relative role of denitrification and immobilization on the fate of inorganic nitrogen in mangrove sediments of Terminos Lagoon, Mexico. Limnology and Oceanography 41: 284–296.

Rivera-Monroy, V. H., J. W. Day, Jr., R. R. Twilley, F. Vera-Herrera, and C. Coronado-Molina. 1995a. Flux of nitrogen and sediment in a fringe mangrove forest in Terminos Lagoon, Mexico. Estuarine, Coastal and Shelf Science 40: 139–160.

Rivera-Monroy, V. H., R. R. Twilley, R. G. Boustany, J. W. Day, Jr., F. Vera-Herrera, and M. C. Ramirez. 1995b. Direct denitrification in mangrove sediments in Terminos Lagoon, Mexico. Marine Ecology Progress Series 126: 97–109.

Roberts, J. K., J. Callis, O. Jardetzky, V. Walbot, and M. Freeling. 1984. Cytoplasmic acidosis as a determinant of flooding intolerance in plants. Proceedings of The National Academy of Science, U.S.A. 81: 6029–6033.

Robertson, A. I. 1986. Leaf-burying crabs:their influence on energy flow and export from mixed mangrove forests (*Rhizophora* spp.) in northeastern Australia. Journal of Experimental Marine Biology and Ecology 102: 237–248.

Robertson, A. I. 1988. Decomposition of mangrove leaf litter in tropical Australia. Journal of Experimental Marine Biology and Ecology 116: 235–247.

Robertson, A. I. and D. M. Alongi. 1992. Tropical Mangrove Ecosystems, Vol. 41. American Geophysical Union, Washington, DC.

Robertson, A. I and S. J. M. Blaber. 1992. Plankton, epibenthos and fish communities. p. 173–224. *In* A. I. Robertson and D. M. Alongi (eds.) Tropical Mangrove Ecosystems. American Geophysical Union, Washington, DC.

Robertson, A. I. and P. A. Daniel. 1989. The influence of crabs on litter processing in high intertidal mangrove forests in tropical Australia. Oecologia 78: 191–198.

Robertson, A. I. and N. C. Duke. 1990. Mangrove fish-communities in tropical Queensland, Australia: spatial and temporal patterns in densities, biomass and community structure. Marine Biology 104: 369–379.

Robertson, A. I., D. M. Alongi, and K. G. Boto. 1992. Food chains and carbon fluxes. p. 293–326. *In* A. I. Robertson and D. M. Alongi (eds.) Tropical Mangrove Ecosystems. American Geophysical Union, Washington, DC.

Robertson, A. I., R. L. Giddins, and T. J. Smith III. 1990. Seed predation by insects in tropical mangrove forests: extent and effects on seed viability and growth of seedlings. Oecologia 83: 213–219.

Robertson, P. A., G. T. Weaver, and J. A. Cavanaugh. 1978. Vegetation and tree species patterns near the northern terminus of the southern floodplain forest. Ecological Monographs 48: 249–267.

Robison, H. W. 1978. Distribution and habitat of the taillight shiner, *Notropis maculatus* Hay, in Arkansas. Arkansas Academy of Science Proceedings 32: 68–70.

Robison, H. W. and T. M. Buchanan. 1988. Fishes of Arkansas. University of Arkansas Press, Fayetteville, AR.

Rodelli, M. R., J. N. Gearing, P. J. Gearing, N. Marshall, and A. Sasekumar. 1984. Stable isotope ratio as a tracer of mangrove carbon in Malaysian ecosystems. Oecologia 61: 326–333.

Rogers, J. and R. Rohli. 1991. Florida citrus freezes and polar anticyclones in the Great Plains. Journal of Climate 4: 1103–1113.

Rogers, M. E. and D. W. West. 1993. The effects of rootzone salinity and hypoxia on shoot and root growth in *Trifolium* species. Annals of Botany 72: 503–509.

Rogers, V. A. 1977. Soil Survey of Barnwell County, South Carolina, Eastern Part. U.S. Department of Agriculture, Soil Conservation Service, Barnwell, SC.

Rogers, V. A. 1985. Soil Survey of Aiken County Area South Carolina. U.S. Department of Agriculture, Soil Conservation Service, Aiken, SC.

Rohde, F. C., R. G. Arndt, D. G. Lindquist, and J. F. Parnell. 1994. Freshwater Fishes of the Carolinas, Virginia, Maryland, and Delaware. University of North Carolina Press, Chapel Hill, NC.

Rojas-Galaviz, J. L., A. Yáñez-Arancibia, J. W. Day, Jr., and F. R. Vera-Herrera. 1992. Estuarine primary producers: Laguna de Terminos-a study case. p. 141–154. *In* U. Seeliger (ed.) Coastal Plant Communities of Latin America. Academic Press, San Diego, CA.

Rosenberg, N. J., B. L. Blad, and S. B. Verma. 1983. Microclimate: The Biological Environment. John Wiley and Sons, New York, NY.

Ross, S. T. and J. A. Baker. 1983. The response of fishes to periodic spring floods in a southeastern stream. American Midland Naturalist 109: 1–14.

Ross, T. E. 1987. A comprehensive bibliography of the Carolina bays literature. Journal of the Elisha Mitchell Scientific Society 103: 28–42.

Rossell, C. R., Jr., I. M. Rossell, J. W. Petranka, and K. K. Moorhead. 1997. Characteristics of a partially-disturbed southern Appalachian forest-gap bog complex, Virginia Museum of Natural History, Martinsville, VA. Special Publication No. 3.

Rosson, J.F., Jr. 1993a. The woody biomass resource of Louisiana, 1991. U.S. Department of Agriculture Forest Service, New Orleans, LA. Resource Bulletin SO-181.

Rosson, J.F., Jr. 1993b. The woody biomass resource of East Texas, 1992. U.S. Department of Agriculture Forest Service, New Orleans, LA. Resource Bulletin SO-183.

Rosson, J. F. Jr. 1995. Forest resources of Louisiana, 1991. U.S. Department of Agriculture Forest Service, Southern Forest Experiment Station, New Orleans, LA. Resource Bulletin SO-192.

Roth, L. C. 1992. Hurricanes and mangrove regeneration: effects of hurricane Juan, October 1988, on the vegetation of Isla del Venado, Bluefields, Nicaragua. Biotropica 24: 375–384.

Rozelle, A. A. 1997. Use of morphological and physiological traits to assess seedling response to flooding of four wetland tree species. M.S. Thesis. Clemson University, Clemson, SC.

Rozema, J. And J. A. C. Verkleij (eds.). 1991. Ecological responses to environmental stresses. Kluwer Academic Publishers, Dordrecht.

Rubenstein, D. I. 1981a. Individual variation and competition in the Everglades pygmy sunfish. Journal of Animal Ecology 50: 337–352.

Rubenstein, D. I. 1981b. Population density, resource patterning, and territoriality in the Everglades pygmy sunfish. Animal Behavior 29: 155–172.

Rubenstein, D. I. 1981c. Combat and communication in the Everglades pygmy sunfish. Animal Behavior 29: 249–258.

Rudolph, D. C. and J. G. Dickson. 1990. Streamside zone width and amphibian and reptile abundance. Southwestern Naturalist 35: 472–476.

Rudolph, H. and J. U. Voigt. 1986. Effects of $N\text{–}NH_4^+$ and $N\text{–}NO_3^-$ on growth and metabolism of *Sphagnum magellanicum*. Physiologia Plantarum 66: 339–343.

Ruffner, J. A. 1978. Climates of the states; with current tables of the National Oceanic and Atmospheric Administration normal 1940 to 1970 and means and extremes to 1975, Vol. I. Gale Research Company, Detroit, MI.

Rummer, R. B., B. J. Stokes, and B. G. Lockaby. 1997. Sedimentation associated with forest road surfacing in a bottomland hardwood ecosystem. Forest Ecology and Management 90: 195–200.

Rumpho, M. E., A. Pradet, A. Khalik, and R. A. Kennedy. 1984. Energy charge and emergence of the coleoptile and radicle at varying oxygen levels in *Echinochloa crus-galli*. Physiologia Plantarum 62: 133–138.

Rutherford, D. A., W. E. Kelso, C. F. Bryan, and G. C. Constant. 1995. Influence of physicochemical characteristics on annual growth increments of four fishes from the lower Mississippi River. Transactions of the American Fisheries Society 124: 687–697.

Rützler, K. and C. Feller. 1988. Mangrove swamp communities. Oceanus 30: 16–24.

Rützler, K. and C. Feller. 1996. Carribbean mangrove swamps. Scientific American 274: 94–99.

Ryan, D. R. and F. H. Bormann. 1982. Nutrient resorption in northern hardwood forests. BioScience 32: 29–32.

Rykiel, E. 1984. Okefenokee Swamp watershed: water balance and nutrient budgets. p. 374–385. *In* K. C. Ewel and H. T. Odum (eds.) Cypress Swamps. University Presses of Florida, Gainesville, FL.

S

Sabo, M. J. and W. E. Kelso. 1991. Relationship between morphometry of excavated floodplain ponds along the Mississippi River and their use as fish nurseries. Transactions of the American Fisheries Society 120: 552–561.

Sabo, M. J., W. E. Kelso, C. F. Bryan, and D. A. Rutherford. 1991. Physiochemical factors affecting larval fish densities in Mississippi River floodplain ponds, Louisiana (U.S.A.). Regulated Rivers Research and Management 6: 109–116.

Saenger, P. and S. C. Snedaker. 1992. Pantropical trends in mangrove above-ground biomass and annual litterfall. Oecologia 96: 293–299.

Saenger, P., E. J. Hegerl, and J. D. S. Davie (eds.). 1983. Global status of mangrove ecosystems. International Union for Conservation of Nature and Natural Resources (IUCN). Commission on Ecology Papers No. 3. The Environmentalist 3(Supplement No. 3): 1–88.

Saglio, P. H. 1985. Effect of path or sink anoxia on sugar translocation in roots of maize seedlings. Plant Physiology 77:285–290.

Saglio, P. H. and A. Pradet. 1980. Soluble sugars, respiration, and energy charge during aging of excised maize root tips. Plant Physiology 66: 516–519.

Saglio, P. H., P. Raymond, and A. Pradet. 1980. Metabolic activity and energy charge of excised maize root tips under anoxia. Plant Physiology 66: 1053–1057.

Sah, R. N. and D. S. Mikkelsen. 1986. Sorption and bioavailability of phosphorus during the drainage period of flooded-drained soils. Plant and Soil 92: 265–278.

Salinas, L. M., R. D. DeLaune, and W. H. Patrick, Jr. 1986. Changes occurring along a rapidly submerging coastal area: Louisiana, U.S.A. Journal of Coastal Research 2(3): 269–284.

Salisbury, F. B. and C. W. Ross. 1978. Plant Physiology. Wadsworth Publishing Company, Inc., Belmont, CA.
Sallabanks, R., J. R. Walters, and J. A. Collazo. In preparation. Breeding bird abundance in bottomland hardwood forests: habitat, edge, and patch size effects. Submitted to The Auk.
Sargent, C. S. 1884. Forests of the United States. 10th Census, Vol. 9.
Saucier, J. R. and N. D. Cost. 1988. The wetland timber resource of the Southeast. American Pulpwood Association, Southeastern Technical Division, Wetland Logging Task Force, Washington, DC. APA Technical Bulletin 88-A-11.
Saucier, R. T. 1994 Evidence of late glacial runoff in the lower Mississippi Valley, Quaternary Science Reviews 13: 973–981.
Sauer, J. D. 1962. Effects of recent tropical cyclones on the coastal vegetation of Mauritius. Journal of Ecology 50: 275–290.
Sauer, J. R. and S. Droege. 1992. Geographic patterns in population trends of Neotropical migrants in North America. p. 26–42. *In* J. M. Hagan, III and D. W. Johnston (eds.) Ecology and Conservation of Neotropical Migrant Landbirds. Smithsonian Institution Press, Washington, DC.
Savage, H., Jr. 1956. Rivers of the Carolinas. Rinehart & Co., New York, NY.
Savage, H., Jr. 1982. The Mysterious Carolina Bays. University of South Carolina Press, Columbia, SC.
Scarnecchia, D. L. 1992. A reappraisal of gars and bowfins in fishery management. Fisheries 17: 6–12.
Schafale, M. P. and A. S. Weakley. 1990. Classification of the natural communities of North Carolina: third approximation. North Carolina National Heritage Program, Division of Parks and Recreation, NC. Department of Environment, Health, and Natural Resources, Raleigh, NC.
Schaffer, B., P. C. Andersen, and R. C. Ploetz. 1993. Responses of fruit crops to flooding. Horticultural Review 13: 257–313.
Schalles, J. F. 1989. Comparative limnology of Carolina bay wetlands on the Upper Coastal Plain of South Carolina. p. 89–111. *In* R. R. Sharitz and J. W. Gibbons (eds.) Freshwater Wetlands and Wildlife Symposium: Perspectives on Natural, Managed and Degraded Ecosystems. U.S. Department of Energy, Office of Scientific and Technical Information, Oak Ridge, TN. CONF-8603101, DOE Symposium Series No. 61.
Schalles, J. F. and D.J. Shure. 1989. Hydrology, community structure, and productivity patterns of a dystrophic Carolina bay wetland. Ecological Monographs 59: 365–385.
Schalles, J. F., R. R. Sharitz, J. W. Gibbons, G. J. Leversee, and J. N. Knox. 1989. Carolina Bays of the Savannah River Plant. U.S. Department of Energy, Aiken, SC. National Environmental Research Park Report.
Scheerer, G. A., W. M. Aust, J. A. Burger, and W. H. McKee, Jr. 1995. Skid trail amelioration following timber harvests on wet flat pines in South Carolina. p. 236–243. *In* M. B. Edwards (compiler) Proceedings of the Eighth Annual Southern Silvicultural Research Conference. U.S. Department of Agriculture Forest Service, Southern Research Station, Asheville, NC. General Technical Report SRS-1.
Schitoskey, F., Jr. and R. L. Linder. 1978. Use of wetlands by upland wildlife. p. 307–311. *In* P. E. Greeson, J. R. Clark, and J. E. Clark (eds.) Wetland Functions and Values: The State of Our Understanding. American Water Resources Association, Minneapolis, MN.
Schlesinger, W. H. 1976. Biogeochemical limits on two levels of plant community organization in the cypress forest of Okefenokee Swamp. Ph.D. Dissertation. Cornell University, Ithaca, NY.
Schlesinger, W. H. 1978. Community structure, dynamics, and nutrient ecology in the Okefenokee cypress swamp-forest. Ecological Monographs 48: 43–65.

Schlesinger, W. H. and B. F. Chabot. 1977. The use of water and minerals by evergreen and deciduous shrubs in Okefenokee Swamp. Botanical Gazette 138: 490–497.

Schluter, U. B., B. Furch, and C.A. Joly. 1993. Physiological and anatomical adaptations by young *Astrocaryum jauari* Mart. (Arecaceae) in periodically inundated biotopes of central Amazonia. Biotropica 25: 384–396.

Schmelz, D. V. and A. A. Lindsey. 1965. Size-class structure of old-growth forests in Indiana. Forest Science 11: 258–264.

Schmidt, K. 1996. Endangered, threatened, and special status fishes in North America, 4th Edition. Special Publication of North American Native Fishes Association, St. Paul, MN.

Schneider, J. P. and J. G. Ehrenfeld. 1987. Suburban Development and cedar swamps: effects on water quality, water quantity and plant community composition. p. 271–288. *In* A. D. Laderman (ed.) Atlantic White Cedar Wetlands. Westview Press, Boulder, CO.

Schneider, R. L. and R. R. Sharitz. 1986. Seed bank dynamics in a southeastern riverine swamp. American Journal of Botany 73: 1022–1030.

Schneider, R. L. and R. R. Sharitz. 1988. Hydrochory and regeneration in a bald cypress-water tupelo swamp forest. Ecology 69: 1055–1063.

Schoeneberger, P. 1996. Soils, geomorphology, and land use of the southeastern United States. p. 59–82. *In* S. Fox and R. A. Mickler (eds.) Impacts of Air Pollutants on Southern Pine Forests. Springer Verlag, New York, NY.

Schoener, T. W. 1983. Field experiments on interspecific competition. American Naturalist 122: 240–285.

Schroder, P. 1989. Aeration of the root system in *Alus glutinosa* L. Gaertn. Annales des Sciences Forestieres 46(suppl.): 310s–314s.

Schroeder, J. G. and M. A. Taras. 1985. Atlantic white-cedar: an American wood. U.S. Department of Agriculture, Washington, DC. Miscellaneous Publication FS-225.

Schumacher, T. E. and A. J. M. Smucker. 1985. Carbon transport and root respiration of split root systems of *Phaseolus vulgaris* subjected to short term localized anoxia. Plant Physiology 78: 359–364.

Schwartz, L. 1996. Personal communication. Grand Strand Water and Sewer Authority, Myrtle Beach, SC.

Schwertmann, U. and R. M. Taylor. 1989. Iron oxides. p. 380–438. *In J. B.* Dixon and S. B. Weed (eds.) Minerals in Soil Environments, 2nd Edition. Soil Science Society of America, Madison, WI.

SCS. 1981. Land resource regions and major land resource areas of the United States. U.S. Department of Agriculture Soil Conservation Service, Washington, DC. Agriculture Handbook 296.

SCS. 1991. Hydric Soils of the United States. U.S. Department of Agriculture, Soil Conservation Service, Washington, DC. Miscellaneous Publication Number 1491.

Sell, M. G. 1977. Modelling the response of mangrove ecosystems to herbicide spraying, hurricane, nutrient enrichment and economic development. Ph. D. dissertation. University of Florida, Gainesville, FL.

Semeniuk, V. 1994. Predicting the effect of sea-level rise on mangroves in northwestern Australia. Journal of Coastal Research 10: 1050–1076.

Semlitsch, R. D., D. E. Scott, J. H. K. Pechmann, and J. W. Gibbons. 1996. Structure and dynamics of an amphibian community: evidence from a 16-year study of a natural pond. p. 217–248. *In* M. L. Cody and J. Smallwood (eds.) Long-term Studies of Vertebrate Communities. Academic Press, Inc., San Diego, CA.

Sena Gomes, A. R. and T. T. Kozlowski. 1980a. Growth responses and adaptations of *Fraxinus pennsylvanica* seedlings to flooding. New Phytologist 66: 267–271.

Sena Gomes, A. R. and T. T. Kozlowski. 1980b. Responses of *Melaleuca quinquenervia* seedlings to flooding. Physiologia Plantarum 49: 373–377.

Sena Gomes, A. R. and T. T. Kozlowski. 1980c. Responses of *Pinus halepensis* seedlings to flooding. Canadian Journal of Forest Research 10: 308–311.

Sena Gomes, A. R. and T. T. Kozlowski. 1988. Physiological and growth responses to flooding of seedlings of *Hevea brasiliensis*. Biotropica 20: 286–293.

Shafer, D. S. 1986. Flat Laurel Gap Bog, Pisgah Ridge, North Carolina: late Holocene development of a high-elevation heath bald. Castanea 51: 1–9.

Shafer, D. S. 1988. Late quaternary landscape evolution at Flat Laurel Gap, Blue Ridge Mountains, North Carolina. Quaternary Research 30: 7–11.

Shaler, N. S. 1890. Fresh-water morasses of the United States. U.S. Geological Survey Annual Report 10: 255–339.

Sharitz, R. R. and J. W. Gibbons. 1982. The ecology of southeastern shrub bogs (pocosins) and Carolina bays: a community profile. U.S. Fish and Wildlife Service, Biological Services Program, Washington, DC. FWS/OBS-82/04.

Sharitz, R. R. and L. C. Lee. 1985a. Limits on regeneration processes in southeastern riverine wetlands. p. 139–143. *In* Riparian Ecosystems and Their Management: Reconciling Conflicting Uses. U.S. Department of Agriculture Forest Service, Ft. Collins, CO. General Technical Report RM-120.

Sharitz, R. R. and L.C. Lee. 1985b. Recovery processes in southeastern riverine wetlands. p. 449–501. *In* Riparian Ecosystems and Their Management: Reconciling Conflicting Uses. U.S. Department of Agriculture Forest Service, Ft. Collins, CO. General Technical Report RM-120.

Sharitz, R. R. and W. J. Mitsch. 1993. Southern floodplain forests. p. 311–372. *In* W. H. Martin, S. G. Boyce, and A. C. Esternacht (eds.) Biodiversity of the Southeastern United States: Lowland Terrestrial Communities. John Wiley and Sons, Inc., New York, NY.

Sharitz, R. R., R. L. Schneider, and L. C. Lee. 1990. Composition and regeneration of a disturbed river floodplain forest in South Carolina. p. 195–218. *In* J. G. Gosselink, L. C. Lee, and T. A. Muir (eds.) Ecological Processes and Cumulative Impacts: Illustrated by Bottomland Hardwood Wetland Ecosystems. Lewis Publishers, Inc., Chelsea, MI.

Shaw, S. P. and C. G. Fredine. 1956. Wetlands of the United States. U.S. Fish and Wildlife Service, Washington, DC. Circular 39.

Shear, T. H., M. Young, and R. C. Kellison. In press. Older growth red river bottom forest in the eastern United States. Torrey Botanical Society.

Sheffield, R. M. and H. A. Knight. 1986. North Carolina's Forests. U.S. Department of Agriculture Forest Service, Southern Research Station, Asheville, NC. Resource Bulletin SE-88.

Sheldon, A. L. and G. K. Meffe. 1995. Path analysis of collective properties and habitat relationships of fish assemblages in coastal plain streams. Canadian Journal of Fisheries and Aquatic Sciences 52: 23–33.

Shepard, J. P. 1992. Effects of forest management on water quality in wetland forests. p. 88–92. *In* M. C. Landin (ed.) Proceedings of the 13th Annual Conference of the Society of Wetland Scientists, New Orleans, LA. South Central Chapter, Society of Wetland Scientists, Utica, MS.

Shepard, J. P. 1993. A statistical description of hydric soils of the Southeastern United States. p. 27–33. *In* J. C. Brissette (ed.) Proceedings of the Seventh Biennial Silvicultural Conference. U.S. Department of Agriculture Forest Service, Southern Forest Experiment Station, New Orleans, LA. General Technical Report SO-93.

Shepard, M. E. and M. T. Huish. 1978. Age, growth, and diet of the pirate perch in a coastal plain stream of North Carolina. Transactions of the American Fisheries Society 107: 457–459.

Sheridan, P. F. 1992. Comparative habitat utilization by estuarine macrofauna within the mangrove system of Rookery Bay, Florida. Bulletin of Marine Science 50: 21–39.

Sherrod, C. L. and C. McMillan. 1981. Black mangrove, *Avicennia germinans*, in Texas: past and present distribution. Contributions in Marine Science 24: 115–131.

Sherrod, C. L. and C. McMillan. 1985. The distribution history and ecology of mangrove vegetation along the northern Gulf of Mexico coastal region. Contributions in Marine Science 28: 129–140.

Sherrod, C. L., D. L. Hockaday, and C. McMillan. 1986. Survival of red mangrove, *Rhizophora mangle*, on the Gulf of Mexico coast of Texas. Contributions in Marine Science 29: 27–36.

Shropshire, F. W. 1980. Willow oak-water oak-diamondleaf (laurel) oak. p. 63. *In F. H.* Eyre (ed.) Forest Cover Types of the United States and Canada. Society of American Foresters, Washington, DC.

Shuey, J .A. 1985. Habitat associations of wetland butterflies near the glacial maxima in Ohio, Indiana, and Michigan. Journal of Research on the Lepidoptera 24: 176–186.

Shure, D. J., M. R. Gottschalk, and K. A. Parsons. 1986. Litter decomposition processes in a floodplain forest. American Midland Naturalist 115: 314–327.

Sietz, L. C. and D. A. Segers. 1993. Nest predation in adjacent deciduous, coniferous and successional habitats. Condor 95: 297–304.

Silker, T. H. 1948. Planting of water-tolerant trees along margins of fluctuating-level reservoirs. Iowa State College Journal of Science 22: 431–448.

Simberloff, D., and J. Cox. 1987. Consequences and costs of conservation corridors. Conservation Biology 1: 63–71.

Simberloff, D. S. and E. O. Wilson. 1969. Experimental zoogeography of islands: the colonization of empty islands. Ecology 50: 278–289.

Simberloff, D., J. A. Farr, J. Cox, and D. W. Mehlman. 1992. Movement corridors: conservation bargains or poor investments? Conservation Biology 6: 493–504.

Simons, R. W., S. W. Vince, and S. R. Humphrey. 1989. Hydric hammocks: a guide to management. U.S. Fish and Wildlife Service, Research and Development, Washington, DC. Biological Report 85 (7.26 supplement).

Simpson, R. H. and H. Riehl. 1981. The Hurricane and Its Impact. Louisiana State University Press, Baton Rouge, LA.

Singer, J. H. 1996. Personal communication. Department of Botany, University of Georgia, Athens, GA.

Sjors, H. 1950. On the relation between vegetation and electrolytes in North Swedish mire waters. Oikos 2: 241.

Skaggs, R. W. 1976. Determination of the hydraulic conductivity — drainable porosity ratio from water table measurements. Transactions of the American Society of Agricultural Engineers 19: 73–84.

Skaggs, R. W. 1978. A water management model for shallow water table soils. Water Resources Research Institute on University of North Carolina, North Carolina State University, Raleigh, NC. Technical Report No. 134.

Skaggs, R. W. 1996. Hydrology and drainage water quality in the North Carolina Coastal Plain. p. 48–52. *In* Solutions: A Technical Conference on Water Quality, North Carolina State University, Raleigh, NC.

Skaggs, R. W., M. A. Breve, and J. W. Gilliam. 1994. Hydrologic and water quality impacts of agricultural drainage. Critical Reviews in Environmental Science and Technology 24: 1–32.

Skaggs, R. W., J. W. Gilliam, and R. O. Evans. 1991. A computer simulation study of pocosin hydrology. Wetlands 11: 399–416.

Sklar, F. H. 1983. Water budget, benthological characterization, and simulation of aquatic material flows in a Louisiana freshwater swamp. Ph.D. Dissertation. Louisiana State University, Baton Rouge, LA.

Sklar, F. H. 1985. Seasonality and community structure of the backswamp invertebrates in a Louisiana cypress-tupelo swamp. Wetlands 5: 69–86.

Sklar, F. H. and W. H. Conner. 1979. Effects of altered hydrology on primary production and aquatic animal population in a Louisiana swamp forest. p. 191–208. *In* J. W. Day, Jr., D. D. Culley, Jr., R. E. Turner, and A. T . Humphrey, Jr. (eds.) Proceedings of the 3rd Coastal Marsh and Estuary Management Symposium. Division of Continuing Education, Louisiana State University, Baton Rouge, LA.

Sklash, M. G. and R. N. Farvolden. 1979. The role of groundwater in storm runoff. Journal of Hydrology 43: 45–65.

Skoglund, S. J. 1990. Seed dispersing agents in two regularly flooded river sites. Canadian Journal of Botany 68: 754–760.

Smale, M. A. and C. F. Rabeni. 1995. Hypoxia and hyperthermia tolerances of headwater stream fishes. Transactions of the American Fisheries Society 124: 698–710.

Smit, B., M. Stacowiak, and E. Van Volkenburgh. 1989. Cellular processes limiting leaf growth in plants under hypoxic roots stress. Journal of Experimental Botany 40(210): 89–94.

Smith, A. M. and T. ap Rees. 1979. Effects of anaerobiosis on carbohydrate oxidation by roots of *Pisum sativum*. Phytochemistry 18: 1453–1458.

Smith, D. A. and M. W. Tome. 1992. Wetland restoration: will we make a difference? p. 61–69. *In* M. C. Landin (ed.) Proceedings of the 13th Annual Conference of the Society of Wetland Scientists, New Orleans, LA. South Central Chapter, Society of Wetland Scientists, Utica, MS.

Smith, D. M. 1986. The Practice of Silviculture, 8th ed. Wiley and Sons, New York, NY.

Smith, D. W. and N. E. Linnartz. 1980. The southern hardwood region. p. 145–230. *In* J. W. Barrett (ed.) Regional Silviculture of the United States. John Wiley and Sons, New York, NY.

Smith, G. S., M. J. Judd, S. A. Miller, and J. G. Buwalda. 1990. Recovery of kiwifruit vines from transient waterlogging of the root system. New Phytologist 115: 325–333.

Smith, H. D., W. L. Hafley, D. L. Holley, and R. C. Kellison. 1975. Yields of mixed hardwood stands occurring naturally in a variety of sites in the southern United States. College of Forest Resources, North Carolina State University, Raleigh, NC. Technical Report Number 55.

Smith, M. W. and P. L. Ager. 1988. Effects of soil flooding on leaf gas exchange of seedling pecan trees. HortScience 23: 370–372.

Smith, M. W., F. K. Wazir, and S. W. Akers. 1989. The influence of soil aeration on growth and elemental absorption of greenhouse-grown seedling pecan trees. Communications in Soil Science and Plant Analysis 20(3&4): 335–344.

Smith, R. D., A. Ammann, C. Bartoldus, and M. M. Brinson. 1995. An approach for assessing wetland functions using hydrogeomorphic classification, reference wetlands, and functional indices. U.S. Army Corps of Engineers Waterways Experiment Station, Vicksburg, MS. Technical Report WRP-DE-9.

Smith, T. J., III. 1987a. Effects of seed predators and light level on the distribution of *Avicennia marina* (Forsk.) Vierh. in tropical, tidal forests. Estuarine, Coastal and Shelf Science 25: 43–51.

Smith, T. J., III. 1987b. Seed predation in relation to tree dominance and distribution in mangrove forests. Ecology 68: 266–273.

Smith, T. J., III. 1992. Forest structure. p. 101–136. *In* A.I. Robertson and D.M. Alongi (eds.) Tropical Mangrove Ecosystems. American Geophysical Union, Washington, DC.
Smith, T. J., III, K. G. Boto, S. D. Frusher, and R. L. Giddins. 1991. Keystone species and mangrove forest dynamics: the influence of burrowing by crabs on soil nutrient status and forest productivity. Estuarine, Coastal and Shelf Science 33: 419–432.
Smith, T. J., III, H. T. Chan, C. C. McIvor, and M. B. Robblee. 1989. Comparisons of seed predation in tropical, tidal forests on three continents. Ecology 70: 146–151.
Smith, T. J., III, M. B. Robblee, H. R. Wanless, and T. W. Doyle. 1994. Mangroves, hurricanes, and lightning strikes. BioScience 44: 256–262.
Smith, W. P., P. B. Hamel, and R. P. Ford. 1993. Mississippi Alluvial Valley forest conversion: implications for eastern North American avifauna. Proceedings of the Annual Conference of the Southeastern Association of Fish and Wildlife Agencies 47: 460–469.
Snedaker, S. C. 1982. Mangrove species zonation: why? p. 111–125. *In* D. N. Sen and K.S. Rajpurohit (eds.) Tasks for Vegetation Science, Vol. 2. Dr W. Junk Publishers, The Hague, The Netherlands.
Snedaker, S. C. 1986. Traditional uses of South American mangrove resources and the socio-economic effect of ecosystem changes. p. 104–112. *In* P. Kunstadter, E. C. F. Bird, and S. Sabhasri (eds.) Proceedings, Workshop on Man in the Mangroves. United Nations University, Tokyo.
Snedaker, S. C. 1989. Overview of ecology of mangroves and information needs for Florida Bay. Bulletin of Marine Science. 44: 341–347.
Snedaker, S. C. And P. D. Biber. 1996. Restoration of mangroves in the United States of America: a case study in Florida. p. 170–188. *In* C. Field (ed.) Restoration of Mangrove Ecosystems. International Society for Mangrove Ecosystems, Okinawa, Japan. South China Printing Co. Ltd., Hong Kong.
Snedaker S. C. and J. G. Snedaker. 1984. The Mangrove Ecosystem: Research Methods. UNESCO, United Kingdom.
Snedaker, S. C., M. S. Brown, E. J. Lahmann, and R. J. Araujo. 1992. Recovery of a mixed-species mangrove forest in South Florida following canopy removal. Journal of Coastal Research 8: 919–925.
Snedaker, S. C., J. F. Meeder, M. S. Ross, and R. G. Ford. 1994. Discussion of Mangrove ecosystem collapse during predicted sea-level rise: holocene analogues and implications. Journal of Coastal Research 10:497–498.
Sneddon, L. 1996. Personal communication. The Nature Conservancy, Eastern Regional Office, Boston, MA.
Snodgrass, J. W., A. L. Bryan, Jr., R. F. Lide, and G. M. Smith. 1996. Factors affecting the occurrence and structure of fish assemblages in isolated wetlands of the upper coastal plain, U.S.A. Canadian Journal of Fisheries and Aquatic Sciences 53: 443–454.
Snyder, J. R. 1980. Analysis of coastal plain vegetation, Croatan National Forest, North Carolina. Veroeffentlichungen des Geobotanischen Institutes der Eidgenoessische Technische Hochschule Stiftung Ruebel in Zuerich 69:40–113.
Soil Survey Staff. 1975. Soil taxonomy: a basic system of soil classification for making and interpreting soil surveys. U.S. Department of Agriculture, Washington, DC. Agricultural Handbook 436.
Soil Survey Staff. 1987. Keys to Soil Taxonomy (third printing). Soil Management Support Services, Ithaca, NY. Technical Monograph No. 6.
Soil Survey Staff. 1975. Soil Taxonomy. Natural Resources Conservation Service, U.S. Department of Agriculture, Washington, DC. Agricultural Handbook 436.
Soil Survey Staff. 1993. Soil Survey Manual. U.S. Department of Agriculture, Washington, DC. Agricultural Handbook 18.

Sojka, R. E. 1992. Stomatal closure in oxygen-stressed plants. Soil Science 154: 269–278.

Solomon, J. D. 1995. Guide to insect borers of North American broadleaf trees and shrubs. U.S. Department of Agriculture Forest Service, Washington, DC. Agriculture Handbook 706.

Somers, A. 1996. Personal communication. Department of Biology, University of North Carolina at Greensboro, Greensboro, NC.

Sorrell, B. K. 1991. Transient pressure gradients in the lacunar system of the submerged macrophyte *Egeria densa* Planch. Aquatic Botany 39: 99–108.

Sorrell, B. K. and W. Armstrong. 1994. On the difficulties of measuring oxygen release by root systems of wetland plants. Journal of Ecology 82: 177–183.

Soule, M. E. 1983. What do we really know about extinction? p. 111–124. *In* C. M. Schonewald-Cox, S. M. Chambers, B. MacBryde, and W. L. Thomas (eds.) Genetics and Conservation. Benjamin/Cummings Publ. Co., Inc., London, England.

South Carolina Forestry Commission. 1994. South Carolina's Best Management Practices for forestry. Columbia, SC.

Southern Forest Soils Council. 1990. Soils and forest productivity of the lower coastal plain of Alabama. Field Guide for the 13th Southern Forest Soils Workshop, Auburn University, Auburn, AL.

Spangler, D. P. 1984. Geologic variability among six cypress domes in north-central Florida. p. 60–66. *In* K. C. Ewel and H. T. Odum (eds.) Cypress Swamps. University Presses of Fla., Gainesville, FL.

Speake, D. W., T. E. Lynch, W. J. Fleming, G. A. Wright, W. J. Hamrick. 1975. Habitat use and seasonal movements of wild turkeys in the Southeast. Proceedings of the National Wild Turkey Symposium 3: 122–130.

Spruill, T. 1996. The results of the Albemarle-Pamlico drainage NAWQA: nutrients and pesticides in surface water and ground water. p. 24–30. *In* SOLUTIONS: A Technical Conference on Water Quality, North Carolina State University, Raleigh, NC.

Stahle, D. W., R. B. VanArsdale, and M. K. Cleveland. 1992. Tectonic signal in baldcypress trees at Reelfoot Lake, Tennessee. Seismological Research Letters 63: 439–447.

Stegman, J. L. 1976. U.S. Fish and Wildlife Service. p. 102–120. *In* J.H. Sather (ed.) Proceedings of the National Wetland Classification and Inventory Workshop. U.S. Fish and Wildlife Service, Washington, DC. FWS/OBS-76/09.

Steinberg, R. E. and M. G. Dowd. 1988. Economic considerations in the Section 404 wetland permit process. Virginia Journal of Natural Resources Law 7: 277–305.

Sternitzke, H. S. 1972. Baldcypress: endangered or expanding species? Economic Botany 26: 130–134.

Stewart, C. N., Jr. and E. T. Nilsen. 1993. Association of edaphic factors and vegetation in several isolated Appalachian peat bogs. Bulletin of the Torrey Botanical Club 120: 128–135.

Stewart, G. R. and M. Popp. 1987. The ecophysiology of mangroves. p. 333–345. *In* R.M.M. Crawford (ed.) The Series Analytic: Plant Life in Aquatic and Amphibious Habitats. Blackwell Scientific Publications, Oxford, England. Special Publications of the British Ecological Society, No. 5.

Stine, J. K. 1983. Regulating wetlands in the 1970s: U.S. Army Corps of Engineers and the environmental organizations. Journal of Forest History 27: 60–75.

Stockard, C. R. 1907. Observations on the natural history of *Polyodon spathula*. American Naturalist 41: 753–766.

Stolt, M. H. and J. C. Baker. 1995. Evaluation of National Wetland Inventory maps to inventory wetlands in the southern Blue Ridge of Virginia. Wetlands 15: 346–353.

Stolt, M. H. and M. C. Rabenhorst. 1987a. Carolina bays on the eastern shore of Maryland. I. Soil characteristics and classification. Soil Science Society of America Journal 51: 394–398.

Stolt, M. H. and M. C. Rabenhorst. 1987b. Carolina bays on the Eastern Shore of Maryland. II. Distribution and origin. Soil Science Society of America Journal 51: 399–405.

Stone, J. H., L. M. Bahr, Jr., J. W. Day, Jr., and R. M. Darnell. 1982. Ecological effects of urbanization on Lake Pontchartrain, Louisiana, between 1953 and 1978, with implications for management. p. 243–253. *In* R. Bornkamm, J. A. Lee and M. R. D. Seaward (eds.) Urban Ecology, The Second European Ecological Symposium. Blackwell Scientific Publications, Oxford, England.

Stout, I. J. and W. R. Marion. 1993. Pine flatwoods and xeric pine forests of the southern (lower) coastal plain. p. 373–446. *In* W. H. Martin, S. G. Boyce, and A. C. Esternacht (eds.) Biodiversity of the Southeastern United States: Lowland Terrestrial Communities. John Wiley and Sons, Inc., New York, NY.

Stowe, J. 1996. Personal communication. South Carolina Heritage Trust, Department of Natural Resources, Columbia, SC.

Strahler, A. H. and A. N. Strahler. 1992. Modern Physical Geography, 4th Edition. Wiley, New York, NY.

Strand, M. N. 1993. Federal wetlands law. p. 5–102. *In* D. Wallenberg and D. Adams (eds.)Wetlands Deskbook. Environmental Law Institute, Washington, DC.

Street, M. and J. McClees. 1981. North Carolina's coastal fishing industry and the influence of coastal alterations. p. 238–254. *In* C. J. Richardson (ed.) Pocosin Wetlands: An Integrated Analysis of Coastal Plain Freshwater Bogs in North Carolina. Hutchinson Ross Publication Co., Stroudsburg, PA.

Streever, W. J. and T. L. Crisman. 1993. A comparison of fish populations from natural and constructed freshwater marshes in central Florida. Journal of Freshwater Ecology 8: 149–153.

Stubbs, J. 1962. Wetland forests. Forest Farmer 21: 6–7, 10–13.

Stubbs, J. 1966. Hardwood type-site relationships on the coastal plain. p. 10–12. *In* Proceedings: 1966 Symposium on Hardwoods of the Piedmont and Coastal Plain. Georgia Forest Research Council, Macon, GA.

Stubbs, J. 1973. Atlantic oak-gum-cypress. p. 89–93. *In* Silvicultural Systems for the Major Forest Types of the United States. U.S. Department of Agriculture Forest Service, Washington, DC. Agricultural Handbook No. 445.

Stuckey, B. N. 1982. Soil Survey of Georgetown County. U.S. Department of Agriculture, Soil Conservation Service, Georgetown, SC.

Stunzi, J. T. and H. Kende. 1989. Gas composition in the internal air spaces of deepwater rice in relation to growth induced by submergence. Plant and Cell Physiology 30: 49–56.

Sun, G. 1995. Measurement and modeling of the hydrology of cypress wetlands-pine uplands ecosystems in Florida flatwoods. Ph.D. Dissertation. University of Florida, Gainesville, FL.

Sutherland, J. P. 1980. Dynamics of the epibenthic community on roots of the mangrove *Rhizophora mangle*, at Bahia de Buche, Venezuela. Marine Biology 58: 75–84.

Sutter, R.D. and R. Kral. 1994. The ecology, status and conservation of two non-alluvial wetland communities in the South Atlantic and Eastern Gulf Coastal Plain, U.S.A. Biological Conservation 68: 235–243.

Suttkus, R. D. 1963. Order Lepisostei. p. 61–88. *In* Y. H. Olsen (ed.) Fishes of the Western North Atlantic, No. 1, Part 3. Sears Foundation for Marine Research, New Haven, CT.

Sutton, R. F. and R. W. Tinus. 1983. Root and root system terminology. Forest Science Monograph 24. Forest Science (S)29(4):15 p.

Swift, M. J., O. W. Heal, and J. M. Anderson. 1979. Decomposition in Terrestrial Ecosystems. Studies in Ecology 5. University of California Press, Los Angeles, CA.

Swindel, B. F., W. R. Marion, L. D. Harris, L. A. Morris, W. L. Pritchett, L. F. Conde, H. Riekerk, and E. T. Sullivan. 1983. Multi-resource effects of harvest, site preparation and planting in pine flatwoods. Southern Journal of Applied Forestry 7: 6–15.

Swinford, K. R. 1948. Standard volume tables for pond cypress. M.S. Thesis. University of Florida, Gainesville, FL.

Switzer, G. L. 1980. Loblolly pine-hardwood. p. 61. *In* F. H. Eyre, (ed.) Forest Cover Types of the United States and Canada. Society of American Foresters, Washington, DC.

Switzer, G. L. and L. E. Nelson. 1972. Nutrient accumulation and cycling in a loblolly pine (*Pinus taeda* L.) plantation ecosystem: the first twenty years. Soil Science Society of America Proceedings 36: 143–147.

Symbula, M. and F. P. Day. 1988. Evaluation of two methods for estimating belowground production in a freshwater swamp forest. American Midland Naturalist 120: 405–415.

Syvertsen, J. P., R. M. Zablotowicz, and M. L. Smith, Jr. 1983 . Soil temperature and flooding effects on two species of citrus. I. Plant growth and hydraulic conductivity. Plant and Soil 72: 3–12.

T

Tang, Z. C. and T. T. Kozlowski. 1982. Some physiological and morphological responses of *Quercus macrocarpa* seedlings to flooding. Canadian Journal of Forest Research 12: 197–202.

Tang, Z. C. and T. T. Kozlowski. 1983. Further studies in flood tolerance of *Betula papyrifera* seedlings. Physiologia Plantarum 59: 218–222.

Tang, Z. C. and T. T. Kozlowski. 1984a. Water relations, ethylene production, and morphological adaptation of *Fraxinus pennsylvanica* seedlings to flooding. Plant and Soil 77: 183–192.

Tang, Z. C. and T. T. Kozlowski. 1984b. Ethylene production and morphological adaptation of woody plants to flooding. Canadian Journal of Botany 62: 1659–1664.

Tansey, J. B. and N. D. Cost. 1990. Estimating the forested wetland resources in the southeastern United States with forest survey data. Forest Ecology and Management 33/34: 193–213.

Tarplee, W. H. 1979. Estimates of fish populations in two northeastern North Carolina swamp streams. Brimleyana 1: 99–112.

Tate, R. L., III. 1980. Microbial oxidation of organic matter in Histosols. Advances in Microbial Ecology 4: 169–201.

Taylor, C. M. and S. M. Norris. 1992. Notes on the reproductive cycle of *Notropis hubbsi* (bluehead shiner) in southeastern Oklahoma. Southwestern Naturalist 37: 89–92.

Taylor, D. S. 1990. Adaptive specializations of the cyprinodontid fish *Rivulus marmoratus*. Florida Scientist 53: 239–248.

Taylor, T. S. 1977. Avian use of moist soil impoundments in southeastern Missouri. M.S. Thesis. University of Missouri, Columbia, MO.

Teas, H. J. 1981. Restoration of mangrove ecosystems. p. 95–103. *In* R. C. Carey, P.S. Markovits, and J.B. Kirkwood (eds.) Proceedings of Workshop on Coastal Ecosystems of the Southeastern United States. U.S. Fish and Wildlife Service, Office of Biological Services, Washington, DC. FWS/OBS-80/59.

Templeton, A. R., Shaw, K., Routman, E. and Davis, S. K. 1990. The genetic consequences of habitat fragmentation. Annals of the Missouri Botanical Garden 77: 13–27.

Tenney, F. G. and S. A. Waksman. 1929. Composition of natural organic materials and their decomposition in the soil. IV. The nature and rapidity of decomposition of the various organic complexes in different plant materials under aerobic conditions. Soil Science 28: 55–84.

Terazawa, K. and K. Kikuzawa. 1994. Effects of flooding on leaf dynamics and other seedling responses in flood-tolerant *Alnus japonica* and flood-intolerant *Betula platyphylla* var. *japonica*. Tree Physiology 14: 251–261.

Terrell, T. L. 1972. The swamp rabbit (*Sylvilagus aquaticus*) in Indiana. American Midland Naturalist 87: 283–295.

Terry, T. A. and J. H. Hughes. 1975. The effects of intensive management on planted loblolly pine (*Pinus taeda*) growth on poorly drained soils of the Atlantic coastal plain. p. 351–379. *In* B. Bernier and C. H. Winget (eds.) Forest Soils and Forest Land Management. Laval University Press, Quebec, Canada.

Terwilliger, K. 1987. Breeding birds of two Atlantic white cedar stands in the Great Dismal Swamp. p. 215–227. *In* A. D. Laderman (ed.) Atlantic White Cedar Wetlands. Westview Press, Boulder, CO.

Terwilliger, V. J. and K. C. Ewel. 1986. Regeneration and growth after logging Florida pondcypress domes. Forest Science 32: 493–506.

Thayer, G. W., D. R. Colby, and W. F. Hettler, Jr. 1987. Utilization of the red mangrove prop root habitat by fishes in south Florida. Marine Ecology Progress Series 35: 25–38.

Theim, G. 1906. Hydrologishe Methoden. Gebhardt, Leipzig.

Theis, C. V. 1935. The relation between lowering of the piezometric surface and the rate and duration of discharge of a well using groundwater storage. American Geophysical Union Transactions 16: 519–524.

Thom, B. G. 1967. Mangrove ecology and deltaic morphology: Tabasco, Mexico. Journal of Ecology 55: 301–343.

Thom, B. G. 1970. Carolina bays in Horry and Marion Counties, South Carolina. Bulletin of the Geological Society of America 81: 783–813.

Thom, B. G. 1982. Mangrove ecology — a geomorphological perspective. p 3–17. *In* B.F. Clough (ed.) Mangrove Ecosystems in Australia: Structure, Function and Management. Australian National University Press, Canberra, Australia.

Thom, B. G. 1984. Coastal landforms and geomorphic processes. p. 3–17. *In* S. C. Snedaker and J.G. Snedaker (eds.) The Mangrove Ecosystem: Research Methods. Unesco, United Kingdom.

Thomas, B. 1976. The Swamp. W. W. Norton & Company, Inc., New York, NY.

Thompson, C. J., W. Armstrong, I. Waters, and H. Greenway. 1990. Aerenchyma formation and associated oxygen movement in seminal and nodal roots of wheat. Plant, Cell and Environment 13: 395–403.

Thompson, F. R., III and E. K. Fritzell. 1990. Bird densities and diversity in clearcut and mature oak-hickory forest. U.S. Department of Agriculture Forest Service. Research Paper NC-293.

Thompson, F. R., III, W. D. Dijak, T. G. Kulowiec, and D. A. Hamilton. 1992. Breeding bird populations in Missouri Ozark forests with and without clearcutting. Journal of Wildlife Management 56: 23–30.

Thompson, M. T. 1989. Forest statistics for Georgia. U.S. Department of Agriculture Forest Service, Southeastern Forest Experiment Station, Asheville, NC. Resource Bulletin SE-109.

Thornbury, W. D. 1965. Regional Geomorphology of the United States. John Wiley and Sons, New York, NY.

Thornthwaite, C. W. and J. R. Mather. 1955. The water balance. Publications in Climatology 8: 9–86.
Thornthwaite, C. W. and J. R. Mather. 1957. Instructions and tables for computing potential evapotranspiration and the water balance. Publications in Climatology 10: 185–311.
Thorpe, J. H., E. M. McEwan, M. F. Flynn, and F. R. Haver. 1985. Invertebrate colonization of submerged wood in a cypress-tupelo swamp and blackwater stream. American Midland Naturalist 113: 56–68.
Tiner, R. W., Jr. 1984. Wetlands of the United States: current status and recent trends. U.S. Fish and Wildlife Service, Washington, DC.
Tiner, R. W., Jr. 1990. Use of high-altitude aerial photography for inventorying forested wetlands in the United States. Forest Ecology and Management 33/34: 593–604.
Tiner, R. W., Jr. 1991. The concept of a hydrophyte for wetland identification: individual plants adapt to wet environments. BioScience 41: 236–247.
Tiner, R. W. and D. B. Foulis. 1994a. Wetland trends for selected areas of the Chickahominy River watershed of Virginia (1982–84 to 1989–90). U.S. Fish and Wildlife Service, Washington, DC.
Tiner, R. W. and D. B. Foulis. 1994b. Wetland trends in selected areas of the Norfolk/Hampton region of Virginia (1982 to 1989–90). U.S. Fish and Wildlife Service, Washington, DC.
Tiner, R. W. and D. B. Foulis. 1994c. Wetland trends in selected areas of northern Virginia (1980–81 to 1988–91). U.S. Fish and Wildlife Service, Washington, DC.
Tiner, R. W., I. Kenenski, T. Nuerminger, J. Eaton, D. B. Foulis, G. S. Smith, and W. E. Frayer. 1994. Wetland status and trends in the Chesapeake watershed (1982–1989): technical report. U.S. Fish and Wildlife Service, Washington, DC.
Toliver, J. R. and B. D. Jackson. 1989. Recommended silvicultural practices in southern wetland forests. p. 72–77. *In D. D.* Hook and R. Lea (eds.) Proceedings of the Symposium: The Forested Wetlands of the Southern United States. U.S. Department of Agriculture Forest Service, Southeastern Forest Experiment Station, Asheville, NC. General Technical Report SE-50.
Toliver, J. R., S. G. Dicke, and R. S. Prenger. 1987. Response of a second-growth natural stand of baldcypress to various intensities of thinning. p 462–465. *In* D. R. Phillips (compiler) Proceedings of the Fourth Biennial Southern Silvicultural Conference. U.S. Department of Agriculture Forest Service, Southeastern Forest Experiment Station, Asheville, NC. General Technical Report SE-42.
Tolley, M. D., R. D. DeLaune, and W. H. Patrick, Jr. 1986. The effect of sediment redox potential and soil acidity on nitrogen uptake, anaerobic root respiration, and growth of rice (*Oryza sativa*). Plant and Soil 93: 323–331.
Tomlinson, P. B. 1986. The Botany of Mangrove. Cambridge University Press, New York, NY.
Toole, R. E. and W. M. Broadfoot. 1959. Sweetgum blight as related to alluvial soils of the Mississippi River floodplain. Forest Science 5: 2–10.
Tooley, M. J. 1994. Sea-level response to climate. p. 172–189. *In* N. Roberts (ed.) The Changing Global Environment. Blackwell, Cambridge, MA.
Topa, M. A. and J. M. Cheeseman. 1992. Effects of root hypoxia and a low P supply on relative growth, carbon dioxide exchange rates and carbon partitioning in *Pinus serotina* seedlings. Physiologia Plantarum 86: 136–144.
Topa, M. A. and K. W. McLeod. 1986a. Responses of *Pinus clausa*, *Pinus serotina* and *Pinus taeda* seedlings to anaerobic solution culture. I. Changes in growth and root morphology. Physiologia Plantarum 68: 523–531.
Topa, M. A. and K. W. McLeod. 1986b. Responses of *Pinus clausa, Pinus serotina* and *Pinus taeda* seedlings to anaerobic solution culture. II. Changes in tissue nutrient concentration and net acquisition. Physiologia Plantarum 68: 532–539.

Topa, M. A. and K. W. McLeod. 1986c. Aerenchyma and lenticel formation in pine seedlings: A possible avoidance mechanism to anaerobic growth conditions. Physiologia Plantarum 68: 540–550.

Topa, M. A. and K. W. McLeod. 1986d. Effects of anaerobic growth conditions on phosphorus tissue concentrations and absorption rates of southern pine seedlings. Tree Physiology 2: 327–340.

Topa, M. A. and K. W. McLeod. 1988. Promotion of aerenchyma formation in *Pinus serotina* seedlings by ethylene. Canadian Journal of Forest Research 18: 276–280.

Trewolla, W. P. and R. W. McDermid. 1969. The feasibility of balloon logging in a Louisiana swamp. Louisiana State University School of Forestry and Wildlife Management, Baton Rouge, LA. Forestry Notes # 86.

Trimble, S. W. 1970. The Alcovy River swamps: the result of culturally accelerated sedimentation. Bulletin of the Georgia Academy of Science 28: 131–144.

Tripepi, R. R. and C. A. Mitchell. 1984. Metabolic responses of river birch and European birch roots to hypoxia. Plant Physiology 76: 31–35.

Trousdell, K. B. and M. D. Hoover. 1955. A change in groundwater level after clearcutting of loblolly pine in the Coastal Plain. Journal of Forestry 53: 493–498.

Tryon, B. W. 1990. Bog turtles (*Clemmys muhlenbergii*) in the South — a question of survival. Bulletin of the Chicago Herpetological Society 25: 57–66.

Turner, J. 1977. Effect of nitrogen availability on nitrogen cycling in a Douglas-fir stand. Forest Science 23: 307–316.

Turner, T. F., J. C. Trexler, G. L. Miller, and K. E. Toyer. 1994. Temporal and spatial dynamics of larval and juvenile fish abundance in a temperate floodplain river. Copeia 1994: 174–183.

Twilley, R. R. 1985. The exchange of organic carbon in basin mangrove forests in a southwest Florida estuary. Estuarine, Coastal and Shelf Science 20: 543–557.

Twilley, R. R. 1988. Coupling of mangroves to the productivity of estuarine and coastal waters. p. 155–180. *In* B.O. Jansson (ed.) Coastal-Offshore Ecosystem Interactions. Springer–Verlag, Berlin, Germany.

Twilley, R. R. 1995. Properties of mangrove ecosystems related to the energy signature of coastal environments. p. 43–62. *In* C.A.S. Hall (ed.) Maximum Power: The Ideas and Applications of H. T. Odum. University Press of Colorado, Niwot, CO.

Twilley, R. R., R. H. Chen, and T. Hargis. 1992. Carbon sinks in mangroves and their implications to carbon budget of tropical coastal ecosystems. Water, Air, and Soil Pollution 64: 265–288.

Twilley, R. R., A. E. Lugo, and C. Patterson-Zucca. 1986. Production, standing crop, and decomposition of litter in basin mangrove forests in southwest Florida. Ecology 67: 670–683.

Twilley, R. R., S. C. Snedaker, A. Yáñez-Arancibia, and E. Medina. 1996. Biodiversity and ecosystem processes in tropical estuaries: perspectives from mangrove ecosystems. p. 327–370. *In* H. Mooney, H. Cushman, and E. Medina (eds.) Biodiversity and Ecosystem Functions: A Global Perspective. John Wiley and Sons, New York, NY.

Twilley, R.R., M. Pozo, V.H. Garcia, V.H. Rivera-Monroy, R. Zambrano, and A. Bodero. 1997. Litter dynamics in riverine mangrove forests in the Guayas River estuary, Ecuador. Oecologia 111: 109–122.

Tyndall, R. W., K. A. McCarthy, J. C. Ludwig, and A. Rome. 1990. Vegetation of six Carolina bays in Maryland. Castanea 55: 1–21.

U

USACE. 1986. Section 404(f)(1) Exemption of farm and forest roads. Regulatory Guidance Letter Number 86-3. U.S. Army Corps of Engineers, Washington, DC.

USACE. 1987. Corps of Engineers Wetlands Delineation Manual. Waterways Experiment Station, Vicksburg, MS. Technical Report Y-87-1.

USACE. 1992. Section 404(q) Memorandum of Agreement. U.S. Army Corps of Engineers, Washington, DC.

USACE. 1996a. Guidance on the application of best management practices to mechanical silvicultural site preparation activities for the establishment of pine plantations in the Southeast. Federal Register 61(39): 7242–7245.

USACE. 1996b. National action plan to develop the hydrogeomorphic approach to assessing wetland functions. Federal Register 61(160): 42593–42603.

USDA. 1996. 1996 Farm bill conservation provisions summary. U.S. Department of Agriculture, Washington, DC.

USEPA. 1993a. Constructed wetlands for wastewater treatment and wildlife habitat. 17 case studies, Washington, DC. EPA 832-R-93-005.

USEPA. 1993b. Wetlands enforcement. Wetlands Fact Sheet # 13. EPA843-F-93-001 m.

USEPA. 1994. Final determination: U.S. EPA Region IV on the Weyerhauser Company's activities in regards to CWA Section 404(f). U.S. Environmental Protection Agency, Washington, DC.

USEPA. 1995. Memorandum to the field: application of best management practices to mechanical silvicultural site preparation activities for the establishment of pine plantations in the Southeast. U.S. Environmental Protection Agency, Washington, DC.

USFS. 1967. National Forest Survey Handbook. U.S. Department of Agriculture Forest Service, Washington, DC.

USFS. 1980. Wildlife management handbook: southern region. U.S. Department of Agriculture Forest Service. FSH 2609.23R.

USFS. 1989. Insects and diseases of trees in the South. U.S. Department of Agriculture Forest Service, Atlanta, GA. Protection Report R8-PR 16.

USFWS. 1981. Habitat evaluation procedures. Washington, DC. FWS/DES-102.

USWFS. 1990. Wetlands: Meeting the President's Challenge. 1990 Wetland Action Plan. U.S. Department of the Interior Fish and Wildlife Service, Washington, DC.

USFWS. 1992. 1991 National survey of fishing, hunting and wildlife-related recreation. U.S. Department of the Interior Fish and Wildlife Service, Washington, DC.

USFWS. 1994a. Cartographic conventions for the National Wetlands Inventory. U.S. Department of the Interior Fish and Wildlife Service, Washington, DC.

USFWS. 1994b. Digitizing photointerpretation conventions for the National Wetlands Inventory. U.S. Department of the Interior Fish and Wildlife Service, Washington, DC.

USFWS. 1995. Photointerpretation conventions for the National Wetlands Inventory. U.S. Department of the Interior Fish and Wildlife Service, Washington, DC.

USFWS. 1996. Voluntary private lands habitat restoration accomplishments, April, 1996. U.S. Department of the Interior Fish and Wildlife Service, Washington, DC. Fact Sheet.

USGS. 1994. United States Geologic Survey Daily Values. Earthinfo Inc., Boulder, CO. CD-ROM.

V

Van der Moezel, P. G., L. E. Watson, and D. T. Bell. 1989. Gas exchange responses of two *Eucalyptus* species to salinity and waterlogging. Tree Physiology 5: 251–257.

van der Pijl, L. 1982. Principles of Dispersal in Higher Plants. Springer–Verlag, New York, NY.

van der Valk, A. G. and P. M. Attiwill. 1984. Acetylene reduction in an *Avicennia marina* community in southern Australia. Australian Journal of Botany 32: 157–164.

Van Diggelen, J. 1991. Effects of inundation stress on salt marsh halophytes. p. 62–73. *In* J. Rozema, and J. A. C. Verkleij (eds.) Ecological Responses to Environmental Stresses. Kluwer Academic Publishers, The Netherlands.

Van Doren, M. (ed.). 1955. Travels of William Bartram. Dover Publications, New York, NY.

Van Genutchen, M. T. and W. J. Alves. 1982. Analytical solutions of the one dimensional convective–dispersive solute transport equation. U.S. Department of Agriculture. Technical Bulletin 1661.

Van Holmes, J. 1954. Loggers of the unknown swamp. Saturday Evening Post 226: 32–33, 102–105.

Van Toai, T. T. 1993. Field performance of abscisic acid induced flood-tolerant corn. Corn Science 33: 344–346.

Van Toai, T. T., N. R. Fausey, and M. B. McDonald, Jr. 1987. Anaerobic metabolism enzymes as markers of flooding stress in maize seeds. Plant and Soil 102: 33–39.

Vartapetian, B.B. and M. B. Jackson. 1997. Plant adaptations to anaerobic stress (review). Annals of Botany January (Special Issue): 3–16.

Vartapetian, B. B., I. N. Andreeva, and N. Nutritdinov. 1978. Plant cells under oxygen stress. p. 13–88. *In* D. D. Hook and R. M. M. Crawford (eds.) Plant Life in Anaerobic Environments. Ann Arbor Science Publishers, Inc., Ann Arbor, MI.

Vermeer, D. E. 1963. Effects of Hurricane Hattie, 1961, on the cays of British Honduras. Zeitschrift fur Geomorphologie 7: 332–354.

Vepraskas, M. J. 1995. Redoximorphic features for identifying Aquic conditions. North Carolina Agricultural Research Service, North Carolina State University, Raleigh, NC. Technical Bulletin 301.

Vickers, C. R., L. D. Harris, and B. F. Swindel. 1985. Changes in herpetofauna resulting from ditching of cypress ponds in Coastal Plain flatwoods. Forest Ecology and Management 11: 17–29.

Vince, S. W. 1996. Unpublished data. School of Forest Resources and Conservation, University of Florida, Gainesville, FL.

Vince, S. W., S. R. Humphrey, and R. W. Simons. 1989. The ecology of hydric hammocks: A community profile. U.S. Fish and Wildlife Service, Washington, DC. Biological Report 85(7.26).

Vissage, J. S. and P. E. Miller. 1991. Forest statistics for Alabama counties — 1990. U.S. Department of Agriculture Forest Service, Southern Forest Experiment Station, New Orleans, LA. Resource Bulletin SO-158.

Vitousek, P. M. 1982. Nutrient cycling and nutrient use efficiency. American Naturalist 119: 553–572.

Vitousek, P. M. and W. A. Reiners. 1975. Ecosystem succession and nutrient retention: a hypothesis. Bioscience 25: 376–381.

Voesenek, L. A. C. J., M. Banga, R. H. Thier, C. M. Mudde, F. J. M. Harren,G. W. M. Barendse, and C. W. P. M. Blom. 1993. Submergence-induced ethylene synthesis, entrapment, and growth in two plant species with contrasting flooding resistances. Plant Physiology 103: 783–791.

Vogel, R. J. 1980. The ecological factors that produce perturbation-dependent ecosystems. p. 63–94. *In* J. Cairns, Jr. (ed.) The Recovery Process in Damaged Ecosystems. Ann Arbor Science, MI.

Vogt, K. A., C. C. Grier, and D. J. Vogt. 1986. Production, turnover, and nutrient dynamics of above- and belowground detritus of world forests. Advances in Ecological Research 15: 303–377.

W

Waddington, J. M., N. T. Roulet, and A. R. Hill. 1993. Runoff mechanisms in a forested groundwater discharge wetland. Journal of Hydrology 147: 37–60.

Wade, D., J. Ewel, and R. Hofstetter. 1980. Fire in south Florida ecosystems. U.S. Department of Agriculture Forest Service. General Technical Report SE-17.

Wadsworth, F. H. 1959. Growth and regeneration of white mangrove in Puerto Rico. Caribbean Forester 20: 59–69.

Waisel, Y. 1972. Biology of Halophytes. Academic Press, New York, NY.

Wakeley, J. S., and T. H. Roberts. 1996. Bird distributions and forest zonation in a bottomland hardwood wetland. Wetlands 16: 296–308.

Walbridge, M. R. 1991. Phosphorus availability in acid organic soils of the lower North Carolina coastal plain. Ecology 72: 2083–2100.

Walbridge, M. R. 1993. Functions and values of forested wetlands in the southern United States. Journal of Forestry 91: 15–19.

Walbridge, M. R. 1994. Plant community composition and surface water chemistry of fen peatlands in West Virginia's Appalachian Plateau. Water, Air and Soil Pollution 77: 247–269.

Walbridge, M. R. and G. E. Lang. 1982. Major plant communities and patterns in four high-elevation headwater wetlands of the unglaciated Appalachian region. p. 131–142. *In* B. R. McDonald (ed.) Proceedings of the Symposium on Wetlands of the Unglaciated Appalachian Region. West Virginia University, Morganton, WV.

Walbridge, M. R. and B. G. Lockaby. 1994. Effects of forest management on biogeochemical functions in southern forested wetlands. Wetlands 14: 10–17.

Walbridge, M. R. and C. J. Richardson. 1991. Water quality of pocosins and associated wetlands of the Carolina coastal plain. Wetlands 11: 417–439.

Walbridge, M. R. and J. P. Struthers. 1993. Phosphorus retention in non-tidal palustrine forested wetlands of the Mid-Atlantic region. Wetlands 13: 84–94.

Walbridge, M. R., R. G. Knox, and P. A. Harcombe. 1992. Phosphorus availability along a landscape gradient in the Big Thicket National Preserve (abstract). Bulletin of the Ecological Society of America 73: 377.

Walbridge, M. R., C. J. Richardson, and W. T. Swank. 1991. Vertical distribution of biological and geochemical phosphorus subcycles in two southern Appalachian forest soils. Biogeochemistry 13: 61–85.

Walker, L. C. 1991. The Southern Forest: A Chronicle. University of Texas Press, Austin, TX.

Walker, M. D. 1985. Fish utilization of an inundated swamp-stream floodplain. U.S. Environmental Protection Agency, Corvalis, OR. 600/3-85/046.

Walsh, G. E. 1974. Mangroves: a review. p. 51–174. *In* R. Reimold and W. Queen (eds.) Ecology of Halophytes. Academic Press, New York, NY.

Wample, R. L. and D. M. Reid. 1979. The role of endogenous auxins and ethylene in the formation of adventitious roots and hypocotyl hypertrophy in flooded sunflower plants (*Helianthus annuus*). Physiologia Plantarum 45: 227–234.

Wang, H. F. and M. P. Anderson. 1982. Introduction to Ground Water Modeling: Finite Difference and Finite Element Methods. W. H. Freeman, San Francisco, CA.

Wanless, H. R., R. W. Parkinson, and L. P. Tedesco. 1994. Sea level control on stability of Everglades wetlands. p. 199–223. *In* Everglades: The Ecosystem and Its Restoration. St. Lucie Press, Delray Beach, FL.

Want, W. L. 1996. Law of Wetlands Regulation. Clark, Boardman, Callaghan, Deerfield, IL. Release No. 7.

Ward, B. J. 1989. Soil Survey of Williamsburg County, South Carolina. U.S. Department of Agriculture, Soil Conservation Service, Kingstree, SC.

Ware, S., C. Frost, and P. D. Doerr. 1993. Southern mixed hardwood forest: the former longleaf pine forest. p. 447–493. *In* W. H. Martin, S. G. Boyce, and A. C. Echternacht (eds.) Biodiversity of the Southeastern United States. John Wiley and Sons, New York, NY.

Waring, R. H. and W. H. Schlesinger. 1985. Forest Ecosystems — Concepts and Management. Academic Press, New York, NY.

Warner, J. H. 1990. Successional patterns in a mangrove forest in southwestern Florida, U.S.A. M.S. Thesis. University of Southwestern Louisiana, Lafayette, LA.

Watson, F. D. 1985. The nomenclature of pondcypress and baldcypress (Taxodiaceae). Taxon 34: 506–509.

Watson, J. 1928. Mangrove forests of the Malay Peninsula. Malayan Forest Records 6. Fraser & Neave, Ltd., Singapore.

Watts, F. C., G. W. Hurt, and V. W. Carlisle. 1993. Identifying hydric soils and estimating the seasonal high water table in Florida based on the ecological communities of Florida. p. 926–939. *In* M.C. Landin (ed.) Proceedings of the 13th Annual Conference of the Society of Wetland Scientists. South Central Chapter, Society of Wetland Scientists, Utica, MS.

Weakley, A. S. and K. K. Moorhead. 1991. Unpublished data. Environmental Studies, University of North Carolina — Asheville, Asheville, NC.

Weakley, A. S. and M. P. Schafale. 1991a. Classification of the Natural Areas of North Carolina. Third Approximation. North Carolina Natural Heritage Program, Raleigh, NC.

Weakley, A. S. and M. P. Schafale. 1991b. Classification of pocosins of the Carolina coastal plain. Wetlands 11: 355–375.

Weakley, A. S. and M. P. Schafale. 1994. Non-alluvial wetlands of the southern Blue Ridge — diversity in a threatened ecosystem. Water, Air and Soil Pollution 77: 359–383.

Weakley, A. S., K. D. Patterson, S. Landaal, and M. Gallyoun. 1996. International classification of ecological communities: terrestrial vegetation of the southeastern United States. Working Draft (April 1996). Southeast Regional Office, The Nature Conservancy, Chapel Hill, NC.

Weaver, K. M. and M. R. Pelton. 1994. Denning ecology of black bears in the Tensas River Basin of Louisiana. International Conference on Bear Research and Management 9: 427–433.

Weaver, K. M., D. K. Tabberer, L. U. Moore, Jr., G. A. Chandler, J. C. Posey, and M. R. Pelton. 1990. Bottomland hardwood forest management for black bears in Louisiana. Proceedings of the Annual Conference of the Southeastern Association of Fish and Wildlife Agencies 44: 342–350.

Webb, P. W. 1988. Simple physical principles and vertebrate aquatic locomotion. American Zoology 28: 709–725.

Webb, P. W. 1994. The biology of fish swimming. p. 45–62. *In* L. Maddock, Q. Bone, and J. M. V. Rayner (eds.) Mechanics and Physiology of Animal Swimming. Cambridge University Press, Cambridge.

Wehrle, B. W., R. M. Kaminski, B. D. Leopold, and W. P. Smith. 1995. Aquatic invertebrate resources in Mississippi forested wetlands during winter. Wildlife Society Bulletin 23(4): 774–783.

Welcome, R. L. and D. Hagborg. 1977. Towards a model of a floodplain fish population and its fishery. Environmental Biology of Fishes 2: 7–24.

Wells, B. W. 1928. Plant communities of the coastal plain of North Carolina and their successional relations. Ecology 9: 230–242.

Wells, B. W. and S. G. Boyce. 1953. Carolina bays: additional data on their origin, age and history. Journal of the Elisa Mitchell Scientific Society 69: 119–141.

Wells, C. L. and I. M. Rossell. 1996. Unpublished data. University of North Carolina — Asheville, Asheville, NC.

Welsh, C. J. E. and W. M. Healy. 1993. Effect of even-aged timber management on bird species diversity and composition in northern hardwoods of New Hampshire. Wildlife Society Bulletin 21: 143–154.

Wesley, D. E., C. J. Perkins, and A. D. Sullivan. 1981. Wildlife in cottonwood plantations. Southern Journal of Applied Forestry 5: 37–42.

West, R. C. 1977. Tidal salt-marsh and mangal formations of middle and south America. p. 193–214. *In* V. J. Chapman (ed.) Wet Coastal Ecosystem. Vol. 1. Ecosystems of the World. Elsevier Scientific Publishing Company, Amsterdam, The Netherlands.

Westveld, R. H. 1949. Applied Silviculture in the United States. John Wiley and Sons, Inc. New York, NY.

Wetzel, R. G. 1991. Extracellular enzymatic interactions: storage, redistribution, and interspecific communication. p. 6–28. *In* R. J. Chrost (ed.) Microbial Enzymes in Aquatic Environments. Science Tech Publishers, Madison, WI.

Wharton, C. H. 1978. The Natural Environments of Georgia. Geologic and Water Resources Division and Resource Planning Section, Office of Planning and Research, Georgia Department of Natural Resources, Atlanta, GA.

Wharton, C. H. and M. M. Brinson. 1979. Characteristics of southeastern river systems. p. 32–40. *In* R. R. Johnson and J. F. McCormick (tech. coords.) Strategies for Protection and Management of Floodplain Wetlands and Other Riparian Ecosystems. U.S. Department of Agriculture Forest Service, Washington, DC. Publication GTR-WO-12.

Wharton, C. H., W. M. Kitchens, E. C. Pendelton, and T. W. Sipe. 1982. The ecology of bottomland hardwood swamps of the southeast: a community profile. U.S. Fish and Wildlife Service, Biological Services Program, Washington, DC. FWS/OBS-81/37.

Wharton, C. H., V. W. Lambou, J. Newsome, P. V. Winger, L. L. Gaddy, and R. Mancke. 1981. The fauna of bottomland hardwoods in southeastern United States. p. 87–176. *In* J. R. Clark and J. Benforado (eds.) Wetlands of Bottomland Hardwood Forests. Elsevier Scientific Publishing Company, Amsterdam, The Netherlands.

Wharton, C. H., H. T. Odum, K. Ewel, M. Duever, A. Lugo, R. Boyt, J. Bartholomew, E. DeBellevue, S. Brown, M. Brown, and L. Duever. 1976. Forested wetlands of Florida — their management and use. Center for Wetlands, University of Florida, Gainesville, FL.

Whetstone, B. H. 1982. Techniques for estimating magnitude and frequency of floods in South Carolina. U.S. Geologic Survey, Columbia, SC. Water Resources Investigations Report 82-1.

Whigham, D. F. and M. M. Brinson. 1990. Wetland value impacts. p. 401–421. *In* B. C. Patten et al. (eds.) Wetlands and Shallow Continental Water Bodies, Volume 1. SPB Academic Publishing, The Hague, The Netherlands.

Whigham, D. F. and C. J. Richardson. 1988. Soil and plant chemistry of an Atlantic white cedar wetland on the Inner Coastal Plain of Maryland. Canadian Journal of Botany 66: 568–576.

Whitaker, J. O., Jr. 1995. National Audubon Society Field Guide to North American Mammals. Alfred A. Knopf, Inc., New York, NY.

White, D. A. 1983. Plant communities of the lower Pearl River basin, Louisiana. American Midland Naturalist 110: 381–396.

White, D. C. 1985. Lowland hardwood wetland invertebrate community and production in Missouri. Archiv fur Hydrobiologia 103: 509–533.

White, E. H. and M. C. Carter. 1970. Properties of alluvial soils supporting young stands of eastern cottonwood in Alabama. U.S. Department of Agriculture Forest Service, Southern Forest Experiment Station, New Orleans, LA. Research Note SO-111.

White, F. 1992. History of forest practice guidelines in North Carolina. North Carolina Division of Forest Resources, Raleigh, NC.

Whitehead, J. C. 1992–93. Economic valuation of wetland resources: a review of the value estimates. Environmental Systems 22: 151–161.

Whitman, R. 1989. Clean water or multiple use? Best management practices for water quality control in the national forests. Ecology Law Quarterly 16: 909.

Whittaker, J. O., Jr. 1996. National Audubon Society Field Guide to North American Mammals. Alfred A. Knopf, Inc., New York, NY.

Wiedenroth, E. M. 1993. Responses of roots to hypoxia — their structural and energy relations with the whole plant. Environmental and Experimental Botany 33: 41–51.

Wieder, R. K. 1985. Peat and water chemistry at Big Run Bog, a peatland in the Appalachian mountains of West Virginia, U.S.A. Biogeochemistry 1: 277–302.

Wieder, R. K. and G. E. Lang. 1983. Net primary productivity of the dominant bryophytes in a *Sphagnum*-dominated wetland in West Virginia. Bryologist 86: 280–286.

Wieder, R. K. and G. E. Lang. 1984. Influence of wetlands and coal mining on stream water chemistry. Water, Air and Soil Pollution 23: 381–396.

Wieder, R. K. and G. E. Lang. 1988. Cycling of inorganic and organic sulfur in peat from Big Run Bog, West Virginia. Biogeochemistry 5: 221–242.

Wieder, R. K., C. A. Bennett, and G. E. Lang. 1984. Flowering phenology at Big Run Bog, West Virginia. American Journal of Botany 71: 203–209.

Wieder, R. K., A. M. McCormick, and G. E. Lang. 1981. Vegetational analysis of Big Run Bog, a nonglaciated *Sphagnum* bog in West Virginia. Castanea 46: 16–29.

Wieder, R. K., J. B. Yavitt, and G. E. Lang. 1990. Methane production and sulfate reduction in two Appalachian peatlands. Biogeochemistry 10: 81–104.

Wieder, R. K., M. Novak, W. R. Schell, and T. Rhodes. 1994. Rates of peat accumulation over the past 200 years in five Sphagnum-dominated peatlands in the Unites States. Journal of Paleolimnology 12: 35–47.

Wieder, R. K., J. P. Yavitt, G. E. Lang, and C. A. Bennett. 1989. Aboveground net primary productivity at Big Run Bog, West Virginia. Castanea 54: 209–216.

Wigley, T. B., Jr. and T. H. Filer, Jr. 1989. Characteristics of greentree reservoirs: a survey of managers. Wildlife Society Bulletin 17: 136–142.

Wilber, C. G. 1983. Turbidity in the Aquatic Environment. C. C. Thomas Publishers, Springfield, IL.

Wilbur, H. M. 1981. Pocosin fauna. p. 62–68. *In* C. J. Richardson (ed.) Pocosin Wetlands: An Integrated Analysis of Coastal Plain Freshwater Bogs in North Carolina. Hutchinson Ross Publication Co., Stroudsburg, PA.

Wilbur, R. B. and N. L. Christensen. 1983. Effects of fire on nutrient availability in a North Carolina coastal plain pocosin. American Midland Naturalist 110: 54–61.

Wilcove, D. S. 1989. Forest fragmentation as a wildlife management issue in the eastern United States. p. 1–6. *In* R. M. DeGraaf and W. M. Healy (comps.) Is Forest Fragmentation a Management Issue in the Northeast? U.S. Department of Agriculture Forest Service. General Technical Report NE-140.

Wilen, B. O. and H. R. Pywell. 1992. Remote sensing of the nation's wetlands: the National Wetlands Inventory. p. 6–17. *In* J.D. Greer (ed.) Proceedings of the Fourth Biennial Forest Service Remote Sensing Applications Conference. American Society for Photogrammetry and Remote Sensing, Bethesda, MD.

Wilhite, L. P. and J. R. Toliver. 1990. *Taxodium distichum* (L.) Rich. — baldcypress. p. 563–572. *In* Silvics of North America: 1. Conifers. U.S. Department of Agriculture Forest Service, Washington, DC. Agriculture Handbook 654.

Willett, I. R., W. A. Muirhead, and M. L. Higgins. 1978. The effects of rice-growing on soil phosphorus immobilization. Australian Journal of Experimental Agriculture and Animal Husbandry 18: 270–275.

Williams, C. G. 1996. Personal communication. Texas A&M University, College Station, TX.

Williams, T. M. 1978. Response of shallow water tables to rainfall. p. 363 — 370. *In* W. E. Balmer (ed.) Proceedings: Soil Moisture — Site Productivity Symposium. U.S. Department of Agriculture Forest Service, State and Private Forestry, Atlanta, GA.

Williams, T. M. 1996. Personal communication. Belle W. Baruch Forest Science Institute, Clemson University, Georgetown, SC.

Williams, T. M. and D. J. Lipscomb. 1981. Water table rise after cutting on coastal plain soils. Southern Journal of Applied Forestry 5: 46–48.

Williamson, M. H. 1989. The MacArthur and Wilson theory today: true but trivial. Journal of Biogeography 16: 3–4.

Willingham, P. W. 1989. Wetlands harvesting: Scott Paper Company. p. 63–66. *In* D. D. Hook and R. Lea. (eds.) Proceedings of the Symposium: Forested Wetlands of the Southern United States. U.S. Department of Agriculture Forest Service, Southeastern Forest Experiment Station, Asheville, NC. General Technical Report SE-50.

Williston, H. L., F. W. Shropshire, and W. E. Balmer. 1980. Cypress management: a forgotten opportunity. U.S. Department of Agriculture Forest Service, Southeastern Area, Atlanta, GA. Forestry Report SA-FR 8.

Wilson, E. O. 1975. Sociobiology: The New Synthesis. Harvard University Press, Cambridge, MA.

Wilson, E. O., and E. O. Willis. 1975. Applied biogeography. p. 522–534. *In* M. L. Cosy and J. M. Diamond (eds.) Ecology and Evolution of Communities. Harvard University Press, Cambridge, MA.

Wilson, G. V., P. M. Jardine, R. J. Luxmore, L. W. Zelazny, D. E. Todd, and D. A. Lietzke. 1991. Hydrogeochemical processes controlling subsurface transport from an upper subcatchment of Walker Branch watershed during storm events. 2. Solute transport processes. Journal of Hydrology 123: 317–336.

Wimme, K. J. 1987. Site disturbance and machine performance from tree-length skidding with a rubber tired machine. M.S. Thesis. Virginia Polytechnic Institute and State University, Blacksburg, VA.

Wolf, C., Jr. 1979. A theory of market failures. The Public Interest 55: 114–133.

Woodroffe, C. D. 1985. Studies of a mangrove basin, Tuff Crater, New Zealand. III: The flux of organic and inorganic particulate matter. Estuarine, Coastal and Shelf Science 20: 447–462.

Woodroffe, C. D. 1990. The impact of sea-level rise on mangrove shoreline. Progress in Physical Geography 14: 483–520.

Woodroffe, C. D. 1992. Mangrove sediments and geomorphology. p.7–41. *In* A.I. Robertson and D.M. Alongi (eds.) Coastal and Estuarine Studies. American Geophysical Union, Washington, DC.

Woodroffe, C. D., J. Chappell, B. G. Thom, and E. Wallensky. 1986. Geomorphological dynamics and evolution of the South Alligator tidal river and plains. North Australia Research Unit Monograph 3: 1–190.

Woodwell, G. M. 1958. Factors controlling growth of pond pine seedlings in organic soils of the Carolinas. Ecological Monographs 28: 219–236.

Woolhouse, M. E. J. 1983. The theory and practice of the species-area effect, applied to breeding birds of British woods. Biological Conservation 27: 315–332.

Workman, T. 1996. Personal communication. School of Forest Resource Conservation, University of Florida, Gainesville, FL.

Workman, T. 1996. Bird populations in north-central Florida cypress ponds. M.S. Thesis. University of Florida, Gainesville, FL.

X

Xu, L. 1996. The effect of water table management practices on nutrient losses. p. 106–110. *In* Solutions: A Technical Conference on Water Quality. North Carolina State University, Raleigh, NC.

Y

Yamamoto, F. 1992. Effects of depth of flooding on growth and anatomy of stems and knee roots of *Taxodium distichum*. IAWA Bulletin n.s. 13: 93–104.

Yamamoto, F. and T. T. Kozlowski. 1987a. Effect of soil on growth, stem anatomy, and ethylene production of *Cryptomeria japonica* seedlings. Scandinavian Journal of Forest Research 2: 45–58.

Yamamoto, F. and T. T. Kozlowski. 1987b. Effects of flooding, tilting of stems, and ethrel application of *Acer plantanoides* seedlings. Scandinavian Journal of Forest Research 2: 141–156.

Yamamoto, F. and T. T. Kozlowski. 1987c. Effects of flooding on growth, stem anatomy, and ethylene production of *Pinus halepensis* seedlings. Canadian Journal of Forest Research 17: 69–79.

Yamamoto, F. and T. T. Kozlowski. 1987d. Regulation by auxin and ethylene of responses of *Acer negundo* seedlings to flooding of soil. Environmental and Experimental Botany 27(3): 329–340.

Yáñez-Arancibia, A. 1985. Fish Community Ecology in Estuaries and Coastal Lagoons: Towards an Ecosystem Integration. UNAM Press, Mexico DF.

Yáñez-Arancibia, A. and J. W. Day, Jr. 1982. Ecological characterization of Terminos Lagoon, a tropical lagoon-estuarine system in the Southern Gulf of Mexico. Oceanologica Acta SP: 431–440.

Yáñez-Arancibia, A. and J. W. Day, Jr. 1988. Ecology of Coastal Ecosystems in the Southern Gulf of Mexico: The Terminos Lagoon Region. Universidad Nacional Autonoma de Mexico, Ciudad Universitaria, Mexico.

Yáñez-Arancibia, A., A. L. Lara-Domínguez, and J. W. Day, Jr. 1993. Interactions between mangrove and seagrass habitats mediated by estuarine nekton assemblages: coupling of primary and secondary production. Hydrobiologia 264: 1–12.

Yáñez-Arancibia, A., A. L. Lara-Domínguez, J. L. Rojas-Galaviz, P. Sánchez-Gil, J. W. Day, Jr., and C. J. Madden. 1988. Seasonal biomass and diversity of estuarine fishes coupled with tropical habitat heterogeneity (southern Gulf of Mexico). Journal of Fisheries Biology 33 (Suppl. A): 191–200.

Yang, S. F. 1980. Regulation of ethylene biosynthesis. HortScience 15: 238–243.

Yang, S. F. and N. E. Hoffman. 1984. Ethylene biosynthesis and its regulation in higher plants. Annual Review of Plant Physiology 35: 155–189.

Yarbro, L. A. 1979. Phosphorus cycling in the Creeping Swamp floodplain ecosystem and exports from the Creeping Swamp watershed. Ph.D. Dissertation. University of North Carolina, Chapel Hill, NC.

Yarbro, L. A. 1983. The influence of hydrologic variations on phosphorus cycling and retention in a swamp stream ecosystem. p. 223–245. *In* T. D. Fontaine and S. M. Bartell (eds.) Dynamics of Lotic Ecosystems. Ann Arbor Science, Ann Arbor, MI.

Yarnal, B. 1993. Synoptic Climatology in Environmental Analysis. Belhaven Press, London.

Yavitt, J. B. 1994. Carbon dynamics in Appalachian peatlands of West Virginia and western Maryland. Water, Air and Soil Pollution 77: 271–290.

Yavitt, J. B., G. E. Lang, and R. K. Wieder. 1987. Control of carbon mineralization to CH_4 and CO_2 in anaerobic, Sphagnum-derived peat from Big Run Bog, West Virginia. Biogeochemistry 4: 141–157.

Yavitt, J. B., D. M. Downey, E. Lancaster, and G. E. Lang. 1990. Methane consumption in decomposing Sphagnum-derived peat. Soil Biology and Biochemistry 22: 441–447.

Yeiser, J. L. and J. L. Paschke. 1987. Regenerating wet sites with bare-root and containerized loblolly pine seedlings. Southern Journal of Applied Forestry 11: 52–56.

Yoshida, S. and T. Tadano. 1978 . Adaptation of plants to submerged soils. p. 233–256. *In* Crop Tolerance to Subtropical Land Conditions. Agronomy Society of America, Crop Science Society of America, Soil Science Society of America, Madison, WI.

Young, G. L., G. L. Karr, B. D. Leopold, and J. D. Hodges. 1995a. Effect of greentree reservoir management on Mississippi bottomland hardwoods. Wildlife Society Bulletin 23:525–531.

Young, P. J., B. D. Keeland, and R. R. Sharitz. 1995b. Growth response of baldcypress [*Taxodium distichum* (L.) Rich.] to an altered hydrologic regime. American Midland Naturalist 133: 206–212.

Z

Zaerr, J. B. 1983. Short-term flooding and net photosynthesis is seedlings of three conifers. Forest Science 29: 71–78.

Zsuffa, L. 1976. Vegetative propagation of cottonwood by rooting cuttings. p. 99–108. *In* Proceedings of the Symposium on Eastern Cottonwood and Related Species. U.S. Department of Agriculture Forest Service, New Orleans, LA.

Zuberer, D. A. and W. S. Silver. 1978. Biological nitrogen fixation (acetylene reduction) associated with Florida mangroves. Applied and Environmental Microbiology 35: 567–575.

Zucca, C. 1978. The effects of road construction on a mangrove ecosystem. M.S. Thesis. University of Puerto Rico, Rio Piedras, PR.

Index

Note Page numbers in this index for animal species refer to chapters other than Chapter 9 – *Wildlife Communities*, which offers extensive coverage of wildlife species, their habitats, and management.

B

C

D

E

F

G

H

L

N

O

P

Q

R

S

T

U

V

W

X

Y

Z